Bacterial Toxins

Handbook of
Natural Toxins
volume 8

and Virulence Factors in Disease

HANDBOOK OF NATURAL TOXINS

Series Editor

Anthony T. Tu
Department of Biochemistry
Colorado State University
Fort Collins, Colorado

1. Plant and Fungal Toxins, *edited by Richard F. Keeler and Anthony T. Tu*
2. Insect Poisons, Allergens, and Other Invertebrate Venoms, *edited by Anthony T. Tu*
3. Marine Toxins and Venoms, *edited by Anthony T. Tu*
4. Bacterial Toxins, *edited by M. Carolyn Hardegree and Anthony T. Tu*
5. Reptile Venoms and Toxins, *edited by Anthony T. Tu*
6. Toxicology of Plant and Fungal Compounds, *edited by Richard F. Keeler and Anthony T. Tu*
7. Food Poisoning, *edited by Anthony T. Tu*
8. Bacterial Toxins and Virulence Factors in Disease, *edited by Joel Moss, Barbara Iglewski, Martha Vaughan, and Anthony T. Tu*

ADDITIONAL VOLUMES IN PREPARATION

Bacterial Toxins and Virulence Factors in Disease

Handbook of Natural Toxins
volume 8

edited by

Joel Moss
National Institutes of Health
Bethesda, Maryland

Barbara Iglewski
University of Rochester
  School of Medicine and Dentistry
Rochester, New York

Martha Vaughan
National Institutes of Health
Bethesda, Maryland

Anthony T. Tu
Colorado State University
Fort Collins, Colorado

CRC Press
Taylor & Francis Group
Boca Raton London New York

CRC Press is an imprint of the
Taylor & Francis Group, an **informa** business

CRC Press
Taylor & Francis Group
6000 Broken Sound Parkway NW, Suite 300
Boca Raton, FL 33487-2742

First issued in paperback 2019

© 1995 by Taylor & Francis Group, LLC
CRC Press is an imprint of Taylor & Francis Group, an Informa business

No claim to original U.S. Government works

ISBN-13: 978-0-8247-9381-4 (hbk)
ISBN-13: 978-0-367-40183-2 (pbk)

**Visit the Taylor & Francis Web site at
http://www.taylorandfrancis.com**

**and the CRC Press Web site at
http://www.crcpress.com**

Library of Congress Cataloging-in-Publication Data

Bacterial toxins and virulence factors in disease / edited by Joel
  Moss . . . [et al.].
      p. cm. — (Handbook of natural toxins ; v. 8)
    Includes bibliographical references and index.
    ISBN 0-8247-9381-1 (hbk. : acid-free paper)
    1. Bacterial toxins. 2. Virulence (Microbiology) I. Moss, Joel.
  II. Series.
  QP632.B3833   1995
  615.9'5299—dc20
                                                    94-47040
                                                      CIP

Preface to the Handbook

Natural toxins are unique toxins which possess some common properties, whether they are obtained from plants, microorganisms, or animals. One common characteristic is that they exert a pronounced effect on the metabolism and biological functions of the intoxicated animals with just a minute quantity. Since ancient times human beings have pondered the physiological effects of various toxins and venoms. How do these natural poisons work? Despite possessing some common nature, each toxin, however, has its unique mode of action and its own characteristic structure.

Drugs are compounds that have specific beneficial effects with a minute quantity. Usually, natural toxins also have very specific effects. Therefore, it is not surprising that many natural toxins are potentially good drugs.

Heretofore, the study of each field of toxins has been taking an independent pathway. Scientists in a specific toxin field are often unaware of the activity in other toxin fields. It is thus desirable to have a primary source of information on all natural toxins so that scientists in a specific discipline of toxin research can easily obtain useful information from other toxin researchers.

The editor expresses sincere thanks to Maurits Dekker for initiating this project.

Anthony T. Tu

Preface to Volume 8

The past decade has seen remarkable growth in our understanding of the structure and function of bacterial toxins as well as the recognition of some hitherto unknown members of this remarkable family. The contributors to this volume have attempted to provide, in relatively concise chapters, detailed and up-to-date information on many aspects of the structure and function of some of these fascinating toxic proteins.

Diphtheria toxin occasioned the first description of a mono-ADP-ribosyltransferase reaction, although it was considerably later that the ADP-ribose acceptor in elongation factor 2 was identified as diphthamide, a histidine that has undergone complex posttranslational modification. The molecule that serves as the diphtheria toxin receptor on the cell surface is still not completely characterized, although at least one part of it appears to be a heparin-binding EGF-like protein. Like many of the potent protein toxins, diphtheria toxin has two distinct domains. The A, or active, moiety acts catalytically within the cell, whereas the B, or binding, portion interacts with a specific receptor molecule on the cell surface and facilitates the entry of A. In some toxins, the A and B functions are carried out by separate proteins; in diphtheria toxin they are in different parts of the same polypeptide.

The ADP-ribosyltransferase activity of cholera toxin was first described in 1976; ganglioside G_{M1} was soon identified as the cell surface receptor and $G_{\alpha s}$, the stimulatory guanine nucleotide-binding protein of adenylyl cyclase, as the cellular substrate. It took somewhat longer, however, to identify the specific arginine residue that is modified. Many similarities among cholera toxin and the several *E. coli* heat-labile enterotoxins have been noted. Some of the most exciting current work in this area is elucidating the molecular mechanisms of assembly of these toxins and three-dimensional structures based on X-ray crystallography.

Pertussis toxin, an ADP-ribosyltransferase that modifies a cysteine near the carboxy terminus of several G protein α subunits, is only one of several *Bordetella pertussis* virulence factors, which include the pertussis adenylyl cyclase as well as the tracheal cytotoxin (TCT). The latter has recently been identified as a muramyl peptide that is released during bacteria growth and is toxic to ciliated epithelial cells. It has been suggested that TCT stimulates macrophages to release IL-1, which induces production of nitric oxide (NO) with resulting inhibition of ribonucleotide reductase and ATP synthesis.

Several clostridial toxins are also ADP-ribosyltransferases. The substrate for C2, for example, is non-muscle actin in the monomeric form, which cannot polymerize after modification. A related clostridial transferase, C3, ADP-ribosylates a ~20-kDa GTP-binding protein termed rho. This toxin and similar C3-like toxins apparently lack a B component, and for many studies they have been artificially introduced into cells. Leukocidins, which are 30- to 40-kDa ADP-ribosyltransferases, have been purified and crystallized from *Staphylococcus aureus* and *Pseudomonas aeruginosa*. Their cytocidal effects appear to involve cell protein kinase C as well as phospholipases A and C and calcium influx, although molecular mechanisms, including the proteins that are ADP-ribosylated, remain to be precisely defined.

Like *Bordetella pertussis*, *Bacillus anthracis* releases an adenylyl cyclase that increases host cell cAMP and thereby interferes with phagocytosis. The cyclase requires another protein termed PA (protective antigen) for activity, as does another product of *B. anthracis* called lethal factor (LF), which together with PA causes release of TNF and IL-1 from macrophages, with resulting cytotoxicity.

Several bacterial toxins increase cell cAMP content, because the toxin itself either catalyzes cAMP synthesis from cell ATP or, like cholera toxin, activates the endogenous adenylyl cyclase or, like pertussis toxin, blocks its inhibition. *E. coli* heat-stable toxin, on the other hand, increases cGMP by acting at the surface of the intestinal cell to increase guanylyl cyclase activity. The toxin structure and effect on cGMP were known for almost a decade before the receptor-cyclase with which it interacts and then guanylin, the physiological peptide ligand for that receptor, were described.

Other toxins produced by certain strains of *E. coli* include Shiga and Shiga-like (Vero) toxins. The A portion (like ricin) has RNA-*N*-glycosidase activity; the B moiety binds to and defines the specificity of interaction with a globoside in the plasma membrane. In addition to its well-known ADP-ribosyltransferase, *Vibrio cholerae* secretes two other interesting toxins. One, named accessary cholera enterotoxin (Ace), causes fluid accumulation in the ileal loop model and ion secretion in vitro. Its deduced structure has similarities to known ion-transport ATPases. The other, termed Zot for zona occludens toxin, binds to the cell surface and reversibly alters actin microfilaments, thus loosening intercellular tight junctions.

The two *Clostridium difficile* toxins that cause pseudomembranous colitis apparently alter tight junctions by acting irreversibly on the cytoskeleton and cause other effects indirectly as a result of activation of host immune responses. Similarly, the devastating effects of endotoxin, which is a ubiquitous component of gram-negative organisms, result from activation of the host immune system, which then releases proinflammatory as well as anti-inflammatory mediators. Endotoxins differ somewhat in structure, but a glycophospholipid called lipid A, which reproduces all of their effects, is found in all, hence the interest in these molecules as targets for therapy or prevention of endotoxic shock.

The dire effects of tetanus toxin have been known for a very long time. It is only relatively recently, however, that its mechanisms of cell entry and processing have become known. Several neurotoxins produced by *Clostridium botulinum* resemble this product of *Clostridium tetani* in structure and function. In 1992, it was found that tetanus toxin, a metallo-protease, cleaves synaptobrevin, a critical synaptic vesicle protein, and thereby blocks neurotransmitter release. The ability of the toxin to undergo retrograde axonal transport suggested its use to deliver other molecules (e.g., enzymes, antibodies) to sites in the central nervous system. In the past few years, a similar botulinum A toxin has found diverse applications in the treatment of muscle spasm or dystonia by intramuscular injection, although it is unclear just how localized the toxin remains during the period of therapeutic effectiveness. Other potential biomedical applications of toxins include the use of liposomes as carriers in vaccines and toxin-fusion proteins for delivery of a toxin to specific cells.

This volume offers, we hope, a summary of much (not all, of course) of the new information that is rapidly accumulating concerning several bacterial toxins that act in diverse and intriguing ways to subvert cellular function (metabolism/signaling). In many instances, unraveling a mechanism of toxin action has provided new information concerning cell physiology or biochemistry, as well as, not infrequently, new toxin tools for experimental studies or clinical therapy.

Joel Moss
Barbara Iglewski
Martha Vaughan
Anthony T. Tu

Contents

PART VII. DIVERSITY OF ACTIONS OF OTHER TOXIC BACTERIAL PRODUCTS

Contributors

Klaus Aktories, M.D., Ph.D.   Professor, Institut für Pharmakologie und Toxikologie der Universität des Saarlandes, Homburg, Germany

Carl R. Alving, M.D.   Chief, Department of Membrane Biochemistry, Walter Reed Army Institute of Research, Washington, D.C.

James W. Bodley, Ph.D.   Professor, Department of Biochemistry, University of Minnesota, Minneapolis, Minnesota

Patrice Boquet, M.D.   Department of Microbiology, Institut Pasteur, Paris, France

Drusilla L. Burns, Ph.D.   Chief, Laboratory of Pertussis, Center for Biologics Evaluation and Research, Food and Drug Administration, Bethesda, Maryland

R. John Collier, Ph.D.   Maude and Lillian Presley Professor, Department of Microbiology and Molecular Genetics, Harvard Medical School, Boston, Massachusetts

Laurie E. Comstock, Ph.D.   Department of Medicine, University of Maryland at Baltimore, Baltimore, Maryland

Terry D. Connell, Ph.D.* Department of Microbiology and Immunology, Uniformed Services University of the Health Sciences, Bethesda, Maryland

Robert L. Danner, M.D.   Head, Infectious Diseases Section, Critical Care Medicine Department, National Institutes of Health, Bethesda, Maryland

---

Current affiliation: Assistant Professor, Department of Microbiology, The University at Buffalo, State University of New York, Buffalo, New York

Mario Domenighini, Ph.D. Biocomputing Unit, Department of Biocomputing, Immunobiological Research Institute, Siena, Italy

William E. Goldman, Ph.D. Associate Professor, Department of Molecular Microbiology, Washington University School of Medicine, St. Louis, Missouri

Richard L. Guerrant, M.D. Thomas H. Hunter Professor of Internal Medicine, Divisions of Geographic Medicine and Infectious Diseases, Department of Medicine, University of Virginia School of Medicine, Charlottesville, Virginia

Mark Hallett, M.D. Clinical Director, National Institute of Neurological Disorders and Stroke, National Institutes of Health, Bethesda, Maryland

Jane L. Halpern, Ph.D. Division of Bacterial Products, Center for Biologics Evaluation and Research, Food and Drug Administration, Bethesda, Maryland

Linda Nixon Heiss, Ph.D. Department of Molecular Microbiology, Washington University School of Medicine, St. Louis, Missouri

Erik L. Hewlett, M.D. Associate Dean for Research; Chief, Division of Clinical Pharmacology; and Professor of Medicine and Pharmacology, Departments of Internal Medicine and Pharmacology, University of Virginia School of Medicine, Charlottesville, Virginia

Toshiya Hirayama, Ph.D. Professor, Department of Bacteriology, Institute of Tropical Medicine, Nagasaki University, Sakamoto, Nagasaki, Japan

Timothy R. Hirst, D. Phil. Research School of Biosciences, University of Kent at Canterbury, Canterbury, Kent, United Kingdom

Wim G. J. Hol, Ph.D. Professor, Department of Biological Structure, Biomolecular Structure Center, and Howard Hughes Medical Institute, University of Washington School of Medicine, Seattle, Washington

Randall K. Holmes, M.D., Ph.D. Professor and Chairman, Department of Microbiology and Immunology, Uniformed Services University of the Health Sciences, Bethesda, Maryland

Chi-Ting Huang, B.A. Research Assistant, Department of Biochemistry, University of Minnesota, Minneapolis, Minnesota

Taroh Iiri, M.D., Ph.D. Fellow in the Department of Life Science, Tokyo Institute of Technology, Yokohama, Japan

Michael G. Jobling, Ph.D. Research Assistant Professor, Department of Microbiology and Immunology, Uniformed Services University of the Health Sciences, Bethesda, Maryland

Yasunori Kanaho, Ph.D. Associate Professor, Department of Life Science, Tokyo Institute of Technology, Yokohama, Japan

James B. Kaper, M.D.   Professor of Medicine, Department of Medicine, University of Maryland at Baltimore, Baltimore, Maryland

Barbara Illowsky Karp, M.D.   National Institute of Neurological Disorders and Stroke, National Institutes of Health, Bethesda, Maryland

Toshiaki Katada, Ph.D.   Professor, Department of Physiological Chemistry, Faculty of Pharmaceutical Sciences, University of Tokyo, Tokyo, Japan

Gertrud Koch, Ph.D.   Institut für Pharmakologie und Toxikologie der Universität des Saarlandes, Homburg, Germany

Emmanuel Lemichez, M.D.   Department of Microbiology, Institut Pasteur, Paris, France

Stephen H. Leppla, Ph.D.   Research Chemist, Laboratory of Microbial Ecology, National Institute of Dental Research, National Institutes of Health, Bethesda, Maryland

Kathryn E. Luker, Ph.D.   Post-doctoral Fellow, Department of Molecular Microbiology, Washington University School of Medicine, St. Louis, Missouri

Nancy J. Maloney, M.S.* Departments of Medicine and Pharmacology, University of Virginia School of Medicine, Charlottesville, Virginia

Eisuke Mekada, Ph.D.   Division of Cell Biology, Institute of Life Science, Kurume University, Kurume, Fukuoka, Japan

Ethan A. Merritt, Ph.D.   Research Assistant Professor, Department of Biological Structure, Biomolecular Structure Center, University of Washington School of Medicine, Seattle, Washington

Joel Moss, M.D., Ph.D.   Chief, Pulmonary/Critical Care Medicine, National Heart, Lung, and Blood Institute, National Institutes of Health, Bethesda, Maryland

John R. Murphy, Ph.D.   Professor of Medicine (Molecular Genetics), Evans Department of Clinical Research and Department of Medicine, Boston University Medical Center Hospital, Boston, Massachusetts

Charles Natanson, M.D.   Senior Investigator, Critical Care Medicine Department, National Institutes of Health, Bethesda, Maryland

Hiroshi Nishina, Ph.D.   Assistant Professor, Department of Life Science, Tokyo Institute of Technology, Yokohama, Japan

Masatoshi Noda, Ph.D.   Professor and Chairman, The Second Department of Microbiology, School of Medicine, Chiba University, Chiba, Japan

---

Current affiliation: Medical College of Virginia Commonwealth University, Richmond, Virginia

Lon D. Phan, B.S. Graduate Assistant, Department of Biochemistry, University of Minnesota, Minneapolis, Minnesota

Mariagrazia Pizza, Ph.D. Senior Scientist, Department of Molecular Biology, Immunobiological Research Institute, Siena, Italy

Rino Rappuoli, Ph.D. Scientific Director, Immunobiological Research Institute, Siena, Italy

David A. Relman, M.D. Assistant Professor, Departments of Medicine, and Microbiology and Immunology, Stanford University School of Medicine, Stanford, and Palo Alto Department of Veterans Affairs Medical Center, Palo Alto, California

Titia K. Sixma, Ph.D. Department of Molecular Carcinogenesis, Netherlands Cancer Institute, Amsterdam, The Netherlands

Katsunobu Takahashi, Ph.D. Assistant Professor, Department of Life Science, Tokyo Institute of Technology, Yokohama, Japan

Yoshifumi Takeda, M.D., D.Med. Sci.* Professor and Chairman, Department of Microbiology, Faculty of Medicine, Kyoto University, Yoshida, Kyoto, Japan

Nathan M. Thielman, M.D., M.P.H. Fellow in Infectious Diseases, Divisions of Geographic Medicine and Infectious Diseases, Department of Medicine, University of Virginia School of Medicine, Charlottesville, Virginia

Michele Trucksis, M.D., Ph.D. Assistant Professor, Department of Medicine, University of Maryland at Baltimore, Baltimore, Maryland

Johanna vanderSpek, Ph.D. Evans Department of Clinical Research and Department of Medicine, Boston University Medical Center Hospital, Boston, Massachusetts

Martha Vaughan, M.D. Deputy Chief, Pulmonary/Critical Care Medicine Branch, National Heart, Lung, and Blood Institute, National Institutes of Health, Bethesda, Maryland

Catherine F. Welsh, M.D. Staff Fellow, Laboratory of Cellular Metabolism, National Heart, Lung, and Blood Institute, National Institutes of Health, Bethesda, Maryland

---

Current affiliation: Director-General, Research Institute, International Medical Center of Japan, Toyama, Shinjuku-ku, Tokyo, Japan

Contents of Previous Volumes

Volume 5: Reptile Venoms and Toxins

PART I. SNAKES

1

Therapeutic Effects of Botulinum Toxins

Barbara Illowsky Karp and Mark Hallett

National Institute of Neurological Disorders and Stroke, National Institutes of Health, Bethesda, Maryland

I. INTRODUCTION

Bacterial toxins were first used medicinally during the late 19th century. Small amounts were injected into animals to produce antiserum to counteract the toxin's own poisonous effects. It is only recently, however, that the toxic properties of these compounds have been exploited to directly treat human disease. This latter use of botulinum toxin was pioneered by Scott. Following extensive animal experimentation, he injected the oculomotor muscles of strabismic patients in an attempt to selectively weaken the muscle that was causing misalignment (Scott, 1981). Botulinum toxin was chosen because of its known mechanism of action, selectivity, low antigenicity, and long-lasting effect. The safety and success of botulinum toxin treatment for strabismus led to trials of its use in blepharospasm and subsequently other focal dystonias and hemifacial spasm (Jankovic and Brin, 1991).

Botulinum toxin now appears to be useful for many disorders characterized by excessive muscle contraction. This chapter discusses the preparation of botulinum toxin for injection into humans, the physiologic effects of injection, the method and therapeutic effects of treatment, and currently approved and experimental clinical applications.

II. PREPARATION FOR USE IN HUMANS

Botulinum toxin is the first microbial toxin to be used for therapeutic injection. It must be carefully prepared to ensure purity, sterility, stability, and consistent biological activity (Hambleton, 1992; Schantz and Johnson, 1992). Botulinum toxin A, the only serotype currently approved by the Food and Drug Administration for clinical use, is crystallized in a complex with nontoxic proteins including hemagglutinin proteins. The additional proteins contribute to the potency of the native toxin by enhancing intestinal absorption of ingested toxin. The proteins may also maintain the three-dimensional configuration necessary for receptor binding (Schantz and Johnson, 1992). After isolation and purification without contaminating synthetic solvents or resins, the toxin is stabilized by the addition of human serum albumin.

The weight of the crystallized toxin does not reflect its potency. The dosage for botulinum toxin is therefore usually not standardized by weight. Rather, potency is expressed in mouse units, with 1 mouse unit equal to the LD_{50} for 18–20 g female Swiss-Webster mice injected intraperitoneally. The LD_{50} for intramuscularly injected botulinum toxin is 40 U/kg in monkeys (Herrero et al., 1967; Scott and Suzuki, 1988). The human LD_{50} is believed to be similar. Commonly available preparations differ in relative potency. The Porten product, Dysport, contains 40 U/ng hemagglutinin-toxin complex. The Allergan product, Botox, contains 2.5 U/ng (Quinn and Hallett, 1989; Schantz and Johnson, 1990). To avoid misunderstanding, reports on the use of botulinum toxin should specify the source of the toxin and the potency (in mouse units) (Quinn and Hallett, 1989). Botox units are used throughout this chapter.

III. PHYSIOLOGIC AND MORPHOLOGIC EFFECTS

The neuromuscular junction is a specialized area of contact between a presynaptic cholinergic neuron and the postsynaptic muscle fiber. Botulinum toxin interferes with the release of acetylcholine from the presynaptic neuron at the neuromuscular junction and causes functional denervation of the muscle. Signs of denervation and subsequent recovery can be detected by electromyography (EMG). Cohen et al. found fibrillations, positive sharp waves, and low-amplitude, incomplete interference patterns on EMG of the injected muscles following treatment for writer's cramp (Cohen et al., 1989). Patients with blepharospasm studied 3 months after injection had low-amplitude, polyphasic motor units during maximal contraction, demonstrating both denervation and reinnervation (Valls-Sole et al., 1991).

Although muscles distant from the site of injection do not become weak, subclinical effects of the toxin can be detected electrophysiologically. Increased jitter and increased fiber density, reflecting instability of the neuromuscular junction and reinnervation, have been reported in single-fiber EMG studies of arm or leg muscles in patients who received cervical or facial injections of botulinum toxin (Garner, 1993; Girlanda et al., 1992; Lange et al., 1991; Olney et al., 1988).

The normal preterminal axon is well myelinated and innervates a single muscle fiber, with infrequent collaterals innervating nearby endplates. Noncollateral sprouts, which do not terminate on endplates, are not present. Morphologic changes in muscle and nerve fibers following botulinum toxin injection have been examined in animals (Angaut-Petit et al., 1990; Capra et al., 1991; Duchen, 1970; Duchen and Stritch, 1968) and in orbicularis oculi muscles of patients treated for blepharospasm who subsequently underwent myectomy (Alderson et al., 1991; Borodic and Ferrante, 1992; Harris et al., 1991). Structural changes can be detected as early as 6 days after botulinum toxin exposure. Axons grow numerous myelinated and nonmyelinated noncollateral and collateral branches that spread to innervate both near and distant motor endplates. Acetylcholinesterase staining becomes dispersed throughout the muscle fiber rather than being confined to the neuromuscular junction, and as fibers atrophy the variability in muscle fiber diameter increases. Axonal structural changes persist, while those in muscle are not permanent. The muscle regains normal appearance by about 14 weeks after injection (Alderson et al., 1991; Duchen, 1970; Duchen and Stritch, 1968; Harris et al., 1991). Repeated botulinum toxin injections will lead to muscle atrophy but do not cause permanent alteration in muscle structure (Alderson et al., 1991; Harris et al., 1991).

Axon sprouting probably underlies recovery from botulinum toxin injection. An increase in the number of muscle fibers innervated by a single preterminal axon and expansion of motor endplates are seen between 1 and 6 weeks after injection. The new neuromuscular junctions become functional several weeks after development (Angaut-Petit et al., 1990; Borodic and Ferrante, 1992; Schantz and Johnson, 1992).

IV. METHOD AND OUTCOME OF TREATMENT

Current clinical applications of botulinum toxin require intramuscular injection to weaken muscles selectively. Identification and choice of muscles for injection are an important first step in treatment. For illnesses such as strabismus, blepharospasm, and hemifacial spasm, the abnormal muscles are visually apparent. For many dystonias, however, electromyography is necessary to determine the pattern of muscle involvement and if spasm is present in small or deep muscles.

Similarly, injection of large, superficial muscles can be accomplished under direct visualization, whereas accurate injection of small or deep muscles requires EMG guidance. An EMG needle, teflon-coated except for the tip, is mounted on a syringe containing the toxin, and attached to an EMG machine. After placing the needle into the muscle, proper localization is confirmed by recording EMG activity with contraction of the selected muscle. The toxin is then injected without moving the needle. An alternative method for confirming localization in hand or finger muscles is to observe movement of the needle with passive stretch of the target muscle.

In early trials of botulinum toxin treatment of dystonia, attempts were made to target injection to the neuromuscular junction by locating the motor endpoint electromyographically (Cohen et al., 1989; Jankovic et al., 1990). Such precise placement is not necessary. Animal experiments have found histologic signs of denervation in muscle fibers 30 mm from the injection site (Borodic and Ferrante, 1992), and the toxin can cross fascial planes (Shaari et al., 1991). Clinical experience has shown that muscles as far as 1 cm from the injection site may be weakened (Cohen et al., 1989) and that injections into the approximately defined midportion of the muscle belly are effective (Jankovic et al., 1990).

Muscle weakness and improvement in symptoms are usually not apparent for 5–7 days after injection even though the toxin is almost completely bound to presynaptic receptors within minutes. The delay in clinical effect may occur because of the time required for the multistep process of toxin internalization and neuromuscular blockade. Benefit from a single injection lasts an average of 3–4 months, but some patients have only a brief response, while others notice a much longer effect. Patient age and duration of illness do not influence the response to botulinum toxin. Women are more likely than men to become weak with injection, perhaps due to their smaller muscle mass (Cole et al., 1991; Comella et al., 1992b).

In some conditions, benefit derives directly from muscle weakening. In strabismus, temporary muscle weakness allows proper ocular alignment, which can be stabilized permanently by other mechanisms (Scott et al., 1990). In hemifacial spasm, weakening of overactive facial muscles does not prevent the underlying ectopic generation of impulses but weakens the muscles so that spasms do not occur (Geller et al., 1989). In dystonia, however, the relationship between weakness and benefit is not as clear. Although many patients have best relief of dystonia with injections that cause significant weakness, some dystonia patients have no improvement in their symptoms despite weakening of the involved muscles. Weakness without improvement in dystonia can occur if deep, inaccessible muscles are not treated or if contractures are present. Also, dystonia is refractory to botulinum toxin treatment in approximately 10% of patients (Jankovic et al., 1990; Poewe et al., 1992). In some patients, dystonia improves without detectable loss of strength, although it is likely that subclinical weakness is present.

The lack of association between benefit and weakness suggests that botulinum toxin may work through a different, perhaps central, mechanism in treating dystonia. Muscle weakness cannot account for all changes after injection, such as the reduced frequency of blinking and the photophobia sometimes seen after treatment of blepharospasm (Valls-Sole et al., 1991), a change in the pattern of dystonic muscles after treatment of torticollis (Gelb et al., 1991), or the improvement in abnormal muscle activation on the noninjected side following unilateral injection for spasmodic dysphonia (Ludlow, 1990). Intramuscularly injected botulinum toxin undergoes retrograde transport and can be detected in the spinal cord half-segments innervating the injected muscles (Wiegand et al., 1976), and botulinum toxin receptors are present in the central nervous system (Black and Dolly, 1987). However, botulinum toxin does not affect abnormalities of the blink reflex in patients with blepharospasm (Valls-Sole et al., 1991) or cocontraction of antagonist forearm muscles in patients with hand cramp (Cohen et al., 1989). There is no clear evidence that botulinum toxin acts centrally in treating dystonia.

The dose of toxin needed depends on the size of the muscle and the severity of spasm.

Small muscles, such as the medial rectus, respond to doses as low as 1.0 U. Large muscles, such as trapezius, may require more than 100 U.

The dose of botulinum toxin that can be given is limited by its inherent toxicity. The maximum total dose used in an individual injection session at our institution is 400 U. This dose is inadequate for the injection of all involved muscles in patients with widespread disease such as generalized dystonia or spasticity. However, botulinum toxin may still be useful in such patients for focal treatment of particular muscles that are painful or cause functional disability.

For strabismus and blepharospasm, a standard dose is effective in most patients. There is, however, substantial individual variability in the effective dose for other dystonias. For example, some patients with writer's cramp improve with 5 U injected into flexor carpi ulnaris. Others require 50 U into the same muscle to attain similar benefit.

Similarly, the degree of improvement in dystonia is highly variable. Despite high doses and repeated injections, some patients may not improve at all or never have more than a mild response. Once an individual's maximum level of benefit is reached, increasing the dose of toxin produces more severe weakness but no additional improvement in the dystonia.

Excessive weakness is the main side effect of botulinum toxin injection and may limit treatment. Jankovic et al. reported disabling weakness with 5% of injections (Jankovic et al., 1990). Although weakness is more likely with higher toxin doses, the extent of weakness varies from person to person and from injection to injection. The site of weakness depends on the site of injection. Patients being treated for blepharospasm develop ptosis, while those injected for spasmodic dysphonia or torticollis may have swallowing difficulties. Weakness may actually cause more functional impairment than the dystonia itself. The dose of toxin used in each disorder should therefore be carefully titrated to the lowest that produces benefit.

The use of a small injection volume to limit toxin diffusion can minimize weakness in noninjected muscles. The volume can be kept low by preparing the toxin in more concentrated solutions for large-dose injections. Concentrations commonly used range from 1.25 to 20 U/0.1 cm^3. It has been suggested that excessive spread of botulinum toxin might also be limited by injecting antitoxin immediately after the toxin to bind and deactivate any toxin not already bound to the presynaptic membrane at the neuromuscular junction (Schantz and Johnson, 1992).

Other side effects of treatment include generalized fatigue, malaise, muscle stiffness, pain, paresthesias, nausea, headache, hives, bruising at the injection site, and atrophy of the injected muscles (Brin, 1991; Cohen et al., 1989; Elston, 1990; Jankovic and Brin, 1991; Jankovic et al., 1990). There is no apparent cumulative toxicity for injections given at least 3 months apart. Two cases of possible immunologically mediated demyelination have been reported in patients who received botulinum toxin. Glanzman et al. (1990) reported brachial plexopathy beginning 2 days after injection for torticollis; Haug et al. (1990) described a patient with Guillain-Barré syndrome 11 months after treatment of blepharospasm. Side effects specific to treatment of particular conditions are discussed below.

Anti-botulinum toxin antibodies develop in some patients and are more likely in those receiving high doses, as for torticollis (Hambleton et al., 1992; Jankovic et al., 1990). Of patients with focal hand dystonia reported by Cohen et al. (1989), 11% developed antibodies. Of patients with diminution of response to injection for torticollis, 36–60% had antibodies (Hambleton et al., 1992; Jankovic and Brin, 1991). Antibody formation blocks all response to the toxin; further injections cause neither improvement nor weakness.

Antibodies are presently measured by a mouse neutralization assay (Hambleton et al., 1992; Jankovic and Brin, 1991). The lack of detectable antibodies in some patients with loss of response to treatment may be because of the insensitivity of this bioassay. Enzyme-linked immunosorbent assays (ELISA) are available, but antibody levels measured by ELISA do not correlate as well as the mouse bioassay with clinical response (Dezfulian and Bartlett, 1984; Jankovic and Brin, 1991; Siegel, 1988; Tsui et al., 1988).

Antibodies to different botulinum toxin serotypes do not cross-react (Siegel, 1989). It may be possible to regain response in patients with antibodies to botulinum toxin type A by the use of another toxin serotype. Ludlow et al. reported a preliminary trial of botulinum toxin type F for torticollis, focal hand dystonia, and spasmodic dysphonia in patients with antibodies to type A. In comparison to type A toxin, type F toxin produced a benefit of similar magnitude but shorter duration (Ludlow et al., 1992; Rhew et al., 1994).

Information on the long-term outcome of botulinum toxin treatment is presently available for patients followed as long as 7 years (Jankovic et al., 1990; Karp et al., 1994; Mauriello and Aljian, 1991). Whereas strabismus may be cured by a single injection, benefit in most conditions is maintained by repeated injections. Some studies have noted that later injections may bring less relief than that initially achieved (Alderson et al., 1991; Elston, 1992; Lees et al., 1992; Poewe et al., 1992). A diminution in response could be due to low levels of circulating antibodies (Hambleton et al., 1992) or an increase in the number of neuromuscular junctions formed in response to earlier injections (Alderson et al., 1991). No additional side effects arise with continued treatment.

Botulinum toxin has been used in patients of all ages, including children as young as 2 months (Scott, 1989; Scott et al., 1990). Patients with amyotrophic lateral sclerosis or known neuromuscular junction defects, such as myasthenia gravis or Lambert-Eaton myasthenic syndrome, should be treated only with extreme care. Although use during pregnancy is not recommended, several pregnant patients have been injected without ill effect on the mother or fetus (Scott, 1989).

V. THERAPEUTIC APPLICATIONS

The use of botulinum toxin has been best established for treatment of strabismus, hemifacial spasm, and focal dystonia. These disorders are discussed next.

A. Strabismus

Strabismus, misalignment of the visual axes, can be caused by weakness in an individual extraocular muscle (paralytic strabismus) or an imbalance in the mechanisms that normally maintain ocular alignment (nonparalytic strabismus). Botulinum toxin injection can lead to permanent improvement in both types of strabismus (Scott, 1989; Scott et al., 1990). Paralysis of a shortened or contracted extraocular muscle for several weeks is sufficient to change ocular alignment and allow compensatory contraction of the nonweakened antagonist muscle. Alignment is maintained even after the effect of the toxin has worn off.

Patients ranging in age from 2 months to 90 years have been treated for strabismus. The usual dose is 1.0–2.5 U, but doses may be as high as 12.5 U in some patients. About one-half of patients require more than one injection. Sixty-one percent of patients achieve an excellent outcome with a residual deviation of less than 10 prism diopters. Botulinum toxin is useful as a secondary treatment for over- or undercorrection following strabismus surgery as well as for the primary treatment of strabismus.

Botulinum toxin treatment is most successful in patients with strabismic angles of less than 50 prism diopters and when there is fusion to stabilize the alignment. Esotropia responds better than exotropia, and horizontal strabismus responds better than vertical strabismus. To achieve a permanent change, the antagonist muscle must be intact, the target muscle must be well-paralyzed, and the paralysis must last 2–6 weeks (AAO, 1989). Botulinum toxin is an especially important option for patients who are not surgical candidates.

Chronic paralytic strabismus due to oculomotor or abducens nerve palsies can also be treated with botulinum toxin (Lee, 1992; Saad and Lee, 1992). As for congenital strabismus, botulinum toxin works best if the paretic muscle retains some strength and if there is visual fusion.

Side effects of injection for strabismus include temporary ptosis, induction of temporary vertical strabismus, diplopia, spatial disorientation, and overcorrection. Puncture of the eye by the injection needle and retrobulbar hemorrhage have been reported rarely (Balkan and Poole, 1991; AAO, 1989; Scott et al., 1990).

B. Hemifacial Spasm

Hemifacial spasm is a disorder of the facial nerve. Abnormal activity of the nerve, including ectopic excitation, ephaptic transmission, and lateral spread of impulses, cause twitching, spasm, and synkinesias in the facial muscles. Facial movement is irregular and repetitive and fluctuates in intensity.

Spasm often begins in the orbicularis oculi but, in contrast to blepharospasm, is limited to one side unless the other side is independently affected. Involvement of the lower face, also unilateral, can affect any muscle innervated by the facial nerve, including the platysma and stapedius. Hemifacial spasm begins in adulthood and is rarely physically disabling, but the facial distortion it causes is bothersome and socially unacceptable to many patients.

Hemifacial spasm arises in association with irritation or compression of the facial nerve. Tumors, bony overgrowth in Paget's disease, or other mass lesions anywhere along the course of the facial nerve can cause hemifacial spasm. In many cases previously believed to be idiopathic, microvascular compression of the nerve can be found in the area of nerve root exit from the brainstem (Janetta et al., 1977).

Carbamazepine, clonazepam, and baclofen are used to treat hemifacial spasm. Surgical ablation of the facial nerve can help many patients but permanently weakens the facial muscles. Microvascular decompression of the nerve as it leaves the brainstem can permanently cure the spasm but is a major surgical procedure requiring posterior fossa craniotomy (Janetta et al., 1977).

Botulinum toxin is free of the drawbacks associated with medications and surgery and is now a reasonable first-line treatment for hemifacial spasm. It is injected into multiple sites of the muscles with the most obvious contraction in doses ranging from 1.25 to 10 U. Benefit lasts an average of 5 months, longer than for many focal dystonias (Geller et al., 1989; Jankovic et al., 1990). Treatment improves 92–100% of patients, and benefit can be maintained for at least 7 years (Mauriello and Aljian, 1991).

The only side effect of treatment for hemifacial spasm is transient excessive weakness, which may cause obvious facial asymmetry. Injection of zygomaticus major in particular may cause visible facial droop and is sometimes avoided (Geller et al., 1989). Injection into levator labii superioris or the superior part of orbicularis oris should similarly be avoided if it causes drooping and biting of the upper lip.

C. Dystonia

General Aspects

Dystonia is characterized by sustained abnormal muscle contraction that can cause twisting, turning, and distortion of normal posture. Dystonia is classified as idiopathic or symptomatic and as generalized or focal. Idiopathic, generalized dystonia (torsion dystonia, formerly called dystonia musculorum deformans) is a progressive hereditary disease of childhood onset. Beginning with in-turning of a foot, the dystonia progresses over several years to involve multiple muscles, causing deformed posture and disability. Focal dystonia commonly begins in adulthood and affects a limited area of the body, such as an arm, the face, or the neck. Although focal dystonia can progress to involve more proximal muscles, it rarely generalizes. Focal dystonia may be a forme fruste of generalized dystonia, as elements of generalized dystonia resemble focal dystonia affecting the same body part, generalized dystonia frequently begins with focal symptoms, and similar sensory gestures can be briefly effective in relieving the dystonia in each. Also, in some families with dystonia, some members have focal dystonia and others have generalized dystonia (Sheehy and Marsden, 1982; Zeman et al., 1960).

A family history of similar disorders is seen in 5–25% of dystonia patients (Marsden and Sheehy, 1990; Waddy et al., 1991). The inheritance is most consistent with autosomal dominant transmission with incomplete penetrance. The gene for idiopathic generalized dystonia, DYT1, has been localized to chromosome 9 (Kramer et al., 1990; Ozelius et al., 1989). Nutt et al. (1988) found the prevalence of generalized dystonia to be 34 per million in Rochester, Minnesota; the prevalence of focal dystonia was 295 per million.

The pathophysiology of dystonia is unknown. Dystonia can be symptomatic of brain damage, especially to the brainstem or basal ganglia. It occurs in disorders such as Parkinson's disease and Wilson's disease and can arise following stroke, perinatal hypoxia, head trauma, neck trauma, or neuroleptic medication. A central etiology for idiopathic dystonia is also suggested by electrophysiologic studies. Nakashima et al. (1989) and Panizza et al. (1989) found decreased reciprocal inhibition in the dystonic arm of patients with focal hand dystonia, suggesting a defect in descending supraspinal influences on 1A inhibitory interneurons.

In some cases, dystonia appears to have a peripheral origin. About 10% of patients with focal dystonia have a history of recent trauma to the dystonic area (Marsden, 1986). Dystonia has been reported in association with peripheral neuropathy, brachial plexopathy, or radiculopathy (Scherokman et al., 1986). Charness et al. (1987) and Scherokman et al. (1986) reported patients with otherwise typical hand dystonia who later developed an overt entrapment neuropathy with sensory loss or weakness. Surgical nerve release relieved the cramp as well as the sensorimotor symptoms. The possibility of a peripheral lesion, which may be amenable to surgical treatment, should be carefully evaluated in patients with focal dystonia. Near-nerve recording may be more sensitive in the detection of subtle neuropathies than routine EMG and surface nerve conduction studies.

Dystonia is characterized physiologically by excessive muscle contraction. Cohen et al. (1988) found five aspects of abnormal muscle activation in patients with writer's cramp: (1) cocontraction of antagonist muscles, (2) prolongation of EMG bursts, (3) tremor, (4) overflow or lack of selectivity in muscle activation, and (5) failure of willed activity to occur. Loss of normal reciprocal inhibition of antagonists has also been reported (Nakashima et al., 1989; Panizza et al., 1989). These abnormalities lead to loss of coordinated movement as well as spasm.

Focal dystonias fluctuate in severity and are often exquisitely task-specific, raising suspicions of a psychogenic origin. However, Grafman et al. (1991) found no psychopathology in hand cramp patients. Jahanshahi and coworkers found that depression may improve with treatment of torticollis, suggesting that psychological dysfunction is secondary to, rather than the cause of, dystonia (Jahanshahi and Marsden, 1992).

Dystonia is difficult to treat. Previous approaches have included prolonged rest, relaxation therapy, biofeedback, acupuncture, psychotherapy, physical therapy, and hypnosis. Surgery, including peripheral nerve release or section and muscle stripping, brings many patients only temporary relief. Medications commonly tried include anticholinergics, dopamine agonists, dopamine antagonists, baclofen, benzodiazepines, anticonvulsants, clonidine, alcohol, steroids (local or systemic), beta blockers, tricyclic antidepressants, and nonsteroidal anti-inflammatories. Anticholinergics are usually the most effective medications for dystonia but often must be used in such high doses that side effects become intolerable.

Botulinum toxin is a symptomatic treatment for both idiopathic and secondary dystonias. It weakens the overcontracting muscles sufficiently to relieve pain, cramping, and abnormal posture. Botulinum toxin cannot correct the loss of coordinated muscle activation. As noted above, the toxicity of botulinum toxin limits the dose that can be used and the number of muscles that can be injected. Focal dystonia is therefore amenable to botulinum toxin treatment, whereas injection of all the muscles involved in generalized dystonia would require a larger dose than can be given safely.

Botulinum toxin injection brings partial or complete relief of cramping and amelioration of pain in approximately 80–90% of all patients with focal dystonia (Jankovic et al., 1990). Improvement begins about 1 week after injection and typically lasts 3–4 months. Weakness often accompanies injection but is rarely disabling. Weakness lasts about 2 months, so the overall period of benefit is generally longer than that of overt weakness.

Blepharospasm

Patients with blepharospasm have involuntary forced eye closure due to spasm of orbicularis oculi. The initial symptoms are often a feeling of eye irritation, photophobia, or excessive blinking. Both eyes are usually affected, although the disease can be asymmetric. Eye closure may be sustained or intermittent with spasm precipitated by attempted eye use. About two-thirds of patients are rendered functionally blind. Constant involuntary contraction of orbicularis oculi leads to hypertrophy of the muscle. Attempts to resist the forced eye closure can cause detachment of the antagonist muscles: the levator palpebrae superioris, Mueller's muscle, and inferior lid retractors. Eye irritation, corneal abrasion, scarring, and permanent loss of vision can occur (Borodic and Cozzolino, 1989).

Like other focal dystonias, this disease begins in middle age. It may be associated with dystonia of the orofacial muscles (Meige's syndrome). Blepharospasm responds poorly to medication. Surgical facial nerve avulsion or stripping of the orbicularis oculi are used but can cause permanent weakness.

Botulinum toxin (10–20 U) is injected into multiple sites in the pretarsal orbicularis oculi bilaterally for treatment of blepharospasm. Injections are given laterally and medially into the superior portion of the muscle, and laterally only into the inferior portion. To prevent excessive weakening of the levator palpebrae, which could cause ptosis, the midportion of the upper lid is not injected. The medial inferior portion of the muscle is similarly avoided to prevent weakness of the nasolacrimal pump and the inferior oblique muscle.

Botulinum toxin is remarkably successful for blepharospasm, with 69–100% of patients improving with injection (Balkan and Poole, 1991; Borodic and Cozzolino, 1989; Brin et al., 1987; Elston, 1992; Jankovic et al., 1990), and the response can be maintained over 7 years with repeated injections (Elston, 1992; Mauriello and Aljian, 1991). Although not clearly related to muscle weakening, blink rate and photophobia can improve with treatment. Remission in blepharospasm has occasionally followed botulinum toxin injection (Mauriello and Aljian, 1991).

Apraxia of eyelid opening, impairment of voluntary eye opening, occurs in patients with Parkinson's disease and supranuclear palsy. Botulinum toxin is less effective in these patients than in those with typical blepharospasm (Brin, 1991; Elston, 1992).

Side effects of injection for blepharospasm include ptosis, tearing, lid edema, diplopia, entropion, ectropion, dry eyes, exposure keratitis, conjunctivitis, blurred vision, and pain or hematoma at the injection site (Borodic and Cozzolino, 1989; Brin, 1991; Elston, 1992; Jankovic et al., 1990). A single case of acute angle closure glaucoma precipitated by mydriasis due to botulinum toxin injection has been reported and suggests that botulinum toxin should be used with caution in patients with narrow angle glaucoma (Corridan et al., 1990).

Oromandibular Dystonia

Patients with oromandibular dystonia have complex spasms and movements of multiple facial muscles that cause pain and distorted facial expression. Blepharospasm, laryngeal dystonia, or torticollis frequently accompany this dystonia. Oromandibular dystonia is functionally and socially disabling. Forced mouth closure or opening impairs eating, talking, and oral hygiene.

Oromandibular dystonia requires the injection of multiple muscles such as the pterygoid, masseter, temporalis, platysma, or glossal muscles. Seventy percent of patients with jaw closure dystonia respond; botulinum toxin is less helpful for jaw opening (Clarke, 1992). Half of injections for tongue involvement are complicated by significant dysphagia, which can lead to aspiration (Brin, 1991).

Spasmodic Dysphonia

The laryngeal muscles are overly active in spasmodic dysphonia. Laryngeal adductor spasm causes a hoarse, strangled voice with pitch breaks and voiceless pauses. Laryngoscopy and EMG reveal increased activation of the thyroarytenoid and cricothyroid muscles during quiet respiration and phonation (Brin, 1991; Ludlow, 1990). Spasmodic dysphonia is generally refractory to medical management; unilateral section of the recurrent laryngeal nerve can relieve the symptoms, but respiratory complications are frequent and the symptoms typically recur after several years.

Adductor spasmodic dysphonia is treated by unilateral injection of 15–35 U or bilateral injection of 1.25–2.5 U of botulinum toxin into the thyroarytenoid (TA) muscle of the larynx. Injection into multiple sites of the TA is more effective than a single injection (Ludlow, 1990). The muscle can be reached periorally, but injection is more commonly given percutaneously with EMG guidance following local anesthesia of the skin of the neck. Almost all patients improve with injection and have fewer phonatory breaks, louder voice volume, increased sentence length, and easier speech production (Brin, 1991; Jankovic et al., 1990; Ludlow, 1990). Benefit from injection lasts 3–4 months, with the best voice present during the first 6 weeks following injection. Botulinum toxin can be used in patients who have undergone nerve resection but may not produce as much improvement as in nonoperated patients.

Side effects of TA injection include hematoma at the injection site, transient breathiness, decreased voice volume, and dysphagia for liquids. Aspiration may occur, but no cases of pneumonia have been reported. Even though local anesthesia is used, some patients have pain with injection that may be referred to the ipsilateral ear.

Abductor spasm, which causes a soft, breathy voice, can be treated with botulinum toxin injection into the posterior cricoarytenoid muscle. Treatment of this disorder is not as satisfactory as adductor spasm; only about one-half of patients respond. Side effects such as dysphagia and respiratory compromise are more common than from adductor injection.

Cervical Dystonia

Cervical dystonia often begins as a feeling of neck stiffness. Eventually, the head begins to turn and twist due to abnormal contraction of the neck muscles. The contraction may be sustained, distorting head position, or intermittent, causing jerky head movements or dystonic tremor. The head can be twisted in any direction. It may be pulled back (retrocollis), pulled forward (anterocollis), turned (torticollis), or tilted sideways (laterocollis). Shoulder elevation is frequently present. The most frequently involved muscles are the sternocleidomastoids, trapezii, scalenes, and platysma (Jankovic et al., 1991). The muscles may hypertrophy from the constant contraction.

Torticollis is the most common focal dystonia (Nutt et al., 1988). Women are more frequently affected than men. About 75% of patients have a specific gesture ("geste antagoniste"), such as lightly touching the chin or the back of the neck, which temporarily relieves the spasm (Jankovic et al., 1991).

Few therapies for cervical dystonia are successful. Cervical collars, biofeedback, or physical therapy are occasionally useful. Anticholinergic, benzodiazepine, or monoamine-depleting medications help about 50% of patients but usually require the use of such high doses that side effects become intolerable. Surgery for torticollis has included thalamotomy, cervical rhizotomy with section of the C1–C3 nerve roots, or myotomy. Although helpful in many patients, the benefit from surgery is not always sustained.

Treatment of cervical dystonia with botulinum toxin requires careful selection of the muscles for injection. EMG is useful both in selecting the muscles and in guiding injection (Comella et al., 1992a; Dubinsky et al., 1991). Since many large muscles are affected, large doses of botulinum toxin are needed. The total dose may reach 300–400 U divided among several muscles. Antibody formation is more likely in patients with torticollis than in those with other kinds of dystonia because of the need for such high doses.

Following botulinum toxin treatment 61–92% of patients have better head position and control (Brin et al., 1987; Greene et al., 1990, Jankovic et al., 1990; Lees et al., 1992; Poewe et al., 1992). A slightly higher percent of patients have relief of pain. Remission following injection has been reported (Poewe et al., 1992). Some patients actually have worse pain or spasm following treatment, which may be due to weakness of the treated muscles causing or unmasking excessive contraction in noninjected muscles.

Side effects of injection for torticollis include neck weakness with head drop, dysphagia, transient paresthesias, generalized weakness, nausea, rash, abdominal pain, hives, and hand swelling (Clarke, 1992; Jankovic et al., 1990, 1991). Lethargy and malaise have been reported and may be more common than in other focal dystonias because of the large doses used. Dysphagia is more frequent in women, possibly because of the smaller mass of the neck muscles in women than in men. It is especially likely following injections into the anterior neck muscles, the scalenes and sternocleidomastoids (Comella et al., 1992b).

Focal Hand Dystonia

Hand cramp was first described in the 18th century in writers, musicians, and craftsmen (Gowers, 1888; Ramazzini, 1713; Solly, 1864). It remains a common problem today (Nutt et al., 1988). Patients initially complain of hand or forearm tightness and fatigue while writing or performing fine motor tasks. Hand use becomes effortful; fluency of movement is lost. The hand and arm muscles contract excessively, causing abnormal movement and distortion of posture.

Hand cramp develops between the ages of 20 and 50. It may evolve from a "simple dystonia" that is elicited only by a specific task to "dystonic cramp" that is present during additional activities. Marsden found that 17% of patients ceased writing entirely (Marsden and Sheehy, 1990). Although symptoms often begin during a period of excessive hand use, the role overuse plays in precipitating dystonia is uncertain.

Patients with hand dystonia may have tremor, myoclonic jerks, decreased arm swing, or slight increase in tone of the affected arm (Jedynak et al., 1991; Sheehy and Marsden, 1982). Hand cramp is easily confused with other conditions such as nerve entrapment or muscle strain. Careful examination, however, will exclude sensory loss, weakness, muscle tenderness, and swelling, which are not found in dystonia. Although some hand cramp patients have aching or tightness in the hand or forearm, a primary complaint of pain should suggest another diagnosis.

Medications, splints, physical therapy, biofeedback, and peripheral nerve surgery do not help most patients with hand dystonia. Many patients with writer's cramp will try to write with the nonaffected hand. However, 25% of patients who switch hands will develop dystonia in the second hand (Marsden and Sheehy, 1990).

Botulinum toxin is most frequently injected into the finger and wrist extensors and flexors for the treatment of hand cramp. Some patients require injection of pronators, supinators, or muscles in the upper arm or shoulder as well. EMG is useful in both muscle selection and needle localization, since different combinations of abnormal muscle contraction can give rise to similar patterns of abnormal posture and the muscle bodies in the forearm are thin and overlapping. The dose of botulinum toxin needed depends on the number and size of muscles involved and can range from 2.5 U into a single finger fascicle to 100 U into multiple muscles.

Of patients with hand cramp, 67–93% improve with botulinum toxin injection (Cohen et al., 1989; Jankovic et al., 1990; Poungvarin, 1991; Rivest et al., 1991; Tsui et al., 1993; Yoshimura et al., 1992). The average duration of benefit is 3–4 months, but some patients require injection as infrequently as every 9 months. Some patients have been treated for as long as 6 years with continued response (Karp et al., 1994).

Excessive weakness of the injected or nearby muscles is the most common side effect of treatment. At times, this excessive weakness actually makes hand performance worse than it was prior to injection. When this occurs, a lower dose of botulinum toxin is used for the next injection.

Musicians may be especially prone to hand dystonia because of the long hours of intense practice required to master an instrument (Hochberg et al., 1990; Jankovic and Shale, 1989). Dystonia can destroy the ability to play professionally. Cole et al. treated 18 musicians with botulinum toxin (Cole et al., 1991). While 83% of patients improved with at least one injection, only two patients (11%) continued treatment. Most of the musicians found that the degree of benefit was not sufficient to allow continued professional performance. It is likely that, despite treating the spasm, botulinum toxin did not help the impaired coordination associated with dystonia.

Lower Limb Dystonia

Leg dystonia is frequently the presenting sign of idiopathic, generalized dystonia but may occur later in life as a focal dystonia or symptomatic of another illness, such as Parkinson's disease. The most common abnormal posture is inversion of the foot, which interferes with gait and balance. Some patients have spontaneous extension or flexion of the toes. Successful treatment of several patients with leg or foot dystonia has been reported (Brin et al., 1987; Jankovic et al., 1990). We have treated three patients with dystonias involving the lower limb. Two patients, one with generalized dystonia and one with Parkinson's disease, were treated for foot inversion. One patient had sustained abduction of the fifth toe, which caused great discomfort. The patient with Parkinson's disease did not improve with injection; the other two patients had moderate relief. The patient with generalized dystonia required the maximum allowable dose to achieve even partial improvement. There were no adverse effects from injection.

Segmental Dystonia

Segmental dystonia, the involvement of contiguous groups of muscles, can be treated with botulinum toxin. Successful treatment may require injection of a number of large, proximal muscles. As in generalized dystonia, treatment of segmental dystonia may be inadequate if the required toxin dose exceeds that which can be safely injected.

VI. OTHER APPLICATIONS

In this section we discuss disorders for which botulinum toxin has been tried in a small number or small series of patients. These studies reveal the wide range of possible therapeutic applications. Since only a few patients with each condition have been treated, further trials are needed before botulinum toxin can be recommended for these conditions.

A. Tremor

Tremor, rhythmic oscillating movement, is classified according to the eliciting situation as resting, postural, or kinetic tremor. Resting tremor is commonly a symptom of Parkinson's disease. Essential tremor, the most common type, is present during sustained posture or movement. Kinetic tremor is characteristic of cerebellar dysfunction. EMG studies show two general patterns of abnormal muscle activity in tremor: synchronous or alternating contraction of antagonist muscles.

There appears to be a pathogenetic relationship between dystonia and tremor. Twenty-five to forty-eight percent of dystonia patients also have tremor (Jedynak et al., 1991; Sheehy and Marsden, 1982). In some patients, the tremor is similar to essential tremor. In others, the tremor amplitude varies with position and may be brought out by the same actions that elicit dystonia ("dystonic tremor") (Jedynak et al., 1991; Rosenbaum and Jankovic, 1988). Family members of dystonic patients may similarly have essential or dystonic tremor (Cohen et al., 1987; Rosenbaum and Jankovic, 1988).

Medications successfully control tremor in many cases. Anticholinergic drugs or L-dopa treat resting tremor in Parkinson's disease. Beta-blockers, such as propranolol, are used for essential tremor. Thalamotomy can be very effective but is reserved for patients with debilitating tremor unresponsive to medication.

Following his observation that tremor improved following botulinum toxin injection for dystonia in patients with both conditions, Jankovic tried botulinum toxin to directly treat tremor (Jankovic and Schwartz, 1991). Patients with tremor of the hand or head received injections into antagonist muscles of the arm or neck. Two-thirds of these patients improved. Several patients with EMG studies after injection had complete absence of tremor bursts; others had decreased muscle burst amplitude. The pattern and frequency of EMG activity were not otherwise altered. Weakness, present after 8% of the injections, was well tolerated.

B. Palatal Tremor

Palatal tremor, previously called palatal myoclonus, consists of rhythmic contractions of the soft palate (Deuschl et al., 1990). The palatal movement occurs at rates of up to 260 per minute. Palatal tremor can be idiopathic (essential palatal tremor) or symptomatic of brainstem damage. In essential palatal tremor, the movement is often accompanied by an audible click that causes discomfort and interferes with sleep. There is presently no adequate treatment for this condition.

Deuschl et al. (1991) injected botulinum toxin into the tensor veli palatini muscle of a patient with palatal tremor. An initial right-sided injection followed 4 weeks later by injection of the left side of the palate eliminated audible, bilateral ear click for a 3-month period. Mild transnasal regurgitation of fluids was the only side effect of treatment.

C. Vocal Tremor

Ludlow et al. (1989) reported unilateral or bilateral botulinum toxin injection of the thyroarytenoid muscle in 13 patients with voice tremor. All patients had improved voice quality for approximately 3 months after injection.

D. Tardive Dyskinesia

Patients treated with phenothiazine medications such as neuroleptics or metoclopramide can develop a movement disorder, tardive dyskinesia (TD). Choreiform movements of the face and hands are typical of TD, but some patients develop other types of abnormal movements, including dystonia.

TD may improve if neuroleptics can be stopped quickly after the abnormal movements first appear. However, it is not possible to discontinue treatment in many patients with severe psychiatric illness. After longer treatment, the movements usually persist even if the medication is stopped. Although movements can be temporarily suppressed by raising the dose of the offending medication, there is no effective treatment for TD.

Five TD patients no longer taking neuroleptics received botulinum toxin injections (Stip et al., 1992; Truong et al., 1990). Although TD affected many muscles, only those causing the most distress were injected. The patients were treated for severe cervical dystonia (torticollis or retrocollis) and orofacial dystonia with forced mouth closure and bruxism. All five patients improved. Although the results should be viewed as preliminary, botulinum toxin appears useful in alleviating some disabling symptoms of tardive dyskinesia.

E. Spasticity

Spasticity, an abnormal increase in muscle tone due to corticospinal tract damage, can interfere with functional use of the limbs even if adequate strength is present. Three studies of botulinum toxin treatment of spasticity have been reported. Das and Park (1989) treated the arm and forearm muscles of six stroke patients with hemiplegia. All patients had better ease of movement, range of motion, and function for at least 3 months after a single injection. Snow et al. (1990) injected spastic leg adductor muscles in two patients with multiple sclerosis. Decreased spasticity resulted in better patient hygiene in both patients and allowed healing of a perineal ulcer in one case. Dengler et al. (1992) were similarly able to improve some aspects of spastic foot drop by botulinum toxin injection of calf muscles.

These results are promising and suggest that botulinum toxin may be useful in spasticity. Despite the generalized nature of spasticity, treatment was focused on an area of particular disability. While mobility and function may not improve in patients with severe spasticity, increased ease and quality of care is an important indirect gain.

F. Urinary Retention

In normal urination, the urethral sphincter must relax while the detrusor muscle in the bladder wall contracts to expel urine. Spasticity and loss of coordinated action of these muscles following spinal cord injury impairs bladder emptying. The resulting urinary retention can lead to repeated urinary tract infections, hydronephrosis, and renal damage. Management of detrusor-sphincter dyssynergia currently relies on catheterization, surgical ablation of the sphincter, or medications. None of these treatments is entirely successful or without adverse effects.

Dykstra et al. (1988) injected botulinum toxin percutaneously or cystoscopically into the urethral sphincter in 11 men who had suffered spinal cord injury 2–15 years earlier. Bladder emptying improved for 50 days in eight patients, with a decrease in both postvoid residual urine volume and urethral pressure profile. Normal urination was not restored; the patients still required urinary catheters. Patients with poor detrusor function before injection did not improve. Side effects of injection included hematuria and discomfort associated with cystoscopy.

Fowler et al. (1992) found no significant symptomatic improvement in six women with chronic urinary retention due to urethral sphincter spasm following botulinum toxin injection, even though weakness sufficient to cause stress incontinence was produced.

G. Anismus

Anal sphincter contraction must be inhibited to allow normal passage of stool. In anismus, sphincter inhibition is lost, leading to sustained contraction with pain and chronic constipation. Current therapies include long-term laxative administration and surgical sphincterotomy. As an alternative to these, Hallan et al. (1988) injected botulinum toxin into the puborectalis muscle, which forms part of the internal anal sphincter, to relieve anismus without loss of voluntary bowel control. Four of seven treated patients had an excellent response. There was one treatment failure, and two patients became temporarily incontinent due to excess sphincter laxity.

H. Stuttering

Stuttering is common in children but rarely persists past adolescence. Stutterers have involuntary prolongations, repetitions, and pauses in the production of speech sounds. Increased and inappropriate bursts of muscle activity are seen on EMG during stuttering. Ludlow and coworkers reported a trial of botulinum toxin injection of the intrinsic laryngeal muscles for stuttering (Ludlow, 1990; Stager and Ludlow, 1994). Although there were fewer word repetitions and longer periods of fluent speech, the number of sound prolongations was unchanged. Overall, patients did not consider injections beneficial and did not continue treatment.

I. Ophthalmic Uses Other Than for Strabismus

Botulinum toxin has been tried in oculomotor disorders other than strabismus. Newman and Lambert (1992) treated two patients whose brainstem strokes caused diplopia. One patient with bilateral internuclear ophthalmoplegia had permanent symptomatic relief from a single 2.5-U injection into a lateral rectus muscle. The second patient had permanent relief of skew deviation with a single 2.5-U injection into the inferior rectus muscle.

In nystagmus, rhythmic oscillations of the eyes interfere with visual fixation and cause oscillopsia, the illusion of environmental movement. Retroorbital and direct extraocular muscle botulinum toxin injections have been tried in attempts to improve vision in patients with acquired nystagmus. Although two studies reported successful treatment, a third study reported inadequate response in two patients who received unilateral injection (Crone et al., 1984; Helveston and Pogrebniak, 1988; Leigh et al., 1992).

Complete ptosis from oculomotor nerve palsy can be treated by using a portion of the frontalis muscle to replace the paralyzed levator palpebrae. A complication of this procedure is unwanted lid retraction during active facial expression. Borodic (1992) found botulinum toxin useful in eliminating excessive eye opening in two patients following frontalis sling.

Botulinum toxin has been used to evaluate lateral rectus function prior to strabismus surgery (Riordan-Eva and Lee, 1992), and temporary botulinum toxin-induced ptosis allowed healing of refractory corneal epithelial defects (Wuebbolt and Drummond, 1991).

J. Orthopedic Uses

Although most of the currently accepted uses for botulinum toxin require repeated injections for long-term benefit, some newer uses for the toxin take advantage of the temporary nature of the induced weakness.

Gasser used the toxin to transiently weaken the shoulder girdle muscles of a Parkinson's disease patient with L-dopa-induced segmental dystonia. A torn rotator cuff was repaired 2 weeks after injection. The temporary weakness not only allowed more thorough repair than had been possible prior to injection but also aided in immobilizing the limb to permit recovery after surgery (Gasser et al., 1991). Botulinum toxin was similarly used for postoperative neck immobilization in a torticollis patient who required cervical spinal fusion (Traynelis et al., 1992).

K. Miscellaneous Uses

Unilateral botulinum toxin injection for hemifacial spasm can produce undesirable facial asymmetry. Borodic et al. (1992) injected botulinum toxin into the unaffected facial muscles as well as those with spasm to restore facial symmetry. Aberrant reinnervation following facial nerve (Bell's) palsy can cause synkinesias of the facial muscles. Mountain et al. (1992) relieved synkinesias with botulinum toxin injection in four patients. Benign eyelid fasciculation (Chong et al., 1991) and facial myokymia (Ruusuvaara and Setala, 1990) have also been reported to respond to botulinum toxin. Bruxism associated with traumatic brain damage was successfully treated with bilateral masseter injections (Van Zandijcke and Marchau, 1990).

Nix et al. (1992) described two patients with an unusual syndrome of focal muscle hypertrophy with complex repetitive discharges on EMG. Muscle size decreased and pain lessened with botulinum toxin injection in both cases.

There have been anecdotal reports of botulinum toxin used to treat facial wrinkles, spasm associated with back pain, esophageal sphincter spasm, contraction headache, tics, and temperomandibular joint syndrome.

VII. CONCLUSIONS

Starting from its use to relieve strabismus, the role of botulinum toxin in treating human disorders has been expanding. It is of proven efficacy and safety for strabismus, blepharospasm, and hemifacial spasm and appears similarly useful in the treatment of other focal dystonias.

Botulinum toxin directly relieves muscle contraction or spasm in disorders such as dystonia, hemifacial spasm, spasticity, and sphincter spasm. Botulinum toxin-induced weakness permits easier operation and improved personal care, hygiene, and healing. In strabismus and some dystonia patients, permanent improvement is achieved with one or two injections. Other illnesses require repeated injections to maintain benefit. Botulinum toxin may be especially useful in spastic or dystonic patients who require surgery, when transient focal weakness is desired.

Future research should focus on methods of optimizing treatment outcome as well as broadening indications for botulinum toxin use. Ways should be sought to achieve a higher degree of improvement, more consistent response, better control of the duration of toxin effect, and avoidance of excessive weakness. These goals can be accomplished by better understanding of the physiology of muscle contraction, dystonia, and spasticity and of the physiologic effects of botulinum toxin injection.

REFERENCES

AAO (American Academy of Ophthalmology). (1989). Ophthalmic procedures assessment: botulinum toxin therapy of eye muscle disorders. *Ophthalmology 2*: 37–41.

Alderson, K., Holds, J. B., and Anderson, R. L. (1991). Botulinum-induced alterations of nerve-muscle interactions in the human orbicularis oculi following treatment for blepharospasm. *Neurology 41*: 1800–1805.

Angaut-Petit, D., Molgo, J., Comella, J. X., Faille, L., and Tabti, N. (1990). Terminal sprouting in mouse neuromuscular junctions poisoned with botulinum type A toxin: morphological and electrophysiological features. *Neuroscience 37*: 799–808.

Balkan, R. J., and Poole, T. (1991). A five-year analysis of botulinum toxin type A injections: some unusual features. *Ann. Ophthalmol. 23*: 326–333.

Black, J. D., and Dolly, J. O. (1987). Selective location of acceptors for botulinum neurotoxin A in the central and peripheral nervous systems. *Neuroscience 23*: 767–779.

Borodic, G. E. (1992). Botulinum A toxin for (expressionistic) ptosis overcorrection after frontalis sling. *Ophthalmol. Plast. Reconst. Surg. 8*: 137–142.

Borodic, G. E., and Cozzolino, D. (1989). Blepharospasm and its treatment, with emphasis on the use of botulinum toxin. *Plast. Reconstr. Surg. 83*: 546–554.

Borodic, G. E., and Ferrante, R. (1992). Effects of repeated botulinum toxin injections on orbicularis oculi muscle. *J. Clin. Neuroophthalmol. 12*: 121–127.

Borodic, G. E., Cheney, M., and McKenna, M. (1992). Contralateral injections of botulinum A toxin for the treatment of hemifacial spasm to achieve increased facial symmetry. *Plast. Reconstr. Surg. 90*: 972–979.

Brin, M. F. (1991). Interventional neurology: treatment of neurological conditions with local injection of botulinum toxin. *Arch. Neurobiol. 54*: 173–188.

Brin, M. F., Fahn, S., Moskowitz, C., Friedman, A., Shale, H., Greene, P. E., Blitzer, A., List, T., Lange, D., Lovelace, R. E., and McMahon, D. (1987). Localized injections of botulinum toxin for the treatment of focal dystonia and hemifacial spasm. *Movement Disord. 2*: 237–254.

Capra, N. F., Bernanke, J. M., and Porter, J. D. (1991). Ultrastructural changes in the masseter muscle of macaca fascicularis resulting from intramuscular injections of botulinum toxin type A. *Arch. Oral Biol. 36*: 827–836.

Charness, M. E., Barbaro, N. M., Olney, R. K., and Parry, G. J. (1987). Occupational cubital tunnel syndrome in instrumental musicians (abstract). *Neurology 37*(Suppl. 1): 115.

Chong, P. N., Ong, B., and Chan, R. (1991). Botulinum toxin in the treatment of facial dyskinesias. *Ann. Acad. Med. Singapore 20*: 223–227.

Clarke, C. E. (1992). Therapeutic potential of botulinum toxin in neurological disorders. *Q. J. Med. 82*: 197–205.

Cohen, L. G., Hallett, M., and Sudarsky, L. (1987). A single family with writer's cramp, essential tremor, and primary writing tremor. *Movement Disord. 2*: 109–116.

Cohen, L. G., Hallett, M. (1988). Hand cramps: Clinical features and electromyographic patterns in a focal dystonia. *Neurology 38:* 1005–1012.

Cohen, L. G., Hallett, M., Geller, B. D., and Hochberg, F. H. (1989). Treatment of focal dystonias of the hand with botulinum toxin injections. *J. Neurol. Neurosurg. Psychiatry 52*:: 355–363.

Cole, R. A., Cohen, L. G., and Hallett, M. (1991). Treatment of musician's cramp with botulinum toxin. *Med. Prob. Perform. Art 6*: 137–143.

Comella, C. L., Buchman, A. S., Tanner, C. M., Brown-Toms, N. C., and Goetz, C. G. (1992a). Botulinum toxin injection for spasmodic torticollis: increased magnitude of benefit with electromyographic assistance. *Neurology 42*: 878–882.

Comella, C. L., Tanner, C. M., DeFoor-Hill, L., and Smith, C. (1992b). Dysphagia after botulinum toxin injections for spasmodic torticollis: clinical and radiologic findings. *Neurology 42*: 1307–1310.

Corridan, P., Nightingale, S., Mashoudi, N., and Williams, A. C. (1990). Acute angle-closure glaucoma following botulinum toxin injection for blepharospasm. *Br. J. Ophthalmol. 74*: 309–310.

Crone, R. A. deJong, P. T. V. M., and Notermans, G. (1984). Behandlung des Nystagmus durch Injektion von Botulinustoxin in die Augenmuskeln. *Klin. Monatsbl. Augenheilkd. 184*: 216–217.

Das, T. K., and Park, D. M. (1989). Effect of treatment with botulinum toxin on spasticity. *Postgrad. Med. J. 65*: 208–210.

Dengler, R., Neyer, U., Wohlfarth, K., Bettig, U., and Janzik, H. H. (1992). Local botulinum toxin in the treatment of spastic foot drop. *J. Neurol. 239*: 375–378.

Deuschl, G., Mischke, G., Schenck, E., Schulte-Monting, J., and Lucking, C. H. (1990). Symptomatic and essential rhythmic palatal myoclonus. *Brain 113*: 1645–1672.

Deuschl, G., Lohle, E., Heinen, F., and Lucking, C. (1991). Ear clinic in palatal tremor: its origin and treatment with botulinum toxin. *Neurology 41*: 1677–1679.

Dezfulian, M., and Bartlett, J. G. (1984). Detection of *Clostridium botulinum* type A toxin by enzyme-linked immunosorbent assay with antibodies produced in immunologically tolerant animals. *J. Clin. Microbiol. 19*: 645–648.

Dubinsky, R. M., Gray, C. S., Vetere-Overfield, B., and Koller, W. C. (1991). Electromyographic guidance of botulinum toxin treatment in cervical dystonia. *Clin. Neuropharmacol. 14*: 262–267.

Duchen, L. W. (1970). Changes in motor innervation and cholinesterase localization induced by botulinum toxin in skeletal muscle of the mouse: differences between fast and slow muscles. *J. Neurol. Neurosurg. Psychiatry 33*: 40–54.

Duchen, L. W., and Stritch, S. J. (1968). The effects of botulinum toxin in the pattern of innervation of skeletal muscle in the mouse. *Q. J. Exp. Physiol. 53*: 84–89.

Dykstra, D. D., Sidi, A. A., Scott, A. B., Pagel, J. M., and Goldish, G. D. (1988). Effects of botulinum A toxin on detrusor-sphincter dyssynergia in spinal cord injury patients. *J. Urol. 139*: 919–922.

Elston, J. S. (1990). Botulinum toxin A in clinical medicine. *J. Physiol. Paris 84*: 285–289:

Elston, J. S. (1992). The management of blepharospasm and hemifacial spasm. *J. Neurol. 239*: 5–8.

Fowler, C. J., Betts, C. D., Christmas, T. J., Swash, M., and Fowler, C. G. (1992). Botulinum toxin in the treatment of chronic urinary retention in women. *Br. J. Urol. 70*: 387–389.

Garner, C. G. (1993). Time course of distant effects of local injections of botulinum toxin. *Movement Disord. 8*:

Gasser, T., Gritsch, K., Arnold, G., and Oertel, W. (1991). Botulinum toxin A in orthopaedic surgery. *Lancet 338*: 761.

Gelb, D. J., Yoshimura, D. M., Olney, R. K., Lowenstein, D. H., and Aminoff, M. J. (1991). Change in pattern of muscle activity following botulinum toxin injections for torticollis. *Ann. Neurol. 29*: 370–376.

Geller, B. D., Hallett, M., and Ravits, J. (1989). Botulinum toxin therapy in hemifacial spasm: clinical and electrophysiologic studies. *Muscle Nerve 12*: 716–722.

Girlanda, P., Vita, G., Micolosi, C., Milone, S., and Messina, C. (1992). Botulinum toxin therapy: distant effects on neuromuscular transmission and autonomic nervous system. *J. Neurol. Neurosurg. Psychiatry 55*: 844–845.

Glanzman, R. L., Gelb, D. J., Drury, I., Bromberg, M. B., and Truong, D. D. (1990). Brachial plexopathy after botulinum toxin injections. *Neurology 40*: 1143.

Gowers, W. R. (1888). *A Manual of Disease of the Nervous System,* Blakiston, Philadelphia.

Grafman, J., Cohen, L. G., and Hallett, M. (1991). Is focal hand dystonia associated with psychopathology? *Movement Disord. 6*: 29–35.

Greene, P., Kang, U., Fahn, S., Brin, M., Moskowitz, C., and Flaster, E. (1990). Double-blind, placebo-controlled trial of botulinum toxin injections for the treatment of spasmodic torticollis. *Neurology 40*: 1213–1218.

Hallan, R. I., Melling, J., Womack, N. R., Williams, N. S., Waldron, D. J., and Morrison, J. F. B. (1988). Treatment of anismus in intractable constipation with botulinum A toxin. *Lancet ii*: 714–717.

Hambleton, P. (1992). *Clostridium botulinum* toxins: a general review of involvement in disease, structure, mode of action and preparation for clinical use. *J. Neurol. 239*: 16–20.

Hambleton, P., Cohen, H. E., Palmer, B. J., and Melling, J. (1992). Antitoxins and botulinum toxin treatment. *Br. Med. J. 304*: 959–60.

Harris, C. P., Alderson, K., Nebeker, J., Holds, J. B., and Anderson, R. L. (1991). Histologic features of human orbicularis oculi treated with botulinum A toxin. *Arch. Ophthalmol. 109*: 393–395.

Haug, B. A., Dressler, D., and Prange, H. W. (1990). Polyradiculoneuritis following botulinum toxin therapy. *J. Neurol. 237*: 62–63.

Helveston, E. M., and Pogrebniak, A. E. (1988). Treatment of acquired nystagmus with botulinum A toxin. *Am. J. Ophthalmol. 106*: 584–586.

Herrero, B. A., Ecklund, A. E., Street, C. S., et al. (1967). Experimental botulism in monkeys—a clinical pathological study. *Exp. Mol. Pathol. 6*: 84–95.

Hochberg, F. H., Harris, S. U., and Blattert, T. R. (1990). Occupational hand cramps: professional disorders of motor control. *Hand Clin. 6*: 417–428.

Jahanshahi, M., and Marsden, C. D. (1992). Psychological functioning before and after treatment of torticollis with botulinum toxin. *J. Neurol. Neurosurg. Psychiatry 55*: 229–231.

Janetta, P. J., Abbasy, M., Maroon, J. C., Ramos, F. M., and Albin, M. S. (1977). Etiology and definitive microsurgical treatment of hemifacial spasm. *J. Neurosurg. 47*: 321–328.

Jankovic, J., and Brin, M. F. (1991). Therapeutic uses of botulinum toxin. *N. Engl. J. Med. 324*: 1186–1194.

Jankovic, J., and Schwartz, K. (1991). Botulinum toxin treatment of tremors. *Neurology 41*: 1185–1188.

Jankovic, J., and Shale, H. (1989). Dystonia in musicians. *Semin. Neurol. 9*: 131–135.

Jankovic, J., Schwartz, K., and Donovan, D. T. (1990). Botulinum toxin treatment of cranial-cervical dystonia, spasmodic dysphonia, other focal dystonias and hemifacial spasm. *J. Neurol. Neurosurg. Psychiatry 53*: 633–639.

Jankovic, J., Leder, S., Warner, D., and Schwartz, K. (1991). Cervical dystonia: clinical findings and associated movement disorders. *Neurology 41*: 1088–1091.

Jedynak, C. P., Bonnet, A. M., and Agid, Y. (1991). Tremor and idiopathic dystonia. *Movement Disord. 6*: 230–236.

Karp, B. I., Cohen, L. G., Cole, R., Grill, S., Lou, J. S., and Hallett, M. (1994). Long-term botulinum toxin treatment of focal hand dystonia. *Neurology 44*: 70–76.

Kramer, P. L., deLeon, D., Ozelius, L., Risch, N., Bressman, S. B., Brin, M. F., Schuback, D. E., Burke, R. E., Kwiatkowski, D. J., Shale, H., Gusella, J. F., Breakefield, X. O., and Fahn, S. (1990). Dystonia gene in Ashkenazi Jewish population is located on chromosome 9q32-34. *Ann. Neurol. 27*: 114–120.

Lange, D. J., Rubin, M., Greene, P. E., Kang, U. J., Moskowitz, C. B., Brin, M. F., Lovelace, R. E., and Fahn, S. (1991). Distant effects of locally injected botulinum toxin: a double-blind study of single fiber EMG changes. *Muscle Nerve 14*: 672–675.

Lee, J. (1992). Modern management of sixth nerve palsy. *Aust. N.Z. J. Ophthalmol. 20*: 41–46.

Lees, A. J., Turjanski, N., Rivest, J., Whurr, R., Lorch, M., and Brookes, G. (1992). Treatment of cervical dystonia, hand spasms and laryngeal dystonia with botulinum toxin. *J. Neurol. 239*: 1–4.

Leigh, R. J., Tomsak, R. L., Grant, M. P., Remler, B. F., Yaniglos, S. S., Lystad, L., and Dell'Osso, L. F. (1992). Effectiveness of botulinum toxin administered to abolish acquired nystagmus. *Ann. Neurol. 32*: 633–642.

Ludlow, C. L. (1990). Treatment of speech and voice disorders with botulinum toxin. *J. Am. Med. Assoc. 264*: 2671–2675.

Ludlow, C. L., Sedory, S. E., Fujita, M., and Naunton, R. F. (1989). Treatment of voice tremor with botulinum toxin injection (abstract). *Neurology 39*: 353.

Ludlow, C. L., Hallett, M., Rhew, K., Cole, R., Shimizu, T., Sakaguchi, G., Bagley, J. A., Schulz, G. M., Yin, S. G., and Koda, J. (1992). Therapeutic use of type F botulinum toxin [letter]. *N. Engl. J. Med. 326*: 349–50.

Marsden, C. D. (1986). The focal dystonias. *Clin. Neuropharmacol. 9*: S49–S60.

Marsden, C. D., and Sheehy, M. P. (1990). Writer's cramp. *TINS 13*: 148–153.

Mauriello, J. A., and Aljian, J. (1991). Natural history of treatment of facial dyskinesias with botulinum toxin: a study of 50 consecutive patients over seven years. *Br. J. Ophthalmol. 75*: 737–739.

Mountain, R. E., Murray, J. A., and Quaba, A. (1992). Management of facial synkinesis with *Clostridium botulinum* toxin injection. *Clin. Otolaryngol. 17*: 223–224.

Nakashima, K., Rothwell, J., Day, B., Thompson, P., Shannon, K., and Marsden, C. (1989). Reciprocal inhibition between forearm muscles in patients with writer's cramp and other occupational cramps, symptomatic hemidystonia and hemiparesis due to stroke. *Brain 112*: 681–697.

Newman, N. J., and Lambert, S. R. (1992). Botulinum toxin treatment of supranuclear ocular motility disorders. *Neurology 42*: 1391–1393.

Nix, W. A., Butler, I. J., Roontga, S., Gutmann, L., and Hopf, H. C. (1992). Persistent unilateral tibialis anterior muscle hypertrophy with complex repetitive discharges and myalgia. *Neurology 42*: 602–606.

Nutt, J. G., Muenter, M. D., Aronson, A., Kurland, L. T., and Melton, L. J. (1988). Epidemiology of focal and generalized dystonia in Rochester, Minn. *Movement Disord. 3*: 188–194.

Olney, R. K., Aminoff, M. J., Gelb, D. J., and Lowenstein, D. H. (1988). Neuromuscular effects distant from the site of botulinum neurotoxin injection. *Neurology 38*: 1780–1783.

Ozelius, L., Kramer, P. L., Moskowitz, C. B., Kwiatkowski, D. J., Brin, M. F., Bressman, S. B., Schuback, D. E., Falk, C. T., Risch, N., and de Leon, D. (1989). Human gene for torsion dystonia located on chromosome 9q32–34. *Neuron 2*: 1427–1434.

Panizza, M., Hallett, M., and Nilsson, J. (1989). Reciprocal inhibition in patients with hand cramps. *Neurology 39*: 85–89.

Poewe, W., Schelosky, L., Kleedorfer, B., Heinen, F., Wagner, M., and Geuschl, G. (1992). Treatment of spasmodic torticollis with local injections of botulinum toxin. *J. Neurol. 239*: 21–25.

Poungvarin, N. (1991). Writer's cramp: the experience with botulinum toxin injections in 25 patients. *J. Med. Assoc. Thai. 74*: 239–247.

Quinn, N., and Hallett, M. (1989). Dose standardisation of botulinum toxin. *Lancet i*: 964.

Ramazzini, B. (1713). Diseases of scribes and notaries. In *Diseases of Scribes and Notaries*, (Ed.), Hafner, New York, pp. 421–425.

Rhew, K., Ludlow, C. L., Karp, B. I., and Hallett, M. (1994). Clinical experience with botulinum toxin F. In *Therapy with Botulinum Toxin*, J. Jankovic and M. Hallett (Eds.), Marcel Dekker, New York, Chapter 23, pp. 323–328.

Riordan-Eva, P., and Lee, P. (1992). Management of VIth nerve palsy—avoiding unnecessary surgery. *Eye 6*: 386–390.

Rivest, J., Lees, A. J., and Marsden, C. D. (1991). Writer's cramp: treatment with botulinum toxin injections. *Movement Disord. 6*: 55–59.

Rosenbaum, F., and Jankovic, J. (1988). Focal task-specific tremor and dystonia: categorization of occupational movement disorders. *Neurology 38*: 522–527.

Ruusuvaara, P., and Setala, K. (1990). Long-term treatment of involuntary facial spasm using botulinum toxin. *Acta Ophthalmol. 68*: 331–338.

Saad, N., and Lee, J. (1992). The role of botulinum toxin in third nerve palsy. *Aust. N. Z. J. Ophthalmol. 20*: 121–127.

Schantz, E. J., and Johnson, E. A. (1990). Dose standardisation of botulinum toxin. *Lancet 335*: 421.

Schantz, E. J., and Johnson, E. A. (1992). Properties and use of botulinum toxin and other microbial neurotoxins in medicine. *Microbiol. Rev. 56*: 80–99.

Scherokman, B., Husain, F., Cuetter, A., Jabbari, B., and Maniglia, E. (1986). Peripheral dystonia. *Arch. Neurol. 43*: 830–832.

Scott, A. B. (1981). Botulinum toxin injection of eye muscles to correct strabismus. *Trans. Am. Ophthalmol. Soc. 79*: 734–770.

Scott, A. B. (1989). Clostridial toxins as therapeutic agents. In *Clostridial Toxins as Therapeutic Agents*, L. L. Simpson (Ed.), Academic, San Diego, pp. 399–412.

Scott, A. B., and Suzuki, D. (1988). Systemic toxicity of botulinum toxin by intramuscular injection in the monkey. *Movement Disord. 3*: 333–335.

Scott, A. B., Magoon, E. H., McNeer, K. W., and Stager, D. R. (1990). Botulinum treatment of childhood strabismus. *Ophthalmology 97*: 1434–1438.

Shaari, C. M., George, E., Wu, B. L., Biller, H. F., and Sanders, I. (1991). Quantifying the spread of botulinum toxin through muscle fascia. *Laryngoscope 101*: 960–964.

Sheehy, M. P., and Marsden, C. D. (1982). Writer's cramp—a focal dystonia. *Brain 105*: 461–480.

Siegel, L. S. (1988). Human immune response to botulinum pentavalent (ABCDE) toxoid determined by a neutralization test and by an enzyme-linked immunosorbent assay. *J. Clin. Microbiol. 26*: 2351–2356.

Siegel, L. S. (1989). Evaluation of neutralizing antibodies to Type A, B, E, and F botulinum toxins in sera from human recipients of botulinum pentavalent (ABCDE) toxoid. *J. Clin. Microbiol. 27*: 1906–1908.

Snow, B. J., Tsui, J. K. C., Bhatt, M. H., Varelas, M., Hashimoto, S. A., and Caine, D. B. (1990). Treatment of spasticity with botulinum toxin: a double-blind study. *Ann. Neurol. 28*: 512–515.

Solly, S. (1864). Scrivener's palsy, or the paralysis of writers. *Lancet 2*: 709–711.

Stager, S. V., and Ludlow, C. L. (1994). Responses of stutterers and vocal tremor patients to treatment with botulinum toxin. In *Therapy with Botulinum Toxin*, J. Jankovic and M. Hallett (Eds.), Marcel Dekker, New York, Chapter 37, pp. 481–490.

Stip, E., Faughnan, M., and Desjardin, I. (1992). Botulinum toxin in a case of severe tardive dyskinesia mixed with dystonia. *Br. J. Psychiatry 161*: 867–868.

Traynelis, V. C., Ryken, T., Rodnitzky, R. L., and Menezes, A. H. (1992). Botulinum toxin enhancement of postoperative immobilization in patients with cervical dystonia. Technical note. *J. Neurosurg. 77*: 808–809.

Truong, D. D., Hermanowicz, N., and Rontal, M. (1990). Botulinum toxin in treatment of tardive dyskinetic syndrome. *J. Clin. Psychopharmacol. 10*: 438–439.

Tsui, J. K., Wong, N. L. M., Wong, E., and Calne, D. B. (1988). Production of circulating antibodies to botulinum-A toxin in patients receiving repeated injections for dystonia (abstract). *Ann. Neurol. 23*: 181.

Tsui, J. K. C., Bhatt, M., Calne, S., and Calne, D. B. (1993). Botulinum toxin in the treatment of writer's cramp: a double blind study. *Neurology 43*: 183–185.

Valls-Sole, J., Tolosa, E., and Ribera, G. (1991). Neurophysiological observations on the effects of botulinum toxin treatment in patients with dystonic blepharospasm. *J. Neurol. Neurosurg. Psychiatry 54*: 310–313.

Van Zandijcke, M., and Marchau, M. M. B. (1990). Treatment of bruxism with botulinum toxin injection. *J. Neurol. Neurosurg. Psychiatry 53*: 530.

Waddy, H. M., Fletcher, N. A., Harding, A. E., and Marsden, C. D. (1991). A genetic study of idiopathic focal dystonias. *Ann. Neurol. 29*: 320–324.

Wiegand, H., Erdmann, G., and Wellhoner, H. H. (1976). 125-I-labelled botulinum A neurotoxin: pharmacokinetics in cats after intramuscular injection. *Naunyn Schmiedebergs Arch. Pharmacol. 292*: 161–165.

Wuebbolt, G. E., and Drummond, G. (1991). Temporary tarsorrhaphy induced with type A botulinum toxin. *Can. J. Ophthalmol. 26*: 383–385.

Yoshimura, D. M., Aminoff, M. J., and Olney, R. K. (1992). Botulinum toxin therapy for limb dystonias. *Neurology 42*: 627–630.

Zeman, W., Kaebling, R., and Pasaminick, R. (1960). Idiopathic dystonia musculorum deformans II. The formes frustes. *Neurology 10*: 1068–1075.

2

Diphtheria Toxin and Related Fusion Proteins: Autonomous Systems for the Delivery of Proteins and Peptides to the Cytosol of Eukaryotic Cells

John R. Murphy and Johanna vanderSpek

Boston University Medical Center Hospital, Boston, Massachusetts

Emmanuel Lemichez and Patrice Boquet

Institut Pasteur, Paris, France

I. INTRODUCTION

The specific delivery of proteins or peptides to the cytosol of target eukaryotic cells is now the subject of intense investigation. While the concept of targeted delivery of toxic materials to disease-causing cells dates back to the turn of the century (Erlich, 1956), it has been only in recent years that the confluence of cell biology, DNA, and protein technologies has allowed this goal to be realized. The application of recombinant DNA methodologies, DNA sequence analysis, X-ray crystal structure determination, protein engineering, and the understanding of the cell biology of receptor-mediated endocytosis and membrane trafficking has led to the design and genetic assembly of fusion toxins with extraordinary cytotoxic potency and selectivity. Indeed, the first of the cell surface receptor specific "designer" fusion proteins that is based upon the diphtheria toxin platform has been studied in phase I/II human clinical trials for the past several years. It is now clear that the administration of the interleukin-2 receptor (IL-2R)-targeted fusion toxin $DAB_{486}IL-2$ to patients is safe and may result in the induction of durable clinical remission in diseases ranging from hematologic malignancies to autoimmune disease (LeMaistre et al., 1992). Since this area has been well reviewed in recent years (Murphy and Strom, 1990; Murphy and Williams, 1991; Strom et al., 1993a,b), we focus this review on the emerging understanding of the molecular events involved in the diphtheria toxin mediated delivery of proteins and peptides into target cells.

One must begin with the fundamental question, Why is it important to deliver proteins or peptides into cells? The answer is simple and clear. The design of fusion proteins based on the diphtheria toxin structural platform will continue to provide important new reagents for (1) the cell surface receptor specific destruction of target cells, (2) the potential therapy of a wide array of genetic diseases by protein complementation, (3) the eliciting of specific MHC class I restricted immune response by the delivery of peptides to the cytosol of antigen-presenting cells, and (4) the study of fundamental questions in cell biology. Although these areas of investigation are new and exciting, it should be remembered that the targeted delivery of proteins across biological membranes is a normal cellular process. Membrane translocation, however, most often occurs cotranslationally from the inside (i.e., endoplasmic reticulum) to the outside of the cell. In addition, a number of polypeptides have been shown to be translocated across membranes posttranslationally. For example, the mechanism of protein transport into mitochondria has been extensively studied (Hartl and Neupert, 1990). This process involves the binding of a recognition sequence to a receptor on the surface of the organelle, unfolding of the protein, and delivery across the membrane and into the lumen of the organelle, where the polypeptide refolds into an active conformation. The delivery of these polypeptides into the mitochondria, peroxisome, or chloroplast requires a complex delivery apparatus consisting of several organelle-associated proteins. Protein toxins are examples of large proteins that are able to translocate across a biological membrane from the outside to the inside of the eukaryotic cell. They do so by following a mechanism almost identical to that evolved for cellular proteins that are targeted to intracellular organelles. However, protein toxins, like parasites, have evolved a unique molecular structure that facilitates the delivery of their active components across the eukaryotic cell membrane.

A.   Diphtheria Toxin

Diphtheria toxin (DT) is produced by toxigenic strains of *Corynebacterium diphtheriae*. Uchida et al. (1971) demonstrated by the isolation of mutants that directed the expression of nontoxic cross-reacting material (CRM) that the structural gene for diphtheria toxin was carried by corynebacteriophage β. Diphtheria toxin is synthesized in precursor form on membrane-bound polysomes (Smith et al., 1980). The mature form of the toxin is a single polypeptide chain of 535 amino acids in length (Kaczorek et al., 1983; Greenfield et al., 1983; Pappenheimer, 1977). It is widely known that DT is readily cleaved into two polypeptide fragments after mild digestion with trypsin or other serine proteases (Gill and Dinius, 1971; Drazin et al., 1971). After cleavage, the toxin may be separated under denaturing conditions in the presence of a thiol into two polypeptides. The N-terminal polypeptide (fragment A, M_r 21,167) is an enzyme that catalyzes the NAD$^+$-dependent ADP-ribosylation of elongation factor 2 (EF-2), thus blocking protein synthesis. The C-terminal polypeptide (fragment B, M_r 37,199) is required for binding the toxin to its cell surface receptor on sensitive cells and for facilitating the transport of fragment A into the cytosol (Pappenheimer, 1977). The nucleic acid sequences of the native diphtheria toxin structural genes from different strains of corynebacteriophage have been shown to be identical (Greenfield et al., 1983; Ratti et al., 1983).

Early biochemical and genetic studies clearly demonstrated that diphtheria toxin was structurally organized as a three-domain protein in which each domain played an essential role in the intoxication process (Gill and Pappenheimer, 1971; Drazin et al., 1971; Uchida et al., 1973; Boquet et al., 1976). The structural/functional organization of diphtheria toxin has been confirmed and greatly extended by the recent determination of its X-ray crystal structure at 2.5 Å resolution (Choe et al., 1992). In general, the mechanism of diphtheria toxin intoxication of a eukaryotic cell can be divided into three main steps: (1) the binding of the native toxin to its cell surface receptor, (2) the insertion of the channel forming helices into the membrane and the facilitated delivery of the catalytic domain to the cytosol, and (3) the inhibition of cellular protein synthesis. Remarkably, each of these steps is represented by a specific structural domain of the toxin (Choe et al., 1992). The receptor-binding domain (R), which is localized to the carboxy-terminal half of fragment B, has a flattened β barrel structure, the transmembrane domain (T) is composed of nine α helices, and the catalytic domain (C) corresponding to fragment A is composed of several α helices and β pleated sheet elements.

Since the structural and functional organization of the diphtheria toxin molecule reflects its mechanism of action, one might imagine that each domain (i.e., catalytic, transmembrane, and receptor binding) of this protein can be used in different combinations or replaced by other polypeptide domains in fusion proteins to create an almost infinite number of new molecules with novel properties. In this respect, diphtheria toxin with the help of recombinant DNA technology is comparable to the popular children's game Legos. This game consists of a small number of plastic pieces that can be joined to one another to build an almost infinite number of different constructions. We must point out, however, that unlike expert Lego players, who clip together their plastic pieces according to a set of instructions, with the "diphtheria toxin game" we are only amateurs who build fusion proteins of almost random structure, ever hopeful of attaining a structure that will be active and thereby add to our understanding of either the structure–function relationships of the toxin and the mechanism of cellular intoxication or serve as a probe for a fundamental aspect of cell biology.

We detail below those aspects that have been useful properties (i.e., *instructions*) in the different domains of diphtheria toxin in order to create new fusion proteins (i.e., *playing the game*) in the hope that others will become intrigued and join in.

B. Diphtheria Toxin Interaction with Sensitive Eukaryotic Cells: Instructions for Playing the Game

1. Role of the Receptor Binding Domain

The interaction between diphtheria toxin and sensitive eukaryotic cells follows a complex series of events, which ultimately lead to the irreversible inhibition of cellular protein synthesis and death of the cell. The first step in the intoxication process is the specific binding of the toxin molecule to its cell surface receptor. The apparent sensitivity of a given cell line appears to be directly related to the number of functional diphtheria toxin receptors (DTr) on the cell surface (Middlebrook et al., 1978).

What is the topology of the diphtheria toxin R domain that is involved in the binding of the toxin to its specific receptor? The initial localization of the receptor-binding domain within the carboxy-terminal portion of DT fragment B was based upon the findings that crm45, a nontoxic immunologically cross-reactive protein that lacks the C-terminal 149 amino acids (Giannini et al., 1984), failed to block the toxic activity of DT on cells (Uchida et al., 1971, 1973). A number of observations made with different experimental designs (e.g., monoclonal antibodies, hybrid toxins, point mutants of DT, chemical modification, and synthetic peptides) clearly indicate that the main structural motif implicated in the binding of DT with its cell surface receptor lies within the last 50 amino acids (Hayakawa et al., 1983; Murphy et al., 1986; Greenfield et al., 1987; Myers and Villemez, 1988; Rolf and Eidels, 1993) corresponding to three β strands (RB8, RB9, RB10) in the DT crystal structure (Choe et al., 1992).

African green monkey Vero cells express large numbers of diphtheria toxin receptors and are consequently exquisitely sensitive to the action of the toxin (Middlebrook et al., 1978). Recently, Naglich et al. (1982b) identified and reported the molecular cloning of the DT receptor from a Vero cell cDNA expression library. The toxin receptor has been shown to be a 20-kDa membrane protein whose amino acid sequence was identical to the heparin-binding epidermal growth factor–like precursor (Naglich et al., 1992b). Scatchard analysis of the binding of DT to Vero cells revealed that these cells have approximately 17,000 receptors per cell with a single class of high-affinity receptors with an apparent K_d of 1.2×10^{-9} M (Brown et al., 1993). When the cDNA encoding the DT receptor was stably transfected into DT-resistant mouse LM(TK$^-$) cells, they became sensitive to the toxin (DT^{S}II) (Naglich et al., 1992a). DT^{S}II cells were found to express approximately 10^6 DT receptors per cell with an apparent K_d of 1.5×10^{-8} M. Interestingly, these cells were not more sensitive to DT than Vero cells (Brown et al., 1993). It has been shown that in addition to the HB-EGF-like growth factor precursor, a 27-kDa cell membrane protein modulates DT binding to the cell surface (Iwamoto et al., 1991; Mitamura et al., 1992). This 27-kDa protein (DAP-27) is the monkey homologue of the human CD9 antigen (Boucheix et al., 1991; Lanza et al., 1991). The transfection of DT^{S}II cells with a plasmid encoding the CD9 antigen resulted in the isolation of a cell line that is hypersensitive to DT, DT^{S}III cells (Brown et al., 1993). Most interestingly, DT^{S}III cells were found to be tenfold more sensitive to DT than Vero cells even though the K_d for the toxin was tenfold lower than that found for the Vero receptor (Brown et al., 1993).

Apparently the role of CD9 in the process of diphtheria intoxication is to maintain a large number of DTr at the cell surface. As yet, there is no indication that there is an interaction between CD9 and DTr in either the presence or absence of diphtheria toxin (Brown et al., 1993; Rubinstein and Boucheix, personal communication). The process by which CD9 modulates the level of DTr on the cell surface remains unknown, as does the physiologic role played by CD9. The structure of CD9 with its five transmembrane-spanning domains could suggest that this protein is involved in transmembrane signaling, and the possible link between CD9 and a small G-protein has been reported in platelets (Seehafer and Shaw, 1991). One possible function for CD9 in elevating the number of DTr at the cell surface would be to control the repartition of HB-EGF-like growth factor from an intracellular pool to its delivery at the plasma membrane. Increasing the number of CD9 at the cell surface would lead to an enhanced detection of an external putative CD9 binding ligand and hence to an intracellular signal allowing the discharge of HB-EGF-like growth factor to the plasma membrane.

At neutral pH, diphtheria toxin interacts with negatively charged liposomes, and the interaction has been shown to involve both the A and B fragments (Montecucco et al., 1991). Accordingly, the removal of negatively charged phosphate groups by phospholipase C treatment of cells results in a decrease in the cytotoxicity of diphtheria toxin (Olsnes et al., 1985). On the basis of these studies, it was proposed that DT first binds to the negative moiety of the plasma membrane with low affinity and then moves laterally to encounter its specific receptor (Montecucco et al., 1991). At this stage DT would be bound to the cell surface by both lipid and protein interactions. According to this model there would be a contribution of lipids to the final value of the affinity between DT and DTr. It is possible that the difference in the DT-binding affinity between DT^{S}II and DT^{S}III cells versus Vero cells is related to the difference in phospholipid composition of mouse and monkey cells, especially with respect to their relative composition of negatively charged phospholipids. A close interaction between DT bound to its receptor and the surface of the phospholipid bilayer may be a crucial step for the rapid and efficient translocation of the C domain from DT through the membrane. This hypothesis is supported by the observations of Stenmark et al. (1988), which indicate that only DT that is bound to its receptor can be efficiently translocated to the cytosol. It is important to note, however, that either conjugate toxins prepared by cross-linking fragments of DT to targeting ligands such as transferrin (Gray Johnson et al., 1988; O'Keefe and Draper, 1985) or genetically linked fusion toxins such as DAB$_{486}$IL-2 (Williams et al., 1987; Bacha et al., 1988), which use an alternative receptor, lead to the assembly of highly potent molecules that apparently translocate through the cell membrane by the DT-specific mechanism. These observations suggest that the most likely role of DTr in the intoxication process is restricted to determining the concentration of DT bound to the surface of the cell and allowing for receptor-mediated endocytosis.

Melby et al. (1993) localized the DTr to a particular membrane domain on polarized epithelial cells. Diphtheria toxin inhibited protein synthesis most efficiently when it was applied to the basolateral side of MDCK, MBBK, PtK2, and Caco2 cells grown in a polarized fashion. In accordance with the toxicity data, DT bound specifically to the basolateral side of MDCK-1 cells but not to the apical side. This localization of DTr is in accordance with DTr being a growth factor precursor (Naglich et al., 1992b) that has to be released into the blood rather than to the luminal fluid to reach specific receptors. In addition, the basolateral localization of DTr may have some importance for the subsequent delivery of DT into a specific cellular compartment. Indeed, Bomsel et al. (1989) showed that receptor-mediated endocytosis of molecules from the basolateral side of polarized MDCK-1 cells results in preferent (77%) routing to late versus early endocytic compartments, whereas 90% of the ligand taken up from the apical side was recycled back (45%) or transcytosed (45%). These observations are in agreement with the finding that DTr does not recycle when DT is bound (Rönnberg and Middlebrook, 1989; Lemichez, unpublished observations).

Once bound to its cell surface receptor, DT in its intact form must be processed into an active di-chain protein through a mechanism known as "nicking." In vivo in the disease state, DT released from toxigenic *C. diphtheriae* are single-chain proteins of 535 amino acids in length. As shown in Figure 1, DT has an exposed serine protease-sensitive 14-amino-acid loop that is subtended by a disulfide bond between Cys-185 and Cys-201. DT purified from culture filtrates of *C. diphtheriae* is usually a mixture of intact and "nicked" (i.e., proteolytically cleaved at a site in the exposed loop) molecules in which the A and B fragments are held together by the disulfide bond and weak interactions. On cultivated cells the intact and "nicked" forms of the toxin are equally active (Middlebrook and Leatherman, 1982). Since the A fragment needs to be separated from the B fragment for delivery to the eukaryotic cell cytosol as well as to be enzymatically active, a protease must perform this cleavage. "Nicking" of the toxin molecule may occur either before the

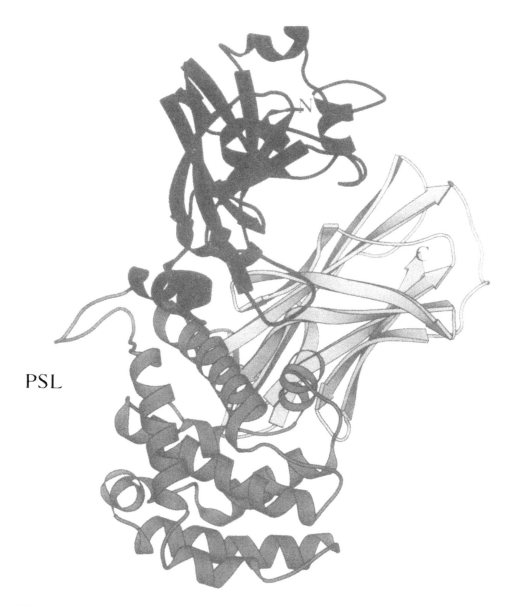

PSL

Fig. 1   Ribbon diagram of the X-ray crystal structure of native diphtheria toxin (Choe et al., 1992). The catalytic (dark gray), transmembrane (medium gray), and receptor binding (light gray) domains of the native toxin are shown. N, N-terminal end of diphtheria toxin; PSL, the exposed protease-sensitive loop; C, C-terminal end of diptheria toxin. The ribbon diagrams were generated using the software program MOLESCRIPT (Kraulis, 1991).

toxin binds to its cell surface receptor through the action of serum serine proteases or at the level of the cell surface. In either case, this reaction must occur before the translocation of the catalytic domain to the cytosol. The protease recognition sequence of DT, Arg-190–Val-191–Arg-192–Arg-193, has been shown to be a site for the endoprotease furin (Schalken et al., 1987; Barr, 1991). Further support for the endoproteolytic role of furin in the intoxication process comes from a site-directed mutational analysis of the fusion toxin $DAB_{486}IL$-2, in which destruction of the furin recognition site in the exposed loop between fragments A and B resulted in a marked loss of cytotoxic activity (Williams et al., 1990b). In addition, it has been shown that intact DT can be specifically nicked by furin in vitro (Klimpel et al., 1992). Tsuneoka et al. (1993) have shown that LoVo cells (a human carcinoma cell line), which do not produce furin, are not sensitive to intact diphtheria toxin. Upon transfection of LoVo cells with the structural gene for furin, the cells became sensitive to the intact toxin. Furin, a subtilisin-like protease like the KEX-2 protein of *Saccharomyces cerevisiae* (Barr, 1991) is localized mainly in the trans-Golgi and late endosomal cell compartments (Bresnahan et al., 1990). As we shall see, it now seems clear that DT facilitates the translocation of its C domain before reaching the late endocytic compartment. Thus, furin molecules or related proteases must be sent from the trans-Golgi to the early endosomal compartment and from there to the membrane in order to cleave intact DT. It has been recently shown that furin cycles from the trans-Golgi to the membrane (Molloy et al., 1994).

2. Role of the Transmembrane Domain

Once DT is bound to the cell surface receptor, it is internalized into the cells by receptor-mediated endocytosis into clathrin-coated pits (Moya et al., 1985; Morris et al., 1985). The $t_{1/2}$ for DT to be internalized into coated vesicles is unknown; however, it is anticipated that it should be close to the classical value of approximately 5 min as reported by Larkin et al. (1986) for other proteins that are internalized by the coated pit–coated vesicle pathway. Diphtheria toxin should then be located in a clathrin-uncoated vesicle and rapidly transported in a microtubule-independent fashion to a site of fusion with an acidic endosomal compartment defined as "early" (Schmid et al., 1988). Early endosomal compartments are acidified by specific vesicular ATPases. The pH value inside early endosomes is limited to an average of 6.2 by the vesicular uptake of the plasma membrane ouabain-sensitive Na/K-dependent ATPase, which pumps positive charges into the endosomal compartment (Fuchs et al., 1989; Cain et al., 1989).

The requirement of an acidic pH for the translocation of the C domain of DT is well documented (Kagan et al., 1981; Donovan et al., 1981; Draper and Simon, 1980; Sandvig and Olsnes, 1980). A pH of less than 6.0 is required to induce a conformational change in the T domain such that membrane insertion and translocation of the C domain occurs (reviewed by Montecucco et al., 1991; London, 1992). In fact, we do not know the exact pH at which the T domain changes from a hydrophilic to a hydrophobic membrane-inserting structure in vivo. Indeed, the values that have been reported in the literature have been obtained under nonphysiologic conditions either with the use of artificial planar lipid membranes (Montecucco et al., 1985; Blewitt et al., 1985; Dumont and Richards, 1988; Zhao and London, 1988; Defrise-Quertain et al., 1989) or by direct interaction of DT with the cell plasma membrane during an acidic pH pulse (Sandvig and Olsnes, 1986; Papini et al., 1988). The threshold acidic pH required for C-domain translocation into the cytosol is most likely to be between 5.5 and 4.8, with an average value of 5.3 (reviewed by Montecucco et al., 1991). It seems apparent that the requirement of a pH value of 5.3 excludes the entry of the C domain from an early endosomal compartment and suggests that the translocation step is likely to occur in a prelysosomal late endocytic compartment as suggested by London (1992).

An amino-terminal extension with a small peptide of 11 residues corresponding to the immunodominant epitope of the c-*myc* antigen (Evans et al., 1985) has been recently used to label DT (c-*myc*-DT) (Lemichez et al., unpublished). The N-terminal extension was not found to alter the toxicity of the labeled DT. By using Vero cells transfected with cDNA encoding DTr and c-*myc*-DT, the specific entry pathway of diphtheria toxin was followed by immunofluorescence microscopy after incubation of the cells with a monoclonal mouse anti-c-*myc* antibody followed by Texas-red labeled anti-rabbit antibodies. After 5 min at 37°C, the labeled DT was found associated with very small vesicles. After 15 and 20 min, the c-*myc*-DT was concentrated in large perinuclear compartments representing late endosomal/lysosomal structures.

Recent data clearly indicate that the C-domain translocation step occurs in an early endosomal compartment (Papini et al., 1993; Lemichez et al., unpublished). Indeed, Papini et al. (1993) showed by cell fractionation experiments that the C domain is released into the cytosol from an endosomal compartment that was identified by specific markers to be an early endocytic organelle. Lemichez et al. (unpublished) found that preventing the transfer of molecules from early to late endocytic compartments did not alter the cytotoxic activity of DT. As a result of these observations, we speculate that there is an early endosomal compartment that has an internal pH of less than 6.2, and it is only when DT molecules reach this subpopulation of vesicles that the translocation of the C domain occurs. The existence of early endosomal compartments with a pH value of less than 6.2 was demonstrated by Schmid et al. (1989) and Kielian et al. (1986) using mutant Semliki Forest virus having different pH values for their entry into the cell cytosol via the limiting membrane of the endosome. From these studies the existence of a small population (ca. 10%) of early endosomes with a pH of 5.3 was found. It is likely that only when DT encounters the membrane of this subpopulation of early endosomes does the T domain insert its hydrophobic channel, forming α helices into the phospholipid bilayer.

Most of the DT molecules that are internalized by cells appear to rapidly cross the early endosomal compartment without encountering a sufficiently low pH to insert into the membrane and as a result should be transferred to the late endosomal compartment (Gruenberg and Howell, 1989). In the prelysosomal/late endosomal compartment, before degradation by proteases, many DT molecules should translocate their C domain to the cytosol as the pH of this compartment is suitable for T domain insertion. Apparently, there is no translocation of the C domain from DT molecules in prelysosomal/late endocytic compartments that contribute to intoxication (Papini et al., 1993). In agreement with these observations, Lemichez et al. (unpublished) found by immunofluorescence microscopy that in Vero cells that have endocytosed c-*myc*-tagged DT molecules there is no colocalization of DT and rab7. rab7 has been shown to be a marker of the prelysosomal/late endocytic compartment (Chavrier et al., 1990). Indeed, as soon as 10 min after receptor binding, c-*myc*-DT molecules were found to be transferred to very large acidic perinuclear compartments. From these observations, we postulate that a large number of DT molecules that have not inserted into the membrane of very acidic early vesicles are transferred directly to lysosomes where degradation occurs.

What are the transmembrane domain conformational changes brought about by the lowering of the pH? It was recognized more than a decade ago, using planar lipid bilayers, that under acidic conditions diphtheria toxin will spontaneously insert into the plane of the membrane and form channels (Kagan et al., 1981; Donovan et al., 1981). Moreover, the diameter of these channels has been reported to be 18 Å (Kagan et al., 1981; Hoch et al., 1985), which is large enough for fragment A in extended form to pass through the channel and into the cytosol. The formation of channels by DT has also been observed in Vero and CHO cell membranes following a low-pH pulse (Papini et al., 1988; Sandvig and Olsnes, 1988).

The channel model of C domain delivery across the membrane is not clear. Several groups using model membrane systems have reported that upon lowering of the pH, the insertion of DT into the membrane resulted in both fragment B and fragment A sequences associating with the hydrophobic core of the phospholipid bilayer (Hu and Holmes, 1984; Zalman and Wisnieski, 1984; Montecucco et al., 1985; Papini et al., 1987; Zhao and London, 1988). It is important to note that these data show that the C domain does not translocate, at least in model membrane systems, through the hydrophilic environment of the channel that is formed by the insertion of the transmembrane helices of the T domain. While the C domain is not translocated in these model systems, it is clear that channel formation in the cell membrane is essential for intoxication. Indeed, mutations or deletions in the T domain that either alter or abolish channel formation activity have also been found to block cytotoxic activity (Stenmark et al., 1991, 1992; Falnes et al., 1992; Cabiaux et al., 1993; O'Keefe et al., 1992; vanderSpek et al., 1993, 1994).

In the presence of liposomes, the T domain of DT, at pH 4.0, interacts with the surface of the phospholipid bilayer with its N-terminal region (i.e., α helices TH1–TH5 according to Choe et al., 1992). The C-terminal region of the T domain (i.e., α helices TH6–TH9) interacts with the lipid core by inserting into the membrane forming transmembrane structures. Moskaug et al. (1991) have attempted to define the topology of DT fragment B following insertion into the plasma membrane of Vero cells by protease protection experiments. These experiments have demonstrated that DT residues 272–378 (i.e., the region containing TH5–TH9) were embedded into the plane of the membrane at low pH. Cabiaux et al. (1993) recently showed that these changes are accomplished by a reduction of the β-sheet content and an increase in the α-helix content of the T domain.

Thus, in both artificial and natural membranes the formation of transmembrane structures, most likely channels, seems to involve the insertion of T-domain α helices TH5–TH9. In support of this hypothesis, O'Keefe et al. (1992) found that the introduction of a site-directed mutation, Glu-349→K, in the short loop connecting TH8 and TH9 resulted in a decrease in the cytotoxic activity and channel-forming activity of the mutant toxin-related protein. These investigators proposed that the role of Glu-349, and possibly Asp-325, after protonation at an acidic pH would be to induce the insertion of the transmembrane helices of the T domain into the membrane. Further, after reaching the neutral pH environment of the cytosol, these residues would become deprotonated (i.e., charged) and thereby lock, or anchor, the transmembrane helices in the membrane. In an analogous fashion, Falnes et al. (1992) reported that the site-directed mutations Asp-295→K (which is positioned in the loop between TH5 and TH6) and Asp-318→K (which is positioned in the loop between TH7 and TH8) also gave rise to mutant forms of DT with reduced cytotoxic activity and impaired ability to form channels in lipid bilayers. In addition to the site-directed point mutations, vanderSpek et al. (1994a) demonstrated that an intact transmembrane helix 9 is required for the selective cytotoxic activity of the fusion toxin DAB$_{389}$IL-2. In-frame deletion mutants that lack an intact TH9 were also found to have a reduced ability to form stable channels in planar lipid membranes.

The amino-terminal portion of the T domain, which encompasses TH1–TH4, is amphipathic and has been shown to be structurally homologous to apolipoprotein A1 (Lambotte et al., 1980). The introduction of site-directed mutations in this portion of the

DT molecule results in the formation of mutants with a reduced ability to bind to the DTr (Stenmark et al., 1992). In the case of the DT-related fusion toxin $DAB_{389}IL-2$, the introduction of site-directed mutations into this region of the fusion toxin results in the formation of nontoxic mutants that both bind with high affinity to the IL-2 receptor and have unimpaired ability to form channels in planar lipid bilayers (vanderSpek et al., 1993). Indeed, since these mutants bound with high affinity to the receptor and were also capable of forming channels in artificial membranes, they defined a hitherto unknown step in the intoxication process.

We can now conclude that upon delivery to an acidic pH environment, the T domain of DT undergoes a conformational change that provokes channel formation in the endocytic vesicle membrane. Moreover, transmembrane helices TH5–TH9 appear to be involved in channel formation, and while channel formation is absolutely necessary for the cytotoxic activity of DT, it is not by itself sufficient for intoxication. While the role of the T domain in the intoxication process is becoming increasingly clear, there are still many unanswered questions regarding the facilitated delivery of the C domain to the cytosol of eukaryotic cells.

3. Role of the Catalytic Domain

At this time, relatively little is known of the molecular details involving the translocation of the catalytic (C) domain through the membrane and into the cytosol of eukaryotic cells. At least three major questions concerning this translocation need to be addressed:

1. Is unfolding of the C domain required for translocation?
2. Is the transfer of the C domain through the membrane initiated by its carboxy-terminal end or its amino-terminal end?
3. Does reduction of the disulfide bond between Cys-185 on fragment A and Cys-201 in fragment B occur before or after translocation of the C domain?

There are two lines of indirect evidence that suggest that the C domain must be unfolded in order to be translocated across the eukaryotic cell membrane. Wiedlocha et al. (1992) showed that the extension of the N terminus of DT with sequences corresponding to the acidic fibroblast growth factor (aFGF) gave rise to a fusion protein that was fully toxic for Vero cells. The addition of heparin, which is known to induce a tight folding of aFGF, to the aFGF-DT fusion protein resulted in complete loss of cytotoxic activity. While other interpretations are possible, these investigators reasoned that heparin-induced tight folding of the aFGF component of the fusion protein caused a block in the translocation of the modified C domain. In the second instance, Falnes et al. (1994) showed that the introduction of paired cysteine residues in the C domain, which were likely to form disulfide bonds, resulted the formation of mutant toxins with decreased cytotoxic activity. Thus, it appears that the C domain must be capable of extensive unfolding if it is to be translocated across the cell membrane.

As yet, there is little information regarding the initiation of C domain translocation into the cytosol. Indirect evidence suggests that the initiation of C domain delivery is mediated through the C-terminal end of fragment A. This evidence is based on the observation that the interchain disulfide bond is reduced very rapidly upon the insertion of fragment B into the membrane (Moskaug et al., 1987, 1989). In addition, relatively large proteins that have been genetically fused to the N terminus of fragment A can be translocated into the cytosol, apparently by the DT-specific mechanism (Madhus et al.,

1991; Wiedlocha et al., 1992; Aullo et al., 1993). At present, there does not appear to be a specific amino acid sequence in either the amino- or carboxy-terminal end of DT fragment A that is necessary for the initiation of C domain translocation into the cytosol (Ariansen et al., 1993; Aullo et al., 1993).

The precise cellular location of the reduction of the interchain disulfide bond between fragments A and B is also unknown. From the work of Moskaug et al. (1987) it appears that the reduction of the disulfide bond occurs either when the T domain becomes buried in the lipid core or when that bond is exposed to the cytosol, since only membrane-permeant sulfhydryl blockers were able to prevent the release of the C domain into the interior of the cell. Recently, Papini et al. (1993) obtained evidence that the reduction of the disulfide between fragments A and B occurs after the low-pH insertion of the T domain in an early endosomal compartment. Moreover, the reduction of the disulfide bond represents the rate-limiting step in the DT intoxication process. These results further suggest that the reduction of the disulfide bond is likely to occur either within the endosomal membrane or in the cytosol. As pointed out by Naglich et al. (1992a) and Papini et al. (1993), the transmembrane domain of the DTr contains a cysteine residue that may be involved in the reduction of the disulfide bond, provided that there are two receptors per DT molecule that are involved in the entry process.

Alternatively, it has been proposed that the reduction of the disulfide bond between fragments A and B could happen either at the level of the cell surface or in the lumen of the endocytic vesicle, as the intoxication of Vero cells by DT was blocked by membrane-impermeant sulfhydryl blockers (Ryser et al., 1991). Moreover, evidence that protein disulfide isomerase (PDI) plays a major role in the DT intoxication process has been recently reported (Mandel et al., 1993). It should be pointed out that the PDI is a resident protein of the endoplasmic reticulum (Freedman, 1989) and therefore may not be operative in the lumen of the early endosome where the entry of the C domain is initiated.

The NAD^+-dependent ADP-ribosyltransferase activity of DT fragment A, which ultimately leads to cell death, has been extensively reviewed (see Collier, 1990) and will not be discussed in this chapter.

C. Fusion Proteins Derived from Diphtheria Toxin: Playing the Game

Now that some of the structural and functional rules concerning the role played by each of the DT domains has been defined, we can begin to use the DT platform to assemble new fusion proteins either for the purpose of directing the toxicity intrinsic to the C domain to eukaryotic cells with unique receptors on their surface or to introduce proteins and peptides to the cytosol.

1. Diphtheria Toxin Receptor Binding Domain Substitution

While the first instance in which the native receptor-binding domain of DT was replaced genetically with a polypeptide hormone was with the construction of $DAB_{486}\alpha$-MSH (Murphy et al., 1986), most of our understanding of the DT-related fusion toxins comes from DAB_{486}IL-2 and DAB_{389}IL-2 (Williams et al., 1987, 1990; Bacha et al., 1988; Waters et al., 1990). The interleukin-2 receptor targeted fusion toxin DAB_{486}IL-2 has been the focus of many recent reviews (Strom et al., 1993a,b; vanderSpek et al., 1994; Murphy et al., 1992). As shown in Table 1, within the past 5 years the native DT receptor-binding domain has been successfully replaced with a number of growth factors,

Table 1 Diphtheria Toxin-Based Cell Receptor-Targeted Fusion Proteins

Fusion toxin	Receptor	IC_{50}	Reference
$DAB_{486}\alpha$-MSH	α-MSH	n.d.	Murphy et al., 1986
$DAB_{389}\alpha$-MSH	α-MSH	3×10^{-10} M	Wen et al., 1991
DAB_{486}IL-2	IL-2	1×10^{-11} M	Williams et al., 1987
DAB_{389}IL-2	IL-2	1×10^{-12} M	Kiyokawa et al., 1991
DAB_{389}mIL-4	IL-4	2×10^{-10} M	Lakkis et al., 1991
DAB_{389}IL-6	IL-6	2×10^{-11} M	Jean and Murphy, 1991
DAB_{389}EGF	EGF	1×10^{-11} M	Shaw et al., 1991
DAB_{389}CD4	HIV gp120	1×10^{-9} M	Aullo et al., 1992

giving rise to a family of fusion toxins that are selectively toxic toward eukaryotic cells that display the targeted receptor (Williams et al., 1987, 1990a; Aullo et al., 1992; Lakkis et al., 1991; Jean and Murphy, 1992; Wen et al., 1991; Shaw et al., 1991).

It is now clear that the most cytotoxic of the DT-based fusion proteins are made by replacing the native DT receptor-binding domain by the genetic fusion of ligand sequences at the end of the random coil that separates the T domain from the R domain (Fig. 2). Williams et al. (1990a) found that the genetic fusion of IL-2 sequences to amino acid 389 of mature DT gave rise to the second-generation form of this fusion toxin, which was tenfold more potent than the prototype DAB_{486}IL-2. A further two- to fivefold increase in the cytotoxic potency of the IL-2 receptor target fusion toxin could be made by the insertion of a 10-amino-acid flexible linker at the fusion junction between DT and IL-2 related sequences (Kiyokawa et al., 1991). In this study, the insertion of a 40-amino-acid flexible linker was found to give rise to a variant fusion toxin with decreased cytotoxic potency. More recently, the genetic insertion of an 11-amino-acid epitope from the VSV G protein (YTDIEMNRLGK; Kreis, 1986) between residue 389 of DT and the N-terminal end of IL-2 sequences was found to increase both the affinity of the modified DAB_{389}IL-2 fusion protein for the high-affinity form of the IL-2 receptor and its cytotoxic potency (vanderSpek et al., unpublished). As suggested by Kiyokawa et al. (1991), the random coil between the T domain and the substitute R domain in the fusion toxins might serve as a "hinge," which allows some degree of flexibility, which may be necessary for either the efficient insertion of the T domain into the membrane or the delivery of the C domain into the cytosol of target cells.

vanderSpek et al. (1994a) recently demonstrated that the maintenance of an intact transmembrane helix 9 (TH9) in the fusion toxins is essential for cytotoxic activity. The genetic fusion of IL-2 sequences to the middle of TH9 resulted in a fusion protein, $DAB(\Delta369\text{-}387)_{389}$IL-2, that bound to the IL-2 receptor with high affinity but was $\geq$2000-fold less cytotoxic and had a lower ability to form channels in planar lipid membranes than the parental DAB_{389}IL-2 form of the fusion toxin. Channel formation by the T domain in both the native DT and the DT-based fusion toxin is indispensable. As described above, although channel formation is essential for cytotoxic activity, channel formation itself is not sufficient, and the N-terminal region of the T domain plays an important role in facilitating the delivery of the C domain into the cytosol of target cells.

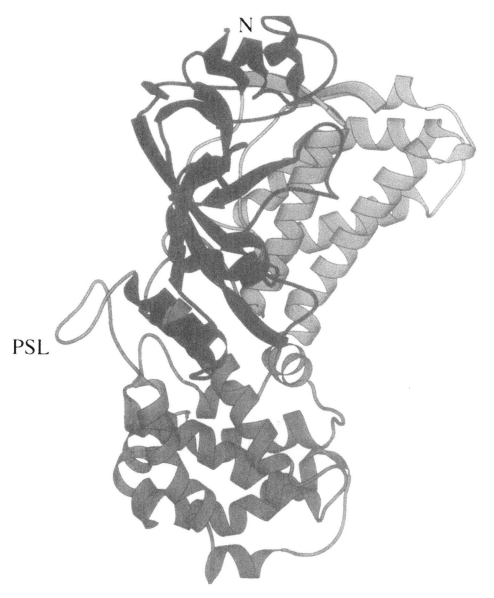

Fig. 2 Ribbon diagram of a molecular model of the interleukin-2 receptor-targeted fusion toxin DAB$_{389}$IL-2. The model was built with a portion of the X-ray crystal structure of native diphtheria toxin (Choe et al., 1992) and the corrected X-ray crystal structure of human IL-2 (McKay, 1992). Following the computational joining of the diphtheria toxin-related sequence with amino acids 2–133 of IL-2, the resulting structure was minimized using CHARMM software (Brooks et al., 1983). The catalytic (dark gray), transmembrane (medium gray), and receptor binding (light gray) domains of DAB$_{389}$IL-2 are shown. N, N-terminal end of the fusion toxin; PSL, the exposed protease-sensitive loop; C, C-terminal end of the fusion toxin. The ribbon diagrams were generated using the software program MOLESCRIPT (Kraulis, 1991).

2. *Clinical Evaluation of the IL-2 Receptor Targeted Fusion Toxin DAB₄₈₆ IL-2 in Hematologic Malignancies*

The high-affinity form of the IL-2 receptor is an attractive target for cytotoxic therapy in both cancer and autoimmune diseases (Waldmann, 1986, 1990). Although the high-affinity receptor for IL-2 is transiently expressed on T and B cells, it is not found on other normal tissue. More important, since the diphtheria toxin-based fusion proteins mediate their cytotoxic effect through the classic ADP-ribosylation of elongation factor 2, these chimeric proteins represent a new prototypical biologic agent for cancer therapy.

A series of phase I/II clinical studies were developed to determine the safety, tolerability, and pharmacokinetics of $DAB_{486}IL-2$ in patients with refractory hematologic malignancies (LeMaistre et al., 1992; Schwartz et al., 1992; Hesketh et al., 1993). This fusion toxin was the first single-chain chimeric toxic protein to be administered to humans. In order to be eligible for study entry, patients had to present with a hematologic malignancy that could not be treated effectively with either standard or, in some cases, experimental therapy. In addition, histologic confirmation of the malignancy and evidence of CD25 expression (i.e., p55, or α-chain of the IL-2 receptor) was also measured. Patients of either sex over the age of 17 years could be treated if they had adequate hepatic and renal function and a Karnofsky performance score of 70% or greater.

The studies were designed as a three-patient cohort dose escalation in which single and multiple doses were administered intravenously either as a bolus or as a 90-min infusion. The initial dose was 700 ng/kg per day and, in the absence of adverse reactions, was followed by three daily doses, followed 1 week later by seven daily doses. The dose was slowly escalated over eight cohort groups to 400 μg/kg per day. Importantly, patients who showed evidence of decreased tumor burden could be re-treated following a minimum of 4 weeks after the first day of therapy. At each dose level, patients were evaluated for adverse effects and response. Toxicities and responses were scored according to the criteria established by the National Cancer Institute. Patients with no evidence of disease for at least 4 weeks were classified as complete responders (CR); patients whose tumor burden decreased by $\geq$50% for at least 4 weeks were classified as partial responders (PR), and patients whose tumor burden decreased by 25–50% were classified as minor responders.

The intravenous administration of $DAB_{486}IL-2$ was found to be well tolerated at all dose levels in this group of patients with refractory malignancies. The adverse reactions were generally mild and included nausea/vomiting, hypersensitivity, fever/malaise/chills, and elevations in serum hepatic transaminases. Renal insufficiency defined the maximum tolerated dose at levels above 400 μg/kg per day. In all instances, the adverse effects were transient, not cumulative, and did not preclude repeated administration of the fusion toxin in this patient population. Most important, no changes in lymphocyte function assays (e.g., blastogenesis), lymphocyte subsets, or opportunistic infections were observed in any of the patients who were treated.

The time-course analysis of serum concentrations of biologically active $DAB_{486}IL-2$ following bolus administration, using a nonlinear open mathematical model, showed that the clearance of the fusion toxin most closely followed a one-component model with a $t_{1/2}$ of approximately 11 min at dose levels of 200, 250, 300, and 400 μg/kg (LeMaistre et al., 1993). The pharmacokinetics of the fusion toxin did not change in a consistent fashion following multiple courses of administration.

In many patients, circulating proteins that could potentially bind $DAB_{486}IL-2$ and thereby decrease its potential effectiveness through enhanced clearance were detected.

Increased serum levels of soluble IL-2 receptor (sIL-2R) were noted in a number of patients; however, there was no correlation between clearance rates of the fusion toxin in patients with high versus low levels of sIL-2R. Since the presence of up to 50,000 units of sIL-2R per milliliter failed to inhibit the cytotoxic action of DAB$_{486}$IL-2 in vitro (P. Bacha, personal communication), the appearance of elevated sIL-2R in serum was not expected to adversely affect the potential action of the fusion toxin in vivo. In addition, some patients had detectable levels of anti-diphtheria toxoid antibodies prior to entry into the clinical study. Following intravenous administration of DAB$_{486}$IL-2, approximately 60% of patients had an anamnestic response to the diphtheria toxin component of the fusion toxin and developed significant titers. In contrast, few patients had titers of anti-IL-2 antibodies in serum prior to study entry; however, after one or more courses of fusion toxin, approximately one-half of the patients developed a low level of anti-IL-2 antibody. As with sIL-2R, the presence of antibodies to either the diphtheria toxin-related or IL-2 components of the fusion toxin did not appear to prevent an antitumor response from being noted. Based upon the clinical experience with the monoclonal antibody OKT3 and the analysis of neutralizing and nonneutralizing antitoxin antibodies, it was not anticipated that preexisting anti-diphtheria toxoid antibodies would interfere with the action of the fusion toxin in vivo (Chatenoud et al., 1986; Bacha et al., 1992; Zucker and Murphy, 1983).

The results from the phase I/II studies of the intravenous administration of DAB$_{486}$IL-2 to patients with refractory hematologic malignancies has now proved in principle the feasibility of fusion toxin therapy in humans. Indeed, the cell surface receptor specific intoxication of neoplastic cells through the catalytic ADP-ribosylation of elongation factor 2 is the prototype of a new class of biologic response modifiers that may be generally applicable. In those circumstances where either the de novo expression or up-regulation of a cell surface receptor can be associated with human disease [e.g., the up-regulation of the epidermal growth factor (EGF) receptor on breast cancer], it should be possible to construct genetically a diphtheria toxin-related growth factor to produce an experimental biologic for the treatment of that malignancy. Indeed, the EGF receptor-targeted fusion toxin DAB$_{389}$EGF has within the last month begun human phase I clinical trials.

3. Diphtheria Toxin Facilitates the Delivery of Peptides and Proteins to the Eukaryotic Cell Cytosol

The ability of DT to deliver its C domain to the eukaryotic cell cytosol has prompted several groups to exploit this property to introduce heterologous proteins to the cytosol. Aullo et al. (1993) reported the first instance in which the native C domain of DT was genetically replaced with a heterologous enzyme. In this construct, Aullo et al. (1993) fused in-frame the gene encoding the *Clostridium botulinum* C3 transferase (Popoff et al., 1991) upstream of Cys-186 of native DT. The *C. botulinum* C3 transferase is an ADP-ribosyltransferase of 25 kDa that is specific for the small GTP-binding protein rho (Rubin et al., 1988; Aktories et al., 1987; Chardin et al., 1989). Aullo et al. (1993) show that although the resulting fusion protein, DC3B, was toxic for Vero cells, the mechanism of C3 entry into the cytosol was different from that used by native DT. These experiments suggest that fragment A of DT, or at least part of fragment A, is required for its efficient delivery into the eukaryotic cell cytosol.

The first attempt to transport peptides as "fusion passengers" of the C domain of DT was reported by Stenmark et al. (1991). These investigators clearly demonstrated that peptides as much as 30 amino acids in length could be delivered to the cytosol as long as they were fused to the N-terminal end of fragment A. The DT-mediated delivery of large proteins to the eukaryotic cell cytosol has been reported (Madhus et al., 1991; Wiedlocha et al., 1992; Aullo et al., 1993). Madhus et al. (1991) showed that the fusion of a second DT fragment A, as an N-terminal extension of the C domain, led to its specific translocation into the cytosol. In a similar fashion, Weidlocha et al. (1992) reported the specific delivery of acidic fibroblast growth factor to the cytosol following its fusion to the N terminus of fragment A of DT. As discussed above, the delivery of aFGF was accomplished as long as the growth factor was allowed to remain in a relaxed conformation. Aullo et al. (1993) showed that the fusion of *C. botulinum* C3 to either the N terminus of fragment A or amino acid 15 of fragment A resulted in the delivery of C3 activity to the cytosol by the DT-specific mechanism. Thus, in the cases that have been described to date, the delivery of peptides and large proteins to the eukaryotic cell cytosol can be accomplished by constructing a fusion gene in which the heterologous peptide is fused to the N-terminal end of DT fragment A. However, it is important to note that the fusion of heterologous peptides to the N terminus of fragment A invariably leads to a fusion protein that binds less avidly to the DTr.

It is clear that the use of DT-based fusion proteins to deliver heterologous peptides to the cytosol of target cells is in its infancy. We have much to learn of the precise molecular events that are required for the entry of the C domain. Nonetheless, this is an exciting area of molecular toxicology, and it will almost inevitably lead to the generation of new reagents for the study of fundamental aspects of cell biology as well as prototypes of new biological response modifiers that may have a therapeutic impact on human disorders that remain refractory to conventional therapies.

D. Conclusions

The considerable effort that has been expended during the past 15 years on the molecular genetics, structure and function, and cell biology of diphtheria toxin has allowed us to exploit this protein as either a "magic bullet" for the specific elimination of target cells or a "molecular syringe" for the introduction of heterologous peptides and proteins into the eukaryotic cell cytosol. Within a short time, by the genetic fusion of amino acid sequences to either the C-terminal or N-terminal end of DT-related proteins, we have seen the development of designer proteins whose unique properties have extended our knowledge of cell biology and hold the promise of a new class of human therapeutic proteins. In summary, the general rules for the genetic construction of these new fusion proteins are as follows:

1. Exchange of the DT-R domain can be made with many different ligands that bind to cell membrane receptors. The resulting fusion proteins have been found to be highly potent as long as they retain sufficient T-domain structure to facilitate the delivery of the C domain into the cytosol of their respective target cell. The optimal fusion junction has been found to be at amino acid 389 of recombinant DT. The insertion of a short flexible linker (i.e., 10–15 amino acid residues) between the T domain and the new receptor-binding moiety may give rise to a fusion toxin with increased receptor-binding affinity and cytotoxic potency.

2. The new binding site incorporated into the fusion toxin must bind to a cell surface receptor that is internalized into a cellular compartment that becomes acidified to a pH of 5.5 or lower.

3. Fusion proteins for the aim of introducing peptides or proteins into the eukaryotic cell cytosol should be made at the amino-terminal portion of the DT fragment A and should not contain internal disulfide bonds.
4. For maximal activity of the recombinant protein, the N-terminal extension of DT fragment A should not result in a fusion protein that binds with lower affinity than DT to the DTr.

Acknowledgments

We thank Larry Cosenza for preparing the ribbon diagrams of native diphtheria toxin and the fusion toxin DAB$_{389}$IL-2.

REFERENCES

Aktories, K., Weller, U., and Chatwal, G. S. (1987). *Clostridium botulinum* type C produces a novel ADP-ribosyltransferase distinct from botulinum C2 toxin. *FEBS Lett.* *212*:109–113.

Ariansen, S., Afanasieu, B. N., Moskaug, J. O., Stenmark, H., Madhus, I. H., and Olsnes, S. (1993). Membrane translocation of diphtheria toxin A-fragment: role of the carboxy-terminal region. *Biochemistry 32*:83–90.

Aullo, P., Alcani, J., Popoff, M. R., Klatzmann, D. R., Murphy, J. R., and Boquet, P. (1992). *In vitro* effects of a recombinant diphtheria–human CD4 fusion toxin on acute and chronically HIV-1 infected cells. *EMBO J.* *11*:575–583.

Aullo, P., Giry, M., Olsnes, S., Popoff, M. R., Kocks, C., and Boquet, P. (1993). A chimeric toxin to study the role of the 21 kDa GTP binding protein rho in the control of actin microfilament assembly. *EMBO J.* *12*:921–931.

Bacha, P., Waters, C., Williams, J., Murphy, J. R., and Strom, T. B. (1988). Interleukin-2 targeted cytotoxicity: selective action of a diphtheria toxin-related interleukin-2 fusion protein. *J. Exp. Med.* *167*:612–622.

Bacha, P., Forte, S. E., Perper, S. J., Trentham, D. E., and Nichols, J. C. (1992). Anti-arthritic effects demonstrated by an interleukin-2 receptor-targeted cytotoxin (DAB$_{486}$IL-2) in rat adjuvant arthritis. *Eur. J. Immunol.* *22*:1673–1679.

Barr, P. J. (1991). Mammalian subtilisins: the long sought dibasic processing endoproteases. *Cell 66*:1–3.

Blewitt, M. G., Chung, L. A., and London, E. (1985). Effect of pH on the conformation of diphtheria toxin and its implications for membrane penetration. *Biochemistry 24*:5458–5464.

Bomsel, M., Prydz, K., Parton, R. G., Gruenberg, J., and Simons, K. (1989). Endocytosis in filter-grown Madin-Darby canine kidney cells. *J. Cell. Biol.* *109*:3243–3258.

Boquet, P., Silverman, M. S., Pappenheimer, A. M., Jr., and Vernon, W. B. (1976). Binding of Triton X-100 to diphtheria toxin, crossreacting material 45, and their fragments. *Proc. Natl. Acad. Sci. U.S.A.* *73*:4449–4453.

Boucheix, C., Benoît, P., Frachet, P., Billard, M., Worthington, R. E., Gagnon, J., and Uzan, G. (1991). Molecular cloning of the CD9 antigen. *J. Biol. Chem.* *266*:117–122.

Bresnahan, P. A., Ledue, R., Thomas, L., Thorner, J., Gibson, H. L., Brake, A. J., Barr, P. J., and Thomas, G. (1990). Human fur gene encodes a yeast KEX2-like endoprotease that cleaves pro-beta-NGF *in vivo*. *J. Cell. Biol.* *111*:2851–2859.

Brooks, B. R., Bruccoleri, R. E., Olafson, B. D., States, D. J., Swaminathan, S., and Karplus, M. (1983). CHARMM: A program for macromolecular energy, minimization, and dynamics calculations. *J. Comp. Chem.* *4*:187–217.

Brown, J. G., Almond, B. D., Naglich, J. G., and Eidels, L. (1993). Hypersensitivity to diphtheria toxin by mouse cells expressing both diphtheria toxin receptor and CD9 antigen. *Proc. Natl. Acad. Sci. U.S.A.* *90*:8184–8188.

Cabiaux, V., Mindell, J., and Collier, R. J. (1993). Membrane translocation and channel-forming activities of diphtheria toxin are blocked by replacing isoleucine 364 with lysine. *Infect. Immun. 61*:2200–2202.

Cain, C. C., Sipe, D. M., and Murphy, R. F. (1989). Regulation of endocytic pH by the Na$^+$, K$^+$-ATPase in living cells. *Proc. Natl. Acad. Sci. U.S.A. 86*:544–548.

Chardin, P., Boquet, P., Madaule, P., Popoff, M. R., Rubin, E. J., and Gill, D. M. (1989). The mammalian protein rhoC is ADP-ribosylated by *Clostridium botulinum* exoenzyme C3 and affects actin microfilaments in Vero cells. *EMBO J. 8*:1087–1092.

Chatenoud, L., Jonker, M., Villemain, F., Goldstein, G., and Bach, J. F. (1986). The human immune response to the OKT3 monoclonal antibody is oligoclonal. *Science 232*:1406–1408.

Chavrier, P., Parton, R. G., Hauri, H. P., Simons, K., and Zerial, M. (1990). Localization of low molecular weight GTP binding proteins to exocytic and endocytic compartments. *Cell 62*:317–329.

Choe, S., Bennett, M. J., Fujii, G., Curmi, P. M. G., Kantardjieff, K. A., Collier, R. J., and Eisenberg, D. (1992). The crystal structure of diphtheria toxin. *Nature 357*:216–222.

Collier, R. J. (1990). Diphtheria toxin: structure function of a cytocidal protein. In *ADP-Ribosylating Toxins and G Proteins*, J. Moss, and M. Vaughan, (Eds.), American Society for Microbiology, Washington, DC, pp. 3–9.

Defrise-Quertain, E., Cabiaux, V., Vandenbranden, M., Wattiez, R., Falmagne, P., and Ruysschaert, J. M. (1989). pH-dependent bilayer destabilization and fusion of phospholipid large unilamellar vesicles induced by diphtheria toxin and fragments A and B. *Biochemistry 28*:3406–3413.

Donovan, J. J., Simon, M. I., Draper, R. K., and Montal, M. (1981). Diphtheria toxin forms transmembrane channels in planar lipid bilayers. *Proc. Natl. Acad. Sci. U.S.A. 78*:172–176.

Draper, R. K., and Simon, M. I. (1980). The entry of diphtheria toxin into the mammalian cell cytoplasm: evidence for lysosomal involvement. *J. Cell Biol. 87*:849–854.

Drazin, R., Kandel, J., and Collier, R. J. (1971). Structure and activity of diphtheria toxin. II. Attack by trypsin at a specific site within the intact toxin molecule. *J. Biol. Chem. 246*:1504–1510.

Dumont, M. E., and Richards, F. M. (1988). The pH-dependent conformational change of diphtheria toxin. *J. Biol. Chem. 263*:2087–2097.

Erlich, P. (1956). The relationship existing between chemical constitution, distribution, and pharmacological action. In *The Collected Papers of Paul Erlich*, Vol. I., F. Himmelweite, M. Marquardt, and Sir H. Dale (Eds.), Pergamon, New York, pp. 596–618.

Evans, G. I., Lewis, G. K., Ramsay, G., and Bishop, J. M. (1985). Isolation of monoclonal antibodies specific for human c-*myc* protooncogene product. *Mol. Cell. Biol. 5*:3610–3616.

Falnes, P. O., Madhus, I. H., Sandvig, K., and Olsnes, S. (1992). Replacement of negative by positive charges in the presumed membrane inserted part of diphtheria toxin B fragment: effect on membrane translocation and on formation of channels. *J. Biol. Chem. 267*:12284–12290.

Falnes, P. O., Choe, S., Madhus, I. H., Wilson, B. A., and Olsnes, S. (1994). Unfolding is required for membrane translocation of diphtheria toxin. *J. Biol. Chem.* (in press).

Freedman, R. B. (1989). Protein disulfide isomerase multiple role in the modification of nascent secretory proteins. *Cell 57*:1069–1072.

Fuchs, R., Schmid, S., and Mellman, I. (1989). A possible role for Na$^+$, K$^+$-ATPase in regulating ATP-dependent endosome acidification. *Proc. Natl. Acad. Sci. U.S.A. 86*:539–543.

Giannini, G., Rappuoli, R., and Ratti, G. (1984). The amino-acid sequence of two non-toxic mutants of diphtheria toxin: CRM45 and CRM197. *Nucleic Acids Res. 12*:4063–4069.

Gill, D. M., and Dinius, L. L. (1971). Observations on the structure of diphtheria toxin. *J. Biol. Chem. 246*:1485–1491.

Gill, D. M., and Pappenheimer, A. M., Jr. (1971). Structure-activity relationships in diphtheria toxin. *J. Biol. Chem. 246*:1492–1495.

Gray Johnson, V., Wilson, D., Greenfield, L., and Youlet, R. J. (1988). The role of the diphtheria toxin receptor in cytosol translocation. *J. Biol. Chem. 263*:1295–1300.

Greenfield, L., Bjorn, M. J., Horn, G., Fong, D., Buck, G. A., Collier, R. J., and Kaplan, D. (1983). Nucleotide sequence of the structural gene for diphtheria toxin carried by corynebacteriophage β. *Proc. Natl. Acad. Sci. U.S.A. 80*:6853–6857.

Greenfield, L., Johnson, V. G., and Youle, R. J. (1987). Mutations in diphtheria toxin separate binding from entry and amplify immunotoxin selectivity. *Science 238*:536–539.

Gruenberg, J., and Howell, K. E. (1989). Membrane traffic in endocytosis: insights from cell-free assays. *Annu. Rev. Cell. Biol. 5*:453–481.

Hartl, F. V., and Neupert, W. (1990). Protein sorting to mitochondria: evolutionary conservations of folding and assembly. *Science 247*:930–938.

Hayakawa, S., Uchida, T, Mekada, E., Moynihan, M. R., and Okada, Y. (1983). Monoclonal antibody against diphtheria toxin: effect on toxin binding and entry into cells. *J. Biol. Chem. 258*:4311–4317.

Hesketh, P., Caguioa, P., Koh, H., Dewey, H., Facada, A., McCaffrey, R., Parker, K., Nylen, P., and Woodworth, T. (1993). Clinical activity of a cytotoxic fusion protein in the treatment of cutaneous T cell lymphoma. *J. Clin. Oncol. 11*:1682–1690.

Hoch, D. H., Romero-Mira, M., Ehrich, B. E., Finkelstein, A., DasGupta, B. R., and Simpson, L. L. (1985). Channels formed by botulinum, tetanus, and diphtheria toxins in planar lipid bilayers: relevance to translocation of proteins across membranes. *Proc. Natl. Acad. Sci. U.S.A. 82*:336–343.

Hu, V. W., and Holmes, R. K. (1984). Evidence of direct insertion of fragments A and B of diphtheria toxin into model membranes. *J. Biol. Chem. 259*:12226–12233.

Iwamoto, R., Senoh, H., Okada, Y., Uchida, T., and Mekada, E. (1991). An antibody that inhibits the binding of diphtheria toxin to cells revealed the association of a 27-kDa membrane protein with the diphtheria toxin receptor. *J. Biol. Chem. 266*:20463–20469.

Jean, L.-F., and Murphy, J. R. (1992). Diphtheria toxin receptor binding domain substitution with interleukin-6: genetic construction and interleukin-6 receptor specific action of a diphtheria toxin-related interleukin-6 fusion protein. *Protein Eng. 4*:989–994.

Kaczorek, M., Delpeyroux, F., Chenciner, N., Streeck, R. E., Murphy, J. R., Boquet, P., and Tiollais, P. (1983). Nucleotide sequence and expression of the diphtheria *tox*228 gene in *Escherichia coli*. *Science 221*:855–858.

Kagan, B. L., Finkelstein, A., and Colombini, M. (1981). Diphtheria toxin fragment forms large pores in phospholipid bilayer membranes. *Proc. Natl. Acad. Sci. U.S.A. 78*:4950–4954.

Kielian, M. C., Marsh, M., and Helenius, A. (1986). Kinetics of endocytic acidification detected by mutant and wild type Semliki Forest virus. *EMBO J. 5*:3103–3109.

Kiyokawa, T., Williams, D. P., Snider, C. E., Strom, T. B., and Murphy, J. R. (1991). Protein engineering of diphtheria toxin-related interleukin-2 fusion toxins to increase biologic potency for high affinity interleukin-2 receptor bearing target cells. *Protein Eng. 4*:463–468.

Klimpel, K. R., Molloy, S. S., Thomas, G., and Leppla, S. (1992). Anthrax toxin protective antigen is activated by a cell surface protease with the sequence specificity and catalytic properties of furin. *Proc. Natl. Acad. Sci. U.S.A. 89*:10277–10281.

Kraulis, P. J. (1991). MOLESCRIPT: a program to produce both detailed and schematic plots of protein structures. *J. Appl. Crystallogr. 24*:946–950.

Kreis, T. E. (1986). Microinjected antibodies against the cytoplasmic domain of vesicular stomatitis virus glycoprotein blocks its transport to the cell surface. *EMBO J. 5*:931–941.

Lakkis, F., Steele, A., Pacheco-Silva, A., Kelley, V. E., Strom, T. B., and Murphy, J. R. (1991). Interleukin-4 receptor targeted cytotoxicity: genetic construction and properties of diphtheria toxin-related interleukin-4 fusion toxins. *Eur. J. Immunol. 21*:2253–2258.

Lambotte, P., Falmagne, P., Capiau, C., Zanen, J., Ruysschaert, J.-M., and Dirkx, J. (1980). Primary structure of diphtheria toxin fragment B: structural similarities with lipid-binding domains. *J. Cell. Biol. 87*:837–840.

Lanza, F., Wolf, D., Fox, C. F., Kieffer, N., Seyer, J. M., Fried, V. A., Coughlin, S. R., Phillips, D. R., and Gennings, L. K. (1991). cDNA cloning and expression of platelet p24/CD9. *J. Biol. Chem. 266*:10638–10645.

Larkin, J. M., Donzell, W. C., and Anderson, R. G. W. (1986). Potassium-dependent assembly of coated pits: new coated pits form as planar clathrin lattices. *J. Cell. Biol. 103*:2619–2627.

LeMaistre, C. F., Meneghetti, C., Rosenblum, M., Reuben, J., Parker, K., Shaw, J., Woodworth, T., and Parkinson, D. (1992). Phase I trial of an interleukin-2 receptor (IL-2R) fusion toxin (DAB$_{486}$IL-2) in hematologic malignancies expressing the IL-2 receptor. *Blood 79*:2547–2554.

LeMaistre, C. F., Craig, F. E., Meneghetti, C., McMullin, B., Parker, K., Reuben, J., Boldt, D. H., Rosenblum, M., and Woodworth, T. (1993). Phase I trial of a 90-minute infusion of the fusion toxin DAB$_{486}$IL-2 in hematological cancers. *Cancer Res. 53*:3930–3934.

London, E. (1992). Diphtheria toxin: membrane interaction and membrane translocation. *Biochim. Biophys. Acta 1113*:25–51.

Madhaus, I. H., Stenmark, H., Sandvig, K., and Olsnes, S. (1991). Entry of diphtheria toxin-protein A chimera into cells. *J. Biol. Chem. 266*:17446–17453.

Mandel, R., Ryser, H.-P., Ghani, F., Wu, M., and Peak, D. (1993). Inhibition of a reductive function of the plasma membrane by bacitracin and antibodies against protein disulfide-isomerase. *Proc. Natl. Acad. Sci. U.S.A. 90*:4112–4116.

Melby, E. L., Jacobsen, J., Olsnes, S., and Sandvig, K. (1993). Entry of protein toxins in polarized epithelial cells. *Cancer Res. 53*:1755–1780.

Middlebrook, J. L., and Leatherman, D. L. (1982). Differential sensitivity of reticulocytes to nicked and unnicked diphtheria toxin. *Exp. Cell Res. 138*:175–182.

Middlebrook, J. L., Dorland, R. B., and Leppla, S. H. (1978). Association of diphtheria toxin with Vero cells: demonstration of a receptor. *J. Biol. Chem. 253*:7325–7330.

Mitamura, T., Iwamoto, R., Umata, T., Yomo, T., Urabe, I., Tsuneoka, M., and Mekada, E. (1992). The 27-kD diphtheria toxin receptor-associated protein (DRAP27) from Vero cells is the monkey homologue of human CD9 antigen: expression of DRAP27 elevates the number of diphtheria toxin receptors on toxin-sensitive cells. *J. Cell Biol. 118*:1389–1399.

Molloy, S. S., Thomas, L., van Slyke, J. K., Stenberg, P. E., and Thomas, G. (1994). Intracellular trafficking and activation of the furin proprotein convertase: localization to the TGN. *EMBO J. 13*:18–33.

Montecucco, C., Schiavo, G., and Tomasi, M. (1985). pH-dependence of the phospholipid interaction of diphtheria toxin. *Biochem. J. 231*:123–128.

Montecucco, C., Papini, E., and Schiavo, G. (1991). Molecular models of toxin membrane translocation. In *Sourcebook of Bacterial Protein Toxins*, J. Alouf, and J. Freer (Eds.), Academic, London, pp. 45–56.

Morris, R. E., Gerstein, A. S., Bonventre, P. F., and Saelinger, C. B. (1985). Receptor-mediated entry of diphtheria toxin into monkey kidney (Vero) cells: electron microscopic evaluation. *Infect. Immun. 50*:721–727.

Moskaug, J. O., Sandvig, K., and Olsnes, S. (1987). Cell-mediated reduction of the interfragment in nicked diphtheria toxin. *J. Biol. Chem. 262*:10339–10345.

Moskaug, J. O., Sandvig, K., and Olsnes, S. (1989). Role of anions in low-pH-induced translocation of diphtheria toxin. *J. Biol. Chem. 264*:11367–11372.

Moskaug, J. O., Stenmark, H., and Olsnes, S. (1991). Insertion of diphtheria toxin B-fragment into the plasma membrane at low pH. *J. Biol. Chem. 266*:2652–2659.

Moya, M., Dautry-Versat, A., Goud, B., Louvard, D., and Boquet, P. (1985). Inhibition of coated pit formation in Hep$_2$ cells blocks the cytotoxicity of diphtheria toxin but not that of ricin toxin. *J. Cell Biol. 101*:548–559.

Murphy, J. R., and Strom, T. B. (1990). Diphtheria toxin-peptide hormone fusion proteins: protein engineering and selective action of a new class of recombinant biological response modifiers. In *ADP-Ribosylating Toxins and G Proteins*, J. Moss, and M. Vaughan (Eds.), American Society for Microbiology, Washington, DC, pp. 141–160.

Murphy, J. R., and Williams, D. P. (1991). Protein engineering of microbial toxins: design and properties of novel fusion proteins with therapeutic and immunogenic potential. In *Structure, Regulation and Activity of Bacterial Toxins*, J. E. Alouf and J. H. Freer (Eds.), Academic, London, pp. 491–505.

Murphy, J. R., Bishai, W., Borowski, M., Miyanohara, A., Boyd, J., and Nagle, S. (1986). Genetic construction, expression, and melanoma-selective cytotoxicity of a diphtheria toxin α-melanocyte stimulating hormone fusion protein. *Proc. Natl. Acad. Sci. U.S.A. 83*:8258–8262.

Murphy, J. R., Lakkis, F., vanderSpek, J., Anderson, P., and Strom, T. B. (1992). Protein engineering of diphtheria toxin: development of receptor-specific cytotoxic agents for the treatment of human disease. In *Immunotoxins*, Vol. II, A. Frankel (Ed.), Martinus Nijhoff, Boston, MA, pp. 365–382.

Myers, D. A., and Villemez, C. L. (1988). Specific chemical cleavage of diphtheria toxin with hydroxylamine: purification and characterization of the modified proteins. *J. Biol. Chem. 263*:17122–17127.

Naglich, J. G., Rolf, J. M., and Eidels, L. (1992a). Expression of functional diphtheria toxin receptors on highly toxin-sensitive mouse cells that specifically bind radioiodinated toxin. *Proc. Natl. Acad. Sci. U.S.A. 89*:2170–2174.

Naglich, J. G., Metherall, J. E., Russell, D. W., and Eidels, L. (1992b). Expression cloning of a diphtheria toxin receptor: identity with a heparin-binding EGF-like growth factor precursor. *Cell 69*:1051–1061.

O'Keefe, D. O., and Draper, R. K. (1985). Characterization of a transferrin-diphtheria toxin conjugate. *J. Biol. Chem. 260*:932–937.

O'Keefe, D. O., Cabiaux, V., Choe, S., Eisenberg, D., and Collier, R. J. (1992). pH-dependent insertion of proteins into membranes: B-chain mutation of diphtheria toxin that inhibits membrane translocation, Glu349 → Lys. *Proc. Natl. Acad. Sci. U.S.A. 89*:6202–6206.

Olsnes, S., Carvajal, E., Sundan, A., and Sandvig, K. (1985). Evidence that membrane phospholipids and protein are required for binding of diphtheria toxin in Vero cells. *Biochim. Biophys. Acta 846*:334–341.

Papini, E., Schiavo, G., Tomasi, M., Colombatti, M., Rappuoli, R., and Montecucco, C. (1987). Lipid interaction of diphtheria toxin and mutants with altered fragment B. 2. Hydrophobic photolabeling and cell intoxication. *Eur. J. Biochem. 169*:637–644.

Papini, E., Sandoná, D., Rappuoli, R., and Montecucco, C. (1988). On the membrane translocation of diphtheria toxin: at low pH the toxin induces ion channels in cells. *EMBO J. 7*:3353–3359.

Papini, E., Rappuoli, R., Murgia, M., and Montecucco, C. (1993). Cell penetration of diphtheria toxin. Reduction of the interchain disulfide bridge is the rate-limiting step of translocation into the cytosol. *J. Biol. Chem. 268*:1567–1574.

Pappenheimer, A. M., Jr. (1977). Diphtheria toxin. *Annu. Rev. Biochem. 46*:69–94.

Popoff, M. R., Hauser, D., Boquet, P., Eklund, M. W., and Gill, D. M. (1991). Characterization of the C3 gene of *Clostridium botulinum* types C and D and its expression in *Escherichia coli*. *Infect. Immun. 59*:3673–3679.

Ratti, G., Rappuoli, R., and Giannini, G. (1983). The complete nucleotide sequence of the gene coding for diphtheria toxin in the corynephage omega (*tox+*) genome. *Nucleic Acids Res. 11*:6589–6595.

Rolf, J. M., and Eidels, L. (1993). Structure-function analyses of diphtheria toxin by use of monoclonal antibodies. *Infect. Immun. 61*:994–1003.

Rönnberg, B. J., and Middlebrook, J. L. (1989). Cellular regulation of diphtheria toxin cell surface receptors. *Toxicon 27*:1377–1388.

Rubin, E. J., Gill, D. M., Boquet, P., and Popoff, M. R. (1988). Modification of a 21-kilodalton G protein when ADP-ribosylated by C3 of *Clostridium botulinum*. *Mol. Cell. Biol. 8*:418–426.

Ryser, H.-J., Mandel, R., and Ghani, F. (1991). Cell surface sulfhydryls are required for the cytotoxicity of diphtheria toxin but not of ricin in Chinese hamster ovary cells. *J. Biol. Chem. 266*:18439–18442.

Sandvig, K., and Olsnes, S. (1980). Diphtheria toxin entry into cells is facilitated by low pH. *J. Biol. Chem. 245*:828–832.

Sandvig, K., and Olsnes, S. (1986). Interactions between diphtheria toxin entry and anion transport in Vero cells. IV. Evidence that entry of diphtheria toxin is dependent on efficient anion transport. *J. Biol. Chem. 261*:1570–1575.

Sandvig, K., and Olsnes, S. (1988). Diphtheria toxin-induced channels in Vero cells selective for monovalent cations. *J. Biol. Chem. 263*:12352–12359.

Schalken, J. A., Roebuck, A. J. M., Oomen, P. P. C. A., Wagenaar, S. S., Debruyne, F. M., Bloemers, H. P. J., and van de Ven, W. J. M. (1987). Furin gene expression as a discriminating marker for small cell and non small cell lung carcinomas. *J. Clin. Invest. 8*:1545–1549.

Schmid, S., Fuchs, R., Male, P., and Mellman, I. (1988). Two distinct subpopulations of endosomes involved in membrane recycling and transport to lysosomes. *Cell 52*:73–83.

Schmid, S., Fuchs, R., Kielian, M., Helenius, A., and Mellman, I. (1989). Acidification of endosome subpopulations in wild-type Chinese hamster ovary cells and temperature sensitive acidification-defective mutants. *J. Cell. Biol. 108*:1291–1300.

Schwartz, G., Tepler, I., Charette, J., Kadin, L., Parker, K., Woodworth, T., and Schnipper, L. (1992). Complete response of a Hodgkin's lymphoma in a phase I trial of DAB$_{486}$IL-2. *Blood 79*:175a.

Seehafer, J. G., and Shaw, A. R. E. (1991). Evidence that the signal-initiating membrane protein CD9 is associated with small GTP-binding proteins. *Biochem. Biophys. Res. Commun. 179*:401–406.

Shaw, J. P., Akiyoshi, D. E., Arrigo, D. A., Rhoad, A. E., Sullivan, B., Thomas, J., Genbauffe, F. S., Bacha, P., and Nichols, J. C. (1991). Cytotoxic properties of DAB$_{486}$EGF and DAB$_{389}$EGF, epidermal growth factor (EGF) receptor-targeted fusion toxins. *J. Biol. Chem. 266*:21118–21124.

Smith, W. P., Tai, P. C., Murphy, J. R., and Davis, B. D. (1980). A precursor in the cotranslational secretion of diphtheria toxin. *J. Bacteriol. 141*:184–189.

Stenmark, H., Olsnes, S., and Sandvig, K. (1988). Requirement of specific receptors for efficient translocation of diphtheria toxin A fragment across the plasma membrane. *J. Biol. Chem. 263*:13449–13455.

Stenmark, H., Moskaug, J. O., Madhus, I. H., Sandvig, K., and Olsnes, S. (1991). Peptides fused to the amino-terminal end of diphtheria toxin are translocated to the cytosol. *J. Cell Biol. 113*:1025–1032.

Stenmark, H., Ariansen, S., Afanasieu, B. N., and Olsnes, S. (1992). Interactions of diphtheria toxin B fragment with cells. Role of amino and carboxy-terminal regions. *J. Biol. Chem. 281*:619–625.

Strom, T. B., Rubin-Kelley, V. E., Murphy, J. R., Nichols, J., and Woodworth, T. G. (1993a). Interleukin-2 receptor-directed therapies: antibody or cytokine based targeting molecules. *Annu. Rev. Med. 44*:343–353.

Strom, T. B., Rubin-Kelley, V. E., Woodworth, T., and Murphy, J. R. (1993b). Interleukin-2 receptor-directed immunosuppressive therapies: antibody or cytokine-based targeting molecules. In *T-Cell-Directed Immunointervention*, J. F. Bach (Ed.), Blackwell, Oxford, UK, pp. 221–255.

Tsuneoka, M., Nakagama, K., Hatsuzawa, K., Komada, M., Kitamura, N., and Mekada, E. (1993). Evidence for involvement of furin in cleavage and activation of diphtheria toxin. *J. Biol. Chem. 268*:26461–26465.

Uchida, T., Gill, D. M., and Pappenheimer, A. M., Jr. (1971) Mutation in the structural gene for diphtheria toxin carried by temperate phage β. *Nature 233*:8–11.

Uchida, T., Pappenheimer, A. M., Jr., and Greany, R. (1973). Diphtheria toxin and related proteins: isolation and properties of mutant proteins serologically related to diphtheria toxin. *J. Biol. Chem. 248*:3838–3844.

vanderSpek, J. C., Mindel, J., Finkelstein, A., and Murphy, J. R. (1993). Structure function analysis of the transmembrane domain of the interleukin-2 receptor target fusion toxin DAB$_{389}$-IL-2: the amphipathic helical region of the transmembrane domain is essential for the efficient delivery of the catalytic domain to the cytosol of target cells. *J. Biol. Chem. 268*:12077–12082.

vanderSpek, J., Cassidy, D., Genbauffe, F., Huynh, P., and Murphy, J. R. (1994a). An intact transmembrane helix 9 is essential for the efficient delivery of the diphtheria toxin catalytic domain to the cytosol of target cells. *J. Biol. Chem.* (in press).

vanderSpek, J., Howland, K., Friedman, T., and Murphy, J. R. (1994b). Maintenance of the hydrophobic face of the diphtheria toxin amphipathic transmembrane helix 1 is essential for the efficient delivery of the catalytic domain to the cytosol of target cells. *Protein Eng.* (submitted for publication).

vanderSpek, J., Cosenza, L., Woodworth, T., Nichols, J. C., and Murphy, J. R. (1994c). Diphtheria toxin-related cytokine fusion proteins: elongation factor 2 as a target for the treatment of neoplastic and autoimmune disease. *Mol. Cell. Biochem.* (in press).

Waldmann, T. A. (1986). The structure, function, and expression of interleukin-2 receptors on normal and malignant T cells. *Science 232*:727–732.

Waldmann, T. A. (1990). The multichain interleukin-2 receptor. A target for immunotherapy in lymphoma, autoimmune disorders, and organ allografts. *J. Am. Med. Assoc. 263*:272–274.

Waters, C. A., Schimke, P., Snider, C. E., Itoh, K., Smith, K. A., Nichols, J. C., Strom, T. B., and Murphy, J. R. (1990). Interleukin-2 receptor targeted cytotoxicity: receptor binding requirements for entry of IL-2-toxin into cells. *Eur. J. Immunol. 20*:785–791.

Wen, Z., Tao, X., Lakkis, F., Kiyokawa, T., and Murphy, J. R. (1991). Expression, purification, and α-melanocyte stimulating hormone receptor-specific toxicity of DAB-α-MSH fusion toxins. *J. Biol. Chem. 266*:12289–12293.

Weidlocha, A., Madhus, I. H., Mach, H., Middaugh, C. R., and Olsnes, S. (1992). Tight folding of acidic fibroblast growth factor prevents its translocation to the cytosol with diphtheria toxin as vector. *EMBO J. 11*:4835–4842.

Williams, D., Parker, K., Bishai, W., Borowski, M., Genbauffe, F., Strom, T. B., and Murphy, J. R. (1987). Diphtheria toxin receptor-binding domain substitution with interleukin-2: genetic construction and properties of a diphtheria toxin-related interleukin-2 fusion protein. *Protein Eng. 1*:493–498.

Williams, D. P., Snider, C. E., Strom, T. B., and Murphy, J. R. (1990a). Structure function analysis of IL-2-toxin (DAB$_{486}$-IL-2): fragment B sequences required for the delivery of fragment A to the cytosol of target cells. *J. Biol. Chem. 265*:11885–11889.

Williams, D. P., Wen, Z., Watson, R. S., Boyd, J., Strom, T. B., and Murphy, J. R. (1990b). Cellular processing of the fusion toxin DAB$_{486}$-IL-2 and efficient delivery of diphtheria toxin fragment A to the cytosol of target cells requires Arg194. *J. Biol. Chem. 265*:20673–20677.

Zalman, L. S., and Wisnieski, B. J. (1984). Mechanism of insertion of diphtheria toxin: peptide entry and pore size determinations. *Proc. Natl. Acad. Sci. U.S.A. 81*:3341–3345.

Zhao, J.-M., and London, E. (1988). Localization of the active site of diphtheria toxin. *Biochemistry 27*:3398–3403.

Zucker, D., and Murphy, J. R. (1983). Monoclonal antibody analysis of diphtheria toxin. I. Localization of epitopes and neutralization of cytotoxicity. *Mol. Immunol. 21*:785–793.

3

Liposomes as Vehicles for Vaccines: Induction of Humoral, Cellular, and Mucosal Immunity

Carl R. Alving

Walter Reed Army Institute of Research, Washington, D.C.

I. INTRODUCTION

Liposome research has been a major beneficiary of a recent resurgence of interest in vaccine adjuvants (reviewed by van Rooijen and Su, 1989; Gregoriadis, 1990; Alving, 1991, 1992; Phillips, 1992). Although liposomes were originally developed as models of efferent mechanisms exhibited by the immune response, it has now become evident that antigens that are presented or reconstituted in liposomes can provide desirable properties that promote effective humoral and cellular immune responses in many vaccines.

Any substance that can increase the immunogenicity of an antigen has generally been considered to be an adjuvant, and hundreds of such substances have been proposed. The mechanisms of action of adjuvants and adjuvant formulations are frequently very complex and are also often poorly understood. Among many mechanisms that have been identified for different immunostimulating substances are the following: depot effect for slow release of antigen, binding or adsorption of antigen, targeting of antigen to antigen-presenting cells, reconstitution of antigen and presentation of T and B epitopes, recruitment of immune cells, activation of complement, induction of cytokine production, and modulation of MHC class I or class II expression (Alving et al., 1992, 1993).

II. ADJUVANTS, CARRIERS, AND VEHICLES

One useful strategy that has been employed in adjuvant formulation is to combine two or more materials having different sources of origin in the hope of gaining additive or synergistic advantage by using multiple mechanisms of action. An interesting way of visualizing this concept has been to classify immunostimulatory materials into three large categories: adjuvants, carriers for haptens and antigens, and vehicles (Edelman and Tacket, 1990; Edelman, 1992). Thirty adjuvants, 13 carriers, and six vehicles are listed by the authors just cited, but the specific list could be expanded greatly. Examples of adjuvants include aluminum salts, saponin, muramyl di- and tripeptides, monophosphoryl lipid A, *Bordetella pertussis*, various cytokines, and many others. Carriers include bacterial toxoids, fatty acids, living vectors, and others. The vehicles, one or more of which are proposed components of most modern synthetic vaccines, consist of mineral oil emulsions (as in Freund's adjuvant), vegetable oil emulsions (e.g., peanut oil), nonionic block polymer surfactants, squalene or squalane, liposomes, and biodegradable polymer microspheres. As pointed out before, vehicles by themselves often have independent immunostimulatory effects, and mixtures of adjuvants with suitable vehicles can therefore be referred to as "adjuvant formulations" (Edelman and Tacket, 1990).

Although it has been demonstrated that liposomal antigens can induce high levels of humoral immunity that could be useful for vaccines, emerging areas of vaccine development are also focusing on induction of mucosal immunity and generation of cytotoxic T lymphocytes (CTLs). Both of these types of immunity, as described below, are promoted by liposomes.

III. INDUCTION OF CYTOTOXIC T LYMPHOCYTES BY LIPOSOMAL ANTIGENS

As noted above, humoral and cellular pathways are both major elements in the generation of immune responses. Induction of CTLs has been proposed as a major strategy for vaccines against intracellular antigens such as viral, parasitic, or tumor antigens (Rouse et al., 1988). Exogenous antigens, such as synthetic soluble peptides or proteins, or for that matter any extracellular antigen, usually must enter the cytoplasm of a CTL in order to participate in the processing pathway leading to presentation with MHC-I molecules to induce $CD8^+$ CTLs. This principle was illustrated by a well-known study in which purified antigen was introduced directly into the cytoplasm of cells by osmotic lysis of pinosomes (Moore et al., 1988).

Although certain protein carrier sequences apparently can promote class I CTLs in combination with an adjuvant (Dillon et al., 1992), any clear-cut method resulting in delivery of antigen to the cytoplasm of appropriate antigen-presenting cells (APCs) would presumably result in CTL induction. It has been established that macrophages can serve as APCs for Class I MHC-restricted immune responses (Debrick et al., 1991) and, as illustrated in Figure 1, phagocytosed liposomal antigen readily gains access to the cytoplasm of macrophages. It is therefore theoretically possible that liposomes could be

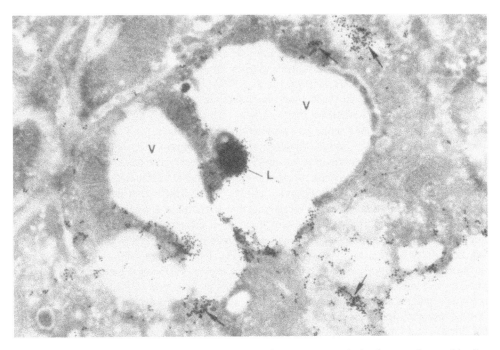

Fig. 1 Immunogold electron microscopy of cultured bone marrow-derived macrophages 6 h after phagocytosis of liposomes containing malaria antigen (R32NS$_{181}$). The malaria antigen was detected by a specific monoclonal antibody to the antigen followed by treatment with gold-labeled second antibody. V, vacuole; L, liposome containing antigen. Four arrows indicate examples of locations of liposomal antigen that has entered the cytoplasm by fusion with the endosomal membrane. See Verma et al. (1991) and Alving (1992) for further details.

used as vaccine carriers for presentation via the class I MHC pathway, thereby generating CTLs.

Numerous reports have now described class I presentation and induction of CTLs by liposomal antigens. These have included both in vitro studies (Hale, 1980; Hale and McGee, 1981; Raphael and Tom, 1982, 1984; Harding et al., 1991; Reddy et al., 1991; Zhou et al., 1991) and in vivo studies with many different antigens (Nerome et al., 1990; White et al., 1991, 1993; Reddy et al., 1992; Collins et al., 1992; Lopes and Chain, 1992; Defoort et al., 1992; Nardelli et al., 1992; Walker et al., 1992; Chen et al., 1993; Miller et al., 1992).

Although cytoplasmic delivery of antigen can be facilitated in cultured cells by so-called pH-sensitive or acid-sensitive liposomes (Harding et al., 1991; Reddy et al., 1991; Zhou et al., 1991), in vivo studies have shown that "acid-insensitive" liposomes can also readily induce CTLs (Hale, 1980; Hale and McGee, 1981; White et al., 1991, 1993; Collins et al., 1992; Reddy et al., 1992; Zhou et al., 1992). CTL induction was facilitated by using liposomes containing a fusion protein to introduce liposomal antigen directly into cells (Miller et al., 1992). Recently, in an in vivo murine model, CTLs were even generated by encapsulation of an extremely small (15-amino-acid) unconjugated peptide in liposomes containing lipid A (Cassatt et al., 1994).

IV. INDUCTION OF MUCOSAL IMMUNITY BY LIPOSOMAL ANTIGENS

In recent years there has been a considerable amount of interest in the possibility of delivering subunit vaccines for immunization by mucosal routes (McGhee and Mestecky, 1990; Gilligan and Po, 1991; McGhee et al., 1992; Husband, 1993). These efforts have been aided by advances in theoretical understanding of the mucosal immune system. It is now believed that numerous mucosal sites are connected by a "common mucosal immune system" in which antigen presentation at one site (e.g., Peyer's patches in the small intestine) can result in the induction of effector lymphocytes that produce secretory IgA both at the local immunization site and, through bloodborne transport of lymphocytes from lymphatic channels via the thoracic duct, at mucosal surfaces throughout the body (McGhee and Mestecky, 1990; Gilligan and Po, 1991; McGhee et al., 1992).

A major theoretical basis for believing that particulate antigens can gain access to the mucosal immune system is derived from the discovery and investigation of a novel intestinal antigen sampling cell known as a "microfold" or M cell (Owen and Jones, 1974; Owen, 1977; Owen et al., 1986; Gilligan and Po, 1991). These cells, which are located in the epithelium overlying the lymphoid dome regions of Peyer's patches, can actively phagocytose particulate material (Owen et al., 1986; Sass et al., 1990). The M cells are not macrophages and, although they contain numerous intracellular vesicles, they have limited capacity for lysosomal degradation and are thought to serve mainly as conduits for introducing antigens to the lymphoreticular tissue under the epithelium. Cells that are analogous to the intestinal M cells are also thought to exist in other mucosal tissues.

From an experimental standpoint, the feasibility of delivering particulate antigens by the oral route has been strengthened by positive results with two major vehicles: liposomes (see below), and poly(D,L-lactide-*co*-glycolide) microspheres (Eldridge et al., 1989, 1990, 1991; Moldoveanu et al., 1989; O'Hagan et al., 1991, 1993; Reid et al., 1993; Edelman et al., 1993).

Liposomes have been used successfully with numerous antigens for induction of mucosal or systemic immune responses after oral immunization (Genco et al., 1983; Pierce and Sacci, 1984; Wachsmann et al., 1985, 1986; Gregory et al., 1986; Ogawa et al., 1986; Jackson et al., 1990; Michalek et al., 1992; Alpar et al., 1992; Thibodeau et al., 1992). Nasal immunization with liposomes has also been achieved (Alpar et al., 1992; Abraham, 1992; Abraham and Shah, 1992). Parenteral (e.g., intramuscular or subcutaneous) immunization with liposomes reportedly does not induce a localized secretory IgA immune response (Genco et al., 1983; Clarke and Stokes, 1992a,b). Liposomal adjuvants, such as lipid A, avridine (a lipoidal amine), or IL-2, can greatly enhance the immune response to liposomal antigen administered orally or nasally (Pierce and Sacci, 1984; Pierce et al., 1984, 1985; Abraham, 1992; Abraham and Shah, 1992). A promising and useful application of oral immunization with liposomes lies in the development of a vaccine against dental caries. Several unique liposomal constructs apparently have achieved the goal of protecting experimental animals against dental caries (Gregory et al., 1986; Jackson et al., 1990; Michalek et al., 1992).

One interesting strategy for liposomal vaccines would be to administer the vaccine orally or nasally in order to obtain a humoral IgG immune response. Although some reports of relatively poor results, or failure, have been reported with liposomes lacking adjuvants (Genco et al., 1983; Wachsmann et al., 1985, 1986; Clarke and Stokes, 1992a,b), this mode of immunization has reportedly been used quite effectively with liposomes containing an adjuvant (Ogawa et al., 1986). In addition, very high IgG antibody titers to orally administered liposomal tetanus toxoid, titers that were equivalent to those obtained after intramuscular injection, were achieved when much larger (tenfold higher) doses of liposomal vaccine were given compared with lower doses administered by the intramuscular route (Alpar et al., 1992).

In summarizing the research on delivery of liposomes to mucosal surfaces for local or systemic immunization, it appears that this route of immunization is relatively inefficient. However, when large doses of liposomal antigen are used, and when the liposomes also contain either a potent adjuvant or certain novel formulations, mucosal delivery of liposomal antigen can be a feasible, effective, and even protective vaccine strategy.

V. PROTECTIVE LIPOSOMAL VACCINES IN ANIMAL DISEASE MODELS

As shown in Table 1, preclinical studies in animals have demonstrated that liposomal antigens can provide protective immunity in numerous infectious disease models. Depending on the model, protection can be obtained after intraperitoneal, intravenous, intracardial, intramuscular, subcutaneous, or oral immunization with liposomal antigen (Table 1).

VI. EXPERIMENTAL LIPOSOMAL VACCINE TRIALS IN HUMANS

Fries et al. (1992) reported a very promising malaria vaccine that employed monophosphoryl lipid A (MPL) as a liposomal adjuvant. The liposomes containing antigen and MPL were adsorbed to aluminum hydroxide to further heighten the immune response and were injected intramuscularly. At the highest doses employed, the liposomes contained an extremely large amount of MPL (2.2 mg)—representing the highest dose of endotoxin ever administered experimentally to humans. Despite the high dose of liposomal endotoxin, the vaccine was nonpyrogenic and had a low level of reactogenicity. The vaccine induced levels of specific IgG antibodies against an antigenic construct derived from the repeat region of the circumsporozoite protein of *Plasmodium falciparum* that were up to 15-fold higher than the levels previously observed in humans when the same antigen was adsorbed to aluminum hydroxide (Fries et al., 1992) (Fig. 2).

The same liposomal malaria vaccine formulation was recently administered in a second phase I trial with new volunteers, and the antibody titers were as high as or higher than those observed previously (unpublished data). In addition, in the latter trial, when experimentally challenged with mosquitoes infected with *P. falciparum*, a substantial fraction of the volunteers were found to be protected against malaria.

Liposomes consisting of a reconstituted influenza virus virosome were used as promising carriers of an inactivated hepatitis A intramuscular vaccine in humans (Glück, 1992; Glück et al., 1992; Just et al., 1992). The antigen, consisting of formalin-inactivated hepatitis A, was coupled to virosomes that contained influenza virus hemagglutinin and neuraminidase. The virosome vaccine formulation exhibited less reactogenicity than an aluminum hydroxide-adsorbed formulation and resulted in higher antibody titers and greater seroconversion (100%, compared to 44% for aluminum hydroxide). In addition, protective levels of antibodies were achieved within 2 weeks after a single immunization.

Another influenza virosome, also containing hemagglutinin and neuraminidase, was combined with a lipophilic muramyl dipeptide (MDP) derivative (B30-MDP) and tested

Table 1 Protective Liposomal Vaccines in Animal Disease Models

Disease or etiologic agent	Antigen	Route of immunization	Animal	Reference
Malaria	Killed *Plasmodium falciparum* merozoite antigen	i.m.	Owl monkey	Siddiqui et al., 1978, 1981
Streptococcus pneumoniae	Hexasaccharide–lipid conjugate	i.p. or i.v.	BALB/c mouse	Snippe et al., 1983
EB virus-induced lymphoma	MA gp340	i.p.	Cottontop tamarin	Epstein et al., 1985
Rabies virus	Glycoprotein	i.m.	Hamster	Perrin et al., 1985
Snake venom				
Carpet viper	Whole venom	s.c.	TFW mouse	New et al., 1985
Carpet viper	Whole venom	i.v.	Sheep	Theakston et al., 1985
Rattlesnake	Whole venom	s.c. or i.v.	TFW mouse	Freitas et al., 1989
Scorpion toxin	Toxic fraction	s.c.	C57Bl/6 mouse	Chaves-Olortegui et al., 1991
Dental caries induced by *Streptococcus sobrinus*	Ribosomal protein	p.o.	Germ-free rat	Gregory et al., 1986
S. mutans	Anti-idiotypic antibodies	p.o.	Germ-free rat	Jackson et al., 1990
S. mutans	Serotype-specific carbohydrate	p.o.	Germ-free rat	Michalek et al., 1992
Melanoma	B16 tumor-associated antigens	s.c.	C57Bl/6 mouse	Phillips et al., 1989
Influenza A virus	Free inactivated virus (mixed with liposomes containing IL-2)	i.p.	BALB/c mouse	Mbawuike et al., 1990
Herpes simplex virus	Glycoprotein D	i.c.	Guinea pig (female Hartley)	Ho et al., 1989, 1990
Herpes simplex virus	Glycoprotein D	i.p.	C3H/HeN mouse	Brynestad et al., 1990
Herpes simplex virus	Free glycoprotein D (mixed with liposomes containing IL-2)	s.c.	Guinea pig (female Hartley)	Ho et al., 1992
Leishmania major (cutaneous leishmaniasis)	Lipophosphoglycan and gp63	i.v.	BALB/c mouse	Kahl et al., 1990
Toxoplasma gondii	Membrane antigen p30	i.p.	Swiss-Webster mouse	Bülow and Boothroyd, 1991
Thymoma (transfected with ovalbumin)	Ovalbumin	i.v.	C57BL/6 mouse	Zhou et al., 1992
Simian immunodeficiency virus	Four SIV envelope-β-galactosidase fusion proteins	i.m.	Rhesus macaque monkey	Alving et al., 1993

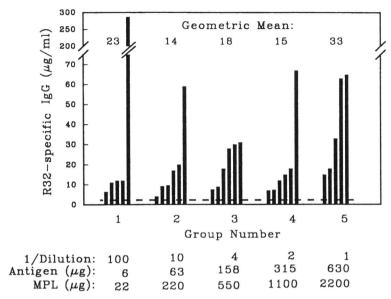

1/Dilution:	100	10	4	2	1
Antigen (μg):	6	63	158	315	630
MPL (μg):	22	220	550	1100	2200

Fig. 2 R32-specific IgG levels in human subjects 8 weeks after the last of three doses of monophosphoryl lipid A (MPL)-bearing liposome-encapsulated R32NS1$_{81}$. Each vertical bar represents a single volunteer studied by R32-specific ELISA. The groups and doses injected were as follows: group 1, 1:100 dilution of liposomes, containing 6.3 μg of R32NS1$_{81}$ and 22 μg of MPL; group 2, 1:10 dilution of liposomes, containing 63 μg of R32NS1$_{81}$ and 220 μg of MPL; group 3, 1:4 dilution of liposomes, containing 158 μg of R32NS1$_{81}$ and 550 μg of MPL; group 4, 1:2 dilution of liposomes, containing 315 μg of R32NS1$_{81}$ and 1100 mg of MPL; group 5, undiluted liposomes containing 630 μg of R32NS1$_{81}$ and 2200 μg of MPL. Dashed line indicates the geometric mean peak R32-specific IgG level (2.3 μg/mL) attained in a previous trial after four injections of 1150 μg of R32NS1$_{81}$ adsorbed to Al(OH)$_3$ alone. The geometric mean value for each dose group in the present study is indicated above the bar graph. (From Fries et al., 1992.)

as a subcutaneous vaccine in humans against three different influenza strains (Kaji et al., 1992). The vaccine exhibited moderate reactogenicity in some of the volunteers, with symptoms including fever, headache, arthralgia, malaise, and mild to moderate local reactions (pain, tenderness, induration). Laboratory analysis revealed the occurrence of significant leukocytosis attributed to the B30-MDP component of the vaccine. Compared with current influenza hemagglutinin vaccines, after only a single dose much higher antibody titers were achieved against two of the three influenza strains tested.

In an immunotherapy study in melanoma patients, after injection of liposome-encapsulated tumor-associated antigens, NK cell activity did not correlate with peripheral blood lymphocyte proliferation or clinical response status. However, there appeared to be some correlation of clinical responsiveness with peripheral blood lymphocyte cytostatic activity against heterologous melanoma tumor cells (Phillips et al., 1990).

On the basis of the apparent successes that are emerging in human applications of liposomal vaccines, I anticipate that this field will have considerable growth in the next few years. It is likely that clinical applications will owe their success to the unique attributes of liposomes that permit depot effects, carrier effects for targeting to APCs and for delivery of adjuvants to APCs, and cellular presentation properties of reconstituted proteins or peptides that allow induction of a combination of humoral and cellular immunity.

REFERENCES

Abraham, E. (1992). Intranasal immunization with bacterial polysaccharide containing liposomes enhances antigen-specific pulmonary secretory antibody response. *Vaccine 10*: 461–468.

Abraham, E., and Shah, S. (1992). Intranasal immunization with liposomes containing IL-2 enhances bacterial polysaccharide antigen-specific pulmonary secretory antibody response. *J. Immunol. 149*: 3719–3726.

Alpar, H. O., Bowen, J. C., and Brown, M. R. W. (1992). Effectiveness of liposomes as adjuvants of orally and nasally administered tetanus toxoid. *Int. J. Pharmacol. 88*: 335–344.

Alving, C. R. (1991). Liposomes as carriers of antigens and adjuvants. *J. Immunol. Methods 140*: 1–13.

Alving, C. R. (1992). Immunologic aspects of liposomes: presentation and processing of liposomal protein and phospholipid antigens. *Biochim. Biophys. Acta 1113*: 307–322.

Alving, C. R., Glass, M., and Detrick, B. (1992). Summary: adjuvants/clinical working group. *AIDS Res. Hum. Retrovir. 8*: 1427–1430.

Alving, C. R., Detrick, B., Richards, R. L., Lewis, M. G., Shafferman, A., and Eddy, G. A. (1993). Novel adjuvant strategies for experimental AIDS vaccines. *Ann. N.Y. Acad. Sci. 690:* 265–275.

Brynestad, K., Babbitt, B., Huang, L., and Rouse, B. T. (1990). Influence of peptide acylation, liposome incorporation, and synthetic immunomodulators on the immunogenicity of a 1–23 peptide glycoprotein D of herpes simplex virus: implications for subunit vaccines. *J. Virol. 64*: 680–685.

Bülow, R., and Boothroyd, J. C. (1991). Protection of mice from fatal *Toxoplasma gondii* infection by immunization with p30 antigen in liposomes. *J. Immunol. 147*: 3496–3500.

Cassatt, D., White, W. I., Madsen, J., Wassef, N. M., Alving, C. R., and Koenig, S. (1994). (Submitted for publication.)

Chavez-Olortegui, B., Amara, D. A., Rochat, H., Diniz, C., and Granier, C. (1991). *Vaccine 9*: 907–910.

Chen, W., Carbone, F. R., and McCluskey, J. (1993). Electroporation and commercial liposomes efficiently deliver soluble protein into the MHC class I presentation pathway. *J. Immunol. Methods 160*: 49–57.

Clarke, C. J., and Stokes, C. R. (1992a). The intestinal and serum humoral immune response of mice to systemically and orally administered antigens in liposomes: I. The response to liposome-entrapped soluble proteins. *Vet. Immunol. Immunopathol. 32*: 125–138.

Clarke, C. J., and Stokes, C. R. (1992b). The intestinal and serum humoral immune response of mice to orally administered antigens in liposomes: II. The response to liposome entrapped bacterial proteins. *Vet. Immunol. Immunopathol. 32*: 139–148.

Collins, D. S., Findlay, K., and Harding, C. V. (1992). Processing of exogenous liposome-encapsulated antigens in vivo generates class I MHC-restricted T cell responses. *J. Immunol. 148*: 3336–3341.

Debrick, J. E., Campbell, P. A., and Staerz, U. D. (1991). Macrophages as accessory cells for class I MHC-restricted immune responses. *J. Immunol. 147*: 2846–2851.

Defoort, J., Nardelli, B., Huang, W., and Tam, J. P. (1992). A rational design of synthetic peptide vaccine with a built-in adjuvant. *Int. J. Peptide Protein Res. 40*: 214–221.

Dillon, S. B., Demuth, S. G., Schneider, M. A., Weston, C. B., Jones, C. S., Young, J. F., Scott, M., Bhatnaghar, P. K., LoCastro, S., and Hanna, N. (1992). Induction of protective class I MHC-restricted CTL in mice by a recombinant influenza vaccine in aluminum hydroxide adjuvant. *Vaccine 10*: 309–318.

Edelman R. (1992). An update on vaccine adjuvants in clinical trial. *AIDS Res. Hum. Retrovir.* 8: 1409–1411.

Edelman, R., and Tacket, C. O. (1990). Adjuvants. *Int. Rev. Immunol.* 7: 51–66.

Edelman, R., Russell, R. G., Losonsky, G., Tall, B. D., Tacket, C. O., Levine, M. M., and Lewis, D. H. (1993). Immunization of rabbits with enterotoxigenic *E. coli* colonization factor antigen (CFA/I) encapsulated in biodegradable microspheres of poly(lactide-*co*-glycolide). *Vaccine 11*: 155–158.

Eldridge, J. H., Gilley, R. M., Staas, J. K., Moldoveanu, Z., Meulbroek, J. A., and Tice, T. R. (1989). Biodegradable microspheres: vaccine delivery system for oral immunization. *Curr. Top. Microbiol. Immunol. 146*: 59–66.

Eldridge, J. H., Hammond, C. J., Meulbroek, J. A., Staas, J. K., Gilley, R. M., and Tice, T. R. (1990). Controlled vaccine release in the gut-associated lymphoid tissues. I. Orally administered biodegradable microspheres target the Peyer's patches. *J. Controlled Release 11*: 205–214.

Eldridge, J. H., Staas, J. K., Meulbroek, J. A., McGhee, J. R., Tice, T. R., and Gilley, R. M. (1991). Biodegradable microspheres as a vaccine delivery system. *Mol. Immunol. 28*: 287–294.

Epstein, M. A., Morgan, A. J., Finerty, S., Randle, B. J., and Kirkwood, J. K. (1985). Protection of cottontop tamarins against Epstein-Barr virus-induced malignant lymphoma by a prototype subunit vaccine. *Nature 318*: 287–289.

Freitas, T. V., Tavares, A. P., Theakston, R. D. G., Laing, G., and New, R. R. C. (1989). Use of liposomes for protective immunisation against *Crotalus durissus* (tropical rattlesnake) venom. *Toxicon 27*: 341–347.

Fries, L. F., Gordon, D. M., Richards, R. L., Egan, J. E., Hollingdale, M. R., Gross, M., Silverman, C., and Alving, C. R. (1992). Liposomal malaria vaccine in humans: a safe and potent adjuvant strategy. *Proc. Natl. Acad. Sci. U.S.A. 89*: 358–362.

Genco, R. J., Linzer, R., and Evans, R. T. (1983). Effect of adjuvants on orally administered antigens. *Ann. N.Y. Acad. Sci. 409*: 650–668.

Gilligan, C. A., and Po, A. L. W. (1991). Oral vaccines: design and delivery. *Int. J. Pharmaceut. 75*: 1–24.

Glück, R. (1992). Immunopotentiating reconstituted influenza virosomes (IRIVs) and other adjuvants for improved presentation of small antigens. *Vaccine 10*: 915–919.

Glück, R., Mischler, R., Brantschen, S., Just, M., Althaus, B., and Cryz, S. J. (1992). Immunopotentiating reconstituted influenza virus virosome vaccine delivery system for immunization against hepatitis A. *J. Clin. Invest. 90*: 2491–2495.

Gregoriadis, G. (1990). Immunological adjuvants: a role for liposomes. *Immunol. Today 11*: 89–97.

Gregory, R. L., Michalek, S. M., Richardson, G., Harmon, C., Hilton, T., and McGhee, J. R. (1986). Characterization of immune response to oral administration of *Streptococcus sobrinus* ribosomal preparations in liposomes. *Infect. Immun. 54*: 780–786.

Hale, A. H. (1980). H-2 antigens incorporated into phospholipid vesicls elicit specific allogeneic cytotoxic T lymphocytes. *Cell. Immunol. 55*: 328–341.

Hale, A. H., and McGee, M. P. (1981). A study of the inability of subcellular fractions to elicit primary anti-H-2 cytotoxic T lymphocytes. *Cell. Immunol. 58*: 277–285.

Harding, C. V., Collins, D. S., Kanagawa, O., and Unanue, E. R. (1991). Liposome-encapsulated antigens engender lysosomal processing for class II MHC presentation and cytosolic processing for class I presentation. *J. Immunol. 9*: 2860–2863.

Ho, R. J. Y., Burke, R. L., and Merigan, T. C. (1989). Antigen-presenting liposomes are effective in treatment of recurrent herpes simplex virus genitalis in guinea pigs. *J. Virol. 63*: 2951–2958.

Ho, R. J. Y., Burke, R. L., and Merigan, T. C. (1990). Physical and biological characterization of antigen presenting liposome formulations: relative efficacy for the treatment of recurrent genital HSV-2 in guinea pigs. *Antiviral Res. 13*: 187–200.

Ho, R. J. Y., Burke, R. L., and Merigan, T. C. (1992). Liposome-formulated interleukin-2 as an adjuvant of recombinant HSV glycoprotein gD for the treatment of recurrent genital HSV-2 in guinea pigs. *Vaccine 10*: 209–213.

Husband, A. J. (1993). Novel vaccination strategies for the control of mucosal infection. *Vaccine 11*: 107–112.

Jackson, S., Mestecky, J., Childers, N. K., and Michalek, S. M. (1990). Liposomes containing anti-idiotypic antibodies: an oral vaccine to induce protective secretory immune responses specific for pathogens of mucosal surfaces. *Infect. Immun. 58*: 1932–1936.

Just, M., Berger, R., Drechsler, H., Brantschen, S., and Glück, R. (1992). A single vaccination with an inactivated hepatitis A liposome vaccine induces protective antibodies after only two weeks. *Vaccine 10*: 737–739.

Kahl, L. P., Lelchuk, R., Scott, C. A., and Beesley, J. (1990). Characterization of *Leishmania major* antigen-liposomes that protect BALB/c mice against cutaneous leishmaniasis. *Infect. Immun. 58*: 3233–3241.

Kaji, M., Kaji, Y., Kaji, M., Ohkuma, K., Honda, T., Oka, T., Sakoh, M., Nakamura, S., Kurachi, K., and Sentoku, M. (1992). Phase 1 clinical tests of influenza MDP-virosome vaccine (KD-5382). *Vaccine 10*: 663–667.

Lopes, L. M., and Chain, B. M. (1992). Liposome-mediated delivery stimulates a class I-restricted cytotoxic T cell response to soluble antigen. *Eur. J. Immunol. 22*: 287–290.

Mbawuike, I. N., Wyde, P. R., and Anderson, P. M. (1990). Enhancement of the protective efficacy of inactivated influenza A virus vaccine in aged mice by IL-2 liposomes. *Vaccine 8*: 347–352.

McGhee, J. R., and Mestecky, J. (1990). In defense of mucosal surfaces. Development of novel vaccines for IgA responses protective at the portals of entry of microbial pathogens. *Infect. Dis. Clin. N. Am. 4*: 315–341.

McGhee, J. R., Mestecky, J., Dertzbaugh, M. T., Eldridge, J. H., Hirasawa, M., and Kiyono, H. (1992). The mucosal immune system: from fundamental concepts to vaccine development. *Vaccine 10*: 75–88.

Michalek, S. M., Childers, N. K., Katz, J., Dertzbaugh, M., Zhang, S., Russell, M. W., Macrina, F. L., Jackson, S., and Mestecky, J. (1992). *Adv. Exp. Med. Biol. 327*: 191–198.

Miller, M. D., Gould-Fogerite, S., Shen, L., Woods, R. M., Koenig, S., Mannino, R. J., and Letvin, N. L. (1992). Vaccination of rhesus monkeys with synthetic peptide in a fusogenic proteoliposome elicits simian immunodeficiency virus-specific CD-8[+] cytotoxic T lymphocytes. *J. Exp. Med. 176*: 1739–1744.

Moldoveanu, Z., Staas, J. K., Gilley, R. M., Ray, R., Compans, R. W., Eldridge, J. H., Tice, T. R., and Mestecky, J. (1989). Immune responses to influenza virus in orally and systemically immunized mice. *Curr. Top. Microbiol. Immunol. 146*: 91–99.

Moore, M. W., Carbone, F. R., and Bevan, M. J. (1988). Introduction of soluble protein into the class I pathway of antigen processing and presentation. *Cell 54*: 777–785.

Nardelli, B., Defoort, J., Huang, W., and Tam, J. P. (1992). Design of a complete synthetic peptide-based AIDS vaccine with a built-in adjuvant. *AIDS Res. Hum. Retrovir. 8*: 1405–1407.

Nerome, K., Yoshioka, Y., Ishida, M., Okuma, K., Oka, T., Kataoka, T., Inoue, A., and Oya, A. (1990). Development of a new type of influenza subunit vaccine made by muramyldipeptide-liposome: enhancement of humoral and cellular immune responses. *Vaccine 8*: 503–509.

New, R. R. C., Theakston, R. D. G., Zumbuehl, O., Iddon, D., and Friend, J. (1985). Liposomal immunisation against snake venoms. *Toxicon 23*: 215–219.

Ogawa, T., Kotani, S., and Shimauchi, H. (1986). Enhancement of serum antibody production in mice by oral administration of lipophilic derivatives of muramylpeptides and bacterial lipopolysaccharides with bovine serum albumin. *Method Find. Exp. Clin. Pharmacol. 8*: 19–26.

O'Hagan, D. T., Rahman, D., McGee, J. P., Jeffery, H., Davies, M. C., Williams, P., Davis, S. S., and Challacombe, S. J. (1991). Biodegradable microparticles as controlled release antigen delivery systems. *Immunology 73*: 239–242.

O'Hagan, D. T., McGee, J. P., Holmgren, J., Mowat, A. MCl., Donachie, A. M., Mills, K. H. G., Gaisford, W., Rahman, D., and Challacombe, S. J. (1993). Biodegradable microparticles for oral immunization. *Vaccine 11*: 149–154.

Owen, R. L. (1977). Sequential uptake of horseradish peroxidase by lymphoid follicle epithelium of Peyer's patches in the normal unobstructed mouse intestine: an ultrastructural study. *Gastroenterology 72*: 440–451.

Owen, R. L., and Jones, A. L. (1974). Epithelial cell specialization within human Peyer's patches: an ultrastructural study of intestinal lymphoid follicles. *Gastroenterology 66*: 189–203.

Owen, R. L., Pierce, N. F., Apple, R. T., and Cray, W. C., Jr. (1986). M cell transport of *Vibrio cholerae* from the intestinal lumen into Peyer's patches: a mechanism for antigen sampling and for microbial transepithelial migration. *J. Infect. Dis. 153*: 1108–1118.

Perrin, P., Thibodeau, L., and Sureau, P. (1985). Rabies immunosome (subunit vaccine) structure and immunogenicity. Pre- and post exposure protection studies. *Vaccine 3*: 325–332.

Phillips, N. C. (1992). Liposomal carriers for the treatment of acquired immune deficiency syndrome. *Bull. Inst. Pasteur 90*: 205–230.

Phillips, N. C., Loutfi, A., Kareem, M., Shibata, H., and Baines, M. (1989). Experimental and clinical evaluation of liposome-tumor antigen immunotherapy. In *Liposomes in the Therapy of Infectious Diseases and Cancer*, G. Lopez-Berestein and I. J. Fidler (Eds.), Alan R. Liss, New York, pp. 15–24.

Phillips, N. C., Loutfi, A., A-Kareem, A. M., Shibata, H. R., and Baines, M. G. (1990). Clinical evaluation of liposomal tumor antigen vaccines in patients with stage-III melanoma. *Cancer Detect. Prev. 14*: 491–496.

Pierce, N. F., and Sacci, J. B., Jr. (1984). Enhanced mucosal priming by cholera toxin and procholeragenoid with a lipoidal amine adjuvant (avridine) delivered in liposomes. *Infect. Immun. 44*: 469–473.

Pierce, N. F., Sacci, J. B., Jr., Alving, C. R., and Richardson, E. C. (1984). Enhancement by lipid A of mucosal immunogenicity of liposome-associated cholera toxin. *Rev. Infect. Dis. 6*: 563–566.

Pierce, N. F., Alving, C. R., Richardson, E. C., and Sacci, J. B., Jr. (1985). Enhancement of specific mucosal antibody responses by locally administered adjuvants. In *Advances in Research on Cholera and Related Diarrheas*, S. Kuwahara and N. F. Pierce (Eds.), KTK Scientific Publishers, Tokyo, pp. 163–170.

Raphael, L., and Tom, B. H. (1982). *In vitro* induction of primary and secondary xenoimmune responses by liposomes containing human colon tumor cell antigens. *Cell. Immunol. 71*: 224–240.

Raphael, L., and Tom, B. H. (1984). Liposome facilitated xenogeneic approach for studying human colon cancer immunity: carrier and adjuvant effect of liposomes. *Clin. Exp. Immunol. 55*: 1–13.

Reddy, R., Zhou, F., Huang, L., Carbone, F., Bevan, M., and Rouse, B. T. (1991). pH sensitive liposomes provide an efficient means of sensitizing target cells to class I restricted CTL recognition of a soluble protein. *J. Immunol. Methods 141*: 157–163.

Reddy, R., Zhou, F., Nair, S., Huang, L., and Rouse, B. T. (1992). In vivo cytotoxic T lymphocyte induction with soluble proteins administered in liposomes. *J. Immunol. 148*: 1585–1589.

Reid, R. H., Boedeker, E. C., McQueen, C. E., Davis, D., Tseng, L.-Y., Kodak, J., Sau, K., Wilhelmsen, C. L., Nellore, R., Dalal, P., and Bhagat, H. R. (1993). Preclinical evaluation of microencapsulated CFA/II oral vaccine against enterotoxigenic *E. coli*. *Vaccine 11*: 159–167.

Rouse, B. T., Norley, S., and Martin, S. (1988). Antiviral cytotoxic T lymphocyte induction and vaccination. *Rev. Infect. Dis. 10*: 16–32.

Sass, W., Dreyer, H.-P., and Seifert, J. (1990). Rapid insorption of small particles in the gut. *Am. J. Gastroenterol. 85*: 255–260.

Siddiqui, W. A., Taylor, D. W., Kan, S., Kramer, K., Richmond-Crum, S. M., Kotani, S.,

Shiba, T., and Kusumoto, S. (1978). Vaccination of experimental monkeys against *Plasmodium falciparum*: a possible safe adjuvant. *Science 201*: 1237–1240.

Siddiqui, W. A., Kan, S., Kramer, K., Case, S., and Palmer, K. (1981). Use of a synthetic adjuvant in an effective vaccination of monkeys against malaria. *Nature 289*: 64–66.

Snippe, H., van Dam, J. E. G., van Houte, A. J., Willers, J. M. N., Kamerling, J. P., and Vliegenthart, J. F. G. (1983). *Infect. Immun. 42*: 842–844.

Theakston, R. D. G., Zumbuehl, O., and New, R. R. C. (1985). Use of liposomes for protective immunisation in sheep against *Echis carinatus* snake venom. *Toxicon 23*: 921–925.

Thibodeau, L., Tremblay, C., and Lachapelle, L. (1992). Oral priming followed by parenteral immunization with HIV-immunosomes induce HIV-1-specific salivary and circulatory IgA in mice and rabbits. *AIDS Res. Hum. Retrovir. 8*: 1379.

van Rooijen, N., and Su, D. (1989). Immunoadjuvant action of liposomes: mechanisms. In *Immunological Adjuvants and Vaccines*, G. Gregoriadis, A. C. Allison, and G. Poste (Eds.), Plenum, New York, pp. 95–106.

Verma, J. N., Wassef, N. M., Wirtz, R. A., Atkinson, C. T., Aikawa, M., Loomis, L. D., and Alving, C. R. (1991). Phagocytosis of liposomes by macrophages: intracellular fate of liposomal malaria antigen. *Biochim. Biophys. Acta 1066*: 229–308.

Wachsmann, D., Klein, J. P., Schöller, M., and Frank, R. M. (1985). Local and systemic immune response to orally administered liposome-associated soluble *S. mutans* cell wall antigens. *Immunology 54*: 189–193.

Wachsmann, D., Klein, J. P., Schöller, M., Ogier, J., Ackermans, F., and Frank, R. M. (1986). Serum and salivary antibody responses in rats orally immunized with *Streptococcus mutans* carbohydrate protein conjugate associated with liposomes. *Infect. Immun. 52*: 408–413.

Walker, C., Selby, M., Erickson, A., Cataldo, D., Valensi, J., and Van Nest, G. (1992). Cationic lipids direct a viral glycoprotein into the class I major histocompatibility complex antigen-presentation pathway. *Proc. Natl. Acad. Sci. U.S.A. 89*: 7915–7918.

White, K., Gordon, D., Gross, M., Richards, R. L., Alving, C. R., Ballou, W. R., and Krzych, U. (1991). Induction of cytolytic T cell responses by the repeatless *Plasmodium falciparum* circumsporozoite protein molecule incorporated into liposomes. *Am. J. Trop. Med. Hyg. 45*(3, Suppl): 284.

White, K., Krzych, U., Gordon, D. M., Porter, T. G., Richards, R. L., Alving, C. R., Deal, C. D., Hollingdale, M., Silverman, C., Sylvester, D. R., Ballou, W. R., and Gross, M. (1993). Induction of cytolytic and antibody responses using *P. falciparum* repeatless circumsporozoite protein encapsulated in liposomes. *Vaccine 11*: 1341–1346.

Zhou, F., Rouse, B. T., and Huang, L. (1991). An improved method of loading pH-sensitive liposomes with soluble proteins for class I restricted antigen presentation. *J. Immunol. Methods 145*: 143–152.

Zhou, F., Rouse, B. T., and Huang, L. (1992). Prolonged survival of thymoma-bearing mice after vaccination with a soluble protein antigen entrapped in liposomes: a model study. (1992). *Cancer Res. 52*: 6287–6291.

4

Bacterial ADP-Ribosyltransferases

Mario Domenighini, Mariagrazia Pizza, and Rino Rappuoli

Immunobiological Research Institute, Siena, Italy

I. INTRODUCTION

ADP-ribosylation is a mechanism widely used in nature to modify the properties of proteins and modulate their function. The prokaryotic enzymes that are known to have ADP-ribosylating activity are reported in Table 1. In bacteria, most of the proteins with ADP-ribosylating activity are toxins that have profound effects on eukaryotic organisms and are often the primary cause of severe diseases. For this reason, bacterial toxins with ADP-ribosylating activity have been widely studied and are the best-known

Table 1 ADP-Ribosyltransferases

Bacterial toxins	Eukaryotic acceptors	Target amino acids	Effects	Reference
Diphtheria toxin (DT)	EF-2	Diphthamide 715	Inhibition of protein synthesis	Collier, 1982
Pseudomonas ETA (PAETA)	EF-2	Diphthamide 715	Inhibition of protein synthesis	Allured et al., 1986
Pertussis toxin (PT)	Gi, Go, and T	Cysteine 352	Uncoupling of G protein and receptor	Tamura et al., 1982
Cholera toxin (CT)	Gs, T, (Gi, Go)	Arginine 201	Inhibition of GTP hydrolysis	Mekalanos et al., 1983
E. coli LT1 and LT2	Gs, T, (Gi, Go)	Arginine 201	Alteration of transmembrane signal transduction	Dallas and Falkow, 1980; Pickett et al., 1987
Bacillus sphaericus mosquitocidal toxin (MTX)	Cell proteins and MTX	Unknown	Mosquito larval mortality	Thanabalu et al., 1993
C. botulinum C3 exoenzyme	Rho, Rac	Asparagine 41	Reorganization of actin fibers	Aktories et al., 1989
Staphylococcal EDIN	Rho	Unknown	Disassembly of Golgi apparatus	Sugai et al., 1992
C. botulinum C2 toxin, C. perfringens iota toxin, C. spiriforme toxin, C. difficile transferase	Actin	Arginine 177	Inhibition of actin polymerization	Aktories et al., 1986, 1990
Pseudomonas exoenzyme S	Ras, vimentin	Arginine	Unknown	Coburn et al., 1989a,b
Rhodospirillum rubrum	Dinitrogenase reductase	Arginine 101	Enzyme regulation	Fitzmaurice et al., 1989
Azospirillum spp.	Dinitrogenase reductase	Arginine 101	Enzyme regulation; induced by NH_4^+ and glutamine	Fu et al., 1989
Escherichia coli	Bacterial proteins			Skorko and Kur, 1982
Streptomyces griseus	Bacterial proteins	Arginine	Regulation in differentiation and sporulation	Penyge et al., 1992
Legionella pneumophila	GTP-binding proteins		Unknown	Belyi et al., 1991
Bacteriophage T4 ALT and MOD proteins	Bacterial RNA polymerase, α subunit		Unknown	Hilse et al., 1989.
N4 phage	Bacterial RNA polymerase, α subunit		Unknown	Skorko and Kur, 1982

ADP-ribosylating proteins. However, a great deal of evidence is being accumulated on many enzymes that control crucial pathways of eukaryotic cells through ADP-ribosylation. In this chapter we focus mainly on bacterial toxins with ADP-ribosylating activity. Since each of these toxins is separately described in this book, in this chapter we concentrate on the features that are common to this family of toxins and point out the common properties that underlie their activity.

II. BACTERIAL ADP-RIBOSYLATING TOXINS

Bacterial ADP-ribosylating toxins are proteins, usually secreted into the extracellular medium by pathogenic bacteria, that cause disease by killing or altering the metabolism of eukaryotic cells. Classically, bacterial toxins are considered to be composed of two functionally different domains: the enzymatically active domain, which has ADP-ribosylating activity and is responsible for the toxicity (A subunit), and the binding domain, which interacts with the receptors on the surface of eukaryotic cells (B subunit). A schematic representation of the main structural features of ADP-ribosylating toxins is given in Figure 1. Although all the A subunits have similar enzymatic activities, which is reflected in a similarity of size and primary and tertiary structure, the B subunits bind different receptors on eukaryotic cells and have different strategies to translocate the A subunit into the cell. These differences are reflected in the variety of structures and compositions of the B subunits. Most of the bacterial ADP-ribosylating toxins can be classified into three major families.

A. Diphtheria Toxin and *Pseudomonas* Exotoxin A

The first family comprises diphtheria toxin (DT) and *Pseudomonas* exotoxin A (PAETA). These are composed of a single polypeptide chain containing both the A and B subunits, which in diphtheria toxin are also linked by a disulfide bridge. The primary sequences (Ratti et al., 1983; Greenfield et al., 1983; Gray et al., 1984) and the three-dimensional structures of the two proteins (Choe et al., 1992; Allured et al., 1986) have a very similar organization that is schematically shown in Figure 1a. Each protein can be divided into three domains. One is located in the catalytic A subunit; the other two domains, the translocator (T) and the receptor-binding domain (R), form the B subunit. The A subunit is at the amino terminus of DT (amino acids 1–193) and at the carboxy terminus of PAETA (amino acids 405–613) and has several homologous domains in the two proteins. The most conserved part is a cavity that binds NAD; in DT it faces the R domain, and in PAETA it faces the T domain (Fig. 1a). The R domains are involved in receptor binding and do not have any primary structural similarity. The R domain of PAETA comprises two regions spanning amino acids 1–252 and 365–404, whereas the R domain of DT contains the carboxy terminal region of the molecule (amino acids 386–535).

The T domains are membrane-spanning peptides that assist in translocation of the A domain across the cell membrane. They comprise amino acids 205–378 of DT (Choe et al., 1992) and 253–364 of PAETA (Hwang et al., 1987; Siegall et al., 1989) and do not have any primary structural similarity but do have a strong similarity in secondary structure that reflects their homologous function. They are composed entirely of alpha helices, a property typical of transmembrane proteins. In DT, the external loops between alpha helices are charged and water-soluble at neutral pH, whereas they become uncharged and membrane-soluble at acidic pH (Choe et al., 1992).

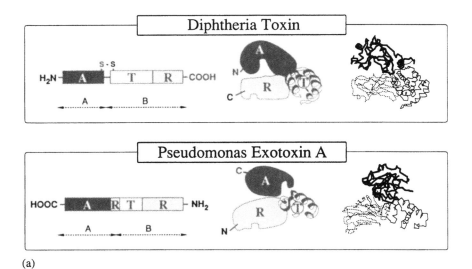

(a)

Fig. 1 Structural features of ADP-ribosylating toxins. (a) Scheme of the structures of DT and PAETA showing on the left the features of the primary structure, which can be divided into three domains: the enzymatically active subunit (A), the translocator domain (T), and the receptor-binding domain (R). The backbone of the X-ray structure of the proteins is shown on the right and is schematically represented in the center of the figure. The cavity in the A subunit shows the location of the NAD-binding site. (b) Scheme of the primary structure of CT, LT1, LT2, and PT (left), showing the enzymatically active A1 subunit, the linker A2 fragment, and the B oligomer, which is made up of five identical subunits in CT, LT1 and LT2, and five subunits of different sequence and size in PT. On the right is shown the backbone of the X-ray structure of LT1 (top right) and of the B oligomer seen from the top (lower right). A schematic representation of the structures is shown in the center. The structural organization shown for LT1 is identical to that of CT and is likely to be very similar also for LT2 and PT. (c) Schematic representation of the structural organization of the ADP-ribosylating toxins not described in Figure 1a and 1b.

B. Cholera, *E. coli* Heat-Labile and Pertussis Toxins

The second family of toxins comprises cholera toxin (CT), *E. coli* heat-labile enterotoxins 1 and 2 (LT1 and LT2), and pertussis toxin (PT). These proteins have the general structure A_1B_5 (Van Heyningen, 1985; Betley et al., 1986). One molecule of the A subunit is associated with a pentameric B subunit (composed of five monomers) (Fig. 1b). The A subunit contains an enzymatically active domain (A1) of approximately 190 amino acids and a linker comprising the remaining 45–50 amino acids (A2) connecting the A and B subunits. The A1 domains of CT, LT1 (Dallas and Falkow, 1980; Mekalanos et al., 1983), and LT2 (Pickett et al., 1987) are almost identical to primary structure, whereas the PT A1 domain (subunit S1) is less similar but contains several blocks of sequences homologous to those in CT and LT (Nicosia et al., 1986; Locht and Keith, 1986). The A2 domains are less conserved at the amino acid level but have a common feature. They contain a long alpha helix that is connected to the A1 domain by a disulfide bridge. The B subunits are composed of five identical monomers in CT, LT1 and LT2, and four different monomers in PT (S2, S3, S4, and S5), present in a 1:1:2:1 ratio, respectively (Tamura et al., 1982, 1983). The three-dimensional structures of LT1 and CT (Sixma et

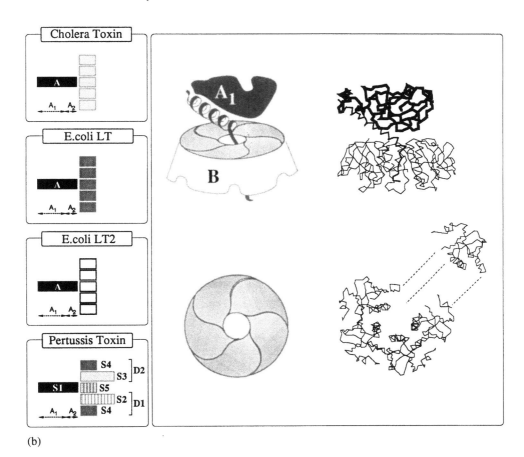

(b)

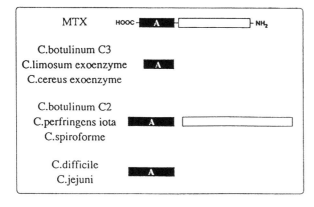

(c)

al., 1991; Gibbons, 1991) (Fig. 1b) are almost identical. The A1 domain has a globular structure with an NAD-binding cavity similar to that of DT and PAETA, whereas the five monomers of the B subunit associate in a ringlike structure containing an internal canal flanked by five alpha helices, each in one of the monomers (see Fig. 1b for details). The A2 domain is a long alpha helix, covalently linked to the A1 domain and inserted into the canal of the B oligomer, thus connecting the A and B subunits. Although the structures of LT2 and PT have not yet been determined, given the high degree of similarity they are predicted to be very similar to the one shown in Figure 1b for LT1. The B_5 ringlike structure of the B oligomer of LT1 and CT is common also to verotoxin, a bacterial toxin that is totally different in amino acid sequence and mechanism of action (Stein et al., 1992). This suggests that the structure of the B subunits of this group of toxins is so efficient that it has emerged more than once during evolution.

C.   *Clostridia, Pseudomonas,* and *Bacillus* Toxins

The third family of toxins comprises the remaining toxins, such as *Clostridium botulinum* C2 toxin (Popoff et al., 1988, 1989; Aktories et al., 1990) and C3 toxin (Aktories et al., 1989, 1992; Sekine et al., 1989), *Clostridium perfringens* iota toxin and related proteins, *Pseudomonas* exoenzyme S (Sokol et al., 1990; Coburn et al., 1989a,b; Kulich et al., 1993), and the mosquitocidal toxin (MTX) from *Bacillus sphaericus* (Thanabalu et al., 1992, 1993). As shown in Figure 1c, this is a heterogeneous group of toxins not well characterized. C3 is an enzyme produced by *C. botulinum* that does not seem to have a B subunit and therefore cannot intoxicate eukaryotic cells. C2, related proteins, and MTX have a typical A-B structure with one A and one B subunit. The information available is, however, too little to draw general conclusions. One important feature of the MTX fragment A is its high degree of homology to the A subunits of CT and PT (Thanabalu et al., 1993).

III.   THE ENTRY PROCESS

The different mechanisms of translocation of the A subunits into cells is reflected in the different structures of the B subunits in each family of toxins. In fact, the B fragments of DT and PAETA, containing the receptor binding and translocator domains, are quite different in architecture from the B domains of CT, LT, and PT, which have a ringlike oligomer composed of five subunits (see Fig. 1a and b). The different mechanisms of entry of the A subunit into the cytosol of eukaryotic cells are summarized in Figure 2a and b. DT, following binding to the receptor on the surface of the cells, is internalized into endosomes that acquire an acidic pH. The low pH neutralizes the charges of the T domain, which becomes hydrophobic and enters into the membrane, facilitating the translocation of the A subunit across the cell membrane to a compartment where the disulfide bridge is reduced and the A subunit is released into the cytosol (Fig. 2a) (Sandvig and Olsnes, 1980; Montecucco et al., 1985; Papini et al., 1987, 1988, 1993). A similar mechanism is used by PAETA, but in this case the role of pH is less critical. It has been shown that the T domains of both DT and PAETA are able to translocate the A subunits or other peptides even when bound to different R domains, suggesting that they should be considered pure translocator domains able to support the transmembrane passage of polypeptides independently of the structure of the upstream and downstream sequences

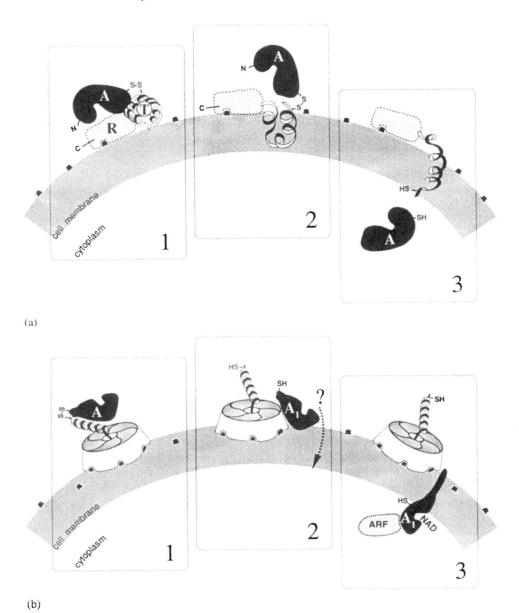

(a)

(b)

Fig. 2 Schematic representation of the translocation of the A subunit across the eukaryotic cell membrane. (a) shows the process occurring in DT and PAETA, while (b) shows the process used by CT and the other toxins having a pentameric B oligomer. In the case of DT and PAETA, the translocator domain (T) enters the cell membrane and facilitates entry of the A subunit. The T domain is not present in CT, which must use a different mechanism to translocate the A subunit.

(Walz et al., 1990; FitzGerald and Pastan, 1992; Williams et al., 1987; Chaudhary et al., 1987, 1988; Prior et al., 1991, 1992; Arora et al., 1992).

In CT and related toxins, the pentameric B oligomer has multiple receptor binding sites in the side opposite the A subunit. Following receptor binding, the A subunit is close to the membrane but not directly in contact with it (part 1 of Fig. 2b), so that it is difficult to visualize how it can reach and cross the membrane in the absence of a translocator domain. It is very likely, however, that it is sufficient for the A subunit to be in the vicinity of the membrane to be able to interact with and translocate across it, even in the absence of a real translocator domain. This hypothesis is supported by experiments showing that following reduction of the disulfide bridge that links the A1 and A2 domains, the A1 subunit becomes hydrophobic and enters the lipid bilayer (Tomasi and Montecucco, 1981; Tomasi et al., 1986). Recent studies suggest that following membrane interaction the A1 subunit is processed by the Golgi apparatus and then reaches the substrate G_s protein (Donta et al., 1993; Lencer et al., 1992).

IV. THE TARGET PROTEINS

Bacterial toxins have evolved to be toxic for animals and therefore have as targets proteins that are essential for the life of eukaryotic cells or for the function of animal tissues. The proteins that are ADP-ribosylated by bacterial toxins have a common feature: they are all GTP-binding proteins that belong to the GTPase superfamily, a class of proteins with conserved structure and molecular mechanism (Bourne et al., 1991). These proteins are molecular switches that are turned on by binding GTP and off by hydrolyzing GTP to GDP, thus changing their affinity for other molecules. These GTPase switches control crucial circuits of eukaryotic cells. They sort and amplify transmembrane signals, control the synthesis and translocation of proteins, guide the vesicular traffic through the cytoplasm, and control the differentiation and proliferation of animal cells and cyto-skeleton structure. Therefore alteration of function of these proteins caused either by bacterial toxins or by somatic mutation may result in severe tissue intoxication or oncogenesis, respectively (Landis et al., 1989; Lyons et al., 1990). The only known exception to this rule is actin, a protein that binds ATP instead of GTP.

The similarity in function of these proteins is reflected by homologies in their primary sequences and three-dimensional structures. The crystal structure of the bacterial EF-Tu and Ras (Jurnak, 1985; La Cour et al., 1985; de Vos et al., 1988; Pai et al., 1989, 1990; Milburn et al., 1990; Brunger et al., 1990) has shown that their guanine nucleotide-binding domains are remarkably similar (Jurnak et al., 1990). The amino acid sequence, although less conserved than the three-dimensional structure, contains several domains of homol-ogy—G1, G2, G3, G4, and G5 (Bourne et al., 1991)—that are involved in GTP binding and are common to most GTP-binding proteins (see Fig. 3). The proteins that are ADP-ribosylated by bacterial toxins are described below and are shown schematically in Figure 3.

A. Elongation Factor 2 (EF-2)

Elongation factor 2 (EF-2) is a GTP-binding protein of 97.5 kDa that is involved in protein synthesis (Kohno et al., 1986). It contains a histidine residue in position 715 that is posttranslationally modified to diphthamide (Ness et al., 1980a,b). This modified residue is ADP-ribosylated by diphtheria toxin and *Pseudomonas* exotoxin A. Following

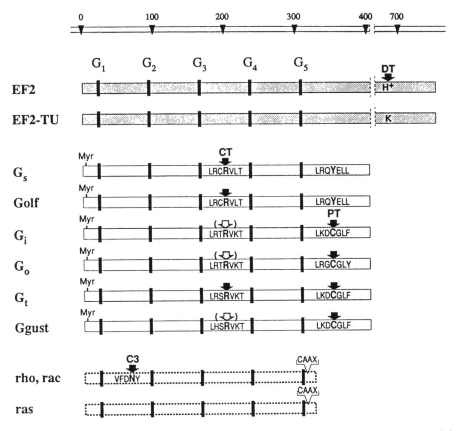

Fig. 3 Schematic representation of the proteins that are ADP-ribosylated by bacterial protein toxins. The top line shows the numbers of the amino acids. G_1–G_5 represent the regions of homology within the target G proteins. Myr and CAAX indicate sites that are modified by the addition of myristic acid or a farnesyl residue, respectively. Amino acids are indicated in the one-letter code; those that are ADP-ribosylated by the toxins are indicated by arrows. Ggust stands for Gustducin. (From McLaughlin et al., 1992.)

ADP-ribosylation, EF-2 becomes unable to carry out protein synthesis, and this results in cell death (Honjo et al., 1968). In bacteria, there is a homologous protein also involved in protein synthesis, EF-Tu, containing a lysine in position 715, that is not a substrate for DT and PAETA. Therefore, bacteria are not susceptible to DT and PAETA (Honjo et al., 1968). In evolution, the first organisms that contain an EF-2 that is susceptible to DT and PAETA ADP-ribosylation are the archaebacteria.

B. G Proteins

G proteins are a family of GTP-binding proteins composed of three subunits (alpha, beta, and gamma) that are involved in signal transduction across the membrane of eukaryotic cells (Linder and Gilman, 1992; Hepler and Gilman, 1992). The alpha subunit is usually anchored to the cell membrane by a 14-carbon fatty acid (myristic acid) (Jones et al.,

1990; Mumby et al., 1990; Spiegel et al., 1991). Typically, a signal that is sensed by a receptor on the surface of eukaryotic cells is received by the alpha subunit of the G protein, which consequently binds GTP, dissociates from the beta and gamma subunits, and transmits the signal to the enzymes that release the second messengers such as adenylate cyclase, phospholipase C, and cyclic GMP phosphodiesterase. Adenylate cyclase, for instance, is regulated by two classes of receptors that can transmit their signals to two different GTP binding proteins: G_s and G_i. G_s receives signals from the stimulatory receptors and activates the adenylate cyclase, whereas G_i receives signals from the inhibitory receptors and inhibits the activity of adenylate cyclase. In addition to G_s and G_i, the family of G proteins contains several other proteins, including G_o, G_t, G_g, Golf (Hepler and Gilman, 1992; Jones and Reed, 1989), and several other G proteins with similar function that have not yet been fully characterized.

As shown in Figure 3, cholera toxin ADP-ribosylates Arg-201 of the alpha subunit of G_s, G_t, and Golf (Van Dop et al., 1984), whereas it ADP-ribosylates the corresponding Arg in G_i and G_o only when these are activated by the receptor (Iiri et al., 1989).

Pertussis toxin ADP-ribosylates Cys-352 residue in G_i, G_o, G_t, Ggust, and other G proteins but is not able to ADP-ribosylate G_s and Golf, which have a tyrosine in position 352 (Katada and Ui, 1982; Bokoch et al., 1983; West et al., 1985; Fong et al., 1988). ADP-ribosylation of G_s by cholera or LT toxins causes constitutive activation of adenylate cyclase, which results in intracellular accumulation of the second messenger cAMP. ADP-ribosylation of G_i by pertussis toxin uncouples this protein from the receptor so that it is unable to transmit signals to inhibit adenylate cyclase.

C. Ras, Rho, and the Small GTP-Binding Proteins

Ras, rho, and the small GTP-binding proteins belong to a large family of regulatory proteins that are known to be involved in controlling diverse essential cellular functions including growth, differentiation, cytoskeletal organization, intracellular vesicle transport, and secretion. They usually contain a carboxy terminal CAAX box that is modified by addition of a farnesyl lipid moiety that increases the hydrophobicity and localizes them to the plasma membrane or to the membranes of intracellular vesicles (Hall, 1990). The oncoprotein ras, which is localized to the plasma membrane, controls cell growth and differentiation (Egan et al., 1993) and is a target of ADP-ribosylation by *Pseudomonas* exoenzyme S (Coburn et al., 1989b). Following ADP-ribosylation, the GTPase activity of ras is decreased (Tsai et al., 1985). Rho and rac are structurally similar proteins involved in the organization of actin (Ridley et al., 1992; Ridley and Hall, 1992). They are ADP-ribosylated on Asn-41 by *Clostridium botulinum* C3 (Aktories et al., 1989; Sekine et al., 1989). This causes reorganization of actin into stress fibers so that cells lose their shape and round up.

D. Actin

Actin, the protein involved in cell shape and movement, is an ATP-binding protein that is ADP-ribosylated on Arg-177 by *Clostridium botulinum* C2 toxin (Aktories et al., 1986, 1990), *Clostridium perfringens* iota toxin, and related toxins from *C. spiriforme* and *C. difficile* (Popoff et al., 1988, 1989) (Table 1). Actin ADP-ribosylation results in disruption of the microfilaments causing rounding up of the cells, which is followed by cell lysis.

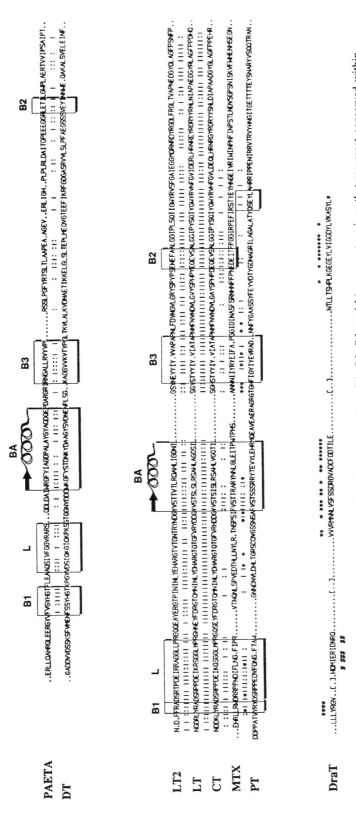

Fig. 4 Homology of toxin amino acid sequences. Boxed sequences (BA, B1, B2, B3, and L) represent regions that are most conserved within each group of toxins. These regions make the NAD-binding and catalytic site, conformation of which is shown in Figures 5 and 6, and are conserved in all toxins. * and # with DraT indicate homology to sequences of group 2 and 1 toxins, respectively.

69

V. THE ACTIVE SITE: A COMMON STRUCTURE AND ACTIVITY

A. Primary Structure Homology

Eight amino acid sequences of bacterial ADP-ribosylating toxins are available in public databases: diphtheria toxin (DT) (Ratti et al., 1983; Greenfield et al., 1983; Kaczoreck et al., 1983), *Pseudomonas aeruginosa* exotoxin A (PAETA) (Gray et al., 1984), pertussis toxin (PT) (Nicosia et al., 1986; Locht and Keith, 1986), cholera toxin (CT) (Mekalanos et al., 1983), the plasmidic heat-labile toxin from enterotoxigenic *E. coli* strains (LT1) (Dallas and Falkow, 1980), the chromosomal equivalent (LT2) (Pickett et al., 1989), the C3 toxin from *Clostridium botulinum* (Nemoto et al., 1991; Popoff et al., 1990), and the mosquitocidal toxin produced by *Bacillus sphaericus* (MTX) (Thanabalu et al., 1991). Analysis of the amino acid sequences revealed homologies only among the enzymatically active A subunits. The B subunits are totally unrelated, with the exception of CT and LT1, which are highly homologous. Based on the degree of homology, seven of the eight A subunit sequences fall into two groups (Fig. 4). One contains DT and PAETA, with very modest homology, and the other contains CT, LT1, LT2, PT, and MTX, which share significant homology in five regions. There is very high homology along the entire sequence among CT, LT1, and LT2, which appear to be just variants of the same molecule (Fig. 4).

No sequence homology was detected between the two groups of proteins or between them and the C3 enzyme. A remarkable similarity is found also in the ADP-ribosylating enzyme from *Rhodospirillum rubrum* DraT (see Fig. 4).

B. Tertiary Structure Homology

In spite of the low degree of homology at the amino acid level, the enzymatically active domains of ADP-ribosylating toxins have several common features. As shown in Figure 1, most of them have a size of approximately 21,000 Da. Larger A fragments (approximately 50,000 Da) are found only in *Clostridium botulinum* C2, *Clostridium perfringens* iota, and *Pseudomonas* exoenzyme S.

The minimal fragment bearing ADP-ribosylating activity contained 179 amino acids in PT (Bartoloni et al., 1988), a size very similar to that of the A1 domains of LT and CT and to that of A fragments from DT and PAETA, which contain 193 and 208 amino acids, respectively. Following the determination of tertiary structure of DT, PAETA, CT, and LT, numerous studies of site-directed mutagenesis and experiments on the purified proteins allowed identification of the amino acids that are important for NAD binding and catalytic activity (Carroll and Collier, 1984, 1987, 1988; Carroll et al., 1985; Brandhuber et al., 1988; Burnette et al., 1988; Pizza et al., 1988; Barbieri et al., 1989). In a model trying to accommodate present knowledge, the amino acids involved in NAD binding and catalysis give rise to a cavity open to the exterior (see Figs. 1 and 5) that, in spite of the low sequence homology, is remarkably similar in structure in all ADP-ribosylating enzymes. This NAD binding and catalytic domain is made up of five noncontiguous regions, which in Figure 4 are named BA, B1, B2, B3, and L, that are equivalent in the two groups of toxins and whose spatial arrangement is schematically represented in Figures 5 and 6. These regions represent the most conserved segments between DT and PAETA and among LT, CT, PT, and MTX (Fig. 4); however, often there is no homology of primary sequence between the DT group and the LT group. In spite of this apparent lack of homology, the secondary and tertiary folding of these regions are identical in all toxins. As shown in Figure 5, the nicotinamide ring of NAD is housed in a cavity formed by a beta sheet followed by an alpha helix (box BA in Fig. 4) and in the case of DT is stacked between two tyrosine rings, one from the beta sheet and the other coming from the alpha helix. Remarkably, these two aromatic amino acids that firmly hold the nicotinamide ring are present in DT, PAETA, and PT, which bind NAD very avidly, whereas they are absent in CT and LT, which bind NAD with an affinity constant 1000-fold lower than those of the other enzymes.

The adenine ring of NAD is housed in a second broad cavity having as floor a beta strand (B3 region in Figs. 4 and 6) and delimited by a surface exposed loop (L region in Figs. 4 and 6). Two amino acids that are essential for catalysis and present in all toxins are located on two beta strands (B1 and B2 in Figs. 4 and 6) flanking the cavity binding the nicotinamide. These are a negatively charged glutamic acid (C2) present in all enzymes and a positively charged amino acid (C1) that is a histidine in DT and PAETA and an arginine in CT, LT, PT, and MTX (Sixma et al., 1993). The spatial location of these

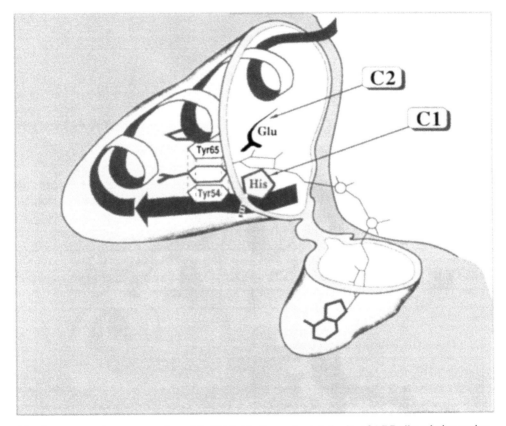

Fig. 5 Schematic representation of the NAD-binding and catalytic site of ADP-ribosylating toxins. The cavity on the left is made by the BA region shown in Figure 4, while the cavity on the right is made by the L region of Figure 4. NAD is shown in the binding site with the nicotinamide ring packed between two tyrosines, one from the alpha helix and the other from the beta strand. The adenine ring is housed in the cavity made by the L loop. The two catalytic amino acids (Glu and His) are located on each side of the nicotinamide and ribose rings. In CT, LT, and PT, the catalytic histidine is replaced by an arginine.

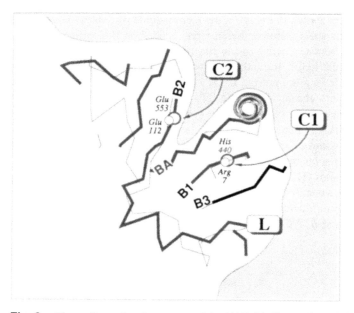

Fig. 6 Three-dimensional structure of the NAD-binding and catalytic site of PAETA (thick line) and of LT (thin line). As shown, the regions BA, B1, B2, B3, and L (shown in Figs. 5 and 6) that make the structures of the NAD-binding and catalytic sites have a remarkably similar spatial configuration in spite of the lack of amino acid homology. C1 and C2 represent the catalytic amino acids that come from the B1 and B2 beta strands.

two amino acids is strictly conserved in the two groups of proteins (see Fig. 6). The positive and negative charges of both catalytic amino acids point toward the nicotinamide ring, and this must cause a rearrangement of the electrons within the ring, resulting in destabilization of the bonds with the ribose and the phosphates that are exposed to the solvent on the external surface of the toxin. Numerous studies have shown that replacement of amino acids C1 and C2 abolishes enzymatic activity (Burnette et al., 1988; Barbieri and Cortina, 1988). A mutant protein in which C1 and C2 were replaced by noncatalytic amino acids was found to be totally devoid of enzymatic activity and toxicity and is being used for vaccination against pertussis (Pizza et al., 1988, 1989; Rappuoli et al., 1992).

C.   The ADP-Ribosylation Reaction

During ADP-ribosylation, the catalytic enzymes bind NAD and transfer the ADP-ribose moiety to the target protein by a mechanism that may be summarized by the model shown in Figure 7. The NAD molecule in solution is mostly in a folded conformation in which the two rings, adenine and nicotinamide, are stacked over each other (part 1 of Fig. 7) (Oppenheimer et al., 1971; McDonald et al., 1972; Lee et al., 1973). To enter the binding site, the compact NAD structure is partially unfolded (part 2 of Fig. 7) so that the adenine and nicotinamide bases can enter into two adjacent cavities forming the catalytic site of the ADP-ribosylating enzymes (part 3 of Fig. 8), leaving phosphates and ribose groups exposed at the surface of the cavity, where they can interact with and be transferred to the target protein that comes in contact with the enzyme (part 4b). In the absence of the

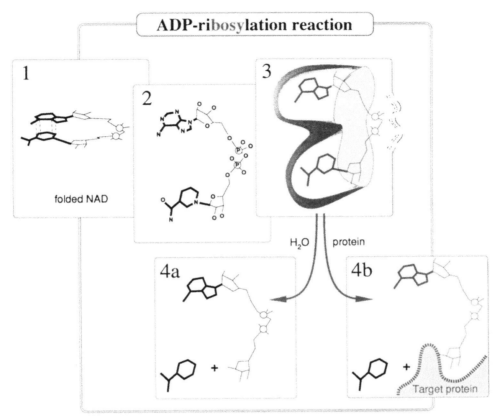

Fig. 7 Schematic representation of the ADP-ribosylation reaction. The NAD that in solution is mostly in the folded conformation (1), opens up (2) and binds the two cavities of the binding site that house the nicotinamide and the adenine rings, leaving the phosphates in contact with the external environment (3). The bound NAD can then interact with water, giving rise to ADP-ribose and nicotinamide (4a), or with the target protein, which is ADP-ribosylated (4b).

target protein, the metastable groups are exposed to the attack of water molecules, and NAD is slowly hydrolyzed (part 4a). This glycohydrolytic activity is only a consequence of NAD binding in the active site and has no biological role.

In nature there are two types of ADP-ribosyltransferases (Ueda and Hayaishi, 1985; Althaus and Richter, 1987): the mono-ADP-ribosyltransferases, which transfer a single ADP-ribose group to the target proteins, and the poly-ADP-ribosyltransferases, which transfer many ADP-ribose groups sequentially to the target proteins (Simonin et al., 1991; Thibodeau et al., 1990). The latter enzymes are found mainly in the nucleus of eukaryotic cells, where they poly-ADP-ribosylate histones and other nuclear proteins. Poly-ADP-ribosylation increases the negative charge of the histones and decreases their interaction with DNA, possibly with a role in controlling transcription.

Poly-ADP-ribosylation involves two steps. In the first a nuclear mono-ADP-ribosyltransferase transfers the ADP-ribose group to a carboxylic group in aspartic acid or glutamic acid or a carboxy terminal amino acid. The first ADP-ribose group is a primer for the second step, in which the nuclear poly-ADP-ribosyltransferase elongates the chain by adding additional ADP-ribose groups.

Mono-ADP-ribosylation is a reaction in which a single ADP-ribose is transferred to a side chain of amino acids such as cysteine, asparagine, and arginine or to a posttranslationally modified amino acid such as diphthamide. Mono-ADP-ribosylation is carried out by the bacterial toxins described in this chapter, phage and *E. coli* proteins, and many cytoplasmic and membrane-associated enzymes in eukaryotic cells.

Acknowledgments

We thank Claudia Magagnoli for help in assembling the manuscript, Giorgio Corsi for the drawings, and Joseph Alouf, Cesare Montecucco, and John Collier for critical reading of the manuscript.

REFERENCES

Aktories, K., Barmann, M., Ohishi, I., Tsuyama, S., Jakobs, K. H., and Habermann, E. (1986). Botulinum C2 toxin ADP-ribosylates actin. *Nature 322*: 390–392.

Aktories, K., Braun, U., Rosener, S., Just, I., and Hall, A. (1989). The rho gene product expressed in *E. coli* is a substrate of botulinum ADP-ribosyltransferase C3. *Biochem. Biophys. Res. Commun. 158*: 209–213.

Aktories, K., Geipel, U., Wille, M., and Just, I. (1990). Characterization of the ADP-ribosylation of actin by *Clostridium botulinum* C2 toxin and *Clostridium perfringens* iota toxin. *J. Physiol. (Paris) 84*: 262–266.

Aktories, K., Mohr, C., and Koch, G. (1992). *Clostridium botulinum* C3 ADP-ribosyltransferase. *Curr. Top. Microbiol. Immunol. 175*: 115–131.

Allured, V. S., Collier, R. J., Carroll, S. F., and McKay, D. B. (1986). Structure of exotoxin A of *Pseudomonas aeruginosa* at 3.0-angstrom resolution. *Proc. Natl. Acad. Sci. U.S.A. 83*: 1320–1324.

Althaus, F. R., and Richter, C. (1987). ADP-ribosylation of proteins. Enzymology and biological significance. *Mol. Biol. Biochem. Biophys. 37*: 1–237.

Arora, N., Klimpel, K. R., Singh, Y., and Leppla, S. H. (1992). Fusions of anthrax toxin lethal factor to the ADP-ribosylation domain of *Pseudomonas* exotoxin A are potent cytotoxins which are translocated to the cytosol of mammalian cells. *J. Biol. Chem. 267*: 15542–15548.

Barbieri, J. T., and Cortina, G. (1988). ADP-ribosyltransferase mutations in the catalytic S-1 subunit of pertussis toxin. *Infect. Immun. 56*: 1934–1941.

Barbieri, J. T., Mende-Mueller, L. M., Rappuoli, R., and Collier, R. J. (1989). Photolabeling of Glu-129 of the S-1 subunit of pertussis toxin with NAD. *Infect. Immun. 57*: 3549–3554.

Bartoloni, A., Pizza, M., Bigio, M., Nucci, D., Manclark, C., Sato, H., and Rappuoli, R. (1988). Mapping of a protective epitope of pertussis toxin by in vitro refolding of recombinant fragments. *Biotechnology 6*: 709–712.

Belyi, Yu F., Tartakovskii, I. S., Vertiev, Yu V., and Prosorovskii, S. V. (1991). ADP-ribosyltransferase activity of *Legionella pneumophila* is stimulated by the presence of macrophage lysates. *Biomed. Sci. 2*: 94–96.

Betley, M. J., Miller, V. L., and Mekalanos, J. J. (1986). Genetics of bacterial enterotoxins. *Annu. Rev. Microbiol. 40*: 577–605.

Bokoch, G. M., Katada, T., Northup, J. K., Hewlett, E. L., and Gilman, A. G. (1983). Identification of the predominant substrate for ADP-ribosylation by islet activating protein. *J. Biol. Chem. 258*: 2072–2075.

Bourne, H. R., Sanders, D. A., and McCormick, F. (1991). The GTPase superfamily: conserved structure and molecular mechanism. *Nature 349*: 117.

Brandhuber, B. J., Allured, V. S., Falbel, T. G., and McKay, D. B. (1988). Mapping the enzymatic active site of *Pseudomonas aeruginosa* exotoxin A. *Proteins 3*: 146–154.

Brunger, A. T., Milburn, M. V., Tong, L., de Vos, A. M., Jancarik, J., Yamaizumi, Z., Nishimura, S., Ohtsuka, E., and Kim, S. H. (1990). Crystal structure of an active form of RAS protein, a complex of a GTP analog and the HRAS p21 catalytic domain. *Proc. Natl. Acad. Sci. U.S.A. 87*: 4849–4853.

Burnette, W. N., Cieplak, W., Mar, V. L., Kaljot, K. T., Sato, H., and Keith, J. M. (1988). Pertussis toxin S1 mutant with reduced enzyme activity and a conserved protective epitope. *Science 242*: 72–74.

Carroll, S. F., and Collier, R. J. (1984). NAD binding site of diphtheria toxin: identification of a residue within the nicotinamide subsite by photochemical modification with NAD. *Proc. Natl. Acad. Sci. U.S.A. 81*: 3307–3311.

Carroll, S. F., and Collier, R. J. (1987). Active site of *Pseudomonas aeruginosa* exotoxin A. Glutamic acid 553 is photolabeled by NAD and shows functional homology with glutamic acid 148 of diphtheria toxin. *J. Biol. Chem. 262*: 8707–8711.

Carroll, S. F., and Collier, R. J. (1988). Amino acid sequence homology between the enzymic domains of diphtheria toxin and *Pseudomonas aeruginosa* exotoxin A. *Mol. Microbiol. 2*: 293–296.

Carroll, S. F., McCloskey, J. A., Crain, P. F., Oppenheimer, N. J., Marschner, T. M., and Collier, R. J. (1985). Photoaffinity labeling of diphtheria toxin fragment A with NAD: structure of the photoproduct at position 148. *Proc. Natl. Acad. Sci. U.S.A. 82*: 7237–7241.

Chaudhary, V. K., FitzGerald, D. J., Adhya, S., and Pastan, I. (1987). Activity of a recombinant fusion protein between transforming growth factor type alpha and *Pseudomonas* toxin. *Proc. Natl. Acad. Sci. U.S.A. 84*: 4538–4542.

Chaudhary, V. K., Mizukami, T., Fuerst, T. R., FitzGerald, D. J., Moss, B., Pastan, I., and Berger, E. A. (1988). Selective killing of HIV-infected cells by recombinant human CD4-*Pseudomonas* exotoxin hybrid protein. *Nature 335*: 369–372.

Choe, S., Bennett, M. J., Fujii, G., Curmi, P. M., Kantardjieff, K. A., Collier, R. J., and Eisenberg, D. (1992). The crystal structure of diphtheria toxin. *Nature 357*: 216–222.

Coburn, J., Dillon, S. T., Iglewski, B. H., and Gill, D. M. (1989a). Exoenzyme S of *Pseudomonas aeruginosa* ADP-ribosylates the intermediate filament protein vimentin. *Infect. Immuno. 57*: 996–998.

Coburn, J., Wyatt, R. T., Iglewski, B. H., and Gill, D. M. (1989b). Several GTP-binding proteins, including p21c-H-ras, are preferred substrates of *Pseudomonas aeruginosa* exoenzyme S. *J. Biol. Chem. 264*: 9004–9008.

Collier, R. J. (1982). Structure and activity of diphtheria toxin. In *ADP-Ribosylation Reactions*, D. Hayashi and K. Ueda (Eds.), Academic, New York, pp. 575–592.

Dallas, W. S., and Falkow, S. (1980). Amino acid homology between cholera toxin and *Escherichia coli* heat labile toxin. *Nature 288*: 499–501.

de Vos, A. M., Tong, L., Milburn, M. V., Matias, P. M., Jancarik, J., Noguchi, S., Nishimura, S., Miura, K., Ohtsuka, E., and Kim, S. H. (1988). Three-dimensional structure of an oncogene protein: catalytic domain of human c-H-ras p21. *Science 239*: 888–893.

Donta, S. T., Beristain, S., and Tomicic, T. K. (1993). Inhibition of heat-labile cholera and *E. coli* enterotoxins by brefeldin A. *Infect. Immun. 61*: 3282–3286.

Egan, S. E., Giddings, B. W., Brooks, M. W., Buday, L., Sizeland, A. M., and Weinberg, R. A. (1993). Association of Sos Ras exchange protein with Grb2 is implicated in tyrosine kinase signal transduction and transformation. *Nature 363*: 45.

FitzGerald, D. J. and Pastan, I. (1992). *Pseudomonas* exotoxin: recombinant conjugates as therapeutic agents. *Biochem. Soc. Trans. 20*: 731–734.

Fitzmaurice, W. P., Saari, L. L., Lowery, R. G., Ludden, P. W., and Roberts, G. P. (1989). Genes coding for the reversible ADP-ribosylation system of dinitrogenase reductase from *Rhodospirillum rubrum. Mol. Gen. Genet. 218*: 340–347.

Fong, H. K., Yoshimoto, K. K., Eversole-Cire, P., and Simon, M. I. (1988). Identification of a GTP-binding protein alpha subunit that lacks an apparent ADP-ribosylation site for pertussis toxin. *Proc. Natl. Acad. Sci. U.S.A. 85*: 3066–3070.

Fu, H. A., Hartmann, A., Lowery, R. G., Fitzmaurice, W. P., Roberts, G. P., and Burris, R. H. (1989). Posttranslational regulatory system for nitrogenase activity in *Azospirillum* spp. *J. Bacteriol. 171*: 4679–4685.

Gibbons, A. (1991). New 3-D protein structures revealed. The shape of cholera [news]. *Science 253*: 382–383.

Gray, G. L., Smith, D. H., Baldridge, J. S., Harkins, R. N., Vasil, M. L., Chen, E. Y., and Heyneker, H. L. (1984). Cloning, nucleotide sequence, and expression in *Escherichia coli* of the exotoxin A structural gene of *Pseudomonas aeruginosa. Proc. Natl. Acad. Sci. U.S.A. 81*: 2645–2649.

Greenfield, L., Bjorn, M. J., Horn, G., Fong, D., Buck, G. A., Collier, R. J., and Kaplan, D. A. (1983). Nucleotide sequence of the structural gene for diphtheria toxin carried by corynebacteriophage beta. *Proc. Natl. Acad. Sci. U.S.A. 80*: 6853–6857.

Hall, A. (1990). The cellular functions of small GTP-binding proteins. *Science 249*: 635–640.

Hepler, J. R., and Gilman, A. G. (1992). G proteins. *Trends Biochem. Sci. 17*: 383–387.

Hilse, D., Koch, T., and Ruger, W. (1989). Nucleotide sequence of the alt gene of bacteriophage T4. *Nucleic Acids Res. 17*: 6731.

Honjo, T., Nishizuka, Y., Hayaishi, O., and Kato, I. (1968). Diphtheria toxin-dependent adenosine diphosphate ribosylation of aminoacyl transferase II and inhibition of protein synthesis. *J. Biol. Chem. 243*: 3553–3555.

Hwang, J., FitzGerald, D. J., Adhya, S., and Pastan, I. (1987). Functional domains of *Pseudomonas* exotoxin identified by deletion analysis of the gene expressed in *E. coli*. *Cell 48* :129–136.

Iiri, T., Tohkin, M., Morishima, N., Ohoka, Y., Ui, M., and Katada, T. (1989). Chemotactic peptide receptor-supported ADP-ribosylation of a pertussis toxin substrate GTP-binding protein by cholera toxin in neutrophil-type HL-60 cells. *J. Biol. Chem. 264*: 21394–21400.

Jones, D. T., and Reed, R. R. (1989). Golf: an olfactory neuron specific-G protein involved in odorant signal transduction. *Science 244*: 790–795.

Jones, T. L., Simonds, W. F., Merendino, J. J., Jr., Brann, M. R., and Spiegel, A. M. (1990). Myristoylation of an inhibitory GTP-binding protein alpha subunit is essential for its membrane attachment. *Proc. Natl. Acad. Sci. U.S.A. 87*: 568–572.

Jurnak, F. (1985). Structure of the GDP domain of EF-Tu and location of the amino acids homologous to ras oncogene proteins. *Science 230*: 32–36.

Jurnak, F., Heffron, S., and Bergmann, E. (1990). Conformational changes involved in the activation of ras p21: implications for related proteins. *Cell 60*: 525–528.

Kaczorek, M., Delpeyroux, F., Chenciner, N., Streeck, R. E., Murphy, J. R., Boquet, P., and Tiollais, P. (1983). Nucleotide sequence and expression of the diphtheria tox-228 gene in *Escherichia coli. Science 221*: 855–858.

Katada, T., and Ui, M. (1982). ADP ribosylation of the specific membrane protein of C6 cells by islet-activating protein associated with modification of adenylate cyclase activity. *J. Biol. Chem. 257*: 7210–7216.

Kohno, K., Uchida, T., Ohkubo, H., Nakanishi, S., Nakanishi, T., Fukui, T., Ohtsuka, E., Ikehara, M., and Okada, Y. (1986). Amino acid sequence of mammalian elongation factor 2 deduced from the cDNA sequence: homology with GTP-binding proteins. *Proc. Natl. Acad. Sci. U.S.A. 83*: 4978–4982.

Kulich, S. M., Frank, D. W., and Barbieri, J. T. (1993). Purification and characterization of exoenzyme S from *Pseudomonas aeruginosa* 388. *Infect. Immun. 61*: 307–13.

La Cour, T. F., Nyborg, J., Thirup, S., and Clark, B. F. (1985). Structural details of the binding of guanosine diphosphate to elongation factor Tu from *E. coli* as studied by X-ray crystallography. *EMBO J. 4*: 2385–2388.

Landis, C. A., Masters, S. B., Spada, A., Pace, A. M., Bourne, H. R., and Vallar, L. (1989). GTPase inhibiting mutations activate the alpha chain of Gs and stimulate adenylyl cyclase in human pituitary tumours. *Nature 340*: 692–696.

Lee, C., Eichner, R. D., and Kaplan, N. O. (1973). Conformations of diphosphopyridine coenzymes upon binding to dehydrogenases. *Proc. Natl. Acad. Sci. U.S.A. 70*: 1593–1597.

Lencer, W. I., Delp, C., Neutra, M. R., and Madara, J. L. (1992). Mechanism of cholera toxin action on a polarized human intestinal epithelial cell line: role of vesicular traffic. *J. Cell Biol. 117*: 1197–1209.

Linder, M. E., and Gilman, A. G. (1992). G. proteins. *Sci. Am.* July, pp. 56–65.

Locht, C., and Keith, J. M. (1986). Pertussis toxin gene: nucleotide sequence and genetic organization. *Science 232*: 1258–1264.

Lyons, J., Landis, C. A., Harsh, G., Vallar, L., Grunewald, K., Feichtinger, H., Duh, Q. Y., Clark, O. H., Kawasaki, E., Bourne, H. R., and McCormick, F. (1990). Two G protein oncogenes in human endocrine tumours. *Science 249*: 655–659.

McDonald, G., Brown, B., Hollis, D., and Walter, C. (1972). Some effects of environment on the folding of nicotinamide-adenine dinucleotides in aqueous solutions. *Biochemistry 11*: 1920.

McLaughlin, S. K., McKinnon, P. J., and Margolskee, R. F. (1992). Gustducin is a taste-cell-specific G protein closely related to the transducins. *Nature 357*: 563–569.

Mekalanos, J. J., Swartz, D. J., Pearson, G. D., Harford, N., Groyne, F., and de Wilde, M. (1983). Cholera toxin genes: nucleotide sequence, deletion analysis and vaccine development. *Nature 306*: 551–557.

Milburn, M. V., Tong, L., de Vos, A. M., Brunger, A., Yamaizumi, Z., Nishimura, S., and Kim. S. H. (1990). Molecular switch for signal transduction: structural differences between active and inactive forms of protooncogenic ras proteins. *Science 247*: 939–945.

Montecucco, C., Schiavo, G., and Tomasi, M. (1985). pH-dependence of the phospholipid interaction of diphtheria-toxin fragments. *Biochem. J. 231*: 123–128.

Mumby, S. M., Heukeroth, R. O., Gordon, J. I., and Gilman, A. G. (1990). G-protein alpha-subunit expression, myristoylation, and membrane association in COS cells. *Proc. Natl. Acad. Sci. U.S.A. 87*: 728–32.

Nemoto, Y., Namba, T., Kozaki, S., and Narumiya, S. (1991). *Clostridium botulinum* C3 ADP-ribosyltransferase gene. Cloning, sequencing, and expression of a functional protein in *Escherichia coli. J. Biol. Chem. 266*: 19312–19319.

Ness, B. G. van, Howard, J. B., and Bodley, J. W. (1980a). ADP-ribosylation of elongation factor 2 by diphtheria toxin. NMR spectra and proposed structure of ribosyl-diphthamide and its hydrolysis products. *J. Biol. Chem. 255*: 10710–10716.

Ness, B. G. van, Howard, J. B., and Bodley, J. W. (1980b). ADP-ribosylation of elongation factor 2 by diphtheria toxin. Isolation and properties of the novel ribosylamino acid and its hydrolysis products. *J. Biol. Chem. 255*: 10717–10720.

Nicosia, A., Perugini, M., Franzini, C., Casagli, M. C., Borri, M. G., Antoni, G., Almoni, M., Neri, P., Ratti, G., and Rappuoli, R. (1986). Cloning and sequencing of the pertussis toxin genes: operon structure and gene duplication. *Proc. Natl. Acad. Sci. U.S.A. 83*: 4631–4635.

Oppenheimer, N. J., Arnold, L. J., and Kaplan, N. O. (1971). A structure of pyridine nucleotides in solution. *Proc. Natl. Acad. Sci. U.S.A. 68*: 3200.

Pai, E. F., Kabsch, W., Krengel, U., Holmes, K. C., John, J., and Wittinghofer, A. (1989). Structure of the guanine-nucleotide-binding domain of the Ha-ras oncogene product p21 in the triphosphate conformation. *Nature 341*: 209–214.

Pai, E. F., Krengel, U., Petsko, G. A., Goody, R. S., Kabsch, W., and Wittinghofer, A. (1990). Refined crystal structure of the triphosphate conformation of H-ras p21 at 1.35 Å resolution: implications for the mechanism of GTP hydrolysis. *EMBO J. 9*: 2351–2359.

Papini, E., Schiavo, G., Tomasi, M., Colombatti, M., Rappuoli, R., and Montecucco, C. (1987). Lipid interaction of diphtheria toxin and mutants with altered fragment B. 2. Hydrophobic photolabelling and cell intoxication. *Eur. J. Biochem. 169*: 637–644.

Papini, E., Sandona, D., Rappuoli, R., and Montecucco, C. (1988). On the membrane translocation of diphtheria toxin: at low pH the toxin induces ion channels on cells. *EMBO J. 7*: 3353–3359.

Papini, E., Rappuoli, R., Murgia, M., and Montecucco, C. (1993). Cell penetration of diphtheria toxin. Reduction of the interchain disulfide bridge is the rate-limiting step of translocation in the cytosol. *J. Biol. Chem. 268*: 1567–1574.

Penyige, A., Vargha, G., Ensign, J. C., and Barabas, G. (1992). The possible role of ADP ribosylation in physiological regulation of sporulation in *Streptomyces griseus. Gene 115*: 181–185.

Pickett, C. L., Weinstein, D. L., and Holmes, R. K. (1987). Genetics of type IIa heat-labile enterotoxin of *Escherichia coli*: operon fusions, nucleotide sequence, and hybridization studies. *J. Bacteriol. 169*: 5180–5187.

Pickett, C. L., Twiddy, E. M., Coker, C., and Holmes, R. K. (1989). Cloning, nucleotide sequence, and hybridization studies of the type IIb heat-labile enterotoxin gene of *Escherichia coli. J. Bacteriol. 171*: 4945–4952.

Pizza, M., Bartoloni, A., Prugnola, A., Silvestri, S., and Rappuoli, R. (1988). Subunit S1 of pertussis toxin: mapping of the regions essential for ADP-ribosyltransferase activity. *Proc. Natl. Acad. Sci. U.S.A. 85*: 7521–7525.

Pizza, M., Covacci, A., Bartoloni, A., Perugini, M., Nencioni, L., Manetti, R., Bugnoli, M., Barbieri, J. T., Sato, H., and Rappuoli, R. (1989). Mutants of pertussis toxin suitable for vaccine development. *Science 246*: 497–500.

Popoff, M. R., Rubin, E. J., Gill, D. M., and Boquet, P. (1988). Actin-specific ADP-ribosyltransferase produced by a *Clostridium difficile* strain. *Infect. Immun. 56*: 2299–2306.

Popoff, M. R., Milward, F. W., Bancillon, B., and Boquet, P. (1989). Purification of the *Clostridium spiroforme* binary toxin and activity of the toxin on HEp-2 cells. *Infect. Immun. 57*: 2462–2469.

Popoff, M., Boquet, P., Gill, D. M., and Eklund, M. W. (1990). DNA sequence of exoenzyme C3, an ADP-ribosyltransferase encoded by *Clostridium botulinum* C and D phages. *Nucleic Acids Res. 18*: 1291.

Prior, T. I., FitzGerald, D. J., and Pastan, I. (1991). Barnase toxin: a new chimeric toxin composed of *Pseudomonas* exotoxin A and barnase. *Cell 64*: 1017–1023.

Prior, T. I., FitzGerald, D. J., and Pastan, I. (1992). Translocation mediated by domain II of *Pseudomonas* exotoxin A: transport of barnase into the cytosol. *Biochemistry 31*: 3555–3559.

Rappuoli, R., Podda, A., Pizza, M., Covacci, A., Bartoloni, A., de Magistris, M. T., and Nencioni, L. (1992). Progress towards the development of new vaccines against whooping cough. *Vaccine 10*: 1027–1032.

Ratti, G., Rappuoli, R., and Giannini, G. (1983). The complete nucleotide sequence of the gene coding for diphtheria toxin in the corynephage omega (tox+) genome. *Nucleic Acids Res. 11*: 6589–6595.

Ridley, A. J., and Hall, A. (1992). The small GTP-binding protein rho regulates the assembly of focal adhesions and actin stress fibers in response to growth factors. *Cell 70*: 389–399.

Ridley, A. J., Paterson, H. F., Johnston, C. L., Diekmann, D., and Hall, A. (1992). The small GTP-binding protein rac regulates growth factor-induced membrane ruffling. *Cell 70*: 401–410.

Sandvig, K., and Olsnes, S. (1980). Diphtheria toxin entry into cells is facilitated by low pH. *J. Cell Biol. 87*: 828–832.

Sekine, A., Fujiwara, M., and Narumiya, S. (1989). Asparagine residue in the rho gene product is the modification site for botulinum ADP-ribosyltransferase. *J. Biol. Chem. 264*: 8602–8605.

Siegall, C. B., Chaudhary, V. K., FitzGerald, D. J., and Pastan, I. (1989). Functional analysis of domains II, Ib, and III of *Pseudomonas* exotoxin. *J. Biol. Chem. 264*: 14256–14261.

Simonin, F., Briand, J. P., Muller, S., and de Murcia, G. (1991). Detection of poly(ADP ribose) polymerase in crude extracts by activity-blot. *Anal. Biochem. 195*: 226–231.

Sixma, T. K., Pronk, S. E., Kalk, K. H., Wartna, E. S., van Zanten, B. A., Witholt, B., and Hol, W. G. (1991). Crystal structure of a cholera toxin-related heat-labile enterotoxin from *E. coli* [see comments]. *Nature 351*: 371–377.

Sixma, T. K., Kalk, K. H., van Zanten, A. M., Dauter, Z., Kingma, J, Witholt, B., and Hol,

W. (1993). Refined structure of *Escherichia coli* heat-labile enterotoxin, a close relative of cholera toxin. *J. Mol. Biol. 230*: 890–918.

Skorko, R., and Kur, J. (1982). ADP-ribosylation of proteins in non-infected *Escherichia coli* cells. *Eur. J. Biochem. 116*: 317–322.

Sokol, P. A., Dennis, J. J., MacDougall, P. C., Sexton, M., and Woods, D. E. (1990). Cloning and expression of the *Pseudomonas aeruginosa* exoenzyme S toxin gene. *Microb. Pathog. 8*: 243–257.

Spiegel, A. M., Backlund, P. S., Jr., Butrynski, J. E., Jones, T. L., and Simonds, W. F. (1991). The G protein connection: molecular basis of membrane association *Trends Biochem. Sci. 16*: 338–341. [Published erratum appears in *Trends Biochem. Sci. 17*(5): 177 (1992).]

Stein, P. E., Boodhoo, A., Tyrrell, G. J., Brunton, J. L., and Read, R. J. (1992). Crystal structure of the cell-binding B oligomer of verotoxin-1 from *E. coli*. *Nature 355*: 748–750.

Sugai, M., Hashimoto, K., Kikuchi, A., Inoue, S., Okumura, H., Matsumoto, K., Goto, Y., Ohgai, H., Moriishi, K., Syuto, B., Yoshikawa, K., Suginaka, H., and Takai, Y. (1992). Epidermal cell differentiation inhibitor ADP-ribosylates small GTP-binding proteins and induces hyperplasia of epidermis. *J. Biol. Chem. 267*: 2600–2604.

Tamura, M., Nogimori, K., Katada, T., Ui, M., and Ishii, S. (1982). Subunit structure of islet-activating protein, pertussis toxin, in conformity with the A-B model. *Biochemistry 21*: 5516–5520.

Tamura, M., Nogimori, K., Yajima, M., Ase, K., and Ui, M. (1983). A role of the B-oligomer moiety of islet-activating protein, pertussis toxin, in development of the biological effects on intact cells. *J. Biol. Chem. 258*: 6756–6761.

Thanabalu, T., Hindley, J., Jackson-Yap, J., and Berry, C. (1991). Cloning, sequencing, and expression of a gene encoding a 100-kilodalton mosquitocidal toxin from *Bacillus sphaericus* SSII-1. *J. Bacteriol. 173*: 2776–2785.

Thanabalu, T., Hindley, J., and Berry, C. (1992). Proteolytic processing of the mosquitocidal toxin from *Bacillus sphaericus* SSII-1. *J. Bacteriol. 174*: 5051–5056.

Thanabalu, T., Berry, C., and Hindley, J. (1993). Cytotoxicity and ADP-ribosylating activity of the mosquitocidal toxin from *Bacillus sphaericus* SSII-1: possible roles of the 27- and 70-kilodalton peptides. *J. Bacteriol. 175*: 2314.

Thibodeau, J., Simonin, F., Favazza, M., and Murcia, G. (1990). Expression in *E. coli* of the catalytic domain of rat poly(ADP-ribose) polymerase. *FEBS Lett. 264*: 81–83.

Tomasi, M., and Montecucco, C. (1981). Lipid insertion of cholera toxin after binding to GM1-containing liposomes. *J. Biol. Chem. 256*: 11177–11181.

Tomasi, M., Montecucco, C., Gallina, A., and D'Agnolo, G. (1986). Interaction of cholera toxin with lipid model membrane. In *Bacterial Protein Toxins*, P. Falmagne, J. E. Alouf, F. J. Fehrenbach, J. Jeljaszewicz, and M. Thelestam (Eds.), Gustav Fischer Verlag, Stuttgart, pp. 19–25.

Tsai, S. C., Adamik, R., Moss, J., Vaughan, M., Manne, V., and Kung, H. F. (1985). Effects of phospholipids and ADP-ribosylation on GTP hydrolysis by *Escherichia coli*-synthesized Ha-ras-encoded p21. *Proc. Natl. Acad. Sci. U.S.A. 82*: 8310–8314.

Ueda, K., and Hayaishi, O. (1985). ADP-ribosylation. *Annu. Rev. Biochem. 54*: 73–100.

Van Dop, C., Tsubokawa, M., Bourne, H. R., and Ramachandran, J. (1984). Amino acid sequence of retinal transducin at the site ADP-ribosylated by cholera toxin. *J. Biol. Chem. 259*: 696–698.

Van Heyningen, S. (1985). Cholera and related toxins. In *Molecular Medicine*, A. D. B. Malcom (Ed.), IRL Press, Oxford.

Walz, G., Zanker, B., Murphy, J. R., and Strom, T. B. (1990). A kinetic analysis of the effects of interleukin-2 diphtheria toxin fusion protein upon activated T cells. *Transplantation 49*: 198–201.

West, R. E., Jr., Moss, J., Vaughan, M., Liu, T., and Liu, T. Y. (1985). Pertussis toxin-catalyzed

ADP-ribosylation of transducin. Cysteine 347 is the ADP-ribose acceptor site. *J. Biol. Chem.* *260*: 14428–14430.

Williams, D. P., Parker, K., Bacha, P., Bishai, W., Borowski, M., Genbauffe, F., Strom, T. B., and Murphy, J. R. (1987). Diphtheria toxin receptor binding domain substitution with interleukin-2: genetic construction and properties of a diphtheria toxin-related interleukin-2 fusion protein. *Protein Eng.* *1*: 493–498.

5

Three-Dimensional Structure of Diphtheria Toxin

R. John Collier

Harvard Medical School, Boston, Massachusetts

I. INTRODUCTION

Diphtheria toxin (DT) has served in recent years as a paradigm for the investigation of toxic bacterial proteins, particularly those that act intracellularly and more specifically those that catalyze ADP-ribosylation reactions. DT was the first toxin shown to act intracellularly and the first identified as catalyzing ADP-ribosylation of its cytoplasmic target protein. The toxin was crystallized in 1982 by two groups (McKeever and Sarma, 1982; Collier et al., 1982), and a 2.5 Å structure was reported in 1992 by Choe et al. (1992). The crystallographic model has provided a basis for interpreting known structure–function relationships and will undoubtedly foster additional advances in our understanding of how this toxin binds to its receptor, penetrates membranes, and catalyzes the covalent modification of its target protein.

Well after structure determination had begun, we found that the form of DT that had crystallized, and whose structure was being analyzed, was a dimer. Later we learned that this form, together with other oligomeric forms, was generated by freezing the toxin under certain conditions, specifically in the presence of buffers such as sodium phosphate, which can undergo a dramatic reduction in pH during slow freezing (Carroll et al., 1986). The dimer is nontoxic but dissociates slowly at neutral pH to yield a fully toxic monomer; the dissociation may be accelerated by treatment with dimethylsulfoxide (Carroll et al., 1986). Recently M. Bennett and D. Eisenberg (personal communication) performed diffraction studies on crystals of monomeric toxin that show that the structure of the monomer corresponds closely to that predicted by Choe et al. (1992) from crystallography of the dimer. This correlates with other data such as the relative affinity constants of dinucleotides for the monomeric and dimeric forms, which imply close similarities of structure (Carroll et al., 1986).

Monomeric diphtheria toxin consists of 4127 non-hydrogen atoms, arranged in three domains (Fig. 1): the N-terminal C domain (residues 1–193); a middle T domain (residues 205–378); and a C-terminal R domain (residues 386–535). The C domain corresponds to fragment A (DTA), and fragment B (DTB) contains domains T and R. The three domains are arranged in the form of a Y (about 90 Å high, 30 Å thick, and 50 Å across at the top), with the T domain serving as the base and the C and R domains as the two arms. The hydrophilic 14-residue loop (residues 187–200) that links the C and T domains (or A and B fragments) and represents the site of nicking by trypsin (arginines 190, 192, and 193) or other proteases is not seen in the crystallographic structure, implying that it extends into solution and is highly mobile. This is consistent with the finding that the loop is highly vulnerable to proteolytic attack.

The structure gives a straightforward explanation of the finding that the C domain within the whole toxin is inactive in ADP-ribosylating elongation factor 2 (EF-2) and must be dissociated from the remainder of the molecule to become enzymatically active. The R domain clearly blocks the active site cleft of the C domain and would be expected to prevent EF-2 ($M_r \sim 100{,}000$) from docking to the cleft. An aqueous passage remains by which small molecules, such as NAD, can diffuse into the cleft, however, which is consistent with the observation that whole DT exhibits NAD-glycohydrolase activity similar to that of DTA (Kandel et al., 1974).

II. THE CATALYTIC DOMAIN

The C domain contains a mixture of α and β structures [eight β strands (CB1–CB8) and seven α helices (CH1–CH7)]. The active site cleft is subtended by two β-sheet subdomains and is similar in folding to that of *Pseudomonas* exotoxin A (ETA) (Allured et al., 1986). Least squares superposition of the α-carbon atoms of the C domain of DT with domain III, the catalytic domain of ETA, yields an rms difference of 1.44 Å between 85 residues. The A chain of *E. coli* heat-labile toxin (LT) has a fold similar to that of ETA and DT (Sixma et al., 1991) (the α-carbon atoms of 44 residues of LT are superimposable with ETA within an rms difference of 1.5 Å) β strands CB2, CB4, and CB8, surrounded by α helices CH2, CH3, CH6, and CH7, form one subdomain of C, while β strands CB1, CB3, and CB5–CB7 surrounded by helices CH1, CH4, and CH5 form the other. Four extended loops connect the two subdomains and may endow the domain with a mode of flexibility that conceivably might function during its enzymatic function or its traversal of the endosomal membrane.

The active-site cleft of C is occupied by ApUp, the endogenous dinucleotide found in many preparations of toxin (Barbieri et al., 1981; Lory and Collier, 1980). ApUp is known to bind competitively with NAD and shows an extraordinarily high affinity for DT ($K_d \sim 0.3$ nM), which may result from multiple contacts with C, complemented by salt bridges between the 3'-terminal phosphate of ApUp and the side chain of Arg-458.

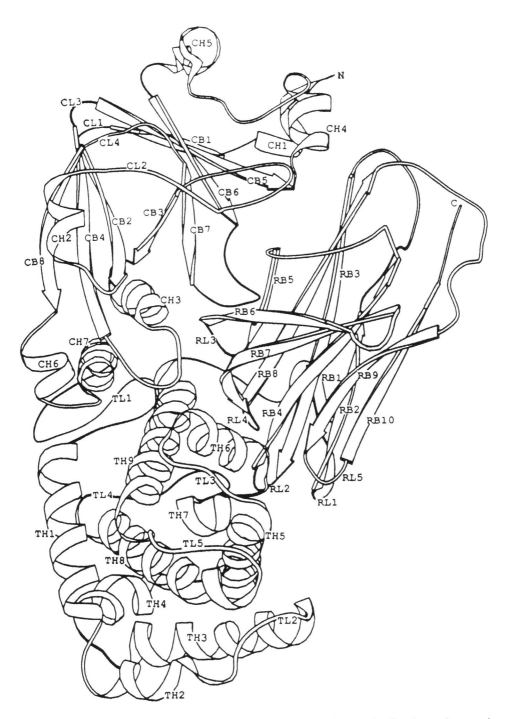

Fig. 1 A ribbon drawing of diphtheria toxin. In the labels shown, the first letter denotes the domain (C for catalytic, T for transmembrane, and R for receptor-binding). The second letter denotes the structure class (H for helix, B for strand, L for loop), and the third symbol is the sequential number of the secondary segment from the N terminus of each domain.

The latter residue is from the R domain, which implies that ApUp forms a noncovalent bridge between the C and R domains, a configuration that had been predicted earlier on the basis of biochemical results. While one would hope to be able to predict the conformation of substrate NAD in the active site cleft from that of ApUp, such predictions remain largely speculative, because of both differences between the structures of NAD and ApUp and effects on ApUp conformation of contacts with the R domain. Despite these qualifications, one finds the uridine ring of ApUp, which represents a quasi-structural analogue of the nicotinamide ring of NAD, located adjacent to the side chains of Glu-148, His-21, Tyr-54, and Tyr-65, all of which have been implicated as residing at or near the catalytic center.

There is increasing evidence that a crucial active-site Glu residue—Glu-148 in DT (Carroll et al., 1985; Carroll and Collier, 1984), Glu-553 in ETA (Carroll and Collier, 1987), Glu-112 in *E. coli* heat-labile toxin (Sixma et al., 1991), and Glu-129 in pertussis toxin (Cockle, 1989; Barbieri et al., 1989)—may prove to be a universal feature of the ADP-ribosylating toxins. We have studied in detail the effects of mutating Glu-148 in DT to Gln, Asp, or Ser (Wilson et al., 1990). These mutations all reduce $k$cat by 100-fold or more but have little effect on the affinity for NAD or EF-2. Analysis of current data (Wilson and Collier, unpublished) has led us to propose that the major function of Glu-148 in catalysis is, in all probability, to deprotonate the attacking diphthamide residue of EF-2. The homologous active-site Glu residues of the other ADP-ribosylating toxins are therefore inferred to deprotonate the attacking nucleophilic residues of their respective target proteins.

The crucial active-site Glu seen in the toxic ADP-ribosyltransferases (and perhaps the active-site fold seen in these toxins) may be present in nontoxic ADP-ribosyltransferases as well. Dominighini et al. (1991) predicted from sequence comparisons that residue Glu-988 of poly(ADP-ribose) polymerase corresponded to Glu-148 of DT, and support for this hypothesis has recently been obtained by Marsischky and Collier, who have shown that conservative mutations of this residue cause dramatic reductions in formation of poly(ADP-ribose) (unpublished results).

The studies of Papini et al. (1989, 1990) implicated His-21 of DT, which corresponds in location to His-440 of ETA, and Arg-7 in LT as important for NAD binding. Blanke et al. (1992) demonstrated that although most substitutions for H21 in DT reduce activity by a large factor (100-fold), substitution of Asn reduces ADP-ribosyltransferase activity by only about three- to fourfold. From the retention of activity by Asn-21 DTA, we infer that the H-bonding capacity of His-21, rather than its potential to ionize, is most important in its ability to promote the ADP-ribosylation of EF-2. Recently, we found that Asn-21 DTA shows a reduction in affinity for NAD of ca. tenfold relative to that of wild-type DTA. Our current model is that His-21 forms hydrogen bonds with the carboxamide moiety of the nicotinamide ring of NAD and thereby serves to orient it optimally within the active-site cleft.

Residues that represent contact points for EF-2 or its diphthamide moiety have not been defined in DT. However, the crystallographic structure of the toxin, noncleavable NAD analogues, synthetic diphthamide, and purified EF-2 should foster studies leading to a detailed description of the interaction of DT with both of its substrates.

III. THE TRANSMEMBRANE DOMAIN

The feature of intracellularly acting toxins that has attracted perhaps the greatest interest is how the catalytic domain crosses a membrane barrier shielding it from its cytoplasmic substrates. In the case of DT, it is well documented that acidic pH, near pH 5, triggers the translocation of the C domain from the endosomal compartment to the cytoplasm. Also, it is clear that functional sites within the amino-terminal half of DTB mediate this process. This region, which contains hydrophobic and amphipathic stretches of sequence, is capable of inserting into membranes at pH near 5 to form ion-conductive channels, as demonstrated by studies in planar lipid bilayers and whole cells (Kagan et al., 1981). The relevance of these channels to the process of translocation is uncertain, but at a minimum their existence is a clear indication that the toxin is capable of being converted from a soluble globular form to an integral membrane protein.

The T domain of DT contains nine α-helices (TH1–TH9), which may be conceived to be formed of three layers, each consisting of two or more antiparallel helices (Fig. 2). A similar α-helical bundle has been found to constitute the channel-forming domain of colicin A (Parker et al., 1989). A helical hairpin containing two long, unusually apolar helices, TH8 and TH9, forms the innermost layer. The second layer is formed of a second hairpin, composed of helices TH5–TH7, two of which, TH6 and TH7, are also hydrophobic. The third, outermost layer consists of helices TH1–TH3, which are hydrophilic and contain large numbers of ionizable residues.

This overall structure—a buried hydrophobic helical hairpin, covered by another hydrophobic hairpin and an outer hydrophilic layer—led Choe et al. (1992) to propose a model of membrane insertion. Perturbation of the tertiary structure of T by acidic pH was assumed to expose the TH8–TH9 hydrophobic hairpin; and this hairpin, and possibly the TH5–TH7 hairpin as well, were postulated to undergo membrane insertion such that the loops at the tips of the hairpins became exposed to solvent at the opposite face of the membrane. This is similar to the model by which the buried helical hairpin of the channel-forming domain of colicin A (helices 9 and 10) is believed to penetrate the membrane of target *Escherichia coli* cells (Parker et al., 1989). However, the loops of the TH8–TH9 and TH5–TH7 hairpins, unlike the loop connecting helices 9 and 10 of colicin A, are acidic and contain two and three acidic side chains, respectively. This

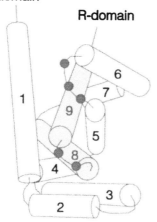

C-domain

R-domain

Fig. 2   A schematic diagram showing the arrangement of the nine helices within the T domain. The filled circles in the loops represent acidic residues in the loops between helices TH5 and TH6 (loop TL3) and between helices TH8 and TH9 (loop TL5).

suggests that acidic pH may, as one role among many possible, serve to protonate these acidic residues and thereby facilitate traversal of the bilayer by the loops.

Recent evidence supports the notion that the TH8–TH9 hairpin of T inserts into bilayers and that this insertion is required for membrane translocation by DTA. Evidence for this came initially from the properties of a translocation-deficient random mutant of DT selected in *E. coli*, containing a charge-reversal mutation located in the TL5 loop (O'Keefe and Collier, 1989, 1990; O'Keefe et al., 1992). This E349K mutation dramatically reduced cytotoxicity, pH-dependent translocation, and the ability to form channels in artificial lipid bilayers. Recently, Silverman and Collier (unpublished results) found that substitution of Lys for the other acidic residue within TL5, Asp-352, causes a similar pattern of reduction in these activities. Also, substitution of Lys for Ile-364, which is centrally located within TH9, caused similar reductions in cytotoxicity, translocation activity, and channel formation (Cabiaux et al., 1993). Inasmuch as a positively charged residue would be required to undergo an energetically unfavorable membrane traversal (in the case of E349K and D352K mutants) or insertion into the hydrophobic core of the membrane (in the case of I364K mutant), the loss of activity by these mutants supports the notion that the TH8–TH9 hairpin undergoes membrane insertion as an essential step in toxin action.

More direct evidence for the proposed topography of insertion of the TH8–TH9 hairpin into the bilayer has come from studies of the conductance of single channels generated by various mutant forms of DT (Mindell et al., 1992). When wild-type DT was inserted into planar lipid bilayers from the cis compartment at acidic pH and the single channel conductance was measured as a function of the pH on the opposite (trans) side of the membrane, a conductance increase over the trans range of ca. pH 4.5–7 was observed, with a midpoint near pH 5.5 (Mindell et al., 1992). These results are consistent with the notion that the flow of ions through the channel (known to be cation-selective) is influenced by the ionization state of a carboxyl group titratable from the trans compartment (analogous to the cytoplasmic compartment of cells). In channels formed by the D352N mutant, this titration behavior was absent; that is, single-channel conductance remained essentially constant over the same pH range. This implies that the ionization state of the side-chain carboxyl of Asp-352 controls the trans pH-dependent conductance changes observed. This residue was titratable only from the trans compartment, under conditions in which the channel is strongly buffered at acidic pH from the cis side. Thus Asp-352, and hence the TL5 loop in which it resides, are localized to a site at or near the trans face of the bilayer (Fig. 3).

To identify other residues besides Asp-352 that affect DT channel conductance, the cis and trans pH-dependence of single-channel conductance was measured with other charge-reversal mutants. All ionizable residues in the TH8–TH9 hairpin and flanking regions were surveyed. The only other residues identified that significantly influenced conductance were Glu-326, Glu-327, and Glu-362. Glu-326 and Glu-327, which lie at the base of TH8, were titratable from the cis side of the membrane, while Glu-362, which is located centrally within TH9, was accessible to titration from both sides. These findings are consistent with the model of the TH8–TH9 hairpin spanning the membrane, with the TL5 loop located at the trans face and the bases of the helices at the cis face. In other studies we have shown that the minimal channel-forming region of DT corresponds with the TH8–TH9 hairpin region, containing those four acidic residues that are the primary determinants of single-channel conductance (E326, E327, D352, and E362).

Recently we completed a survey of all other ionizable residues in and around the

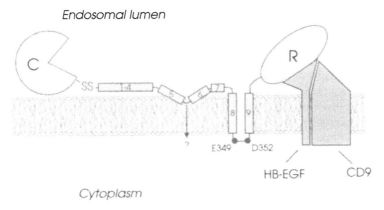

Fig. 3 Schematic model of the insertion of diphtheria toxin into a lipid bilayer. Not represented here is the possibility that channel formation involves oligomerization of the toxin, and particularly of the TH8–TH9 hairpin, in the membrane.

TH8–TH9 and TH5–TH7 hairpins, in which each was mutated to one of opposite charge (Silverman, Mindell, Finkelstein, and Collier, unpublished results). Only one other charge-reversal mutation examined, D352K, located in the same (TL5) loop as D349, caused a major reduction in cytotoxic activity (> 100-fold) similar to that of E349K. A second mutation, D395K, located in the TL3 loop of the TH5–TH7 helical hairpin, caused a reproducible, but less dramatic, effect on toxicity (discussed below). Both the E349K and D352K mutations in TL5 caused a striking loss in pH-dependent formation of channels in artificial lipid bilayers and a loss in ability to permeabilize Vero cells to $^{86}Rb^+$ under acidic conditions. Toxin containing these mutations formed occasional channels in artificial bilayers, however, and the channels had approximately the same lifetime as wild-type channels. This supports the notion that these mutations inhibit, but do not entirely block, insertion of the TH8–TH9 hairpin.

Although the evidence outlined above for insertion of the TH9–TH9 hairpin into membranes at pH 5 is strong, there is no clear indication to date that the TH5–TH7 hairpin undergoes similar insertion. Among the three acidic residues within TL3, the hairpin loop of TH5–TH7, changes in only one had a significant effect on cytotoxic activity. Thus cytotoxicity and channel formation by DT containing the D290K or E292K mutation were normal. In contrast, the D295K substitution reduced the cytotoxicity as assayed at 6 h, but not at 24 h, suggesting that D295K affects the rate of intoxication. The D295K mutation also caused a major reduction of channel lifetime in artificial bilayers, and the ability of toxin to induce low-pH-dependent leakage of $^{86}Rb^+$ leakage from Vero cells was abolished. Further work will be necessary to clarify the basis of these effects and obtain evidence on the status of the TH5–TH7 hairpin during intoxication. In the absence of clear evidence that the TH5–TH7 helices undergo membrane insertion, we are modeling them as remaining on the surface of the bilayer after insertion of TH8–TH9 (Fig. 3).

The helices of the outermost layer of the T domain, TH1–TH4, have a high density of charged residues, implying that this region of the domain does not undergo insertion into the bilayer. At pH near 5 this region should be negatively charged and may bind via electrostatic interactions to the phospholipid head groups of the bilayer. In this capacity, this region may serve as the initial point of contact of the T domain with the bilayer.

Current evidence indicates that a pH-dependent perturbation of tertiary structure of the T domain precedes its insertion into bilayers. The toxin has been proposed to undergo a partial unfolding over a narrow range near pH 5.3 in which secondary structure is largely retained (Dumont and Richards, 1988; Jiang et al., 1991; Kieleczawa et al., 1990; Zhao and London, 1988). Titration in this pH range suggests that the unfolding might depend on the disruption of carboxyl-dependent salt bridges by protonation of Asp and Glu residues, combined with possible charge repulsion resulting from protonation of one or more His residues. We have identified a network of salt bridges in the crystallographic structure of DT involving linkages between various of the helices of T. Helix TH1, which serves as a "backbone" of the T domain, is linked to other helices via salt bridges. Thus, for example, Arg-210 within TH1 forms a bridge with Glu-362 within TH9. Also, both Lys-212 and Lys-216 appear to interact with Glu-327 at the amino-terminal end of TH8. Falnes et al. (1992) showed that mutation of Glu-362 shifts the curve of cell-surface intoxication to higher pH, thus supporting the notion that the R210–E362 bridge is involved in stabilization of T. Two His residues are found in close proximity, His-223 in TH2 and His-257 in TH3; electrostatic repulsion following protonation of these residues may perturb tertiary structure significantly. Systematic mutagenesis should reveal the relative importance of the various bridges and potentially repulsive cationic pairs of residues in inducing unfolding.

Substitution of Lys for either Glu-349 or Asp-352 within the TL5 loop was found not to alter the profile of pH-dependent unfolding of DT, as monitored by trypsin sensitivity or fluorescence methods (Silverman and Collier, unpublished results). This is consistent with the fact that neither of these acidic residues forms a salt bridge in the crystal structure. The ionized acidic residues within TL5 impede the insertion of the TH8–TH9 hairpin into the membrane at neutral pH, and protonation at acidic pH may facilitate penetration. Once the TL5 loop has penetrated the bilayer and been exposed to the neutral pH of the cytoplasm, the acidic tip residues should reionize and anchor the toxin in a transmembrane configuration. This would tether the A fragment directly to the bilayer and presumably facilitate its penetration (Fig. 2). The conformation in which the A fragment traverses the bilayer is unknown and is of major interest in understanding the process of translocation. Evidence from various sources indicates that isolated A fragment is able to insert into the bilayer at acidic pH (Mindell et al., 1992; Jiang et al., 1989; Hu and Holmes, 1984), and it may be that its being anchored to the bilayer by the inserted T domain promotes this insertion. Direct evidence regarding the question of whether the A chain traverses the membrane directly through the bilayer or, alternatively, through a proteinaceous channel or groove will be useful in analyzing this question. The detailed role(s) of the T domain in the process of translocation and the structure of DTA as it passes through the bilayer are now within experimental reach, with the diverse and powerful biochemical, biophysical, and genetic tools now available.

IV. THE RECEPTOR-BINDING DOMAIN

The specific proteinaceous receptor on the surface of sensitive mammalian cells recognized by the R domain was identified recently by Eidels and coworkers as heparin-binding EGF-like growth factor precursor (Naglich et al., 1992). Also, Mekada and coworkers identified a 27-kDa protein that copurifies with the receptor and appears to play a role in intoxication, as CD-9 (Mitamura et al., 1992; Iwamoto et al., 1991). The R domain of DT is constructed of two β sheets, one containing β strands RB2, RB3, RB5, and RB8 and the other containing strands RB4, RB6, RB7, and RB10. The topography of these two sheets is reminiscent of the variable domain of immunoglobulins and is similar to the jelly-roll topology found in viral coat proteins, tumor necrosis factor, and the receptor-binding domain of ETA.

The manner in which the DT monomers are associated within the dimer provides an explanation for the dimer's being nontoxic, due to its inability to bind to cell-surface receptors. The monomers are associated via six hydrogen bonds, between RB1/RB2 of one DT molecule and RB1/RB2 of the other, giving the dimer twofold rotational symmetry. Productive contact with the toxin receptor is thereby apparently blocked. Recent studies by Bennett and Eisenberg (unpublished results) strongly suggest that the R domain seen associated with the C and T domains of one monomer in fact belongs to that of the other monomer within the dimer. Dimer formation occurs during slow freezing of toxin in certain buffers that undergo a decline in pH during freezing. Under such conditions, noncovalent contacts of R with C and T are apparently weakened, allowing the R domain of a given molecule to become associated with C and T of another. The elevation of solute concentrations during slow freezing would be expected to increase the frequency of intermolecular collisions of DT, thereby facilitating oligomerization. There is no evidence that dimer formation of the type described here occurs during intoxication of cells.

After the R domain was defined by Choe et al. (1992), we cloned and expressed this part of the toxin and demonstrated that it alone was capable of blocking the action of DT on sensitive cells (Fu et al., 1993). This demonstrates that the receptor-binding site of the toxin is fully contained within the R domain. Rolf and coworkers (Rolf et al., 1990; Rolf and Eidels, 1993) have shown that a peptide fragment, HA6DT, generated by hydroxylamine cleavage and corresponding to the last 54 residues of DT, blocks the binding of the toxin to sensitive cells and protects them from its action. This peptide contains, approximately, RB8, RB9, RB10, and intervening loops. Toxin mutants with Ser-508 to Phe or Ser-525 to Phe substitutions, all within this peptide, show greatly reduced receptor-binding activity (Greenfield et al., 1987). Systematic mutagenesis studies should permit a detailed mapping of the receptor-binding region of the toxin.

Besides its receptor-binding site, the R domain also contains part or all of a second (and possibly overlapping) site, the P site, occupancy of which has been shown to block binding of the toxin to receptors. Middlebrook and coworkers showed that ATP and certain related phosphate-containing compounds inhibit the action of DT on whole cells and that this inhibition is due to interference with toxin attachment to cell surface receptors (Middlebrook and Dorland, 1979). Lory and coworkers later showed that binding of NAD-site and P-site ligands was competitive (Lory and Collier, 1980; Lory et al., 1980). On the basis of this and other studies they proposed a model in which the NAD site on DTA in whole DT was adjacent to a strongly cationic region within the carboxyl-terminal half of DTB, with a strong affinity for polyanions. Proia et al. (1979, 1980, 1981) incubated DT with periodate-treated ATP, followed by sodium cyanoborohydride reduction, and demonstrated affinity labeling of an 8-kDa CNBr fragment from the carboxyl terminus of DT. Most of the label was attached to Lys-474, near the second disulfide bridge (Cys-461–Cys-471). This and other basic residues in the linear sequence near this disulfide were proposed to contribute to the P site.

The crystallographic structure shows that the face of R domain to the active-site cleft of C contains a high concentration of basic residues, including lysines 447, 456, 460, and 474; arginines 455, 458, 460, 462, and 472; and histidine 449. Histidine 492 is on the same face but somewhat distant from the other residues listed. The side chain of Arg-458 forms a salt bridge with the 3' terminal phosphate of ApUp, as noted earlier. This high concentration of basic residues is likely to constitute, or contain, the P site. Further work should reveal those basic residues within this set that interact directly and most avidly with polyanionic compounds, and the way in which such interactions interfere with the binding of the toxin to its receptor.

V. APPLICATIONS

Besides fostering a deeper understanding of structure–function relationships, knowledge of the crystallographic structure of DT has provided a firmer basis for developing various medical applications of this toxin. Chemically inactivated DT has, of course, traditionally served as the basis for immunization against diphtheria. Antibodies that interfere with receptor binding are most effective in blocking toxin action, and it may well be that unmodified R domain alone, now that its structure and boundaries have been defined, will prove to be a sufficiently good immunogen to be developed into a vaccine (Fu et al., 1993). The R domain may be particularly useful in developing live vaccines in which an immunogenic domain of DT is expressed by a carrier microbe. Expression of the R domain alone should eliminate fear of reversion to a toxic state. Also, DT is one of several cytocidal toxins that have been employed in developing targeted cytotoxic agents. The major aim of these studies has been to substitute the receptor-binding function of some surrogate cell-binding protein for that of the toxin without perturbing those portions of the toxin that serve the catalytic and translocation functions. For example, Murphy and coworkers have constructed a hybrid protein in which C-terminal regions of the toxin were replaced by interleukin-2 (Williams et al., 1988). Knowledge of the precise boundaries of the T domain, together with other details of the structure of DT, should foster development of more effective targeted toxins.

Acknowledgments

Efforts to elucidate the crystallographic structure of diphtheria toxin began when I spent a sabbatical year in the laboratory of Fred Richards, at Yale, in 1980–1981, during which I was able to obtain X-ray grade crystals of diphtheria toxin (and *Pseudomonas aeruginosa* exotoxin A). Returning from Yale to UCLA, my academic home at the time, I contacted my friend and colleague, David Eisenberg, who was eager to undertake the crystallographic structure determination. The final structure determination was the result of a long collaborative effort, in which many individuals from the Eisenberg and Collier laboratories participated in one way or another. Particular thanks to Joe Barbieri, Melanie Bennett, Steve Carroll, Seunghyon Choe, Bauke Dykstra, Kathy Kantardjieff, Gary Fujii, Paul Kurmi, and Ed Westbrook. Seunghyon Choe deserves special credit for bringing a long and difficult structure determination to a successful conclusion.

REFERENCES

Allured, V. S., Collier, R. J., Carroll, S. F., and McKay, D. B. (1986). Structure of exotoxin A of *Pseudomonas aeruginosa* at 3.0-angstrom resolution. *Proc. Natl. Acad. Sci. U.S.A.* **83**: 1320.
Barbieri, J. T., Carroll, S. F., Collier, R. J., and McCloskey, J. A. (1981). An endogenous

dinucleotide bound to diphtheria toxin adenylyl-(3',5')-uridine 3'-monophosphate. *J. Biol. Chem.* 256: 12247.

Barbieri, J. T., Mende Mueller, L. M., Rappuoli, R., and Collier, R. J. (1989). Photolabeling of Glu-129 of the S-1 subunit of pertussis toxin with NAD. *Infect. Immun.* 57: 3549.

Blanke, S. R., Collier, R. J., Covacci, A., Fu, H., Killeen, K., Montecucco, C, Papini, E., Rappuoli, R., and Wilson, B. A. (1992). Mutations affecting ADP-ribosyltransferase activity of diphtheria toxin. In *Bacterial Protein Toxins, Fifth European Workshop*, B. Witholt, J. E. Alouf, G. J. Boulnois, P. Cossart, B. W. Dijkstra, P. Falmagne, F. J. Fehrenbach, J. Freer, H. Niemann, R. Rappuoli, and T. Wadstrom, (Eds.), Gustav Fischer Verlag, New York, pp. 349–354.

Cabiaux, V., Mindell, J., and Collier, R. J. (1993). Membrane translocation and channel-forming activities of diphtheria toxin are blocked by replacing isoleucine 364 with lysine. *Infect. Immun.* 61: 2200.

Carroll, S. F., and Collier, R. J. (1984). NAD binding site of diphtheria toxin: identification of a residue within the nicotinamide subsite by photochemical modification with NAD. *Proc. Natl. Acad. Sci. U.S.A.* 81: 3307.

Carroll, S. F., and Collier, R. J. (1987). Active site of *Pseudomonas aeruginosa* exotoxin A. Glutamic acid 553 is photolabeled by NAD and shows functional homology with glutamic acid 148 of diphtheria toxin. *J. Biol. Chem.* 262: 8707.

Carroll, S. F., McCloskey, J. A., Crain, P. F., Oppenheimer, N. J., Marschner, T. M., and Collier, R. J. (1985). Photoaffinity labeling of diphtheria toxin fragment A with NAD: structure of the photoproduct at position 148. *Proc. Natl. Acad. Sci. U.S.A.* 82: 7237.

Carroll, S. F., Barbieri, J. T., and Collier, R. J. (1986). Dimeric form of diphtheria toxin: purification and characterization. *Biochemistry* 25: 2425.

Choe, S., Bennett, M., Fujii, G., Curmi, P. M. G., Kantardjieff, K. A., Collier, R. J., and Eisenberg, D. (1992). The crystal structure of diphtheria toxin. *Nature* 357: 216.

Cockle, S. A. (1989). Identification of an active-site residue in subunit S1 of pertussis toxin by photocrosslinking to NAD. *FEBS Lett.* 249: 329.

Collier, R. J., Westbrook, E. M., McKay, D. B., and Eisenberg, D. (1982). X-ray grade crystals of diphtheria toxin. *J. Biol. Chem.* 257: 5283.

Domenighini, M., Montecucco, C., Ripka, W. C., and Rappuoli, R. (1991). Computer modelling of the NAD binding site of ADP-ribosylating toxins: active-site structure and mechanism of NAD binding. *Mol. Microbiol.* 5: 23.

Dumont, M. E., and Richards, F. M. (1988). The pH-dependent conformational change of diphtheria toxin. *J. Biol. Chem.* 263: 2087.

Falnes, P. O., Madshus, I. H., Sandvig, K., and Olsnes, S. (1992). Replacement of negative by positive charges in the presumed membrane-inserted part of diphtheria toxin B fragment. Effect on membrane translocation and on formation of cation channels. *J. Biol. Chem.* 267: 12284.

Fu, H., Shen, W. H., and Collier, R. J. (1993). Receptor-binding domain of diphtheria toxin as a potential immunogen. In *Vaccines 93*, H. S. Ginsberg, F. Brown, R. M. Chanock, and R. A. Lerner, (Eds.), Cold Spring Harbor Laboratory Press, Cold Spring Harbor, ME, pp. 379–383.

Greenfield, L., Johnson, V. G., and Youle, R. J. (1987). Mutations in diphtheria toxin separate binding from entry and amplify immunotoxin selectivity. *Science* 238: 536.

Hu, V. W., and Holmes, R. K. (1984). Evidence for direct insertion of fragments A and B of diphtheria toxin into model membranes. *J. Biol. Chem.* 259: 12226.

Iwamoto, R., Senoh, H., Okada, Y., Uchida, T., and Mekada, E. (1991). An antibody that inhibits the binding of diphtheria toxin to cells revealed the association of a 27-kDa membrane protein with the diphtheria toxin receptor. *J. Biol. Chem.* 266: 20463.

Jiang, G., Solow, R., and Hu, V. W. (1989). Fragment A of diphtheria toxin causes pH-dependent lesions in model membranes. *J. Biol. Chem.* 264: 17170.

Jiang, J. X., Abrams, F. S., and London, E. (1991). Folding changes in membrane-inserted diphtheria toxin that may play important roles in its translocation. *Biochemistry* 30: 3857.

Kagan, B. L., Finkelstein, A., and Colombini, M. (1981). Diphtheria toxin fragment forms large pores in phospholipid bilayer membranes. *Proc. Natl. Acad. Sci. U.S.A. 78*: 4950.

Kandel, J., Collier, R. J., and Chung, D. W. (1974). Interaction of fragment A from diphtheria toxin with nicotinamide adenine dinucleotide. *J. Biol. Chem. 249*: 2088.

Kieleczawa, J., Zhao, J.-M., Luongo, C. L., Dong, L.-Y. D., and London, E. (1990). The effect of high pH upon diphtheria toxin conformation and model membrane association: role of partial unfolding. *Arch. Biochem. Biophys. 282*: 214.

Lory, S., and Collier, R. J. (1980). Diphtheria toxin: nucleotide binding and toxin heterogeneity. *Proc. Natl. Acad. Sci. U.S.A. 77*: 267.

Lory, S., Carroll, S. F., and Collier, R. J. (1980). Ligand interactions of diphtheria toxin. II. Relationships between the NAD site and the P site. *J. Biol. Chem. 255*: 12016.

McKeever, B., and Sarma, R. (1982). Preliminary crystallographic investigation of the protein toxin from *Corynebacterium diphtheriae*. *J. Biol. Chem. 257*: 6923.

Middlebrook, J. L., and Dorland, R. B. (1979). Protection of mammalian cells from diphtheria toxin by exogenous nucleotides. *Can. J. Microbiol. 25*: 285.

Mindell, J. A., Silverman, J. A., Collier, R. J., and Finkelstein, A. (1992). Locating a residue in the diphtheria toxin channel. *Biophys. J. 62*: 41.

Mitamura, T., Iwamoto, R., Umata, T., Yomo, T., Urabe, I., Tsuneoka, M., and Mekada, E. (1992). The 27-kD diphtheria toxin receptor-associated protein (DRAP27) from Vero cells is the monkey homologue of human CD9 antigen: expression of DRAP27 elevates the number of diphtheria toxin receptors on toxin-sensitive cells. *J. Cell. Biol. 118*: 1389.

Naglich, J. G., Metherall, J. E., Russell, D. W., and Eidels, L. (1992). Expression cloning of a diphtheria toxin receptor: identity with a heparin-binding EGF-like growth factor precursor. *Cell 69*: 1051.

O'Keefe, D., and Collier, R. J. (1989). Cloned diphtheria toxin within the periplasm of *Escherichia coli* causes lethal membrane damage at low pH. *Proc. Natl. Acad. Sci. U.S.A. 86*: 343.

O'Keefe, D. O., and Collier, R. J. (1990). A genetic approach to studying pH-dependent membrane interactions of diphtheria toxin. In *Bacterial Protein Toxins, Fourth European Workshop, Urbino, July 3–July 6, 1989*, R. Rappuoli, J. E. Alouf, P. Falmagne, F. J. Fehrenbach, J. Freer, R. Gross, J. Jeljaszewicz, C. Montecucco, M. Tomasi, T. Wadstrom, and B. Witholt, (Eds.), Gustav Fischer Verlag, New York, pp. 21–28,

O'Keefe, D. O., Cabiaux, V., Choe, S., Eisenberg, D., and Collier, R. J. (1992). pH-dependent insertion of proteins into membranes: B-chain mutation of diphtheria toxin that inhibits membrane translocation, Glu-349—-Lys, *Proc. Natl. Acad. Sci. U.S.A. 89*: 6202.

Papini, E., Schiavo, G., Sandona, D., Rappuoli, R., and Montecucco, C. (1989). Histidine 21 is at the NAD^+ binding site of diphtheria toxin. *J. Biol. Chem. 264*: 12385.

Papini, E., Schiavo, G., Rappuoli, R., and Montecucco, C. (1990). Histidine-21 is involved in diphtheria toxin NAD^+ binding. *Toxicon 28*: 631.

Parker, M. W., Pattus, F., Tucker, A. D., and Tsernoglou, D. (1989). Structure of the membrane-pore-forming fragment of colicin A. *Nature 337*: 93.

Proia, R. L., Hart, D. A., and Eidels, L. (1979). Interaction of diphtheria toxin with phosphorylated molecules. *Infect. Immun. 26*: 942.

Proia, R. L., Wray, S. K., Hart, D. A., and Eidels, L. (1980). Characterization and affinity labeling of the cationic phosphate-binding (nucleotide-binding) peptide located in the receptor-binding region of the B-fragment of diphtheria toxin. *J. Biol. Chem. 255*: 12025.

Proia, R. L., Eidels, L., and Hart, D. A. (1981). Diphtheria toxin:receptor interaction. Characterization of the receptor interaction with the nucleotide-free toxin, the nucleotide-bound toxin, and the B-fragment of the toxin. *J. Biol. Chem. 256*: 4991.

Rolf, J. M., and Eidels, L. (1993). Characterization of the diphtheria toxin receptor-binding domain, *Mol. Microbiol. 7*: 585.

Rolf, J. M., Gaudin, H. M., and Eidels, L. (1990). Localization of the diphtheria toxin receptor-binding domain to the carboxyl-terminal M_r-6000 region of the toxin. *J. Biol. Chem. 265*: 7331.

Sixma, T. K., Pronk, S. E., Kalk, K. H., Wartna, E. S., van Zanten, B. A. M., Witholt, B., and Hol, W. G. J. (1991). Crystal structure of a cholera toxin-related heat-labile enterotoxin from *E. coli. Nature 351*: 371.

Williams, D. P., Regier, D., Akiyoshi, D., Genbauffe, F., and Murphy, J. R. (1988). Design, synthesis and expression of a human interleukin-2 gene incorporating the codon usage bias found in highly expressed *Escherichia coli* genes. *Nucleic Acids Res. 16*: 10453.

Wilson, B. A., Reich, K. A., Weinstein, B. R., and Collier, R. J. (1990). Active-site mutations of diphtheria toxin: effects of replacing glutamic acid-148 with aspartic acid, glutamine, or serine. *Biochemistry 29*: 8643.

Zhao, J. M., and London, E. (1988). Conformation and model membrane interactions of diphtheria toxin fragment A. *J. Biol. Chem. 263*: 15369.

6

The Diphtheria Toxin Receptor

Eisuke Mekada

Institute of Life Science, Kurume University, Kurume, Fukuoka, Japan

I. INTRODUCTION

Since the discovery by Roux of its role in the disease caused by *Corynebacterium diphtheriae*, diphtheria toxin (DT) has been studied first for the prevention of disease and later to understand the pathogenic mechanism at the molecular level. A number of important discoveries in the field of bacterial pathogenesis and bacterial protein toxins have resulted from these studies; for example, (1) antibodies directed against the toxin can protect against the disease, (2) diphtheria toxin acts within cells by inactivating a target protein by ADP-ribosylation, and (3) the toxin consists of two segments with distinct functions. Because of extensive studies over a long period, diphtheria toxin is one of the best known of the bacterial protein toxins, and discoveries such as those enumerated above have frequently led to increased understanding of the biology of other bacterial toxins.

To understand fully the intoxication process, it is necessary to identify the host cell factors that are involved and to determine their structure and function. Although various cellular factors may be involved in the intoxication, a cell surface receptor for diphtheria toxin, the first molecule with which diphtheria toxin interacts, is a key factor that determines the sensitivity of cells to this toxin. Thus, investigators have been interested in identifying and characterizing the receptor. A number of approaches have been taken, such as (1) treatment of the surface of diphtheria toxin-sensitive cells with specific enzymes (Moehring and Crispell, 1974; Sandvig et al., 1978; Mekada et al., 1979); (2) searches for substances from cell membranes that block diphtheria toxin toxicity (Chin and Simon, 1981; Mekada et al., 1988); (3) binding studies using labeled toxin (Bonventre et al., 1975; Boquet and Pappenheimer, 1976; Chang and Neville, 1978); (4) precipitation of a diphtheria toxin-binding molecule using diphtheria toxin and anti-diphtheria toxin antibody (Proia et al., 1979); and (5) isolation of diphtheria toxin-resistant mutants (Moehring and Moehring, 1979; Didsbury et al., 1983; Kohno et al., 1985). However, in spite of those endeavors, the diphtheria toxin receptor has only recently been characterized at the molecular level.

In 1988, the diphtheria toxin receptor protein was first identified as a protein band in SDS-PAGE (Mekada et al., 1988). A cDNA for the receptor was cloned in 1992 (Naglich et al., 1992). Thus information on the diphtheria toxin receptor at the molecular level has accumulated in the last few years. In this chapter, I emphasize recent advances in our knowledge of the diphtheria toxin receptor. Earlier studies have been reviewed elsewhere (Eidels and Draper, 1988).

II. A BRIEF DESCRIPTION OF DIPHTHERIA TOXIN

Diphtheria toxin is produced in *Corynebacterium diphtheriae* lysogenized with β-phage carrying the *tox* structure gene (Freeman, 1951; Uchida et al., 1971). The toxin is secreted into the extracellular space as a single polypeptide chain (Collier and Kandel, 1971; Gill and Dinius, 1971) with a molecular weight of 58,342 (Greenfield et al., 1983). The single polypeptide form, called intact diphtheria toxin, is easily cleaved by trypsin (Gill and Dinius, 1971), furin (Tsuneoka et al., 1993), and other proteolytic enzymes to yield two functionally different fragments, A and B, linked by a disulfide bridge (Collier and Kandel, 1971; Gill and Dinius, 1971). Fragment A catalyzes ADP-ribosylation of elongation factor 2 (EF-2) in the presence of NAD, a reaction that inactivates elongation factor 2 and inhibits cellular protein synthesis (Collier, 1967; Honjo et al., 1968). Fragment B is required for the binding of the toxin to cell surface receptors (Uchida et al., 1972) and is involved in the translocation of endocytosed toxin molecules from an acidic compartment to the cytosol.

Entry of the toxin, or at least its A fragment, into the cytoplasm is required for the cytotoxic action (Uchida et al., 1977). The generally accepted mechanism for toxin entry into the cell is as follows. In the first step, the toxin binds to a specific receptor on the cell surface of the toxin-susceptible cells. The toxin bound to the receptor is concentrated in coated pits and then internalized by endocytosis (Morris et al., 1985; Moya et al., 1985). A conformational change of the toxin molecule takes place in an acidic compartment, resulting in the exposure of hydrophobic domains (Sandvig and Olsnes, 1980; Blewitt et al., 1985; Cabiaux et al., 1989), which are mainly in the B fragment (Boquet et al., 1976; Greenfield et al., 1983). Finally, the enzymatically active fragment A is translocated to the cytosol, where it exerts its toxicity (Moskaug et al., 1991). The exposure of diphtheria toxin to low pH is essential for its translocation to the cytosol (Kim and Groman, 1965; Draper and Simon, 1980; Sandvig and Olsnes, 1980; Mekada et al., 1981). A vacuolar type H^+-ATPase is responsible for acidification of diphtheria toxin-bearing vesicles (Umata et al., 1990).

The foregoing description is intended to provide a minimal knowledge of diphtheria toxin to facilitate the reading of this chapter. Excellent reviews have been published on diphtheria toxin by Collier (1975), Pappenheimer (1977), Uchida (1983), and Eidels and Draper (1988). In another chapter of this volume, Collier describes the three-dimensional structure of diphtheria toxin. One can refer to these reviews for additional information.

III. DIFFERENCES IN THE SENSITIVITY OF CELLS TO DIPHTHERIA TOXIN

Most animals, including rabbits, guinea pigs, and primates, are highly sensitive to the lethal effect of diphtheria toxin. The lethal dose for susceptible animals is 100 ng/kg body weight or less (Collier, 1975; Pappenheimer, 1977). The only mammals known to be resistant to the toxin are rats and mice. These animals are about 1000 times more resistant than susceptible species (Collier, 1975; Pappenheimer, 1977). In general, cells derived from sensitive species are sensitive to the cytotoxic action of diphtheria toxin, while cells from toxin-resistant species are quite resistant to its action (Lennox and Kaplan, 1957; Placido-Sousa and Evans, 1957; Gabliks and Solotorovsky, 1962; Middlebrook and Dorland, 1977). The exception is lymphocytes; even some human cell lines of lymphocyte origin are quite resistant to diphtheria toxin. There are quite large differences in sensitivity between diphtheria toxin-sensitive cells and resistant cells as shown in Figure 1. For

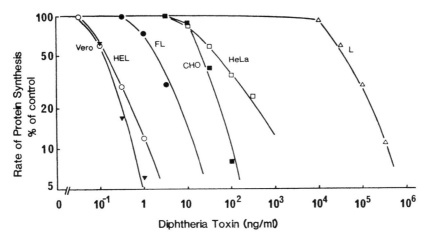

Fig. 1 Sensitivity of various cell lines to diphtheria toxin assayed by measuring the inhibition of protein synthesis. (From Mekada et al., 1982.)

example, Vero cells, one of the cell lines most sensitive to diphtheria toxin, are about 100,000 times more sensitive than L cells (Mekada et al., 1982).

So far, EF-2 from any eukaryote is inactivated by the enzymatically active A fragment of diphtheria toxin in vitro. Thus the resistance of cells to diphtheria toxin is due to the binding and entry process. Indeed, in the case of L cells, one molecule of the A fragment can kill a cell when the fragment is introduced directly into the cytoplasm (Yamaizumi et al., 1978). A number of cellular factors are likely to be involved in the binding and entry of diphtheria toxin and could influence cell sensitivity, but most of these factors are essential for normal cellular functions even in mouse and rat cells. One exception is diphtheria toxin receptor; cells from rats and mice have no functional diphtheria toxin receptor that specifically binds to diphtheria toxin as discussed below. Thus, in many types of cells, cell sensitivity to diphtheria toxin is determined by the presence of functional diphtheria toxin receptors.

IV. BIOCHEMICAL STUDIES OF THE DIPHTHERIA TOXIN RECEPTOR

A. Binding of Diphtheria Toxin to Intact Cells

The existence of a specific receptor for diphtheria toxin in diphtheria toxin-sensitive cells was suggested by competition experiments using CRM197 (Itelson and Gill, 1973). CRM197 is the product of a missense mutation within the fragment A region leading to an enzymatically inactive product that still binds to diphtheria toxin-sensitive cells (Uchida et al., 1972). CRM197 inhibits the toxicity of diphtheria toxin in HeLa cells. Dose-ratio analyses indicated that the inhibition is competitive. Although rat cells can be intoxicated by very high doses of toxin, indicating that DT enters cells inefficiently by a nonspecific route, CRM197 did not inhibit the toxin activity in a rat skin test. These results suggested that a limited number of specific molecules that bind diphtheria toxin confer sensitivity to diphtheria toxin to diphtheria toxin-susceptible cells.

Studies of ^{125}I-labeled diphtheria toxin binding to intact cells were attempted by several groups, but observation of specific binding of diphtheria toxin to cells was hampered by the high background of nonspecifically associated diphtheria toxin (Bonventre et al., 1975; Boquet and Pappenheimer, 1976). This problem was solved by using Vero cells. Middlebrook and Dorland (1977) compared the sensitivity of a number of cultured cells to diphtheria toxin and showed that Vero cells are among the most sensitive. They demonstrated specific binding of ^{125}I-labeled diphtheria toxin to Vero cells with low background levels (Middlebrook et al., 1978). Specific binding was not observed in mouse L cells under the same conditions, so L cells were suggested to have no functional diphtheria toxin receptors.

The number of diphtheria toxin receptors (binding sites) on Vero cells and the affinity for toxin were first reported as 1×10^5 to 2×10^5 sites/cell and a K_a of 5×10^8 to 9×10^8 M^{-1} by Scatchard plot analysis (Middlebrook et al., 1978). However, subsequent studies by our group and others show that the number of diphtheria toxin-binding sites on Vero cells is 0.5×10^4 to 2×10^4 sites/cell, one order of magnitude lower than the earlier report, with K_a values in the range of 0.8×10^9 to 1.7×10^9 M^{-1} (Mekada and Uchida, 1985; Naglich et al., 1992).

Specific binding of diphtheria toxin to Vero cells can be detected using the radiolabeled toxin, but binding to less sensitive cell lines, such as human HeLa cells and CHO cells, is difficult to detect. The following method facilitates observation of specific uptake of diphtheria toxin by less sensitive cells. When Vero cells are incubated with ^{125}I-labeled diphtheria toxin at 37°C, the cell-associated radioactivity increases for 1–2 h and then decreases owing to rapid degradation of internalized diphtheria toxin (Middlebrook et al., 1978). However, in the presence of methylamine or bafilomycin A1, inhibitors of the acidification of intracellular vesicles, degradation of the toxin is blocked, and thus diphtheria toxin accumulates in the cells (Mekada et al., 1981; Umata et al., 1990). We showed that in the presence of methylamine, specific association of ^{125}I-labeled diphtheria toxin with many cell lines can be detected (Mekada et al., 1982; Kohno et al., 1985). Although this method does not directly measure the amount of diphtheria toxin bound to the receptor, a good correlation has been observed between toxin sensitivity and the amount of toxin associated with the cells.

The binding studies described above indicate that diphtheria toxin-sensitive cells possess a functional diphtheria toxin receptor but toxin-insensitive mouse and rat cells do not. However, in some reports a significant association of toxin with toxin-insensitive cells was observed, and the amount of toxin associated with cells was similar for toxin-insensitive and toxin-sensitive cells (Chang and Neville, 1978; Keen et al., 1982). These studies suggested that diphtheria toxin-insensitive cells also bear toxin receptors.

Our current interpretation is that this binding is not due to the receptor but rather to another diphtheria toxin-binding substance in membrane fractions. One candidate is an inhibitory substance we found in membrane preparations of toxin-sensitive and toxin-insensitive cell lines (Mekada et al., 1988). A similar substance was also reported by others (Chin and Simon, 1981). The inhibitor binds to diphtheria toxin, inhibits the binding of toxin to Vero cells, and consequently inhibits the toxicity (Mekada et al., 1988). Since the inhibitor seems to associate with membranes, it is possible to show specific binding of diphtheria toxin if the binding studies are performed with relatively high concentrations of toxin. The inhibitory activity was lost after treatment with RNase, indicating that it contains a ribonucleotide structure. Collier's group demonstrated that some dinucleotides, such as adenyl-(3', 5')-uridine-3'-monophosphate (ApUp), bind tightly to diphtheria toxin (Collins and Collier, 1984). ATP also binds to diphtheria toxin (Lory and Collier, 1980) and inhibits its cytotoxic activity (Middlebrook and Dorland, 1979). Thus the inhibitor seems to inhibit the cytotoxicity of diphtheria toxin by a mechanism similar to that of the nucleotides.

B. Identification and Purification of the Diphtheria Toxin Receptor

The first successful approach to the identification of diphtheria toxin receptor was reported by Cieplak et al. (1987). They incubated surface-iodinated Vero cells or other diphtheria toxin-sensitive cells with DT, followed by solubilization and immunoprecipitation with diphtheria toxin and anti-diphtheria toxin antibody. SDS-PAGE showed that a 10–20-kDa protein is specifically precipitated by this method. In accord with this result, cross-linking experiments resulted in the appearance of an 80-kDa complex, comprising diphtheria toxin and an approximately 20-kDa cellular protein. These results suggest that a 10–20-kDa cell surface protein is, or constitutes a portion of, the diphtheria toxin receptor.

To identify and purify the diphtheria toxin receptor, a suitable assay for detecting the binding of solubilized receptor to diphtheria toxin is essential. As mentioned above, membrane fractions contain an inhibitor that binds to diphtheria toxin. Therefore, it is necessary to distinguish the binding of diphtheria toxin to the diphtheria toxin receptor from binding to the inhibitor. CRM197, a mutant protein of diphtheria toxin, has been used for this purpose. This protein differs from wild-type toxin in one amino acid residue of fragment A (Giannini et al., 1984). CRM197 does not bind the inhibitor (Mekada et al., 1988) but binds to diphtheria toxin receptor with an affinity similar to or greater than that of diphtheria toxin (Mekada and Uchida, 1985).

We demonstrated specific binding of CRM197 to a membrane preparation from Vero cells (Mekada et al., 1988). Such specific binding is not observed in membrane preparations from mouse L cells. Unlabeled diphtheria toxin inhibits the binding of CRM197. Solubilization of diphtheria toxin receptor with diphtheria toxin-binding activity has been achieved with a neutral detergent, octyl-β-D-glucoside (Mekada et al., 1988). The diphtheria toxin-binding activity in the solubilized membrane fraction is assayed by a gel filtration assay (Mekada et al., 1988) or an acetone precipitation assay (Mekada et al., 1991). The acetone precipitation assay allows detection of the binding activity of the solubilized receptor in a rapid and quantitative manner. Figure 2 shows the result of the binding assay using ^{125}I-CRM197 for the crude solubilized receptor fraction from Vero cell membrane. Scatchard analysis showed single class binding with a K_a value of 2.4×10^9 M^{-1}, quite similar to that obtained with intact Vero cells.

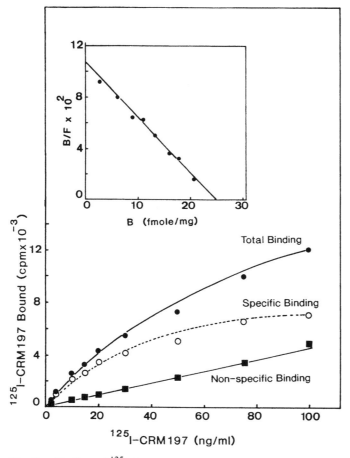

Fig. 2   Binding of ^{125}I-CRM197 to solubilized Vero cell membrane. Inset: Scatchard plot of the specific binding of ^{125}I-CRM197. (From Mekada et al., 1991.)

Using a combination of several chromatographic steps, diphtheria toxin receptor has been purified in the presence of detergent (Mekada et al., 1991). The purified receptor showed essentially a single band of 14.5 kDa by SDS-PAGE. Evidence that the 14.5-kDa protein is the receptor, or at least a part of the receptor, was obtained by immunoprecipitation with diphtheria toxin and anti-diphtheria toxin antibody from partially purified receptor fraction or by blotting using ^{125}I-CRM197 as a probe (Mekada et al., 1988).

Although the 14.5-kDa protein was the major form isolated from Vero cell membrane lysate, other sizes of diphtheria toxin-binding molecules (ca. 20 and 17 kDa) were detected in different fractions from CM-Sepharose ion-exchange chromatography (Mekada et al., 1991). The 14.5-kDa protein and the 17- and 20-kDa proteins may be derived from a single precursor protein, as discussed below, although the possibility that multiple proteins are used as diphtheria toxin receptors has not been excluded.

V. GENETIC APPROACHES TO THE DIPHTHERIA TOXIN RECEPTOR

Genetic mapping of the diphtheria toxin receptor has been done using interspecific hybrid cells. In human-mouse hybrid cells, human chromosomes are selectively lost during cell proliferation. Loss of diphtheria toxin sensitivity was observed when the hybrid cells were cultured. Evidence that hybrid cells that contain human chromosome 5 are sensitive to diphtheria toxin while those that have lost it are resistant indicates that a gene for diphtheria toxin sensitivity is located on human chromosome 5 (Creagan et al., 1975). Further analysis of microcell hybrids with subchromosomal fragments showed that the diphtheria toxin sensitivity gene is located in the q23 region of human chromosome 5 (Hayes et al., 1987).

A diphtheria toxin sensitivity gene has been isolated by expression cloning (Naglich et al., 1992). Mouse L cells were transfected with a cDNA library obtained from monkey Vero cells. Transfectants that were sensitive to diphtheria toxin were isolated using a replica plate assay (Naglich and Eidels, 1990). A cDNA (pDTS) was recovered from the diphtheria toxin-sensitive transfectants. Mouse L cells transfected with this cDNA become sensitive to diphtheria toxin and display diphtheria toxin-binding molecules on the cell surface that have the characteristics of the diphtheria toxin receptor (Naglich et al., 1992). The predicted protein product of pDTS has 185 amino acids, a quite basic isoelectric point, and characteristics of an integral membrane protein. These characteristics are consistent with the characteristics of diphtheria toxin receptor protein isolated by us (Mekada et al., 1991), suggesting that pDTS encodes the diphtheria toxin receptor.

Homology search analysis showed that the product of pDTS is identical to a known heparin-binding growth factor, HB-EGF (Naglich et al., 1992), first identified by Higashiyama et al. (1991). Further evidence that the product of pDTS, HB-EGF, is the diphtheria toxin receptor has been obtained from direct binding experiments with diphtheria toxin and a recombinant human HB-EGF (Iwamoto et al., 1994). A mature form of HB-EGF, produced in *E. coli*, was immobilized on heparin-Sepharose beads, and binding of ^{125}I-diphtheria toxin to the recombinant HB-EGF was studied. The amount of diphtheria toxin bound was a linear function of the amount of immobilized recombinant HB-EGF, and saturation of binding occurred when a fixed amount of immobilized HB-EGF was incubated with increasing amounts of diphtheria toxin. The binding of ^{125}I-diphtheria toxin was inhibited by excess unlabeled diphtheria toxin. Moreover, Scatchard analysis demonstrated that diphtheria toxin binds to HB-EGF with an affinity similar to that for intact Vero cells (K_a of about 1×10^9 M^{-1}). These results clearly show that the diphtheria toxin receptor is HB-EGF.

VI. RECEPTOR STRUCTURE AND REQUIREMENTS FOR DIPHTHERIA TOXIN BINDING

Diphtheria toxin receptor/HB-EGF cDNA encodes a protein of 208 amino acids (Higashiyama et al., 1991; Naglich et al., 1992). This protein consists of a characteristic signal sequence of 23 amino acid residues, a presumed extracellular domain of 136 residues (24–159), a putative transmembrane domain of 25 residues (160–184), and a carboxy-terminal cytoplasmic domain of 24 residues (185–208) (Fig. 3). The mature protein, after cleavage of the signal peptide, seems to be 185 amino acids with a calculated molecular weight of 20,652. The extracellular domain includes two characteristic features: (1) an EGF-like domain with six cysteine residues with highly conserved spacing and (2) a highly hydrophilic stretch of amino acid residues upstream of the EGF-like domain.

Diphtheria toxin receptor/HB-EGF exists not only as a membrane-anchored protein but also as a soluble protein without the transmembrane region. The membrane-anchored form of diphtheria toxin receptor/HB-EGF molecule is cleaved upstream of the transmembrane domain by a cell-associated protease, and the extracellular domain is released into the medium as a soluble growth factor (Higashiyama et al., 1991, 1992). The secreted HB-EGF protein is also processed at multiple sites in the N-terminal portion. In addition, there are potential sites for O-glycosylation in the extracellular domain, and indeed secreted proteins are O-glycosylated (Higashiyama et al., 1992). Thus, multiple forms of HB-EGF with different molecular sizes are secreted.

Which domain of HB-EGF is involved in diphtheria toxin binding? HB-EGF secreted

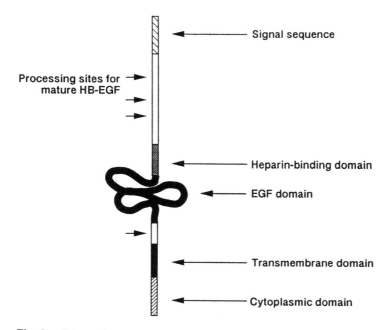

Fig. 3   Schematic structure of the diphtheria toxin receptor.

into culture media binds to diphtheria toxin (Iwamoto et al., 1994). Thus the diphtheria toxin binding site lies wholly or partially within this region. Furthermore, a human recombinant HB-EGF binds to ^{125}I-diphtheria toxin with an affinity similar to that of the diphtheria toxin receptor for diphtheria toxin (Iwamoto et al., 1994). As the recombinant material was made in *E. coli*, involvement of glycosylation of the protein in the binding activity can be ruled out. Because one can now obtain modified HB-EGF protein by modification of the HB-EGF structure gene, more precise information on the binding site will be available soon.

Because HB-EGF is expressed in multiple tissues in rats, mice, and humans with a very similar tissue distribution (Abraham et al., 1993), the question arises as to why cells from rats and mice are resistant to diphtheria toxin. Transfection of human diphtheria toxin receptor/HB-EGF cDNA into mouse L cells confers sensitivity to diphtheria toxin (Naglich et al., 1992), but transfection of mouse HB-EGF cDNA does not (Mitamura et al., submitted). Therefore, it is likely that mouse and rat HB-EGF do not have an optimal diphtheria toxin-binding sequence.

VII. MOLECULES ASSOCIATED WITH THE DIPHTHERIA TOXIN RECEPTOR

Iwamoto et al. (1991) isolated a monoclonal antibody (mAb 007) that inhibits the binding of diphtheria toxin to intact Vero cells. This antibody does not inhibit the binding of diphtheria toxin to solubilized receptor, suggesting that membrane integrity is necessary for the inhibition. Immunoprecipitation and Western blotting analysis revealed that this antibody recognizes a membrane protein of 27 kDa (DRAP27) and does not bind to diphtheria toxin receptor. When diphtheria toxin receptor was passed through an affinity column prepared with this antibody, the receptor was trapped on the column in the presence of DRAP27, whereas in the absence of DRAP27 the receptor was not trapped. It has also been demonstrated recently that either anti-DRAP27 antibody or CRM197 coprecipitates DRAP27 and diphtheria toxin receptor protein from cell lysates prepared from cells overexpressing diphtheria toxin receptor (Iwamoto et al., 1994). These results show that diphtheria toxin receptor forms a complex with DRAP27. The antibody 007 probably binds to DRAP27 closely associated with diphtheria toxin receptor on the cell surface and causes inhibition of the binding of diphtheria toxin to the receptor.

Analysis of the nucleotide sequence of a DRAP27 cDNA has shown that DRAP27 has 228 amino acids containing four putative transmembrane domains and that DRAP27 is the monkey homologue of the human CD9 antigen (Mitamura et al., 1992).

The role of DRAP27 on diphtheria toxin binding and diphtheria toxin sensitivity has been studied by transfection of DRAP27 cDNA to cells. A human-mouse hybrid cell line (3279-10) expressing diphtheria toxin receptor (Hayes et al., 1987) but not DRAP27/CD9 antigen was transfected with cDNA for DRAP27. The transfectants showed increased diphtheria toxin binding and were 3–25 times more sensitive to diphtheria toxin than untransfected cells (Fig. 4) (Mitamura et al., 1992). However, when mouse L cells with no functional diphtheria toxin receptor were transfected with DRAP27 cDNA, neither increased diphtheria toxin binding nor enhancement of diphtheria toxin sensitivity was observed. Thus, DRAP27 serves to enhance the sensitivity of DT-sensitive cell lines.

Cotransfection of DRAP27 cDNA with diphtheria toxin receptor cDNA into mouse L cells showed more clearly the role of DRAP27 (Iwamoto et al., 1994). L cells transfected transiently with both DRAP27 and diphtheria toxin receptor cDNA bound about 10 times

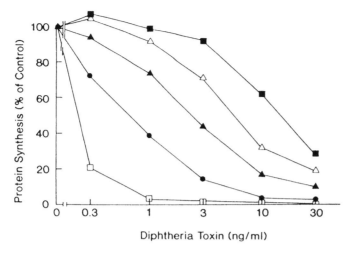

Diphtheria Toxin (ng/ml)

Fig. 4 Enhancement of diphtheria toxin sensitivity by the transfection of DRAP27 cDNA. Inhibition of protein synthesis by diphtheria toxin in stable transfectants of 3279-10 cells expressing DRAP27 (△, ▲, ●, parental 3279-10 cells (■), and monkey Vero cells (□). [Reproduced from the *Journal of Cell Biology, 118*: (1992).]

more diphtheria toxin than cells transfected with diphtheria toxin receptor alone. Transcription of diphtheria toxin receptor mRNA was not increased by cotransfection of DRAP27 cDNA. Stable L cell transfectants expressing both diphtheria toxin receptor and DRAP27 had a 15 times greater cell surface diphtheria toxin receptor number and were 20 times more sensitive to diphtheria toxin than stable L-cell transfectants expressing diphtheria toxin receptor alone, though the cell lines contained similar levels of diphtheria toxin receptor mRNA. Thus, it appears that DRAP27 up-regulates diphtheria toxin receptor number and diphtheria toxin sensitivity by a posttranscriptional mechanism. DRAP27 increases the number of diphtheria toxin receptors on the cell surface but does not change the affinity for diphtheria toxin. Although the mechanism is not clear, physical association of these proteins suggests that DRAP27 may affect the diphtheria toxin receptor through protein–protein interaction.

VIII. THE RECEPTOR AND THE DIPHTHERIA TOXIN ENTRY PROCESS

After binding to the receptor, diphtheria toxin is internalized by receptor-mediated endocytosis. Diphtheria toxin receptor, in particular its carboxy-terminal region located on the cytoplasmic side, would be expected to have a signal for internalization or a site that interacts with endocytosis machinery. Tyrosine residues in the cytoplasmic domain have been shown to be important for rapid endocytosis (Davis et al., 1987; Lazarovits and Roth, 1988). Diphtheria toxin receptor has two tyrosine residues in the cytoplasmic domain, and it has been suggested that one of these, Tyr-192, is in a region that resembles the signal necessary for receptor-mediated endocytosis (Naglich et al., 1992).

Are the diphtheria toxin receptor and the receptor-associated protein involved in the translocation of the A fragment from endosome to cytoplasm? I think that it is likely for the following reasons. We observed that solubilized diphtheria toxin receptor is not eluted from a diphtheria toxin-conjugated column by addition of the solutions down to pH 3.5

(E. Mekada, unpublished observation). The pH of the endosome interior is 5–6, so diphtheria toxin could remain bound to the receptor in the endosomal compartment. In the acidic environment of the endosome, the transmembrane domain of diphtheria toxin, adjacent to the receptor-binding domain, is inserted into the endosome membrane to translocate the A fragment (Collier, 1994). If the receptor-binding domain remains bound to the receptor molecule even at this stage, the transmembrane domain of the toxin may be close to the transmembrane domain of the diphtheria toxin receptor in the endosome bilayer. It is possible that diphtheria toxin receptor affects the translocation step of diphtheria toxin. A more interesting possibility is that DRAP27 is involved in the translocation step. Although it is not clear whether DRAP27 is internalized together with diphtheria toxin receptor, DRAP27 may be located in the endosome in association with the diphtheria toxin receptor. DRAP27 is an integral membrane protein with four transmembrane domains, and it looks like a channel-forming protein. Whether diphtheria toxin utilizes DRAP27 when the toxin penetrates the endosome membrane is an intriguing question for the future.

IX. THE DIPHTHERIA TOXIN RECEPTOR AS A GROWTH FACTOR

Diphtheria toxin receptor is identical to a membrane-anchored form of heparin-binding EGF-like growth factor, HB-EGF (Naglich et al., 1992). HB-EGF was first identified and purified as a soluble factor from conditioned medium of the human monocyte/macrophage-like cell line U-937 (Higashiyama et al., 1991, 1992). Mature HB-EGF polypeptide contains at least 86 amino acid residues, while the precursor molecule has 208 amino acids that include a putative N-terminus signal sequence and a hydrophobic transmembrane domain near the C terminus as mentioned in Section VI (Higashiyama et al., 1992). HB-EGF belongs to the EGF family of growth factors (Carpenter and Wahl, 1990), which includes TGF-α (Lee et al., 1985) and amphiregulin (Shoyab et al., 1989). HB-EGF binds to and gives a mitogenic signal through the EGF receptor. This growth factor is a potent mitogen for fibroblasts, smooth muscle cells, and keratinocytes but not for endothelial cells (Higashiyama et al., 1991). The heparin-binding property of HB-EGF distinguishes this protein from other members of the EGF family of growth factors.

In the case of TGF-α, the cell surface-associated precursor has been shown to be biologically active and to transmit the mitogenic signal to adjacent cells by interaction with EGF receptors, a process referred to as juxtacrine stimulation (Massagué, 1990). By analogy, the membrane-anchored form of HB-EGF, which is a diphtheria toxin receptor, has been postulated to have juxtacrine activity toward neighboring cells. Studies of the diphtheria toxin receptor revealed that DRAP27 associates with the membrane-anchored HB-EGF and increases the number of diphtheria toxin receptors on the cell surface. Thus, it is suggested that DRAP27/CD9 also affects growth factor activity. Since DRAP27/CD9 is a membrane-spanning molecule, as is the diphtheria toxin receptor/HB-EGF precursor, it is possible that DRAP27 plays a role in HB-EGF-mediated juxtacrine activity.

X. CONCLUSION

Diphtheria toxin receptors exist on the surface of diphtheria toxin-susceptible cells, and the number of such receptors on the cell surface essentially determines the sensitivity of the cell to diphtheria toxin. This receptor is a bifunctional protein that acts as a diphtheria toxin-binding receptor and as the precursor of a biologically active growth factor.

Diphtheria toxin receptor associates with another membrane protein (DRAP27) in the cell membrane and forms a protein complex that acts as a functional diphtheria toxin receptor and possibly as a mitogenic growth factor by a juxtacrine mechanism. Future studies on diphtheria toxin receptor and related proteins would further elucidate the diphtheria toxin–diphtheria toxin receptor interaction and intoxication process. These studies would also help to elucidate the function of this receptor as a mitogenic mediator.

Acknowledgments

I thank Dr. Michael R. Moynihan for critical review of this manuscript. The author's studies referred to in this chapter were supported by grants for scientific research from the Ministry of Education, Science, and Culture of Japan.

REFERENCES

Abraham, J. A., Damm, D., Bajardi, A., Miller, J., Klagsbrun, M., and Ezekowitz, R. E. B. (1993). Heparin-binding EGF-like growth factor: characterization of rat and mouse cDNA clones, protein domain conservation across species and transcript expression in tissues. *Biochem. Biophys. Res. Commun.* *190*: 125.

Blewitt, M. G., Chung, L. A., and London, E. (1985). Effect of pH on the conformation of diphtheria toxin and its implications for membrane penetration. *Biochemistry 24*: 5458.

Bonventre, P. F., Saelinger, C. B., Ivins, B., Woscinski, C., and Amorini, M. (1975). Interaction of cultured mammalian cells with [^{125}I]diphtheria toxin. *Infect. Immun. 11*: 675.

Boquet, P., and Pappenheimer, A. M., Jr. (1976). Interaction of diphtheria toxin with mammalian cell membranes. *J. Biol. Chem. 251*: 5770.

Boquet, P., Silverman, M. S., Pappenheimer, A. M. J., and Vernon, W. (1976). Binding of Triton X-100 to diphtheria toxin, crossreacting material 45 and their fragments. *Proc. Natl. Acad. Sci. U.S.A. 73*: 4449.

Cabiaux, V., Brasseur, R., Wattiez, R., Falmagne, P., Ruysschaert, J.-M., and Goormaghtigh, E. (1989). Secondary structure of diphtheria toxin and its fragments interacting with acidic liposomes studied by polarized infrared spectroscopy. *J. Biol. Chem. 264*: 4928.

Carpenter, G., and Wahl, M. I. (1990). The epidermal growth factor family. In *Peptide Growth Factors and Their Receptors*, Vol. 95, M. Sporn and A. Roberts (Eds.), Springer-Verlag, Berlin, p. 69.

Chang, T., and Neville, D. M. J. (1978). Demonstration of diphtheria toxin receptors on surface membranes from both toxin-sensitive and toxin-resistant species. *J. Biol. Chem. 253*: 6866.

Chin, D., and Simon, M. I. (1981). Diphtheria toxin interaction with an inhibitory activity from plasma membranes. In *Receptor-Mediated Binding and Internalization of Toxins and Hormones*, J. L. Middlebrook and L. Kohn (Eds.), Academic, New York, p. 53.

Cieplak, W., Gaudin, H. M., and Eidels, L. (1987). Diphtheria toxin receptor. Identification of specific diphtheria toxin-binding proteins on the surface of Vero and BS-C-1 cells. *J. Biol. Chem. 262*: 13246.

Collier, R. J. (1967). Effect of diphtheria toxin on protein synthesis: inactivation of one of the transfer factors. *J. Mol. Biol. 25*: 83.

Collier, R. J. (1975). Diphtheria toxin: mode of action and structure. *Bacteriol. Rev. 39*: 54.

Collier, R. J. (1994). Three-dimensional structure of diphtheria toxin. In *Handbook of Natural Toxins, Bacterial Toxins and Virulence Factors in Disease*, Vol. 8, Moss, Iglewski, Vaughan, and Tu (Eds.), Marcel Dekker, New York.

Collier, R. J., and Kandel, J. (1971). Thiol-dependent dissociation of a fraction of toxin into enzymically active and inactive fragments. *J. Biol. Chem. 246*: 1496.

Collins, C. M., and Collier, R. J. (1984). Interaction of diphtheria toxin with adenylyl-(3',5')-uridine

3'-monophosphate. II. The NAD-binding site and determinants of dinucleotide affinity. *J. Biol. Chem. 259*: 15159.

Creagan, R. P., Chen, S., and Ruddle, F. H. (1975). Genetic analysis of the cell surface: association of human chromosome 5 with sensitivity to diphtheria toxin in mouse-human somatic cell hybrids. *Proc. Natl. Acad. Sci. U.S.A. 72*: 2237.

Davis, C. G., van Driel, I. R., Russell, D. W., Brown, M. S., and Goldstein, J. L. (1987). The low density lipoprotein receptor. Identification of amino acids in cytoplasmic domain required for rapid endocytosis. *J. Biol. Chem. 262*: 4075.

Didsbury, J. R., Moehring, J. M., and Moehring, T. J. (1983). Binding and uptake of diphtheria toxin by toxin-resistant Chinese hamster ovary and mouse cells. *Mol. Cell. Biol. 3*: 1283.

Draper, R. K., and Simon, M. I. (1980). The entry of diphtheria toxin into the mammalian cell cytoplasm: evidence for lysosomal involvement. *J. Cell. Biol. 87*: 849.

Eidels, L., and Draper, R. K. (1988). Diphtheria toxin. In *Handbook of Natural Toxins*, Vol. 4, C. Hardegree and A. T. Tu (Eds.), Marcel Dekker, New York, p. 217.

Freeman, V. J. (1951). Studies on the virulence of bacteriophage-infected strains of *Corynebacterium diphtheriae*. *J. Bacteriol. 61*: 675.

Gabliks, J., and Solotorovsky, M. (1962). Cell culture reactivity to diphtheria, staphylococcus, tetanus and *Escherichia coli* toxins. *J. Immunol. 88*: 505.

Giannini, G., Rappuoli, R., and Ratti, G. (1984). The amino-acid sequence of two non-toxic mutants of diphtheria toxin: CRM45 and CRM197. *Nucleic Acids Res. 12*: 4063.

Gill, D. M., and Dinius, L. L. (1971). Observations on the structure of diphtheria toxin. *J. Biol. Chem. 246*: 1485.

Greenfield, L., Bjorn, M. J., Horn, G., Fong, D., Buck, G. A., Collier, R. J., and Kaplan, D. A. (1983). Nucleotide sequence of the structural gene for diphtheria toxin carried by corynebacteriophage β. *Proc. Natl. Acad. Sci. U.S.A. 80*: 6853.

Hayes, H., Kaneda, Y., Uchida, T., and Okada, Y. (1987). Regional assignment of the gene for diphtheria toxin sensitivity using subchromosomal fragments in microcell hybrids. *Chromosoma 96*: 26.

Higashiyama, S., Abraham, J. A., Miller, J., Fiddes, J. C., and Klagsbrun, M. (1991). A heparin-binding growth factor secreted by macrophage-like cells that is related to EGF. *Science 251*: 936.

Higashiyama, S., Lau, K., Besner, G. E., Abraham, J. A., and Klagbrun, M. (1992). Structure of heparin-binding EGF-like growth factor. Multiple forms, primary structure, and glycosylation of the mature protein. *J. Biol. Chem. 267*: 6205.

Honjo, T., Nishizuka, Y., Hayaishi, O., and Kato, I. (1968). Diphtheria-toxin-dependent adenosine diphosphate ribosylation of aminoacyl transferase II and inhibition of protein synthesis. *J. Biol. Chem. 243*: 3553.

Itelson, T. R., and Gill, D. M. (1973). Diphtheria toxin: specific competition for cell receptors. *Nature 242*: 330.

Iwamoto, R., Senoh, H., Okada, Y., Uchida, T., and Mekada, E. (1991). An antibody that inhibits the binding of diphtheria toxin to cells revealed the association of a 27-kDa membrane protein with the diphtheria toxin receptor. *J. Biol. Chem. 266*: 20463.

Iwamoto, R., Higashiyama, S., Mitamura, T., Taniguchi, N., Klagsbrun, M., and Mekada, E. (1994). Heparin-binding EGF-like growth factor, which acts as the diphtheria toxin receptor, forms a complex with membrane protein DRAP27/CD9, which up-regulates functional receptors and diphtheria toxin sensitivity. *Embo. J. 13*: 2322.

Keen, J. H., Maxfield, F. R., Hardegree, M. C., and Habig, W. H. (1982). Receptor-mediated endocytosis of diphtheria toxin by cells in culture. *Proc. Natl. Acad. Sci. U.S.A. 79*: 2912.

Kim, K., and Groman, N. B. (1965). In vitro inhibition of diphtheria toxin action by ammonium salts and amines. *J. Bacteriol. 90*: 1552.

Kohno, K., Uchida, T., Mekada, E., and Okada, Y. (1985). Characterization of diphtheria-toxin-resistant mutants lacking receptor function or containing non-ribosylatable elongation factor 2. *Somat. Cell Genet. 11*: 421.

Lazarovits, J., and Roth, M. (1988). A single amino acid change in the cytoplasmic domain allows the influenza virus hemagglutinin to be endocytosed through coated pits. *Cell 53*: 743.

Lee, D. C., Rose, T. M., Webb, N. R., and Todaro, G. J. (1985). Cloning and sequence analysis of a cDNA for rat transforming growth factor-α. *Nature 313*: 489.

Lennox, E. S., and Kaplan, A. S. (1957). Action of diphtheria toxin on cells cultivated in vitro. *Proc. Soc. Exp. Biol. Med. 95*: 700.

Lory, S., and Collier, R. J. (1980). Diphtheria toxin: nucleotide binding and toxin heterogeneity. *Proc. Natl. Acad. Sci. U.S.A. 77*: 267.

Massagué, J. (1990). Transforming growth factor-α. A model for membrane-anchored growth factors. *J. Biol. Chem. 265*: 21393.

Mekada, E., and Uchida, T. (1985). Binding properties of diphtheria toxin to cells are altered by mutation in the fragment A domain. *J. Biol. Chem. 260*: 12148.

Mekada, E., Uchida, T., and Okada, Y. (1979). Modification of the cell surface with neuraminidase increases the sensitivities of cells to diphtheria toxin and *Pseudomonas aeruginosa* exotoxin. *Exp. Cell Res. 123*: 137.

Mekada, E., Uchida, T., and Okada, Y. (1981). Methylamine stimulates the action of ricin toxin but inhibits that of diphtheria toxin. *J. Biol. Chem. 256*: 1225.

Mekada, E., Kohno, K., Ishiura, M., Uchida, T., and Okada, Y. (1982). Methylamine facilitates demonstration of specific uptake of diphtheria toxin by CHO cell and toxin-resistant CHO cell mutants. *Biochem. Biophys. Res. Commun. 109*: 792.

Mekada, E., Okada, Y., and Uchida, T. (1988). Identification of diphtheria toxin receptor and a nonproteinous diphtheria toxin-binding molecule in Vero cell membrane. *J. Cell Biol. 107*: 511.

Mekada, E., Senoh, H., Iwamoto, R., Okada, Y., and Uchida, T. (1991). Purification of diphtheria toxin receptor from Vero cells. *J. Biol. Chem. 266*: 20457.

Middlebrook, J. L., and Dorland, R. B. (1977). Response of cultured mammalian cells to the exotoxins of *Pseudomonas aeruginosa* and *Corynebacterium diphtherieae*: differential cytotoxicity. *Can. J. Microbiol. 23*: 183.

Middlebrook, J. L., and Dorland, R. B. (1979). Protection of mammalian cells from diphtheria toxin by exogenous nucleotides. *Can. J. Microbiol. 25*: 285.

Middlebrook, J. L., Dorland, R. B., and Leppla, S. H. (1978). Association of diphtheria toxin with Vero cells. *J. Biol. Chem. 253*: 7325.

Mitamura, T., Iwamoto, R., Umata, T., Yomo, T., Urabe, I., Tsuneoka, M., and Mekada, E. (1992). The 27-kD diphtheria toxin receptor-associated protein (DRAP27) from Vero cells is the monkey homologue of human CD9 antigen: expression of DRAP27 elevates the number of diphtheria toxin receptors on toxin-sensitive cells. *J. Cell Biol. 118*: 1389.

Moehring, T. J., and Crispell, J. P. (1974). Enzyme treatment of KB cells: the altered effect of diphtheria toxin. *Biochem. Biophys. Res. Commun. 60*: 1446.

Moehring, J. M., and Moehring, T. J. (1979). Characterization of the diphtheria toxin-resistance system in Chinese hamster ovary cells. *Somat. Cell Genet. 5*: 453.

Morris, R. E., Gerstein, A. S., Bonventre, P. F., and Saelinger, C. B. (1985). Receptor-mediated entry of diphtheria toxin into monkey kidney (Vero) cells: electron microscopic evaluation. *Infect. Immun. 50*: 721.

Moskaug, Y. O., Stenmark, H. and Olsnes, S. (1991). Insertion of diphtheria toxin B-fragment into the plasma membrane at low pH. Characterization and topology of inserted regions. *J. Biol. Chem. 266*: 2652.

Moya, M., Datry-Varsat, A., Goud, B., Louvard, D., and Bouquet, P. (1985). Inhibition of coated pit formation in Hep2 cells blocks the cytotoxicity of diphtheria toxin but not that of ricin toxin. *J. Cell Biol. 101*: 548.

Naglich, J. G., and Eidels, L. (1990). Isolation of diphtheria toxin-sensitive mouse cells from a toxin-resistant population transfected with monkey DNA. *Proc. Natl. Acad. Sci. U.S.A. 87*: 7250.

Naglich, J. G., Metherall, J. E., Russell, D. W., and Eidels, L. (1992). Expression cloning of a diphtheria toxin receptor: identity with a heparin-binding EGF-like growth factor precursor. *Cell 69*: 1051.

Pappenheimer, A. M., Jr. (1977). Diphtheria toxin. *Annu. Rev. Biochem. 46*: 69.

Placido-Sousa, C., and Evans, D. G. (1957). The action of diphtheria toxin on tissue cultures and its neutralization by antitoxin. *Br. J. Exp. Pathol. 38*: 644.

Proia, R. L., Hart, D. A., Holmes, R. K., Holmes, K. V., and Eidels, L. (1979). Immunoprecipitation and partial characterization of diphtheria toxin-binding glycoproteins from the surface of guinea pig cells. *Proc. Natl. Acad. Sci. U.S.A. 76*: 685.

Sandvig, K., and Olsnes, S. (1980). Diphtheria toxin entry into cells is facilitated by low pH. *J. Cell Biol. 87*: 828.

Sandvig, K., Olsnes, S., and Pihl, A. (1978). Binding, uptake and degradation of the toxic proteins abrin and ricin by toxin-resistant cell variants. *Eur. J. Biochem. 82*: 13.

Shoyab, M., Plowman, G. D., McDonald, V. L., Bradley, J. G., and Todaro, G. J. (1989). Structure and function of human amphiregulin: a member of the epidermal growth factor family. *Science 243*: 1074.

Tsuneoka, M., Nakayama, K., Hatsuzawa, K., Komada, M., Kitamura, N., and Mekada, E. (1993). Evidence for involvement of furin in cleavage and activation of diphtheria toxin. *J. Biol. Chem. 268:* 26461.

Uchida, T. (1983). Diphtheria toxin. *Pharm. Ther. 19*: 107.

Uchida, T., Gill, D. M., and Pappenheimer, A. M., Jr. (1971). Mutation in the structural gene for diphtheria toxin carried by temperate phage β. *Nature New Biol. 233*: 8.

Uchida, T., Pappenheimer, A. M., Jr., and Harper, A. A. (1972). Reconstitution of diphtheria toxin from two nontoxic cross-reacting mutant proteins. *Science 175*: 901.

Uchida, T., Yamaizuma, M., and Okada, Y. (1977). Reassembled HVJ (Sendai virus) envelopes containing non-toxic mutant proteins of diphtheria toxin show toxicity to mouse L cell. *Nature 266*: 839.

Umata, T., Moriyama, Y., Futai, M., and Mekada, E. (1990). The cytotoxic action of diphtheria toxin and its degradation in intact Vero cells are inhibited by bafilomycin A1, a specific inhibitor of vacuolar-type H^+-ATPase. *J. Biol. Chem. 265*: 21940.

Yamaizumi, M., Mekada, E., Uchida, T., and Okada, Y. (1978). One molecule of diphtheria toxin fragment A introduced into a cell can kill the cell. *Cell 15*: 245.

7

Diphthamide: The Toxin Target in Protein Synthesis Elongation Factor 2

Lon D. Phan, Chi-Ting Huang, and James W. Bodley

University of Minnesota, Minneapolis, Minnesota

I. INTRODUCTION

Diphtheria toxin and *Pseudomonas* exotoxin A are produced as single peptide chains.[1] Both proteins are organized into three physically and functionally distinct domains. Two of these domains are concerned with the binding of the toxins to receptors on cell surfaces and penetration of the cell membrane. The third domain enters the cytoplasm and functions catalytically to inhibit protein synthesis. It is this inhibition of protein synthesis that is responsible for the lethal effects of these toxins (see Perentesis et al., 1992a).

The catalytic domains of the two toxins each contain about 200 amino acids, but

---

[1]The properties of diphtheria toxin and *Pseudomonas* exotoxin A are reviewed in accompanying chapters by R. J. Collier and B. Iglewski, respectively. References to work reviewed by them will not be repeated here.

their structures are quite different. The catalytic fragment of diphtheria toxin occurs at the N terminus of the protein, whereas the catalytic fragment of *Pseudomonas* exotoxin A occurs at its C terminus. Little sequence homology exists between the catalytic domains of these proteins, but careful alignment and comparison of the three-dimensional structures suggests that critical catalytic residues are conserved in what appears to be their catalytic clefts. Both contain a catalytically essential glutamate residue in this cleft that can be covalently cross-linked to the nicotinamide moiety of the NAD^+ that serves as cofactor.

Despite their structural differences, the two toxins catalyze identical enzymatic reactions. Both inactivate protein synthesis elongation factor 2 (EF-2) through the covalent attachment of ADP-ribose. NAD^+ serves as the ADP-ribosyl donor. The specificity of the catalytic fragment of the toxins is remarkable, since in contrast to the ADP-ribosyl-transfering toxins discovered subsequently, these toxins do not modify any protein other than EF-2. Moreover, the toxins recognize EF-2 from any eukaryotic or archaebacterial source.

At least a partial explanation of the basis of the unusual specificity of diphtheria toxin and *Pseudomonas* exotoxin A was provided by the discovery that ADP-ribose is attached to an unusual amino acid present in EF-2. Subsequent purification and chemical analysis of this amino acid indicated that it is 2-[3-carboxyamido-3-(trimethylammonio)propyl]his-tidine. It was given the trivial name diphthamide. Here we review what is known of the biosynthesis and function of diphthamide in EF-2 and in the recognition of the factor by their ADP-ribosylating toxins.

II.   THE CHEMISTRY OF DIPHTHAMIDE

Robinson et al. (1974) first suggested that diphtheria toxin modifies an unusual amino acid in EF-2 that might be responsible for the recognition of the factor by the toxin. Subsequently, Van Ness et al. (1980a), following enzymatic digestion of ADP-ribosylated EF-2, purified the amino acid attached to ribose by the glycosidic linkage formed by the toxin. Because it contained an amide and originated at the site of diphtheria toxin modification, the amino acid was given the trivial name diphthamide. The deaminated amino acid was given the trivial name diphthine.

Analysis of diphthamide by NMR (Van Ness et al., 1980b) and fast atom bombardment mass spectrometry (Bodley et al., 1984) led to the conclusion that it is a posttranslational derivative of histidine with the proposed structure 2-[3-carboxyamino-3-(trimethylammonio)propyl]histidine, shown in Figure 1. Confirmation of histidine as the translational precursor of diphthamide was obtained by biosynthetic labeling (Dunlop and Bodley, 1983) and by sequence analysis of EF-2 cDNA (Kohno et al., 1986) and subsequently of the EF-2 genes in yeast and other organisms (Perentesis et al., 1992b). The nature of the posttranslational modification of histidine in diphthamide was confirmed by biosynthetic labeling (Dunlop and Bodley, 1983) that showed the incorporation of $[\beta^{-3}H]$histidine, $[\alpha^{-3}H]$methionine, and $[Me^{-3}H]$methionine in a ratio of 1:1:3. Also in agreement with the proposed structure, $[^{35}S]$methionine was not incorporated.

Nuclear Overhauser enhancement NMR spectroscopy of ribosyl-diphthamide indicated that the glycosidic linkage formed by diphtheria toxin is of the α configuration with imidazole N_1 (Oppenheimer and Bodley, 1981). This result suggests that ADP-ribosylation of EF-2 is accomplished by direct transfer from NAD^+ with inversion of configuration of the glycosidic bond.

The structures proposed for diphthamide and diphthine were confirmed in 1992, when

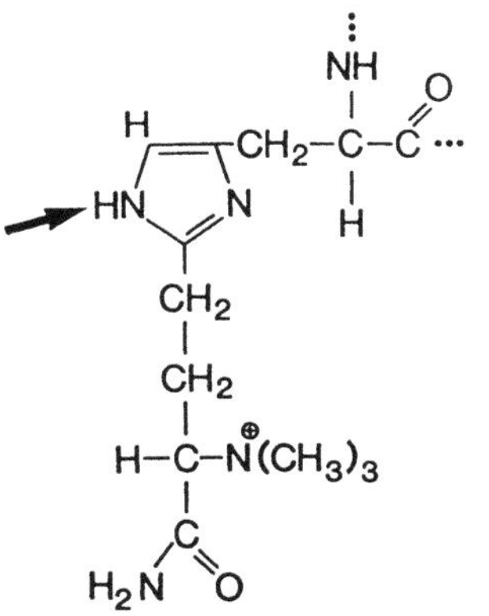

Fig. 1 The structure of diphthamide. Diphthamide, 2-[3-carboxyamido-3-(trimethylammonio)pro-pyl]histidine, is shown as it occurs in peptide linkage in the backbone of EF-2. Diphthamide is derived by posttranslational modification of a histidine that is residue 715 in mammalian EF-2 and residue 699 in yeast EF-2. The arrow designates the point of covalent attachment of ADP-ribose from NAD^+ by diphtheria toxin and *Pseudomonas* exotoxin A that inactivates EF-2.

the chemical synthesis of the structures proposed for both compounds was accomplished by Evans and Lundy (1992). Considerable difficulty was encountered in forming the carbon–carbon bond joining the imidazole C-2 to the propyl side chain, illustrating the resistance of the imidazole carbon to modification and hence the unusual nature of this histidine modification. Successful synthesis was achieved by creating the imidizole ring with this carbon–carbon bond preformed. This was accomplished by condensing two glutamate derivatives. Diphthamide contains two asymmetric centers, and Evans and Lundy (personal communication) showed that in the natural compound both are of the L configuration expected of natural amino acids.

III. THE GENETICS OF DIPHTHERIA TOXIN RESISTANCE

There are two general types of host mutations that can influence the killing action of diphtheria toxin: those that alter access of fragment A to its target in the cytoplasm and those that prevent modification of the target by fragment A. Thus, mutations that block binding or internalization of the toxin and mutations that produce a non-ADP-ribosylatable form of EF-2 can both confer resistance to mutant cells (Moehring and Moehring, 1979; Kohno et al., 1985). In vitro ADP-ribosylation by the A fragment can readily distinguish between these two types of mutations.

There are also two types of mutations that yield forms of EF-2 that are resistant to the in vitro action of the toxin: mutations in the enzymes of diphthamide biosynthesis and mutations in the structural gene(s) for EF-2. Both types of mutations were recognized very early by Moehring and coworkers (Moehring et al., 1979, 1980), who have investigated the basis of diphtheria toxin resistance in cultured CHO cells. Mutations in the enzymes of diphthamide biosynthesis are also readily observed in yeast cells (Chen et al., 1985), where-as toxin-resistant mutations in the yeast structural genes for EF-2 have thus far been obtained only by site-specific mutagenesis (Phan et al., 1993).

A. Mutants with Defects in the Pathway of Diphthamide Biosynthesis

The existence of mutations in the enzymes of diphthamide biosynthesis that resulted in diphtheria toxin resistance was first suggested by Moehring et al. (1980), and these were designated as MOD⁻ mutations. The MOD⁻ mutations were characterized by a complete loss of ADP-ribose acceptor activity in vitro and were genetically recessive, as would be expected of mutations in genes for transacting factors. Similar mutations were subsequently obtained in the yeast *Saccharomyces cerevisiae*, and these were designated as *dph* mutations (Chen et al., 1985).

The diphthamide biosynthetic mutants were selected by treating mutagenized cells with lethal levels of toxin and screening survivors for the inability of their EF-2 to be ADP-ribosylated in vitro by diphtheria toxin fragment A. Intact yeast cells are refractory to toxin, but they are rendered susceptible by creating spheroplasts through the removal of their cell walls. Thus, preparation of spheroplasts and their subsequent regeneration were necessary intermediate steps in selecting yeast *dph* mutants (Murakami et al., 1982). To eliminate unspheroplasted cells, the spheroplasts were simultaneously transformed with a plasmid bearing a selectable metabolic marker (Chen et al., 1985). *dph* mutants have also been selected by the conditional expression of fragment A-encoding plasmids (Perentesis et al., 1988; Parentesis and Bodley, unpublished observations).

Hybridization of the MOD⁻ strains of CHO cells with a normal toxin-sensitive strain yielded hybrids that were toxin-sensitive. Hence the MOD⁻ mutants are recessive. Hybridization of independently isolated MOD⁻ mutants with each other revealed that they belong to three complementation groups. These complementation groups were designated CG-1, CG-2, and CG-3 (Moehring et al., 1984).

In a similar manner, 31 independent toxin-resistant haploid yeast mutants were analyzed (Chen et al., 1985). Tetrad analysis indicated that each resulted from a single chromosomal mutation. Each isolate was crossed with a toxin-sensitive strain of the opposite mating type. The heterozygous diploids resulting from these crosses were in each case sensitive to diphtheria toxin, indicating that each mutation was recessive. Pairwise mating of the mutants and analysis of the diploids by in vitro ADP-ribosylation with diphtheria toxin showed that the mutants fall into five genetic complementation groups, revealing that at least five gene products participate in the biosynthesis of diphthamide. It remains to be determined whether the products of these five genes participate directly in diphthamide synthesis. Biochemical evidence (Chen and Bodley, 1988) suggested that in addition to the products of these five genes, at least one other gene product is required for the final step of ATP-dependent amidation of diphthine to diphthamide. Mutants with a defect in this last step have not been isolated, because amidation of diphthine is not required for ADP-ribosylation by diphtheria toxin. In addition, preliminary screening by plasmid expression suggests the existence of at least one additional complementation group of toxin-resistant mutants (Perentesis and Bodley, unpublished observations).

Neither the yeast mutants (Chen et al., 1985) nor the mammalian mutants (Moehring et al., 1984) exhibited any detectable growth defects when the cells were grown in culture. Thus, the only phenotype known to be associated with the failure to complete the synthesis of diphthamide in both yeast and cultured animal cells is the resistance to diphtheria and *Pseudomonas* toxins. Moreover, these results demonstrate that neither yeast nor mammalian cells require diphthamide-containing EF-2 for growth in culture.

B. Cloning of the *DPH* Genes

Isolation of yeast mutants defective in the pathway of diphthamide biosynthesis provided a means of cloning the genes involved in this unusual posttranslational modification reaction. Yeast is an ideal organism for this purpose because it is amenable to genetic manipulation and to the application of cloning procedures. Two screening procedures are available to detect the presence of a plasmid containing the complementing wild-type *DPH* genes in a toxin-resistant *dph* strain.

We have reported a screening procedure that is capable of directly distinguishing yeast colonies that contain ADP-ribosylatable EF-2 (Donovan et al., 1992). For this procedure, a *dph* strain of yeast is simply transformed with a yeast genomic library plasmid and the transformants are replica-plated onto nitrocellulose membranes. The colonies are converted to spheroplasts, the spheroplasts are lysed in situ, and the released intracellular proteins are drawn through and absorbed to the membrane. The released EF-2 is retained by the filter and can be ADP-ribosylated on it. Colonies containing the complementing *DPH* gene can be identified by autoradiographing the filters after incubating them with A fragment and [^{32}P]NAD$^+$. This method has not yet been used to obtain the relevant genes.

Screening can also be accomplished by transforming a toxin-resistant *dph* strain containing a plasmid library with the gene for the catalytic component of either diphtheria toxin or *Pseudomonas* exotoxin under control of the GAL1 promoter (Mattheakis et al., 1992). These strains are resistant to modification when grown on galactose. Thus, transformants are screened for complementation of the *dph* mutant by their *failure* to grow following replication onto galactose. Mattheakis and coworkers employed this procedure to obtain two of the genes involved in diphthamide biosynthesis.

1. The DPH5 Gene

The *DPH5* gene and its gene product are the best characterized participants in diphthamide biosynthesis. As will be described later, biochemical evidence suggests that the *DPH5* gene product corresponds functionally to the product of the MOD$^-$ mammalian CG-1 gene.

Mattheakis et al. (1992), cloning by complementation in a *dph5* strain as described above, obtained a genomic fragment that contained an open reading frame encoding a 300-amino-acid translation product. This fragment complemented the *dph5* mutant in vivo, restoring the ability of EF-2 to be ADP-ribosylated by toxin. Expression of this gene in *E. coli* yielded a protein that was capable of converting *dph5* EF-2 into an ADP-ribosylatable form and of methylating *dph5* EF-2 in vitro (see below). These findings provide definitive proof that this translation product is the complete *DPH5* gene product. In addition, Mattheakis and coworkers have shown that a *DPH5* deletion mutant is viable, thus conclusively demonstrating that complete diphthamide synthesis is not essential for growth.

Analysis of the *DPH5* gene product shows that it has sequence similarity to the bacterial *S*-adenosylmethionine:uroporphyrinogen III methyltransferases. These proteins are involved in cobalamin (vitamin B$_{12}$) biosynthesis. These observations are in complete agreement with the independently observed properties of the enzyme that complements the *dph5* mutation in vivo (see below).

2. The DPH2 Gene

The *dph2* mutant has an uncharacterized defect that occurs prior to the point of *DPH5* function in the diphthamide biosynthetic pathway. Mattheakis and coworkers (1993) have also cloned the gene that complements the *dph2* mutation by the same method employed in the cloning of *DPH5*. The genomic fragment with the *DPH2* gene contains an open reading frame of 534 amino acids. The predicted sequence of this protein does not exhibit significant homology with any sequence in the databank. In addition, the *E. coli* expression product does not complement a yeast *dph2* extract in vitro. Thus, the function of the *DPH2* gene product in diphthamide biosynthesis is unknown.

C. Mutations in the EF-2 Genes That Confer Toxin Resistance

The molecular genetic analysis of EF-2 is complicated by the fact that its central role in translation, which makes it a lethal target for toxins, also makes the presence of a functional gene essential for cell viability. Thus, the analysis of EF-2 gene mutations must either be limited to functionally active mutations or be conducted in the presence of a functional gene.

Mutations in a gene for EF-2 that confer toxin resistance were first described in cultured CHO cells by Moehring et al. (1979). Mammalian cells possess only a single gene for EF-2 per haploid genome (Nakanishi et al., 1988). Moehring and coworkers conducted their analysis with diploid cells that were heterozygous for the toxin resistance-conferring muta-tion. They found that these cells contained approximately equal amounts of the two forms of EF-2: wild-type toxin-sensitive and toxin-resistant. Treatment of these cells with diph-theria toxin reduced the rate of protein synthesis but did not kill the cells. Thus, mutations in the EF-2 gene of CHO cells that confer toxin resistance are codominant. Cells bearing these mutations are also distinguishable from the recessive MOD⁻ mutations because their extracts contain a significant amount of ADP-ribosylatable EF-2. Subsequent studies of mammalian EF-2 mutants were done in a similar manner with cells that contain both wild-type and mutant forms of the protein.

The yeast *Saccharomyces cerevisiae* possesses two genes for EF-2 (*EFT1* and *EFT2*) per haploid genome that encode identical proteins (Perentesis et al., 1992b). Gene disruption experiments have demonstrated that either *EFT1* or *EFT2* is able to satisfy the cellular requirement for EF-2 and that the required gene may be either on the chromosome or on a plasmid. As a consequence, the technique of plasmid shuffling can be used to investigate mutations in a single yeast EF-2 gene. For this purpose, the chromosomal copies of *EFT* are disrupted, a wild-type *EFT* is provided on a *URA*⁺ plasmid, and the cells are transformed with a second plasmid containing a mutant form of *EFT*. When these cells are grown in the presence of uracil, the *URA*⁺ plasmid can be lost in those cells that contain a mutant *EFT* that is functional. Cells containing such genes can be easily selected as a consequence of their growth in the presence of 5′-fluoroorotic acid, which is lethal to *URA*⁺ cells. We are using this plasmid-shuffling system to investigate the features of EF-2 that are recognized by toxins and by the enzymes of diphthamide biosynthesis.

Saturation mutagenesis of the *EFT* codon specifying histidine 699 in *EFT* and plasmid shuffling were used to determine if the precursor of diphthamide is essential to the function of EF-2 in protein synthesis (Phan et al., 1993). This possibility was tested with the assumption that functional replacements for this residue would also confer toxin resistance to the factor. Thus, toxin-resistant mutants were selected following shuffling with a plasmid containing *EFT* in which all codons replaced the native CAC for histidine 699. The sequence of *EFT* in the surviving mutants was determined following PCR amplifi-cation. Analysis of randomly selected mutants revealed that histidine 699 had been replaced with one of four amino acids: asparagine, glutamine, leucine, or methionine.

All 11 possible codons for these four amino acids were recovered. Yeasts bearing these mutant forms of *EFT* grew slightly slower than the corresponding wild-type strains but, except for their toxin resistance, exhibited no other phenotype. We conclude that despite its strict conservation, the histidine precursor of diphthamide is not essential to the function of EF-2 in yeast.

Omura et al. (1989) failed to find an active form of mammalian EF-2 following site-specific replacement of histidine 715 with either arginine, aspartate, or lysine. They concluded from these experiments that the precursor of diphthamide is functionally essential. However, since the replacements they tested are probably not active in yeast EF-2 and because of the high degree of sequence homology between the yeast and mammalian proteins in the region of diphthamide (Perentesis et al., 1992b), it seems unlikely that histidine 715 is required for the function of mammalian EF-2.

Kohno and Uchida (1987) described a mutagenic hot spot in the gene for mammalian EF-2, near but not involving histidine 715, that confers toxin resistance. This mutation results from a G-to-A transition in the first position of codon 717 of the EF-2 gene that causes the substitution of an arginine for a universally conserved glycine. Foley et al. (1992) demonstrated that this mutation produces toxin resistance indirectly by preventing step A (Fig. 2) of diphthamide synthesis. Apparently, the one or more enzymes involved in this reaction require glycine in position 717 in order to recognize histidine 715 for modification. The analogous mutation in yeast *EFT*, changing glycine 701 to arginine, was also found to produce toxin resistance. Circumstantial evidence (see below) suggests that this resistance is also caused by a defect in diphthamide synthesis, further emphasizing the similarity in diphthamide synthesis between yeast and mammalian systems.

Other mutations in the mammalian EF-2 gene that do not involve either histidine 715 or glycine 717, and are as yet uncharacterized, have also been observed to produce resistance to diphtheria toxin (Aonuma et al., 1991).

D. ADP-Ribosylation of Toxin-Resistant EF-2

As described above, yeast cells either possessing mutations in the *DPH* genes or sustained by plasmids encoding EF-2s with mutations in positions 699 or 701 are refractory to the intracellular expression of diphtheria toxin fragment A. However, Mattheakis et al. (1992) have demonstrated that a yeast *dph5* null mutant is killed by the high-level expression of a very active form of diphtheria toxin A fragment. They also demonstrated that EF-2 in extracts of these cells is capable of undergoing a small amount of ADP-ribosylation in the presence of massive amounts of fragment A (approximately 10^6 times that required to ADP-ribosylated wild-type EF-2). More recently, Mattheakis and coworkers have shown (L. C. Mattheakis et al., unpublished) that the extent of this ADP-ribosylation is limited by the competing hydrolysis of NAD$^+$ that is catalyzed by these high levels of A fragment. Under suitable conditions they observed that stoichiometric and specific ADP-ribosylation of what was previously thought of as completely toxin-resistant EF-2 can be achieved in vitro.

Subsequently, Mattheakis et al. (unpublished) showed that EF-2 in all other *dph* mutants, as well as the protein possessing the arginine substitution at position 701, can be ADP-ribosylated in vitro and that cells containing these mutations are killed by high-level fragment A expression in vivo. The in vitro rates of ADP-ribosylation of EF-2 derived from all six types of mutants were indistinguishable. In contrast, EF-2 containing the four functional substitutions for histidine 699, described above, were resistant to all

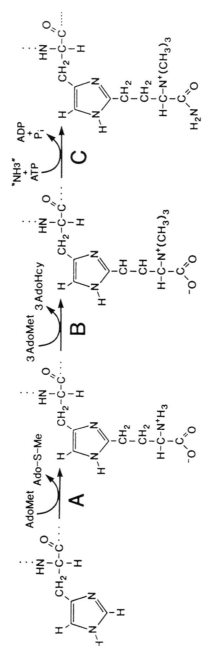

Fig. 2 The proposed posttranslational pathway of diphthamide biosynthesis. Reaction A is presumably catalyzed by the action of some or all of the products of the yeast genes *DPH1*, *DPH2*, *DPH3*, and *DPH4* or the mammalian MOD⁻ genes CG-2 and CG-3. Possibly these gene products function as a complex in the catalysis of this unusual transfer reaction. Reaction B is probably catalyzed by a single specific S-adenosylmethionine:EF-2 methyltransferase that is the product of the yeast *DPH5* gene and the mammalian MOD⁻ gene CG-1. Reaction C is not essential for toxin sensitivity in vivo. As a consequence, neither yeast nor mammalian mutations that are defective in this step have been obtained. The reaction appears to require ATP, but the amino donor is unknown, as is the enzyme that catalyzes the reaction.

levels of diphtheria toxin both in vivo and in vitro. Thus, it is clear that diphtheria toxin fragment A is capable of recognizing and specifically ADP-ribosylating the histidine precursor of diphthamide, probably without any prior modification, albeit at a slow rate. The capability to achieve complete and stoichiometric ADP-ribosylation of the histidine precursor of diphthamide in EF-2 from the *dph* mutants provides a convenient way to label, purify, and chemically analyze the state of its modification in these mutants. This analysis is currently under way.

IV. THE PATHWAY OF DIPHTHAMIDE BIOSYNTHESIS

The stepwise addition reactions that are involved in converting histidine 699 of yeast EF-2 or histidine 715 of mammalian EF-2 into diphthamide are shown in Figure 2.

The initial step in this pathway (step A) involves the addition of the four-carbon 3-carboxy-3-aminopropyl moiety to carbon-2 of the imidizole ring. The product of this step is presumably 2-(3-carboxy-3-aminopropyl)histidine, although the chemical identity of this compound has not been investigated. Methionine is the metabolic precursor of this moiety because label from $[\alpha\text{-}^3\text{H}]$methionine is incorporated into diphthamide in yeast. Ado-Met (S-adenosylmethionine) is the likely donor of the 3-carboxy-3-aminopropyl moiety because of the precedent provided by the biosynthesis of 3-(3-carboxy-3-aminopropyl)uridine during the posttranscriptional modification of tRNA (Nishimura et al., 1974). Circumstantial evidence supporting this conclusion is provided by the observation that the expected product of the reaction, Ado-Me (S-methylthioadenosine), is an inhibitor of diphthamide biosynthesis under some in vivo conditions (Yamanaka et al., 1986). The more readily formed carbon–nitrogen bond joins the side chain to uridine in 3-(3-carboxy-3-aminopropyl)uridine, whereas the more difficult to form carbon–carbon bond joins the side chain to the imidizole carbon-2 in diphthamide. It is thus possible that some type of prior activation step is a prerequisite to this joining.

Presumably all four yeast genes *DPH1*, *DPH2*, *DPH3*, and *DPH4* (or in the mammalian system the MOD⁻ genes CG-2 and CG-3) or their products are either directly or indirectly involved in step A. It may be that all four yeast gene products (or both mammalian gene products) are directly involved in a single step, perhaps as a multisubunit complex. Instability of such a hypothetical complex in the absence of one component would explain the inability of mutant extracts involving these genes to complement in vitro (see below).

Step B is most likely catalyzed solely by the product of the yeast *DPH5* gene and the mammalian MOD⁻ gene CG-1. All three methyl groups are derived from Ado-Met and added to 2-(3-carboxy-3-aminopropyl)histidine to form diphthine. The enzymology of this reaction appears to have been highly conserved because Moehring and Moehring (1988) have shown that extracts of a wide range of organisms, including *S. cerevisiae*, complement extracts of mammalian CG-1 extracts in vitro.

The last step that converts diphthine to diphthamide, step C, is not required to generate in vivo sensitivity to diphtheria toxin. Hence, there are no mutations in the corresponding gene(s). Studies in both yeast and mammalian cells demonstrate that this step cannot occur until diphthine is formed and suggest that ATP hydrolysis is required. The identity of the amino donor is unknown.

V. ENZYMES OF DIPHTHAMIDE BIOSYNTHESIS

The in vitro complementation of mutations involving the reaction(s) (step A) in the diphthamide pathway prior to the methylation step B has not been achieved. For this reason there is no direct information about the nature of the enzymes involved in this interesting step. As mentioned above, it is possible that this failure to complement could be the result of instability of an essential macromolecular complex in the absence of one of its components. It is likely that the *DPH2* gene product is involved in this step, but the cloning of this gene has not provided insight into the possible function of its translation product.

The specific *S*-adenosylmethionine:elongation factor 2 methyltransferase (Chen and Bodley, 1988) that is the product of the *DPH5* gene (Mattheakis et al., 1992) is the best characterized enzyme that is involved in diphthamide biosynthesis. The *DPH5* methyltransferase has been partially purified, and it is clear that diphthine is the final product of its action (Chen and Bodley, 1988). The available data are consistent with the view that diphthine is the minimal substrate recognized by diphtheria toxin (under the normal in vitro reaction conditions). This methyltransferase adds at least two of the three methyl groups present in diphthine, but it is not yet certain that it adds the first methyl group. Diphthine is also the final product of the in vitro complementation of the mammalian MOD⁻ CG-1 mutation (Moehring and Moehring, 1988). In this case it is clear that complementation involves the addition of all three methyl groups and that this addition is nonprocessive as mono- and dimethylated intermediates were found. Because the complementing enzyme has not been purified, the defect could reside in an enzyme that adds only the first methyl group. However, the simplest explanation for all of the observations is that there is only one methyltransferase, but this possibility has not yet been conclusively proven. The determination of the state of chemical modification of histidine 699 in the EF-2 of the *dph5* mutant should clarify this point.

The *DPH5* methyltransferase lacks the sequence motifs that are common to the *S*-adenosylmethionine-dependent methyltransferases that act on DNA, RNA, and small molecules, but it exhibits considerable homology with *S*-adenosylmethionine:uroporphyrinogen III methyltransferase (SUMT). These two enzymes share a unique consensus: VXXLXXGDPF. This homology is especially interesting because *DPH5* catalyzes a nitrogen methylation, whereas SUMT catalyzes a carbon methylation.

The final step in diphthamide biosynthesis is catalyzed by an amino transferase. Evidence in both the yeast (Chen and Bodley, 1988) and mammalian (Moehring and Moehring, 1988) systems suggests that ATP is required for this reaction and that it has only a modest effect on the recognition of EF-2 by diphtheria toxin. The amino donor is unknown, and the transferase has not been identified or purified.

REFERENCES

Aonuma, S., Ushijima, T., Nakayasu, M., Shima, H., Sugimura, T., and Nagao, M. (1991). Mutation induction by okadaic acid, a protein phosphatase inhibitor, in CHL cells, but not in *S. typhimurium. Mutat. Res. 250*:375–381.

Bodley, J. W., Upham, R., Crow, W., Tomer, K. B., and Gross, M. L. (1984). Ribosyl-diphthamide: confirmation of structure by fast atom bombardment mass spectrometry. *Arch. Biochem. Biophys. 230*:590–593.

Chen, J. C., and Bodley, J. W. (1988). Biosynthesis of diphthamide in *Saccharomyces cerevisiae.* Partial purification and characterization of a specific *S*-adenosylmethionine:elongation factor 2 methyltransferase. *J. Biol. Chem. 263*:11692–11696.

Chen, J. C., Bodley, J. W., and Livingston, D. M. (1985). Diphtheria toxin-resistant mutants of *Saccharomyces cerevisiae. Mol. Cell. Biol. 5*:3357–3360.

Donovan, M. G., Veldman, S. A., and Bodley, J. W. (1992). A screening procedure for the intracellular expression of native proteins by *Saccharomyces cerevisiae*: discrimination of diphtheria toxin-resistant mutants. *Yeast 8*:629–633.

Dunlop, P. C., and Bodley, J. W. (1983). Biosynthetic labeling of diphthamide in *Saccharomyces cerevisiae*. *J. Biol. Chem. 258*:4754–4758.

Evans, D., and Lundy, K. (1992). Synthesis of diphthamide: the target of diphtheria-toxin catalyzed ADP-ribosylation in protein-synthesis elongation factor 2. *J. Am. Chem. Soc. 114*(4):1495–1496.

Foley, B., Moehring, J., and Moehring, T. (1992). A mutation in codon-717 of the CHO-K1 elongation factor-2 gene prevents the 1st step in the biosynthesis of diphthamide. *Somatic Cell Mol. Genet. 18*(3):227–231.

Kohno, K., and Uchida, T. (1987). Highly frequent single amino acid substitution in mammalian elongation factor 2 (EF-2) results in expression of resistance to EF-2-ADP-ribosylating toxins. *J. Biol. Chem. 262*:12298–12305.

Kohno, K., Uchida, T., Mekada, E., and Okada, Y. (1985). Characterization of diphtheria-toxin-resistant mutants lacking receptor function or containing nonribosylatable elongation factor 2. *Somatic Cell Mol. Genet. 11*:421–431.

Kohno, K., Uchida, T., Ohkubo, H., Nakanishi, S., Nakanishi, T., Fukui, T., Ohtsuka, E., Ikehara, M., and Okada, Y. (1986). Amino acid sequence of mammalian elongation factor 2 deduced from the cDNA sequence: homology with GTP-binding proteins. *Proc. Natl. Acad. Sci. U.S.A. 83*:4978–4982.

Mattheakis, L. C., Shen, W. H., and Collier, R. J. (1992). *DPH5*, a methyltransferase gene required for diphthamide biosynthesis in *Saccharomyces cerevisiae*. *Mol. Cell. Biol. 12*(9): 4026–4037.

Mattheakis, L. C., Sor, F., and Collier, R. J. (1993). Diphthamide synthesis in *Saccharomyces cerevisiae*: structure of the DPH2 gene. *Gene 132:* 149–154.

Moehring, J., and Moehring, T. (1979). Characterization of the diphtheria toxin-resistance system in Chinese hamster ovary cells. *Somatic Cell Genet. 5*:453–468.

Moehring, J., and Moehring, T. (1988). The post-translational trimethylation of diphthamide studied in vitro. *J. Biol. Chem. 263*:3840–3844.

Moehring, T., Danley, D., and Moehring, J. (1979). Codominant translational mutants of chinese hamster ovary cells selected with diphtheria toxin. *Somatic Cell Genet. 5*:469–480.

Moehring, J., Moehring, T., and Danley, D. (1980). Posttranslational modification of elongation factor 2 in diphtheria toxin-resistant mutants of CHO-K1 cells. *Proc. Natl. Acad. Sci. U.S.A. 77*:1010–1014.

Moehring, T., Danley, D., and Moehring, J. (1984). *In vitro* biosynthesis of diphthamide, studied with mutant Chinese hamster ovary cells resistant to diphtheria toxin. *Mol. Cell. Biol. 4*:642–650.

Murakami, S., Bodley, J. W., and Livingston, D. M. (1982). *Saccharomyces cerevisiae* spheroplasts are sensitive to the action of diphtheria toxin. *Mol. Cell. Biol. 2*:588–592.

Nakanishi, T., Kohno, K., Ishiura, M., Ohashi, H., and Uchida, T. (1988). Complete nucleotide sequence and characterization of the 5′-flanking region of mammalian elongation factor 2 gene. *J. Biol. Chem. 263*:6384–6391.

Nishimura, S., Taya, Y., Kuchino, Y., and Ohasi, Z. (1974). Enzymatic synthesis of 3-(3-amino-3-carboxypropyl)uridine in *Escherichia coli* phenylalanine transfer RNA:transfer of the 3-amino-3-carboxypropyl group from *S*-adenosylmethionine. *Biochem. Biophys. Res. Commun. 57:* 702–708.

Omura, F., Kohno, K., and Uchida, T. (1989). The histidine residue of codon 715 is essential for function of elongation factor 2. *Eur. J. Biochem. 180*:1–8.

Oppenheimer, N. J., and Bodley, J. W. (1981). Diphtheria toxin: site and configuration of ADP-ribosylation of diphthamide in elongation factor 2. *J. Biol. Chem. 256*:8579–8581.

Perentesis, J., Genbauffe, F., Veldman, S., Galeotti, C., Livingston, D., Bodley, J. W., and Murphy, J. R. (1988). Expression of diphtheria toxin fragment A and hormone-toxin fusion proteins in toxin-resistant yeast mutants. *Proc. Natl. Acad. Sci. U.S.A. 85*:8386–8390.

Perentesis, J. P., Miller, S. P., and Bodley, J. W. (1992a). Protein toxin inhibitors of protein synthesis. *BioFactors 3*:173–184.

Perentesis, J. P., Phan, L. D., Gleason, W., LaPorte, D., Livingston, D., and Bodley, J. W. (1992b). *Saccharomyces cerevisiae* elongation factor 2. Genetic cloning, characterization of expression, and G-domain modeling. *J. Biol. Chem. 267*:1190–1197.

Phan, L. D., Perentesis, J. P., and Bodley, J. W. (1993). *Saccharomyces cerevisiae* elongation factor 2: mutagenesis of the histidine precursor of diphthamide yields a functional protein that is resistant to diphtheria toxin. *J. Biol. Chem. 268:8665–8668.*

Robinson, E. A., Henriksen, O., and Maxwell, E. S. (1974). Elongation factor 2:amino acid sequence at the site of adenosine diphosphate ribosylation. *J. Biol. Chem. 249*:5088–5093.

Van Ness, B. G., Howard, J. B., and Bodley, J. W. (1980a). ADP-ribosylation of elongation factor 2 by diphtheria toxin: isolation and properties of the novel ribosyl-amino acid and its hydrolysis products. *J. Biol. Chem. 255*:10717–10720.

Van Ness, B. G., Howard, J. B., and Bodley, J. W. (1980b). ADP-ribosylation of elongation factor 2 by diphtheria toxin. NMR spectra and proposed structures of ribosyl-diphthamide and its hydrolysis products. *J. Biol. Chem. 255*:10710–10716.

Yamanaka, H., Kajander, E., and Carson, D. (1986). Modulation of diphthamide synthesis by 5'-deoxy-5'-methylthioadenosine in murine lymphoma cells. *Biochim. Biophys. Acta 888*:157–162.

8

Biogenesis of Cholera Toxin and Related Oligomeric Enterotoxins

Timothy R. Hirst

Research School of Biosciences, University of Kent at Canterbury, Canterbury, Kent, United Kingdom

I. INTRODUCTION

Virtually all of the classical virulence factors produced by pathogenic microorganisms, such as surface-associated adhesins, capsular polysaccharides, lipopolysaccharides, siderophores, and toxins, are *secreted* molecules. Thus, they must undergo a process of translocation from their site of synthesis (usually in the cytoplasm or on the cytoplasmic membrane), through the microbial envelope to a location where they can exhibit their functional virulence properties. Incorporation of virulence factors in the microbial cell surface or their secretion into the surrounding milieu depends on multiple cellular processes, which include the events of macromolecular synthesis, translocation, processing, modification, folding, assembly, and secretion or integration. A term that usefully encompasses such a wide range of events is macromolecular *biogenesis.*

In this chapter I focus specifically on the biogenesis of toxins responsible for causing cholera and related diarrheal diseases (for an earlier review, see Hirst, 1991). These toxins are among the best characterized of all the virulence factors produced by pathogenic microorganisms, and their biogenesis serves as an excellent paradigm of coordinate gene regulation and protein secretion by bacteria. Cholera toxin (produced by *Vibrio cholerae*) and heat-labile enterotoxin (produced by certain enterotoxigenic strains of *Escherichia coli*) are structurally and functionally related proteins. Both toxins are macromolecular complexes consisting of one A subunit (M_r 28,000) and five B subunits (each of M_r ~11,600). The A subunit is an ADP-ribosyltransferase, whereas the B subunits are lectins that bind the holotoxin complex to G_{M1} ganglioside receptors ubiquitously present on the surface of eukaryotic cells (for a more detailed description of the structure and mode of action of these toxins, see Section III and Chapters 4 and 9 of this volume). The genes encoding the A and B subunits are organized into an operon of overlapping cistrons that encode a polycistronic mRNA, the synthesis of which is subject to regulatory control. The subunits are synthesized as separate precursor polypeptides that are exported, processed, and assembled into an AB_5 holotoxin complex. In *E. coli* the assembled toxin is located within the periplasm of the bacterial envelope, suggesting that its capacity to interact with the host gut and induce a diarrheagenic response must involve bacterial lysis or nonspecific release. In contrast, *V. cholerae* efficiently and selectively secretes the assembled toxin across its outermost membrane. This latter phenomenon challenges the current prevailing paradigm that proteins translocate across biological membranes as unfolded molecules. Thus, the biogenesis of cholera toxin and related enterotoxins involves a complex sequence of events, some of which remain challenging enigmas awaiting a molecular description.

II. CHOLERA AND RELATED DIARRHEAL DISEASES

A. Cholera

In 1884 Robert Koch established that a "comma-bacillus," now called *Vibrio cholerae* serogroup O1, was the causative agent of Asiatic cholera (Koch, 1884) (Fig. 1). The organism, which is transmitted in contaminated water, colonizes the surface of the small bowel and produces a potentially life-threatening secretory diarrhea. The rapid spread of the organism and its capacity to reduce a previously healthy individual to the point of death in as little as 6–8 h (with a mortality rate as high as 60%) made cholera a disease to be feared. However, with the widespread introduction of intravenous and oral rehydration therapies, the mortality rate can be reduced to less than 1%. The present

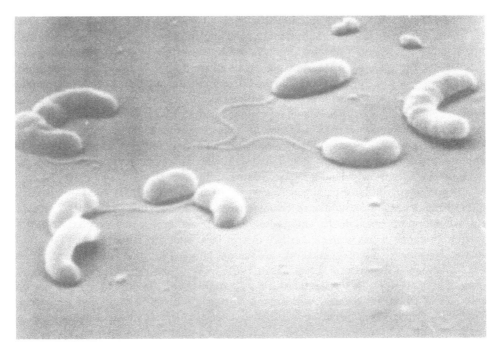

Fig. 1 Scanning electronmicrograph of *V. cholerae* O1. (Magnification 22,400×). (From D. K. Banergee, *Microbiology of Infection Diseases*, Mosby-Year Book Europe, Ltd., used by permission.)

number of cases of cholera in the world each year has been estimated to be approximately 8 million, including 124,000 deaths (Black, 1986).

In Asia, and in particular the Ganges river delta, cholera is endemic and has probably been causing disease and death for hundreds if not thousands of years. In contrast, the first well-documented experience of cholera in Europe was in 1817, when the first of a series of seven pandemics saw the spread of cholera along routes of trade from its ancestral home to large areas of Asia, Africa, Europe, and North America (Glass, 1986). The sixth pandemic, which began in the early 1900s, was caused by what is now referred to as the *classical* biotype of *V. cholerae* O1. These strains exhibit characteristic sensitivities to vibriophages and antibiotics and characteristic hemagglutination patterns. The seventh pandemic, which began in Sulawesi in 1961, was due to the emergence of a somewhat less virulent El Tor biotype of *V. cholerae* O1, which quickly spread through the Pacific Islands and into most of Southeast Asia. Since then the El Tor biotype has caused major cholera epidemics in large areas of Africa and was responsible for the spectacular cholera epidemic in Latin America in 1991 in which over 1 million people were infected (Swerdlow and Mintz, 1992).

A poignant reminder of the enormous pathogenic potential of *V. cholerae* came with emergence of a new serogroup of *V. cholerae* at the end of 1992 that is presently causing a large number of cholera cases and deaths in India and Bangladesh, with over 100,000 cases of cholera and 1473 deaths reported in Bangladesh alone during the first three months of 1993 (Albert and Ansaruzzaman, 1993). This strain, which does not react with antibodies directed against the O1 antigen or any of the other 137 non-O1 serogroups, has been designated O139. The clinical features of patients infected with *V. cholerae* O139 are indistinguishable from those of typical cholera, and this strain produces a toxin that is indistinguishable from that produced by *V. cholerae* O1 strains (Bhattacharya et al., 1993). The majority of the infections caused by the O139 strain in Bengal are in adults, suggesting that prior exposure to the O1 sergroup did not elicit a protective immune response to the new strain (Albert and Ansaruzzaman, 1993). A recent report that the O139 strain has spread to Thailand has prompted some speculation that this could be the beginning of the eighth cholera pandemic (Chongsa-nguan and Chaicumpa, 1993).

Non-O1/non-O139 strains of *V. cholerae* are commonly found in marine and brackish environments and are generally considered avirulent, although they have been reported to cause sporadic (nonepidemic) cases of diarrheal disease.

In common with most pathogens, the capacity of *V. cholerae* O1 and O139 to cause disease requires multiple virulence characteristics that enable this organism to not only survive within the host but also successfully colonize and proliferate and produce factors toxic to the host. It has been established that surface-associated adhesins of *V. cholerae* (particularly the toxin-coregulated pili) are essential for colonization and hence pathogenesis in humans (Taylor et al., 1987; Herrington et al., 1988). Nevertheless, it is quite clear that Robert Koch's speculation that "comma-bacilli produce a special poison" responsible for causing diarrhea was an example of great scientific intuition (Koch, 1884). This was eventually confirmed some 75 years later in independent experiments by S. N. De and N. K. Dutta, who demonstrated that sterile culture filtrates from *V. cholerae* contained a toxin that fully accounted for the enterotoxicity of cholera vibrios (De, 1959; Dutta et al., 1959). Levine and coworkers subsequently tested the potency of cholera toxin (CT) in healthy adult American volunteers and showed that as little as 5 μg of pure toxin, given in bicarbonate to neutralize gastric acidity, was able to cause significant diarrhea, while a 25 μg dose elicited more than 20 liters of diarrhea, sufficient to kill the individual in the absence of fluid and electrolyte replacement therapy (Levine et al., 1983).

B. Enterotoxinogenic *E. coli*

In addition to De's important contributions to the identification of CT, he noted that the common gut bacterium *E. coli* was also capable of causing a cholera-like disease (De et al., 1956). By the early 1970s it had become clear that enterotoxigenic *E. coli* (ETEC) strains were of global significance as agents of diarrheal disease in both humans and domestic animals. ETEC have a worldwide distribution, but their significance as human pathogens is most evident in developing countries. In Asia, Africa, and Latin America, the incidence of ETEC infection in patients with diarrhea attending hospitals or clinics has been reported to range from 10 to 50% (Evans et al., 1977; Black et al., 1980; Agbonlahor and Odugbemi, 1982; DeMol et al., 1983; Echeverria et al., 1985; Steffen, 1986). It has also been estimated that approximately 20% of all acute (life-threatening) diarrheal cases in children <5 years of age are due to ETEC, and in the population as a whole ca. 5% of acute diarrhea is due to these microorganisms. A community-based study of children under 5 years of age in rural villages in Bangladesh revealed an incidence of one to two ETEC episodes per child per year (Black et al., 1981). For travelers to developing countries, ETEC is the most commonly identified cause of diarrhea: in 19 studies in Latin America, 28–72% of all diarrheal episodes in travelers were shown to be due to ETEC (Black, 1986; Steffen, 1986).

The secretory diarrhea produced by ETEC is due to its ability to produce either one or both of two enterotoxins, an oligomeric heat-labile enterotoxin (Etx) that is structurally and functionally homologous to CT (see Section III) or a heat-stable enterotoxin (ST), an 18- or 19-amino-acid hydrophobic peptide that activates guanylate cyclase (for a review of ST, see Chapter 12 of this volume). Strains producing either ST alone or both toxins have been found to cause severe symptoms, whereas strains producing only Etx cause mild diarrhea. This is somewhat surprising given that CT and Etx exhibit the same mode of action. The level of Etx produced by ETEC in the intestine is difficult to estimate, but it is likely to exceed 5–25 μg, the amount of CT that has been sufficient to induce profuse diarrhea in volunteers. An explanation for the difference in the pathogenicity of *V. cholerae* and ETEC strains is not immediately obvious; it may be a result of factors unrelated to the capacity to produce oligomeric enterotoxins, such as the chemotactic response of *V. cholerae* strains that enables them to swim through mucus and efficiently colonize the surface of the intestine. Alternatively, it may be due to features directly related to toxin biogenesis, notably the failure of *E. coli* to release Etx from the periplasm compared with the active secretion of CT from *V. cholerae* (for a more detailed discussion, see Section V.E).

Enterotoxigenic strains of *E. coli* also cause diarrheal disease (colibacillosis) in neonatal pigs, calves, lambs, and chickens with concomitant losses to the farming community (Smith and Halls, 1967; Tsuji et al., 1988; Inoue et al., 1993; Moon and Bunn, 1993). Most of the detailed studies of the toxins produced by veterinary ETEC strains have been from pigs. This has revealed that *porcine* heat-labile Etx (pEtx) shares 98% sequence identity to hEtx produced by ETEC strains of human origin (Leong et al., 1985; Yamamoto et al., 1987; Inoue et al., 1993). The A and B subunits of chicken Etx (cEtx) were found to have primary sequences identical to those of hEtx (Inoue et al., 1993).

Holmes and coworkers have characterized an *E. coli* strain isolated from a water buffalo in Thailand that produces a toxin (designated LT-II) that resembles Etx in several respects, in that it forms an AB$_5$ structure possessing ADP-ribosyltransferase activity, but it differs in its glycolipid binding specificity from CT and Etx and is not neutralized by antisera raised against CT, hEtx, or pEtx (Green et al., 1983; Holmes et al., 1986; Pickett et al., 1987).

C. Other Enteropathogens

Several other enteric microorganisms, including certain *Salmonella Aeromonas*, and *Campylobacter* sp. as well as *Vibrio mimicus*, have been reported to occasionally cause watery cholera-like diarrhea (Ruiz-Palacious et al., 1983; Spira and Fedorka-Cray, 1984; Chopra et al., 1987; Fernadez et al., 1988). However, in the case of *Salmonella*, *Aeromonas*, and *Campylobacter* sp., further studies are required to confirm whether or not toxins are produced that are both structurally and functionally related to CT.

III. CHOLERA TOXIN AND RELATED ENTEROTOXINS

A. Toxin Structure

The structural analysis of CT was initiated by Finkelstein and LoSpalluto with the successful purification of CT from sterile culture filtrates of *V. cholerae* (Finkelstein et al., 1966; Finkelstein and LoSpalluto, 1969). An analysis of the purified protein revealed that it contained two factors of differing biological activity. One of the factors, termed choleragen (now known as cholera toxin), was capable of inducing fluid accumulation in guinea pig ileal loops and was found to contain two different subunits, A and B. The second factor, termed choleragenoid, was devoid of toxic activity and consisted exclusively of B subunits (Finkelstein and LoSpalluto, 1969). The stoichiometry and arrangement of subunits within the CT molecule was investigated by Gill using cross-linking reagents and SDS-polyacramide gel electrophoresis (Gill, 1976). This led to the accurate prediction, subsequently verified by X-ray crystallography, that CT was a heterohexamer composed of one A subunit arranged on a ring of five B subunits (Gill, 1976). When the A subunit of CT is analyzed by SDS-PAGE under nonreducing conditions, a single polypeptide is observed with an apparent electrophoretic mobility of ca. 28,000 Da. However, under reducing conditions, it separates into two polypeptides, A1 ($M_r \approx 22,000$) and A2 ($M_r \approx 5,500$), corresponding to the N-terminal and C-terminal fragments, respectively, of the A subunit (Gill, 1976).

The purification of Etx from ETEC of human (hEtx) and porcine (pEtx) origin confirmed that these proteins also consisted of A and B subunits (Clements and Finkelstein, 1979; Kunkel and Robertson, 1979; Geary et al., 1982) arranged with an AB_5 stoichiometry homologous to that of CT (Gill et al., 1981).

1. Primary Sequence

The first sequence analysis of CT was performed on purified B subunits by direct sequencing of the polypeptide (Kurosky et al., 1977; Lai, 1977). However, with the advent of gene cloning, the protein sequences of the A and B subunits for hEtx, pEtx, chicken Etx (cEtx), and CT (from classical and El Tor strains) have been deduced from the DNA sequence (Dallas and Falkow, 1980; Gennaro et al., 1982; Spicer and Noble, 1982; Dallas, 1983; Gennaro and Greenaway, 1983; Lockman and Kaper, 1983; Mekalanos et al., 1983; Yamamoto and Yokota, 1983; Lockman et al., 1984; Yamamoto et al., 1984, 1987; Dykes et al., 1985; Leong et al., 1985; Takao et al., 1985; Sanchez and Holmgren, 1989; Brickman et al., 1990; Dams et al., 1991; Inoue et al., 1993). Figure 2 shows a comparison of the complete amino acid sequences of the A and B subunits of CT from the classical V. cholerae O1 strain 569B and those of hEtx from E. coli strain H10407 of human origin (Yamamoto et al., 1987; Dams et al., 1991; Inoue et al., 1993). From the DNA sequence it was apparent that both the A and B subunits of CT and Etx are synthesized as longer precursors with amino-terminal signal sequence extensions of 18 and 21 amino acids, respectively (denoted by negative numbers in Fig. 2). Such signal sequences are involved in targeting precursors to the cytoplasmic membrane and are cleaved off during translocation to yield a mature polypeptide (for a fuller description of the role of signal sequences, see Section V.C).

The mature A subunits of the CT (CT-A) and Etx (EtxA), shown in Figure 2, are 240 amino acids in length ($M_r \approx 27,200$) and share considerable sequence identity, with 195 identical residues (81.2%). Similarly, the mature B subunits (CT-B and EtxB) are composed of 103 amino acids ($M_r \approx 11,800$), of which 85 (82.5%) are identical. The only region where there is little homology between Etx and CT is between amino acid residues 190 and 212 of the A subunits, where the sequence identity is 34.7%. This region contains an exposed loop formed by the generation of a disulfide bond between two cysteine residues at positions 187 and 199. The loop can be easily "nicked" by exogenous proteases, thereby cleaving the A subunit to its A1/A2 polypeptides. Commercially available preparations of CT contain A subunits that have already undergone proteolytic nicking, with cleavage occurring between residues 192 and 193 (Xia et al., 1984). Extensive studies by Finkelstein and coworkers have demonstrated that V. cholerae secrete

A-subunit

```
                                                     ↓
                        CtxA -18 MVKIIFVFFIFLSSFSYA  0
                                 | .|.|:|||:|.|  ||
                        EtxA -18 MKNITFIFFILLASPLYA  0

      .              .              .              .
  1 NDDKLYRADSRPPDEIKQSGGLMPRGQSEYFDRGTQMNINLYDHARGTQT  50
    |:|:||||||||||||||.||||||||:.||||||||||||||||||||||
  1 NGDKLYRADSRPPDEIKRSGGLMPRGHNEYFDRGTQMNINLYDHARGTQT  50

      .              .              .              .
 51 GFVRHDDGYVSTSISLRSAHLVGQTILSGHSTYYIYVIATAPNMFNVNDV  100
    |||||.|||||||||:|||||||.|| ||||.|||||||||||||||||||
 51 GFVRYDDGYVSTSLSLRSAHLAGQSILSGYSTYYIYVIATAPNMFNVNDV  100

       .              .              .              .
101 LGAYSPHPDEQEVSALGGIPYSQIYGWYRVHFGVLDEQLHRNRGYRDRYY  150
    ||.|||||  |||||||||||||||||||||||:||.||||||:||||||
101 LGVYSPHPYEQEVSALGGIPYSQIYGWYRVNFGVIVERLHNNREYRDRYY  150

       .              .              .              .
151 SNLDIAPAADGYGLAGFPPEHRAWREEPWIHHAPPGCGNAPRSSMSNTCD  200
    .||:||||.||| ||||||:|.||||||||||||||.||||..|.  ::||:
151 RNLNIAPAEDGYRLAGFPPDHQAWREEPWIHHAPQGCGNSSRTITGDTCN  200

       .              .              .
201 EKTQSLGVKFLDEYQSKVKRQIFSGYQSDIDTHNRIKDEL  240
    |.||.|:. :| ||||||||||||||:|||::|..|||.|||
201 EETQNLSTIYLRKYQSKVKRQIFSDYQSEVDIYNRIRNEL  240
```

B-subunit

```
                                                    ↓
                        CtxB -21    MIKLKFGVFFTVLLSSAYAHG  0
                                    | | |   | ||:||||..| |
                        EtxB -21    MNKVKCYVLFTALLSSLCAYG  0

      .              .              .              .
  1 TPQNITDLCAEYHNTQIHTLNDKIFSYTESLAGKREMAIITFKNGATFQV  50
    .||.||:||.||.|.||||.|:||||||||||:||||||.|||||.|||||
  1 APQSITELCSEYRNTQIYTINDKILSYTESMAGKREMVIITFKSGATFQV  50

      .              .              .              .
 51 EVPGSQHIDSQKKAIERMKDTLRIAYLTEAKVEKLCVWNNKTPHAIAAIS  100
    ||||||||||||||||||||||||||.||||.|::||||||||||:.||||
 51 EVPGSQHIDSQKKAIERMKDTLRITYLTETKIDKLCVWNNKTPNSIAAIS  100
```

```
101 MAN 103
    |.|
101 MEN 103
```

Fig. 2   Primary amino acid sequence alignments of the A and B subunits of cholera toxin and *E. coli* heat-labile enterotoxin. Cholera toxin sequences are from the classical biotype *V. cholerae* O1 strain 569B (Dams et al., 1991); heat-labile enterotoxin sequences are derived from *E. coli* strain H10407 of human origin (Yamamoto et al., 1984; Inoue et al., 1993). Amino acids composing the signal sequences of the A and B subunits are denoted by negative numbers. The arrows indicate the positions of cleavage between the signal sequences and mature polypeptides.

a zinc metalloprotease (hemagglutinin/protease) that can nick the A subunit of CT in vitro (Booth et al., 1984). In contrast, Etx purified from *E. coli* is unnicked, although the A subunit can be easily cleaved by exogenously added trypsin, yielding A1 and A2 polypeptides similar to those of CT (Clements and Finkelstein, 1979; Clements et al., 1980). The nicking of the A subunit activates the ADP-ribosylating activity of the A1 polypeptide (see below) and is thus important for toxin action. However, it remains to be established whether it is trypsin, a bacterially encoded protease, or proteases present on the surface of intestinal epithelia that nick the A subunits in vivo.

The primary sequences of the B subunits of CT and Etx reveal a number of other distinctive features. First, there are a pair of conserved cysteine residues (at positions 9 and 86) that have been shown to form a single intramolecular disulfide bridge (Hardy et al., 1988; Sixma et al., 1991), the formation of which is essential for toxin assembly (see section V.D). Second, they have a high content of charged residues, which for each B subunit of CT include four Asp, seven Glu, nine Lys, and three Arg residues (Fig. 2). Third, each B subunit possesses a single conserved tryptophan residue (at position 88).

The analysis of the *ctx* genes from different strains of *V. cholerae* has revealed a number of differences in the B-subunit sequence compared to that shown in Figure 2 (see Table 1). Similarly, the subunits of Etx from human and porcine origin also show slight sequence heterogeneity (Table 2). These minor differences have no known effect on the toxin three-dimensional structure or functional properties, but they do appear to give rise to significant novel antigenic specificities.

2. *Quaternary Structure*

An enormous stride forward in our understanding of the tertiary and quaternary structure of these toxins was made in 1991, when Hol and coworkers successfully determined the three-dimensional structure of porcine heat-labile enterotoxin from *E. coli* (Sixma et al., 1991, 1993). This provided a glimpse of a molecule with a quite spectacular architecture in which the A2 polypeptide (composed of a 23-residue α helix) abuts one of the faces of the triangular A1 polypeptide and then continues as an extended chain down into the highly charged central pore of a doughnut-shaped ring of five B subunits

TABLE 1 Amino Acid Sequence Heterogeneity in B Subunits of Cholera Toxin

V. cholerae strain	Biotype	Amino acid number[a]			Reference
		B subunit			
		+18	+47	+54	
569B	Classical	His	Thr	Gly	Dams et al., 1991
41	Classical	His	Thr	Gly	Dams et al., 1991
O395	Classical	His	Thr	Gly	Dams et al., 1991
2125	El Tor	Tyr	Ile	Gly	Dams et al., 1991
62746	El Tor	Tyr	Ile	Gly	Dams et al., 1991
3083	El Tor	Tyr	Ile	(Ser)	Dams et al., 1991; Brickman et al., 1990

[a]Corresponds to the amino acid numbers in the mature subunit sequence.

TABLE 2 Amino Acid Sequence Heterogeneity in A and B Subunits of Etx from *E. coli* of Human and Porcine Origin

E. coli strain	Origin	Amino acid number[a]				Reference[b]
		A subunit				
		+4	+212	+213	+238	
H10407[c]	Human	Lys	Arg	Lys	Asn	1,2
H74-114	Human	Lys	Lys	Glu	Asp	3
P307	Porcine	Arg	Arg	Glu	Asp	4
		B subunit				
		+4	+13	+46	+102	
H10407[c]	Human	Ser	Arg	Ala	Glu	2,5
H74-114	Human	Ser	His	Ala	Glu	6
P307	Porcine	Thr	Arg	Glu	Lys	6

[a]Corresponds to the amino acid number in the mature subunit sequence.
[b]1, Yamamoto and Yokota (1983); 2, Inoue et al. (1993); 3, Webb and Hirst, unpublished results; 4, Dykes et al. (1985); 5, Yamamoto and Yokota (1983); 6, Leong et al. (1985).
[c]The amino acid sequences of the A and B subunits of cEtx from chicken ETEC isolate were found to be identical to those of strain H10407 (Inoue et al., 1993).

(Fig. 3). A full description of the three-dimensional structure of pEtx can be found in Chapter 9 of this volume.

Some of the features to emerge from the structural analysis of pEtx are that each B-subunit monomer is composed of two three-stranded antiparallel β sheets, a small amino-terminal α helix, and a large central helix. In the pentameric structure the central helix is part of the wall of the central pore, with the five helices from each subunit forming a pentagonal helix barrel. This creates a pore 30 Å long, with a diameter ranging from ca. 11 Å near the surface at which the A subunit is positioned to ca. 15 Å on the lower surface (see Fig. 3). There are six β strands in each B subunit, which form a pair of three-stranded antiparallel β sheets (sheets I and II) on each side of the molecule. These sheets interact directly with the β sheets of adjacent B subunits (via multiple hydrogen bonds)—a feature that contributes to the remarkable stability of the B pentamer. Visual inspection of the B pentamer reveals that the top surface (at which the A subunit is located) is rather flat, whereas the lower surface is somewhat convoluted. The site of receptor binding to the B subunit occurs at this lower convoluted surface by interaction with residues from two adjacent B subunits (Sixma et al., 1992).

Of the three proline residues in each B subunit of Etx and CT, one (Pro-93) is in the cis configuration, which may have important implications for the pathway of subunit folding (see Section V.D).

The pEtx used for the crystallographic determination was purified from *E. coli*, and as such the A subunit had not undergone proteolytic nicking. However, the two fragments of the A subunit corresponding to the A1 and A2 polypeptides were revealed as two separately structured domains, with a disulfide bridge between Cys-187 and Cys-199. A

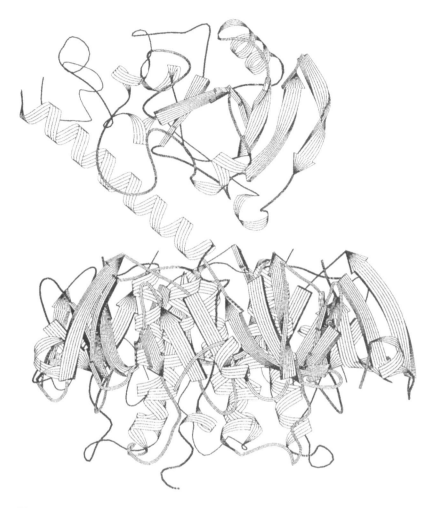

Fig. 3 Ribbon diagram of the crystal structure of porcine heat-labile enterotoxin from *E. coli*. The A subunit (comprising the A1 and A2 fragments) forms a wedge-shaped structure; the long α helix of the A2 fragment is clearly visible. The B subunits form a pentameric ring with a central pore into which the C-terminal portion of the A2 fragment is situated. (From T. K. Sixma and W. G. J. Hol, used with permission.)

number of regions within the A subunit were not visible within the refined structure, with no density apparent for the three amino-terminal and four carboxy-terminal amino acids, or for residues 189–195 corresponding to the exposed loop between the A1 and A2 polypeptides.

The A1 polypeptide (from amino acids 1–194) is a single-domain protein, with numerous secondary structure elements. It contains the enzyme active site capable of ADP-ribosylation (for a fuller description of the enzymatic activity of Ctx and related enterotoxins, see Chapters 4 and 9 of this volume). When the crystal structure of Etx was compared with that of another ADP-ribosylating toxin, exotoxin A (ETA) from *Pseudo-*

monas aeruginosa, it was apparent that 44 residues in the A1 polypeptide could be superimposed with residues in the enzymatic domain of ETA (Sixma et al., 1993). Of these, only three amino acids were identical: Tyr-6, Ala-69, and Glu-102 in Etx, with the latter corresponding to the active site residue Glu-553 of ETA.

The A2 polypeptide (from amino acids 195–240) has a kink near the C terminus of A2, at Ser-224, at approximately the site at which it enters the central pore of the B-subunit pentamer. One side of the long α helix of A2 (amino acids 197–226) makes numerous contacts with the A1 polypeptide. The major contacts between the A and B subunits are localized at the region where the A2 polypeptide enters the B_5 pore as well as within the central pore, where Glu-229 and Arg-235 of the A2 polypeptide form salt bridge interactions with three different B subunits.

Other toxins also adopt AB_5 structures, although they do not necessarily show primary sequence homology with Etx and CT. One of the most striking comparisons is with the Verotoxins (also known as Shiga-like toxins) produced by certain pathogenic strains of *E. coli*. These toxins have a B-pentamer moiety that binds to Gb3 and Gb4 glycolipids and an A-subunit moiety that, like the A subunit of Shiga toxin and ricin, inhibits protein synthesis by cleavage of a specific adenine residue in the 28 S RNA of eukaryotic ribosomes (for a review, see Brunton, 1990). Given that there is no sequence identity whatsoever between the B subunits of Verotoxin and Etx, it came as somewhat of a surprise to find that 52 out of the 69 amino acids of the B subunit of Verotoxin can be superimposed onto the structure of EtxB. Moreover, seven of the eight secondary structure elements present in Etx are also found in Verotoxin, with only the amino-terminal helix of EtxB being absent from the latter (Stein et al., 1992). The structural fold of EtxB has also been found in two other proteins, staphylococcal nuclease and the anticodon binding domain of Asp-tRNA synthase, which bind oligonucleotides (Arnone et al., 1971; Murzin, 1993; Sixma et al., 1993). Thus, the B subunits of Etx (and CT) are members of a family of oligosaccharide and oligonucleotide binding proteins that have the same folding motif, even though they do not share any homology in their primary sequences.

B. Toxin Action

The mode of action of CT is described in detail in Chapters 4 and 11 of this volume. Briefly, the biological activity of CT and related enterotoxins depends on the B subunits acting to bind the toxin to specific receptors and initiate internalization into the cell, whereas the A subunit possesses the "toxic" activity responsible for initiating the biochemical events that lead to the activation of adenylate cylase, electrolyte efflux, and ultimately diarrhea (Holmgren, 1981). The B subunits of CT bind with high affinity and virtually exclusively to G_{M1} ganglioside, a ubiquitous glycolipid found in the membranes of all eukaryotic cells (van Heyningen et al., 1971; Cuatrecasas, 1973; Holmgren et al., 1973; King and van Heyningen, 1973). The B subunits of Etx have a somewhat broader spectrum of receptor recognition, binding not only to G_{M1} but also to certain other receptors that possess a terminal galactosyl moiety (Holmgren, 1973; Holmgren et al., 1985; Griffiths and Critchley, 1991). Upon binding, the toxin appears to be internalized via non-coated vesicles, followed by translocation of the A1 polypeptide across the vesicle membrane (Kassis et al., 1982; Montesano et al., 1982; Janicot et al., 1991; Lencer et al., 1992). The A subunit then catalyzes the NAD-dependent ADP-ribosylation of $G_{s\alpha}$, a component of the heterotrimeric signal transducing guanine-nucleotide binding protein (G protein) that stimulates adenylate cyclase and regulates Ca^{2+} channels (Moss and

Vaughan, 1991). The A subunit's activity is normally latent, requiring reduction of the disulfide bond between the two cysteines (and proteolytic nicking to the A1 and A2 polypeptides) to yield the enzymatically active A1 moiety. Disulfide reduction is believed to occur after toxin receptor binding and internalization (Kassis et al., 1982; Janicot et al., 1991; Lencer et al., 1992). The A1 fragment catalyzes the ADP-ribosylation of a specific arginine residue at position 201 of $G_{s\alpha}$ (Moss and Vaughan, 1977; Van Dop et al., 1984; Rappuoli and Pizza, 1991). This inhibits the capacity of $G_{s\alpha}$ to hydrolyze and release bound GTP, thereby maintaining $G_{s\alpha}$ in a [$G_{s\alpha}$ GTP]-activated state capable of prolonged stimulation of adenylate cyclase (Cassel and Selinger, 1977; Moss and Richardson, 1978; Moss and Vaughan, 1991). The ensuing rise in cytosolic cAMP levels causes a down-regulation of neutral Na^+ adsorption by enterocytes and an increase in Cl^- secretion from crypt cells. Since water movement across the intestinal mucosa passively follows the osmotic gradients established by the transport of charged ions, the increased Cl^- secretion and inhibition of Na^+ adsorption causes that rapid movement of water into the intestinal lumen, resulting in the characteristic voluminous watery diarrhea of cholera.

One of the most intriguing and most poorly defined aspects of toxin action on eukaryotic cells comprises the events involved in toxin uptake, intracellular trafficking, and A-subunit (or A1-polypeptide) translocation. In this regard, it is particularly interesting that the last four carboxy-terminal amino acid residues of the A2 polypeptide of CT are Lys-Asp-Glu-Leu (KDEL) [or RD(N)EL for the A subunit of Etx] (Mekalanos et al., 1983; Leong et al., 1985; Yamamoto et al., 1987), homologous to sequences that occur at the carboxy termini of eukaryotic proteins located in the endoplasmic reticulum (ER) (Pelham, 1991). When ER resident proteins move by a process of bulk flow from the ER into the Golgi complex, the KDEL sequence interacts with a KDEL receptor that recycles such proteins back into the ER (Pelham, 1991). The presence of a KDEL sequence at the carboxy terminus of the A subunit is suggestive of the possibility that vesicles containing receptor-bound AB_5 toxin move through the cell and eventually fuse with the Golgi membrane, permitting the direct interaction of the A subunit with the KDEL-receptor protein. This might either aid the movement of the toxin into the ER, whereupon the A1 fragment translocates and activates $G_{s\alpha}$, or alternatively it might destabilize A/B subunit interaction and promote the release and translocation of the A1 polypeptide directly across the Golgi membrane. Recent evidence has emerged for partial involvement of the KDEL (RDEL) sequence in enterotoxin action from studies of the kinetics and magnitude of Cl^- ion efflux in polarized T84 epithelial cells (Lencer et al., 1992) treated with *E. coli* Etx or a mutant in which the terminal leucine residue of the A subunit had been replaced by valine (W. I. Lencer et al., unpublished results). The lag period, before elevation of cAMP and concomitant activation of Cl^- efflux, was significantly increased for the mutant Etx, supporting the view that this region of the A2 polypeptide contributes to the efficiency of toxin action. Findings by Pastan and coworkers on the intoxication by *Pseudomonas* exotoxin A indicate that KDEL–receptor interaction may play a role in ETA translocation to the cytosol (Seethaharam et al., 1991). By altering the native carboxy-terminal sequence of ETA from REDLK to an authentic KDEL sequence, its cytotoxicity increased two- to three-fold (Seethaharam et al., 1991). Other toxins, such as Shiga toxin and Shiga-like toxin (Verotoxin), also appear to traffic to the ER, but since these lack a KDEL sequence, they must do so via a KDEL-independent route (Jackson et al., 1987; Strockbine et al., 1988; Weinstein et al., 1988). In summary, it seems likely that uptake and trafficking of toxins is determined by both the receptor to which a toxin initially binds and components of the vesicles or intracellular compartments (such as the KDEL-receptor protein) that affect the efficiency of toxin translocation to its intracellular target. In all cases, however, the mechanism by which the "toxic A domains" are able to translocate across an intracellular membrane remains to be elucidated.

IV. TOXIN GENE ORGANIZATION

In this and following sections the genetic and biochemical basis of toxin expression in *V. cholerae* and ETEC are considered.

A. The *ctx* Genes Encoding Cholera Toxin

The *ctxA* and *ctxB* genes, encoding the A and B subunits of CT, are organized into an operon (*ctxAB*) located in a 4.5-kb "virulence cassette" on the chromosome of *V. cholerae* (Fig. 4) (Mekalanos, 1983; Mekalanos et al., 1983; Goldberg and Mekalanos, 1986; Pearson et al., 1993; Trucksis et al., 1993). In addition to the *ctxAB* operon, the cassette contains genes encoding zonula occludins toxin (Zot), accessory cholera enterotoxin (Ace), a core-encoded pilin (Cep), and an open reading frame (orfU) of unknown function (Fig. 4) (Fasano et al., 1991; Baudry et al., 1992; Pearson et al., 1993; Trucksis et al., 1993). For *V. cholerae* El Tor strains the *ctx*-containing cassette maps to the same chromosomal location in all strains that have so far been examined (Mekalanos, 1983; Pearson et al., 1993), whereas in classical strains of *V. cholerae* the *ctx* operon has been shown to be located at two distinct chromosomal loci (Mekalanos, 1983).

In most strains of *V. cholerae*, the 4.5-kb virulence cassette is flanked by one or more copies of a 2.7-kb directly repeated sequence called RS1 (Fig. 4) (Mekalanos, 1983; Goldberg and Mekalanos, 1986). These RS1 sequences are capable of undergoing RecA-dependent homologous recombination, to give tandem amplification of the entire virulence cassette and flanking RS1 repeats. The high tendency of this cassette to undergo such amplification was convincingly demonstrated by Goldberg and Mekalanos (1986). These workers replaced the *ctx* operon with a DNA fragment coding for resistance to kanamycin and demonstrated that the single copy of the $\Delta ctxAB$ Kan^r allele in the resultant strain conferred resistance to approximately 150 μg of kanamycin per milliliter. They then isolated derivatives capable of growing on media containing 3 mg of kanamycin per milliliter at a frequency of 9.4×10^{-8} and showed that these derivatives had 20 amplified genetic elements (Goldberg and Mekalanos, 1986). By the same token, spontaneous deletion of the *ctx* virulence cassette can also occur, presumably by events reciprocal to those leading to amplification. Various models can be envisaged for these events, the most likely being that amplification or deletion occurs as a result of unequal crossing over between RS1 sequences of daughter strands during DNA replication (Goldberg and Mekalanos, 1986). The relevance of such events becomes clear when it is considered that they appear to occur naturally during infection (at least in animal models), resulting in organisms with more copies of the *ctx* virulence cassette and presumably enhanced pathogenicity (Mekalanos, 1983; Mekalanos et al, 1983). The presence on the virulence cassette of a gene encoding a core-encoded pilin that appears to confer enhanced intestinal colonization (Pearson et al., 1993) may provide sufficient selective advantage for tandem duplication to occur in vivo.

The *ctx* virulence cassette, with its "core" region flanked by direct RS1 repeats, is structurally analogous to composite transposons, such as Tn9, where IS*1* insertion sequences are directly duplicated at the termini of the transposon. Until recently there

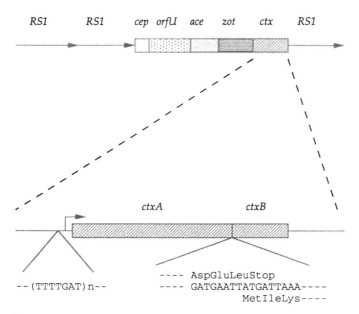

Fig. 4 Schematic representation of the *ctx* virulence cassette showing the core region and flanking RS1 elements. Open reading frames; *cep*, core-encoded pilin; *orfU*, open reading frame of unknown function; *ace*, accessory cholera enterotoxin; *zot*, zonula occludens toxin; and *ctx*, cholera toxin operon. The lower portion represents the *ctx* operon, including the *ctxA* and *ctxB* genes. The promoter, the upstream TTTTGAT repeat motif involved in ToxR activation of *ctx* transcription, and the amino acid and nucleotide sequences at the site of overlap between the *ctxA* and *ctxB* genes are shown.

was no evidence that the *ctx* element could undergo direct transposition, as opposed to amplification and deletion. However, Mekalanos and coworkers have recently shown in an elegant series of experiments that RS1 encodes a site-specific recombination system that allows it to integrate into an 18-bp sequence, termed *attRS1*, located on the chromosome of nontoxinogenic strains of *V. cholerae* (Pearson et al., 1993). They cloned a single copy of RS1 into a conditionally replication-defective plasmid (that replicates in suitable *E. coli* strains but not in *V. cholerae*) and introduced this plasmid into a nontoxinogenic El Tor isolate that lacked the entire *ctx* genetic element. Southern blot and sequence analysis revealed that the plasmid integrated into the same chromosomal location at which the toxin genes are normally found in natural toxinogenic isolates of El Tor strains. Comparison of the DNA sequence of the site of RS1 insertion in the recipient, revealed an 18-bp CCTTAGTGCGTATTTATGT identity with the left end of RS1. This sequence exists at the junction between the *ctx* genetic element and chromosomal DNA in all *V. cholerae* strains so far examined and also marks the junction between tandem copies of the RS sequence (Pearson et al., 1993).

These findings thus signify that the *ctx* genes are located on a potentially highly mobile genetic element and raise the specter that *V. cholerae* acquired the entire virulence cassette from a hitherto unidentified ancestral organism by a process of horizontal gene transfer. This would be consistent with the fact that the GC content and codon usage of the *ctx* genes are not those expected for genes of *V. cholerae* origin (Yamamoto et al., 1987).

B.   The *etx* Genes Encoding *E. coli* Heat-Labile Enterotoxin

In contrast to the chromosomal location of the CT genes, the genes specifying the production of Etx are found on large naturally occurring plasmids, called *ENT* plasmids, present in enterotoxinogenic strains of *E. coli* (Gyles et al., 1977). In many instances these plasmids have been shown to be transmissible by conjugation and often to possess additional genes for drug resistance and colonization antigens (Smith and Linggood, 1971; Gyles et al., 1977). Some *ENT* plasmids, such as pCG86 and P307, encoding hEtx and pEtx, respectively, have been extensively investigated. For example, Gyles et al. (1977) showed that pCG86 contained genes for Etx, ST, conjugal transfer, and resistance to several antibiotics. The considerable diversity of naturally occurring *ENT* plasmids, with some possessing resistance determinants and others not, and some encoding both Etx and ST whereas others encode only one toxin determinant, suggests that *ENT* plasmids may undergo frequent recombination events.

The gene for ST is flanked by IS*1* inverted repeats and has been shown to be capable of transposition (So et al., 1979). A study by Yamamoto on an *ENT* plasmid of human origin, using alkali-denaturation and self-annealing followed by electron microscopy, revealed several double-stranded stem loops, due to the presence of repeated sequences (Yamamoto and Yokota, 1981). The Etx DNA from this plasmid was shown to be flanked by three repeat sequences, two in the same orientation and one in the reverse orientation. Moreover, an ST gene that was also present on this plasmid was located within the DNA bounded by the repeats of the Etx DNA and was itself flanked by two small inverted repeat sequences (Yamamoto and Yokota, 1981). This *ENT* plasmid would therefore appear to have arisen by transposition of an ST transposon into an Etx transposon. No DNA sequence homology exists between the Etx repeats and the RS1 repeats that flank the *ctx* virulence cassette (Mekalanos et al., 1983).

C.   Operon Structure

Because the *E. coli* Etx genes were plasmid-located, Falkow and coworkers were able to use the emerging techniques of genetic engineering in the mid-1970s to isolate and clone the pEtx genes from *E. coli* P307. This pioneering work made Etx (and ST) the first virulence factors to be subjected to such a molecular analysis (So et al., 1978). The genes encoding the A and B subunits of Etx have been variously designated by early investigators as either *eltA / eltB* or *toxA / toxB*, with, in some instances, the use of subscripts H or P to denote the origin of the gene as deriving from a human or porcine strain of *E. coli*. Because of these disparities in nomenclature, I have adopted the general mnemonic, *etxA* and *etxB*, for the genes encoding the A and B subunits of *E. coli* heat-labile enterotoxin, irrespective of the source of the infection.

The *etx* genes from P307, and two human ETEC isolates, H10407 and H74-114, have been sequenced (Dallas and Falkow, 1980; Spicer and Noble, 1982; Yamamoto and Yokota, 1982, 1983; Yamamoto et al., 1984, 1987; Leong et al., 1985). This revealed that the *etxA* and *etxB* genes overlap one another by four nucleotides, with the last four nucleotides of *etxA* (including the stop codon for the A subunit) forming the first four nucleotides of *etxB*. When Mekalanos et al. (1983) sequenced the *ctxAB* operon, an identical 4-bp overlap between the A and B genes was found, as shown in Figure 4 (Lockman and Kaper, 1983; Mekalanos et al., 1983).

Proximal to *etxA* is a sequence of nucleotides characteristic of promoters involved in core RNA polymerase binding. No similar sequence can be found in the nucleotides proximal to *etxB*, supporting the view that the enterotoxin genes are transcribed as a polycistronic message from a single promoter proximal to *etxA*. This view is also consistent with early cloning and deletion analyses, which indicated that the expression of the B subunit was dependent upon the presence of the upstream *etxA* gene (Dallas and Falkow, 1979; Dallas et al., 1979).

Upstream of the −35 promoter site of the *ctxA* gene are a series of three to eight tandemly repeated copies of a TTTTGAT motif, involved in ToxR-mediated activation of CT expression (see Section V.A; Mekalanos et al., 1983; Miller and Mekalanos, 1984; Miller et al., 1987).

V. BIOGENESIS OF CHOLERA TOXIN AND RELATED ENTEROTOXINS

Certain of the steps in toxin biogenesis have been particularly well studied, including the coordinate regulation of toxin gene expression and the folding and assembly of the toxin subunits in vivo. However, other steps have received less attention, such as targeting of the enterotoxin subunits to the export apparatus in the bacterial cytoplasmic membrane and the genetic basis of toxin secretion across the outer membrane of *V. cholerae*. In such instances, I draw upon the data that have emerged from the study of other exported and secreted proteins, thereby providing some insight into the likely events that occur at these stages of toxin biogenesis.

A. Regulation of Toxin Expression

1. *Coordinate Expression of Virulence Genes*

Bacteria possess numerous regulatory systems that enable them to rapidly respond to changes in their environment. In the case of pathogenic microorganisms, such systems enable them to adapt to the changes encountered when leaving a free-living environment and entering a host (DiRita and Mekalanos, 1989; Dorman and Ni Bhriain, 1992; Mekalanos, 1992). This permits the organism not only to resist the differing hostile conditions within the host, but also to switch on the production of specific virulence factors that contribute to its pathogenicity. Over the past few years many regulatory systems have been identified and found to be widespread among bacterial species. They can be grouped into superfamilies of bacterial transcriptional regulators, including the two-component regulatory systems, the LysR group of regulatory elements and AraC group (Albright et al., 1989; Dorman and Ni Bhriain, 1992). The two-component transcriptional regulators usually consist of a sensor kinase protein that detects particular environmental signals and phosphorylates a response regulator DNA-binding protein that coordinately regulates the transcription of a number of genes. All of the genes under the control of a common regulator are defined as a regulon.

Induction of virulence gene expression can involve the sensing of "host" molecules, as occurs in the case of phytopathogens (Ankenbauer and Nester, 1990; Mo and Gross, 1991). In animal and human pathogens, however, the regulatory systems appear to sense environmental cues, such as the transition from ambient low temperatures to body temperature, or low to starvation levels of iron in host tissues, or changes in osmolarity, pH, or redox conditions (for a review see Mekalanos, 1992).

2. *Control of Cholera Toxin Expression by ToxR*

In *V. cholerae*, several genes required for virulence, including CT gene expression, are under the control of ToxR (Fig. 5) (for an excellent review, see DiRita, 1992). This protein influences the transcription of the *ctx* operon in response to changes in several environmental factors, including aeration, osmolarity, pH, temperature, and the availability of several amino acids (Miller and Mekalanos, 1988; DiRita, 1992; Mekalanos, 1992). ToxR is a 32.5-kDa integral cytoplasmic membrane protein that is essential for pathogenicity of the organism in humans (Miller and Mekalanos, 1984; Miller et al., 1987; Herrington et al., 1988). It has been shown to directly regulate CT production as well as that of another control protein ToxT, which in turn regulates the expression of toxin-coregulated pili (TCP), various outer membrane proteins, and an accessory colonization factor (ACF) (Miller and Mekalanos, 1984; Taylor et al., 1987). The ToxR protein has been found to increase directly the transcription of the *ctxAB* promoter, by binding to the TTTTGAT repeat motif found upstream of the −35 region (see Figs. 4 and 5) (Miller et al., 1987).

Approximately one-third of the ToxR protein is located in the periplasm, where it is presumed to sense environmental signals, such as pH and temperature, while the remaining two-thirds of the molecule is located in the cytoplasm, where it acts as a DNA-binding protein (Miller et al., 1987).

The *tox*R gene is transcribed in an operon containing a second gene, *toxS*, which enhances the ToxR-mediated activation of the CT promoter (Miller et al., 1989). ToxS is also an integral membrane protein with a large periplasmic domain and is thought to stabilize ToxR as homodimer in the membrane. This conclusion was serendipitously drawn from the observation that ToxR–PhoA fusion proteins, lacking the C-terminal periplasmic domain of ToxR, dimerize as a result of the PhoA moiety and activate the *ctx* promoter in a ToxS-independent manner (DiRita and Mekalanos, 1991). When the periplasmic domain of ToxR was replaced by PhoA, the fusion was no longer able to respond to certain environmental signals, supporting the view that this portion of the protein is important for sensing the environment. Miller et al. (1987) also showed that isolated membranes from *E. coli* expressing a ToxR–PhoA fusion can interact with radiolabeled DNA containing the *ctx* promoter, resulting in a retardation of the DNA in agarose gels. This was abolished when the TTTTGAT repeat motif was deleted, which supports the view that the amino-terminal domain of ToxR is a DNA-binding protein that triggers an increase in *ctxAB* transcription by direct interaction with the TTTTGAT motif (Miller et al., 1987).

The amino-terminal sequence of ToxR shows some homology with the DNA-binding proteins of the OmpR subclass of the two-component regulators, namely OmpR, VirG, and PhoP, involved in transcriptional activation (Miller et al., 1987; Albright et al., 1989). However, ToxR should be viewed as an atypical member of the OmpR subclass of the two-compartment regulators, because these are usually cytoplasmic proteins that are phosphorylated upon activation of a sensor kinase protein. Since ToxR does not appear to have the phosphoacceptor domain characteristic of members of this regulator family, it may not be activated by phosphorylation. Indeed, the fact that ToxR is a transmembrane protein may have circumvented the need to transduce the environmental signal by phosphorylation, because long-range conformational changes in the molecule may be sufficient to alter its promoter-binding activity.

In contrast to the direct activation of the *ctxAB* promoter by ToxR, the promoters of other ToxR-regulated genes, such as TCP and ACF, were not activated when cloned into *E. coli* in the presence of ToxR and ToxS. This observation suggested that ToxR might

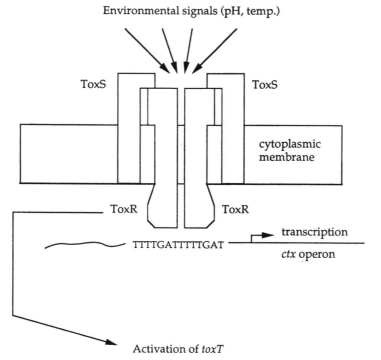

Fig. 5 Schematic model of the activation of CT expression by ToxR. Activation of ToxR by environmental signals is presumed to cause conformational changes (that could involve dimerization and association with ToxS), which lead to the interaction of the N-terminal domain of ToxR with the TTTTGAT$_n$ motif upstream of the *ctx* promoter. The ensuing increase in *ctx* transcription causes an increase in the level of CT production. ToxR also activates the transcription of *toxT*, which in turn regulates the expression of a number of other genes involved in the pathogenesis of *V. cholerae* infection.

mediate activation of these promoters in an indirect fashion that required an additional regulatory protein. This subsequently led to the identification of ToxT (DiRita et al., 1991). This protein has been shown to share sequence homology with the AraC family of transcriptional activators found in gram-negative bacteria, which possess a characteristic helix-turn-helix motif typical of DNA binding proteins (Higgins et al., 1992). The *toxT* gene is located in the *tcp* gene cluster, and it thus not surprising that it is involved in regulating TCP biogenesis. Transcriptional studies have shown that at lower pH, where ToxR-regulated virulence genes are expressed, there is a rise in *toxT* mRNA levels and that this is dependent on ToxR (Higgins et al., 1992). This is not mediated by an increase in ToxR synthesis but through a pH-induced effect on the activity of ToxR. It is not exactly clear how *toxT* expression is controlled by ToxR, since, unlike the ToxR activation of the *ctxAB* operon, which requires ToxR to bind to the tandemly repeated heptameric sequence TTTTGAT, no such sequence is present immediately upstream of *toxT* (Higgins et al., 1992). At this stage, it cannot be ruled out that *toxT* is part of a larger operon and that ToxR binds further upstream to activate transcription in some way.

Interestingly, constitutive expression of ToxT in a *toxR* mutant strain of *V. cholerae* resulted in CT synthesis, regardless of the culture pH (DiRita, 1992). This suggests either that ToxT as well as ToxR modulates the activity of the *ctx* promoter, presumably by direct DNA binding within the *ctx* promoter region, or alternatively that ToxT induces expression of hitherto unidentified regulatory factors that influence toxin expression. Footprint studies or gel retardation assays to map the possible site(s) of ToxT binding to the *ctx* promoter, and thereby establish that ToxT directly modulates *ctx* promoter activity, have as yet not been reported.

It should also be borne in mind that the environmental conditions that elevate the expression of the ToxR-regulated genes in vitro (pH 6.5 and 30°C), may be important only during the initial stages of infection in vivo. Thus, as the organism is ingested, the low-pH conditions in the stomach may serve as an appropriate signal to produce TCP to aid initial colonization of the jejunum. However, once the organism has entered the small intestine and is proliferating at 37°C and at alkaline pH, ToxR-regulated genes should be repressed and presumably other virulence factors produced. These may include new colonization factors, such as *cep* (although its regulation has not yet been studied), as well as changes in aspects of the biogenesis and secretion of virulence factors. In this regard, we have obtained evidence to suggest that environmental conditions that activate ToxR have an unexpected adverse effect on the efficiency of toxin secretion (J. Yu and T. R. Hirst, unpublished observations).

3. Other Factors Influencing Toxin Gene Expression

ToxR, S, and T may represent only the beginning of our understanding of what could be multiple, overlapping regulatory cascades that influence toxin and other virulence factor expression in *V. cholerae*.

An emerging area of study is the influence of DNA topology on the transcription of certain genes (for a review, see Dorman and Ni Bhriain, 1992). Bacteria maintain their genomic DNA in an underwound, negatively supercoiled state, the extent of which is controlled by the activities of various enzymes, namely DNA gyrase and DNA topoisomerases, and by histone-like proteins that bind to DNA. The activities of some promoters are strongly influenced by the extent of DNA supercoiling. Some show elevated transcription after an increase in DNA supercoiling (e.g., the *lac* promoter); others prefer relaxed DNA (e.g., *gyrA*, *gyrB*, and *tonB* promoters), while a third class appear to be unaffected by changes in DNA supercoiling (e.g., the *trp* promoter) (Sanzey, 1979; Menzel and Gellert, 1987; Tse-Dinh and Beran, 1988; Dorman and Ni Bhriain, 1992). DNA supercoiling fluctuates dramatically in response to changes in environmental conditions, such as changes in osmolarity, temperature, and anaerobiosis, which are likely to be encountered when *V. cholerae* or ETEC infect a host.

The expression of *E. coli* heat-labile enterotoxin has been observed to be influenced by growth temperatures, with an approximate fourfold elevation in Etx production at 37°C compared with 30°C (T. R. Hirst and B.-E. Uhlin, unpublished observations). Although this could be due to an influence on any one of the multiple steps in toxin biogenesis, recent evidence suggests that it is mediated at the level of *etx* mRNA transcription as a result of changes in the local topology of DNA caused by a histone-like protein H1 (also called H-NS). Mutations in the gene *osmZ*, encoding H1, were found to alter the transcription of osmoregulated and thermoregulated genes in *E. coli* and to attenuate the virulence of microbial pathogens (Goransson et al., 1990; Dorman and Ni Bhriain, 1992). Goransson et al. (1990) isolated a mutant allele of the *osmZ* gene (designated *drdX*), which showed deregulated expression of PAP pili at 30°C. When a plasmid carrying an *etxA'-cat* operon fusion was introduced into the *drdX* mutant strain, CAT expression was found to be derepressed at 30°C and considerably elevated at 37°C (Uhlin et al., 1991). This suggests that the *etx* promoter is probably thermoregulated by H1. No studies have yet been reported on the identification of a *Vibrio* homologue of *osmZ* or its possible role in the control of toxin expression.

B. Translational Control of Subunit Stoichiometry

Translation of the polycistronic *ctx* or *etx* mRNAs that arise from transcriptional initiation upstream of the *ctxA* or *etxA* cistrons is the start of events that lead to the synthesis of the A and B subunits. Initially, the subunits are produced as precursor polypeptides (pre-A and pre-B), which then undergo export across the cytoplasmic membrane (see Section V.C). Since the subunit structure of the assembled toxins is one A to five B subunits, this implies that the synthesis of the A and B subunits may be stoichiometrically controlled in some way. Indeed, in minicells harboring a recombinant plasmid carrying the *etx* operon, the level of expression of the Etx subunits is significantly different, with more B subunits than A subunits being produced (Yamamoto and Yokota, 1981; Hirst and Hardy, unpublished observations). A careful examination of the relative levels of A- and B-subunit synthesis in *E. coli* expressing either porcine or human Etx indicated that one to two A subunits are synthesized for every five B subunits (Hofstra and Witholt, 1984; Hirst and Hardy, unpublished observations). The molecular mechanism(s) responsible for this have not as yet been comprehensively investigated, and several possible explanations may be considered. The fact that both subunits are expressed from a polycistronic mRNA indicates that the relative levels of A- and B-subunit synthesis must be governed by events that occur after transcriptional initiation. These could include (1) differential stability of the mRNA transcript, with the 5' EtxA-encoding region being subject to a higher rate of turnover; (2) differences in the efficiency of the Shine-Dalgarno (SD) sequences just upstream of the start codons of the A and B cistrons, which influence translational initiation; (3) the presence of local stable secondary mRNA structures that either interfere with or promote ribosomal recognition of the SD sequence; (4) translational reinitiation or coupling at the SD sequence for B-subunit expression (see below), thereby elevating B-subunit synthesis; (5) different stable mRNA secondary structures that affect ribosome movement along the mRNA and consequently alter the translation rate for the A and B subunits; and (6) events that occur posttranslationally, such as the relative susceptibility of A and B subunits to proteolytic turnover, which results in a steady-state level of subunits that approaches the 1:5 stoichiometry that is experimentally observed.

The Shine-Dalgarno sites of ribosome binding and translational initiation are usually found seven nucleotides upstream of the initiation codon and comprise the consensus sequence 5'-AGGAGG-3', which base-pairs with the 3' end of 16 S ribosomal RNA (Shine and Dalgarno, 1974; Gold and Stormo, 1987). An analysis of the nucleotides found upstream of the initiation codons of the A and B subunits of CT and Etx is presented in Table 3, with the similarities to the *E. coli* SD consensus sequence underlined. There are clear differences in the numbers of nucleotides of A and B cistrons that share complementarity with the SD consensus, consistent with the higher level of B-subunit synthesis being due, at least in part, to the presence of a more efficient ribosome binding site. When the majority of the *ctxA* cistron was deleted by Mekalanos and coworkers to give an in-frame fusion between amino acid 17 (−2) of the A-subunit signal sequence and amino acid 19 (−3) of the B-subunit signal sequence, the normal cleavage site for the B subunit was maintained, but B-subunit expression became dependent on the translational initiation sequences of the A cistron, and it was found that B-subunit pentamers were produced at one-ninth the level observed with the wild-type operon (Mekalanos et al., 1983). Later it was found that mutations introduced into the upstream region of *etxB*, which alter the nucleotides of the *etxB* SD sequence, alter the yield of EtxB (Streatfield et al., 1992). For example, an increase in the complementarity from GGA to AGGA increased the level of EtxB production by ca. 80% (Table 2) (Streatfield et al., 1992). These findings are consistent with an important role for the SD sequence in determining the stoichiometry of subunit expression, although the possibility that such deletions or substitutions give rise to altered mRNA stability or secondary structures was not excluded.

Translational reinitiation may also be a key feature in the translational control of subunit stoichiometry. In general, if either an initiation codon or a Shine-Dalgarno sequence, or both, are found within 10 or so nucleotides of a termination codon, a given ribosome undergoing translational termination may diffuse momentarily on the mRNA and find a place to begin translation again (Oppenheim and Yanofsky, 1980; Normark et al., 1984; Gold and Stormo, 1987). These criteria are met by both the *ctxB* and *etxB* cistrons, although the formal demonstration that such translational coupling may be involved in CT or Etx biogenesis has not been investigated.

The possibility that mRNA secondary structure may also be involved in translational control of toxin subunit synthesis was suggested by Yamamoto et al. (1985). After initiation of translation, ribosomes move along the mRNA, with gaps of approximately 100 bases between adjacent ribosomes. This affords the possibility that transient secondary structures may form in the mRNA that impede ribosome movement and the rate of translation (Yamamoto et al., 1985). Based on free-energy calculations of the likelihood of stable stem-loop structures being formed, Yamamoto et al. (1985) concluded that such structures would be more abundant in the A-subunit mRNA and could therefore contribute to a lower level of A-subunit translation than that of the B subunit. Since the extent and position of such putative stem-loop structures in the *ctxA* and *etxA* cistrons appear to have been conserved during the evolution of the toxin genes, they may be of functional significance.

TABLE 3 Influence of Upstream Shine-Dalgarno[a] (SD) Sequences on the Translation of the A and B Subunits of CT and Etx

Operon	Putative Shine-Dalgarno sequences	Reference
	Met - - - - - - CT A subunit	Mekalanos et al., 1983
ctx	5'-AAGGGAGCATTATATG - - - - - -	
	Met - - - - - - CT B subunit	Mekalanos et al., 1983
ctx	5'-TTAAGGATGAATTATG - - - - - -	
	Met - - - - - - Etx A subunit	Yamamoto et al., 1987
etx	5'-TAAGTTTTCCTCGATG - - - - - -	
	Met - - - - - - Etx B subunit	Leong et al., 1985
etx	5'-TTCGGGATGAATTATG	
	Met - - - - - - Etx B subunit	Streatfield, 1992
Mutant *etx* SD	5'-TTTAGGATGAATTATG	

[a]The consensus Shine-Dalgarno sequence of *E. coli* is 5'-AGGAGG-3' (Gold and Stormpo, 1987).

C. Toxin Export Across the Cytoplasmic Membrane

Protein export across the cytoplasmic membranes of bacteria, especially in *E. coli*, has been extensively investigated during the past 20 years and has recently been summarized in an excellent review by Pugsley (1993). Protein translocation is dependent on both distinct structural properties of exported proteins, such as the presence of amino-terminal signal peptides (see below), and a highly efficient translocation machinery that not only delivers exported proteins to the membrane but also achieves their translocation across it. Our current perception of how the A and B subunits of Etx and CT cross the cytoplasmic membranes of *E. coli* and *V. cholerae* is based primarily on the understanding of protein export in general. This includes the finding that protein components exist in the cytoplasmic membrane (see below) that form a specific translocation channel through which exported proteins are transported, resulting in their liberation into the aqueous environment of the periplasm where the subsequent events of protein biogenesis, such as folding and assembly, can occur. The precise molecular mechanism by which polypeptides translocate across the cytoplasmic membrane remains to be elucidated, but on the basis of current concepts it seems most likely that the polypeptide is threaded like a "string of beads" through the translocation channel in an energy-dependent process requiring both ATP and proton motive force (see below).

1. The Role of Signal Peptides in Protein Export

The overwhelming majority of exported proteins, including the subunits of CT and related enterotoxins, are synthesized as precursors with characteristic amino-terminal extensions called signal peptides (also referred to as leader peptides) (Gierasch, 1989). Each residue in the signal sequence is designated by a negative number, starting from the cleavage site between the signal peptide (-1) and the mature protein ($+1$) (Fig. 2). Signal peptides, which range in size from 15 to 30 amino acids, play an essential role in the export process both by maintaining precursors in a partially unfolded conformation competent for export (Park et al., 1988; Liu et al., 1989) and by engaging the membrane/export apparatus (Gierasch, 1989; Pugsley, 1993). During translocation of precursors, signal peptides are cleaved off, yielding the *mature* polypeptide. The processing of precursors is performed by one of two signal peptidases. Signal peptidase I (leader peptidase, LepB) is responsible for the cleavage of most precursors, including the toxin subunits; and signal peptidase II (LspA), for cleaving lipoproteins (Table 4) (Wolfe et al., 1983; Tokunaga et al., 1984). Signal peptidases I and II have cytoplasmic membrane-spanning segments and have a *protease* active site situated on the periplasmic face of the membrane (Dalby, 1991). Proteolytic removal of signal peptides from precursor proteins is taken as an indicator that translocation has occurred, and it provides a convenient experimental assay for studying the export process.

An authoritative analysis of signal sequences, carried out by von Heinje, identified three distinctive regions or domains characteristic of all signal sequences (von Heijne, 1985). Amino acids adjacent to the mature polypeptide, that is, the cleavage site (the c region) comprise 5–7 residues, with those occupying the -1 and -3 positions being amino acids with a small R group, usually Ala, Ser, or Gly. Adjacent to this region are 7–13 hydrophobic amino acids (the h region), which appear to be able to adopt an α-helical structure when they partition into and span a lipid bilayer. At the amino terminus (the n region) are a variable number of residues of differing composition, containing one or more Lys or Arg residues that give this portion a net positive charge; this appears to be important, since it orients the signal peptide with respect to the membrane bilayer, by means of electrostatic interactions with the negatively charged phospho-head groups of phosphatidylglycerol and phosphatidylethanolamine. Consequently a signal peptide will insert "tail-first" into a membrane, with the n region remaining in association with the cytoplasmic face, while the h region partitions into the bilayer and drags the c region and mature portion of the precursor across the membrane. The cleavage site of the signal peptide thus becomes accessible to the periplasmic face of the membrane and to signal peptidase action.

The signal peptides present at the amino termini of the precursor A and B subunits of CT/Etx are 18 and 21 amino acids long, respectively (Fig. 2). Each of them possesses the characteristic n, h, and c regions. A comparison of the signal peptides encoded by the *ctx* and *etx* genes, and between *etx* genes of human and porcine origin, shows that the signal sequences are less well conserved than the mature sequences (Fig. 2) (Mekalanos et al., 1983; Yamamoto et al., 1987; Dams et al., 1991). This is not too unexpected since there is considerable sequence diversity among signal peptides in general, even though they perform a conserved function (Randall and Hardy, 1989). Changes in signal sequences can, nonetheless, improve or hinder the efficacy of translocation of an attached polypeptide (Morioka-Fujimoto et al., 1991), which raises the possibility that the differences in the signal peptides that have arisen between the A and B subunits and between the *ctx* and *etx* genes may contribute to subtle differences in rates of subunit entry into, and translocation through, the export machinery.

To understand the function of signal peptides it is important to consider the sequence of events that occur during polypeptide translation and the initiation of translocation. Relatively soon after synthesis of the polypeptide begins, the nascent chain with its amino-terminal signal peptide will emerge from the ribosome and begin to fold spontaneously, with the formation of secondary structural elements and loosely associated domains that sequester hydrophobic residues away from the aqueous environment (Jaenicke, 1987; Randall et al., 1987). It would appear that one important role for signal peptides is to slow down the acquisition of a compacted tertiary structure and in so doing maintain precursor polypeptides in a looser nonnative conformation that can successfully engage the export apparatus (Park et al., 1988; Liu et al., 1989). As a probable consequence of this, many precursor proteins also readily associate with a class of cytoplasmic proteins, known collectively as molecular chaperones, which recognize nonnative protein conformations (Gething and Sambrook, 1992). Thus, precursor proteins may be maintained in a conformation suitable for export as a result of contributions from both the signal peptide and the binding of exogenous chaperones (see below).

In an elegant series of experiments on the export of maltose-binding protein (MBP) to the periplasm of *E. coli*, Josefsson and Randall showed that the proteolytic removal of the signal peptide from preMBP could occur either posttranslationally or cotranslationally, with the relative proportion processed in each mode capable of being altered by changing the rate of polypeptide translation (Josefsson and Randall, 1981; Randall, 1983). This provided the first unequivocal evidence that translocation in vivo may take place either while the polypeptide is still attached to the ribosome or after the precursor has been completely translated. Thus the mechanisms of protein export in bacteria must be able to accommodate polypeptides at various stages of translation and presumably with quite diverse nonnative conformations. No systematic evaluation of whether the precursor A and B subunits of either CT or Etx are co- or posttranslationally processed has been undertaken. However, given the relatively small size of the B-subunit precursor (only

TABLE 4 Components Involved in the Translocation of Precursor Proteins Across the Cytoplasmic Membrane of *E. coli*

Component	M_r (kDa)	Location	Function/comments	Reference[a]
I. Well-characterized components				
SecA (PrlD)	102	Cytoplasm, cytoplasmic membrane	Translocation ATPase	1–3
SecB	16.6	Cytoplasm	Chaperone (antifolding) factor	4–6
SecD	67	Cytoplasmic membrane with a ca. 400-amino-acid periplasmic domain	Unknown, possibly involved in protein folding or release	7–9
SecE (PrlG)	14	Integral cytoplasmic membrane	Part of the translocation channel	10, 11
SecF	39	Cytoplasmic membrane with a ca. 200-amino-acid periplasmic domain	Unknown, possibly involved in protein folding or release; cotranscribed with *secD*	8
SecG (Band I)	15	Integral cytoplasmic membrane	Part of the translocation channel. Copurifies with SecE/SecY	19, 27
SecY	48	Integral cytoplasmic membrane	Part of the translocation channel	11–13
LepB	36	Cytoplasmic membrane	Signal (leader) peptidase	14, 15
LspA	18	Cytoplasmic membrane	Signal peptidase II; removal of signal sequences from lipoproteins	16
DsbA	21	Periplasm	Catalyst of disulfide bond formation in exported and secreted proteins	17, 18
II. Components whose participation in protein translocation has not been fully established				
Ydr	19	Integral cytoplasmic membrane	Suppresses *secY* (*ts*)	11
DnaK	69	Cytoplasm	Chaperone; member of Hsp70	20
GroEL	62	Cytoplasm	Chaperone; member of Hsp60 class of chaperone	20, 21
GroES	11	Cytoplasm	Chaperone component of the GroEL/ES complex	20, 21
Trigger factor	63	Cytoplasm	Chaperone for OmpA export	22, 23
Other heat-shock proteins	?	Cytoplasm?	Substitute for SecB function	24
4.5 S RNA (*ffs*)		Cytoplasm	Structure similar to the 7 S RNA of signal recognition particle (SRP) in eukaryotes	25

TABLE 4 (*Continued*)

Component	M_r (kDa)	Location	Function/comments	Reference[a]
Ffh	48	Cytoplasm	Shares homology with the 54-kDa protein of SRP	25, 26
FtsY		Cytoplasmic membrane	Shares homology with the α subunit of the SRP receptor	26

[a]1, Oliver (1993); 2, Schmidt et al. (1988); 3, Lill et al. (1989); 4, Kumamoto and Beckwith (1983); 5, Kumamoto (1991); 6, Randall (1992); 7, Gardel et al. (1987); 8, Gardel et al. (1990); 9, Pogliano and Beckwith (1994); 10, Schatz et al. (1989); 11, Ito (1992); 12, Ito and Beckwith (1981a); 13, Akiyama and Ito (1987); 14, Wolfe et al. (1983); 15, Dalby (1991); 16, Yu et al. (1984); 17, Bardwell et al. (1991); 18, Kamitani et al. (1992); 19, Brundage et al. (1992); 20, Phillips and Silhavy (1990); 21, Kusukawa et al. (1989); 22, Crooke and Wickner (1987); 23, Crooke et al. (1988); 24, Altman et al. (1991); 25, Poritz et al. (1990); 26, Romisch et al. (1989); and 27, Nishiyama et al. (1994).

124 amino acids in length) it would not be surprising if its translation had finished before translocation commenced, whereas the relative mode of A-subunit translocation could be either co- or posttranslationally or a mix of both.

2. Cytoplasmic and Membrane Components Involved in Precursor Translocation Across the Cytoplasmic Membrane

To date, six Sec proteins (shown in Table 4) are known to be involved in the targeting and translocation of precursor proteins across the cytoplasmic membrane in *E. coli* (reviewed in Schatz and Beckwith, 1990; Pugsley, 1993). Similar proteins have been identified in other gram-negative and gram-positive bacterial species (Pugsley, 1993; Simonen and Palva, 1993), although as yet the respective homologues in the *Vibrionaceae*, including *V. cholerae*, have not been described. Two of the proteins, SecA and SecB, are found in the cytoplasm, although SecA also appears to be associated with cytoplasmic membrane (see below). The SecB protein possesses the characteristics of a molecular chaperone, in that it interacts with certain precursor polypeptides while they are in still in a nonnative, partially unfolded state (Randall, 1992). In so doing, it would appear to retard precursor folding, thereby maintaining a conformation suitable for translocation across the membrane (Kumamoto, 1989, 1991; Randall, 1992). Only a subset of exported proteins, such as MBP and LamB, appear to be dependent on SecB for their export, suggesting either that other chaperones (see below) or the signal sequence alone may be sufficient to maintain SecB-independent proteins in conformations suitable for export (Kumamoto and Beckwith, 1985; Watanabe and Blobel, 1989a,b; Randall, 1992). Unlike all other *sec* genes, *secB* (*null*) mutations are still viable, at least for growth in minimal media (Kumamoto and Beckwith, 1985). Our studies have shown that the CT/EtxB subunits are members of the Sec-independent class of exported proteins, since a *secB* (*null*) mutant strain of *E. coli* harboring a plasmid encoding EtxB produced exactly the same amount of assembled EtxB pentamers as an isogenic wild-type strain (J. Yu and T. R. Hirst, unpublished observations).

Other cytoplasmic chaperones, such as GroEL and DnaK of the Hsp-60 and Hsp-70 families of heat-shock proteins, respectively (Gething and Sambrook, 1992), as well as several newly identified but unclassified heat-shock proteins (Altman et al., 1991) may prevent selected precursors from folding and thereby contribute to an improvement in the efficiency of their export (Kusukawa et al., 1989; Lecker et al., 1989; Phillips and Silhavy, 1990; Kumamoto, 1991). For some proteins, such as the precursor of the outer membrane protein LamB, all of the major chaperones, namely DnaK, GroEL, and SecB, have been implicated in contributing to the efficient targeting of the protein to the membrane translocation apparatus (Phillips and Silhavy, 1990; Randall, 1992). Since DnaK and GroEL normally act sequentially in catalyzing protein folding (Langer et al., 1992) as a result of their recognition of different nonnative conformations in proteins, it is likely that their roles in protein export involve a sequential series of interactions whereby a precursor is transferred from one chaperone to another, finishing with transfer to SecB, or that each chaperone functions by interacting with a different subset of conformational intermediates, to promote initial folding into an appropriate nonnative conformation, to prevent aggregation, or to unfold precursors that have already attained a compacted conformation incompatible with export. The sum of all of these events would be that a higher proportion of the molecules of any given precursor would be able to engage the translocation machinery. The relative requirement for a particular chaperone would then depend on the propensity of a particular precursor or its nascent chain intermediates to fold or aggregate into states recognized by different chaperones. Preliminary investigations in our laboratory suggest that the export of hEtxB in *E. coli* may be partially influenced by GroEL, since expression of the B subunit in a GroEL (*ts*) mutant resulted in an approximately threefold reduction in the level of assembled EtxB pentamers at 37°C, compared with the amount produced by an isogenic wild-type strain (J. Yu and T. R. Hirst, unpublished observations). As yet, it is not clear whether this finding indicates a direct involvement of GroEL in EtxB export or whether the adverse consequences of growth at the nonpermissive temperature results in an accumulation of other precursors that leads to a partial dysfunction of protein translocation in general.

The SecA protein, also called translocation ATPase (see Table 4), is essential for protein translocation and for cell viability (Oliver, 1993). It binds and hydrolyzes ATP, with the ATPase activity being elevated upon interaction with the cytoplasmic membrane (Lill et al., 1989, 1990; Oliver, 1993). Wickner and coworkers have used in vitro reconstitution experiments with inner membrane vesicles to establish that SecA interacts with both acidic phospholipids and the integral membrane proteins SecE and SecY, which are thought to compose a *translocation channel* through which precursors are exported (Brundage et al., 1990; Lill et al., 1990). The fact that mutations in *secA(prlD)* can lead to the export of precursors with defective signal peptides and that signal peptides can be cross-linked to SecA in vitro supports the view that signal peptides interact directly with SecA at some stage of the export process, although it does not exclude the possibility that SecA also interacts with other parts of nascent or fully translated precursors (Fikes and Bassford, 1989; Akita et al., 1990). The SecA/precursor complex is then thought to interact with phospholipids and the SecE/SecY translocase, leading to membrane insertion of the signal peptide. It also conceivable that the ATPase activity of SecA plays a role in the dissociation of precursors bound to chaperones, thereby ensuring their transfer into the SecE/SecY translocase (Oliver, 1993; Pugsley, 1993).

The SecD, E, F, and Y proteins are all located in the cytoplasmic membrane. Since SecY possesses multiple membrane-spanning segments, and SecE/SecY can be purified as a complex together with a third protein band I (secG), these proteins are the prime candidates for formation of the translocation channel (see Table 4). In an elegant series of experiments,

Silhavy and coworkers obtained compelling evidence that precursors interact first with SecE and then with SecY during transit through the membrane (Bieker and Silhavy, 1990; Bieker-Brady and Silhavy, 1992). The fact that mutations in *secE* and *secY* permit the export of precursor proteins with defective signal sequences suggests that these proteins interact with the signal peptide in some way during the export process (Emr et al., 1981; Stader et al., 1989; Schatz and Beckwith, 1990). Whether such an interaction triggers the opening of the translocation channel for the subsequent entry of the remainder of the polypeptide chain has yet to be established. The SecD and SecF proteins, which have large periplasmic loops, are thought to be involved in late events of export, such as the catalysis of folding or release of proteins as they emerge from the translocation channel (see below).

Various cytoplasmic and membrane components have been identified in *E. coli* that show sequence homology with eukaryotic proteins and RNA involved in protein translocation into the endoplasmic reticulum of higher eukaryotes, namely, the signal recognition particle (SRP) and the SRP-receptor protein (Table 4) (Gill et al., 1986; Poritz et al., 1988, 1990; Romisch et al., 1989; Lutcke et al., 1992). This has led to some speculation that *E. coli* could possess a rudimentary SRP-dependent export pathway that functions alongside the Sec pathway. However, the only indication to date that these components may play a role in protein export in *E. coli* has come from the finding that pre-β-lactamase accumulates when 4.5 S RNA is depleted or when Ffh is produced at abnormally high levels (Ribes et al., 1990). No studies have yet been undertaken to assess whether these components exert any influence on toxin subunit translocation.

3. Bioenergetics of Protein Export

Protein export is an energy-dependent process, requiring both ATP and proton motive force (pmf) (for a review, see Geller, 1991). Early studies on the biogenesis of Etx in *E. coli* showed that in the presence of uncouplers of pmf, unprocessed precursor B subunits accumulate (Palva et al., 1981). Since the addition of such uncouplers not only causes a collapse in pmf but also depletes cytoplasmic ATP, due to the futile attempt by ATP-synthase to restore the pmf by ATP hydrolysis, the use of uncouplers cannot satisfactorily distinguish between a requirement for pmf, a requirement for ATP, or both. Randall and coworkers directly addressed this issue by showing that an *uncA* mutant of *E. coli* carrying a nonfunctional ATP-synthase exported and processed precursor proteins only under aerobic conditions; under such conditions the pmf is provided by oxidative phosphorylation (Enequist et al., 1981). In contrast, under anaerobic conditions, where the *uncA* strain has no pmf and ATP levels remain high because ATP from glycolysis cannot be used by ATP-synthase to generate an electrochemical potential, precursor processing and export did not occur (Enequist et al., 1981). This provided unequivocal in vivo evidence for the importance of pmf in protein export.

The requirement for ATP is presumed to be due solely to the activity of SecA (see Section IV.C). In vitro experiments using inner membrane vesicles have revealed that protein translocation can be driven by ATP alone, which is a somewhat surprising result given the observations for a role for pmf in vivo (Chen and Tai, 1985, 1986, 1987; Yamada et al., 1989). However, subsequent findings have shown that by imposing a pmf across such inner membrane vesicles, the level of SecA and ATP required for translocation can be reduced (Yamada et al., 1989; Shiozuka et al., 1990). It is now accepted that ATP and pmf function at different stages of the translocation process, with ATP playing an essential role in the early events of precursor insertion and translocation, while pmf completes the process (Geller, 1990, 1991). However, the exact function of pmf in translocation remains to be precisely determined, especially in light of the findings that the translocation complex may be a proton antiporter and that different precursors appear to require different levels of pmf for efficient export (Daniels et al., 1981; Driessen and Wicker, 1991; Driessen, 1992). It should be possible to unravel further the roles of ATP and pmf by studying the kinetics of insertion, translocation, processing, and release of polypeptides under conditions where ATP and pmf levels are strictly controlled. It would not be surprising if the level of the pmf were found to correlate with the rate of transport through the translocation channel, as this would explain why the initial studies of protein export in vivo, including toxin export, had highlighted the importance of pmf. Thus, the in vivo rate of translocation must far exceed the rates that have currently been achieved in vitro.

4.   Release from the Membrane: The Late Event in Protein Export

The last step in the export of the toxin subunits across the cytoplasmic membrane of both *E. coli* and *V. cholerae* involves their release from the translocation channel into the aqueous environment of the periplasm. This step coincides with the important events of subunit folding and the acquisition of tertiary structure (see Section V.D). Interactions between secondary structural elements are likely to begin as soon the amino-terminal portion of the polypeptide emerges from the translocation channel, and will lead to the formation of loosely packed tertiary domains. If the polypeptide is exported cotranslationally, it is conceivable that a significant degree of folding may have occurred on the periplasmic face of the cytoplasmic membrane *before* the carboxy-terminal portion of the polypeptide has been synthesized. The release of polypeptides from the membrane may in fact be dependent on the acquisition of a soluble tertiary-like structure, since parameters such as low temperature (which slows down folding) or the use of truncated proteins that do not fold correctly can result in the failure to release the polypeptide from the membrane (Ito and Beckwith, 1981b; Koshland and Botstein, 1982; Hengge and Boos, 1985; Minsky et al., 1986; Fitts et al., 1987; Sandkvist et al., 1987, 1990).

It is probable that the close proximity of the exported polypeptide to proteins of the translocation channel during the time at which it begins to fold could influence its folding pathway and the efficiency of its release to the periplasm. There is clear genetic evidence that SecD and SecF play a distinct role in the late events of polypeptide export, and it has been reported that the addition of anti-SecD antibodies to *E. coli* spheroplasts inhibits the release of exported proteins (Bieker-Brady and Silhavy, 1992; Pugsley, 1993). In addition to being involved in promoting the folding of exported proteins, SecD and SecF may also play a role in clearing the translocation channel for another round of protein export. Importantly clearance of the channel does not depend on the proteolytic removal of the signal peptide, since export of a mutant maltose-binding protein with an altered cleavage site resulted in preMBP being tethered by its signal peptide to the periplasmic face of the membrane, without interfering with the export of other polypeptides (Fikes and Bassford, 1987).

Pulse-chase experiments, using [^{35}-S]methionine, have been used to follow the kinetics of toxin subunit export and release into the periplasm of *E. coli* and *V. cholerae* (Hirst et al., 1983; Hofstra and Witholt, 1984, 1985; Hirst and Holmgren, 1987b; Wilholt et al., 1988). This revealed that, in common with other proteins, export of the B subunits of porcine and human Etx was very fast; release was shown to be complete within 10 s. In contrast, Hofstra and Witholt (1984) showed that release of mature A subunits occurred much more slowly, with up to a third remaining membrane-associated 3 min after the initiation of the pulse-chase. This could possibly be explained by the fact that the recombinant plasmid EWD299 used in this study produced porcine A and B subunits in a molar ratio of approximately 2:5, and so only about half of the A subunits produced could assemble into an AB$_5$ structure (Hofstra and Witholt, 1984). Thus, retention of a proportion of the A subunits on the membrane may be due to a requirement for toxin assembly to be initiated at the membrane surface and thereby trigger A subunit release. Alternatively the A subunits may be expelled from the channel, and those that fail to associate with B subunits may irreversibly reassociate with the membrane. This latter explanation could account for more recent observations on the export behavior of mutant hEtx B subunits (with minor deletions or substitutions at their carboxy termini), which fail to assemble and appear to remain associated with the cytoplasmic membrane (Sandkvist et al., 1990) (see Section V.D). In a series of semi-in vivo experiments, Sandkvist (1992) showed that *E. coli* spheroplasts were able to efficiently secrete the mutant B subunits into the supernatant, indicating that they are not defective in their release from the cytoplasmic membrane. Indeed, under such conditions, all of the mutant B subunits were exported and released as efficiently as the wild-type B subunits (Sandkvist, 1992). When pulse-chase experiments were performed on intact cells to follow the fate of the mutant B subunits, it was found that the B subunits were rapidly exported and released into the periplasm, but upon continuation of the chase they disappeared and an increasing amount were found associated with the membrane (Sandkvist, 1992).

In summary, polypeptide release is a key event in protein translocation, and its further study holds the prospect of providing an understanding of the importance of membrane surfaces and the role of translocation proteins or other as yet unidentified factors in modulating the folding of exported proteins.

D. Toxin Folding and Assembly

The folding and assembly of the A and B subunits to form a final toxin structure clearly represent key molecular processes in toxin biogenesis. Although the events of folding and assembly can be thought of as distinct processes, it is perhaps more appropriate to view them as a pathway of events involving both intra- and intermolecular protein interactions

It has been observed that the denaturation of both purified CT and Etx by various denaturants (under nonreducing conditions) resulted in the reversible disassembly and unfolding of the toxins, which reassembled into biologically active AB$_5$ complexes upon their return to nondenaturing conditions (Finkelstein et al., 1974; Hardy et al., 1988). This is consistent with the generally accepted model of Anfinsen that all information necessary to achieve correct folding (and in this case assembly) of a protein resides in the primary amino acid sequence (Epstein et al., 1963). However, as mentioned in Section V.C and discussed further below, the folding and assembly of proteins within cells can be markedly affected by interactions with other proteins such as chaperones or enzymes that catalyze distinct folding steps, as well as by membrane surfaces. Thus, a study of toxin folding and assembly must consider not only the role of amino acids in the primary sequences of the A and B subunits but also the role of intermolecular interactions with cellular folding and assembly factors.

The solving of the Etx crystal structure by Sixma et al. (1991) provided a three-dimensional image of the final outcome of all of the folding and assembly events (Fig. 2). In the crystal structure each B subunit is shown to consist of an N-terminal helix (α 1), followed by four β strands (β1, β2, β3, and β4), a long α helix (α 2), and two further β strands (β5 and β6) (Sixma et al., 1991, 1993). As discussed in Section III.A, the β strands fold to form two three-stranded antiparallel sheets, with $\beta 2$, $\beta 3$, and $\beta 4$ forming sheet I and $\beta 1$, $\beta 5$, and $\beta 6$ forming sheet II. These sheets interact with each other via hydrogen-bond interactions between $\beta 4$ and $\beta 6$ to form a β barrel (Sixma et al., 1993). This interaction is possible because the proline at position 93 in the loop connecting $\beta 5$ and $\beta 6$ is in a cis configuration (Sixma et al., 1993). Since *cis* prolines are rare in proteins, and cis-trans isomerization of peptidyl-prolyl residues is recognized as a slow event in protein folding (Jaenicke, 1991), the trans to cis isomerization of Pro-93 in the B subunit is likely to be a crucial step in the formation of the correct tertiary fold. Similarly, formation of the disulfide bonds between Cys-9 in the N-terminal α 1 helix and Cys-86 in $\beta 5$ are also essential for toxin biogenesis; such a step is generally subject to catalysis by cellular enzymes (see Section V.D). The crystal structure also revealed that there are five internal salt bridges in each B subunit (Asp-22–Lys-81 or Asp-22–Lys-43; Lys-23–Glu-79; Glu-51–Lys-91; Glu-66–Lys-69; and Asp-70–Arg-73), which are likely to contribute to stabilizing the folded structure (Sixma et al., 1993).

The X-ray crystallographic image of the A subunit revealed that A1 and A2 fragments were separate domains (Fig. 3) (Sixma et al., 1993). Whereas these may fold independently of one another, the fact that the long α helix of A2 abuts A1 may mean that this interaction effects the conformational stability of the A2 polypeptide. The A1 fragment contains 10 β strands and 12 α helices that fold into a wedge-shaped structure, partly stabilized by 12 internal salt bridges (Sixma et al., 1993).

During assembly the intramolecular interactions that give rise to tertiary structure in the individual toxin subunits will create interfaces that allow specific intermolecular interactions that ultimately lead to stable quaternary complexes. As will be discussed below, the nature of the interactions between B subunits, and between the A and B subunits, affords the possibility that these associations may affect late folding events. It is also important to consider that the folded B-subunit monomer contains two surfaces that extensively pack against adjacent B subunits in the assembled toxin (Fig. 2) (Sixma et al., 1993). Since these surfaces are particularly hydrophobic, the folded monomer may require "shielding" by chaperone-like proteins to prevent either aggregation or nonspecific association with the membrane (discussed in Section V.D).

Several experimental techniques have been developed to allow the study of toxin assembly in vivo. The most important is based on the finding that the B_5 oligomer is stable in solutions containing up to 1% SDS at pH > 5.5 and so will migrate as a stable pentamer on SDS-polyacrylamide gels (Hirst and Holmgren, 1987b). This allows discrimination between unassembled and assembled B subunits in cell fractions. Thus, the kinetics of B-subunit assembly can be studied by sampling radiolabeled cells at different times and immediately treating them with SDS. However, since the presence of SDS causes the dissociation of the A subunit from the assembled holotoxin complex, this technique cannot be used to study A and B subunit interaction. Therefore an alternative approach has been used based on enzyme-linked immunosorbent assays (Svennerholm and Holmgren, 1978; Hardy et al., 1988). This technique uses G_{M1} ganglioside-coated microtiter plates to capture assembled B subunits, which are then probed with antibodies directed against determinants on the B oligomer or the A subunit (Hirst and Holmgren, 1987a; Hardy et al., 1988; Streatfield et al., 1992).

1. Cellular Location of Toxin Folding and Assembly

The A- and B-subunit polypeptides of CT and Etx undoubtedly begin to fold as soon as they emerge from the translocation channel, although it is not yet clear to what extent they achieve a folded tertiary structure before being fully released into the periplasm. Recently it was demonstrated that the formation of the disulfide bond in the B subunit was critically dependent on a periplasmic enzyme, DsbA (see Section V.D): the periplasmic location of this enzyme suggests that this step in folding of the B subunit may occur in the periplasm. Furthermore, it has been shown by pulse-chase experiments that the release of toxin subunits into the periplasm *precedes* the appearance of assembled, SDS-stable oligomers (Hirst et al., 1983; Hofstra and Witholt, 1984; Hirst and Holmgren, 1987b). As a result of these observations, it has been proposed that assembly occurs largely within the periplasmic compartment of the bacterial envelope, although the possibility remains that certain subunit–subunit interactions may start at the membrane (Hirst, 1991).

2. Cellular Factors Involved in Toxin Folding and Assembly

Disulfide bond formation. In common with many other exported and secreted polypeptides, the A and B subunits of Etx and CT possess cysteine residues that must oxidize to form specific intrachain disulfide bonds during folding. Each of the toxin subunits possesses a single disulfide bond, between Cys-187 and Cys-199 in the A subunit and between Cys-9 and Cys-86 in each B subunit (Sixma et al., 1993). The importance of the formation of disulfide bonds in toxin biogenesis was first demonstrated by Hardy et al. (1988), who showed that the addition of the sulfhydryl reducing reagent dithiothreitol (DTT) to an EtxB-producing strain of *E. coli* immediately stopped the assembly of new EtxB pentamers, although it had no effect on the pentamers that had already assembled. The subsequent demonstration that replacement of Cys-9 or Cys-86 by Ser in the B subunit of CT (Jobling and Holmes, 1991) or Etx (S. J. S. Hardy, personal communication) abolished B-subunit assembly into stable oligomers confirmed that disulfide bond formation in the B subunit is an essential step in toxin biogenesis.

Until recently it was presumed that the more oxidizing environment of the periplasm would allow the spontaneous (air) oxidation of juxtaposed cysteine residues in secretory proteins. However, it is now clear that the formation of disulfide bonds in the enterotoxin molecule is catalyzed by a periplasmic thiol-disulfide oxidoreductase (Yu et al., 1992). Genes encoding analogous enzymes responsible for disulfide bond formation (*dsbA*) have been identified by others in *E. coli*, *V. cholerae*, and *Haemophilus influenzae* (where the mnemonics *ppfA*, *tcpG*, and *por* have also been used) (Bardwell et al., 1991; Kamitani et al., 1992; Peek and Taylor, 1992; Tomb, 1992; Yu et al., 1992).

Peek and Taylor (1992) and Yu et al. (1992) independently identified the *V. cholerae* periplasmic thiol-disulfide oxidoreductase by characterizing transposon mutants (Tn*phoA*) that were defective in the biogenesis of TCP and the assembly of enterotoxin B subunits. Yu et al. (1992, 1993) used an El Tor strain of *V. cholerae* that had been engineered to secrete large quantities of the heterologous *E. coli* EtxB subunit into the medium. Two transposon derivatives of this strain were obtained and found to produce EtxB at a level of only 2% of that found in the isogenic wild-type strain. When the fate of newly synthesized B subunits in one such mutant was examined by radioactive pulse labeling, it was apparent that the B subunits were efficiently exported and released into the periplasm in both the mutant and wild-type strains, but only in the latter strain were they assembled and secreted to the medium (Findlay and Hirst, 1993). The profiles of other periplasmic and extracellular proteins in the mutant and wild-type strains were equivalent, with the exception of a major 24-kDa periplasmic protein that was absent from the mutant strain (Yu et al., 1993). This protein was subsequently shown to be a thiol-disulfide oxidoreductase that (1) shares 40% sequence identity with PpfA(DsbA), the *E. coli* protein involved in the formation of disulfide bonds (Bardwell et al., 1991; Kamitani et al., 1992; Peek and Taylor, 1992; Yu et al., 1992); (2) is identical to the *tcpG* gene required for TCP synthesis in *V. cholerae* 0395 (Peek and Taylor, 1992); (3) contains a characteristic Phe-X-X-X-X-Cys-X-X-Cys motif similar to the active sites of enzymes that catalyze thiol-disulfide interchange reactions, such as thioredoxin and eukaryotic protein disulfide isomerase (Fig. 6); and (4) exhibits disulfide isomerase activity in vitro (Yu et al., 1993). Expression of Etx subunits in a *dsbA* mutant of *E. coli* confirmed that toxin biogenesis in both *E. coli* and *V. cholerae* is dependent on DsbA (Yu et al., 1992).

The exact enzymatic mechanism by which DsbA catalyzes disulfide bond formation in vivo remains to be elucidated. However, based upon the mechanism of action of thioredoxin and eukaryotic protein disulfide isomerase (Holmgren, 1985; Freedman, 1989), it is reasonable to speculate that the cysteines in the active site undergo thiol–disulfide interchange reactions in which a transient mixed disulfide intermediate forms between the enzyme and the target protein (Fig. 7). Thus, the folding of the B subunit into its "correct" tertiary structure will involve the formation of a transient covalent bond with DsbA. The consequences of this for the formation of other secondary and tertiary structural elements during subunit folding and the prospect that DsbA may function as a chaperone shielding the B subunit interfaces from nonspecific aggregation (in addition to its role as a catalyst of disulfide bond formation) have yet to be investigated. The crystal structure of *E. coli* DsbA was recently reported, providing scope to evaluate which surface features of DsbA might interact with target proteins during their folding (Martin et al., 1993).

Evidence has been obtained from studies in *E. coli* that the reoxidation of DsbA after it has catalyzed disulfide bond formation in a target protein is mediated by DsbB (Bardwell et al., 1993; Missiakis et al., 1993), an integral protein of the cytoplasmic membrane that possesses an active-site motif similar to that of DsbA. It has been suggested that the electron transport system or an as yet unidentified oxidant may in turn be responsible for DsbB oxidation (Bardwell et al., 1993).

Recently, another periplasmic enzyme, DsbC, was identified in *E. coli* that also catalyzes disulfide bond formation (Missiakis et al., 1994). Although it possesses an active-site motif similar to that of DsbA (Fig. 6), it is not dependent on DsbB for reoxidation (Missiakis et al., 1994). Thus, *E. coli* appears to possess two independent and parallel pathways for disulfide bond formation in the periplasm. No studies have yet been reported on whether there are differences in substrate specificity between the two pathways or whether the A or B subunits can use the second DsbC pathway as a means of disulfide bond formation.

V.cholerae DsbA	N	E	F	F	S	F	Y	C	P	H C[33]
E.coli DsbA	L	E	F	F	S	F	F	C	P	H C[33]
E.coli DsbC	T	V	F	T	D	I	T	C	G	Y C[121]
H.influenzae Por	I	E	F	F	S	F	Y	C	P	H C[32]
Human PDI	V	E	F	Y	A	P	W	C	G	H C[39]
E.coli thioredoxin	V	D	F	W	A	E	W	C	G	P C[35]
Human thioredoxin	V	D	F	S	A	I	W	C	G	P C[35]

Fig. 6 Comparison of the putative active-site regions of *V. cholerae* and *E. coli* DsbAs to the active sites of DsbC, protein disulfide isomerase, and thioredoxin. Identical amino acids are shaded.

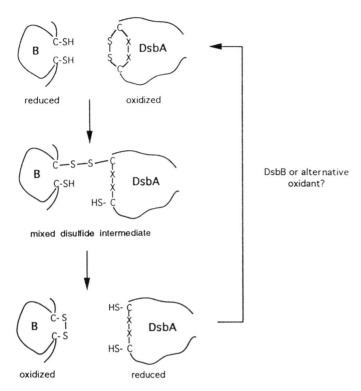

Fig. 7 Model of the oxidation/reduction cycle of DsbA during the formation of an interchain disulfide bond in the B subunit. The model predicts a reaction between oxidized DsbA and a reduced B subunit to form a mixed disulfide intermediate. At least one of the two cysteine residues in the B subunit must undergo ionization to give a thiolate anion (S⁻) prior to reaction with DsbA, since only the thiolate ion is known to participate in thiol/disulfide exchange reactions. The subsequent resolution of the mixed disulfide intermediate will yield oxidized B subunit and reduced DsbA, with the latter being reoxidized by DsbB or an alternative (unknown) oxidant. (Adapted from Yu et al., 1992.)

Prolyl isomerization. The cis-trans isomerization of peptidylprolyl bonds is recognized to be a slow, rate-limiting step in the refolding of proteins in vitro (Jaenicke, 1987, 1991; Lang et al., 1987). Peptidylprolyl *cis-trans* isomerases are located in both the cytoplasm and periplasm of bacteria and are thought to accelerate this folding step in vivo (Liu and Walsh, 1990; Hayano et al., 1991). Since the folded pEtx B subunit, revealed by X-ray crystallography, was found to contain a single *cis*-Pro at position 93, it is tempting to speculate that the B subunit interacts with a peptidylprolyl *cis-trans* isomerase as it folds in the periplasm. It would be of interest to identify whether *V. cholerae* possesses a periplasmic peptidylprolyl *cis-trans* isomerase and to evaluate the consequences of mutations in this enzyme on toxin production in either *E. coli* or *V. cholerae*.

Are general molecular chaperones involved? The bacterial periplasm may be considered to be structurally and functionally analogous to the lumen of the endoplasmic reticulum (ER) of eukaryotic cells, since both represent sites in which folding and assembly of secretory proteins take place. Extensive studies have shown that the ER contains a number of molecular chaperones that mediate the folding and assembly of secretory proteins (Gething et al., 1986; Gething and Sambrook, 1992). For example, the chaperone, BiP, first characterized as an immunoglobulin heavy-chain binding protein, is an ATPase that transiently interacts with newly ported polypeptides to prevent their aggregation and to assist their assembly (Gething et al., 1986; Gething and Sambrook, 1992). A homologue of BiP (designated *KAR2*), has been identified in the ER of yeast, where it again plays an essential role in polypeptide folding (Normington et al., 1989). However, the bacterial periplasm is unlikely to contain a functional homologue of BiP since this compartment is presumed to lack ATP, not only because of the absence of nucleotide transport systems, but chiefly because high phosphatase activity would ensure the spontaneous hydrolysis of ATP. This view may, however, need to be rectified if the recently discovered *E. coli* member of the ATP-dependent Clp family of proteases proves to have a periplasmic location as suggested by Squires and Squires (1992).

The possibility that toxin folding and assembly may be aided by interaction with a general molecular chaperone was raised by the studies of Schonberger et al. (1991), who investigated the heterologous expression of hEtxB in *Saccharomyces cerevisiae*. Of interest was their finding that B-subunit assembly in the yeast ER was dependent on a functional *KAR2* gene, the homologue of BiP (Schonberger et al., 1991). Thus, the folding of the B subunit in this heterologous environment appears to involve noncovalent interaction with a well-recognized chaperone.

It is also noteworthy that pulse-chase experiments on the kinetics of Etx assembly in *E. coli* showed that when B subunits were immunoprecipitated from radiolabeled cells an unknown 8-kDa polypeptide coprecipitated (Hofstra and Witholt, 1985). This coprecipitation event was observed only during the first 2 min of a pulse-chase, that is, during the period in which EtxB oligomerization occurs. This makes the 8-kDa polypeptide a possible candidate for a periplasmic chaperone in *E. coli*.

It remains to be established, however, whether the folding of secretory proteins in the periplasm of bacteria is assisted by the interaction with general chaperones and whether such interactions could depend on the hydrolysis of ATP for their function.

3. Subunit Assembly

The precise pathway of subunit interactions and the identity of assembly intermediates that lead to the formation of an AB$_5$ holotoxin complex are not known. However, given that the B subunits of CT and Etx readily assemble (in the absence of concomitant A-subunit synthesis) to form stable pentamers (Hirst et al., 1984), let us first consider B subunit–B subunit interactions and second, the interaction of A and B subunits. This does not imply that assembly involves the formation of the B pentamer before association with the A subunit, since there is compelling evidence, as discussed below, that the A subunit interacts with B subunits before pentamerization is complete.

The kinetics of assembly of radiolabeled toxin subunits in vivo has been investigated in several laboratories. It has emerged that the attainment of an SDS-stable B-pentamer structure (the most convenient measure of toxin assembly) occurs with a half-time of approximately 15–60 s, depending on the bacterial strain, growth temperature, Etx plasmid, and level of toxin expression (Hirst et al., 1983, 1984; Hofstra and Witholt, 1985; Hirst and Holmgren, 1987a; Hardy et al., 1988). In all of these studies the assembly process was quenched by rapidly chilling the cells on ice before fractionation and analysis. By such analyses it is clear that there is the rapid appearance of an exported pool of unassembled toxin subunits, which disappear from the periplasm concomitantly with the appearance of SDS-stable pentamers (Hardy et al., 1988). It is, of course, conceivable that these quenching procedures are not adequate to prevent certain assembly intermediates from progressing through to an SDS-stable structure. This could perhaps be resolved by the use of improved technology. For example, protein–protein interactions could be studied in situ as they occur in living cells. Alternatively, rapid-acting cross-linkers could be used to stabilize assembly intermediates and so permit more stringent quenching.

B subunit–B subunit interactions. While the exact tertiary conformation of B-subunit monomers at the time they begin to interact with one another isn't known, the monomer intermediate probably bears at least some resemblance to the B subunit revealed in the X-ray crystallographic analysis of the AB_5 complex. Consequently, much of the tertiary fold and the interfaces of the putative B monomer are likely to have formed before assembly begins. However, the attainment of the final B-monomer conformation may depend on structural rearrangements that occur when the subunit interfaces combine. From the X-ray structure of the holotoxin it is clear that in its final state each B subunit forms extensive hydrophobic interactions with its neighbor via the β sheets (I and II) as well as interacting by means of multiple intersubunit salt bridges (between Glu-11 and Arg-35, the C-terminal carboxylate and Lys-23, Lys-63 and Glu-66, Arg-67 and Glu-66, and Arg-67 and Asp-70) (Sixma et al., 1993). It is not known which, if any, of these interactions are necessary to initiate assembly or to stabilize assembly intermediates.

Obviously, the first productive interaction during Etx or CT assembly must be dimerization of two B-subunit monomers (at least in cells that do not produce the A subunit). Thereafter a number of pathways can be envisaged involving association of monomers, dimers, or oligomers to yield the B pentamer; which pathway predominates has yet to be elucidated. The simplest could involve the sequential addition of B monomers to the B dimer.

$$B_1 + B_1 \rightarrow B_2 + B_1 \rightarrow B_3 + B_1 \rightarrow B_4 + B_1 \rightarrow B_5 \tag{1}$$

Alternatively, a dimer may associate with a monomer to give a trimer followed by incorporation of another dimer,

$$B_1 + B_1 \rightarrow B_2 + B_1 \rightarrow B_3 + B_2 \rightarrow B_5 \tag{2}$$

or two dimers may associate followed by incorporation of a monomer,

$$B_1 + B_1 \rightarrow B_2 + B_2 \rightarrow B_4 + B_1 \rightarrow B_5 \tag{3}$$

Our current working model favors the hypothesis illustrated by Eq. (2), which draws upon in vivo studies of the kinetics of assembly of AB_5 as well as the nature of the interactions between the A and B subunits in the crystal structure. Both aspects are discussed further below.

The use of mutant B subunits that have an altered capacity to fold or assemble provides a particularly powerful approach for studying the pathway of B pentamer and AB_5 formation. This approach was first employed by Sandkvist et al. (1987, 1990), who showed that when three or more amino acids were deleted from the C terminus of hEtxB, the B subunits failed to assemble into oligomers. This suggested that this region either stabilizes the folded B-subunit monomer in a conformation suitable for assembly or contains residues essential for subunit interactions (Sandkvist et al., 1990). The crystal structure revealed that this region contributes extensively to intersubunit interaction, forming part of strand β6, which has seven hydrogen-bond interactions with strand β2 (of the adjacent subunit); the carboxylate of Asn-103 (the last amino acid) undergoes both salt-bridge and hydrogen-bond interactions with the neighboring B subunit (Sixma et al., 1993). However, because Asn-103 can be deleted, and amino acids can be readily added onto the C terminus without perturbing B-subunit assembly, it seems likely that it is the propensity of this region to form a stable intersubunit β sheet that is most important (Sandkvist et al., 1987; Schodel and Will, 1989; Fergusson et al., 1990; Aitken and Hirst, 1993; Nashar et al., 1993). Certain modifications to the C-terminal region render B subunits temperature-sensitive (*ts*) for assembly, with stable pentamers forming when cells are cultured at 30°C but not at or above 37°C (Sandkvist and Bagdasarian, 1993). Once assembled, however, such mutant B subunits exhibit the same SDS stability properties that characterize wild-type B-subunit pentamers. An example is EtxB191.5 (Sandkvist et al., 1990), in which the last four residues of the B subunit (Ser-Met-Glu-Asn) were replaced by three different ones (Gly-Leu-Asn). EtxB191.5 readily assembles when produced at 30°C, while at 42°C no pentamers are detected at all (Sandkvist and Bagdasarian, 1993). This is consistent with the view that the folding pathway that leads to the formation of a crucial secondary structure element is *ts*.

Intragenic suppressor mutations of EtxB191.5 have been isolated that allow this protein to assemble into oligomers at elevated temperatures (Sandkvist and Bagdasarian, 1993). This was achieved by subjecting the *etxB191.5* gene to chemical mutagenesis with hydroxylamine and screening for phenotypic revertants that produced B pentamers at 37°C. One such suppressor mutation was a substitution Thr-75 for Ile (Sandkvist and Bagdasarian, 1993). The final crystal structure of the native holotoxin showed that Thr-75 is positioned toward the end of the α2 helix and is in contact with C^δ of Met-101 in the β6 strand by means of van der Waals interactions. However, in EtxB191.5, Met-101 has been replaced by Leu; thus residue 101 has been shortened by one atom. This substitution probably results in the loss of van der Waals contact with Thr-75. The suppressor mutation (in which Thr-75 is replaced by Ile) extends this residue by one atom and should reestablish the van der Waals interaction with Leu-101. This suggests that the interaction between the α2 helix and the β6 strand during the folding of B monomers is likely to stabilize a conformation that is essential for B-subunit assembly (Sandkvist and Bagdasarian, 1993; Sixma et al., 1993).

A second suppressor mutation was isolated in which Met-31 had been replaced with a leucine residue. Again analysis of the holotoxin structure revealed that Met-31 is located at the end of the β2 strand and forms part of a β hairpin in the B subunit. This residue interacts with the C-terminal β6 strand of the adjacent B subunit in the final assembled structure (Sixma et al., 1993), forming part of a hydrophobic pocket that is completely buried when assembly takes place. It is conceivable that the Met-31 to Ile mutation affects monomer folding and stability by altering the ease with which this region forms a β hairpin. Alternatively, the substitution may increase the propensity for association of the B-subunit interfaces, promoting more stable subunit–subunit interaction (Sandkvist and Bagdasarian, 1993). It would be of interest to examine the consequences of these substitutions on the assembly of B subunits with a wild-type C terminus.

Met-31 is positioned so that its R group extends into the pocket facing Ala-64 (in the α2 helix) of the neighboring B subunit (Sixma et al., 1993). The replacement of Ala-64 with Val in pEtx was reported by Iida and coworkers to disrupt the formation of B oligomers and to prevent holotoxin formation (Iida et al., 1989). This mutation probably interferes with the packing of the C^β of Ala-64 and the C^β of Met-31 (of the adjacent B subunit). Recently, we have shown that the substitution of either Met-31 for Asp or Ala-64 for Asp renders the B subunit *ts* for assembly; notably, the double mutant failed to assemble *at all temperatures* (H. Webb et al., unpublished observations). Replacement of Met-31 with Trp or Arg did not interfere with B-pentamer formation (H. Webb and T. R. Hirst, unpublished results). It should be possible to use the mutant in which Met-31 has been replaced by Trp to investigate the environment of the subunit interface, by monitoring tryptophan fluorescence during assembly.

A subunit–B subunit interactions. The association of the A and B subunits to form an AB_5 holotoxin adds another level of complexity to the assembly process. The crystal structure of the holotoxin gives a picture of the final outcome of these interactions. The major contacts between the A and B subunits are clustered toward the C-terminal portion of the A2 polypeptide, which is inserted into the central pore of the B pentamer, where there are five salt-bridge interactions (Sixma et al., 1993) (see Figs. 2 and 8). Within the central pore there are surprisingly few specific contacts, just two regions of hydrophobic contact at each end of the pore and three salt bridges between Arg-235 of A2 and Glu-66 (in B#2) and Asp-70 (in B#1) and a salt bridge in the middle of the pore between Glu-229 of A2 and Arg-73 (in B#3) (Sixma et al., 1993).

Studies on the in vivo assembly of the Etx in *E. coli* using pulse-labeling experiments revealed that the rate at which the B subunits attained an SDS-stable pentameric structure in the periplasm increased by about fourfold when the A subunit was coexpressed (Hardy et al., 1988). This suggested that the A subunit stabilizes an intermediate in B-subunit assembly. Furthermore, in vitro experiments on the reassembly of acid-denatured B subunits showed that holotoxin assembly could occur if the A subunit was added to B subunits before pentamerization was complete, whereas the A subunit failed to associate with B subunits that had already assembled (Hardy et al., 1988). Taken together, these results indicate that the A subunit must interact with assembling B subunits.

Examination of the holotoxin suggests that the alternating positively and negatively charged residues in the central pore will prevent the insertion of A2 into a preformed pore. Alternatively, it could be suggested that the short C-terminal α helix near the end of A2 may not fit into the narrow end of the preformed pore.

Several pathways for A/B-subunit assembly can be envisaged (see below). However, given that the major contacts between the A2 fragment and B subunits observed in the crystal structure are exclusively between three adjacent B subunits (B#1, B#2, and B#3), with salt-bridge interactions involving all three subunits and hydrophobic contacts with two of them (B#1 and B#3), it seems plausible to hypothesize that assembly of the A and B subunits proceeds via the formation of an AB_3 intermediate. Indeed, it may be the stabilization of the B trimer by the A subunit that leads to the enhancement in the rate of B-subunit pentamerization observed by Hardy et al. (1988). This intermediate could be formed in the following ways:

$$B_1 + B_1 \rightarrow B_2 + B_1 \rightarrow B_3 + A \rightarrow AB_3 \tag{4}$$

$$B_1 + B_1 \rightarrow B_2 + A \rightarrow AB_2 + B_1 \rightarrow AB_3 \tag{5}$$

$$B_1 + A \rightarrow AB_1 + B_1 \rightarrow AB_2 + B_1 \rightarrow AB_3 \tag{6}$$

The use of in vitro spectroscopic and cross-linking studies may be necessary to unravel the pathway of these interactions. If the generation of the AB_3 intermediate is part of the normal pathway of toxin assembly, then its sequential association with two B monomers

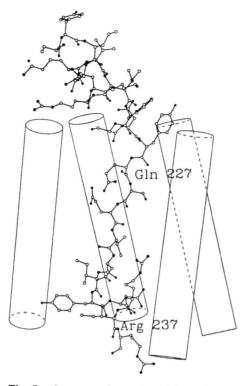

Fig. 8 Structure of a portion of the carboxy-terminal fragment of the A subunit within the central pore of the B pentamer. The A1 fragment (not shown) would be positioned at the top of the figure. A region of the A2 fragment (residues 213–237) is shown (with the positions of Gln-227 and Arg-237 indicated); four out of the five B-subunit α helices lining the central pore are shown as cylinders.

[Eq. (7)] or its association with a preformed B dimer [Eq. (8)] would form the holotoxin complex.

$$AB_3 + B_1 \rightarrow AB_4 + B_1 \rightarrow AB_5 \tag{7}$$

$$AB_3 + B_2 \rightarrow AB_5 \tag{8}$$

The crystal structure revealed that five salt bridges as well as hydrophobic interactions are found between the A1 polypeptide and B#2, B#3, and B#4 in the B pentamer. This suggests a means by which the AB_4 intermediate may be stabilized during holotoxin assembly.

A recent study by Jobling and Holmes (1992) provided unequivocal evidence that only the A2 polypeptide of CT was necessary and sufficient for association with the B subunit to form a stable oligomer; thus, structurally speaking, the A2 polypeptide serves to anchor A1 to the B pentamer. They constructed translational fusions in which the entire A2 fragment (from amino acid residues ca. 197–240) was fused to the C terminus of either alkaline phosphatase (PhoA-A2), β-lactamase (Bla-A2), or maltose-binding protein

(MBP-A2). In all cases, the A2 component of the fusion retained its capacity to assemble with CT B subunits to form holotoxin-like chimeras (Jobling and Holmes, 1992). The effect of alterations in the A2 fragment on its capacity to assemble with B subunits was also tested. This revealed that changing the last four carboxy-terminal amino acids from Lys-Asp-Glu-Leu ($_{-237}$KDEL$_{240}$) to Gln-Asp-Glu-Leu ($_{-237}$QDEL$_{240}$) did not affect the capacity of the otherwise native A subunit to assemble with B subunits. Likewise, the replacement of the KDEL sequence by Arg-Gly-Gly-Ala-Arg ($_{-237}$RGGAR$_{241}$) in a PhoA-A2 fusion did not prevent its assembly into a holotoxin-like chimera (Jobling and Holmes, 1992). Thus, the *identity* of the last four amino acids in the A2 fragment does not appear to be important in ensuring the formation of stable A/B-subunit interactions. However, the *presence* of amino acids at those positions is essential for holotoxin formation, since the deletion of the last four residues in *E. coli* EtxA resulted in an A subunit [designated EtxA(R237*)] that was unable to form a stable AB$_5$ complex (Streatfield et al., 1992). Interaction of the A and B subunits has also been investigated using the mutant B subunit EtxB191.5 (see Section V.D). Not only does EtxB191.5 exhibit a *ts* assembly defect but it also shows a striking dependence on the A subunit for its assembly (Streatfield et al., 1992). When EtxA(R237*) was coexpressed with EtxB191.5, the mutant A subunit promoted the assembly of EtxB191.5 almost as efficiently as the wild-type A subunit. Thus, the defect in EtxA(R237*) assembly into stable AB$_5$ complexes cannot be due to its failure to be released from the membrane and interact with assembling B subunits (Streatfield et al., 1992). Rather, it would seem that the deletion of the last four amino acids prevented the A2 fragment from maintaining a stable interaction with B subunits. However, since this region can be readily replaced by heterologous amino acids without perturbing holotoxin formation, specific contacts cannot be important. Instead, these C-terminal amino acids are likely to stabilize the small α helix from residues 232–236 in A2. Perhaps without this α helix the A subunit simply slips out of the pore.

Other deletions and substitutions in the A2 fragment have been generated that affect holotoxin assembly (Jobling and Holmes, 1992; Streatfield et al., 1992). For example, when an amber mutation was introduced at Gln-227 (corresponding to the 14th residue from the C terminus of the A2 fragment of EtxA), the mutant A subunit [designated EtxA(Q227*)] failed to either assemble with wild-type B subunits or promote the assembly of EtxB191.5 (Streatfield et al., 1992). This region must therefore normally contribute to stabilizing A/B-subunit interaction during toxin assembly.

The crystal structure of pEtx also revealed a number of salt-bridge interactions at the entrance to the pore between the A2 fragment and the B pentamer, involving Glu-213, Lys-217, Lys-219, and Arg-220 of the A subunit (Sixma et al., 1993) (Section II.A). With the exception of Glu-213, which has mutated to Arg in hEtx, these residues are conserved between CT and pEtx (Fig. 2). Overall, the sequence identity between CT and pEtx in the A2 fragment residues 213–240 is 78%. It should therefore come as no surprise that hybrid holotoxins form in vitro when the A subunits of CT and the B subunits of Etx (and vice versa) are mixed together under appropriate conditions (Takeda et al., 1981).

A number of mutations in the B subunits have been obtained that prevent A and B subunit association during assembly but do not appear to inhibit B-pentamer formation (Sandkvist et al., 1987; Jobling and Holmes, 1991). One example is the introduction of extensions at the C terminus of EtxB (Sandkvist et al., 1987). This is probably due to steric hindrance, since both the C terminus of the B subunit and the A subunit are positioned on the upper face of the B pentamer. Another mutant in which Arg-35 in the B subunit of CT was replaced by negatively charged residues (either Glu or Asp) produced pentamers (Jobling and Holmes, 1991). This finding is more difficult to explain since the crystal structure did not show any interaction between Arg-35 and the A subunit. Perhaps the loss of the intersubunit salt bridge between Arg-35 and Glu-11 of the neighboring B subunits destabilizes the B subunit–B subunit interaction to such a degree that nonnative B-subunit assembly intermediates are produced that are unable to interact appropriately with the A subunit.

Several site-specific point mutations in the A1-fragment of porcine EtxA have resulted in the A subunit failing to assemble into a stable holotoxin complex (M.-G. Pizza and R. Rappuoli, personal communication). The majority of these substitutions introduce either a large hydrophobic amino acid or Pro or Gly, which is likely to affect the folding of the A1 fragment. Thus it could be suggested that these mutants fail to assemble because they are unable to attain a soluble folded conformation, which is required for release from the membrane. Alternatively, these mutant A subunits may be rapidly degraded by periplasmic proteases before or after assembly.

These various studies have provided a significant insight into the assembly of cholera toxin and related enterotoxins in vivo. Nevertheless the precise pathway of subunit–subunit interactions and the relative importance of particular residues in ensuring the formation an AB_5 holotoxin complex remain to be determined.

E. Toxin Secretion Across the Outer Membrane

The final step of toxin biogenesis in *V. cholerae* is the translocation of the periplasmically located holotoxin complex across the bacterial outer membrane. In contrast, in enterotoxinogenic strains of *E. coli* the assembled toxin remains entrapped within the periplasm (Hirst et al., 1984a; Hofstra and Witholt, 1984; Hunt and Hardy, 1991). ETEC strains may therefore have to rely on damage to their outer membranes by host intestinal factors, such as by bile salts and proteases, to cause the *nonspecific release* of toxin to the gut milieu. Hunt and Hardy (1991) showed that when *E. coli* $286C_2$ (an ETEC isolate of human origin) was cultured in a defined low-phosphate medium, the addition of bile salts caused the leakage of Etx and other periplasmic proteins (e.g., alkaline phosphatase) into the medium. This leakage could be augmented by growing the cells under iron-restricted conditions in the presence of trypsin, thereby liberating up to 70% of the toxin into the external medium (Hunt and Hardy, 1991). Ironically, therefore, the level of Etx delivered to the intestinal epithelium during infection appears to be dependent on the host's natural physiological defense mechanisms. It must nonetheless be borne in mind that the efficacy of intestinal factors in causing the leakage of Etx may be somewhat reduced if the bacteria grow as microcolonies on the gut mucosa. The lack of an efficient mechanism to secrete Etx across the *E. coli* outer membrane may explain in part the reduced severity of Etx-mediated ETEC infections compared with those caused by *V. cholerae*.

By contrast, *V. cholerae* is able to efficiently and selectively secrete CT across its outer membrane (Neill et al., 1983; Hirst et al., 1984). This is not due to differences in the properties of CT and Etx per se since the expression of the cloned *ctx* genes in *E. coli* resulted in CT remaining in the periplasm whereas expression of Etx in *V. cholerae* resulted in its efficient secretion to the medium (Pearson and Mekalanos, 1982; Mekalanos et al., 1983; Neill et al., 1983; Hirst et al., 1984; Hirst, 1991). Because the secretory apparatus of *V. cholerae* appears to be able to translocate a fully folded and assembled protein (see below), it must function in a fundamentally different manner from the "string of beads" mechanism postulated to be involved in protein export across the bacterial cytoplasmic membrane.

1. Secretion of Assembled Holotoxin Across the Outer Membrane of V. cholerae

The roles of the A and B subunits in secretion. The roles of the various subunits in toxin secretion have been investigated by studying engineered bacterial strains carrying either a wild-type operon encoding both the A and B subunits or mutant operons in which the A or B subunits are expressed separately (Hirst et al., 1984). This led to the conclusion that the B subunit (or the assembled B pentamer) contains the relevant structural determinants to permit secretion across the *V. cholerae* outer membrane. Thus, *V. cholerae* strains expressing *E. coli* Etx or the individual EtxA or EtxB subunits were found to be able to secrete the assembled EtxB subunits as efficiently as the fully assembled AB_5 toxin. In contrast, when expressed alone, the A subunit remained cell-associated (Hirst et al., 1984). This finding led to the hypothesis that the A subunits must be associated with the B pentamer prior to the step at which the B subunits engage the secretory machinery if the holotoxin is to be translocated across the outer membrane. It also implicated a need for the assembly of the A and B subunits to be well coordinated in order to avoid secretion of the B pentamer (a potential competitive inhibitor of toxin action in the gut). As indicated in Section V.D, the ability of the A subunit to stabilize a *B-subunit intermediate* during assembly should ensure production of AB_5 complexes rather than B pentamers and thereby avoid this problem.

Kinetics of toxin efflux from the periplasm. A direct examination of the structure of the toxin prior to translocation across the outer membrane of *V. cholerae* was achieved by monitoring the kinetics of formation of SDS-stable pentamers compared with the kinetics of toxin efflux from the periplasm (Hirst and Holmgren, 1987a,b). This provided unequivocal evidence that the toxin subunits rapidly assemble into holotoxin complexes ($t_{1/2} < 1$ min) and then undergo relatively slow efflux from the periplasm ($t_{1/2}$ ca. 5–13 min) (Hirst and Holmgren, 1987a,b and unpublished results).

An analysis of the rate of efflux of labeled holotoxin from the periplasm showed that it resembled a simple first-order kinetic process. Thus, once the toxin molecules have assembled in the periplasm, all of them have an equal probability of engaging the secretory machinery and being successfully translocated to the medium (Hirst and Holmgren, 1987b). Since the size of the periplasmic pool of holotoxin, as well as the rate of toxin efflux across the outer membrane, can be measured experimentally, it is possible to estimate the number of toxin molecules secreted per minute by each *V. cholerae* cell. Hirst and Holmgren (1987b) used this approach in studies of a *V. cholerae* strain that had been engineered to express the A and B subunits of *E. coli* Etx. They calculated that the apparent first-order rate constant for efflux of radiolabeled toxin from the periplasm was 0.053 min^{-1} (representing a half-time for efflux of 13 min) and calculated the size of the periplasmic pool of holotoxin as 640 molecules per cell (assuming that all cells in the culture had an equivalent amount of toxin in the periplasm). Given the estimated size of the periplasmic pool and the rate of toxin efflux, it was calculated that 34 holotoxin molecules were secreted per minute per cell (Hirst and Holmgren, 1987b). This figure agrees closely with the estimated number of molecules that had to be secreted in order to account for the requisite rise in extracellular toxin concentration (Hirst and Holmgren, 1987b). Thus, the rate of toxin flux through the periplasm can fully account for all of the secreted toxin that subsequently appears in the medium.

The entry of secreted proteins into the periplasm prior to their secretion across the bacterial outer membrane is now recognized as a common step in the secretion of most (but not all) extracellularly located proteins that are produced by gram-negative bacteria (for reviews, see Hirst and Welch, 1988; Pugsley, 1993). However, a few bacterially secreted proteins (such as α-hemolysin of *E. coli*, the adenylate cyclase of *Bordetella pertussis*, and various proteases from *P. aeruginosa*) use transport mechanisms that are similar to the multidrug resistance (MDR) transporters of eukaryotic cells. These proteins are not synthesized with amino-terminal signal peptides and do not depend on the Sec proteins for their translocation across the bacterial cytoplasmic membrane (see Section V.C); instead they use specific transport proteins that are usually encoded in the same operon as the secreted protein. Most notably, these proteins appear to be translocated through the bacterial envelope by a mechanism that bypasses the periplasm and delivers the secreted protein directly into the medium [for a review of these MDR-like transport systems in bacteria, see Holland et al. (1990)]. However, the overwhelming majority of secreted proteins of gram-negative bacteria are synthesized with amino-terminal signal peptides and, like the toxin A and B subunits, appear to enter the periplasm prior to translocation across the bacterial outer membrane. Thus, the secretion of such proteins may be characterized as involving two distinct, kinetically separable translocation events, the first involving rapid export across the cytoplasmic membrane and the second involving secretion across the outer membrane.

Studies by Pugsley (1992, 1993) on pullulanase, a 117-kDa starch-debranching enzyme that is secreted by such a two-step mechanism, revealed that the enzyme must at least partially achieve its native folded structure in the periplasm in order to be translocated across the bacterial outer membrane. Pugsley reported that strains lacking DsbA, the enzyme responsible for catalyzing disulfide bond formation (see Section V.D), accumulated pullulanase in the periplasm, implying that the formation of the correctly folded disulfide-bonded pullulanase structure was necessary for secretion (Pugsley, 1992). Similarly, secretion of aerolysin from *Aeromonas hydrophila* involves the transient entry of the protein into the periplasm, where it appears to dimerize before being translocated across the bacterial outer membrane (G. van der Goot, personal communication). Thus, following the pioneering work on CT and Etx, an increasing number of proteins appear to fold, and even to oligomerize, before translocation across the outer membrane. How the translocation of folded proteins across membranes might be achieved is discussed in a later section.

2. The Genetics of Toxin Secretion from V. cholerae

The first indication that *V. cholerae* contains gene(s) involved in the process of toxin translocation through the outer membrane was obtained nearly 20 years ago, when nontoxinogenic isolates of *V. cholerae* strain 569B were obtained by NTG chemical mutagenesis (Finkelstein et al., 1974; Holmes et al., 1975). This led to identification of several mutants that did not produce CT and a single mutant, designated M14, that failed to secrete CT into the growth medium (Holmes et al., 1975). The defect in M14 was shown not to have been caused by a mutation in the CT molecule itself, because when Etx was heterologously expressed in M14 it was also found to remain cell-associated (Neill et al., 1983). The CT produced by M14 was subsequently shown to react with *oligomer-specific* monoclonal antitoxin antibodies and to be trapped in the periplasm (Hirst and Holmgren, 1987b). Thus, the toxin subunits produced by M14 cross the cytoplasmic membrane and assemble normally, indicating that this strain is defective in one or more genes involved in toxin translocation across the outer membrane.

Evidence that the toxin is likely to use a pleiotropic secretory machinery that functions in the secretion of a variety of extracellular proteins was suggested by a study of the heterologous expression and secretion of the B subunit of *E. coli* enterotoxin (hEtxB) in a range of pathogenic and nonpathogenic vibrios and aeromonads (Hirst and Leece, 1991). None of the species tested in that study, with the exception of *Vibrio mimicus*, normally produce cholera-like enterotoxins, yet they all exhibited a capacity to efficiently secrete hEtxB into the medium. Since it seems unlikely that they would have retained a translocation machinery that could be used only by CT or Etx, oligomeric enterotoxins are likely to use a preexisting secretory machinery. This view was further substantiated by investigations of protein secretion in a marine *Vibrio* and various mutant derivatives of that strain that were pleiotropically defective in the secretion of extracellular proteins (Ichige et al., 1988; Wong et al., 1990; Leece and Hirst, 1992). Unlike the parental *Vibrio* strain, the mutants accumulated several endogenously secreted proteases and amylases in the periplasm and were shown to be unable to secrete heterologously expressed aerolysin (from *Aeromonas hydrophila*) or hEtxB (Wong et al., 1990; Leece and Hirst, 1992). Thus, EtxB, aerolysin, and several other extracellular enzymes appear to use the same machinery for translocation through the bacterial outer membrane of this marine vibrio.

Extensive studies have been undertaken in recent years on the genetic basis of aerolysin secretion from *A. hydrophila*. This has resulted in the identification of an operon containing several genes that are required for secretion (Jiang and Howard, 1992). These genes show extensive sequence homology with genes identified in other gram-negative bacterial species that have been shown to function in the secretion of extracellular proteins across the outer membrane. Pugsley (1993) has termed these secretion systems the *main terminal branch* of the *general secretion pathway* of gram-negative bacteria (for reviews see Pugsley et al., 1990; Pugsley, 1993).

The best-characterized example of this pathway is that involved in production of pullulanase (PulA) by *Klebsiella oxytoca* (Fig. 9). Pullulanase is a lipoprotein that is synthesized as a longer precursor with an amino-terminal signal sequence. It is exported across the cytoplasmic membrane with the aid of the Sec proteins and undergoes proteolytic processing by LspA (Pugsley et al., 1991). However, secretion of pullulanase across the outer membrane of *Klebsiella* (and also when expressed heterologously in *E. coli*) is dependent on 14 out of 15 genes that flank *pulA*, namely *pulC-0* and *pulS* (Fig. 9). Although 13 of these genes encode proteins that are exported across the cytoplasmic membrane, only PulD (and possibly PulS) appear to be integrated into the outer membrane (Pugsley et al., 1990; Pugsley, 1993). Since several of the Pul proteins (PulG–J; Fig. 9) are homologous to type IV pilin proteins, it has been suggested that they may assemble into a pseudopilus that provides a scaffold that spans the periplasm between the cytoplasmic and outer membranes. Uniquely, PulE, one of the Pul proteins that is essential for pullulanase secretion, is not synthesized with a recognizable signal sequence for export and appears to be located in the cytoplasm (Pugsley, 1993). The PulE protein and its various homologues from other gram-negative bacteria have a conserved ATP-binding motif similar to that of ATPases and kinases (Walker et al., 1982; Pugsley, 1993). It is possible that this protein may either transduce energy in the process of protein secretion or provide energy for the assembly of Pul proteins into a complex. When the *exeE* gene of *Aeromonas hydrophila* (which shares extensive sequence homology with *PulE*) was mutated by Jiang and Howard (1992), aerolysin secretion from this organism was prevented. The introduction of a plasmid encoding hEtxB into the *exeE* mutant strain of *A. hydrophila* resulted in the accumulation of correctly assembled B subunits in the periplasm that, unlike the parental strain, failed to be secreted into the medium (Yu and Hirst, 1994).

This further supports the view that toxin secretion by the Vibrionaceae is dependent on genes involved in the secretion of several different extracellular proteins.

Recently, Sandkvist et al. (1993) obtained direct evidence that toxin secretion from *V. cholerae* is dependent on the main terminal branch of the protein secretion pathway of gram-negative bacteria. They identified a fragment of chromosomal DNA from an El Tor strain that complemented the toxin secretion defect in M14. This study took advantage of the observation that the M14 mutant exhibits a pleiotropic defect in several cellular functions, including the secretion of proteases. Thus, screening for restored protease secretion provided a convenient means of testing for transcomplementing DNA (Sandkvist et al., 1993). Various positive clones were identified and tested in a G_{M1}-ELISA to verify that the cloned DNA restored toxin secretion as well as protease secretion. One

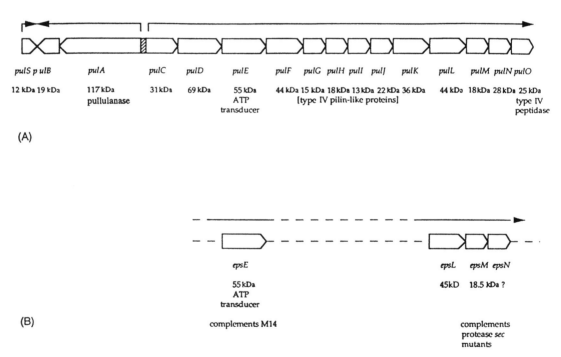

Fig. 9 Genes involved in the secretion of extracellular proteins from *K. oxytoca* and *V. cholerae*. (A) The secretion of the starch debranching enzyme, pullulanase (PulA) across the outer membrane of *K. oxytoca* is dependent on 14 genes (*pulC–O* and *S*). Mutations in any one of these genes prevents the extracellular secretion of pullulanase. Between *pulA* and *pulC* (hatched box) are promoter regions containing four binding sites for the positive transcriptional activator protein, MalT, that induces expression of the *pulAB* and *pulC–O* transcripts in the presence of maltose or maltodextrins. The expression of the *pulS* transcript is not regulated by MalT. PulE is a cytoplasmically located protein with a Walker ATP-binding motif that is likely to catalyze ATP hydrolysis and to transduce energy into the process for protein secretion through the outer membrane. PulG–J show similarities with type IV pilins, each being expressed as a precursor with a short amino-terminal extension that is removed by the prepilin peptidase, PulO. (Adapted from Pugsley, 1993.) (B) Genes of *V. cholerae* involved in extracellular protein secretion (*eps*), which show homology in sequence and organization with the *pul* gene cluster.

complementing clone (containing a DNA fragment of 15 kb) was identified of which the minimum subclone necessary to restore toxin secretion was of 1.6 kb (Sandkvist et al., 1993). Sequence analysis revealed that this subclone contained an open reading frame that showed a high degree of homology (45–60% identity) to the PulE/ExeE proteins involved in the secretion of extracellular proteins by other gram-negative bacteria (Sandkvist et al., 1993). The *V. cholerae* gene (designated *epsE*) was shown to encode a 503-amino-acid (56.3-kDa) protein that exhibited the characteristic features of the PulE-like proteins, the lack an amino-terminal signal peptide and the characteristic ATP-binding motif (Walker et al., 1982; Possot et al., 1992; Sandkvist et al., 1993).

Overbye et al. (1993) recently showed that additional genes located downstream of *epsE* are required for toxin secretion by *V. cholerae* (Fig. 9). Three genes, *epsLMN*, were identified that share sequence homology with *pulLMN* (Overbye et al., 1993). Thus, *V. cholerae* possesses a gene cluster analogous to other outer membrane secretion systems of the main terminal branch of the general secretion pathway.

3. Mechanism of Toxin Secretion Across the Outer Membrane

In common with that of other extracellularly located proteins, the exact mechanism by which CT is secreted remains something of an enigma. The mechanism must take account of the following facts.

1. The toxin attains a stable AB_5 structure before translocation across the outer membrane.
2. The toxin secretion machinery appears to be shared with several proteins.
3. The *eps* gene products of the main terminal branch of the secretion pathway are involved in toxin secretion in *V. cholerae*.

The outer membrane of *V. cholerae*, like that of other gram-negative bacteria, does not allow the free movement of solutes of >500 Da. This observation means that the translocation of CT or Etx, which are nearly 200 times larger than this limit, cannot involve translocation through large open pores, since this would result in the membrane being highly permeable and would lead to the nonspecific loss of resident periplasmic proteins. I shall therefore discuss two speculative models that assume that the limited permeability of the outer membrane is maintained throughout the process of toxin secretion (Fig. 10). The first model is based on the concept that the holotoxin partitions into the outer membrane bilayer and that this is facilitated by the components of the secretory machinery. The second model assumes that the components of the secretory machinery assemble into a gated pore that opens to allow proteins to translocate through it.

The facilitated partitioning model. It is now accepted that proteins destined to be incorporated into the outer membranes of gram-negative bacteria, such as the trimeric porins, can spontaneously insert into the outer membrane when they interact with phospholipids and lipopolysaccharide (LPS) (Sen and Nikaido, 1990). A number of laboratories have produced evidence to suggest that porin proteins transiently enter the periplasm during their export, which indicates that certain proteins possess an inherent capacity to partition from the periplasm into the outer membrane. In the case of the porins, insertion is triggered by interaction with LPS, which results in the formation of SDS-stable trimers. When heated, such trimers dissociate into monomers, in a way similar to the B pentamer (Sen and Nikaido, 1990).

If the toxin is translocated across the outer membrane by a mechanism that involves partitioning into the outer membrane bilayer, it must do so in a manner that permits insertion on one side of the membrane and release on the other. The secretory machinery might in some way act to trigger insertion by causing a conformational change in the toxin that increases its affinity for the membrane. In this context, it has been shown that pentameric CT-B and EtxB can undergo pH-induced conformational changes in vitro at around pH 5.0 that make the protein more hydrophobic without disrupting the pentameric structure of the molecule (L. Ruddock and T. R. Hirst, unpublished observations). If such conformational changes are important during toxin secretion, then it is conceivable that one of the functions of the secretory machinery is to generate a localized environment of low pH. Thus it could be postulated that EpsE may hydrolyze ATP in order to translocate protons into the secretory machinery to achieve such a localized low pH.

Studies on the translocation aerolysin across the outer membrane of *A. hydrophila* have demonstrated that lowering the external pH prevented aerolysin secretion (Wong and Buckley, 1989). Although this appears to contradict the suggestion that translocation is triggered by a pH-induced conformational change, a low external pH might have prevented aerolysin from undergoing the reverse transition to a more native soluble structure for release from the outer membrane. It was also noted in this study that uncouplers of pmf inhibit the efflux of aerolysin fom the periplasm. It could be suggested that pmf may be required to provide a source of protons for a pH-induced transition. Alternatively, the treatment of *A. hydrophila* with an uncoupler may have caused a depletion of the cellular ATP pool and thereby inhibited the function of ExeE.

Adjacent to the *epsE* gene of *V. cholerae* is *epsD*, which shows strong sequence homology with PulD, the outer membrane component of the Pul complex (J. Yu and T. R. Hirst, unpublished observations). It is thus expected that EpsD will also be located in the outer membrane. Although the exact function of PulD-like proteins in secretion remains to be elucidated, their outer membrane location implicates them as the component of the secretory machinery most likely to be involved in interacting with secreted proteins as they traverse the outer membrane. Thus, if toxin translocation involves insertion into the outer membrane bilayer, EpsD might aid the initial insertion event or alternatively may maintain an interaction with the toxin while it translocates across the membrane.

The possibility that facilitated diffusion may play a role in toxin secretion from *V. cholerae* was raised by Hirst and Holmgren (1987a) when it was shown that the concentration of holotoxin in the periplasm of *V. cholerae* was at least 1800-fold higher than that in the external medium. This large toxin concentration gradient between the periplasm and the external milieu would favor the partitioning of the toxin into the periplasmic face of the outer membrane but would almost certainly preclude such an event on the external membrane surface.

The facilitated partitioning model for toxin secretion assumes that the toxin interacts with the outer membrane bilayer, and with protein components of the secretory machinery, during translocation. Entry and release would be triggered by requisite conformational changes that alter the relative hydrophobicity and affinity of the toxin for the membrane.

The gated-pore model. This model assumes that components of the secretory machinery form a channel or pore through which toxin molecules are translocated without interacting with the lipid phase of the outer membrane. Several features common to most of the proteins that use the main terminal branch of the general secretory pathway suggest that these proteins may prefer to translocate via hydrophilic (water-filled) pores. Most of the proteins appear to adopt folded structures prior to translocation, and, more important,

(A)   (B)

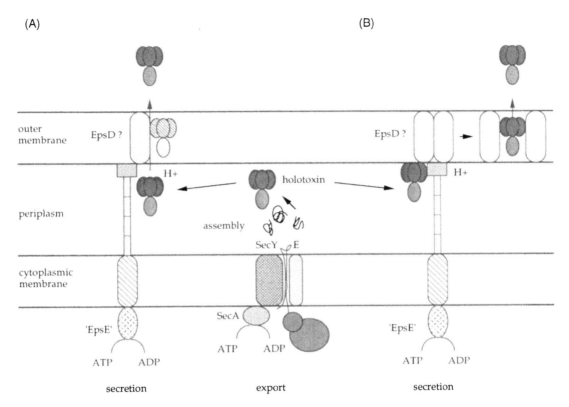

Fig. 10 Schematic model of the pathway of toxin secretion from *V. cholerae*. In this model it is proposed that the precursor A and B subunits are exported across the cytoplasmic membrane via a Sec-dependent pathway involving homologues of SecA, SecE, and SecY (SecD and SecF are not shown), which are cleaved by LepB (not shown) to yield mature subunits that are released into the periplasm, where they assemble into an AB$_5$ holotoxin complex. Factors that are (or may be) involved in folding reactions such as DsbA and peptidyl prolyl *cis-trans* isomerase are not shown. (A) Secretion through the outer membrane by facilitated membrane partitioning. It is proposed that the Eps proteins (or an alternative secretory machinery) mediate translocation of the assembled holotoxin complex across the outer membrane by a mechanism that involves toxin insertion into the lipid phase of this membrane. In this model the function of the secretory machinery is to trigger a conformational change in the toxin (possibly induced by a local low-pH environment) that promotes toxin insertion into the outer membrane. Release from the membrane into the external milieu would occur spontaneously as a result of the corresponding inverse conformational change. (B) Secretion through the outer membrane via a gated pore. It is proposed that the Eps proteins (or an alternative secretory machinery) mediate translocation of the assembled holotoxin complex across the outer membrane by a mechanism that involves the toxin triggering the opening of a pore formed by one or more outer membrane proteins of the secretory machinery. It is expected that such a major reorganization of proteins in the outer membrane would be dependent either on ATP hydrolysis (via EpsE) or on transduction of protons. This model presupposes that pore opening would not render the outer membrane permeable to nonspecific protein release from the periplasm.

the proteins appear to adopt folded structures prior to translocation, and, more important, many of them not only have hydrophilic surfaces but are in fact soluble in aqueous environments. Thus, in the absence of some trigger to expose buried hydrophobic regions these proteins will not insert into the lipid phase. Such pores must be able to exclude not only small solutes but also small resident periplasmic proteins while at the same time permitting the secretion of bulky proteins such as CT, aerolysin, or pullulanase. Although the Pul-like secretion machinery in different gram-negative bacteria appears to be able to translocate several proteins, these homologous secretory systems nonetheless discriminate between different secreted proteins. This argues for some kind of specific recognition between a particular secreted protein and components of the secretory machinery. Such a mechanism would provide a convenient means of discriminating between resident periplasmic proteins and those passing through the periplasm en route to being secreted through the outer membrane. Consequently, secreted proteins may possess some kind of common structural motif, analogous to the specific nuclear localization signal that targets eukaryotic proteins through the pores of the nuclear envelope (Hinshaw et al., 1992). Since no linear sequence motifs have been identified that could fulfill this role, it may be necessary to search for common three-dimensional motifs once a sufficient number of crystal structures of secreted proteins have been determined.

The most likely component of the secretory machinery to form an outer membrane pore to allow toxin secretion would be the EpsD protein. The PulD homologue of EpsD has been extensively investigated, but as yet there is no evidence that it has a pore-forming activity (Pugsley, 1993). It is of interest to note that PulD-like proteins have been identified in a diverse array of systems involved in the transport of macromolecules. These include proteins required for assembly and secretion of a filamentous phage, for secretion of proteins by pathways other than the main terminal branch of the general secretory pathway, and for DNA transformation of *H. influenzae* (Brisette and Russel, 1990; Michels et al., 1991; Tomb et al., 1991; Allaoui et al., 1993; Pugsley, 1993). All of the functions carried out by these proteins could conceivably rely on transport through channels or pores.

If such pores are not permanently occupied by a secretory protein, their opening would presumably have to be regulated in order to maintain the limited permeability properties of the outer membrane. This regulation could operate by secretory proteins triggering the pores to open by direct interaction. This could require a source of energy, either pmf or ATP hydrolysis via EpsE, to ensure that the pore opens. Thus, in this model the pore would act as a gate, opening upon interaction with any secreted protein that possesses the postulated structural motif. To provide an understanding of such a mechanism, it will be necessary to define (1) the architecture and biochemical properties of the postulated pore and (2) the identity (if any) of a putative secretion motif.

Other considerations. Several additional aspects of toxin secretion by *V. cholerae* remain to be resolved. For example, is the *eps* operon of *V. cholerae* the only secretory machinery responsible for CT secretion into the extracellular milieu? The specter that bacteria might possess alternative pathways to the main terminal branch of the general secretory pathway for translocating oligomeric toxins across their outer membranes was raised by studies on pertussis toxin secretion by *Bordetella pertussis* (see Chapter 19 of this volume) (Weiss et al., 1993). Several genes adjacent to the pertussis toxin operon are required for toxin secretion from *B. pertussis*. These *ptl* genes show a striking homology to certain of the *virB* genes from *Agrobacterium tumefaciens*, which play a role in the transport of T-DNA across bacterial and plant membranes (Weiss et al., 1993).

It would also be of interest to establish whether *V. cholerae* possess a *ptl*-like pathway that might function under different environmental conditions.

Another feature of toxin secretion that needs further investigation is the possibility that the activity of the secretory machinery may be regulated by transcriptional or environmental factors. The consequence of such regulation would be that the number of functional "sites" in the outer membrane through which CT and other proteins could be secreted may vary. This could result in the number of functional "sites" being a rate-determining factor for the efficiency of toxin secretion. Such regulation could be seen as a change in the size of the periplasmic pool of toxin subunits. Recently, we obtained evidence that both *toxR* and another gene (currently being characterized at a molecular level) influence the size of the periplasmic pool of toxin (J. Yu et al., manuscript in submitted; unpublished results). While there are a number of explanations for this, the most likely is that these factors regulate the expression of the genes involved in toxin secretion across the outer membrane of *V. cholerae*.

VI. CONCLUDING REMARKS

The biogenesis of cholera toxin and related oligomeric enterotoxins represents a paradigm of the sequence of events involved in the production of secreted virulence factors. Our current understanding of many of these events is well advanced, while other areas remain challenging enigmas awaiting a molecular explanation.

Acknowledgments

I especially thank Gordon Findlay for critically reading the manuscript, and Lloyd Ruddock, Jun Yu, Ed Lowe, and Stephen McLaughlin for their help in its preparation. The extensive financial support of the Wellcome Trust is gratefully acknowledged.

REFERENCES

Agbonlahor, D. E., and Odugbemi, T. O. (1982). Enteropathogenic, enterotoxigenic and enteroinvasive *Escherichia coli* isolated from acute gastroenteritis patients in Lagos, Nigeria. *Trans. Roy. Soc. Trop. Med. Hyg. 76*: 265–267.

Aitken, R., and Hirst, T. R. (1993). Recombinant enterotoxins as vaccines against *Escherichia coli*-mediated diarrhoea. *Vaccine 11*: 227–223.

Akita, M., Sasaki, S., Matsuyami, S.-I, and Mizushima, S. (1990). SecA interacts with secretory proteins by recognizing the positive charge at the amino acid terminus of the signal peptide in *Escherichia coli. J. Biol. Chem. 265*: 8164–8169.

Akiyama, Y., and Ito, K. (1987). Topology analysis of the SecY protein, an integral membrane protein involved in protein export in *Escherichia coli. EMBO J. 6*: 3465–3470.

Albert, M. J., and Ansaruzzaman, M. (1993). Large epidemic of cholera-like disease in Bangladesh caused by *Vibrio cholerae* O139 synonym Bengal. *Lancet 342*: 387–390.

Albright, L. M., Huala, E., and Ausubel, F. M. (1989). Prokaryotic signal transduction mediated by sensor and regulator protein pairs. *Annu. Rev. Genet. 23*: 311–316.

Allaoui, A., Sansonetti, P., and Parsot, C. (1993). MxiD, an outer membrane protein necessary for the secretion of the *Shigella flexneri* Ipa invasins. *Mol. Microbiol. 7*: 59–68.

Altman, E., Kumamoto, C. A., and Emr, S. (1991). Heat-shock proteins can substitute for Sec B function during protein export in *Escherichia coli. EMBO J. 10*: 239–245.

Ankenbauer, R. G., and Nester, E. W. (1990). Sugar-mediated induction of *Agrobacterium tumefaciens* virulence genes: structural specificity and activities of monosaccharides. *J. Bacteriol. 172*: 6442–6446.

Arnone, A., Bier, C. J., Cotton, F. A., Day, V. W., Hazen, E. E. J., Richardson, D. C., Richardson, J. S., and Yonath, A. (1971). A high resolution structure of an inhibitor complex of the extracellular nuclease of *Staphylococcus aureus*. *J. Biol. Chem. 245*: 2302–2316.

Bardwell, J. C. A., McGovern, K., and Beckwith, J. (1991). Identification of a protein required for disulfide bond formation *in vivo*. *Cell 65*: 581–589.

Bardwell, J. C. A., Lee, J.-O., Jander, G., Martin, N., Belin, D., and Beckwith, J. (1993). A pathway for disulphide bond formation *in vivo*. *Proc. Natl. Acad. Sci. U.S.A. 90*: 1038–1042.

Baudry, B., Fasano, A., Ketley, J., and Kaper, J. B. (1992). Cloning of a gene (*zot*) encoding a new toxin produced by *Vibrio cholerae*. *Infect. Immun. 60*: 428–434.

Bhattacharya, S. K., Bhattacharya, M. K., Nair, G. B., Dutta, D., Deb, A., and Ramamaurthy, T., Garg, S., Saha, P. K., Dutta, P., Moitra, A., Mandel, B. K., Shimada, T. Takeda, T., and Deb, B. C. (1993). Clinical profile of acute diarrhoea cases infected with the new epidemic strain of *Vibrio cholerae* O139. *J. Infect. 27*: 11–15.

Bieker, K. L., and Silhavy, T. J. (1990). PrlA (SecY) and PrlG (SecE) interact directly and function sequentially during protein translocation in *E. coli*. *Cell 61*: 833–842.

Bieker-Brady, K., and Silhavy, T. J. (1992). Suppressor analysis suggests a multistep, cyclic mechanism for protein secretion in *Escherichia coli*. *EMBO J. 11*: 3165–3174.

Black, R. E. (1986). The epidemiology of cholera and enterotoxigenic *E. coli* diarrheal disease. In *Development of Vaccines and Drugs Against Diarrhea*, J. Holmgren, A. Lindberg, and R. Mollby (Eds.), Studentlitteratur, Lund pp. 23–32.

Black, R. E., Mersen, M. H., Rahman, A. S. M. M., Yunus, M., Alim, A. R. M. A., Huq, I., Yolken, R. H., and Curlin, G. T. (1980). A two year study of bacterial, viral, and parasitic agents associated with diarrhea in rural Bangladesh. *J. Infect. Dis. 142*: 660–664.

Black, R. E., Merson, M. H., Huq, I., Alim, A. R. M., and Yunus, M. D. (1981). Incidence and severity of rotavirus and *Escherichia coli* diarrhea in rural Bangladesh: implications for vaccine development. *Lancet 1*: 141–143.

Booth, B. A., Boesman-Finkelstein, M., and Finkelstein, R. A. (1984). *Vibrio cholerae* hemagglutinin/protease nicks cholera enterotoxin. *Infect. Immun. 45*: 558–560.

Brickman, T. J., Boesman-Finkelstein, M., Finkelstein, R. A., and McIntosh, M. A. (1990). Molecular cloning and nucleotide sequence analysis of cholera toxin genes of the ctxA⁻ *Vibrio cholerae* strain Texas Star-SR. *Infect. Immun. 58*: 4142–4144.

Brisette, J. L., and Russel, M. (1990). Secretion and membrane integration of a filamentous phage encoded morphogenic protein. *J. Mol. Biol. 211*: 565–580.

Brundage, L., Hendrick, J. P., Schiebel, E., Driessen, A. J. M., and Wickner, W. (1990). The purified *E. coli* integral membrane protein SecY/E is sufficient for reconstitution of SecA-dependent precursor protein translocation. *Cell 62*: 649–657.

Brundage, L., Fimmel, C. J., Mizushima, S., and Wickner, W. (1992). SecY, SecE and band 1 form the membrane-embedded domain of *Escherichia coli* preprotein translocase. *J. Biol. Chem. 267*: 4166–4170.

Brunton, J. L. (1990). The shiga toxin family: molecular nature and possible role in disease. In *The Bacteria*, I. C. Gusalus, and R. Y. Stanier (Eds.) Academic, New York, pp. 377–397.

Cassel, D., and Selinger, Z. (1977). Mechanism of adenylate cyclase activation by cholera toxin: inhibition of GTP hydrolysis at the regulatory site. *Proc. Natl. Acad. Sci. U.S.A. 84*: 3307–3311.

Chen, L., and Tai, P.-C. (1986). Roles of H⁺-ATPase and proton motive force for ATP-dependent protein translocation *in vitro*. *J. Bacteriol. 167*: 389–392.

Chen, L., and Tai, P.-C. (1987). Evidence for the involvement of ATP in cotranslational protein translocation. *Nature (Lond.) 328*: 164–166.

Chen, M. Y., and Tai, P.-C. (1985). ATP is essential for protein translocation into *Escherichia coli* membrane vesicles. *Proc. Natl. Acad. Sci. U.S.A. 82*: 4384–4388.

Chongsa-nguan, M., Chaicumpa, W., Moolasart, P., Kandhasingha, P., Shinmada, T., Kurazono, H., and Takeda, Y. (1993). *Vibrio cholerae* O139 Bengal in Bangkok. *Lancet 342*: 431.

Chopra, A. K., Houston, C. W., Peterson, J. W., Prasad, R., and Mekalanos, J. J. (1987). Cloning and expression of the *Salmonella* enterotoxin gene. *J. Bacteriol. 169*: 5095–5100.

Clements, J. D., and Finkelstein, R. A. (1979). Isolation and characterisation of homogeneous heat-labile enterotoxins with high specific activity from *Escherichia coli* cultures. *Infect. Immun. 24*: 760–769.

Clements, J. D., Yancy, R. J. and Finkelstein, R. A. (1980). Properties of homogeneous heat-labile enterotoxin from *Escherichia coli. Infect. Immun. 29*: 91–97.

Crooke, E., Brundage, L., Rice, M., and Wickner, W. (1988). ProOmpA spontaneously folds in a membrane assembly competent state which trigger factor stabilizes. *EMBO J. 7*: 1831–1835.

Crooke, E., and Wickner, W. (1987). Trigger factor: a soluble protein that folds pro-OmpA into a membrane-assembly competent form. *Proc. Natl. Acad. Sci. U.S.A. 84*: 5216–5220.

Cuatrecasas, P. (1973). Gangliosides and membrane receptors for cholera toxin. *Biochemistry 12*: 3558–3566.

Dalby, R. E. (1991). Leader peptidase. *Mol. Microbiol. 5*: 2855–2860.

Dallas, W. S. (1983). Conformity between heat-labile toxin genes from human and porcine enterotoxigenic *Escherichia coli. Infect. Immun. 40*: 647–652.

Dallas, W. S., and Falkow, S. (1979). The molecular nature of heat-labile enterotoxin from *Escherichia coli. Nature (Lond.) 277*: 406–407.

Dallas, W. S., and Falkow, S. (1980). Amino acid sequence homology between cholera toxin and *Escherichia coli* heat-labile toxin. *Nature (Lond.) 288*: 499–501.

Dallas, W. S., Gill, D. M., and Falkow, S. (1979). Cistrons encoding *Escherichia coli* heat-labile enterotoxin (LT). J. Bacteriol. 139: 850–858.

Dams, E., De Wolf, M., and Dierick, W. (1991). Nucleotide sequence analysis of the CT operon of *Vibrio cholerae* classical strain 569B. *Biochim. Biophys. Acta 1090*: 139–141.

Daniels, C. J., Bole, D. G., Quay, S. C., and Oxender, D. L. (1981). Role of membrane potential in the secretion of protein into the periplasm of *Escherichia coli. Proc. Natl. Acad. Sci. U.S.A. 78*: 5396–5400.

De, S. N. (1959). Enterotoxicity of bacteria-free culture filtrate of *Vibrio cholerae. Nature (Lond.). 183*: 1533–1534.

De, S. N., Bahattacharya, K., and Sarkar, J. K. (1956). A study of the pathogenicity of strains of *Bacterium coli* from acute and chronic enteritis. *J. Pathol. Bacteriol. 71*: 201–209.

DeMol, P., Brasseur, D., Hemelhof, W., Kalala, T., Butzler, J. P., and Vis, H. L. (1983). Enteropathogenic agents in children with diarrhoea in rural Zaire. *Lancet i*: 516–518.

DiRita, V. J. (1992). Coordinate expression of virulence genes by ToxR in *Vibrio cholerae. Mol. Microbiol. 6*: 451–458.

DiRita, V. J., and Mekalanos, J. J. (1989). Genetic regulation of bacterial virulence. *Annu. Rev. Genet. 23*: 455–482.

DiRita, V. L., and Mekalanos, J. J. (1991). Periplasmic interaction between two membrane proteins, ToxR and ToxS, results in signal transduction and transcriptional activation. *Cell 64*: 29–37.

DiRita, V. J., Parsot, C., Jander, G., and Mekalanos, J. J. (1991). Regulatory cascade controls virulence in *Vibrio cholerae. Proc. Natl. Acad. Sci. U.S.A. 88*: 5403–5407.

Dorman, C. J., and Ni Bhriain, N. (1992). Global regulation of gene expression during environmental adaptation. In *Molecular Biology of Bacterial Infection: Current Status and Future Perspectives*, C. Hormaeche (Ed.), Cambridge Univ. Press, Cambridge, pp. 193–230.

Driessen, A. J. M. (1992). Precursor protein translocation by the *Escherichia coli* translocase is directed by the proton motive force. *EMBO J. 11*: 847–853.

Driessen, A. J. M., and Wicker, W. (1991). Proton transfer is rate-limiting for translocation of precursor proteins by the *Escherichia coli* translocase. *Proc. Natl. Acad. Sci. U.S.A. 88*: 2471–2475.

Dutta, N. K., Panse, M. W., and Kulkarni, D. R. (1959). Role of cholera toxin in experimental cholera. *J. Bacteriol. 78*: 594–595.

Dykes, C. W., Halliday, I. J., Hobden, A. N., Read, M. J., and Harford, S. (1985). A comparison of the nucleotide sequence of the A subunit of heat-labile enterotoxin and cholera toxin. *FEMS Microbiol. Lett. 26*: 171–174.

Echeverria, P., Seriwarana, J., Taylor, D. N., Yanggratoke, S., and Tirapat, C. (1985). A comparative study of enterotoxigenic *Escherichia coli*, *Shigella*, *Aeromonas* and *Vibrios* as etiologies of diarrhea in Northeastern Thailand. *Am. J. Trop. Med. Hyg. 35*: 547–554.

Emr, S. D., Hanely-Way, S., and Sihavy, T. J. (1981). Suppressor mutations that restore export of a protein with a defective signal sequence. *Cell 23*: 79–88.

Enequist, H. G., Hirst, T. R., Harayama, S., Hardy, S. J. S., and Randall, L. L. (1981). Energy is required for maturation of exported proteins in *Escherichia coli*. *Eur. J. Biochem. 116*: 227–233.

Epstein, C. J., Goldberger, R. F., and Anfinsen, C. B. (1963). The genetic control of tertiary protein structure: studies with model systems. *Cold Spring Harbor Symp. Quant. Biol. 28*: 439–449.

Evans, D. G., Olarte, J., DuPont, H. L., Evans, D. J. J., Galindo, E., Portnoy, B. L., and Conklin, R. H. (1977). Enteropathogens associated with pediatric diarrhea in Mexico City. *J. Pediatr. 91*: 65–68.

Fasano, A., Baudry, B., Pumplin, D. W., Wasserman, S. S., Tall, B. D., Ketley, J. M., and Kaper, J. B. (1991). *Vibrio cholerae* produces a second enterotoxin, which affects intestinal tight junctions. *Proc. Natl. Acad. Sci. U.S.A. 88*: 5242–5246.

Fergusson, L. F., McDiarmid, N., and Hirst, T. R. (1990). A multivalent carrier for delivery of epitopes and antigens based upon the B subunit enterotoxoid of *Escherichia coli*. In *Bacterial Protein Toxins*, R. Rappuoli et al. (Eds.), Gustav Fischer, Stuttgart, pp. 519–520.

Fernadez, M., Sierra-Madero, J., de la Vega, H., Vazquez, M., Lopez-Vidal, Y., Ruiz-Palacios, G. M., and Clava, E. (1988). Molecular cloning of *Salmonella typhi* LT-like enterotoxin gene. *Mol. Microbiol. 2*: 821–825.

Fikes, J. D., and Bassford, P. J., Jr. (1987). Export of unprocessed maltose-binding protein to the periplasm of *Escherichia coli* cells. *J. Bacteriol. 169*: 2353–2359.

Fikes, J. D., and Bassford, P. J., Jr. (1989). Novel secA alleles improve export of maltose-binding protein synthesized with a defective signal peptide. *J. Bacteriol. 171*: 402–409.

Findlay, G., and Hirst, T. R. (1993). Analysis of enterotoxin secretion in a *Vibrio cholerae* strain lacking DsbA, a periplasmic enzyme involved in disulphide bond formation. *Biochem. Soc. Trans. 21*: 212S.

Finkelstein, R. A., and LoSpalluto, J. J. (1969). Pathogenesis of experimental cholera: preparation and isolation of choleragen and choleragenoid. *J. Exp. Med. 130*: 185–202.

Finkelstein, R. A., Atthasampunu, P., Chulasmya, M., and Charunmethee, P. (1966). Pathogenesis of experimental cholera: biologic activities of purified procholeragen. *J. Immunol. 96*: 440–449.

Finkelstein, R. A., Boseman, M., Neoh, S. H., LaRue, M. K., and Delaney, R. (1974a). Dissociation and recombination of the subunits of the cholera enterotoxin (choleragen). *J. Immunol. 113*: 145–150.

Finkelstein, R. A., Vasil, M. L., and Holmes, R. K. (1974b). Studies on toxinogenesis in *Vibrio cholerae*. I. Isolation of mutants with altered toxinogenicity. *J. Infect. Dis. 129*: 117–123.

Fitts, R., Reuveny, Z., van Amsterdam, J., Mulholland, J., and Botstein, D. (1987). Substitution of tyrosine for either cysteine in β-lactamase prevents release from the membrane during secretion. *Proc. Natl. Acad. Sci. U.S.A. 84*: 8540–8543.

Freedman, R. B. (1989). Protein disulphide isomerase: multiple roles in the modification of nascent secretory proteins. *Cell 57*: 1069–1072.

Gardel, G., Benson, S., Hunt, J., Michaelis, S., and Beckwith, J. (1987). SecD, a new gene involved in protein export in *Escherichia coli*. *J. Bacteriol. 169*: 1286–1290.

Gardel, C., Johnson, K., Jacq, A., and Beckwith, J. (1990). The secD locus of *E. coli* codes for two membrane proteins required for protein export. *EMBO J. 9*: 3209–3216.

Geary, S. J., Marchlewicz, B. A., and Finkelstein, R. A. (1982). Comparison of heat-labile enterotoxins from porcine and human strains of *Escherichia coli*. *Infect. Immun.* 36: 215–220.

Geller, B. L. (1990). Electrochemical potential releases a membrane-bound secretion intermediate of maltose-binding protein in *Escherichia coli*. *J. Bacteriol.* 172: 4870–4876.

Geller, B. L. (1991). Energy requirements for protein translocation across the *Escherichia coli* inner membrane. *Mol. Microbiol.* 5: 2093–2098.

Gennaro, M. L., and Greenaway, P. J. (1983). Nucleotide sequences within the cholera toxin operon. *Nucleic Acids Res.* 11: 3855–3861.

Gennaro, M. L., Greenaway, P. J., and Broadbent, D. A. (1982). Expression of biologically active cholera toxin in *Escherichia coli*. *Nucleic Acids Res.* 10: 4883–4890.

Gething, M. J., and Sambrook, J. (1992). Protein folding in the cell. *Nature (Lond.)* 355: 33–45.

Gething, M.-J., McGammon, K., and Sambrook, J. (1986). Expression of wild-type and mutant forms of influenza haemagglutinin: the role of folding in intracellular transport. *Cell* 46: 939–950.

Gierasch, L M. (1989). Signal sequences. *Biochemistry* 28: 923–930.

Gill, D. M. (1976). The arrangement of subunits in cholera toxin. *Biochemistry* 15: 1242–1248.

Gill, D. M., Clements, J. D., Robertson, D. C., and Finkelstein, R. A. (1981). Subunit number and arrangement in *Escherichia coli* heat-labile enterotoxin. *Infect. Immun.* 33: 677–682.

Gill, D. R., Hatfull, G. F., and Salmond, G. P. C. (1986). A new cell division operon in *Escherichia col. Mol. Gen. Genet.* 205: 134–145.

Glass, R. I. (1986). Cholera. *Clin. Trop. Med. Communicable Dis.* 1: 603–615.

Gold, L., and Stormo, G. (1987). Translational initiation. In *Escherichia coli and Salmonella typhimurium: Cellular and Molecular Biology*, F. C. Neidhardt, J. L. Ingraham, K. B. Low, B. Magasanik, M. Schaecter, and H. E. Umbarger (Eds.) American Society for Microbiology, Washington, DC, pp. 1302–1307.

Goldberg, I., and Mekalanos, J. J. (1986). Effect of recA mutation on cholera toxin gene amplification and deletion events. *J. Bacteriol.* 165: 723–731.

Goransson, M., Sonden, B., Nilsson, P., Dagberg, B., Forsman, K., Emmanuelsson, K., and Uhlin, B.-E. (1990). Transcriptional silencing and thermoregulation of gene expression in *Escherichia coli*. *Nature (Lond.)* 344: 682–685.

Green, B. A., Neill, R. J., Ruyechan, W. T., and Holmes, R. K. (1983). Evidence that a new enterotoxin of *Escherichia coli* which activates adenylate cyclase in eukaryotic target cells is not plasmid mediated. *Infect. Immun.* 41: 383–390.

Griffiths, S. L., and Critchley, D. R. (1991). Characterisation of the binding sites for *Escherichia coli* heat-labile toxin type I in intestinal brush orders. *Biochim. Biophys. Acta* 1075: 154–161.

Gyles, C. L., Palchaudhuri, S., and Maas, W. (1977). Naturally occurring plasmid carrying genes for enterotoxin production and drug resistance. *Science* 198: 198–199.

Hardy, S. J. S., Holmgren, J., Johansson, S., Sanchez, J., and Hirst, T. R. (1988). Coordinated assembly of multisubunit proteins: oligomerization of bacterial enterotoxins *in vivo* and *in vitro*. *Proc. Natl. Acad. Sci. U.S.A.* 85: 7109–7113.

Hayano, T., Takahashi, N., Kato, S., Maki, N., and suzuki, M. (1991). Two distinct forms of peptidyl-*cis-trans*-isomerase are expressed separately in periplasmic and cytoplasmic compartments of *Escherichia coli* cells. *Biochemistry* 30: 3041–3048.

Hengge, R., and Boos, W. (1985). Defective secretion of maltose- and ribose-binding proteins caused by a truncated periplasmic protein in *Escherichia coli*. *J. Bacteriol.* 162: 972–978.

Herrington, D. A., Hall, R. H., Losonsky, D., Mekalanos, J. J., Taylor, R. K., and Levine, M. M. (1988). Toxin, toxin-coregulated pili, and the *toxR* regulon are essential for *Vibrio cholerae* pathogenesis in humans. *J. Exp. Med.* 168: 1487–1492.

Higgins, D. E., Nazareno, E., and DiRita, V. J. (1992). The virulence gene activator ToxT from *Vibrio cholerae* is a member of the AraC family of transcriptional activators. *J. Bacteriol.* 174: 6974–6980.

Hinshaw, J. E., Carragher, B. O., and Milligsan, R. A. (1992). Architecture and design of the nuclear pore complex. *Cell* 69: 1133–1141.

Hirst, T. R. (1991). Assembly and secretion of oligomeric toxins by gram negative bacteria. In

Sourcebook of Bacterial Protein Toxins, J. E. Alouf and J. Freer (Eds.), Academic, London, pp. 75–100.

Hirst, T. R., and Holmgren, J. (1987a). Conformation of protein secreted across bacterial outer membranes: a study of enterotoxin translocation from *Vibrio cholerae*. *Proc. Natl. Acad. Sci. U.S.A. 84*: 7418–7422.

Hirst, T. R., and Holmgren, J. (1987b). Transient entry of enterotoxin subunits into the periplasm occurs during their secretion from *Vibrio cholerae*. *J. Bacteriol. 169*: 1037–1045.

Hirst, T. R., and Leece, R. (1991). The phenomenon of toxin secretion by vibrios and aeromonads. *Experientia 47*: 429–431.

Hirst, T. R., and Welch, R. A. (1988). Mechanisms for secretion of extracellular proteins by gram-negative bacteria. *Trends Biochem. Sci. 13*: 265–269.

Hirst, T. R., Randall, L. L., and Hardy, S. J. S. (1983). Assembly in vivo of enterotoxin from *Escherichia coli*: formation of the B subunit oligomer. *J. Bacteriol. 153*: 21–26.

Hirst, T. R., Randall, L. L., and Hardy, S. J. S. (1984a). Cellular location of heat-labile enterotoxin in *Escherichia coli*. *J. Bacteriol. 157*: 637–632.

Hirst, T. R., Sanchez, J., Kaper, J., Hardy, S. J. S., and Holmgren, J. (1984b). Mechanism of toxin secretion by *Vibrio cholerae* investigated in strains harboring plasmids that encode heat-labile enterotoxins of *Escherichia coli*. *Proc. Natl. Acad. Sci. U.S.A. 81*: 7752–7756.

Hofstra, H., and Witholt, B. (1984). Kinetics of synthesis, processing and membrane transport of heat-labile enterotoxin, a periplasmic protein in *Escherichia coli*. *J. Biol. Chem. 259*: 15182–15187.

Hofstra, H., and Witholt, B. (1985). Heat-labile enterotoxin in *Escherichia coli*: kinetics of association of subunits into periplasmic holotoxin. *J. Biol. Chem. 260*: 16037–16044.

Holland, I. B., Blight, M. A., and Kenny, B. (1990). The mechanism of secretion of hemolysin and other polypeptides from gram negative bacteria. *J. Bioenerget. Biomem. 22*: 473–491.

Holmes, R. K., Vasil, M. L., and Finkelstein, R. A. (1975). Studies on toxigenesis in *Vibrio cholerae*. III. Characterization of nontoxinogenic mutants in vitro and in experimental animals. *J. Clin. Invest. 55*: 551–560.

Holmes, R. K., Widdy, E. M., and Pickett, C. L. (1986). Purification and characterization of type II heat-labile enterotoxin of *Escherichia coli*. *Infect. Immun. 53*: 464–473.

Holmgren, A. (1985). Thioredoxin. *Annu. Rev. Biochem. 54*: 237–271.

Holmgren, J. (1973). Comparison of the tissue receptors for *Vibrio cholerae* and *Escherichia coli* enterotoxins by means of gangliosides and natural cholera toxoid. *Infect. Immun. 8*: 851–859.

Holmgren, J. (1981). Actions of cholera toxin and the prevention and treatment of cholera. *Nature (Lond.) 292*: 413–417.

Holmgren, J., Lonnroth, I., and Svennerholm, L. (1973). Tissue receptor for cholera exotoxin: postulated structure from studies with G_{M1}-ganglioside and related glycolipids. *Infect. Immun. 8*: 208–214.

Holmgren, J., Lindblad, M., Fredman, P., Svennerholm, L., and Myrvold, H. (1985). Comparison of receptors for cholera and *Escherichia coli* enterotoxins in human intestine. *Gastroenterology 89*: 27–35.

Hunt, P. D., and Hardy, S. J. S. (1991). Heat-labile enterotoxin can be released from *Escherichia coli* cells by host intestinal factors. *Infect. Immun. 59*: 168–171.

Ichige, A., Oishi, K., and Mizushima, S. (1988). Isolation and characterization of mutants of a marine *Vibrio* strain that are defective in the secretion extracellular proteins. *J. Bacteriol. 170*: 3537–3542.

Iida, T., Tsuji, T., Honda, T., Miwatani, T., Wakabayashi, S., Wada, K., and Matsubura, H. (1989). A single amino acid substitution in B subunit of *Escherichia coli* enterotoxin affects its oligomer formation. *J. Biol. Chem. 264*: 14065–14070.

Inoue, T., Tsuji, T., Koto, M., Imamura, S., and Miyama, A. (1993). Amino acid sequence of heat-labile enterotoxin from chicken enterotoxigenic *Escherichia coli* is identical to that of human strain H10407. *FEMS Microbiol. Lett. 108*: 157–162.

Ito, K. (1992). SecY and integral membrane components of the *Escherichia coli* protein translocation system. *Mol. Microbiol. 6*: 2423–2428.

Ito, K., and Beckwith, J. (1981a). Protein localization in *E. coli*: is there a common step in the secretion of periplasmic and outer-membrane proteins? *Cell 24*: 707–717.

Ito, K., and Beckwith, J. R. (1981b). Role of the mature protein sequence of maltose binding protein in its secretion across the *E. coli* cytoplasmic membrane. *Cell 25*: 143–150.

Jackson, M. P., Neill, R. J., O'Brien, A. D., Holmes, R. K., and Newland, J. W. (1987). Nucleotide sequence analysis and comparison of the structural genes for Shiga-like toxin I and Shiga-like toxin II encoded by bacteriophages from *Escherichia coli*. *FEMS Lett. 44*: 109–114.

Jaenicke, R. (1987). Folding and association of proteins. *Prog. Biophys. Mol. Biol. 49*: 117–237.

Jaenicke, R. (1991). Protein folding: local structures, domains, subunits and assemblies. *Biochemistry 30*: 3147–3161.

Janicot, M., Fouque, F., and Desbouquois (1991). Activation of rat liver adenylate cyclase by cholera toxin requires internalization and processing in endosomes. *J. Biol. Chem. 226*: 12858–12865.

Jiang, B., and Howard, S. P. (1992). The *Aeromonas hydrophila exeE* gene, required for both protein secretion and normal outer membrane biogenesis, is a member of the general secretion pathway. *Mol. Microbiol. 6*: 1351–1361.

Jobling, M. G., and Holmes, R. K. (1991). Analysis of structure and function of the B-subunit of cholera toxin by use of site-directed mutagenesis. *Mol. Microbiol. 5*: 1755–1767.

Jobling, M. G., and Holmes, R. K. (1992). Fusion proteins containing the A2 domain of cholera toxin assemble with B polypeptides of cholera toxin to form immunoreactive and functional holotoxin-like chimeras. *Infect. Immun. 60*: 4915–4924.

Josefsson, L.-G., and Randall, L. L. (1981). Processing *in vivo* of precursor maltose-binding protein in *Escherichia coli* occurs post-translationally as well as co-translationally. *J. Biol. Chem. 256*: 2504–2507.

Kamitani, S., Akiyama, Y., and Ito, K. (1992). Identification and characterization of an *Escherichia coli* gene required for the formation of correctly folded alkaline phosphatase, a periplasmic enzyme. *EMBO J. 11*: 57–62.

Kassis, S., Hagmann, J., Fishman, P. H., Chang, P. P., and Moss, J. (1982). Mechanism of action of cholera toxin on intact cells: generation of A1 peptide and activation of adenylate cyclase. *J. Biol. Chem. 257*: 12148–12152.

King, C. A., and van Heyningen, W. E. (1973). Deactivation of cholera toxin by sialidase-resistant monosialosylganglioside. *J. Infect. Dis. 127*: 639–647.

Koch, R. (1884). An address on cholera and its bacillus. *Br. Med. J. 2*: 403–407.

Koshland, D., and Botstein, D. (1982). Evidence for post-translational translocation of β-lactamase across the bacterial inner membrane. *Cell 30*: 893–902.

Kumamoto, C. A. (1989). *Escherichia coli* SecB protein associates with exported protein precursors *in vivo*. *Proc. Natl. Acad. Sci. U.S.A. 86*: 5320–5324.

Kumamoto, C. A. (1991). Molecular chaperones and protein translocation across the *Escherichia coli* inner membrane. *Mol. Microbiol. 5*: 19–22.

Kumamoto, C. A., and Beckwith, J. (1983). Mutations in a new gene, *secB*, cause defective protein localization in *Escherichia coli*. *J. Bacteriol. 154*: 254–260.

Kumamoto, C. A., and Beckwith, J. (1985). Evidence for specificity at an early stage in protein export in *Escherichia coli*. *J. Bacteriol. 163*: 267–274.

Kunkel, S. L., and Robertson, D. C. (1979). Purification and chemical characterization of the heat-labile enterotoxin produced by enterotoxigenic *Escherichia coli*. *Infect. Immun. 25*: 586–596.

Kurosky, A., Markel, D. E., and Peterson, J. W. (1977). Covalent structure of the β chain of cholera enterotoxin. *J. Biol. Chem. 252*: 7257–7256.

Kusukawa, N., Yura, T., Ueguchi, C., Akiyama, Y., and Ito, K. (1989). Effects of mutations in heat-shock genes groES and groEL on protein export in *Escherichia coli*. *EMBO J. 8*: 3517–3521.

Lai, C.-Y. (1977). Determination of the primary structure of cholera toxin B subunit. *J. Biol. Chem.* 252: 7249–7256.

Lang, K., Schmid, F. X., and Fischer, G. (1987). Catalysis of protein folding by prolyl isomerase. *Nature (Lond.)* 329: 268–270.

Langer, T., Lu, C., Echols, H., Flanagan, J., Hayer, M. K., and Hartl, F.-U. (1992). Successive action of DnaK, DnaJ and GroEL along the pathway of chaperone-mediated protein folding. *Nature (Lond.)* 356: 683–689.

Lecker, S. H., Lill, R., Ziegelhoffer, T., Georgopoulos, C., Bassford, P. J., Kumamoto, C. A., and Wickner, W. (1989). Three pure chaperones of *Escherichia coli*—SecB, trigger factor and GroEL—form soluble complexes with precursor proteins *in vitro*. *EMBO J.* 8: 2703–2709.

Leece, R., and Hirst, T. R. (1992). Expression of the B subunit of *Escherichia coli* heat-labile enterotoxin in a marine *Vibrio* and in a mutant that is pleiotropically defective in the secretion of extracellular proteins. *J. Gen. Microbiol.* 138: 719–724.

Lencer, W. I., Delp, C., Neutra, M. R., and Madera, J. L. (1992). Mechanism of action on a polarized human epithelial cell line: role of vesicular traffic. *J. Cell. Biol.* 117: 1197–1209.

Leong, J., Vinal, A. C., and Dallas, W. S. (1985). Nucleotide sequence comparison between heat-labile toxin B-subunit cistrons from *Escherichia coli* of human and porcine origin. *Infect. Immun.* 48: 73–77.

Levine, M. M., Kaper, J. B., Black, R. E., and Clements, M. L. (1983). New knowledge on the pathogenesis of bacterial enteric infections as applied to vaccine development. *Microbiol. Rev.* 47: 510–550.

Lill, R., Cunningham, K., Brundage, L. A., Ito, K., Oliver, D., and Wickner, W. (1989). Sec A protein hydrolyzes ATP and is an essential component of the protein translocation ATPase of *Eschericia coli*. *EMBO J.* 8: 961–966.

Lill, R., Dowhan, W., and Wickner, W. (1990). The ATPase activity of SecA is regulated by acidic phospholipids SecY and the leader and mature domains of precursor proteins. *Cell* 60: 271–280.

Liu, G., Topping, T. B., and Randall, L. L. (1989). Physiological role during export for the retardation of folding by the leader peptide of maltose-binding protein. *Proc. Natl. Acad. Sci. U.S.A.* 86: 9213–9217.

Liu, J., and Walsh, C. T. (1990). Peptidyl-prolyl *cis-trans*-isomerase, a periplasmic homologue of cyclophilin that is not inhibited by cyclosporin A. *Proc. Natl. Acad. Sci. U.S.A.* 87: 4028–4032.

Lockman, H., and Kaper, J. B. (1983). Nucleotide sequence analysis of the A2 and B subunits of *Vibrio cholerae* enterotoxin. *J. Biol. Chem.* 258: 13722–13726.

Lockman, H. A., Galen, J. E., and Kaper, J. B. (1984). *Vibrio cholerae* enterotoxin genes: nucleotide sequence of DNA encoding ADP-ribosyl transferase. *J. Bacteriol.* 159: 1086–1089.

Lutcke, H., High, S., Romisch, K., Ashford, A. J., and Dobberstein, B. (1992). The methionine-rich domain of the 54 kDa subunit of signal recognition particle is sufficient for the interaction with signal sequences. *EMBO J.* 11: 1543–1551.

Martin, J. L., Bardwell, J. C. A., and Kuriyan, J. (1993). Crystal structure of the DsbA protein required for disulphide bond formation *in vivo*. *Nature (Lond.)* 365: 464–468.

Mekalanos, J. J. (1983). Duplication and amplification of toxin genes in *Vibrio cholerae*. *Cell* 35: 253–263.

Mekalanos, J. J. (1992). Environmental signal controlling expression of virulence determinants. *J. Bacteriol.* 174: 1–7.

Mekalanos, J. J., Swartz, D. J., Pearson, G. D. N., Harford, N., Groyne, F., and deWilde, M. (1983). Cholera toxin genes: nucleotide sequence deletion analysis and vaccine development. *Nature (Lond.)* 306: 551–557.

Menzel, R., and Gellert, M. (1987). Fusions of the *Escherichia coli* gyrA and gyrB control regions to galactokinase gene are inducible by coumermycin treatment. *J. Bacteriol.* 171: 6206–6212.

Michels, T., Vanooteghen, J.-C., Lambert de Rouvroit, C., China, B., Gustin, A., Boudry, P., and Cornelis, G. R. (1991). Analysis of *virC*, an operon involved in the secretion of YOP proteins by *Yersinia enterocolitica*. *J. Bacteriol. 173*: 4994–5009.

Miller, V. L., and Mekalanos, J. J. (1984). Synthesis of cholera toxin is positively regulated at the transcriptional level by *toxR*. *Proc. Natl. Acad. Sci. U.S.A. 81*: 3471–34575.

Miller, V. L., and Mekalanos, J. J. (1988). A novel suicide vector and its use in construction of insertion mutation: osmoregulation of outer membrane proteins and virulence determinants in *Vibrio cholerae* requires *toxR*. *J. Bacteriol. 170*: 2575–2583.

Miller, V. L., Taylor, R. K., and Mekalanos, J. J. (1987). Cholera toxin transcriptional activator ToxR is a transmembrane DNA binding protein. *Cell 48*: 271–279.

Miller, V. L., DiRita, V. J., and Mekalanos, J. J. (1989). Identification of *toxS*, a regulatory gene whose product enhances ToxR-mediated activation of the cholera toxin promoter. *J. Bacteriol. 171*: 1288–1293.

Minsky, A., Summers, R. G., and Knowles, J. R. (1986). Secretion of β-lactamase into the periplasm of *Escherichia coli*: evidence for a distinct release step associated with a conformational change. *Proc. Natl. Acad. Sci. U.S.A. 83*: 4180–4184.

Missiakis, D., Georgopoulos, C., and Raina, S. (1993). Identification and characterization of the *Escherichia coli* gene *dsbB*, whose gene product is involved in the formation of disulfide bonds *in vivo*. *Proc. Natl. Acad. Sci. U.S.A. 90*: 7084–7088.

Missiakis, D., Georgopoulos, C., and Raina, S. (1994). The *Escherichia coli dsbC* (*xprA*) gene encodes a periplasmic protein involved in disulphide bond formation. *EMBO J.* (in press).

Mo, Y.-Y., and Gross, D. C. (1991). Plant signal molecules activate the *syrB* gene, which is required for syringomycin production by *Pseudomonas syringae* pv. syringae. *J. Bacteriol. 173*: 5784–5792.

Montesano, R., Roth, J., Robert, A., and Orci, L. (1982). Non-coated membrane invaginations are involved in binding and internalization of cholera and tetanus toxins. *Nature (Lond.) 296*: 651–653.

Moon, H. W., and Bunn, T. O. (1993). Vaccines for preventing enterotoxigenic *Escherichia coli* infections in farm animals. *Vaccine 11*: 213–220.

Morioka-Fujimoto, K., Marumoto, R., and Fukuda, T. (1991). Modified enterotoxin signal sequences increase secretion level of the recombinant human epidermal growth factor in *Escherichia coli*. *J. Biol. Chem. 266*: 1728–1732.

Moss, J., and Richardson, S. H. (1978). Activation of adenylate cyclase by *Escherichia coli* enterotoxin. Evidence for ADP-ribosyltransference activity similar to that of choleragen. *J. Clin. Invest. 62*: 281–285.

Moss, J., and Vaughan, M. (1977). Mechanism of choleragen: evidence for ADP-ribosyltransferase activity with arginine as an acceptor. *J. Biol. Chem. 252*: 2455–2457.

Moss, J., and Vaughan, M. (1991). Activation of cholera toxin and *Escherichia coli* heat-labile enterotoxins by ADP-ribosylation factors, a family of 20kDa guanine nucleotide binding proteins. *Mol. Microbiol. 5*: 2621–2627.

Murzin, A. (1993). OB (oligonucleotide/oligosaccharide binding)-fold: common structural and functional solution for nonhomologous sequences. *EMBO J. 12*: 861–867.

Nashar, T. O., Amin, T., Marcello, A., and Hirst, T. R. (1993). Current progress in the development of the B subunits of cholera toxin and *Escherichia coli* heat-labile enterotoxin as carriers for the oral delivery of heterologous antigens and epitopes. *Vaccine 11*: 235–240.

Neill, R. J., Ivins, B. E., and Holmes, R. K. (1983). Synthesis and secretion of the plasmid-encoded heat-labile enterotoxin of *Escherichia coli* in *Vibrio cholerae*. *Science 221*: 289–291.

Nishiyama, K., Hanada, M. and Tokuda, H. (1994). Disruption of the gene encoding P12 (*secG*) reveals the direct involvement and important function of SecG in protein translocation of *Escherichia coli* at low temperature. *EMBO J. 13*: 3272–3277.

Normark, S., Bergstrom, S., Edlund, T., Grundstrom, T., Jaurin, B., Lindberg, F. P. L., and Olsson, O. (1984). Overlapping genes. *Annu. Rev. Genet. 17*: 499–525.

Normington, K., Kohno, K., Kozutsumi, Y., Gething, M.-J., and Sambrook, J. (1989). *S. cerevisiae* encodes an essential protein homologous in sequence and function to mammalian BiP. *Cell 57*: 1223–1236.

Oliver, D. B., and Beckwith, J. (1982). Identification of a new gene (*secA*) and gene product involved in the secretion of envelope proteins. *J. Bacteriol. 150*: 686–691.

Oliver, D. B. (1993). SecA protein: autoregulated ATPase catalyzing preprotein insertion and translocation across the *Escherichia coli* inner membrane. *Mol. Microbiol. 7*: 159–165.

Oppenheim, D. S., and Yanofsky, C. (1980). Translational coupling during expression of the tryptophan operon of *Escherichia coli*. *Genetics 95*: 785–795.

Overbye, L. J., Sandkvist, M., and Bagdasarian, M. (1993). Genes required for extracellular secretion of enterotoxin are clustered in *Vibrio cholerae*. *Gene 132*: 101–106.

Palva, E. T., Hirst, T. R., Hardy, S. J. S., Holmgren, J., and Randall, L. L. (1981). Synthesis of a precursor to the B subunit of heat-labile enterotoxin in *Escherichia coli*. *J. Bacteriol. 146*: 325–330.

Park, S., Liu, G., Topping, T. B., Cover, W. H., and Randall, L. L. (1988). Modulation of folding pathways of exported proteins by the leader sequence. *Science 239*: 1033–1035.

Pearson, G. D. N., and Mekalanos, J. J. (1982). Molecular cloning of *Vibrio cholerae* enterotoxin genes in *Escherichia coli* K-12. *Proc. Natl. Acad. Sci. U.S.A. 79*: 2976–2980.

Pearson, G. D. N., Woods, A., Chiang, S. L., and Mekalanos, J. J. (1993). Ctx genetic element encodes a site-specific combination system and an intestinal colonization factor. *Proc. Natl. Acad. Sci. U.S.A. 90*: 3750–3754.

Peek, J. A., and Taylor, R. K. (1992). Characterization of a periplasmic thiol:disulphide interchange protein required for the functional maturation of secreted virulence factors of *Vibrio cholerae*. *Proc. Natl. Acad. Sci. U.S.A. 89*: 6210–6214.

Pelham, H. R. B. (1991). Recycling of proteins between the endoplasmic reticulum and Golgi complex. *Curr. Opin. Cell Biol. 3*: 585–589.

Phillips, G. J., and Silhavy, T. J. (1990). Heat-shock proteins DnaK and GroEL facilitate export of LacZ hybrid proteins in *E. coli*. *Nature (Lond.) 344*: 882–884.

Pickett, C. L., Weinstein, D. L., and Holmes, R. K. (1987). Genetics of type IIa heat-labile enterotoxin of *Escherichia coli* operon fusions, nucleotide sequence, and hybridisation studies. *J. Bacteriol. 169*: 5180–5187.

Pogliano, K. T., and Beckwith, J. (1994). Genetic and molecular characterization of the *Escherichia coli secD* operon and its products. *J. Bacteriol. 176*: 804–814.

Poritz, M. A., Strub, K., and Walter, P. (1988). Human SRP RNA and *E. coli* 4.5S RNA contain a highly homologous domain. *Cell 55*: 4–6.

Poritz, M. A., Bernstein, H. D., Strub, K., Zopf, D., Wilhelm, H., and Walter, P. (1990). An *E. coli* ribonucleoprotein containing 4.5S RNA resembles mammalian signal recognition particle. *Science 250*: 1111–1117.

Possot, O., d'Enfert, C., Reyss, I., and Pugsley, A. P. (1992). Pullulanase secretion in *Escherichia coli* K-12 requires a cytoplasmic protein and a putative polytopic cytoplasmic membrane protein. *Mol. Microbiol. 6*: 95–105.

Pugsley, A. P. (1992). Translocation of a folded protein across the outer membrane in *Escherichia coli*. *Proc. Natl. Acad. Sci. U.S.A. 89*: 12058–12062.

Pugsley, A. P. (1993). The complete general secretion pathway in gram-negative bacteria. *Microbiol. Rev. 57*: 50–108.

Pugsley, A. P., d'Enfert, C., Reyss, I., and Kornacker, M. G. (1990). Genetics of extracellular protein secretion by gram negative bacteria. *Annu. Rev. Genet. 24*: 67–90.

Pugsley, A. P., Kornacker, M. G., and Poquet, I. (1991). The general secretion pathway is directly required for extracellular pullulanase secretion in *Escherichia coli*. *Mol. Microbiol. 5*: 343–352.

Randall, L. L. (1983). Translocation of domains of nascent periplasmic proteins across the cytoplasmic membrane is independent of elongation. *Cell 33*: 231–240.

Randall, L. L. (1992). Peptide binding by chaperone SecB: implications for recognition by nonnative structure. *Science 257*: 341–345.

Randall, L. L., and Hardy, S. J. S. (1989). Unity in function in the absence of consensus in sequence: role of the leader peptides in export. *Science 243*: 1156–1159.

Randall, L. L., Hardy, S. J. S., and Thorn, J. R. (1987). Export of proteins: a biochemical view. *Annu. Rev. Microbiol. 41*: 507–541.

Rappuoli, R., and Pizza, M. (1991). Structure and evolutionary aspects of ADP-ribosylating toxins. In *Sourcebook of Bacterial Protein Toxins*, J. E. Alouf and J. Freer (Eds.), Academic, London, pp. 1–22.

Ribes, V., Romisch, K., Giner, A., Dobberstein, B., and Tollervey, B. (1990). *E. coli* 4.5S RNA is part of a ribonucleoprotein particle that has properties related to signal recognition particle. *Cell 63*: 591–600.

Romisch, K., Webb, J., Herz, J., Prehn, S., Frank, R., Vingron, M., and Dobberstein, B. (1989). Homology of 54K protein of signal-recognition particle proteins with putative GTP-binding domains. *Nature (Lond.) 340*: 478–482.

Ruiz-Palacious, G. M., Torres, J., Torres, N. I., Escamilla, E., Ruiz-Palacious, B. R., and Tamayo, J. (1983). Cholera-like enterotoxin produced by *Campylobacter jejuni. Lancet ii*: 250–252.

Sanchez, J., and Holmgren, J. (1989). Recombinant system for overexpression of cholera toxin B subunit in *Vibrio cholerae* as a basis for vaccine development. *Proc. Natl. Acad. Sci. U.S.A. 86*: 481–485.

Sandkvist, M. (1992). Assembly and secretion of *E. coli* heat-labile enterotoxin. Ph.D. thesis, University of Umea, Umea.

Sandkvist, M., and Bagdasarian, M. (1993). Suppression of temperature assembly mutants of heat-labile enterotoxin B subunits. *Mol. Microbiol. 10*: 635–645.

Sandkvist, M., Hirst, T. R., and Bagdasarian, M. (1987). Alterations at the carboxyl terminus change the assembly and secretion properties of the B subunit of *Escherichia coli* heat-labile enterotoxin. *J. Bacteriol. 169*: 4570–4576.

Sandkvist, M., Hirst, T. R., and Bagdasarian, M. (1990). Minimal deletion of amino acids from the carboxyl terminus of the B subunit of heat-labile enterotoxin causes defects in its assembly and release from the cytoplasmic membrane of *Escherichia coli. J. Biol. Chem. 265*: 15239–15244.

Sandkvist, M., Morales, V., and Bagdasarian, M. (1993). A protein required for secretion of cholera toxin through the outer membrane of *Vibrio cholerae. Gene 123*: 81–86.

Sanzey, B. (1979). Modulation of gene expression by drugs affecting deoxyribonucleic acid gyrase. *J. Bacteriol. 138*: 40–47.

Schatz, P. J., and Beckwith, J. (1990). Genetic analysis of protein export in *Escherichia coli. Annu. Rev. Genet. 24*: 215–248.

Schatz, P. J., Riggs, P. D., Jacq, A., Fath, M. J., and Beckwith, J. (1989). The *secE* gene encodes an integral membrane protein required for protein export in *Escherichia coli. Genes Dev. 3*: 1035–1044.

Schmidt, M. G., Rollo, E. E., Grodberg, J., and Oliver, D. B. (1988). Nucleotide sequence of the *secA* gene and secA(*ts*) mutations preventing protein export in *Escherichia coli. J. Bacteriol. 170*: 3404–3414.

Schodel, F., and Will, H. (1989). Construction of a plasmid for expression of foreign epitopes as fusion proteins with subunit B of *Escherichia coli* heat-labile enterotoxin. *Infect. Immun. 57*: 1347–1350.

Schonberger, O., Hirst, T. R., and Pines, O. (1991). Targeting and assembly of an oligomeric bacterial enterotoxoid in the endoplasmic reticulum of *Saccharomyces cerevisiae. Mol. Microbiol. 5*: 2663–2672.

Seethaharam, S., Chaudhary, V. K., Fitzgerald, D., and Pastan, I. (1991). Increased cytotoxic activity of *Pseudomonas* exotoxin and two chimeric toxins ending in KDEL. *J. Biol. Chem. 266*: 17376–17381.

Sen, K., and Nikaido, H. (1990). *In vitro* trimerization of OmpF porin secreted by spheroplasts of *Escherichia coli*. *Proc. Natl. Acad. Sci. U.S.A.* 87: 743–747.

Shine, J., and Dalgarno, L. (1974). The 3'-terminal sequence of *Escherichia coli* 16S ribosomal RNA: complementarity to nonsense triplets and ribosome binding sites. *Proc. Natl. Acad. Sci. U.S.A. 71*: 1342–1346.

Shiozuka, K., Tani, K., Mizushima, S., and Tokuda, H. (1990). The proton motive force lowers the level of ATP required for *in vitro* translocation of secretory proteins in *Escherichia coli*. *J. Biol. Chem. 265*: 18843–18847.

Simonen, M., and Palva, I. (1993). Protein secretion in *Bacillus* species. *Microbiol. Rev. 57*: 109–137.

Sixma, T. K., Pronk, S. E., Kalk, K. H., Wartna, E. S., van Zantan, B. A. M., Witholt, B., and Hol, W. G. J. (1991). Crystal structure of cholera toxin-related heat-labile enterotoxin from *E. coli*. *Nature (Lond.) 351*: 371–378.

Sixma, T. K. Pronk, S. E., Kalk, K. H., van Zantan, B. A. M., Berghuis, A. M., and Hol, W. G. J. (1992). Lactose binding to heat-labile enterotoxin revealed by X-ray crystallography. *Nature (Lond.) 355*: 561–564.

Sixma, T., Kalk, K., van Zanten, B., Dauter, Z., Kingma, J., Witholt, B., and Hol, W. (1993). Refined structure of *Escherichia coli* heat-labile enterotoxin, a close relative of cholera toxin. *J. Mol. Biol. 230:* 890–918.

Smith, H. W., and Halls, S. (1967). Observations by the ligated intestinal segment and oral inoculation methods on *Escherichia coli* infections in pigs, calves, lambs and rabbits. *J. Pathol. Bacteriol. 93*: 499.

Smith, H. W., and Linggood, M. A. (1971). The transmissible nature of enterotoxin production in a human enteropathogenic strain of *Escherichia coli*. *J. Med. Microbiol. 4*: 301–305.

So, M., Dallas, W. S., and Falkow, S. (1978). Characterization of an *Escherichia coli* plasmid encoding for synthesis of heat-labile toxin: molecular cloning the toxin determinant. *Infect. Immun. 21*: 405–411.

So, M., Heffron, F., and McCarthy, B. J. (1979). The *E. coli* gene encoding heat-stable enterotoxin is a bacterial transposon flanked by inverted repeats of IS*1*. *Nature (Lond.) 277*: 453–456.

Spicer, E. K., and Noble, J. A. (1982). *Escherichia coli* heat-labile enterotoxin. Nucleoside sequence of the A subunit gene. *J. Biol. Chem. 257*: 5716–5721.

Spira, W. M., and Fedorka-Cray, P. J. (1984). Purification of enterotoxins from *Vibrio mimicus* that appear to be identical to cholera toxin. *Infect. Immun. 45*: 679–684.

Squires, C., and Squires, C. L. (1992). The Clp proteins: proteolysis regulators or molecular chaperones? *J. Bacteriol. 174*: 1081–1085.

Stader, J., Gansheroff, L. J., and Silhavy, T. J. (1989). New suppressors of signal sequence mutations, *prlG*, are linked tightly to the *secE* gene of *Escherichia coli*. *Genes Dev. 3*: 1045–1052.

Steffen, R. (1986). Epidemiologic studies of travellers' diarrhea, severe gastrointestinal infections, and cholera. *Rev. Infect. Dis. 8*: S122–S130.

Stein, P. E., Boodhoo, A., Tyrrell, G. J., Brunton, J. L., and Read, R. J. (1992). Crystal structure of the cell-binding B oligomer of Verotoxin-1 from *E. coli*. *Nature (Lond.) 355*: 748–750.

Streatfield, S. J., Sandkvist, M., Sixma, T. K., Bagdasarian, M., Hol, W. G. J., and Hirst, T. R. (1992). Intermolecular interactions between the A and B subunits of heat-labile enterotoxin from *Escherichia coli* promote holotoxin assembly and stability *in vivo*. *Proc. Natl. Acad. Sci. U.S.A. 89*: 12140–12144.

Strockbine, N. A., Jackson, M. P., Sung, L. M., Holmes, R. K., and O'Brien, A. D. (1988). Cloning and sequencing of the genes for Shiga toxin from *Shigella dysenteriae* type 1. *J. Bacteriol 170*: 1116–1122.

Svennerholm. A.-M., and Holmgren, J. (1978). Identification of *Escherichia coli* heat-labile enterotoxin by means of ganglioside immunosorbent assay (G_{M1}-ELISA) procedure. *Curr. Microbiol. 1*: 19–27.

Swerdlow, D. L., and Mintz, E. D. (1992). Water transmission of epidemic cholera in Truijillo, Peru: lessons for a continent at risk. *Lancet 340*: 28–32.

Takao, T., Watanabe, H., and Shimonishi, Y. (1985). Facile identification of protein sequences by mass spectrometry—B-subunit of *Vibrio cholerae* classical biotype INABA 569B toxin. *Eur. J. Biochem. 146*: 503–508.

Takeda, Y., Honda, T., Taga, S., and Miwantani, T. (1981). In vitro formation of hybrid toxins between subunits of *Escherichia coli* heat-labile enterotoxin and those of cholera enterotoxin. *Infect. Immun. 34*: 341–346.

Taylor, R. K., Miller, V. L., Furlong, D., and Mekalanos, M. M. (1987). The use of *phoA* gene fusions to identify a pilus colonization factor coordinately regulated with cholera toxin. *Proc. Natl. Acad. Sci. U.S.A. 84*: 2833–2837.

Tokunaga, M., Loranger, J., and Wu, H. (1984). Prolipoprotein modification and processing enzymes in *Escherichia coli. J. Biol. Chem. 259*: 3825–3830.

Tomb, J. F. (1992). A periplasmic protein disulphide oxidoreductase is required for transformation of *Haemophilus influenzae* RD. *Proc. Natl. Acad. Sci. U.S.A. 89*: 10252–10256.

Tomb, J.-F., El-Hajj, H., and Smith, H. O. (1991). Nucleotide sequence of a cluster of genes involved in the transformation of *Haemophilus influenzae* Rd. *Gene 104*: 1–10.

Trucksis, M., Galen, J. E., Michalski, J., Fasano, A., and Kaper, J. B. (1993). Accessory cholera enterotoxin (Ace), the third toxin of a *Vibrio cholerae* virulence cassette. *Proc. Natl. Acad. Sci. U.S.A. 90*: 5267–5271.

Tse-Dinh, Y., and Beran, R. K. (1988). Multiple promoters for transcription of the *Escherichia coli* DNA topoisomerase I gene and their regulation by DNA supercoiling. *J. Mol. Biol. 202*: 735–742.

Tsuji, T., Joya, J. E., Yoa, S., Honda, T., and Miwatani, T. (1988). Purification and characterisation of heat-labile enterotoxin isolated from chicken enterotoxigenic *Escherichia coli. FEMS Microbiol. Lett. 52*: 79–84.

Uhlin, B.-E., Dagberg, B., Forsman, K., Goransson, M., Knepper, B., Nilsson, P., and Sonden, B. (1991). Genetics of histone-like protein H-NS/H1 and regulation of virulence determinants in enterobacteria. In *Molecular Pathogenesis of Gastrointestinal Infections*, T. Wadstrom, P. H. Makela, A.-M. Svennerholm, and H. Wolf-Watz (Eds.), Plenum, New York, pp. 55–59.

Van Dop, C., Tsubokawa, M., Bournet, H. R., and Ramachandran, J. (1984). Amino acid sequence of retinal transducin at the site ADP-ribosylated by cholera toxin. *J. Biol. Chem. 259*: 696–698.

van Heyningen, W. E., Carpenter, C. C. J., Pierce, N. F., and Greenough, B. I. (1971). Deactivation of cholera toxin by ganglioside. *J. Infect. Dis. 124*: 415–418.

von Heijne, G. (1985). Signal sequences. The limits of variation. *J. Mol. Biol. 184*: 99–105.

Walker, J. E., Saraste, M., Runswick, M. J., and Gay, N. J. (1982). Distantly related sequences in the α- and β-subunits of ATP-synthase, myosin, kinases and other ATP-requiring enzymes and a common nucleotide binding fold. *EMBO J. 1*: 945–951.

Watanabe, M., and Blobel, G. (1989a). Binding of a soluble factor of *Escherichia coli* to preproteins does not require ATP and appears to be the first step in protein export. *Proc. Natl. Acad. Sci. U.S.A. 86*: 2248–2252.

Watanabe, M., and Blobel, G. (1989b). Cytosolic factor purified from *Escherichia coli* is necessary and sufficient for export of a preprotein and is a heterotetramer of SecB. *Proc. Natl. Acad. Sci. U.S.A. 86*: 2728–2732.

Weinstein, D. L., Jackson, M. P., Samuel, J. E., Holmes, R. K., and O'Brien, A. D. (1988). Cloning and sequencing of a Shiga-like type II from an *Escherichia coli* strain responsible for edema disease in swine. *J. Bacteriol. 170*: 4223–4230.

Weiss, A. A., Johnson, F. D., and Burns, D. L. (1993). Molecular characterization of an operon required for pertussis toxin secretion. *Proc. Natl. Acad. Sci. U.S.A. 90*: 2970–2974.

Wilholt, B., Hofstra, H., Kingma, J., Proak, S. E., Hol, W. G. J., and Drenth, J. (1988). Studies on the synthesis and structure of the heat-labile enterotoxin (LT) of *Escherichia coli*. In *Bacterial Protein Toxins*, B. Witholt, J. E. Alouf, G. J. Boulnois, P. Cossart, B. W.

Dijkstra, P. Falmagne, F. J. Fehrenbach, J. Freer, H. Niemann, R. Rappuoli, and T. Wadstrom (Eds.), Gustav Fischer, Stuttgart, pp. 3–12.

Wolfe, P. B., Wickner, W., and Goodman, J. M. (1983). Sequence of the leader peptidase gene of *Escherichia coli* and the orientation of leader peptidase in the bacterial envelope. *J. Biol. Chem. 258*: 12073–12080.

Wong, K. R., and Buckley, J. T. (1989). Proton motive force involved in protein transport across the outer membrane of *Aeromonas salmonicida*. *Science 246*: 654–656.

Wong, K. R., McLean, D. M., and Buckley, J. T. (1990). Cloned aerolysin of *Aeromonas hydrophila* is exported by a wild-type marine *Vibrio* strain but remains in the periplasm of pleiotropic export mutants. *J. Bacteriol. 172*: 372–376.

Xia, Q.-C., Chang, D., Blacher, R., and Lai, C.-Y. (1984). The primary structure of the COOH-terminal half of cholera toxin A1 containing the ADP-ribosylation site. *Arch. Biochem. Biophys. 234*: 363–370.

Yamada, H., Matsuyama, S.-I., Tokuda, H., and Mitzushima, S. (1989). A high concentration of SecA allows proton motive force-independent translocation of a model secretory protein into *Escherichia coli* membrane vesicles. *J. Biol. Chem. 264*: 18577–18581.

Yamamoto, T., and Yokota, T. (1981). *Escherichia coli* heat-labile enterotoxin genes are flanked by repeated deoxyribonucleic acid sequences. *J. Bacteriol. 145*: 850–860.

Yamamoto, T., and Yokota, T. (1982). Release of heat-labile enterotoxin subunits in *Escherichia coli*. *J. Bacteriol. 150*: 1482–1484.

Yamamoto, T., and Yokota, T. (1983). Sequence of heat-labile enterotoxin of *Escherichia coli* pathogenic for humans. *J. Bacteriol. 155*: 728–733.

Yamamoto, T., Tamura, T., and Yokota, T. (1984). Primary structure of heat-labile enterotoxin produced by *Escherichia coli* pathogenic for humans. *J. Biol. Chem. 259*: 5037–5044.

Yamamoto, T., Suyama, A., Mori, N., Yokota, T., and Wada, A. (1985). Gene expression in the polycistronic operons of *Escherichia coli* heat-labile toxin and cholera toxin: new model of translational control. *FEBS Lett. 181*: 377–380.

Yamamoto, T., Gojobori, T., and Yokota, T. (1987). Evolutionary origin of pathogenic determinants in enterotoxinogenic *Escherichia coli* and *Vibrio cholerae* O1. *J. Bacteriol. 169*: 1352–1357.

Yu, F., Yamada, H., Daishima, K., and Mizushima, S. (1984). Nucleotide sequence of the *lspA* gene, the structural gene for lipoprotein signal peptidase of *Escherichia coli*. *FEBS Lett. 173*: 264–268.

Yu, J., Webb, H., and Hirst, T. R. (1992). A homologue of the *Escherichia coli* DsbA protein involved in disulphide bond formation is required for enterotoxin biogenesis in *Vibrio cholerae*. *Mol. Microbiol. 6*: 1949–1958.

Yu, J., McLaughlin, S., Freedman, R. B., and Hirst, T. R. (1993). Cloning and active site mutagenesis of *Vibrio cholerae* DsbA, a periplasmic enzyme that catalyses disulphide bond formation. *J. Biol. Chem. 268*: 4326–4330.

Yu, J., and Hirst, T. R. (1994). A pleiotropic secretion mutant of *Aeromonas hydrophila* is unable to secrete heterologously expressed *E.coli* enterotoxin: implications for common mechanisms of protein secretion. *Biochem. Soc. Trans. 22*: 34S.

9

Structure and Function of *E. coli* Heat-Labile Enterotoxin and Cholera Toxin B Pentamer

Wim G. J. Hol

Biomolecular Structure Center, and Howard Hughes Medical Institute, University of Washington School of Medicine, Seattle, Washington

Titia K. Sixma

Netherlands Cancer Institute, Amsterdam, The Netherlands

Ethan A. Merritt

Biomolecular Structure Center, University of Washington School of Medicine, Seattle, Washington

I. INTRODUCTION

In 1959 it was recognized for the first time that the major virulence factor of the cholera-causing bacteria *Vibrio cholerae* O1 is a secreted protein, thereafter called cholera toxin (CT) (De, 1959; Dutta et al., 1959). At that time De had already discovered that certain coliform bacteria also can cause a cholera-like disease (De et al., 1956). In the late 1960s it was found that enterotoxigenic *Escherichia coli* (ETEC) produces a toxin (Smith and Halls, 1967) that is immunologically related to cholera toxin (Gyles and Barnum, 1969) and has similar effects (Gorbach et al., 1971). This protein was called heat-labile enterotoxin (LT) and constitutes the major subject of this chapter.

E. coli LT and *V. cholerae* CT are very similar in amino acid sequence, in mode of action, and in biological properties. Cholera toxin was initially purified to homogeneity by Finkelstein and LoSpalluto (1969). Finkelstein and coworkers were also the first to purify LT (Clements and Finkelstein, 1979). Since then a large body of information has been obtained regarding both of these toxins. Cholera toxin has been more extensively studied than LT, and this extra information can often support interpretation of the LT action. Conversely, the details of the LT work can be useful for the cholera toxin field. Hence this chapter discusses the structure–function relationships of both LT and CT.

A. The Two Toxins

Heat-labile enterotoxin and cholera toxin are hexameric AB_5 proteins. The A subunit has 240 amino acids (Table 1), and the five identical B subunits have 103 amino acids each (Table 2). There are some small sequence differences between LT from porcine *E. coli* (p-LT) and that from the two types of human *E. coli* (h-LT-1 and h-LT-2): four amino acids differ in the B subunits (positions 4, 13, 46, and 102) and three in the A subunit (position 4, 213, and 237) [for reviews see Finkelstein et al. (1987) and Finkelstein (1988)]. The B subunit of CT from *V. cholerae* El Tor exhibits sequence variability at positions 18, 25, and 47 (Olsvik et al., 1993); the El Tor and classical biotypes are further distinguished by variation at positions 22 and 70 (review: Finkelstein et al., 1987). The sequence identity of LT with CT is about 80%, and there are no insertions and deletions (Tables 1 and 2).

In *E. coli* the AB_5 toxin remains in the periplasm until released by host intestinal factors (Hirst and Holmgren, 1987a; Hunt and Hardy, 1991). In *V. cholerae*, however, there is an active system for secretion of the mature toxin across the outer cell membrane. This difference is due to the bacterium rather than to structural or sequence differences in the two toxins, as LT is secreted when expressed in *V. cholerae* (Neill et al., 1983; Hirst et al., 1984) and CT is retained in the periplasm when expressed in *E. coli* (Gennaro et al., 1982; Pearson and Mekalanos, 1982).

B. Membrane Binding

1. The Receptor: Ganglioside G_{M1}

The membrane receptor for LT and CT is ganglioside G_{M1}, a glycosphingolipid situated in the outer leaflet of the cell membrane (van Heyningen et al., 1971; Cuatrecasas 1973; reviews: Van Heyningen, 1983; Eidels et al., 1983). G_{M1} is distinguished by the characteristic pentasaccharide attached to the ceramide head group and extending from the cell surface (Fig. 1). The solution structure of the G_{M1} pentasaccharide has been determined by NMR methods (Sabesan et al., 1984; Acquotti et al., 1990). Small-angle

TABLE 1 Alignment of A Subunit Sequences for LT-I, CT, Pertussis Toxin, and LT-II[a]

```
            |----β1---|      |----α1----|  |β2|                              |---α2---|
                       9              19          29              39              49
Porcine LT-I    - - N Q D R L Y R A D S R P P D E I K R S G G L M P R Q H N E Y F D R G T Q M N I N L Y D H A R G T Q T G F V R Y D D Q
Human LT-I      - - . . K . . . . . . . . . . . . . . . . . . . . . . . . . . . . . . . . . . . . . . . . . . . . . . . . . . . . . . .
Cholera toxin   - - . D . K . . . . . . . . . . . . . Q . . . . . . . Q S . . . . . . . . . . . . . . . . . . . . . . . . . . H . . .
Human LT-IIA    - - . D - - F F . . . . . T . . . . R . A . . . L . . . Q Q . A Y E . . . P I . . . . . E . . . . . V . . N T . . N . .
Human LT-IIB    - - . D - - Y F . . . . . T . . . V R . . . . I . . . Q D . A Y E . . . P I . . . . . D . . . . A . N T . . N . .
Pertussis toxin D D P P A T V . . Y . . . . . E D V F Q N . F T A W G N - - - - - - - - - - - - - D . V L . . L T . R S C Q V G S S N S A
```

```
            |--β3---|    |--------α3-------|                        |------β4-----|
            59              69          77              77                  |87              91
Porcine LT-I    Y V S T S L S L R S A H L A G Q S I L - - - - - - - - - - - - - S G Y S T Y Y I Y V I A T A - - - - - - - - - - - - - - - P
Human LT-I      . . . . . . . . . . . . . . . . . . . - - - - - - - - - - - - - . . . . . . . . . . . . . . . - - - - - - - - - - - - - - - .
Cholera toxin   . . . . . I . . . . . . V . . T . . . - - - - - - - - - - - - - . . . H . . . . . . . . . . . - - - - - - - - - - - - - - - .
Human LT-IIA    . . . T V T . . Q . . . I . . N - - - - - - - - - - - - - Q S . N E . . . . V . P . - - - - - - - - - - - - - - - .
Human LT-IIB    . . . T T T . . Q . . . L . . N M - - - - - - - - - - - - - Q Q . N E . . . . V . A . - - - - - - - - - - - - - - - .
Pertussis toxin F . . . . . S . R . Y T E V Y L E H R M Q S A V E A E R A G R G T . H F I Q . . . E V R A D N N F Y G A A S G Y P E Y V D
```

```
            |-β5-|-α4---|-α5-|    |-α6-|  |--β6---|    |β7|-α7-|  |-----β8------|    |β9-|  |β10|      |---α8----|
            93              103              113              123              133              143
Porcine LT-I    N M F N V N D V L G V Y S P H P Y E Q E V S A L G G I P Y S Q I Y G W Y R V N F G V I D E R L H R N R E Y R D R Y Y R N
Human LT-I      . . . . . . . . . . . . . . . . . . . . . . . . . . . . . . . . . . . . . . . . . . M . . . . . . . . . . . . . . . .
Cholera toxin   . . . . . . . . . A . . . . D . . . . . . . . . . . . . . . . . . . . H . . L . . Q . . . . . G . . . . . . S . .
Human LT-IIA    . L . D . . G . . . R . . . . Y . S . N . F A . . . . . . L . . I . . . . S . . A . E G G M Q . . . . D . . G D L P . G
Human LT-IIB    . L . D . . G . . . R . . . Y . S . N . Y A . . . . . . L . . I . . . . S . . A . E G G M . . . . D . . R D L F . G
Pertussis toxin T Y G D N A G R I L A G A L A T . Q S . Y L . H R R . . P E N . R R V T R V Y H N G . T G E T T T T E Y S N A R . V S Q
```

```
            |-α9-|                  |-α10--|  |-α11--|                              |-------------------------|
            153              163              173              183              193              203
Porcine LT-I    L N I A P A E D G Y R L A G F P P D H Q A W R E E P W I E H A P Q G C G N S S R T I T G D T C N E E T Q N L S T I Y L R
Human LT-I      K . . . . . . . . . . . . . . . . . . . . . . . . . . . . . . . . . . . . . . . . . . . . . . . . . N . . . . . . . .
Cholera toxin   . D . . . A . . . G . . . E . R . . . . . . . . . . . . . . P . . . . P . S S M S N . . D . K . . S . G V K F . D
Human LT-IIA    . T V . . N . . . . Q . . . . S N F P . . . . . M . . S T F . . E Q . V P N N K E F K . G V . I S A . N V . . K Y D . M
Human LT-IIB    . S A . . N . . . . . I . . . D G F P . . B . V . . R E F . . N S . L P N N K A S S D T . . A S L . N K . . Q H D . A
Pertussis toxin Q T R . N P N P Y T S R R S V A S I V G T L V R K A P V I G A - - - - - - - - - - - - - - - - M A R Q A E S . E A M A A
```

```
            ----A2|α1------|  |A2|α2|      |-A2|α3--|
            213              223              233
Porcine LT-I    E Y Q S K V K R Q I F S D Y Q S E V D I Y N R I - - - - - R D E L
Human LT-I      K . . . . . . . . . . . . . . . . . . . . . . . . . . . . N . . .
Cholera toxin   . . . . . . . . . . . . Q . . D I . T H . . . . . . . K . . . .
Human LT-IIA    N F K K L L . . R L A L T F F M S E . D F I G V - - H G E . . . .
Human LT-IIB    D F K K Y I . . K F T L M T L L S I N N D G F F S N N G Q K . . . .
Pertussis toxin W S E R A G E A M V L V Y . E . I A Y S F - - - - - - - - - - - -
```

[a]Sequences are given for pLT-I (Dykes et al., 1985; Yamamoto et al., 1987), hLT-I (Yamamoto et al., 1984), CT (Mekalanos et al., 1983; Lockman et al., 1984; Lockman and Kaper, 1983), LT-IIa (Picket et al., 1987), LT-IIb (Picket et al., 1989), and pertussis toxin (Nicosia et al., 1986; Locht and Keith, 1986).

X-ray diffraction studies on model membranes have shown that the G_{M1} pentasaccharide extends to a distance of 21 Å from the hydrocarbon–water surface (McDaniel and McIntosh, 1986). On model membranes the gangliosides have a random distribution (Thompson et al., 1985) and a distinct lateral flexibility (Goins et al., 1986; Song and Rintoul, 1989; Reed et al., 1987). High local concentrations of G_{M1} can induce a destabilization of the membrane (Thompson and Tillack, 1985).

2. *Saccharide Binding: Affinity and Selectivity*

Cholera toxin binding is quite specific for the pentasaccharide of G_{M1}, and, conversely, cells lacking this ganglioside cannot bind the toxin. Addition of G_{M1} can, however, induce sensitivity in cells initially lacking the receptor (Fishman et al., 1976; review: van Heyningen, 1983; Moss and Vaughan, 1988). For LT the specificity is less stringent; it binds to several glycoproteins (Critchley et al., 1981; Holmgren et al., 1982; Donta et al., 1982) in addition to the G_{M1} ganglioside (Osborne et al., 1982). The different binding affinities of LT and CT could be due mainly to better binding of galactose alone by LT (Clements and Finkelstein, 1979; Merritt et al., 1994b). In that case most glycoprotein binding noticed so far is likely to be due to binding of galactoproteins (Critchley et al., 1981). This may also be the reason why LT binds weakly to asialo-G_{M1}

TABLE 2 Alignment of B Subunit Sequences[a]

	|---α1---|	|-----β1-----|	|---β2--|	|---β3--|	|--β4--|	|--
	1	11	21	31	41	51
Porcine LT	A P Q T I T E L C S E	Y R N T Q I Y T I N D	K I L S Y T E S M A	G K R E M V I I T F K S	G E T F Q V E V P G	S Q H I D S
Human LT-1	. . . S	. . . . . . H	. . . .	. . . .	. . A	. . . . . .
Human LT-2	. . . S	. . . .	. . . .	. . . .	. . A	. . . . . .
CT-1a	T . . M . . D . . A .	. H . . . . H . L . N	. . F L	. . . . . . A	. N . A	. . . . . .
CT-1b	T . . M . . D . . A .	. H . . . . H . L . N	. . F L	. . . . . . A	. N . A . . E . . .	. . . . . .
CT-3	T . . M . . D . . A .	. H . . . . . . L . .	. . F L	. . . . . . A	. N . A I	. . . . . .
CT-4	T . . M . . D . . A .	. H . . . . . . L . .	. . F L	. . . . . . A	. N . A I	. . S . . . . . .

	|------α2----------------------|	|----β5-----|	|------β6------|
	61	71	81	91	101
Porcine LT	Q K K A I E R M K D T L R I T Y L T E T K	I D K L C V W N N K T P N	S I A A I S M K N
Human LT-1	. . . .	. . . .	. . . . . E . .
Human LT-2	. . . .	. . . .	. . . . . E . .
CT-1a	. . . . M . . . A . . . . A . V E	. . . . H A .	. . . . A . . .
CT-1b	. . . . M . . . A . . . . A . V E	. . . . H A .	. . . . A . . .
CT-3	. . . . A . . . . A . V E	. . . . H A .	. . . . A . . .
CT-4	. . . . A . . . . A . V E	. . . . H A .	. . . . A . . .

[a]Sequences are given for pLT-I (Dallas and Falkow, 1980; Leong et al., 1985; Yamamoto et al., 1984; Tsuji et al., 1984), hLT-I (Leong et al., 1985). hLT-II (Yamamoto and Yokota, 1983; Yamamoto et al., 1987); classical cholera strains CT-1a (Kurosky et al., 1977) and CT-1b (Lai, 1977); and El Tor cholera strains CT-3 (Mekalanos et al., 1983), and CT-4 (Lockman and Kaper, 1983).

(Fukuta et al., 1988); that is, tighter binding of galactose compensates for loss of binding interactions due to a second sugar.

A number of different values have been reported for the binding constants of CT to G_{M1} (see van Heyningen, 1983). The dissociation constant of CT for G_{M1} on cells was estimated at 10^{-9}–10^{-10} M (Cuatrecasas, 1973; Sattler et al., 1978). Measurements of concentrations needed to inhibit binding of G_{M1}-coated plastic wells gave an inhibition constant of 5.1×10^{-8} M for G_{M1}. Schengrund and Ringler (1989) found a 100-fold stronger binding constant for the complete G_{M1} ganglioside than for the G_{M1} pentasaccharide alone and attributed this to the fact that the intact gangliosides aggregate into

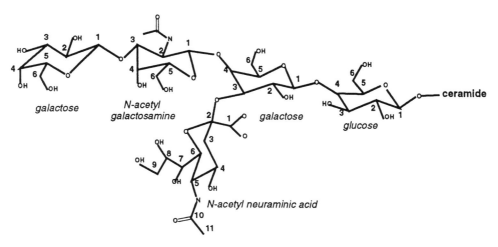

Fig. 1 Schematic representation of G_{M1} ganglioside, the membrane receptor of LT. The characteristic pentasaccharide, Gal($\beta1\rightarrow3$)GalNAc($\beta1\rightarrow4$){ NeuAc($\alpha2\rightarrow3$) } Gal($\beta1\rightarrow4$)Glc, has a negative charge due to the carboxylate group on the sialic acid (NeuAc moiety).

multimeric systems whereas the pentasaccharides remain single entities in solution. Isothermal titration calorimetry of the saccharide portion alone gave an association constant of 1.05×10^6 M^{-1} at 37°C (Schön and Freire, 1989), also weaker than that for the G_{M1} ganglioside in solution, which was determined at 1.9×10^7 M^{-1} (Masserini et al., 1992). However, interaction of the toxins with the fatty acid component of G_{M1} is not specific. The lipid tail can be replaced by an acetyl group without detrimental effect on the binding (Fishman et al., 1980). Replacement of the ceramide by another lipid group even provides better binding in some cases (Pacuszka and Fishman, 1990). Hence the pentasaccharide head group of G_{M1} appears to be the major specificity-determining factor.

C. Enzymatic Activity

1. Reaction Catalyzed

The reaction catalyzed by LT and CT is the following (Moss and Vaughan, 1988):

$$\text{NAD}^+ + \text{ACCEPTOR} \rightarrow \text{ADP-ribose-ACCEPTOR} + \text{nicotinamide} + \text{H}^+$$

The involvement of NAD (Fig. 2) in the CT reaction was first shown by Gill (1975). The acceptor can be either water, resulting in an NAD glycohydrolase reaction (Moss et al., 1976), or a guanidinium group (Moss et al., 1977). The guanidinium group can be part of a small molecular compound, or it can be part of an arginine side chain in a protein (Moss and Vaughan, 1977). The most prevalent in vivo reaction is the ADP-ribosylation of an arginine in the α subunit of G_s, the stimulatory trimeric G protein of the adenylate cyclase pathway (Cassel and Pfeuffer, 1978; Gill and Richardson, 1980). The modified arginine residue is thought to be residue Arg-201 by extrapolation from the modifications of the related protein G_t (transducin), which can also be modified by CT (Van Dop et al., 1984; Bourne et al., 1991). After ADP-ribosylation, the G_s loses its intrinsic GTPase activity and remains permanently in its GTP state (Cassel and Pfeuffer, 1978). This causes a continuous stimulation of adenylate cyclase by the activated G_s and an increase of cyclic AMP levels in the cell. The latter presumably leads to the pathogenic effects of the disease, the massive loss of fluids and ions from the cell, although prostaglandin production (Peterson and Ochoa, 1989) and certain calcium ionophore effects have also been implicated (Maenz and Forsyth, 1986; Maenz et al., 1987). An excellent review on the current debate about these aspects of the action of cholera toxin and heat-labile enterotoxin is given by Kaper et al. (1993).

The enzymatic activity of LT and CT is located solely on the A1 fragment of the A subunit (Gill, 1976). For optimal activity of this fragment it needs to be separated from

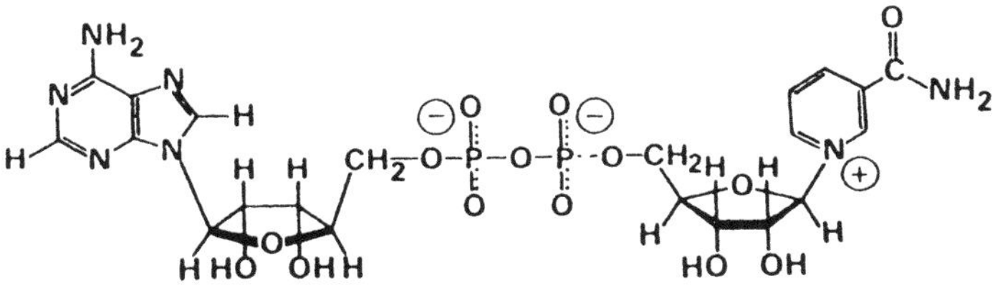

Fig. 2 The structure of β-NAD$^+$, one of the two substrates in the ADP-ribosylation reaction catalyzed by LT and CT. Figure taken from Moss and Vaughan (1988) with permission.

the A2 fragment both by proteolytic cleavage of the main chain and by the reduction of a disulfide bond that links A1 to A2 (Mekalanos et al., 1979b; Moss et al., 1981). The rate of activity after nicking and reduction is similar for LT and CT (Kunkel and Robertson, 1979), or slightly higher for CT than for LT (Moss et al., 1979; Lee et al., 1991).

2.   Binding of Acceptor Compounds

No binding constants have been reported for the protein substrate, but various different small molecular compounds have been studied as acceptors for the enzymatic reaction of LT and CT (Fig. 3). It was shown that the guanidinium group is essential for the ADP-ribose transfer (Moss and Vaughan, 1977; Moss et al., 1979). An additional charged group in the compound seems to be marginally detrimental to the activity, as was seen by the binding constant and the k_{cat}/K_m for the L-arginine methyl ester (LAME, Fig. 3b) ($K_m = 39$ mM) versus the amino acid arginine alone ($K_m = 50$ mM) (Larew et al., 1991). There is no preference for the L- or the D- amino acid in the reaction (Moss and Vaughan, 1977).

An arginine with an ether oxygen instead of the γ-methylene (canaverine, Fig. 3e) gave a reduced binding constant (Larew et al., 1991). This could be due to the lowering of the pK_a of the guanidinium group, which changes from 12.5 for arginine to 7 for canaverine and may indicate that this compound is deprotonated under the assay conditions. This could indicate that the deprotonation of the guanidinium group forms part of the binding process. The best binding constant to CT ($K_m = 44$ μM) has been

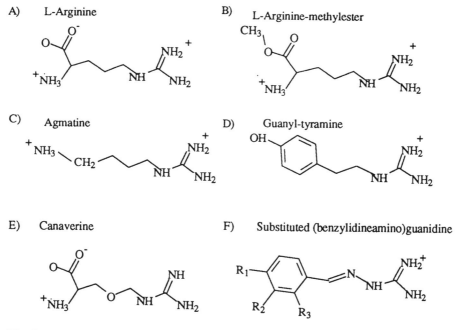

Fig. 3   Various small compounds used as acceptor analogues for LT and CT activity tests. (A) L-arginine; (B) L-arginine methyl ester (LAME) (see, e.g., Larew et al., 1991); (C) agmatine (see, e.g., Osborne et al., 1985; Lee et al., 1991); (D) guanyl-tyramine (Makalanos et al., 1979a); (E) canaverine (Larew et al., 1991); (F) substituted (benzylidineamino)guanidines; R 1, various substi- tutions; R2, R3 usually hydrogen (Soman et al., 1986).

reported for guanyl-tyramine (Fig. 3d) (Mekalanos et al., 1979a), which was also an inhibitor of the reaction with LT (Mekalanos et al., 1979a).

A detailed study of K_m values for CT with a number of similar, highly colored, easily detectable, (benzylidine-amino) guanidine compounds with different substitutions on the benzene group (Fig. 3f) (Soman et al., 1986) resulted in K_m values varying between 0.7 and 5.6 mM and k_{cat}/K_m values ranging between 0.28×10^6 and 1.92×10^6 s^{-1} mol^{-1}.

The stereochemistry for the ADP-ribosylation reaction was studied by Oppenheimer (1978) for CT and by Moss et al. (1979) for LT. These authors showed that β-NAD is converted to α-ADP-ribose, which subsequently anomerizes to a mixture of α- and β-ADP-ribose in solution (Fig. 4).

3. Binding of NAD

NAD binding to cholera toxin has been studied by several groups, who have reported K_m values of 1 mM (Osborne et al., 1985), 4 mM (Moss et al., 1976), and 5.6 mM (Larew et al., 1991). This binding is almost 1000 times weaker than in other ADP-ribosylating toxins such as exotoxin A ($K_m = 2.5$ μM; Lory and Collier, 1980), diphtheria toxin ($K_m = 9$ μM; Lorey et al., 1980), and pertussis toxin ($K_m = 21.6$ μM; Cieplak et al., 1990). Certain fragments of NAD have some inhibitory power. The purine group substituents seem to be important for the binding; AMP shows better inhibition of the intrinsic NAD glycohydrolase activity than GMP, while IMP and CMP show virtually no effect. The measured inhibition values for adenine are somewhat contradictory; Moss et al. (1977) found a value of $K_i = 3$ mM whereas Galloway and van Heyningen (1987) reported no inhibition by adenine and claim that only compounds containing both the adenine and nicotinamide parts of NAD can function as competitive inhibitors. NADP shows a very weak inhibitory power.

The binding constants for both substrates of the reaction follow Michaelis-Menten kinetics. They increase in the presence of the other substrate; for example, K_m for NAD goes from 4 mM to 50 mM for NAD in the presence of artificial small inhibitors (Galloway and van Heyningen, 1987), and K_m for either NAD or agmatine (Fig. 3c) increases by more than 70% in the presence of the other (Osborne et al., 1985). Such behavior of substrates in the presence of inhibitors for the opposite site is indicative of a random sequential mechanism, where either substrate can bind first (Osborne et al., 1985; Larew et al., 1991), although an early report claimed some evidence for an ordered sequential mechanism (Mekalanos et al., 1979b).

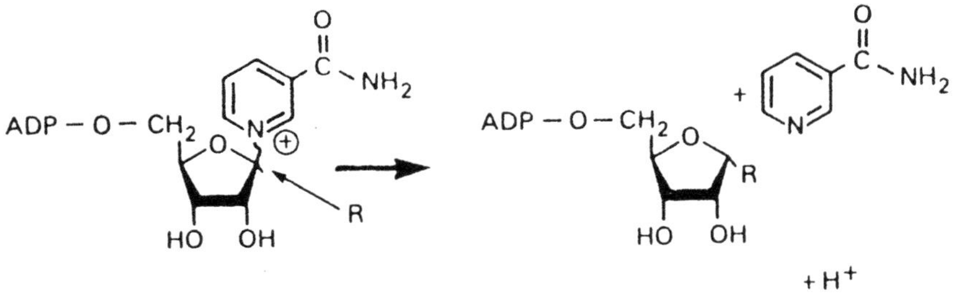

Fig. 4 Stereospecificity of LT/CT-catalyzed ADP-ribosylation of an acceptor R. The β-NAD$^+$ is converted to an α-anomeric product. R is an appropriate guanidinium-containing acceptor (e.g., arginine) or water. Figure taken from Moss and Vaughan (1988) with permission.

4. Involvement of ARFs

LT and CT enzymatic activity in vitro is increased in the presence of ADP-ribosylation factors (ARFs), a class of small GTP-binding proteins (Gill and Meren, 1978; Kahn and Gilman, 1984; Moss et al., 1993; review: Moss and Vaughan, 1991). Several different ARFs have been characterized. The effect is sensitive to the presence of lipids (DMPC/cholate), Mg^{2+}, and detergents and is similar for LT and CT (Lee et al., 1991). It is mediated by direct binding of ARF to A1 (Gill and Coburn, 1987) and was shown to lower the binding constants for the two substrates of the protein (Noda et al., 1990). Not much is known yet about the precise site on the LT/CT A subunit responsible for binding of ARF, but it is apparently obscured prior to activation of the toxin by nicking and disulfide reduction (Moss et al., 1993). It is also not known whether ARF activation occurs in vivo (Moss and Vaughan, 1991). A function in the secretory pathway has recently been found for the class of GTP-binding proteins to which ARF belongs; they assist in the vesicle transport from the Golgi apparatus to the plasma membrane (Serafini et al., 1991; review: Rothman and Orci, 1992). It is therefore an intriguing possibility that ARFs play a double role in intoxication by CT and LT.

II. X-RAY DIFFRACTION STUDIES

Prior to successful determination of the crystal structure of LT, careful and ingenious studies revealed the major characteristics of the LT and CT AB_5 holotoxin assembly: a ring of B subunits to which the A subunit was attached, with a linker fragment (A2) connecting the actual catalytic fragment (A1) to the B pentamer [for a review, see Middlebrook and Dorland (1984)]. As we shall see, the X-ray studies bore out these conclusions and at the same time revealed a molecule of unusual architecture and beauty.

Initial structure determination of heat-labile enterotoxin was far from trivial (Sixma et al., 1991b). Good quality crystals could be grown for a period of only a few months after fresh protein had been purified, and major difficulties were caused by variation in the diffraction pattern of the native crystals. It perhaps suffices to say that the preparation of three isomorphous derivatives of rather poor quality enabled us to obtain phases of equally poor quality, which in turn resulted in a mediocre electron density distribution. Nevertheless, the outline of the molecule became apparent. The presence of a fivefold symmetry axis in the B pentamer proved to be crucial; density averaging of the B-pentamer region dramatically increased the quality of this part of the molecule and allowed the building of a virtually complete molecular model of the B pentamer. The density for the A subunit was, however, still very difficult to interpret. Only by eventual combination of the phase information from the isomorphous derivatives, plus that from fivefold averaging, plus that from a partial molecular model (Sixma et al., 1991a, 1993) were we able to elucidate the three-dimensional structure of the catalytically active A subunit. A complete model of the holotoxin could then be built and crystallographically refined at 1.95 Å resolution (Sixma et al., 1993). The structure amply rewarded the effort required for determination, turning out to have many unique features.

As we had already observed in the early stages of the structure determination effort (Pronk et al., 1987), LT is able to crystallize from a variety of conditions and in a number of different unit cells. With the 1.95 Å structure of the native LT in hand, crystallographic determination of these related structures was relatively straightforward. By the same token, the known structure of LT enabled a fast structure elucidation of the B pentamer of cholera toxin in complex with the receptor G_{M1} pentasaccharide (Merritt et al., 1994a). An

TABLE 3 Crystal Structures for LT and CT

Structure	Resolution (Å)	R factor (%)	Space group	Unit cell a (Å)	b (Å)	c (Å)	β (deg)	Reference
LT	1.95	18.2	$P2_12_12_1$	119.2	98.2	64.8		Sixma et al., 1993
LT (cleaved)	2.6	17.2	$P2_12_12_1$	119.2	98.2	64.8		Merritt et al., 1994c
LT : Sm(NO$_3$)$_3$	3.4	33.7	$P2_12_12_1$	118.8	97.5	65.1		Sixma et al., 1992b
LT	3.4	26.2	$P2_1$	84.1	78.8	123.2	99.5	Sixma et al., 1992c
LT : lactose	2.3	16.4	$P2_12_12_1$	119.8	101.2	64.2		Sixma et al., 1992a
LT : galactose	2.2	17.5	$P2_12_12_1$	70.7	73.5	163.3		Merritt et al., 1994b
LTB (CdCl$_2$)	2.4	16.0	$P2_1$	40.2	94.4	65.4	99.5	Sixma et al., unpublished
CTB : G$_{M1}$ OS	2.2	17.0	$C2$	101.9	67.6	80.5	105.7	Merritt et al., 1994a

overview of the X-ray structures that we have determined up to mid-egg3 is given in Table 3. All this has led to a wealth of structural information that is described in the next few sections.

A. The B Subunit and B Pentamer

The 103 residues of the B subunit, as seen in the B pentamer, adopt a quite simple topology that is schematically depicted in Figure 5. The first 15 residues form a small helix α1, which is connected by a short loop to the strand β1. A second loop brings the

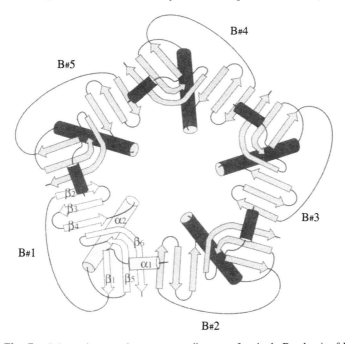

Fig. 5 Schematic secondary structure diagram of a single B subunit of LT (light gray) and the LT B pentamer. The secondary structural elements are α_1, 5–9; β_1,15–22; β_2, 26–30; β_3, 37–41; β_4, 47–50; α_2, 59–78; β_5, 82–88; β_6, 94–102. Figure taken from Sixma et al. (1991a) with permission.

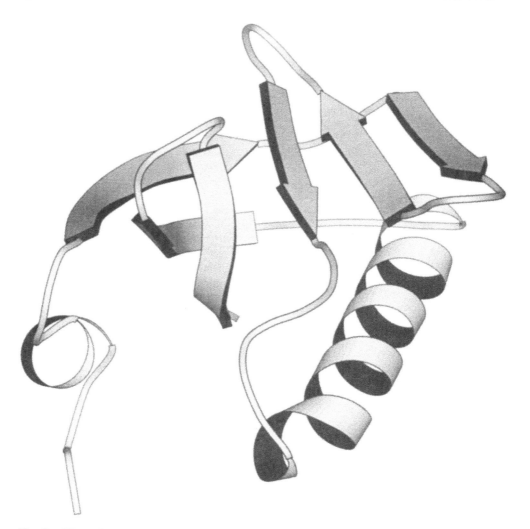

Fig. 6   Three-dimensional representation of secondary structure in a single B subunit of LT.

chain to a three-strand antiparallel beta sheet. Next comes the major helix $\alpha 2$, which in the assembled pentamer forms the inner surface of the central pore. The chain ends in two final strands, $\beta 5$ and $\beta 6$, which lie adjacent to the initial $\beta 1$ to form a second three-strand antiparallel β sheet. The three-dimensional structure of a single B subunit is depicted in Figure 6. Initially, we thought that this constituted a previously unknown fold (Sixma et al., 1991a), but Alex Murzin (1993, and personal communication) has pointed out that this general topology may be a feature common to many oligonucleotide/oligo-saccharide binding proteins and was originally seen for staphylococcal nuclease.

The B pentamer has the shape of a ring or doughnut (Fig. 7) with an approximate radius of 30 Å and a height of 40 Å. A characteristic and very important feature is the pore in the center of the ring. This pore is lined by the $\alpha 2$ helices of the five subunits. These helices are amphiphilic, presenting a very polar and charged surface to the pore while the hydrophobic side chains interact with the interior of the protein. There is a net

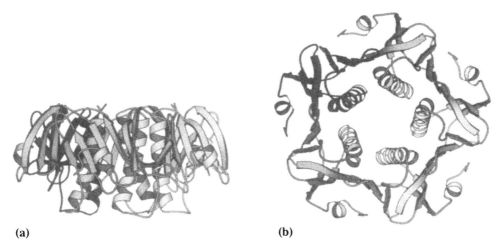

(a)   (b)

Fig. 7   Three-dimensional representation of the assembled LT B pentamer. (a) Viewed perpendicular to the fivefold axis; note the remarkably "flat" upper surface, against which the A subunit lies, and the opposite "convoluted" surface, which contains the five receptor-binding sites. (b) Viewed along the fivefold axis from the receptor-binding surface.

positive change from the side chains facing the interior of the pore. Adjacent subunits associate by juxtaposition of the two three-stranded β sheets to form a six-stranded antiparallel β sheet spanning the subunit interface. Five of these sheets surround the five central $\alpha2$ helices like a kind of "β ring," while the small N-terminal helices sit on the outside, filling up crevices on the surface of the ring.

The pentamer is additionally stabilized at each subunit–subunit interface by at least four salt bridges between each adjacent pair of subunits and approximately 797 Å^2 of buried hydrophobic surface (Sixma et al., 1993). Forty percent of the surface of a free subunit is buried upon pentamer formation. The numerous subunit–subunit interactions are quite consistent with the reported great stability of the B pentamer (Surewicz et al., 1990).

Viewed perpendicular to the fivefold axis (Fig. 7a), it is clear that the two sides of the pentamer are distinctly different. The upper side (Fig. 7b) is quite flat, with the carboxy-terminal residue 103 extending just beyond the surface. This "upper" or "flat" surface also contains a large number of charged residues (Sixma et al., 1993). The "lower" surface of the pentamer (Fig. 7b) is quite "convoluted." Its major feature is the protrusion formed by residues 51–63, starting with the extended chain following strand $\beta4$ in the second beta sheet, extending toward the solvent, and making a turn, thereby exposing in particular residues 55 and 56, and includes the first turn of the central helix $\alpha2$.

These two faces of the pentamer will be discussed many times with respect to a wide variety of properties of LT and CT in the remainder of this review.

B.   The A Subunit

The catalytically active A subunit has a topology totally different from that of the B subunit. As mentioned in the introduction, it is crucial for full activity of the toxin that the A chain be proteolytically "nicked" somewhere between residues 192 and 195 and that the disulfide bridge linking Cys-187 and Cys-199 be reduced. Reduction and nicking create the large A1 fragment containing the ADP-ribosylating activity and the much smaller A2 fragment comprising approximately residues 196–240. The topology of the chain is depicted in Figure 8, showing the numerous elements of secondary structure. An irregular antiparallel β sheet in the A1 fragment is formed by strands $\beta 5$, $\beta 6$, $\beta 3$, and $\beta 1$, $\beta 4$, $\beta 8$, $\beta 9$, with a strong twist between $\beta 3$ and $\beta 1$. The short strands $\beta 2$ and $\beta 7$ are linked by a single hydrogen bond. The A2 fragment has a quite unique conformation. The long α helix containing residues 197–222 of the A2 fragment lies snugly against the surface of the A1 fragment, while the remainder of the A2 fragment forms a tail that anchors the

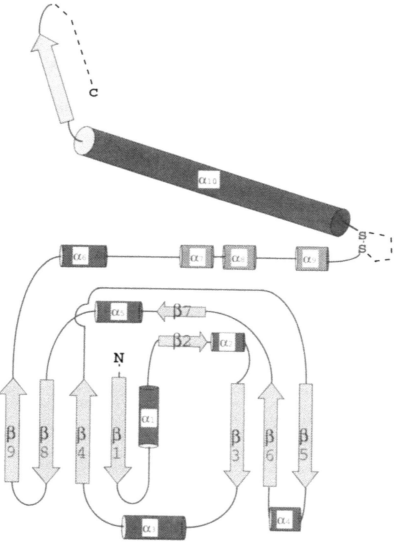

Fig. 8 Schematic drawing of secondary structure elements of the LT A subunit. $\beta 1$, 5–9; $\alpha 1$, 13–19; $\beta 2$, 21–22; $\alpha 2$, 41–45; $\beta 3$, 59–62; $\alpha 3$, 66–76; $\beta 4$, 82–88; $\beta 5$, 94–96; $\alpha 4$, 97–104; $\beta 6$, 113–116; $\beta 7$, 119–120; $\alpha 5$, 121–123; $\beta 8$, 124–131, $\beta 9$, 134–141; $\alpha 6$, 147–150; $\alpha 7$, 158–164; $\alpha 8$, 172–175; $\alpha 9$, 179–182; $\alpha 10$, 200–222. Figure adapted from Sixma et al. (1991a) with permission.

A subunit to the B pentamer as described below. This tail begins with a short 3_{10} helix (residues 223–226); residues 227–231 are very much extended and are followed by an α helix starting at residue 232. The four C-terminal residue, 237–240, with the sequence RDEL in LT and KDEL in CT, are largely invisible in the electron density maps, indicating flexibility of this part of the chain. The 187–199 disulfide bridge connecting the A1 and A2 fragments of the A subunit is quite well defined in the electron density, but the peptide chain in the region of the cleavage site (residues 189–195) has no convincing density. This again indicates considerable flexibility, probably required for easy access by and accommodation in the active site of one or more proteases. The second half of the A2 fragment extends totally away from the remainder of the A chain (Fig. 9), a very important feature of the AB_5 holotoxin, as we shall see in a moment.

C. Architecture of the AB₅ Holotoxin

The association of the catalytic A subunit with the receptor-binding B pentamer is most remarkable (Fig. 10): the second half of the A2 chain extends through the pore of the B pentamer. This is a manner of associating subunits that, to the best of our knowledge, had never been seen before the determination of the *E. coli* LT structure (Sixma et al., 1991a). However, since then this association mode has also been observed in shigatoxin (Fraser et al., 1994), another member of the AB_5 toxin family. It is expected to be a constant and intriguing feature of this entire family of bacterial toxins, including the

Fig. 9   Three-dimensional representation of secondary structure in the LT A subunit.

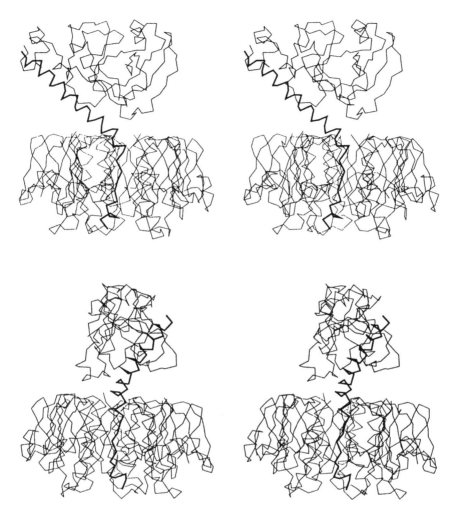

Fig. 10 Stereo pair C$^\alpha$ tracings of LT, with A$_2$ fragment in dark lines, showing the complete AB$_5$ structure. The figure is oriented so that the A subunit lies above the B pentamer. The two views are 90° apart about the vertical axis and show that the bulk of the A subunit is rather like a flat triangular wedge, with base ~ 57 Å, height ~ 35 Å, and thickness ~ 25 Å. Note that the A$_2$ fragment extends through the central pore of the B pentamer. Figure taken from Sixma et al. (1993) with permission.

related LT-II enterotoxin of *E. coli* (Pickett et al., 1987, 1989), the structure determination of which has been initiated in our laboratory.

Unfortunately, we do not as yet have a crystal structure for LT after activation by proteolytic "nicking" of the cleavage peptide and reduction of the disulfide bridge in the A chain. From the available structures however, we see, that the interactions between A1 and A2 are still quite considerable, even without a covalent component, and would not be much affected by the nicking and reduction. It is reasonable to expect that the loss of the two covalent interactions between A1 and A2 will affect the mutual affinity of the two A fragments, but it is difficult to derive quantitative values for the dissociation constant from structural data. Moreover, if separation of A1 and A2 occurs during the membrane translocation process, the possible interactions of the A1–A2 interface regions of both the A1 and A2 fragments with hydrophobic lipid molecules also has to be taken into account. Here we come to the most intriguing, and most mysterious, aspects of the structure and function of LT and CT: the detailed course of events during membrane translocation and activation of the toxin. This is discussed later in a separate section.

D.   Flexibility of the A Subunit with Respect to the B Pentamer

The mode of association of the A subunit with the B pentamer and the small number of direct interactions between these two elements of the toxin make it an interesting question as to how flexible the A subunit is with respect to the B pentamer. There are at present several different crystal structures of the toxin (Table 3) available to address this question. It should be realized that all crystal structures are at least partially affected by having to pack into a specific crystal lattice. Although this may imply that we have not observed the full range of orientations that the A subunit can assume relative to the B pentamer, it nevertheless gives some insight.

The structures of the LT AB_5 molecule in five different crystal environments have been compared by Sixma et al. (1992c) (Fig. 11). Since then the structure of nicked and oxidized LT (Merritt et al., 1994c) and that of the LT: galactose complex have become available. The nicked form had the same cell dimensions as the native LT crystals, and, not surprisingly, the relative orientation of the A subunit was the same as in the native crystals. In the case of the LT: galactose complex, the relative orientation of the A subunit deviates about 5° from that in the native LT crystals (Merritt et al., 1994b). The maximum observed orientational deviation of 7.5° was seen when comparing one of the molecules

Fig. 11   Relative flexibility of LT subunit interactions. The B pentamers from AB_5 complexes in five different crystal structures were superimposed, and the variation in the resulting position of the corresponding A subunits is depicted in the stereo pair. The rotation axes that describe the pairwise variation among these five structures are shown as dashed lines. The five structures shown are LT in an orthorhomic cell (Sixma et al., 1991a), LT:lactose complex (Sixma et al., 1992a), LT:Sm complex (Sixma et al., 1992b), and two independent molecules of LT from a monoclinic crystal form (Sixma et al., 1992c). Figure taken from Sixma et al. (1992c) with permission.

in the $P2_1$ crystal form of native LT with that of native LT in the $P2_12_12_1$ crystal form (Sixma et al., 1992c).

It turned out that the hinge point of the rotation is in all these instances near residue Gln-221, except in the case of the samarium complex. This is, however, quite a special case, as we shall see in the next section, so that the conclusion at present is that the maximum deviation of the A subunit with respect to the B pentamer is on the order of 7–8°. Clearly, this set of structures does not provide information about the conformational changes that may take place upon reduction or reduction plus nicking of the A subunit, let alone about the conformational changes that might take place in the presence of lipid molecules or when the A subunit is interacting with "assisting proteins" like ARF or with target proteins like $G_{s\alpha}$. It is likely that in one or more of these steps in the intoxication process the A subunit undergoes much more spectacular conformational changes than we have observed so far.

E. Interactions of Divalent Cations with LT

Although the physiological importance is unclear, it has been reported that LT and CT have calcium ionophore activity (Knoop and Thomas, 1984; Dixon et al., 1987). Also, lanthanum ions have an inhibitory effect on the toxin-induced secretion of Cl^- and HCO_3^- (Leitch and Amer, 1975) as well as on the calcium ionophore activity (Maenz et al., 1987). Because of this, and because we saw effects of several lanthanides on LT crystals in the course of our attempts to prepare suitable heavy atom derivatives for crystal structure determination, we chose to examine cation-binding sites of LT crystallographically. The structure determination (Sixma et al., 1992b) revealed one samarium site that was due to crystal contacts, and that will be ignored hereafter, and two samarium-binding sites that are at a quite interesting position: between the A subunit and the B pentamer (Fig. 12). One of these samarium-binding sites is equivalent to the single site observed in the erbium soaking experiment. The residues interacting with the lanthanide ions are

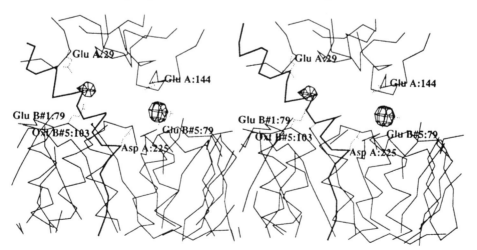

Fig. 12   Stereo pair showing cation-binding sites in LT AB_5 as seen in LT:Sm complex (Sixma et al., 1992b). The ligands on the left side in the figure are conserved in CT and may well be the site responsible for inhibitory lanthanum binding. Figure taken from Sixma et al. (1992b) with permission.

B:79, A1:144, A2:225 for ion site 1, and B 1:79, B 5:103, A1:29 for ion site 2. Note that these sites do not follow the pentamer fivefold symmetry, but, remarkably, in both sites a B-subunit Glu-79 residue is involved in ion coordination. The residues involved in the first samarium-binding sites are not conserved in cholera toxin, but the side chains of the second cation-binding site are all conserved in LT and CT. The implications of these observations, if any, for the processes of LT and CT when affecting the host epithelial cells remain to be established.

It may be relevant to mention here that the use of relatively high concentrations of cadmium chloride during the crystallization of LT AB_5 holotoxin led to crystals that did not appear to contain the full holotoxin (Pronk et al., 1987). The structure of these crystals has been solved in the meantime (Sixma et al., manuscript in preparation), and it appears that the A subunit is absent and the B pentamers in the crystals have virtually the same conformation as in the AB_5 LT holotoxin. A number of cadmium-binding sites could be determined. Apparently a concentration of 300 mM cadmium is able to dissociate the A subunit from the B pentamer. Although unexpected and intriguing, it is not clear what the significance of this observation is for LT under physiological conditions.

F. The Active Site of LT

1. Comparison with Exotoxin A and Diphtheria Toxin

Heat-labile enterotoxin and cholera toxin share the ADP-ribosylation activity with other bacterial toxins such as *Pseudomonas aeruginosa* exotoxin A (ETA) and diphtheria toxin (DT). The architecture and many properties of these latter two toxins are entirely different from that of LT and CT: (1) ETA and DT consist of a single polypeptide chain, which is later nicked to form two chains; (2) the target protein of ETA and DT is elongation factor 2, whereas LT and CT target $G_s\alpha$; (3) ETA and DT transfer the ADP-ribose to a modified histidine residue, diphthamide (Van Ness et al., 1980), while LT and CT transfer the ADP-ribose to an arginine side chain (van Dop et al., 1984). In spite of all these differences, the comparison of LT with ETA proves to be very useful indeed (Sixma et al., 1991a).

Exotoxin A consists of three domains: a receptor-binding domain, a translocation domain, and a catalytic domain. The first two domains have no structural features in common with LT. The comparison of the catalytic domain of ETA, the first structure solved of any ADP-ribosylating toxin (Allured et al., 1986), with that of LT (Sixma et al., 1991a) revealed a similarity between the core of the catalytic domain of ETA and the A subunit of LT: 44 residues can be superimposed with an rms difference of 1.5 Å for the C^α coordinates. The equivalent residues coincide approximately with helix $\alpha 3$ and strands $\beta 1$–$\beta 6$, which form the central part of both enzymes (Fig. 8). Only three out of the 44 equivalent residues are identical: A1:Tyr 6, Ala-69, and Glu-112 of LT correspond with Tyr-439, Ala-478, and Glu-553, respectively, in exotoxin A (Fig. 13). The latter amino acid has been identified as an exotoxin A active-site residue for its ability to be covalently linked to NAD and for the loss of activity after mutation to an aspartate (Carroll and Collier, 1987; Douglas and Collier, 1987). Residue A1:Glu-112 in LT has been shown to be important for activity by mutagenesis studies (Lobet et al., 1991). This has been further confirmed by the recent structure determination of diphtheria toxin (Choe et al., 1992), which confirmed and extended the sequence alignment of DT and ETA given in Fig. 13b and showed this region to be the binding site of ApUp, a competitive inhibitor

(a)

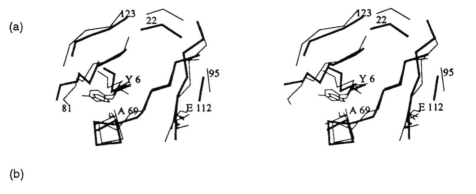

(b)

CT :	LYRADS	LMP	YVSTSISLRSAH	STYYIYVIA	FN	QEVSALG	QIYGW
	5 • 10	22 24	59 • 70	81 89	95 96	111• 117	123 127
LT :	LYRADS	LMP	YVSTSLSLRSAH	STYYIYVIA	FN	QEVSALG	QIYGW
ETA:	GYHGTF	VRA	GFYIAGDPALAY	RNGALLRVY	AI	LETILGW	VVIPS
	438 443	455 457 468	479	494 502	541 542	552 558	565 569
DT :	SYHGTK	QKP	GFYSTDNKYDAA	KAGGVVKVT	LS	VEYINNW	SVELE
	19 24	36 38 52	63	76 84	136 137	147 153	160 164

Fig. 13 Comparison of the putative active site in LT and in *P. aeruginosa* exotoxin A (ETA). (a) Stereo pair superposition of LT residues (thick lines) onto the enzymatic domain of ETA (thin lines). C^α numbering is that of LT; side chains are shown for identical residues (Tyr-6, Ala-69, Glu-112). (b) Sequence alignment of 44 structurally equivalent residues in LT (Yamamoto et al., 1987) and ETA (Gray et al., 1984) as well as the related diphtheria toxin (Ratti et al., 1983) and cholera toxin (Mekalanos et al., 1983). Figure taken from Sixma et al. (1991a) with permission.

of NAD^+ binding. These results clearly indicate that the structural comparison identifies a region crucial for the activity of the A1 fragment.

2. The Architecture of the Active Site

So far our attempts to obtain a picture of the active site with bound NAD, arginine, or various substrate analogues have not been successful, although we continue to pursue this goal. The active-site superposition mentioned in the previous section allows us, nevertheless, to describe the architecture of the active sites of LT and CT.

The crucial glutamic acid 112 is located in a large crevice that runs across the A subunit between strands $\beta1$ and $\beta3$ (Fig. 14). In the X-ray structure this crevice is also lined by helix $\alpha3$ and filled with a number of water molecules. Intriguingly, the shape of the crevice is expected to be different in CT than in LT, since Ala-72 of helix $\alpha3$ in LT is replaced by a larger Leu residue in CT, which would make the crevice narrower. It is nevertheless possible that the crevice is involved in NAD binding (as suggested by Sixma et al., 1991a), but experimental evidence for this would be most welcome.

Arginine A1:7 forms part of the active-site crevice near Glu-112 (Fig. 14), making, at the floor of the cleft, a salt bridge with Asp A1:9 and hydrogen bonds with the main-chain carbonyl groups of Val A1:53, Arg A1:54, and Ser A1:61. The importance of this arginine residue for the integrity of the active site was shown by mutagenesis studies in both LT (Lobet et al., 1991) and CT (Burnette et al., 1991). Even a lysine at this position is not tolerated, probably due to disruption of the intricate hydrogen-bonding network involving the guanidinium group of Arg-7. Characterization of these mutations of Arg-7 had, in fact, been suggested by analogy with mutants in pertussis toxin (Burnette

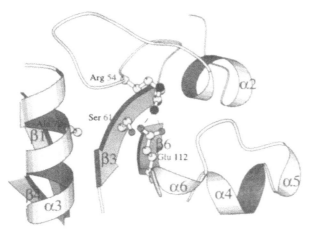

Fig. 14 Schematic representation of the secondary structural environment of the putative active site in LT. This figure shows the active site in the unnicked and oxidized form of the enzyme. Conformational changes might take place upon reduction and nicking, as activity of the toxin increases considerably after these events (Moss et al., 1993). A three-dimensional structure of oxidized but nicked LT has just been completed (Merritt et al., 1994c). It appears from this structure that nicking without reduction does not cause any significant changes on the active site, or anywhere else in the LT holotoxin. Figure taken from Sixma et al. (1993) with permission.

et al., 1988; Pizza et al., 1988). The S1 subunit of pertussis toxin has a high sequence identity with LT and CT in this region (Nicosia et al., 1986; Locht and Keith, 1986), and a mutation of the equivalent Arg-9 to a lysine in pertussis toxin also caused loss of activity. This indicates that the function of this arginine residue might be similar in pertussis toxin, LT, and CT, and hence the architectures of the three active sites may have several features in common.

The equivalent residue of Arg-7 in LT/CT is His-440 in exotoxin A and His-21 diphtheria toxin. The hydrogen-bonding networks are therefore different, but in exotoxin A His-440 makes a hydrogen bond to the carbonyl oxygen of Tyr-470 in strand $\beta3$, which implies that this hydrogen bond is probably equivalent to the one between Arg-7 and Ser-61 in LT/CT. Thus there are apparently common features in the active sites of the entire family of ADP-ribosylating bacterial toxins. The loop formed in LT by residues 49–55, which is located at the "top" of the active-site crevice as depicted in Fig. 14, has no equivalent in ETA or DT. The presence of this loop in LT/CT may be the explanation for the much higher K_m for NAD in LT and CT compared to other ADP-ribosylating toxins. Further structural and mutagenesis experiments may well reveal definitely the reason for this interesting difference among the ADP-ribosylating toxins in the not too distant future.

Although it is difficult to make predictions about conformational changes, we would nevertheless like to point out that the interactions between residues 100–110 and residues 40–54 in the LT A chain (right-hand side of Fig. 14) are virtually all hydrophobic. This hydrophobic interface is at right angles to the active-site cleft. The active site is on the "left" side from the hydrophobic interface between the two segments (Fig. 14). The "right" side of this interface is not too far from the disulfide bridge that is to be cleaved upon activation of LT and CT. Since it is known that the activity of LT and CT increases in the presence of detergents (Tsuji et al., 1991; Gill and Woolkalis, 1988) and also that model compounds with a hydrophobic group next to the guanidinium group become better acceptors (Tait and Nassau, 1984; Soman et al., 1986), we have speculated (Sixma et al., 1993) that this hydrophobic interface opens up to a smaller or larger degree upon nicking, reduction, and substrate binding. However, in the absence of solid facts it is best to defer further speculations on the active site and activation of LT and CT here, and continue with a field where solid data are accumulating rapidly, that is, receptor binding, as will be seen in the next sections.

G. Receptor Recognition by LT and CT

The receptor of LT and CT at the surface of the epithelial cell is the ganglioside G_{M1} [G_{M1} = Gal($\beta 1 \rightarrow 3$)GalNAc($\beta 1 \rightarrow 4$){NeuAc($\alpha 2$-3)}Gal($\beta 1 \rightarrow 4$)Glc($\beta 1 \rightarrow 1$)ceramide]. Figure 1 depicts the five sugars of the G_{M1} branched oligosaccharide. Crystallographic study of LT AB_5 complexed with the G_{M1} pentasaccharide has been hampered by the tendency of pregrown crystals of the toxin to break up when soaked in a solution of the pentasaccharide and by the inability to obtain good quality crystals from LT:pentasaccharide cocrystallization trials. Some success was achieved for cocrystallization of LT with simpler sugars (lactose and galactose), using a three-layer liquid diffusion technique. Recently, in striking contrast with the difficulties in obtaining crystals of LT AB_5:sugar complexes, crystals of the CT B-pentamer with the G_{M1} pentasaccharide grew almost immediately after the initial crystallization experiments were set up (S. Sarfaty, personal communication). Consequently, we now have three well-refined structures of toxin:saccharide complexes available. In addition, the three-dimensional structures of the CT B-pentamer in complex with galactose and sialic acid have also been determined (E. Westbrook, personal communication).

1. LT Complexes with Lactose and Galactose

A first look at the sugar-binding site in LT came from the three-dimensional structure of the LT:lactose complex (Sixma et al., 1992a). Only the galactose moiety of lactose is directly analogous to the actual receptor saccharide, and it was reassuring that the LT:lactose structure revealed extensive protein interactions with the galactose residue and hardly any with the second, "incorrect," residue of the bound lactose. The galactose sugar ring binds directly parallel to the indole ring of Trp-88. This residue had already been implicated as crucial for G_{M1} binding by fluorescence studies by De Wolf et al. (1981), and our investigations completely confirm these results. Secondary hydrophobic interactions also arise from van der Waals contact between the galactose ring and His-57. The specificity of the galactose for its binding pocket is seen to stem mainly from the extensive network of hydrogen bonds between the galactose hydroxyl oxygens and the protein (Fig. 15a,b). These involve the side chains of residues Glu-51, Gln-61, Asn-90, and Lys-91 as well as the main-chain carbonyl oxygen of residue Gln-56. The involvement of Lys-91 is in accordance with studies suggesting the importance of one or more lysine residues for G_{M1} binding (Ludwig et al., 1985). The ring oxygen atom O5 does not participate in any hydrogen-bonding interactions, even though this atom is entirely solvent-inaccessible in the complex.

Analysis of the LT:lactose complex left unclear the extent to which the highly specific galactose: LT interactions were affected by the presence of the incorrect second sugar of lactose. The presence of the Gal($\beta 1 \rightarrow 4$)Glc linkage in lactose, instead of the correct

Gal(β1→3)GalNAc linkage in G_{M1}, might have perturbed the normal binding site stereochemistry. This issue was resolved by structure determination of the LT:galactose complex (Merritt et al., 1994b) and the CT:G_{M1} oligosaccharide complex discussed separately below (Merritt et al., 1994a). The location and conformation of the galactose itself in each of these structures is essentially superimposable onto that seen for the galactose moiety of the LT:lactose complex. After rigid-body superposition of the LT:galactose structure onto the LT:lactose structure to optimize the fit for all protein atoms in the B pentamer, the rms deviation of coordinates for the five galactose moieties in LT:galactose compared to LT:lactose is a mere 0.3 Å.

As in the LT:lactose complex (Fig. 15), the galactose sugar ring in the LT:galactose complex lies parallel to the ring system of Trp-88, which contributes 25% of the total surface area of the binding site seen by the galactose. The stacking arrangement of the hydrophobic surface of the sugar against the tryptophan ring is reminiscent of the galactose-binding site seen for the *E. coli* galactose chemoreceptor protein (Vyas et al., 1988) and is typical of protein–sugar association (Vyas, 1991). The direct hydrogen-bonding interactions between the protein and the bound sugar are the same as those previously observed for the LT:lactose complex (Fig. 15), with the exception of the involvement of Gln-56. The hydrogen bond between the galactose O6 and the backbone carbonyl oxygen of Gln-56 implied by the conformation seen in LT:lactose is not supported by the longer distance found in LT:galactose. The Gln-56 carbonyl oxygen remains involved in sugar binding, but only via solvent-mediated hydrogen bonds.

The buried surface of the galactose is 133 Å^2. This constitutes 79% of the solvent-accessible surface area of the free sugar; that is, the galactose is quite deeply inserted into the binding site. Two solvent molecules are seen to be a conserved feature of the sugar binding site in all of the LT structure determinations; when these are considered as part of the recognition surface, the galactose buried surface becomes 145 Å^2, or 86% of the total sugar surface. The only galactose atoms that remain significantly solvent-accessible in the complex are O-2 and O-1, the latter atom being the attachment site for the remainder of the G_{M1} pentasaccharide.

Apart from a small rotation of the A subunit with respect to the B pentamer, typical of variations in different crystal forms as discussed above, there is minimal change in the conformation of the peptide chains of LT as seen with and without bound sugar. In the LT:galactose structure there is an rms shift of only 0.4 Å for all 515 C^α atoms of the B pentamer after superposition onto the 1.95-Å LT structure. The major effect of galactose binding is decreased flexibility of residues 51–60, as discussed below.

The location of the binding site with respect to the entire AB$_5$ complex (Fig. 16) is of the greatest importance for understanding the molecular events accompanying the membrane translocation process. This is the subject of a separate section below. The conclusion from comparison of the LT:lactose and LT:galactose complexes is that the galactose binding mode is virtually the same and hence is expected to be similar in LT:G_{M1} and LT:galactoprotein interactions.

3. *CT B-Pentamer Complex with* G_{M1} *Pentasaccharide*

The recent structure determination of the cholera toxin B pentamer bound to the pentasaccharide from G_{M1} (Merritt et al., 1994a) allows analysis of the entire receptor-binding surface of the toxin. Each of the five G_{M1} binding sites lies primarily within a single B monomer, on the convoluted side of the B pentamer, in a cleft formed by the loop 51–58 on one side and two loops formed by residues 10–14 and 89–93 on the other

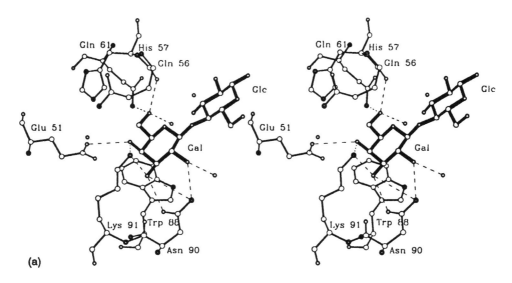

Fig. 15 Galactose-binding site in LT as seen in LT:lactose complex (Sixma et al., 1992a). (a) Stereo pair of lactose binding site, showing the amino acids interacting directly with the galactose moiety. (b) Schematic representation of first- and second-shell hydrogen-bonding interactions involving the bound galactose moiety. Figure taken from Sixma et al. (1992a) with permission.

side. At one end of this cleft, the $\beta2$–$\beta3$ loop (residues 31–36) from an adjacent monomer forms the remainder of the binding site, although these residues do not interact directly with the bound oligosaccharide.

By far the largest number of direct interactions between the toxin and the pentasaccharide involve the two terminal residues of the branched saccharide: galactose and sialic acid. The sialic acid occupies roughly 43% of the binding surface, the terminal galactose 39%, and the GalNAc residue 17%. The association of G_{M1} to the protein may thus be thought of as a two-fingered grip; the Gal($\beta1{\rightarrow}4$)GalNAc forefinger is inserted fairly deeply into the more buried end of the binding cleft, and the sialic acid thumb extends in the other direction along the protein surface.

Exactly as in the LT:lactose and LT:galactose complexes, the sugar ring of the terminal galactose lies directly over the indole ring of Trp-88. The primary hydrophobic interactions at the sialic acid binding site involve association with Tyr-12. The methyl group of the GalNAc residue is accommodated by proximity to C^{β} of His-13. In total, six protein side chains are involved in hydrogen-bonding interactions with the receptor oligosaccharide, and four in hydrophobic interactions (Fig. 17).

The conformation of the G_{M1} pentasaccharide in solution has previously been analyzed by solution NMR studies (Acquotti et al., 1990). The torsion angles of the sugar linkages seen in the CTB:G_{M1} complex are fully consistent with the NMR-derived model for the pentasaccharide in solution. The Gal($\beta1{\rightarrow}4$)Glc linkage has been reported to exhibit great flexibility in solution (Poppe et al., 1990), and this probably implies that the terminal four sugars are relatively free to move with respect to the glucose, which is attached to the membrane-bound ceramide group.

In the crystal structure the conformation of the sialic acid residue is stabilized by an internal hydrogen bond from O7 to O10, and the sugar as a whole is constrained by several intrasaccharide hydrogen bonds. The bound saccharide is positioned so that the

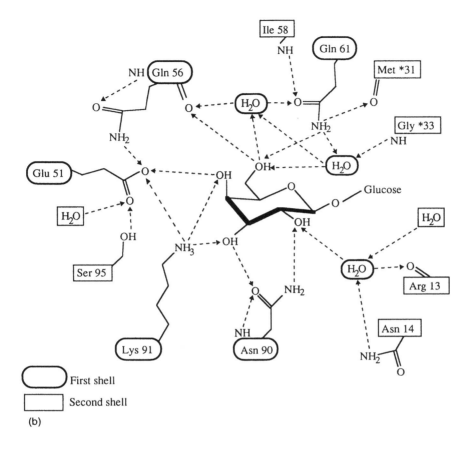

First shell
Second shell

(b)

longest dimension of the sugar chain lies diagonally away from the bottom surface of the toxin. The O1 hydroxyl group of the glucose residue is thus furthest from the convoluted surface of the B pentamer, entirely consistent with its being the point from which the ceramide lipid tail of the intact G_{M1} ganglioside is anchored in the cell membrane.

The only contribution to the binding interactions from the neighboring subunit visible in the X-ray structure is a small decrease in the solvent-accessible surface of the Lys-34 and Arg-35 side chains, due to proximity of the sialic acid acetyl group, and a solvent-mediated hydrogen bond from the backbone nitrogen of Gly-33 to O6 of the terminal galactose.

It is noteworthy that all protein atoms directly involved in the binding of the G_{M1} pentasaccharide are identical in LT and CT (Fig. 17). This means that the reported differences in selectivity between the two toxins are due to second-order effects from residues further away from the binding site, the closest of which are residues 13 and 95. It may be pointed out, however, that the studies of Schengrund and Ringler (1989) revealed remarkably similar affinities of a variety of chemically modified pentasaccharides for LT and CT. The results of these and other authors correspond very well with our studies; in particular, the modifications of the sialic acid that maintain high affinity for the toxins are in excellent agreement with the observed LT–sugar interactions in the crystal structures. For instance, the O8 and O9 of the sialic acid can be replaced with hydroxymethylene or 2-aminoethanol without much effect on the binding properties

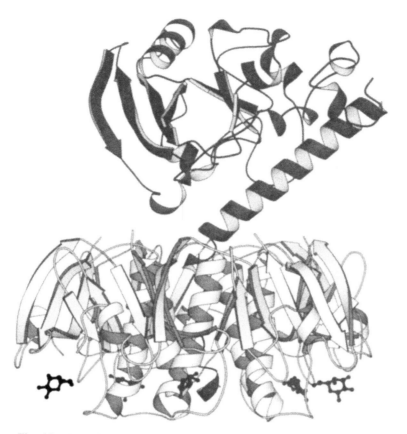

Fig. 16 Location of sugar-binding site on AB$_5$ holotoxin, as seen in LT:galactose complex (Merritt et al., 1994b). The five identical sugar-binding sites are located on the convoluted surface of the B pentamer, implying that the toxin binds to the cell membrane with the catalytic A subunit facing away from the membrane surface. That is, when the galatose constitutes the terminal sugar of the complete G$_{M1}$ ganglioside, the remainder of the receptor will extend downward from the toxin as shown into a membrane surface lying below it.

(Schengrund and Ringler, 1989). The *N*-acetyl is also not essential; it can be replaced by *N*-glycolyl (Fishman et al., 1980). The carboxylate on the sialic acid can be methyl esterified without detrimental effect on the binding (Sattler et al., 1977). All these modifications agree well with the receptor-binding mode depicted in Figure 17. Also, the glucose can be replaced by a (β1→4)sorbitol linkage (Sattler et al., 1977) or by a (β1→3)galactose linkage and extended with more sugar moieties (Nakamura et al., 1987) without change in binding properties of CT. This corresponds well with the fact that the glucose has few, if any, interactions with the toxins in the crystal structure. Extension by another sugar is also possible at the other end, at the O2 of the terminal galactose (Masserini et al., 1992).

3. *Conformational Changes Due to Saccharide Binding*

One notable difference between the free toxin and complexes of the toxin with bound sugar is the degree to which the loop consisting of residues 51–60 in the B pentamer is

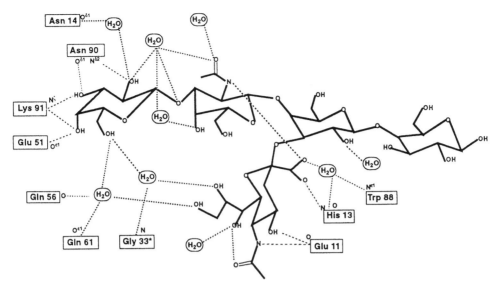

Fig. 17 Schematic representation of hydrogen-bonding interactions at the sugar-binding site in the CT B pentamer:G_{M1} oligosaccharide complex (Merritt et al., 1994a).

well-ordered. This loop connects strand $\beta 4$ to helix $\alpha 2$ of the B monomer and protrudes from the convoluted surface of the assembled AB_5 holotoxin. In the free LT structure this loop exhibited high thermal parameters ($B_{avg} = 55.1$ Å^2 for residues 51–60 as compared to $B_{avg} = 30.2$ Å^2 for the B pentamer as a whole) and also conformational flexibility as evidenced by the variation seen in the five separate B monomers. Residues Glu-51, Gln-56, His-57, and Gln-61 are involved in sugar binding, however, and in the LT:sugar and CT:pentasaccharide complexes the 51–60 loop is much better ordered. Presumably this loop exhibits flexibility in solution, allowing easy entry of the receptor into the binding site. Upon binding, the loop becomes much more strongly ordered. This interpretation is supported by the higher stability noticed for the B pentamer with bound sugar (Surewicz et al., 1990).

It is of interest that a 15-residue synthetic peptide (CTP3) spanning this loop and corresponding in sequence to residues 50–64 has been shown to elicit antibodies that are cross-reactive with both native LT and native CT. Antibodies to the peptide were partially effective in neutralizing biological activity of both toxins (Jacob et al., 1984, 1986). The fact that the CTP3 antigen is derived from the loop defining one side of the receptor-binding site suggests that the binding of anti-CTP3 antibodies may block the toxin's cell-binding ability. This case would be parallel to that of influenza virus hemagglutinin, where a majority of antigenic sites surround the sialic acid-binding pocket on the surface of the protein (Weiss et al., 1988).

Our observation that in the three toxin:saccharide complexes described above the structural changes with respect to unliganded toxin are very small indeed corresponds well with many other studies. For instance:

1. Binding to the G_{M1} pentasaccharide causes a small change in the local environment of Trp-88, according to fluorescence measurements (De Wolf et al., 1981), but no large conformational differences.

2. FT-IR measurements also indicate small conformational changes and no major change of secondary structure (Surewicz et al., 1990) upon G_{M1} ganglioside binding.
3. Most monoclonal antibodies bind equally well in the presence or absence of ganglioside G_{M1}, except for those with an epitope overlapping the G_{M1} binding site. This indicates that the structure does not change its external surface upon binding. One exception is an antibody that binds only to toxin:G_{M1}. The epitope for this antibody seems to contain residue B:46, and it is possible that the epitope is changed by the G_{M1} binding, indicating that there must be some (local) conformational changes (Finkelstein et al., 1987).

III. ORIENTATION OF AB$_5$ BOUND TO THE MEMBRANE

A. Noncrystallographic Investigations

The orientation of CT and LT upon binding to the target cell surface was controversial for a long time. An early model by Gill (1976) proposed the insertion of the B pentamer into the membrane as a pore through which the A subunit could pass. This model was later abandoned when surface pressure measurements on monolayers established that the B pentamer does not enter into the hydrophobic layer of the membrane to any great extent (Mosser and Brisson, 1991). This had already been suggested by photolabeling studies, where the B pentamer was not labeled when the photolabel was present in the hydrophobic core of the membrane (Tomasi and Montecucco, 1981; Wisnieski et al., 1979; Wisnieski and Bramhall, 1981), and by calorimetric studies, where only the A subunit was shown to perturb the hydrophobic core of the bilayer by causing a decrease in the enthalpy of the phospholipid transition from fluid to gel phase (Goins and Freire, 1985). It seemed a justifiable conclusion that the effect on the conformation of the B pentamer induced by receptor binding is minor and the B pentamer does not enter the hydrophobic core of the membrane.

Some of the photolabeling studies were interpreted in terms of binding with the A subunit toward the membrane (Wisnieski and Bramhall, 1981), while others concluded the opposite orientation (Tomasi and Montecucco, 1981). Quasi-elastic laser scattering experiments seemed to indicate that the A subunit pointed toward the membrane (Dwyer and Bloomfield, 1982). It is noteworthy that antibodies with specificity for the A subunit could still bind when the toxin was bound to cells (Janicot et al., 1988), indicating that this subunit was still accessible after binding. In 15-Å electron microscopy studies by Ribi et al. (1988), binding of the holotoxin was shown to give extra density in a sphere around the central pore, which could possibly be interpreted as due to an A subunit partially immersed in the monolayer membrane. The alternative explanation, however— that this density was due to an A subunit on the other side of the pentamer in a fivefold (or more) degenerate position—could not be excluded.

Yang et al. (1993) used atomic force microscopy to obtain 20-Å resolution images of cholera toxin bound to supported synthetic bilayers containing G_{M1}. Bound molecules of the B pentamer were clearly visible as a pentamer resting above the plane of the bilayer and containing a central pore. Bound molecules of the AB$_5$ holotoxin showed instead as a relatively featureless hump protruding somewhat higher above the lipid surface than the B pentamer alone. Although the vertical scale of the atomic force images obtained is very compressed, these images are consistent with the model deduced from X-ray structures as described below, in which the B pentamer binds to the surface of the membrane and the bulk of the A subunit lies on the opposite face of the pentamer.

B. Crystallographic Results

Obviously, crystallographic studies cannot easily provide a movie of the events taking place during membrane translocation. However, we can construct a pretty good picture of the location and orientation of the AB_5 complex until the very moment before translocation takes place. From the X-ray structures of the LT:galactose, LT:lactose, and CTB_5:G_{M1} pentasaccharide complexes it can be concluded that, upon receptor binding, the convoluted side of the B pentamer points to the membrane surface while the A subunit points away from the membrane. If this were not the case, then the attachment point of O-1 of the glucose moiety, being 35 Å away from the "flat" surface of the pentamer, would require that the ganglioside hydrophobic tails extend along the hydrophilic sides of the B-pentamer ring and immerse the pentamer into the lipid bilayer. There is strong evidence, summarized above, that this is not the case. Moreover, the lateral side of the B pentamer is quite hydrophilic, which means that upon immersion into the membrane there would need to be a major conformational change of the B pentamer. As discussed in the previous section, there is evidence only to the contrary. Hence, by combining the X-ray studies with the previous biochemical and biophysical results, it must be concluded that CT and LT bind with their enzymatically active A subunit pointing away from the membrane (Fig. 16).

C. The C-Terminal [K/R]DEL Sequence of the A Subunit

It is most interesting that the C-terminal tetrapeptide of the A subunit, which is largely invisible in the structures we have determined, has a highly intriguing amino acid sequence. This is, in one-letter code, RDEL in the heat-labile enterotoxin and KDEL in cholera toxin. The KDEL sequence has recently been identified as an endoplasmic reticulum retention signal. Proteins with this sequence at their C terminus are recognized by specific receptors in the Golgi network and shuttled back into the endoplasmic reticulum [for reviews, see Pelham (1991) and Rothman and Orci (1992)]. From the glycosidic markers these proteins acquire in this process it was concluded that they reach the cis-Golgi but do not go any further, although the receptors are distributed over the other parts of the Golgi as well (Lewis and Pelham, 1990). Involvement of these receptors in the translocation process of LT/CT might potentially occur if either the receptors or the toxins move along this vesicle pathway.

This would require, at some point, the direct interaction of the KDEL receptor with the A subunit. As has been discussed in considerable detail above, the AB_5 molecule binds to gangliosides on the target cell in such a manner that the convoluted side of the B pentamer is facing the outer membrane surface. This implies that immediately after binding the C-terminal R/KDEL sequence is near the membrane.

It is obvious that this leads to speculation that the KDEL receptor may bind the KDEL sequence while the B pentamer is still attached to the A subunit and the A1 fragment is still facing away from the membrane. If this were the case, then it is possible that the mode of entry of the A subunit into the cell would involve unfolding of the A1 chain, followed by the entire 200 residues threading through the central pore of the B pentamer. As mentioned above, this mode of entry has already been proposed by Gill (1976).

A variant of this model is that the B pentamer somehow disociates from the

AB_5:G_{M1}:KDEL receptor complex after the association of the KDEL sequence with the KDEL receptor is made. The membrane translocation process then would not involve threading the A chain through the pore of the B pentamer. But still the membrane translocation process in this model might require considerable unfolding of the A subunit.

In both variants of the KDEL-receptor-mediated membrane translocation model it is important for the process that the disulfide bridge in the nicked A subunit remain intact for a sufficiently long time; otherwise "pulling" at the C terminus of A2 may not lead to the translocation of A1. However, little information is available regarding the exact sequence of events during membrane translocation of LT and CT.

Obviously, the C-terminal sequence may have no role in the membrane translocation process upon entry into the cell but rather play an important role at some later stages of the intoxication event. Alternatively, the C-terminal tetrapeptide of the A subunit may have no function whatsoever. Further studies are clearly required to reveal the specific events and structures involved in the entry of A1 into the target cell.

IV. THERAPEUTIC APPLICATIONS

A. LT, CT, and Vaccine Development

The ultimate goal of much of the research on LT and CT is to provide better protection against the diseases caused by these toxins than is currently available. Treatment of cholera and related diarrheal diseases is perfectly possible, at least since the development of the oral rehydration therapy (ORS) (see Hirschhorn and Greenough, 1991). This type of therapy, which is very effective, is based on restoring the water and ion levels in the body. To enable the transport of the ions into the cells, addition of carbohydrates is required. In most cases oral rehydration is sufficient as a therapy against dehydration, and only in very severe cases are intravenous supplies necessary. Notwithstanding the incredible success of the method, it is not infallible, and people still die of these diseases. Moreover, those who recover have suffered immense discomfort. Hence there is a tremendous need for more effective treatment, and an even greater need for methods to prevent the disease.

Infection by cholera of the classical type affords protection against recurrence for at least 3 years in volunteers, the longest period tested (Levine et al., 1981), and therefore vaccination should in principle be possible. For the El Tor type the evidence is somewhat less convincing (Levine et al., 1983). Vaccine development against cholera and related diseases is being pursued along a number of different lines (Levine and Pierce, 1992). Unfortunately, none of the various approaches followed has so far resulted in an efficient cholera vaccine that gives sufficient long-term prophylactic protection to merit large-scale application.

Vaccine development against enterotoxigenic Escherichia coli (ETEC) has concentrated very much on the fimbriae and colonization factors (Levine et al., 1983). Other attempts have focused on making a general vaccine against ETEC-mediated disease by combining LT and the heat-stable 18 amino acid long ST in a fusion protein, either by chemical (Klipstein et al., 1982, 1983) or by genetic means (Sanchez et al., 1986, 1988a). The disadvantage of some of these fusion proteins is that they are still toxic. A fusion protein of LT-B with 10 amino acids of ST at the N terminus does not have this problem (Sanchez et al., 1988b), but neutralizing antibodies have not been detected against ST (Clements, 1990). Another attempt with ST at the C terminus of the B subunit results in a fusion protein that no longer forms oligomers and loses most of its G_{M1} binding ability (Clements, 1990). Since this ability is important for the stimulation of the mucosal immunogenic effect of LT (de Aizpurua and Russell-Jones, 1988), it is not surprising that the fusion protein is less immunogenic than the parent protein (Clements, 1990).

Knowledge of the three-dimensional structure of LT and CT may be of use in the development of a vaccine against disease caused by *Vibrio cholerae* and enterotoxigenic *E. coli*. For instance, knowledge of the active site as outlined above allows easy identification of specific residues that should cripple the ADP-ribosylating activity of the toxin and hence remove serious side effect of an AB_5-based, or AB_5-containing, oral vaccine. However, generation of long-lasting and effective immune protection against the intestinal disease caused by these bacteria requires many factors to be interacting properly. Whether the three-dimensional information described in this overview will help lead to a breakthrough in the development of a highly effective vaccine protective against *Vibrio cholerae* or enterotoxigenic *E. coli* remains to be seen.

B. Adjuvant Activity of LT and CT

LT and CT are also increasingly being tested as immune adjuvants because of their exceptional capability to generate a significant mucosal immune response (Elson & Ealding, 1984) and because of their influence on tolerance induction. This effect is most noticeable when the other immunogen is linked to the toxin (McKenzie and Halsey, 1984). Many such fusion proteins have been made, and the method is currently tested with a variety of antigens, among them the *E. coli* heat-stable toxin (ST) but also horseradish peroxidase (McKenzie and Halsey, 1984), ovalbumin (Clements et al., 1988; van der Heijden et al., 1991), and, of more direct medical relevance, epitopes of surface proteins of the cariogenic bacterium *Streptococcus mutans* (Czerkinsky et al., 1989; Dertzbaugh et al., 1990) and hepatitis B virus nucleocapsid (Schödel et al., 1990).

In these hybrid assemblies the B subunits act to enhance the immunogenicity of the mucosally presented antigen, while the A subunit contributes adjuvant activity (van der Heijden et al., 1991; Czerkinsky et al., 1989). Optimal adjuvanticity of the A subunit may require that it not be defective in ADP-ribosylation activity (Lycke et al., 1992). This may constitute a serious obstacle for the development of effective oral LT- or CT-based vaccines that stimulate the mucosal immune system. However, it might be possible to use the knowledge of the three-dimensional structure of LT for the creation of LT and CT variants that have decreased activity of the A subunit but are still able to stimulate the M cells in the Peyer patches for an effective secretory IgA response.

Knowledge of the three-dimensional structure of the toxins can be of assistance in the development of fusion proteins that combine the adjuvant properties of the toxins with epitopes or complete domains of surface proteins of a wide variety of disease-causing organisms. A clear example of this is the fact that the structure allows decisions about the place of insertion or addition of epitopes to be made such that they leave the G_{M1} binding site intact. The latter requirement has been shown to be important for the adjuvanticity of the toxin (de Aizpurua and Russell-Jones, 1988).

C. Prospects for Drug Design

The three-dimensional structures of LT and CT offer marvelous perspectives for the development of a whole series of compounds that should prevent the action of these toxins in the human host. Some of these are listed below.

1. One obvious possibility is the design of molecules that block binding of G_{M1} to the receptor-binding site on the toxin. This prospect was already much on the mind of several early workers in the field when determining the receptor specificity of CT. Now, with detailed structural information available, it should be possible to arrive at receptor-blocking molecules in a rational manner using an interdisciplinary approach as described by Hol (1986) and Verlinde and Hol (1994).

2. The active-site crevice offers good opportunities for a very similar approach. Substrate analogues or molecules designed to block access to the substrate-binding sites would constitute potential drugs.

3. Peptide-mimicking molecules might compete with the binding of the long A2 helix to the A1 fragment and thereby interfere with the membrane translocation or other steps in the disease-causing process.

4. Since we know that relatively mild conditions, such as high concentrations of divalent cations, are sufficient to dissociate the A subunit from the B pentamer, and since we also know that the association of the A subunit with the B pentamer is essentially mediated only by the second half of the A2 chain and is not very extensive, it might be possible to design molecules that fit much tighter in the central pore of the B pentamer than the A2 chain does, thereby removing the toxic A subunit from the holotoxin and rendering the molecule innocuous.

It will be fascinating to see which of these approaches will be successful. A sobering thought here is that the costs of producing and delivering the final product must obviously be extremely low if we are to ensure that the benefits from our scientific investigation will reach those who need them most.

V. CONCLUDING REMARKS

The structural studies of heat-labile enterotoxin and of the B pentamer of cholera toxin described in this review have given an atomic resolution view and a strong structural basis for interpreting the enormous amount of information that has already been accumulated for these fascinating molecules. By no means have all the intriguing questions of the structure and function of CT and LT been settled by these investigations; assembly, excretion, membrane translocation, ARF recognition, target protein interaction, and the catalytic mechanism of the A subunit are still major mysteries, for instance. Also, the scourge of cholera and related diarrheal diseases has certainly not yet been banished from this planet. But it is our hope that by these structural investigations new steps in that direction can be made.

Acknowledgments

It is a great pleasure to thank the many colleagues who participated in our structural studies on LT and CT-B. Sylvia Pronk contributed greatly to the early stages of the project, Kor Kalk has collected an incredible number of data sets on the area detector in Groningen, and Focco van den Akker has processed several data sets more recently. The crystallization and heavy atom derivative preparation has greatly benefited from the enthusiasm and dedication of Hillie Groendijk, Ellen Wartna, Albert Berghuis, Anke Terwisscha, Angel Aguirre, and Ben van Zanten. Protein material of LT was generously provided in large amounts by Jaap Kingma and Bernard Witholt, while Cecile L'Hoir and Joseph Martial provided us with easily crystallized CT-B. The assistance of Steve Sarfaty and Stewart Turley with crystallization and data collection on the CT component, respectively, has been deeply appreciated. Stimulating and critical discussions with Christophe Verlinde, Ingeborg Feil, Tim Hirst, Randy Holmes, Rino Rappuoli, Jan Wilschut, and Piet Herdewijn have been most fruitful. We enjoyed reading the recent review by Jim Kaper prior to publication. Financial support for the crystallographic research described has been provided by the Dutch Chemical Foundation (SON) and The Netherlands Organization for Scientific Research (NWO) the National Institute of Health (NIH), the Murdock Charitable Trust, and by the University of Washington School of Medicine. Finally, we thank all our colleagues over the years for a most enjoyable atmosphere, both in Groningen and in Seattle.

REFERENCES

Acquotti, D., Poppe, L., Dabrowski, J., Lieth, C.-W. von der, Sonnino, S., and Tettamanti, G. (1990). Three dimensional structure of the oligosaccharide chain of G_{M1} ganglioside revealed by a distance-mapping procedure: a rotating and laboratory frame nuclear Overhauser enhancement investigation of native glycolipid in dimethyl sulfoxide and in water-dodecyl-phosphocholine solutions. *J. Am. Chem. Soc. 112*: 7772–7778.

Allured, V. S., Collier, R. J., Carroll, S. F., and McKay, D. B. (1986). Structure of exotoxin A of *Pseudomonas aeruginosa* at 3.0 angstrom resolution. *Proc. Natl. Acad. Sci. U.S.A. 83*: 1320–1324.

Bourne, H. R., Sanders, D. A., and McCormick, F. (1991). The GTPase superfamily: conserved structure and molecular mechanism. *Nature 349*: 117–127.

Burnette, W. N., Cieplak, W., Mar, V. L., Kaljot, K. T., Sato, H., and Keith, J. M. (1988). Pertussis toxin S1 mutant with reduced enzyme activity and a conserved protective epitope. *Science 242*: 72–74.

Burnette, W. N., Mar, V. L., Platler, B. W., Schlotterbeck, J. D., McGinley, M. D., Stoney, K. S., Rohde, M. F., and Kaslow, H. R. (1991). Site-specific mutagenesis of the catalytic subunit of cholera toxin: substituting lysine for arginine 7 causes loss of activity. *Infect. Immun. 59*: 4266–4270.

Carroll, S. F., and Collier, R. J. (1987). Active site of *Pseudomonas aeruginosa* exotoxin A. *J. Biol. Chem. 262*: 8707–8711.

Cassel, D., and Pfeuffer, T. (1978). Mechanism of cholera toxin action: covalent modification of the guanyl nucleotide-binding protein of the adenylate cyclase system. *Proc. Natl. Acad. Sci. U.S.A. 75*: 2669–2673.

Choe, S., Bennett, M. J., Fujii, G., Curmi, P. M. G., Kantardjieff, K. A., Collier, R. J., and Eisenberg, D. (1992). The crystal structure of diphtheria toxin. *Nature 357*: 216–222.

Cieplak, W., Jr., Locht, C., Mar, V. L., Burnette, W. N., and Keith, J. M. (1990). Photolabelling of mutant forms of the S1 subunit of pertussis toxin with NAD^+. *Biochem. J. 268*: 547–551.

Clements, J. D. (1990). Construction of a nontoxic fusion peptide for immunization against *Escherichia coli* strains that produce heat-labile and heat-stable enterotoxins. *Infect. Immun. 58*: 1159–1166.

Clements, J. D., and Finkelstein, R. A. (1979). Isolation and characterization of homogeneous heat-labile enterotoxins with high specific activity from *Escherichia coli* cultures. *Infect. Immun. 24*: 760–769.

Clements, J. D., Hartzog, N. M., and Lyon, F. L. (1988). Adjuvant activity of *Escherichia coli* heat-labile enterotoxin and effect on the induction of oral tolerance in mice to unrelated protein antigens. *Vaccine 6*: 269–277.

Critchley, D. R., Magnani, J. L., and Fishman, P. H. (1981). Interactions of cholera toxin with rat intestinal brush border membranes. *J. Biol. Chem. 256*: 8724–8731.

Cuatrecasas, P. (1973). Interaction of *Vibrio cholerae* enterotoxin with cell membranes. *Biochemistry 12*: 3547–3558.

Czerkinsky, C., Russell, M. W., Lycke, N., Lindblad, M., and Holmgren, J. (1989). Oral administration of a streptococcal antigen coupled to cholera toxin B subunit evokes strong antibody responses in salivary glands and extramucosal tissues. *Infect. Immun. 57*: 1072–1077.

Dallas, W. S., and Falkow, S. (1980). Amino acid sequence homology between cholera toxin and *Escherichia coli* heat-labile toxin. *Nature 288*: 499–501.

De, S. N., Bhattacharya, K., and Sarkar, J. K. (1956). A study of the pathogenicity of strains of *Bacterium coli* from acute and chronic enteritis. *J. Path. Bact. 71*: 201–209.

De, S. N. (1959). Enterotoxicity of bacteria-free culture-filtrate of *Vibrio cholerae. Nature 183*: 1533–1534.

de Aizpurua, H. J., and Russell-Jones, G. J. (1988). Oral vaccination. Identification of classes of proteins that provoke an immune response upon oral feeding. *J. Exp. Med. 167*: 440–451.

De Wolf, M. J. S., Fridkin, M., and Kohn, L. D. (1981). Tryptophan residues of cholera toxin and its A and B protomers; intrinsic fluorescence and solute quenching upon interacting with the ganglioside G_{M1}, oligo-G_{M1} or dansylated oligo-G_{M1}. *J. Biol. Chem. 256*: 5489–5496.

Dertzbaugh, M. T., Peterson, D. L., and Macrina, F. L. (1990). Cholera toxin B-subunit gene fusion: structural and functional analysis of the chimeric protein. *Infect. Immun. 58*: 70–79.

Dixon, S. J., Stewart, D., Grinstein, S., and Spiegel, S. (1987). Transmembrane signaling by the B subunit of cholera toxin: increased cytoplasmic free calcium in rat lymphocytes. *J. Cell Biol. 105*: 1153–1161.

Donta, S. T., Poindexter, N. J., and Ginsberg, B. H. (1982). Comparison of the binding of cholera and *Escherichia coli* enterotoxins to Y1 adrenal cells. *Biochemistry 21*: 660–664.

Douglas, C. M., and Collier, R. J., (1987). Exotoxin A of *Pseudomonas aeruginosa*: substitution of glutamic acid 553 with aspartic acid drastically reduces toxicity and enzymatic activity. *J. Bacteriol. 169*: 4967–4971.

Dutta, N. K., Panse, M. W., and Kulkarni, D. R. (1959). Role of cholera toxin in experimental cholera. *J. Bacteriol. 78*: 594–595.

Dwyer, J. D., and Bloomfield, V. A. (1982). Subunit arrangement of cholera toxin in solution and bound to receptor-containing model membranes. *Biochemistry 21*: 3227–3231.

Dykes, C. W., Halliday, I. J., Hobden, A. N., Read, M. J., and Harford, S. (1985). A comparison of the nucleotide sequence of the A subunit of heat-labile enterotoxin and cholera toxin. *FEMS Microbiol. Lett. 26*: 171–174.

Eidels, L., Proia, R. L., and Hart, D. A. (1983). Membrane receptors for bacterial toxins. *Microbiol. Rev. 47*: 596–620.

Elson, C.O., and Ealding, W. (1984). Generalized systemic and mucosal immunity in mice after mucosal stimulation and cholera toxin. *J. Immunol. 132*: 2736–2741.

Finkelstein, R. A. (1988). Cholera, the cholera enterotoxins, and the cholera enterotoxin-related enterotoxin family. In *Immunochemical and Molecular Genetic Analysis of Bacterial Pathogens*, P. Owen and T. J. Foster (Eds.), Elsevier, New York, pp. 85–102.

Finkelstein, R. A., and LoSpalluto, J. J. (1969). Pathogenesis of experimental cholera. Preparation and isolation of choleragen and choleragenoid. *J. Exp. Med. 130*: 185–202.

Finkelstein, R. A., Burks, M. F., Zupan, A., Dallas, W. S., Jacob, C. O., and Ludwig, D. S. (1987). Epitopes of the cholera family of enterotoxins. *Rev. Infect. Dis. 9*: 544–561.

Fishman, P. H., Moss, J., and Vaughan, M. (1976). Uptake and metabolism of gangliosides in transformed mouse fibroblasts. Relationship of ganglioside structure to choleragen response. *J. Biol. Chem. 251*: 4490–4494.

Fishman, P. H., Pacuszka, T., Hom, B., and Moss, J. (1980). Modification of ganglioside G_{M1}, effect of lipid moiety on choleragen action. *J. Biol Chem. 255*: 7657–7664.

Fraser, M. E., Chernaia, M. M., Kozlov, Y. V., and James, M.N.G. (1994). Crystal Structure of the holotoxin from shigella dysenteriae at 2.5Å resolution. *Nature Struct. Biol. 1*: 59–64.

Fukuta, S., Magnani, J. L., Twiddy, E. M., Holmes, R. K., and Ginsburg, V. (1988). Comparison of the carbohydrate-binding specificities of cholera toxin and *Escherichia coli* heat-labile enterotoxins LTh-I, LT-IIa and LT-IIb. *Infect. Immun. 56*: 1748–1753.

Galloway, T. S., and van Heyningen, S. (1987). Binding of NAD$^+$ by cholera toxin. *Biochem. J. 244*: 225–230.

Gennaro, M. L., Greenaway, P. J., and Broadbent, D. A. (1982). The expression of biologically active cholera toxin in *Escherichia coli*. *Nucleic Acids Res. 10*: 4883–4890.

Gill, D. M. (1975). Involvement of nicotinamide adenine dinucleotide in the action of cholera toxin in vitro. *Proc. Natl. Acad. Sci. U.S.A. 72*: 2064–2068.

Gill, D. M. (1976). The arrangement of subunits in cholera toxin. *Biochemistry 15*: 1242–1248.

Gill, D. M., and Coburn, J. (1987). ADP-ribosylation by cholera toxin: functional analysis of a cellular system that stimulates the enzymic activity of cholera toxin fragment A1. *Biochemistry 26*: 6364–6371.

Gill, D. M., and Meren, R. (1978). ADP-ribosylation of membrane proteins catalyzed by cholera toxin: basis of the activation of adenylate cyclase. *Proc. Natl. Acad. Sci. U.S.A. 75*: 3050–3054.

Gill, D. M., and Richardson, S. H. (1980). Adenosine diphosphate-ribosylation of adenylate cyclase catalyzed by heat-labile enterotoxin of *Escherichia coli*: comparison with cholera toxin. *J. Infect. Dis. 141*: 64–70.

Gill, D. M., and Woolkalis, M. (1988). ^{32}ADP-ribosylation of proteins catalyzed by cholera toxin and related heat-labile enterotoxins. *Methods Enzymol. 165*: 235–245.

Goins, B., and Freire, E. (1985). Lipid phase separation induced by the association of cholera toxin to phospholipid membranes containing ganglioside G_{M1}. *Biochemistry 24*: 1791–1797.

Goins, B., Masserini, M., Barisas, B. G., and Freire, E. (1986). Lateral diffusion of ganglioside G_{M1} in phospholipid bilayer membranes, *Biophys. J. 49*: 849–856.

Gorbach, S. L., Banwell, J. G., Chatterjee, B. D., Jacobs, B., and Sack, R. B. (1971). Acute undifferentiated human diarrhea in the tropics. I. Alterations in intestinal microflora. *J. Clin. Invest. 50*: 881–889.

Gray, G. L., Smith, D. H., Baldridge, J. S., Harkins, R. N., Vasil, M. L., Chen, E. Y., and Heyneker, H. L. (1984). Cloning, nucleotide sequence, and expression in *Escherichia coli* of the exotoxin A structural gene of *Pseudomonas aeruginosa*. *Proc. Natl. Acad. Sci. U.S.A. 81*: 2645–2649.

Gyles, C. L., and Barnum, D. A. (1969). A heat labile enterotoxin from strains of *Escherichia coli* enteropathogenic for pigs. *J. Infect Dis. 120*: 419–426.

Hirschhorn, N., and Greenough, W. B., III. (1991). Progress in oral rehydration therapy. *Sci. Am. 264*: 16–22.

Hirst, T. R., and Holmgren, J. (1987a). Transient entry of enterotoxin subunits into the periplasm occurs during their secretion from *Vibrio cholerae*. *J. Bacteriol. 169*: 1037–1045.

Hirst, T. R., and Holmgren, J. (1987b). Conformation of protein secreted across bacterial outer membranes: a study of enterotoxin translocation from *Vibrio cholerae*. *Proc. Natl. Acad. Sci. U.S.A. 84*: 7418–7422.

Hirst, T. R., Sanchez, J., Kaper, J. B., Hardy, S. J., and Holmgren, J. (1984). Mechanism of toxin secretion by *Vibrio cholerae* investigated in strains harboring plasmids that encode heat-labile enterotoxins of *Escherichia coli*. *Proc. Natl. Acad. Sci. U.S.A. 81*: 7752–7756.

Hol, W. G. J. (1986). Protein crystallography and computer graphics—toward rational drug design. *Angew. Chem. Int. Ed. Engl. 25*: 767–778.

Holmgren, J., Fredman, P., Lindblad, M., Svennerholm, A.-M., and Svennerholm, L. (1982). Rabbit intestinal glycoprotein receptor for *Escherichia coli* heat-labile enterotoxin lacking affinity for cholera toxin. *Infect. Immun. 38*: 424–433.

Hunt, P. D., and Hardy, S. J. S. (1991). Heat-labile enterotoxin can be released from *Escherichia coli* cells by host intestinal factors. *Infect. Immun. 59*: 168–171.

Jacob, C. O., Pines, M., and Arnon, R. (1984). Neutralization of heat-labile toxin of *E. coli* by antibodies to synthetic peptides derived from the B subunit of cholera toxin. *EMBO J.* *3*: 2889–2893.

Jacob, C. O., Arnon, R., and Finkelstein, R. A. (1986). Immunity to heat-labile enterotoxins of porcine and human *Escherichia coli* strains achieved with synthetic cholera toxin peptides. *Infect. Immun. 52*: 562–567.

Janicot, M., Clot, J.-P., and Desbuquois, B. (1988). Interactions of cholera toxin with isolated hepatocytes. Effects of low pH, chloroquine and monensin on toxin internalization, processing and action. *Biochem. J. 253*: 735–743.

Kahn, R. A., and Gilman, A. G. (1984). Purification of a protein cofactor required for ADP-ribosylation of the stimulatory regulatory component of adenylate cyclase by cholera toxin. *J. Biol. Chem. 259*: 6228–6234.

Kaper, J. B., Fasano, A., and Trucksis, M. (1993). Toxins of *Fibrio cholerae*. In *Vibrio cholerae and Cholera*, I. K. Wachsmuth, P. Balke, and Ø. Olsvik (Eds.), Am. Soc. Microbiol., Washington, DC.

Klipstein, F. A., Engert, R. F., and Clements, J. D. (1982). Development of a vaccine of cross-linked heat-stable and heat-labile enterotoxins that protects against *Escherichia coli* producing either enterotoxin. *Infect. Immun. 37*: 550–557.

Klipstein, F. A., Engert, R. F., and Houghten, R. A. (1983). Protection in rabbits immunized with a vaccine of *Escherichia coli* heat-stable toxin cross-linked to the heat-labile toxin B subunit. *Infect. Immun. 40*: 888–893.

Knoop, F. C., and Thomas, D. D. (1984). Effect of cholera enterotoxin on calcium uptake and cyclic AMP accumulation in rat basophilic leukemia cells. *Int. J. Biochem. 16*: 275–280.

Kunkel, S. L., and Robertson, D. C. (1979). Purification and chemical characterization of the heat-labile enterotoxin produced by enterotoxigenic *Escherichia coli. Infect. Immun. 25*: 586–596.

Kurosky, A., Markel, D. E., and Peterson, J. W. (1977). Covalent structure of the b chain of cholera enterotoxin. *J. Biol. Chem. 252*: 7257–7264.

Lai, C.-Y. (1977). Determination of the primary structure of cholera toxin B subunit. *J. Biol. Chem. 252*: 7249–7256.

Larew, J. S.-A., Peterson, J. E., and Graves, D. J. (1991). Determination of the kinetic mechanism of arginine-specific ADP-ribosyltransferases using a high performance liquid chromatographic assay. *J. Biol. Chem. 266*: 52–57.

Lee, C.-M., Chang, P. P., Tsai, S.-C., Adamik, R., Price, S. R., Kunz, B. C., Moss, J., Twiddy, E. M., and Holmes, R. K. (1991). Activation of *Escherichia coli* heat-labile enterotoxins by native and recombinant adenosine diphosphate-ribosylation factors, 20-kD guanine nucleotide-binding proteins. *J. Clin. Invest. 87*: 1780–1786.

Leitch, G. J., and Amer, M. S. (1975). Lanthanum inhibition of *Vibrio cholerae* and *Escherichia coli* enterotoxin-induced enterosorption and its effects on intestinal mucosa cyclic adenosine 3′-5′ monophosphate and cyclic guanosine 3′-5′ monophosphate levels. *Infect. Immun. 11*: 1038–1044.

Leong, J., Vinal, A. C., and Dallas, W. S. (1985). Nucleotide sequence comparison between heat-labile toxin B-subunit cistrons from *Escherichia coli* of human and porcine origin. *Infect. Immun. 48*: 73–77.

Levine, M. M., and Pierce, N. F. (1992). Immunity and vaccine development. In *Cholera*, Barua and Greenough (Eds.), Plenum, New York, pp. 285–327.

Levine, M. M., Black, R. E., Clements, J. L., Cisneros, L., Nalin, D. R., and Young, C. R. (1981). Duration of infection-derived immunity to cholera. *J. Infect. Dis. 143*: 818–820.

Levine, M. M., Kaper, J. B., Black, R. E., and Clements, M. L. (1983). New knowledge on pathogenesis of bacterial enteric infections as applied to vaccine development. *Microbiol. Rev. 47*: 510–550.

Lewis, M. J., and Pelham, H. R. B. (1990). A human homologue of the yeast HDEL receptor. *Nature 348*: 162–163.

Lobet, Y., Cluff, C. W., and Cieplak, W., Jr. (1991). Effect of site-directed mutagenic alterations on ADP-ribosyltransferase activity of the A subunit of *Escherichia coli* heat-labile enterotoxin. *Infect. Immun. 59*: 2870–2879.

Lockman, H., and Kaper, J. B. (1983). Nucleotide sequence analysis of the A2 and B subunits of *Vibrio cholerae* enterotoxin. *J. Biol. Chem. 258*: 13722–13726.

Lockman, H. A., Galen, J. E., and Kaper, J. B. (1984). *Vibrio cholerae* enterotoxin genes: nucleotide sequence analysis of DNA encoding ADP-ribosyltransferase. *J. Bacteriol. 159*: 1086–1089.

Lory, S., and Collier, R. J. (1980). Expression of enzymic activity by exotoxin A from *Pseudomonas aeruginosa. Infect. Immun. 28*: 494–501.

Lory, S., Carrol, S. F., Bernard, P. D., and Collier, R. J. (1980). Ligand interactions of diptheria toxin. II. Ligand interactions between the NAD site and the P site. *J. Biol. Chem. 255*: 12016–12019.

Ludwig, D. S., Holmes, R. K., and Schoolnik, G. K. (1985). Chemical and immunochemical studies on the receptor binding domain of cholera toxin B subunit. *J. Biol. Chem. 260*: 12528–12534.

Lycke, N., Tsuji, T., and Holmgren, J. (1992). The adjuvant effect of *Vibrio cholerae* and *Escherichia coli* heat-labile enterotoxin is linked to their ADP-ribosyltransferase activity. *Eur. J. Immunol. 22*: 2277–2281.

McDaniel, R. V., and McIntosh, T. J. (1986). X-ray diffraction studies of the cholera toxin receptor, G_{M1}. *Biophys. J. 49*: 94–96.

McKenzie, S. J., and Halsey, J. F. (1984). Cholera toxin B subunit as a carrier protein to stimulate a mucosal immune response. *J. Immunol. 133*: 1818–1823.

Maenz, D. D., and Forsyth, G. W. (1986). Cholera toxin facilitates calcium transport in jejunal brush border vesicles. *Can. J. Physiol. Pharmacol. 64*: 568–574.

Maenz, D. D., Gabriel, S. E., and Forsyth, G. W. (1987). Calcium transport affinity, ion competition and cholera toxin effects on cytosolic Ca concentration. *J. Membrane Biol. 96*: 243–249.

Masserini, M., Freire, M., Palestini, P., Calappi, E., and Tettamanti, G. (1992). Fuc-G_{M1} ganglioside mimics the receptor function of G_{M1} for cholera toxin. *Biochemistry 31*: 2422–2426.

Mekalanos, J. J., Collier, R. J., and Romig, W. R. (1979a). Enzymic action of cholera toxin. I. New method of assay and the mechanism of ADP-ribosyl transfer. *J. Biol. Chem. 254*: 5849–5854.

Mekalanos, J. J., Collier, R. J., and Romig, W. R. (1979b). Enzymic action of cholera toxin. II. Relationships to proteolytic processing, disulfide bond reduction, and subunit composition. *J. Biol. Chem. 254*: 5855–5861.

Mekalanos, J. J., Swartz, D. J., Pearson, G. D. N., Harford, N., Groyne, F., and De Wilde, M. (1983). Cholera toxin genes: nucleotide sequence, deletion analysis and vaccine development. *Nature 306*: 551–557.

Merritt, E. A., Sarfaty, S., van den Akker, F., L'hoir, C., Martial, J. A., and Hol, W. G. J. (1994a). Crystal structure of cholera toxin B-pentamer bound to receptor G_{M1} pentasaccharide. *Protein Sci. 3*: 166–175.

Merritt, E. A., Sixma, T. K., Kalk, K. H., van Zanten, B. A. M., and Hol, W. G. J. (1994b). Galactose binding site in *E. coli* heat-labile enterotoxin (LT) and cholera toxin (CT) *Mol. Microbiol. 13*: 745–753.

Merritt, E. A., Pronk, S. E., Sixma, T. K., Kalk, K. H., van Zanten, B. A. M., and Hol, W. G. J. (1994c). 2.6Å structure of partially-activated *E. coli* heat-labile enterotoxin (LT). *FEBS Lett. 307*: 88–92.

Middlebrook, J. L., and Dorland, R. B. (1984). Bacterial toxins: cellular mechanisms of action. *Microbiol. Rev. 48*: 199–221.

Moss, J., and Vaughan, M. (1977). Mechanism of action of choleragen. Evidence for ADP-ribosyltransferase activity with arginine as an acceptor. *J. Biol. Chem. 252*: 2455–2457.

Moss, J., and Vaughan, M. (1988). ADP-ribosylation of guanyl nucleotide-binding regulatory proteins by bacterial toxins. *Adv. Enzymol. 61*: 303–379.

Moss, J., and Vaughan, M. (1991). Activation of cholera toxin and *Escherichia coli* heat-labile enterotoxins by ADP-ribosylation factors, a family of 20 kDa guanine nucleotide-binding proteins. *Mol. Microbiol. 5*: 2621–2627.

Moss, J., Manganiello, and Vaughan, M. (1976). Hydrolysis of nicotinamide adenine dinucleotide by choleragen and its A protomer: possible role in the activation of adenylate cyclase. *Proc. Natl. Acad. Sci. U.S.A. 73*: 4424–4427.

Moss, J., Osborne, J. C., Jr., Fishman, P. H., Brewer, H. B., Jr., Vaughan, M., and Brady, R. O. (1977). Effect of gangliosides and substrate analogues on the hydrolysis of nicotinamide adenine dinucleotide by choleragen. *Proc. Natl. Acad. Sci. U.S.A. 74*: 74–78.

Moss, J., Garrison, S., Oppenheimer, N. J., and Richardson, S. H. (1979). NAD-dependent ADP-ribosylation of arginine and proteins by *Escherichia coli* heat-labile enterotoxin. *J. Biol. Chem. 254*: 6270–6272.

Moss, J., Osborne, J. C., Jr., Fishman, P. H., Nakaya, S., and Robertson, D. C. (1981). *Escherichia coli* heat-labile enterotoxin. Ganglioside specificity and ADP-ribosyltransferase activity. *J. Biol. Chem. 256*: 12861–12865.

Moss, J., Stanley, S., Vaughan, M., and Tsuji, T. (1993). Interaction of ADP-ribosylation factor with *Escherichia coli* enterotoxin that contains an inactivating lysine 112 substitution. *J. Biol. Chem. 268*: 6383–6387.

Mosser, G., and Brisson, A. (1991). Conditions of two-dimensional crystallization of cholera toxin B-subunit on lipid films containing ganglioside G_{M1}. *J. Struct. Biol. 106*: 191–198.

Murzin, A. G. (1993). OB (oligonucleotide/oligosaccharide binding)-fold: common structural and functional solution for non-homologous sequences. *EMBO J. 12*: 861–867.

Nakamura, K., Suzuki, M., Inagaki, F., Yamakawa, T., and Suzuki, A. (1987). A new ganglioside showing choleragenoid-binding activity in mouse spleen. *J. Biochem. 101*: 825–835.

Neill, R. J., Ivins, B. E., and Holmes, R. K. (1983). Synthesis and secretion of the plasmid-coded heat-labile enterotoxin of *Escherichia coli* in *Vibrio cholerae*. *Science 221*: 289–291.

Noda, M., Tsai, S.-C., Adamik, R., Moss, J., and Vaughan, M. (1990). Mechanism of cholera toxin activation by a guanine nucleotide-dependent 19 kDa protein. *Biochim. Biophys. Acta 1034*: 195–199.

Olsvik, Ø., Wahlberg, J., Petterson, B., Uhlén, M., Popovic, T., Wachsmuth, I. K., and Fields, P. I. (1993). Use of automated sequencing of polymerase chain reaction-generated amplicons to identify three types of cholera toxin subunit B in *Vibrio cholerae* O1 Strains. *J. Clin. Microbiol. 31*: 22–25.

Oppenheimer, N. J. (1978). Structural determination and stereospecificity of the choleragen catalyzed reaction of NAD^+ with guanidines. *J. Biol. Chem. 253*: 4907–4910.

Osborne, J. C., Jr., Moss, J., Fishman, P. H., Nakaya, S., and Robertson, D. C. (1982). Specificity in protein–membrane associations: the interaction of gangliosides with *Escherichia coli* heat-labile enterotoxin and choleragen. *Biophys. J. 37*: 168–169.

Osborne, J. C., Jr., Stanley, S. J., and Moss, J. (1985). Kinetic mechanisms of two NAD:arginine ADP-ribosyltransferases: the soluble salt-stimulated transferase from turkey erythrocytes and choleragen, a toxin from *Vibrio cholerae*. *Biochemistry 24*: 5235–5240.

Pacuszka, T., and Fishman, P. H. (1990). Generation of cell surface neoganglioproteins. G_{M1}-Neoganglioproteins are non-functional receptors for cholera toxin. *J. Biol. Chem. 265*: 7673–7678.

Pearson, G. D. N., and Mekalanos, J. J. (1982). Molecular cloning of *Vibrio cholerae* enterotoxin genes in *Escherichia coli* K-12. *Proc. Natl. Acad. Sci. U.S.A. 79*: 2976–2980.

Pelham, H. R. B. (1991). Recycling of proteins between the endoplasmic reticulum and Golgi complex. *Curr. Opinion Cell Biol. 3*: 585–591.

Peterson, J. W., and Ochoa, L. G. (1989). Role of prostaglandins and cAMP in the secretory effects of cholera toxin. *Science 245*: 857–859.

Picket, C. L., Weinstein, D., and Holmes, R. K. (1987). Genetics of type IIa heat-labile enterotoxin of *Escherichia coli*: operon fusions, nucleotide sequence, and hybridization studies. *J. Bacteriol. 169*: 5180–5187.

Picket, C. L., Twiddy, E. M. Coker, C., and Holmes, R. K. (1989). Cloning, nucleotide sequence, and hybridization studies of the type IIb heat-labile enterotoxin gene of *Escherichia coli*. *J. Bacteriol. 171*: 4945–4952.

Pizza, M.-G., Bartoloni, A., Prugnola, A., Silvestri, S., and Rappuoli, R. (1988). Subunit S1 of pertussis toxin: mapping of the regions essential for ADP-ribosyltransferase activity. *Proc. Natl. Acad. Sci. U.S.A. 85*: 7521–7525.

Poppe, L., von der Lieth, C.-W., and Dabrowski, J. (1990). Conformation of the glycolipid globoside head group in various solvents and in the micelle-bound state. *J. Am. Chem. Soc. 112*: 7762–7771.

Pronk, S. E., Hofstra, H., Groendijk, H., Kingma, J., Swarte, M. B. A., Dorner, F., Drenth, J., Hol, W. G. J., and Witholt, B. (1987). Heat-labile enterotoxin of *Escherichia coli*. Characterization of different crystal forms. *J. Biol. Chem. 260*: 13580–13584.

Ratti, G., Rappuoli, R., and Giannini, G. (1983). The complete nucleotide sequence of the gene encoding for diphtheria toxin in the corynephage omega (tox+). genome. *Nucleic Acids Res. 11*: 6589–6595.

Reed, R. A., Mattai, J., and Shipley, G. G. (1987). Interaction of cholera toxin with ganglioside G_{M1} receptors in supported lipid monolayers. *Biochemistry 26*: 824–832.

Ribi, H. O., Ludwig, D. S., Mercer, K. L., Schoolnik, G. K., and Kornberg, R. D. (1988). Three-dimensional structure of cholera toxin penetrating a lipid membrane. *Science 239*: 1272–1276.

Rothman, J. E., and Orci, L. (1992). Molecular dissection of the secretory pathway. *Nature 355*: 409–415.

Sabesan, S., Bock, K., and Lemieux, R. U. (1984). The conformational properties of the gangliosides G_{M2} and G_{M1} based on ^1H and ^{13}C nuclear magnetic resonance studies. *Can. J. Chem. 62*: 1034–1045.

Sanchez, J., Uhlin, B. E., Grundstrøm, T., Holmgren, J., and Hirst, T. R. (1986). Immunoactive chimeric ST-LT enterotoxins of *Escherichia coli* generated by in vitro gene fusion. *FEBS Lett. 208*: 194–198.

Sanchez, J., Hirst, T. R., and Uhlin, B. E. (1988a). Hybrid enterotoxin LTA::STa proteins and their protection from degradation by in vivo association with B-subunits of *Escherichia coli* heat-labile enterotoxin. *Gene 64*: 265–275.

Sanchez, J., Svennerholm, A.-M., and Holmgren, J. (1988b). Genetic fusion of a non-toxic heat-stable enterotoxin-related decapeptide antigen to cholera toxin B-subunit. *FEBS Lett. 241*: 110–114.

Sattler, J., Schwarzmann, G., Staerk, J., Ziegler, W., and Wiegandt, H. (1977). Studies of ligand binding to cholera toxin. II. The hydrophilic moiety of sialoglycolipids. *Hoppe-Seyler's Z. Physiol. Chem. 358*: 159–163.

Sattler, J., Schwarzmann, G., Knack, I., Røhm, K.-H., and Wiegandt, H. (1978). Studies of ligand binding to cholera toxin. III. Cooperativity of oligosaccharide binding. *Hoppe-Seyler's Z. Physiol. Chem. 359*: 719–723.

Schengrund, C.-L., and Ringler, N. J. (1989). Binding of *Vibrio cholerae* toxin and the heat-labile enterotoxin of *Escherichia coli* to G_{M1}, derivatives of G_{M1}, and nonlipid oligosaccharide polyvalent ligands. *J. Biol. Chem. 264*: 13233–13237.

Schödel, F., Enders, G., Jung, M.-C., and Will, H. (1990). Recognition of a hepatitis B virus nucleocapsid T-cell epitope expressed as a fusion protein with the subunit B of *Escherichia coli* heat labile enterotoxin in attenuated salmonellae. *Vaccine 8*: 569–572.

Schön, A., and Freire, E. (1989). Thermodynamics of intersubunit interactions in cholera toxin upon binding to the oligosaccharide portion of its cell surface receptor, ganglioside G_{M1}. *Biochemistry 28*: 5019–5024.

Serafini, T., Orci, L., Amherdt, M., Brunner, M., Kahn, R. A., and Rothman, J. E. (1991).

ADP-ribosylation factor (ARF) is a subunit of the coat of Golgi-derived COP-coated vesicles: a novel role for a GTP-binding protein. *Cell 67*: 239–253.

Sixma, T. K., Pronk, S. E., Kalk, K. H., Wartna, E. S., van Zanten, B. A. M., Witholt, B., and Hol, W. G. J. (1991a). Crystal structure of a cholera toxin-related heat-labile enterotoxin from *E. coli. Nature 351*: 371–378.

Sixma, T. K., Pronk, S. E., Terwisscha van Scheltinga, A. C., Aguirre, A., Kalk, K. H., Vriend, G., and Hol, W. G. J. (1991b). Native non-isomorphism in the structure determination of heat labile enterotoxin (LT) from *E. coli.* In *Isomorphous Replacement and Anomalous Scattering,* W. Wolf, P. R. Evans, and A. G. W. Leslie (Eds.), Daresbury, U. K., pp. 133–140.

Sixma, T. K., Pronk, S. E., Kalk, K. H., van Zanten, B. A. M., Berghuis, A. M., and Hol, W. G. J. (1992a). Lactose binding to heat-labile enterotoxin revealed by X-ray crystallography. *Nature 355*: 561–564.

Sixma, T. K., Terwisscha van Scheltinga, A. C., Kalk, K. H., Zhou, K., Wartna, E. S., and Hol, W. G. J. (1992b). X-ray studies reveal lanthanide binding sites at the A/B$_5$ interface of *E. coli* heat labile enterotoxin. *FEBS Lett 297*: 179–182.

Sixma, T. K., Aguirre, A., Terwisscha van Scheltinga, A. C., Wartna, E. S., Kalk, K. H., and Hol, W. G. J. (1992c). Heat-labile enterotoxin crystal forms with variable A/B$_5$ orientation. *FEBS Lett. 305*: 81–85.

Sixma, T. K., Kalk, K. H., van Zanten, B. A. M., Dauter, Z., Kingma, J., Witholt, B., and Hol, W. G. J. (1993). Refined structure of *Escherichia coli* heat-labile enterotoxin, a close relative of cholera toxin. *J. Mol. Biol. 230*: 890–918.

Smith, H. W., and Halls, S. (1967). *Escherichia coli* enterotoxin. *J. Pathol. Bacteriol. 93*: 531–543.

Soman, G., Narayanan, J., Martin, B. L., and Graves, D. J. (1986). Use of substituted (benzylidineamino) guanidines in the study of guanidino group specific ADP-ribosyl transferase. *Biochemistry 25*: 4113–4119.

Song, W., and Rintoul, D. A. (1989). Synthesis and characterization of *N*-parinaroyl ganglioside G$_{M1}$: effect of choleragen binding on fluorescence anisotropy in model membranes. *Biochemistry 28*: 4194–4200.

Surewicz, W. K., Leddy, J. J., and Mantsch, H. H. (1990). Structure, stability, and receptor interaction of cholera toxin as studied by Fourier-transform infrared spectroscopy. *Biochemistry 29*: 8106–8111.

Tait, M., and Nassau, P. M. (1984). Artificial low-molecular-mass substrates of cholera toxin. *Eur. J. Biochem. 143*: 213–219.

Thompson, T. E., and Tillack, T. W. (1985). Organisation of glycosphingolipids in bilayers and plasma membranes of mammalian cells. *Ann. Rev. Biophys. Biophys. Chem. 14*: 361–386.

Thompson, T. E., Allietta, M., Brown, R. E., Johnson, M. L., and Tillack, T. W. (1985). Organization of ganglioside G$_{M1}$ in phosphatidylcholine bilayers. *Biochim. Biophys. Acta 817*: 229–237.

Tomasi, M., and Montecucco, C. (1981). Lipid insertion of cholera toxin after binding to G$_{M1}$-containing liposomes. *J. Biol. Chem. 256*: 11177–11181.

Tsuji, T., Honda, T., Miwatani, T., Wakabayashi, S., and Matsubara, H. (1984). The amino acid sequence of the b-subunit of porcine enterotoxigenic *Escherichia coli* enterotoxin—analysis and comparison with literature data. *FEMS Microbiol. Lett. 25*: 243–246.

Tsuji, T., Inoue, T., Miyama, A., and Noda, M. (1991). Glutamic acid-112 of the A subunit of heat-labile enterotoxin from enterotoxigenic *Escherichia coli* is important for ADP-ribosyltransferase activity. *FEBS Lett. 291*: 319–321.

van der Heijden, P. J., Bianchi, A. T. J., Dol, M., Pals, J. W., Stok, W., and Bokhout, B. A. (1991). Manipulation of intestinal immune responses against ovalbumin by cholera toxin and its B subunit in mice. *Immunology 72*: 89–93.

Van Dop, C., Tsubokawa, M., Bourne, H. R., and Ramachandran, J. (1984). Amino acid sequence of retinal transducin at the site ADP-ribosylated by cholera toxin. *J. Biol. Chem.* *259*: 696–698.

van Heyningen, S. (1983). The interaction of cholera toxin with gangliosides and the cell membrane. *Curr. Top. Membranes Transp.* *18*: 445–471.

van Heyningen, W. E., Carpenter, C. C., Pierce, N. F., and Greenough, W. C., III. (1971). Deactivation of cholera toxin by ganglioside. *J. Infect. Dis.* *124*: 415–418.

Van Ness, B. G., Howard, J. B., and Bodley, J. W. (1980). ADP-ribosylation of elongation factor 2 by diphtheria toxin: NMR spectra and proposed structures of ribosyl-diphthamide and its hydrolysis products. *J. Biol. Chem.* *255*: 10710–10716.

Verlinde, C.L.M.J., and Hol, W.G.J. (1994). Structure-based drug design: progress, results and challenges. *Structure 2*: 577–587.

Vyas, N. K. (1991). Atomic features of protein–carbohydrate interactions. *Curr. Opinions Struct. Biol.* *1*: 732–740.

Vyas, N. K., Vyas, M. N., and Quiocho, F. A. (1988). Sugar and signal-transducer binding sites of *Escherichia coli* galactose chemoreceptor protein. *Science 242*: 1290–1295.

Weiss, W., Brown, J. H., Cusack, S., Paulson, J. C., Skehel, J. J., and Wiley, D. C. (1988). Structure of the influenza virus haemagglutinin complexed with its receptor, sialic acid. *Nature 333*: 426–431.

Wisnieski, B. J., and Bramhall, J. S. (1981). Photolabelling of cholera toxin subunits during membrane penetration. *Nature 289*: 319–321.

Wisnieski, B. J., Shiflett, M. A., Mekalanos, J., and Bramhall, J. S. (1979). Analysis of transmembrane dynamics of cholera toxin using photoreactive probes. *J. Supramol. Struct.* *10*: 191–197.

Yamamoto, T., and Yokota, T. (1983). Sequence of heat-labile enterotoxin of *Escherichia coli* pathogenic for humans. *J. Bacteriol.* 155: 728–733.

Yamamoto, T., Tamura, T., and Yokota, T. (1984). Primary structure of heat-labile enterotoxin produced by *Escherichia coli* pathogenic for humans. *J. Biol. Chem.* *259*: 5037–5044.

Yamamoto, T., Gojobori, T., and Yokota, T. (1987). Evolutionary origin of pathogenic determinants in enterotoxigenic *Escherichia coli* and *Vibrio cholerae* O1. *J. Bacteriol.* *169*: 1352–1357.

Yang, J., Tamm, L. K., Tillack, T. W., and Shao, Z. (1993). New approach for atomic force microscopy of membrane proteins. *J. Mol. Biol.* *229*: 286–290.

10

Cholera Toxin and Related Enterotoxins of Gram-Negative Bacteria

Randall K. Holmes, Michael G. Jobling, and Terry D. Connell*

Uniformed Services University of the Health Sciences, Bethesda, Maryland

Current affiliation: The University at Buffalo, State University of New York, Buffalo, New York

I. INTRODUCTION

Cholera is a severe and potentially fatal disease that is caused by *Vibrio cholerae* and characterized by massive watery diarrhea. Discoveries during the 1950s showed that products of *V. cholerae* could produce secretory diarrhea when they were injected into ligated ileal loops of adult rabbits or into the intestine of infant rabbits (De, 1959; Dutta et al., 1959). These findings were critical for the subsequent purification of cholera enterotoxin (CT) to apparent homogeneity by Finkelstein and LoSpalluto (1969). The availability of purified CT as a reagent for basic and clinical research was a powerful stimulus for studies both on the pathogenesis of enterotoxin-mediated diarrheal diseases and on the structure, function, immunochemistry, molecular genetics, and regulation of CT and related enterotoxins. Cholera remains the prototype for secretory diarrheal diseases caused by enterotoxin-producing bacteria.

Gyles and Barnum (1969) described a heat-labile enterotoxin (LT), produced by a strain of *Escherichia coli* that causes diarrhea in swine, that is similar to CT. Subsequently, LT-producing strains of *E. coli* were isolated from humans with diarrhea (Sack et al., 1971; Etkin and Gorbach, 1971). Enterotoxigenic *E. coli* (ETEC) that produce LT, heat-stable enterotoxin (ST), or both LT and ST are now recognized as a major cause of diarrhea among infants and young children in developing countries (Huilan et al., 1991), and ETEC is also the single most frequent cause of traveler's diarrhea among visitors to the tropics (Levine, 1987).

Rabbit ileal loop tests, infant rabbit tests, and bioassays using a variety of cultured cell lines (Donta et al., 1974; Guerrant et al., 1974; Maneval et al., 1980) were used by many investigators in the last two decades to test known enteropathogenic or potentially enteropathogenic bacteria for production of heat-labile enterotoxins, and tests for neutralization of enterotoxic activity by specific antibodies against purified CT provided the initial basis for determining which enterotoxins produced by bacteria other than *V. cholerae* are antigenically related to CT (Finkelstein, 1973). Subsequently, genetic and biochemical criteria were used to identify additional enterotoxins, designated type II enterotoxins, that are related to CT but are not neutralized by anti-CT antisera (Holmes and Twiddy, 1983; Pickett et al., 1986; Holmes et al., 1986; Guth et al., 1986a). The type II heat-labile enterotoxins (LT-II) are produced by some strains of *E. coli* that have been isolated from humans, cattle, other livestock, and food (Green et al., 1983; Guth et al., 1986a; Seriwatana et al., 1988), but they have not yet been implicated as causes of diarrheal disease. Gram-negative bacteria of several different genera and species have now been shown to produce CT-related enterotoxins (Table 1).

This chapter summarizes the information currently available about the family of heat-labile enterotoxins related to CT. Its focus is on comparative studies, with emphasis on CT and well-characterized enterotoxins produced by various strains of *E. coli*. Other chapters in this volume provide additional information about the pathobiology, crystallography, and biogenesis of the CT-related enterotoxins, their interactions with ADP-ribosylation factors (ARFs), and their relationships to the superfamily of bacterial toxins that function as ADP-ribosyltransferase enzymes. Several recent reviews also provide useful information on related topics (Nashar et al., 1993; Kaper and Srivastava, 1992; Moon and Bunn, 1993; Spangler, 1992; Gyles, 1992; Acheson, 1992; Moss and Vaughan, 1993b; Holmgren et al., 1989; Sack, 1990; Holmes et al., 1991).

TABLE 1 Bacteria Other Than *Vibrio cholerae* That Produce Cholera Toxin-like Heat-Labile Enterotoxins

Species	Reference
Aeromonas hydrophila	Annapurna and Sanyal, 1977; Honda et al., 1985
Campylobacter jejuni	Johnson and Lior, 1986
Escherichia coli	Gyles and Barnum, 1969; Green et al., 1983
Klebsiella and *Enterobacter*	Klipstein et al., 1979
Non-cholera vibrios	Zinnaka and Carpenter, 1972; Sanyal, 1983
Plesiomonas shigelloides	Gardner et al., 1987
Salmonella typhi	Fernandez et al., 1988
Salmonella typhimurium	Sandefur and Peterson, 1976; Prasad et al., 1992
Vibrio mimicus	Spira and Fedorka-Cray, 1983

II. STRUCTURE

Cholera toxin (CT), also called choleragen, is an 84-kDa heterohexameric protein composed of two subunits, a 27-kDa A subunit (CT-A) consisting of 240 amino acid residues and a pentameric B subunit with five 11.5-kDa B polypeptides (CT-B), each of which has 103 amino acid residues (Olsnes et al., 1990; Spangler, 1992). The primary structure of CT was analyzed first by amino acid sequencing (Lai et al., 1976; Kurosky et al., 1976) and subsequently by nucleotide sequencing of the *ctxA* and *ctxB* genes (Lockman et al., 1984; Lockman and Kaper, 1983; Mekalanos et al., 1983). Each A or B polypeptide has a single intramolecular disulfide bond. The holotoxin (AB$_5$) is held together by very stable noncovalent interactions between the polypeptides.

The A polypeptide in CT can be proteolytically cleaved between Cys-187 and Cys-199 to produce a 22-kDa A1 fragment and a 5.5-kDa A2 fragment linked by the disulfide bond. Such cleavage can be accomplished either by the hemagglutinin/protease (HP) of *V. cholerae* (Booth et al., 1984) or by trypsin (Gill and Rappaport, 1979; Mekalanos et al., 1979b), although the sites of cleavage by these enzymes are different. CT with the cleaved form of polypeptide A is designated as "nicked." Typically, CT recovered after purification from *V. cholerae* culture supernatants is in the nicked form. With some El Tor strains that produce large amounts of HP, however, the A polypeptide in CT recovered from culture supernatants is more extensively degraded (Dubey et al., 1990).

Cross-linking experiments provided direct evidence that the B polypeptides in CT associate to form a pentamer (B$_5$) and that fragment A2, but not fragment A1, interacts closely with the B polypeptides (Gill, 1976). Isolated B$_5$, initially called choleragenoid, is a remarkably stable complex. It does not dissociate in 0.1% SDS at a neutral or alkaline pH (Gill, 1976) or in 6 M urea and 20 mM dithiothreitol (Mekalanos et al., 1978b), although these conditions release fragment A1 from CT, leaving fragment A2 still bound to B$_5$. The heat-lability of CT is due to the irreversible denaturation of CT-A at temperatures above 51°C (Goins and Freire, 1988). At pH 3.2 or below, CT dissociates completely into its constituent A and B polypeptides (Finkelstein et al., 1974).

CT was first crystallized in the 1970s (Finkelstein and LoSpalluto, 1972; Sigler et al., 1977), but crystals suitable for high-resolution X-ray diffraction were not obtained reproducibly. Analysis of two-dimensional crystalline arrays of CT bound to ganglioside G$_{M1}$ receptors in a lipid membrane (Ribi et al., 1988) showed that the B pentamer exists as a planar ring of B polypeptides and that the A subunit is located centrally but asymmetrically with respect to the ring. Elimination of microheterogeneity by using isoelectric focusing to purify a specific charged species of CT made it possible to obtain well-ordered crystals (Spangler and Westbrook, 1989). The crystal structure of CT at a resolution of 2.4 Å is reported (Gibbons, 1991) to be almost identical to that of a closely related LT from *E. coli* described below (Sixma et al., 1991, 1993), but details of the structure for CT are not yet published.

Several LT toxins of *E. coli* have been purified and characterized (Kunkel and Robertson, 1979; Clements and Finkelstein, 1979; Clements et al., 1980; Tsuji et al., 1982; Geary et al., 1982; Holmes et al., 1980, 1983), and the genes that encode them have been cloned and sequenced (Dallas et al., 1979; Dallas and Falkow, 1980; Spicer et al., 1981; Dallas, 1983; Yamamoto et al., 1984a,b; Leong et al., 1985). As described in more detail below, LT toxins are classified into two serogroups designated LT-I and LT-II, with two or more antigenic variants recognized in each serogroup (Finkelstein, 1983; Gilligan et al., 1983; Green et al., 1983; Holmes et al., 1983, 1986; Marchlewicz and Finkelstein, 1983; Takeda et al., 1983; Geary et al., 1982; Honda et al., 1981; Guth et al., 1986b; Pickett et al., 1986). Nevertheless, each LT holotoxin, like CT, is a heterohexameric complex consisting of an A polypeptide and a pentameric array of a B polypeptides (Gill et al., 1981). In contrast to CT, purification of LT-I or LT-II from *E. coli* typically results in recovery of the intact, unnicked holotoxin (Holmes et al., 1986; Clements and Finkelstein, 1979). Treatment of LT-I (Clements and Finkelstein, 1979; Kunkel and Robertson, 1979) or LT-II (Holmes et al., 1986; Guth et al., 1986b) with trypsin cleaves the A polypeptide into A1 and A2 fragments linked by a disulfide bond in a manner similar to that described above for CT (Gill and Rappaport, 1979; Mekalanos et al., 1979b). Characteristics of the A and B polypeptides of representative LT-I and LT-II enterotoxins are summarized in Table 2 and compared with those of CT-A and CT-B.

LT-I from a porcine isolate of *E. coli* was crystallized, and its structure was determined at high resolution by X-ray diffraction analysis (Sixma et al., 1991, 1992a, 1993). These studies, described more fully in the chapter by Hol in this volume, revealed in detail how the five B polypeptides of LT form a planar, donut-shaped ring with a central pore that is penetrated by the carboxyl-terminal segment of A2. A1 forms a globular domain that extends outward from one face of the B pentamer and interacts only weakly with the B polypeptides. The disulfide-linked loop between Cys-187 and Cys-199 of the A polypeptide, which is the target for proteolytic nicking, is flexible and is exposed on the surface of the enterotoxin. Limited conformational flexibility between the A and B subunits was also observed among independently formed crystals of LT-I.

III. IMMUNOCHEMISTRY

The family of CT-related heat-labile enterotoxins is currently divided into two serogroups on the basis of neutralization tests with antisera prepared against highly purified reference toxins (Holmes et al., 1986; Pickett et al., 1986; Guth et al., 1986b). Antiserum raised against each purified type I enterotoxin neutralizes all type I enterotoxins tested (Finkelstein, 1983; Marchlewicz and Finkelstein, 1983; Holmes et al., 1983, 1985; Takeda et al., 1983), and antisera raised against each purified type II enterotoxin neutralizes all type II enterotoxins tested (Holmes et al., 1986; Guth et al., 1986b). When the reference antisera are titrated in neutralization tests against various members of the same toxin

TABLE 2 Comparison of the A and B Polypeptides of Representative Heat-Labile Enterotoxins[a]

Polypeptide	Toxin	Number of amino acid residues	Location of cysteine residues	pI
A	CT	240	187, 199	6.4
	LTh-I	240	187, 199	5.7
	LTp-I	240	187, 199	5.9
	LT-IIa	241	185, 197	6.3
	LT-IIb	243	185, 197	6.5
B	CT	103	9, 86	7.6
	LTh-I	103	9, 86	8.2
	LTp-I	103	9, 86	8.7
	LT-IIa	100	10, 81	7.3
	LT-IIb	99	10, 81	4.8

[a]The number of amino acid residues in the mature A and B polypeptides and the locations of the cysteine residues are deduced from published sequences for the representative enterotoxin genes as cited in Figure 1. The isoelectric points were determined experimentally for the A and B polypeptides of CT, LTh-I, and LTp-I (Holmes et al., 1985) and were predicted for LT-II and LT-IIb from the deduced amino acid sequences.

serogroup, the highest titers are obtained with the specific toxins to which the reference antisera are prepared (Holmes et al., 1983; Finkelstein, 1983; Marchlewicz and Finkelstein, 1983; Guth et al., 1986b).

Serogroup I includes CT, the variants of LT-I produced by ETEC strains from humans and pigs (designated LTh-I and LTp-I, respectively), and enterotoxins produced by several other gram-negative bacteria (Table 1). Serogroup II includes the variants of LT-II from *E. coli* designated LT-IIa and LT-IIb. Genetic evidence indicates that additional variants of LT-II exist among isolates of *E. coli* (Pickett et al., 1989). Production of type II enterotoxins by bacteria other than *E. coli* has not been demonstrated.

When CT, LTh-I, and LTp-I are tested in Ouchterlony gel diffusion tests against antisera prepared against each of the purified toxins, the immunoprecipitates form lines of partial identity, indicating that CT, LTh-I, and LTp-I express both common and unique epitopes (Finkelstein, 1983; Geary et al., 1982; Gilligan et al., 1983; Holmes et al., 1983; Takeda et al., 1983; Tsuji et al., 1982). LTh-I and LTp-I exhibit a closer antigenic relationship to each other than to CT, and CT is related more closely to LTh-I than to LTp-I.

Studies with polyclonal antisera demonstrate that the immunodominant epitopes of CT are associated with the B polypeptides (Holmgren and Svennerholm, 1977). Monoclonal antibodies (mAbs) against CT-B that have potent neutralizing activity interact with conformation-dependent epitopes (Ludwig et al., 1985), but only limited progress has been made in determining the structure of these epitopes (Ludwig et al., 1985; Finkelstein et al., 1987a,b). Antisera raised against synthetic oligopeptides corresponding with specific regions of CT or LT-I can exhibit neutralizing activity, but their titers appear to be low (Jacob et al., 1986a,b, et al., 1990). Monoclonal antibodies raised against oligopeptides of CT-B often do not react with native CT (Spangler, 1991).

Monoclonal antibodies against type I enterotoxins have been characterized by several investigators (Robb et al., 1982; Ludwig et al., 1985; Belisle et al., 1984a,b; Holmes and Twiddy, 1983; Kazemi and Finkelstein, 1990a; Finkelstein et al., 1987a,b; Lindholm et al., 1983). These studies demonstrate that some epitopes of CT, LTh-I, and LTp-I are unique to a single toxin, others are shared by two of the toxins, and a few are common to all three of the toxins. Most mAbs react with epitopes that are determined either by polypeptide A or by polypeptide B, but some epitopes are expressed only by holotoxin (Holmes and Twiddy, 1983). Some mAbs have neutralizing activity and others do not. Neutralizing activity can be expressed by mAbs specific for either polypeptide A or polypeptide B of type I enterotoxins, but the mAbs that react only with CT holotoxin lacked neutralizing activity (Holmes and Twiddy, 1983). The greatest neutralizing activity per unit of immunoglobulin was found among mAbs specific for polypeptide B. No reactivity against type II enterotoxins has been demonstrated with mAbs prepared against type I enterotoxins (unpublished data of R. K. Holmes, E. M. Twiddy, F. M. J. Petitjean, and T. C. Connell).

Monoclonal antibodies against LT-IIa and LT-IIb have been characterized in our laboratory (unpublished data of R.K. Holmes, E. M. Twiddy, F. M. J. Petitjean, and T. Connell). Among collections of 11 mAbs against LT-IIa and 15 mAbs against LT-IIb, none recognized an epitope shared by LT-IIa and LT-IIb, and none cross-reacted with the representative type I enterotoxins CT, LTh-I, and LTp-I.

IV. GENETICS AND REGULATION

The genes for LT-I are located on plasmids in *E. coli* (Gyles et al., 1974; Smith and Linggood, 1971), whereas the genes for CT in *V. cholerae* (Kaper et al., 1982; Sporecke et al., 1984) and for LT-II in *E. coli* (Green et al., 1983; Pickett et al., 1986) are chromosomal. Successful cloning of the genes for LT-I was facilitated by their presence in plasmids, and these genes were among the first bacterial virulence determinants to be cloned (So et al., 1978; Dallas et al., 1979). Subsequently, LT-I genes were used as probes to clone CT genes from the *V. cholerae* chromosome by sequence homology (Pearson and Mekalanos, 1982; Mekalanos et al., 1983; Lockman and Kaper, 1983; Lockman et al., 1984). The strategy used for cloning the LT-II genes was based on expression of their biological activity, since the previously cloned LT-I genes did not hybridize to chromosomal DNA from strains of *E. coli* that produce LT-II (Pickett et al., 1986, 1987, 1989).

A. Organization of Enterotoxin Operons

The structural genes encoding LT-I and CT are highly homologous and are organized into very similar operons. The genes encoding LT-I (designated *tox* or *elt*) or CT (designated *ctx*) are transcribed as a single messenger RNA (Dallas et al., 1979; Moseley and Falkow, 1980; Dallas and Falkow, 1980; Yamamoto and Yokota, 1981; Spicer et al., 1981; Dallas, 1983; Yamamoto et al., 1984; Mekalanos et al., 1983; Lockman and Kaper, 1983). The genes for the A and B polypeptides overlap by four nucleotides. This overlapping organization of the genes suggests translational coupling, but the stoichiometry of the polypeptides in holotoxin requires the B polypeptide to be made in fivefold excess over the A polypeptide. Two mechanisms proposed to account for production of polypeptide B in the required excess over polypeptide A are (1) greater efficiency of the ribosome binding site for the B coding region than that for the A coding region (Mekalanos et al., 1983) and (2) formation of local secondary structures in the polycistronic mRNA that decrease translation of the A coding region (Yamamoto et al., 1985).

CT is produced by strains of *V. cholerae* that belong to serogroup O1 (Finkelstein, 1973) or the recently reported serogroup O139 (Bhattacharya et al., 1993), both of which have the potential to cause epidemic cholera. Strains of *V. cholerae* O1 may be either Ogawa or Inaba serotype as well as El Tor or classical biotype (Finkelstein, 1973). Classical biotype strains have two copies of the *ctx* operon at widely separated loci on the bacterial chromosome, and most El Tor strains have a single copy (Mekalanos, 1983). Some El Tor strains have multiple copies of the *ctx* operon on tandemly repeated genetic elements that are capable of transposition and site-specific recombination (Pearson et al., 1993). These tandemly repeated elements vary in size due to differences within the elements in copy number of a smaller repeated sequence called RS1. Although in some cases the operon for LT-I is flanked by inverted and direct repeats, it has not been demonstrated that these sequences mediate transposition of LT-I genes (Yamamoto and Yokota, 1981). Repetitive elements similar to the RS1 element have not been detected in the sequences flanking LT-I operons (Betley et al., 1986; Mekalanos, 1983).

The genes for LT-IIa and LT-IIb are organized into operons similar to those described above for the type I enterotoxins (Pickett et al., 1987, 1989). The overlap in the coding sequences for the A and B polypeptides of the type II enterotoxins is 11 nucleotides, however, instead of the four nucleotides found in the operons for the type I enterotoxins. Neither the location of the LT-IIa or LT-IIb operon on the chromosome of *E. coli* nor the characteristics of the DNA sequences that flank either operon has been determined.

B. Regulation of Enterotoxin Synthesis

Hypotoxinogenic mutants of *V. cholerae* that produce 1000-fold lower levels of toxin than wild-type classical strains (Holmes et al., 1978; Vasil et al., 1975; Finkelstein et al., 1974b) all mapped to a locus called *tox* (Baine et al., 1978), now called *toxR* (Miller and Mekalanos, 1985). Hypertoxinogenic mutants of *V. cholerae* were also isolated (Mekalanos et al., 1978a) and characterized (Mekalanos et al., 1979; Mekalanos and Murphy, 1980). Expression of the toxin operon is environmentally regulated by the transcriptional activator *toxR* as part of a regulatory cascade that controls the expression of many virulence genes in *V. cholerae* (Miller and Mekalanos, 1984, 1985; Ottemann et al., 1992; Mekalanos, 1992). A tandemly repeated TTTTGAT sequence is located upstream of the *ctx* operon in *V. cholerae* and serves as the recognition sequence for the *toxR* gene product (Miller and Mekalanos, 1984, 1985). This regulatory sequence for *toxR* is not found in association with LT-I and LT-II operons (Pickett et al., 1987, 1989; Yamamoto et al., 1984; Spicer et al., 1981). The *toxR* regulon is discussed in greater detail in the chapter by Mekalanos in this volume.

The regulatory elements that control expression of the LT-I and CT operons differ greatly. Most environmental factors that have significant effects on control of CT production in *V. cholerae* have little effect on synthesis of LT-I in *E. coli*, and mutations in *V. cholerae* that increase or decrease production of CT do not affect the expression of LT-I genes that are introduced into *V. cholerae* (Neill et al., 1983a; Betley et al., 1986). A severalfold stimulation of CT or LT-I production is observed when lincomycin or tetracycline is added to the growth medium at sublethal concentrations (Young and

Broadbent, 1986; Yoh et al., 1983; Levner et al., 1977, 1980). Extracellular cAMP is reported either to simulate (Gibert et al., 1990) or to have no effect on (Neill et al., 1983b) production of LT-I by *E. coli*. Several genetic effects on regulation of LT-I production have also been described. Several independent hypertoxinogenic (*htx*) chromosomal mutations in *E. coli* were found to increase production of LT-I by up to fivefold over the amounts made by the parental strain (Bramucci et al., 1981). Furthermore, when either of two different *ent* plasmids was mobilized into bacteria from several different genera in the family Enterobacteriaceae, the amount of LT-I produced by different bacteria that harbored each plasmid varied over a range of approximately 50-fold (Bramucci et al., 1981; Neill et al., 1983). The regulatory mechanisms that determine these various phenotypes have not yet been established.

Little information is available on regulation of LT-II operons. The amounts of enterotoxin made by different LT-II-producing isolates of *E. coli* vary greatly (Guth et al., 1986a; Holmes et al., 1986), but differences in expression of cloned LT-II genes appear to be determined primarily by the individual LT-II operons and the copy numbers of the plasmids into which they are cloned, rather than by the strains of *E. coli* in which the clones are expressed (Pickett et al., 1986, 1987, 1989; Holmes et al., 1986; Guth et al., 1986b). Addition of sublethal concentrations of lincomycin to the growth medium stimulates production of LT-II by some isolates in a manner similar to that observed for type I enterotoxins (Guth et al., 1986a; Holmes et al., 1986). The mechanisms for the lincomycin effect and for differences in yield of LT-II that are observed in different growth media (Holmes et al., 1986) have not been determined.

V. PROCESSING AND ASSEMBLY

Processing and assembly have been studied in detail with the type I enterotoxins (Hirst and Welch, 1988; Hirst, this volume). The polypeptides of CT or LT-I are synthesized as precursors with amino-terminal signal sequences that direct the preproteins to the periplasm (Hofstra and Witholt, 1984). After removal of signal peptides and delivery of mature A and B polypeptides into the periplasmic space, the polypeptides assemble spontaneously into holotoxin (Hardy et al., 1988). B monomers can also assemble into B_5 in the absence of A polypeptides, but A polypeptides accelerate the assembly of B monomers and favor the formation of holotoxin relative to production of B_5 when both A and B polypeptides are present (Hirst et al., 1983). Recent work using gene fusions demonstrates that the presence of the A2 domain at the carboxyl terminus of several fusion proteins enables them to assemble in vivo with CT-B to form holotoxin-like chimeras in the periplasm of *E. coli* (Jobling and Holmes, 1992).

The A polypeptides of CT and LT-I do not interact spontaneously in vitro with the mature B_5 subunit to form holotoxin (Hardy et al., 1988; Finkelstein et al., 1974a). Assembly of CT in vitro requires prior denaturation of the individual subunits with urea and subsequent dialysis of the mixture (Finkelstein et al., 1974a). LT-I holotoxin and CT/LT-I hybrid holotoxin is also formed under similar conditions in vitro from the appropriate combination of A and B polypeptides (Takeda et al., 1981). The efficient formation of hybrid holotoxin suggests that the differences in amino acid sequence that exist between the homologous A or B polypeptides of LT-I and CT are not critical for the assembly process.

After holotoxin is formed in the periplasm, CT in *V. cholerae* and LT-I in *E. coli* are processed differently. In *V. cholerae*, CT is actively secreted into the extracellular medium by a pathway that involves chromosomally encoded genes (Holmes et al., 1975; Neill et al., 1983a; Hirst and Holmgren, 1987; Sandkvist et al., 1993). The A polypeptide of secreted CT is often nicked between residues 192 and 195, presumably by the HP enzyme of *V. cholerae* (Booth et al., 1984). In contrast, LT-I is not translocated across the outer membrane of *E. coli,* and periplasmic LT-I remains unnicked (Hirst et al., 1984a,b; Clements and Finkelstein, 1979; Holmes et al., 1980). *E. coli* apparently depends either on extraperiplasmic enzymes distinct from the HP of *V. cholerae* or on host proteases for nicking of the A polypeptide of LT-I (Clements and Finkelstein, 1979). When genes encoding LT-I were introduced into *V. cholerae,* LT-I was actively secreted into the culture medium, but nicking of the A polypeptide did not occur (Hirst et al., 1984c). Thus, the A polypeptide of LT-I apparently lacks the target site for HP. The amino acid sequences of the A polypeptides between Cys-187 and Cys-199 of LTp-I and LTh-I are identical, but CT differs from LTp-I or LTh-I at six of the 11 residues (Yamamoto et al., 1984a,b, 1987).

A mutant of classical *V. cholerae* strain 569B, designated M14, is defective in translocation of CT across the outer membrane but synthesizes nearly normal amounts of CT that accumulate in the periplasm (Holmes et al., 1975). Recently a gene designated *epsE* that complements the secretory defect in M14 was cloned from a strain of *V. cholerae* El Tor (Sandkvist et al., 1993). The predicted product of the cloned gene is homologous with a protein that is a component of the pathway for secretion of extracellular proteins in several other gram-negative bacteria.

Processing and assembly have not been studied as thoroughly for type II enterotoxins as for type I enterotoxins. As is the case for type I enterotoxins, however, the type II enterotoxins are formed in *E. coli* as cell-associated, periplasmic proteins (Green et al., 1983; Holmes et al., 1986; Guth et al., 1986b). Genetic complementation tests were performed in *E. coli* using all pairwise combinations of cloned genes for the A and B polypeptides from LTp-I, LT-IIa, and LT-IIb (Connell and Holmes, 1992b). Although the yields of holotoxins with the A polypeptide from LTp-I and the B polypeptide from a type II toxin, or vice versa, were less than those of wild-type LTp-I or of holotoxins with both A and B polypeptides from a type II toxin, the experiment demonstrated that biologically active holotoxin was produced in *E. coli* from every combination of A and B polypeptides tested. This result is striking, since the B polypeptides of LTp-I have very little amino acid sequence homology with those of LT-IIa or LT-IIb. It suggests that LTp-I and the type II enterotoxins have conserved features of tertiary structure that are not evident from their primary structures.

In contrast to the result obtained with mixtures of purified A and B subunits from type I enterotoxins, biologically active type II holotoxins form in vitro when extracts from *E. coli* strains that make only the A or B polypeptide of LT-IIa or LT-IIb are mixed without prior exposure to denaturing agents (Connell and Holmes, 1992b). These findings suggest that the requirements for spontaneous assembly of A and B polypeptides of type II toxins differ significantly from those needed for assembly of type I enterotoxins. However, this inference must still be confirmed with purified A and B polypeptides of the type II toxins when they become available.

Determination of the crystallographic structure of a representative LT-II toxin will make possible direct comparisons of its three-dimensional structure with those of CT and LTp-I. Such an analysis will contribute greatly to understanding the conserved features of tertiary and quaternary structure among members of this enterotoxin family that are important for their biological functions.

VI. FUNCTION OF SUBUNIT A

Cholera toxin and related enterotoxins exert their biological effects by increasing intracellular cAMP concentrations (Field et al., 1989a,b). The A1 fragments of CT, LT-I, and LT-II are very similar NAD-dependent ADP-ribosyltransferases that ADP-ribosylate the Arg-187 residue of $G_s\alpha$, a G protein involved in the regulation of adenylate cyclase (Lee et al., 1991). The GTPase activity of ADP-ribosylated $G_s\alpha$ is greatly decreased, and ADP-ribosylated $G_s\alpha$ remains in the active form bound to GTP, thereby continuously stimulating adenylate cyclase (Gill and Meren, 1978). The resulting high levels of intracellular cAMP cause active secretion of chloride, inhibition of NaCl absorption, and passive loss of water from the small intestinal mucosa, resulting in massive diarrhea (Field et al., 1989a,b). The release of 5-hydroxytryptamine, prostaglandins, or eicosanoids may also contribute to the effects of cholera toxin in the intestine (Rask-Madsen et al., 1990; Peterson and Ochoa, 1989; Beubler et al., 1989).

A. ADP-Ribosyltransferase Activity

Cholera toxin with intact CT-A, intact but reduced CT-A, or nicked but not reduced CT-A has very little enzymatic activity, but reduced and nicked CT, like CT-A1, exhibits both ADP-ribosyltransferase and NAD-glycohydrolase activities (Mekalanos et al., 1979a,b). CT-A1 has broad substrate specificity, being able to use water (NAD-glycohydrolase activity), arginine, other guanidino compounds, several purified proteins, or its normal substrate $G_s\alpha$ as ADP-ribose acceptors (Moss and Vaughan,1977,1979). The most striking difference between the A1 fragments within the family of CT-related enterotoxins is the much lower ADP-ribosyltransferase activity ($<1\%$) of LT-IIa and LT-IIb when agmatine is used as the ADP-ribose acceptor, although ADP-ribosylation of the normal substrate $G_s\alpha$ is only slightly less for LT-IIa and LT-IIb than for CT or LTh-I (Lee et al., 1991). The K_m of CT-A1 for NAD, approximately 5 mM (Mekalanos et al., 1979a; Galloway and van Heyningen, 1987), is much higher than the K_m of several other ADP-ribosylating toxins, which may explain why it has not been possible to demonstrate photolabeling of an active-site residue of CT-A1 with NAD (Galloway et al., 1987).

The A1 fragments of CT, LT-I, and LT-II are also capable of auto-ADP-ribosylation, and up to three ADP-ribose residues can be incorporated per CT-A1 peptide (Moss et al., 1980). The ADP-ribosylated A1 peptide of CT has 30–50% greater activity than native A1. The biological significance of this is unclear, however, since a mutant of LT-I with Arg-146, the preferred site for auto-ADP-ribosylation, changed to Gly-146 retains full toxicity (Okamoto et al., 1988).

NAD-dependent ADP-ribosylation of $G_s\alpha$ by the A1 fragments of CT and related enterotoxins is stimulated by several factors including GTP and ARFs (Lee et al., 1991). ARFs, a family of small GTP-binding proteins of approximately 20 kDa that stimulate the activity of cholera toxin by interacting directly with the enzyme (Moss et al., 1990), are discussed in more detail in the chapter in this volume by Moss.

B. Analysis of Mutant A Polypeptides

Molecular genetic methods are currently being applied to the analysis of the enzymatic activity of the A1 fragments of CT and related heat-labile enterotoxins. The ADP-ribosyltransferase activity of CT-A1 is abolished by replacing Arg-7 with Lys-7 (Burnette et al., 1991), and analogous substitutions in the A1 fragment of LT-I abolish its toxicity

(Lobet et al., 1991). The CT-A1 sequence YRADSRPP (one-letter amino acid code, Arg-7 underlined) is identical at seven of eight positions with a sequence in the S1 polypeptide of pertussis toxin (PT-S1) (Gierschik, 1992), and mutant forms of PT-S1 altered at the homologous Arg residue also have severely decreased ADP-ribosyl-transferase activity (Burnette et al., 1988). Ser-61 in the A1 fragment of LT-I is in a second region (VSTSLSLR, Ser-61 underlined) that is highly homologous with both CT-A1 (seven of eight residues identical) and PT-S1 (six of eight residues identical), and substitution of Phe-61 for Ser-61 in LT-I eliminates its toxicity (Harford et al., 1989). The location of these mutations in regions of homology between CT, LT-I, and PT suggests that Arg-7 and Ser-61 may be involved in NAD binding, since these toxins all recognize NAD but use different molecules as acceptors in their ADP-ribosyltransferase reactions ($G_{s\alpha}$ for CT and LT-I and $G_{i\alpha}$ for PT). A mutant form of the A1 fragment of LT-I with Glu-112 changed to Lys-112 lacks both enzymatic activity and toxicity (Tsuji et al., 1990, 1991). The Glu-112 residue in the A1 fragment of CT-related enterotoxins appears to be homologous with the active-site residues Glu-551 in exotoxin A of *Pseudomonas aeruginosa* and Glu-148 of diphtheria toxin, based on similarities in their crystal structures (Sixma et al., 1991, 1993).

VII. FUNCTION OF SUBUNIT B

Binding of heat-labile enterotoxin to a susceptible cell triggers an endocytic process that results in delivery of fragment A1 into the cytosol, where it catalyzes the ADP-ribosylation of $G_{s\alpha}$ and activation of adenylate cyclase, as summarized above. The B subunits mediate binding of the heat-labile enterotoxins to specific receptors on the plasma membranes of target cells.

A. Type I Enterotoxins

The monosialoganglioside G_{M1} [Galβ1-3GalNAcβ1-4(NANAα1-3)Galβ1-4Glcβ1-ceram-ide], hereafter designated G_{M1}, is a functional receptor both for CT and for LT-I, and exogenous G_{M1} competitively inhibits the binding of CT or LT-I to susceptible cells (Cuatrecasas, 1973; Fishman, 1982b, 1986; Fishman and Atikkan, 1980; Fukuta et al., 1988; Eidels et al., 1983; Moss et al., 1979). Incorporation of exogenous G_{M1} into the plasma membrane of a ganglioside-deficient, enterotoxin-resistant mouse fibroblast cell line restores sensitivity to CT and LT-I (Fishman and Atikkan, 1980; Moss et al., 1979; Galloway et al., 1987; Fishman et al., 1977). LTh-I binds with a higher affinity than LTp-I to G_{M1} in vitro (Olsvik et al., 1983), but the significance of this finding for pathogenicity is uncertain.

Each pentameric B subunit of CT or LT-I can bind five G_{M1} molecules, i.e., one molecule of G_{M1} per B polypeptide (Fishman et al., 1978). Biochemical studies of CT-B initially provided evidence that the binding site for G_{M1} is at the interface between CT-B monomers (DeWolf et al., 1981a,b), but the crystal structure of LTp-I complexed with lactose suggests that each G_{M1} interacts primarily with a single B polypeptide (Sixma et al., 1992b). Determination of the structure of LTp-I or CT complexed with the complete oligosaccharide of G_{M1} will resolve this issue.

The specificity for binding of glycolipids by type I enterotoxins has been examined in considerable detail. Fucosyl G_{M1} (Masserini et al., 1992), G_{M1} containing *N*-glycolylneuraminic acid (G_{M1}[NeuGc]) (Fishman et al., 1980), and a novel ganglioside from mouse spleen (G_{M1}[NeuGc]-Glc-ceramide) (Nakamura et al., 1987) also function as receptors for type I enterotoxins. CT and LT-I bind weakly in vitro to ganglioside G_{D1b}, and LT-I also has low but measurable binding activity for ganglioside G_{M2} (Fukuta et al., 1988). In contrast, CT and LT-I have little or no binding activity for gangliosides G_{D2}, G_{D3}, G_{D1a}, G_{Q1b}, G_{T1b}, G_{M3}, and G_{T3} (Fukuta et al., 1988). The use of chemically modified derivatives of G_{M1} showed that the acetyl groups of N-acetylneuraminic acid (NANA) and GalNAc, and the C8 and C9 residues of NANA, are not essential for binding of type I enterotoxins (Schengrund and Ringler, 1989). The terminal galactose and NANA residues of G_{M1} are essential for binding, since G_{M2} and asialo-G_{M1} do not bind to type I enterotoxins (Fukuta et al., 1988), and binding of toxin to G_{M1} on the cell surface protects these sugars from chemical modification (Moss et al., 1977).

A fucose-containing, blood group A-active, neutral glycosphingolipid that binds CT is present in pig intestinal mucosa (Bennun et al., 1989). If functionally similar compounds are present in the human intestine, they might provide an explanation for epidemiological data showing that people with blood group O who contract cholera are twice as likely to develop severe diarrhea as people with other blood groups (Glass et al., 1985). Glycoproteins that bind CT are found in rat intestinal brush borders (Morita et al., 1980), but not in human (Holmgren et al., 1985) or rabbit intestines (Holmgren et al., 1982; Griffiths et al., 1986).

LT-I, but not CT, interacts with functional receptors other than gangliosides. Pretreatment of cultured Chinese hamster ovary cells (Guerrant and Brunton, 1977) or isolated rat intestinal brush borders (Griffiths and Critchley, 1991) with pentameric CT-B to block receptor sites fully inhibits CT but not LT-I, whereas preincubation with pentameric B subunits of LT-I inhibits both enterotoxins. The molecules that serve as alternative receptors for LT-I but not for CT are cell surface glycoproteins (Holmgren et al., 1982, 1985; Griffiths et al., 1986). Recent studies associated the LT-I-binding activity with partially purified fractions of intestinal brush border membranes enriched for sucrase and isomaltase activities (Griffiths and Critchley, 1991).

Several studies have analyzed the residues of the B subunit that are important for binding to G_{M1}. Several lines of evidence show that Trp-88, the lone tryptophan in the B polypeptide, is closely associated with the G_{M1} binding site. Binding of CT-B to G_{M1} causes a 12-nm blue shift in the fluorescence emission spectrum of Trp-88 (Mullin et al., 1976), and chemical modification of Trp-88 destroys the ability of CT-B to bind G_{M1} (Ludwig et al., 1985; DeWolf et al., 1981b). Mutant forms of CT-B with charged residues replacing Trp-88 are unable to bind G_{M1} (Jobling and Holmes, 1991). Tryptophan 88 appears to have a structural role in establishing or maintaining the native conformation of CT-B, since all mutant forms of CT-B with substitutions of other amino acids for Trp-88 are made in subnormal amounts (Jobling and Holmes, 1991).

Positively charged amino acids in the B subunit are also involved in G_{M1} binding. Modification of arginine residues in CT-B led to the suggestion by one group of investigators that Arg-35 is important for binding to G_{M1} (Duffy and Lai, 1979), but another group was unable to identify which of the three arginine residues in CT-B is involved in G_{M1} binding (Ludwig et al., 1985). The latter group also demonstrated that one or more of the nine lysine residues in CT-B is required for G_{M1} binding and that reduction of the intrachain disulfide bond in CT-B abolishes G_{M1}-binding activity. A mutant of LT-I with Gly-33 in the B subunit replaced by Asp-33 is nontoxic and defective for G_{M1} binding, but the mutant B polypeptide retains the ability to assemble into pentamers (Tsuji et al., 1985). Substitution of Gly-33 in the B subunit of CT with Glu,

Asp, or a bulky hydrophobic residue such as Leu, Ile, or Val also results in loss of G_{M1}-binding activity and toxicity (Jobling and Holmes, 1991). Surprisingly, substitution of a positively charged Lys or Arg residue for Gly-33 in CT-B is tolerated and does not result in loss of G_{M1}-binding activity or the ability of the mutant CT-B to assemble into holotoxin. Mutant CT-Bs with any of several different amino acids substituted for Arg-35 were able to bind to G_{M1}, but the variants in which a glutamate or aspartate residue replaced Arg-35 were unable to form or maintain holotoxin structure (Jobling and Holmes, 1991). These results illustrate the complexity of toxin–receptor interactions. It is remarkable that CT is so specific in its ganglioside-binding activity, and yet many different substitutions are tolerated at positions that are important for G_{M1} binding.

B. Type II Enterotoxins

The B polypeptides of type II enterotoxins differ significantly in receptor-binding activity from those of the type I enterotoxins (Fukuta et al., 1988). Unlike LT-1 or CT, ganglioside G_{M1} is not the preferred receptor. The type II enterotoxins are much less susceptible than either LT-I or CT to competitive inhibition of their biological activities by G_{M1} (Guth et al., 1986b; Holmes et al., 1986). The ganglioside-binding specificities of LT-IIa and LT-IIb were determined in solid-phase radioimmunoassays using purified gangliosides (Fukuta et al., 1988). LT-IIa binds most avidly to ganglioside G_{D1b}, has lower binding activities for gangliosides G_{M1} and G_{D1a}, and has low but measurable affinity for gangliosides G_{T1b}, G_{Q1b}, and G_{D2}. LT-IIb has a more restricted specificity; it binds preferentially to ganglioside G_{D1a} and exhibits lower binding activity for ganglioside G_{T1b}. The binding activity of the type II enterotoxins for glycoproteins is not known. The extent to which differences in ganglioside-binding specificities of the type I and type II enterotoxins are reflected in their cell, tissue, or species tropisms has not yet been studied systematically, but it is striking that the LT-II enterotoxins have equal or greater toxicity than CT or LT-I in mouse Y1 adrenal cell assays but much less activity than CT or LT-I in rabbit ileal loop assays (Guth et al., 1986b; Holmes et al., 1986).

Genetic analysis identified three residues in the B subunit of LT-IIa that are important for binding to gangliosides (Connell and Holmes, 1992a). Replacing any one of the threonine residues at positions 13, 14, or 34 in the B subunit by a variety of amino acids abolishes the binding of LT-IIa to G_{D1b}. Residues involved in the binding of LT-IIa to G_{D1b}, however, are not necessarily important for binding to other gangliosides. For example, replacement of Thr-14 in the B polypeptide of LT-IIa by any of several amino acids abolished binding to ganglioside G_{D1a} as well as G_{D1b} but had little effect on binding to G_{M1}. Mutant forms of the B polypeptide of LT-IIa with serine substituted for threonine at position 13, 14, or 34 retained at least some binding activity for ganglioside G_{D1b}, in contrast to results obtained with other substitutions at these positions. This finding suggests a role for the hydroxyl groups of threonine in the receptor-binding activity of the wild-type LT-IIa enterotoxin.

To test whether the mutant B polypeptides can assemble into biologically active holotoxin, extracts from recombinant strains of *E. coli* that expressed both the wild-type A polypeptide and a mutant B polypeptide of LT-IIa were assayed both for immunoreactivity in solid-phase radioimmunoassays and for biological activity in mouse Y1 adrenal cell assays (Connell and Holmes, 1992a). All of the mutant forms of the B subunit of LT-IIb were incorporated into immunoreactive holotoxins, but the mutant forms of holotoxin were not toxic. Interestingly, the mutant LT-IIa holotoxins that retained G_{M1} binding activity (i.e., those with substitutions for residue Thr-14 in the B subunit) were also nontoxic in the Y1 adrenal cell assay, although the sensitivity of the Y1 cells to CT (Maneval et al., 1980) was confirmed by simultaneous tests. The data are consistent with either of two models. Either the oligosaccharide moiety of G_{M1} is exposed on the surface of Y1 adrenal cells in a manner that is accessible to CT or LT-1 but not LT-IIa, or LT-IIa holotoxins with mutant B polypeptides substituted at Thr-14 have binding activity for G_{M1} that is too low to permit toxicity.

VIII. ENTRY INTO CELLS

Binding of CT and related enterotoxins to cell membrane receptors triggers a process that leads to translocation of fragment A1 into the cytosol, activation of adenylate cyclase, and expression of cAMP-mediated biological effects. The steps involved in internalization of enterotoxins and translocation of their A1 fragments to the cytosol are not well understood. Most studies on entry have used CT; only limited data are available concerning other heat-labile enterotoxins.

Multivalent binding of CT to G_{M1} receptors positions the B pentamer parallel to the plane of the membrane (Ribi et al., 1988). Binding to receptor results in little, if any, change in CT-B secondary structure, but it does cause a marked increase in thermal stability of CT-B (Goins and Freire, 1988; Surewicz et al., 1990). Until recently the orientation of CT-A in membrane-bound CT was controversial. Analysis of two-dimensional crystals of CT bound to G_{M1} in planar lipid membranes indicated that the CT-A1 domain might be located between the planar B pentamer and the membrane (Ribi et al., 1988), but subsequent high-resolution X-ray crystallographic studies of LT-I demonstrated that the A1 domain is situated on the opposite face of the planar B pentamer from the G_{M1}-binding sites (Sixma et al., 1991, 1992b, 1993). The current model, therefore, places the A1 domain of membrane-bound enterotoxin on the opposite side of the B pentamer from the plasma membrane.

Studies using membrane-restricted, photoreactive glycolipids demonstrated that fragment A1, but not CT-B, inserts into the lipid bilayer (Tomasi and Montecucco, 1981; Wisnieski and Bramhall, 1981). These studies reached different conclusions about the necessity to reduce the disulfide bond of CT-A before inserting CT-A1 into the membrane. The presence of free sulfhydryl groups in the membrane of Newcastle disease virus (Patzer et al., 1979), which was used as the source of membrane vesicles for one of the studies, could explain the lack of a requirement for added sulfhydryl reducing agents in that study (Wisneiski and Bramhall, 1981). An early model proposed that pentameric CT-B forms a pore in the plasma membrane through which unfolded CT-A passes into the cytoplasm (Gill, 1976). Although a recent report showed that CT-B can form low-pH-dependent ion channels in bilayer lipid membranes (Krasilnikov et al., 1991), the X-ray crystallographic structures of LT-I and CT do not support the concept that the A1 polypeptide enters the cell through a pore formed by the B pentamer. The central pore in the planar B pentamer interacts in a highly specific manner with the A2 domain and is too narrow in diameter to function as a channel through which an unfolded polypeptide such as the A1 fragment can be translocated (Gibbons, 1991; Sixma et al., 1991, 1993).

Following exposure to CT at 37°C, whole cells exhibit a lag phase of 10–15 min before showing signs of intoxication, even though bound toxin rapidly becomes inaccessible to the inhibitory effects of neutralizing antibodies (Fishman, 1980). This lag phase parallels the generation of free CT-A1 peptide (Kassis et al., 1982), suggesting that reduction of the disulfide bond in CT-A is the rate-limiting step. Cell surface-bound toxin has a half-life of about 2 h (Fishman, 1982a), during which it is internalized via noncoated membrane invaginations (Montesano et al., 1982) and enters a network of tubules and vesicles closely associated with the Golgi apparatus (GERL, Golgi-endoplasmic reticulum-lysosomes) (Joseph et al., 1979; Tran et al., 1987). In rat liver cells, association of CT with an endosomal compartment appears to be an obligatory step in toxin processing, and generation of the A1 peptide and its membrane translocation occur optimally at low pH (Janicot et al., 1991). Brefeldin A, a fungal metabolite that inhibits secretory processes of mammalian cells and causes reversible disintegration of the Golgi apparatus and redistribution of Golgi enzymes to the endoplasmic reticulum (Pelham, 1991a), transiently inhibits the response of cells to cholera and heat-labile enterotoxins (Nambiar et al., 1993; Orlandi et al., 1993; Donta et al., 1993), suggesting that cholera toxin is routed through the Golgi apparatus during entry into the cell. KDEL, a potential endoplasmic reticulum retention sequence (Pelham, 1991b), is present at the carboxy terminus of fragment CT-A2 (Dams et al., 1991; Mekalanos et al., 1983; Lockman and Kaper, 1983), and RDEL or RNEL is present at the same position in the characterized variants of LT-I and LT-II (Pickett et al., 1987, 1989; Yamamoto et al., 1984b, 1987). It is not yet known if this sequence is required for the function of CT, but a similar carboxy-terminal sequence (RDELK), is required for the toxicity of exotoxin A from *Pseudomonas aeruginosa* (Chaudhary et al., 1990). The action of *Pseudomonas* exotoxin A on whole cells is also inhibited by brefeldin A (Yoshida et al., 1991).

Several findings indicate that intoxication of cells by CT is a complex process that reflects, at least in part, both the intracellular trafficking of CT and the compartmentalization of $G_{s\alpha}$, ARFs, and adenylate cyclase. In intestinal epithelial cells the apical and basolateral membranes are separated by circumferential tight junctions through which neither CT nor G_{M1} can pass (Peterson et al., 1972; Spiegel et al., 1985). Although G_{M1} is present on both apical and basolateral membranes, $G_{s\alpha}$ is associated with the apical membrane (Longbottom and van Heyningen, 1989), and the adenylate cyclase activated by CT is located on the cytoplasmic surface of the basolateral membrane (Murer et al., 1976). Furthermore, both cytosolic and membrane-associated ADP-ribosylating factors have been characterized (Moss and Vaughan, 1988, 1993a). This restricted distribution of factors that participate in the action of CT led to the suggestion that $G_{s\alpha}$ might be covalently modified by CT in the brush border membrane and then migrate to the basolateral membrane to activate adenylate cyclase (Longbottom and van Heyningen, 1989). Polarized cells in culture exhibit a temperature-sensitive step, believed to involve vesicular transport, that is essential for the action of CT when it is applied to the apical membrane but not when it is applied to the basolateral membrane (Lencer et al., 1992). Furthermore, the pathway for entry of CT into cells appears to be shared by *Pseudomonas* exotoxin A and ricin, toxins that are very different from CT in both structure and mode of action (Nambiar et al., 1993).

IX. EVOLUTIONARY RELATIONSHPS

Figures 1 and 2 align the deduced amino acid sequences for the mature A and B polypeptides, respectively, of CT, LTh-I, LTp-I, LT-IIa, and LT-IIb reference enterotoxins. Residues are designated as conserved in Figures 1 and 2 only if they are present in at least three of the five enterotoxins, including at least one from each serogroup. Table 3 presents amino acid and nucleotide sequence homologies for the mature A polypeptides

```
             1                                                          50
CT           NDDKLYRADS RPPDEIKQSG GLMPRGQSEY FDRGTQMNIN LYDHARGTQT
LTh-I        NGDKLYRADS RPPDEIKRSG GLMPRGHNEY FDRGTQMNIN LYDHARGTQT
LTp-I        NGDRLYRADS RPPDEIKRSG GLMPRGHNEY FDRGTQMNIN LYDHARGTQT
LT-IIa       ND..FFRADS RTPDEIRRAG GLLPRGQQEA YERGTPININ LYEHARGTVT
LT-IIb       ND..YFRADS RTPDEVRRSG GLIPRGQDEA YERGTPININ LYDHARGTAT
Conserved    ND----RADS R-PDEI-RSG GL-PRGQ-E- --RGT--NIN LYDHARGT-T

             51                                                         100
CT           GFVRHDDGYV STSISLRSAH LVGQTILSGH STYYIYVIAT APNMFNVNDV
LTh-I        GFVRYDDGYV STSLSLRSAH LAGQSILSGY STYYIYVIAT APNMFNVNDV
LTp-I        GFVRYDDGYV STSLSLRSAH LAGQSILSGY STYYIYVIAT APNMFNVNDV
LT-IIa       GNTRYNDGYV STTVTLRQAH LIGQNILGSY NEYYIYVVAP APNLFDVNGV
LT-IIb       GNTRYNDGYV STTTTLRQAH FLGQNMLGGY NEYYIYVVAA APNLFDVNGV
Conserved    G--RY-DGYV ST---LR-AH L-GQ-IL-GY --YYIYV-A- APN-F-VN-V

             101                                                        150
CT           LGAYSPHPDE QEVSALGGIP YSQIYGWYRV HFGVLDEQLH RNRGYRDRYY
LTh-I        LGVYSPHPYE QEVSALGGIP YSQIYGWYRV NFGVIDERLH RNREYRDRYY
LTp-I        LGVYSPHPYE QEVSALGGIP YSQIYGWYRV NFGVIDERLH RNREYRDRYY
LT-IIa       LGRYSPYPSE NEFAALGGIP LSQIYGWYRV SFGAIEGGMQ RNRDYRGDLF
LT-IIb       LGRYSPYPSE NEYAALGGIP LSQIIGWYRV SFGAIEGGMH RNRDYRRDLF
Conserved    LG-YSP-P-E -E--ALGGIP -SQI-GWYRV -FG-I----H RNR-YR----

             151                                                        200
CT           SNLDIAPAAD GYGLAGFPPE HRAWREEPWI HHAPPGCGNA PRSSMSNTCD
LTh-I        RNLNIAPAED GYRLAGFPPD HQAWREEPWI HHAPQGCGNS SRTITGNTCN
LTp-I        RNLNIAPAED GYRLAGFPPD HQAWREEPWI HHAPQGCGNS SRTITGNTCN
LT-IIa       RGLTVAPNED GYQLAGFPSN FPAWREMPWS TFAPEQCVPN NKEFKGGVCI
LT-IIb       RGLSAAPNED GYRIAGFPDG FPAWEEVPWR EFAPNSCLPN NKASSDTTCA
Conserved    R-L--AP-ED GYRLAGFP-- --AWRE-PW- --AP--C--- -----G-TC-

             201                                                        240
CT           EKTQSLGVKF LDEYQSKVKR Q.....IFSG YQSDIDTHNR IKDEL
LTh-I        EETQNLSTIY LRKYQSKVKR Q.....IFSD YQSEVDIYNR IRNEL
LTp-I        EETQNLSTIY LRKYQSKVKR Q.....IFSD YQSEVDIYNR IRDEL
LT-IIa       SATNVLSKYD LMNFKKLLKR RLAL..TFFM SEDDFIGVHG ERDEL
LT-IIb       SLTNKLSQHD LADFKKYIKR KFTLMTLLSI NNDGFFSNNG GKDEL
Conserved    --T--LS--- L-------KR -------FS- --------N- -RDEL
```

Fig. 1 Deduced amino acid sequences for mature A polypeptides of representative heat-labile enterotoxins. Signal sequences are not shown. Residues are indicated by one-letter amino acid codes. Numbers refer to CT-A, and the other polypeptides are aligned to achieve maximal identity with CT-A. Residues designated as conserved are present in at least three of the five reference enterotoxins, including at least one from each serogroup. The disulfide loop that links the A1 and A2 domains extends from Cys-187 to Cys-199 in CT-A, and the proportion of conserved residues is significantly greater in the A1 domain than in the A2 domain. Sources for nucleotide sequences are CT, Dams et al. (1991); LTh-I, Yamamoto et al. (1984b, 1987); LTp-I (Yamamoto et al. (1987); LT-IIa, Pickett et al. (1987); and LT-IIb, Pickett et al., (1989).

and the corresponding genes of the CT, LTh-I, LTp-I, LT-IIa, and LT-IIb reference enterotoxins. For this analysis the signal sequences of the A polypeptides and the corresponding nucleotide sequences are not considered. Table 4 presents a similar analysis of the mature B polypeptides and the corresponding genes.

The classification of enterotoxins into serogroups I and II reflects their genotypic relationships. By nucleotide and amino acid sequence homologies, the type I enterotoxins are related more closely to each other than to the type II enterotoxins, and vice versa (Tables 3 and 4). Among the type I enterotoxins, CT-A shares 81.3% identity with the

```
            1                                                      50
CT          .TPQNITDLCA EYHNTQIHTL NDKIFSYTES LAGKREMAII TFKNGATFQV
LTh-I       .APQSITELCS EYRNTQIYTI NDKILSYTES MAGKREMVII TFKSGATFQV
LTp-I       .APQTITELCS EYRNTQIYTI NDKILSYTES MAGKREMVII TFKSGETFQV
LT-IIa      GVSEHFRNIC. .NQTTADIVA GVQLKKYIAD VNTNTRGIYV VSNTGGVWYI
LT-IIb      GASQFFKDNC. .NRTTASLVE GVELTKYISD INNNTDGMYV VSSTGGVWRI
Conserved   -A-Q-----C- --R-T----- ------Y--- ---------- ----G-----


            51                                                    100
CT          EVPGSQHIDS QKKAIERMKD TLRIAYLTEA KVEKLCVWNN KTPHAIAAIS MAN
LTh-I       EVPGSQHIDS QKKAIERMKD TLRITYLTET KIDKLCVWNN KTPNSIAAIS MEN
LTp-I       EVPGSQHIDS QKKAIERMKD TLRITYLTET KIDKLCVWNN KTPNSIAAIS MKN
LT-IIa      ....PGGRDY PDNFLSGEIR KTAMAAITSD TKVNLCAKTS SSPNHIWAME LDRES
LT-IIb      ....SRAKDY PDNVMTAEMR KIAMAAVLSG MRVNMCASPA SSPNVIWAIE LEAE
Conserved   ----S---D- ---------- ----A--T-- ----LC---- --PN-I-AI- -----
```

Fig. 2   Deduced amino acid sequences for mature B polypeptides of representative heat-labile enterotoxins. Presentation of data for the B polypeptides follows the conventions established for Figure 1. Comparison with Figure 1 reveals the high proportion of conserved residues in the A polypeptides and the low proportion of conserved residues in the B polypeptides within the heat-labile enterotoxin family. Sources for nucleotide sequences are given in Figure 1.

240-amino-acid A polypeptides of LTh-I and LTp-I; and CT-B shares 80.6–81.6% identity with the 103-amino-acid B polypeptides of LTh-I and LTp-I. Furthermore, LTh-I and LTp-I are related much more closely to each other than to CT; they are 99.2% identical in their A polypeptides and 97.1% identical in their B polypeptides. Sequencing of toxin operons from multiple strains of *V. cholerae* demonstrated that the *ctxA* genes encoded identical mature CT-A polypeptides, but three different *ctxB* alleles were shown to encode mature CT-B polypeptides that differed by a maximum of five of the 103 residues (Dams et al., 1991; Brickman et al., 1990; Lockman and Kaper, 1983; Mekalanos et al., 1983). Similar studies demonstrated three variants of LTh-I that differed by a maximum of two residues in their B polypeptides (Leong et al., 1985; Yamamoto and Yokota, 1983; Tsuji et al., 1987), but variability was not observed among independent isolates of LTp-I (Leong et al., 1985). Although some of these CT and LTh-I variants differ in reactivity with specific monoclonal antibodies (Leong et al., 1985; Finkelstein et al., 1987a,b; Kazemi and Finkelstein, 1990a,b; Qu et al., 1991), the significance of minor structural variants

TABLE 3 Nucleotide and Amino Acid Sequence Homologies for Mature A Polypeptides and Corresponding Genes of Representative Type I and Type II Heat-Labile Enterotoxins[a]

		Nucleotide sequence homologies (%)				
		CT	LTh-I	LTp-I	LT-IIa	LT-IIb
Amino	CT	—	78.8	78.9	57.6	59.0
Acid	LTh-I	81.3	—	99.4	57.4	57.1
Sequence	LTp-I	81.3	99.2	—	57.1	56.9
Homologies	LT-IIa	50.4	54.2	54.6	—	72.7
(%)	LT-IIb	51.7	53.4	53.8	75.5	—

[a]Based on data in references in Figure 1. Percentage nucleotide sequence identity for each pair of genes is above the dashes, and percentage amino acid sequence identity for each pair of polypeptides is below the dashes.

TABLE 4 Nucleotide and Amino Acid Sequence Homologies for Mature B Polypeptides and Corresponding Genes of Representative Type I and Type II Heat-Labile Enterotoxins[a]

		Nucleotide sequence homologies (%)[b]				
		CT	LTh-I	LTp-I	LT-IIa	LT-IIb
Amino	CT	—	78.3	77.7	ND	ND
Acid	LTh-I	81.6	—	98.7	ND	ND
Sequence	LTp-I	80.6	97.1	—	ND	ND
Homologies	LT-IIa	11.7	10.7	10.7	—	65.3
(%)[c]	LT-IIb	13.6	13.6	13.6	56.4	—

[a]Based on data in references in Figure 2.
[b]Percentage nucleotide sequence identity for each pair of genes is above the dashes. ND = not determined. Homology for these pairs of genes is sufficiently low that their nucleotide sequences cannot be properly aligned.
[c]Percentage amino acid sequence identity for each pair of polypeptides is below the dashes.

among CTs or LTh-Is for immunity to cholera or travelers' diarrhea has not yet been demonstrated.

The A polypeptides of LT-IIa and LT-IIb exhibit 75.5% identity with each other, but the B polypeptides of LT-IIa and LT-IIb have only 56.4% identity (Tables 3 and 4). The divergence between the LT-IIa and LT-IIb enterotoxins within serogroup II is greater, therefore, than the divergence between the CT and LT-I enterotoxins within serogroup I, particularly with respect to the B polypeptides. The A polypeptides of LT-IIa and LT-IIb are 50.4–54.6% identical with CT-A and the A polypeptides of LTh-I and LTp-I, but the B polypeptides of LT-IIa and LT-IIb are only 10.7–13.6% identical with CT-B and the B polypeptides of LTh-I and LTp-I.

Few of the heat-labile enterotoxins from gram-negative bacteria other than *V. cholerae* or *E. coli* (Table 1) have been characterized in detail at the genetic and biochemical levels. An enterotoxin purified from *V. mimicus* is identical in amino acid sequence to one of the variants of cholera toxin (Spira and Fedorka-Cray, 1983). One report that enterotoxin purified from *Salmonella typhimurium* resembles cholera toxin closely in subunit structure and antigenicity (Finkelstein et al., 1983) has not yet been confirmed by other investigators. Studies in another laboratory led to the cloning of a gene from *S. typhimurium* into *E. coli* that encodes a biologically active enterotoxin that binds ganglioside G_{M1} and is neutralized by antisera against cholera toxin (Prasad et al., 1992; Chopra et al., 1987), but the nucleotide sequence of this cloned *Salmonella* enterotoxin gene is not yet reported and its gene product is not yet purified and characterized. The preliminary data available do not indicate, however, that the cloned enterotoxin gene from *S. typhimurium* is closely related to the genes for cholera toxin and *E. coli* heat-labile enterotoxin.

Examination of Figures 1 and 2 reveals that A1 is the most highly conserved domain of the heat-labile enterotoxins. This finding is consistent with the fact that A1 fragments from all of the type I and type II enterotoxins have ADP-ribosyltransferase activity, use

$G_{s\alpha}$ as the acceptor for ADP-ribose, and exhibit stimulation of activity by ARFs and appropriate cofactors (Chang et al., 1987; Lee et al., 1991). Regions of the A1 peptide that may be involved in its ADP-ribosyltransferase activity are discussed in Section VI of this chapter, but the site(s) for interaction of A1 with ARFs have not been determined. The striking divergence between the B polypeptides of type I and type II enterotoxins is consistent with their different receptor-binding specificities (Fukuta et al., 1988) and their lack of antigenic cross-reactivity (Guth et al., 1986b; Holmes et al., 1986). Nevertheless, the complementation observed in vivo between A and B polypeptides of type I and type II enterotoxins suggests that these toxins have conserved features of three-dimensional structure (Connell and Holmes, 1992b). In this regard, conservation of the two cysteine residues required for formation of the intrachain disulfide bond and the apparent conservation of a sequence between residues 92 and 100 (Fig. 2) that may be important for interaction of adjacent B monomers in formation of the B pentamer (Sixma et al., 1991) are of interest. The A2 domain is intermediate between A1 and B in the extent to which it exhibits conservation among the heat-labile enterotoxins. This is consistent with its important role in the interaction between the A polypeptide and the B polypeptides in holotoxin (Sixma et al., 1991, 1993).

The origin of the genes for the heat-labile enterotoxins is not established. The plasmid-encoded genes for LTp-I and LTh-I have a lower guanosine plus cytosine content (36.8–38.6%) than the genome of *E. coli* (50%), and they use optimal codons for *E. coli* less frequently (0.38–0.44) than genes of *E. coli* that are expressed at high levels (Ikemura, 1981). These findings raise the possibility that the genes for LTh-I and LTp-I were introduced into *E. coli* from another bacterium, possibly *V. cholerae*, at a relatively recent time in the evolutionary history of these bacteria (Yamamoto et al., 1987). The data summarized in Tables 3 and 4 indicate that the chromosomal genes for the type I and type II enterotoxins diverged from each other much earlier than the genes for CT diverged from those for LTh-I and LTp-I. It is interesting, therefore, that the genes for LT-IIa and LT-IIb also have a low average guanosine plus cytosine content (38–41%) and relatively low frequency of usage of optimal codons for *E. coli* (0.38–0.39 for the A genes and 0.54–0.58 for the B genes, respectively), suggesting that they also are not fully adapted as a part of the genome of *E. coli*.

The data summarized above indicate that the heat-labile enterotoxins of gram-negative bacteria demonstrate extensive evolutionary divergence, especially between the B polypeptides of the type I and type II toxins. It is evident that the heat-labile enterotoxins are encoded by an ancient family of genes, but the origin of the gene family and possible roles that the "enterotoxins" may have in microbial physiology during evolution other than to function as virulence factors for present-day enteropathogenic bacteria are unknown.

X. CONCLUSIONS

During the past quarter-century the heat-labile enterotoxins of gram-negative bacteria have become one of the best characterized families of bacterial protein toxins. Cholera toxin is a major virulence factor of *V. cholerae* and has a primary major role in causing the massive watery diarrhea in cholera in humans. Type I heat-labile enterotoxins are important virulence factors for enterotoxigenic *E. coli* infections of humans and animals. Although the more recently discovered type II heat-labile enterotoxins of *E. coli* are potentially important as virulence factors, their roles in pathogenesis are not yet established. Efforts to prevent cholera and travelers' diarrhea by antitoxic immunity, antibacterial immunity, or both have not yet been fully successful, and control of these important infectious diseases by immunization remains an important goal for the future.

Characterization of the heat-labile enterotoxins provided many insights into their structure, function, immunochemistry, genetics, regulation, and evolutionary relationships, as reviewed briefly here. Studies of these toxins contributed greatly to the current understanding of pathophysiological mechanisms in secretory diarrhea and the development of highly effective methods for treatment of cholera and other secretory diarrheas by parenteral and oral replacement of fluid and electrolytes. Cholera toxin and other heat-labile enterotoxins have also been used extensively as exquisitely sensitive and specific reagents to probe the role of G proteins in many different biological systems. The ways in which the heat-labile enterotoxins have contributed to biomedical science are now so numerous and so diverse that no single review can encompass them.

Acknowledgments

Our research on cholera toxin and *E. coli* heat-labile enterotoxins was supported in part by Public Health Service grants AI-14107 and AI-31940 from the National Institute of Allergy and Infectious Diseases and by protocol R07301 from the Uniformed Services University of the Health Sciences. The opinions and assertions contained herein are private views of the authors and should not be construed as official or as necessarily reflecting the views of the Uniformed Services University of the Health Sciences or the Department of Defense.

REFERENCES

Acheson, D. W. (1992). Enterotoxins in acute infective diarrhoea. *J. Infect.* 24:225–245.

Annapurna, E., and Sanyal, S. C. (1977). Enterotoxicity of *Aeromonas hydrophila. J. Med. Microbiol.* 10:317–323.

Baine, W. B., Vasil, M. L., and Holmes, R. K. (1978). Genetic mapping of mutations in independently isolated nontoxinogenic mutants of *Vibrio cholerae. Infect. Immun.* 21:194–200.

Belisle, B. W., Twiddy, E. M., and Holmes, R. K. (1984a). Monoclonal antibodies with an expanded repertoire of specificities and potent neutralizing activity for *Escherichia coli* heat-labile enterotoxin. *Infect. Immun.* 46:759–764.

Belisle, B. W., Twiddy, E. M., and Holmes, R. K. (1984b). Characterization of monoclonal antibodies to heat-labile enterotoxin encoded by a plasmid from a clinical isolate of *Escherichia coli. Infect. Immun.* 43:1027–1032.

Bennun, F. R., Roth, G. A., Monferran, C. G., and Cumar, F. A. (1989). Binding of cholera toxin to pig intestinal mucosa glycosphingolipids: relationship with the ABO blood group system. *Infect. Immun.* 57:969–974.

Betley, M. J., Miller, V. L., and Mekalanos, J. J. (1986). Genetics of bacterial enterotoxins. *Annu. Rev. Microbiol.* 40:577–605.

Beubler, E., Kollar, G., Saria, A., Bukhave, K., and Rask-Madsen, J. (1989). Involvement of 5-hydroxytryptamine, prostaglandin E2, and cyclic adenosine monophosphate in cholera toxin-induced fluid secretion in the small intestine of the rat in vivo. *Gastroenterology* 96:368–376.

Bhattacharya, S. K., Bhattacharya, M. K., Nair, G. B., Dutta, D., Deb, A., Ramamurthy, T., Garg, S., Saha, P. K., Dutta, P., Moitra, A., et al. (1993). Clinical profile of acute diarrhoea cases infected with the new epidemic strain of *Vibrio cholerae* O139: designation of the disease as cholera. *J. Infect.* 27:11–15.

Booth, B. A., Boesman-Finkelstein, M., and Finkelstein, R. A. (1984). *Vibrio cholerae* hemagelutinin/protease nicks cholera enterotoxin. *Infect. Immun. 45*:558–560.

Bramucci, M. G., Twiddy, E. M., Baine, W. B., and Holmes, R. K. (1981). Isolation and characterization of hypertoxinogenic (*htx*) mutants of *Escherichia coli* KL320(pCG86). *Infect. Immun. 32*:1034–1044.

Brickman, T. J., Boesman-Finkelstein, M., Finkelstein, R. A., and McIntosh, M. A. (1990). Molecular cloning and nucleotide sequence analysis of cholera toxin genes of the CtxA-*Vibrio cholerae* strain Texas Star-SR. *Infect. Immun. 58*:4142–4144.

Burnette, W. N., Ceiplak, W., Mar, V. L., Kaljot, K. T., Sato, H., and Keith, J. M. (1988). Pertussis toxin S1 mutant with reduced enzyme activity and a conserved protective epitope. *Science 242*:72–74.

Burnette, W. N., Mar, V. L., Platler, B. W., Schlotterbeck, J. D., McGinley, M. D., Stoney, K. S., Rohde, M. F., and Kaslow, H. R. (1991). Site-specific mutagenesis of the catalytic subunit of cholera toxin: substituting lysine for arginine 7 causes loss of activity. *Infect. Immun. 59*:4266–4270.

Chang, P. P., Moss, J., Twiddy, E. M., and Holmes, R. K. (1987). Type II heat-labile enterotoxin of *Escherichia coli* activates adenylate cyclase in human fibroblasts by ADP ribosylation. *Infect. Immun. 55*:1854–1858.

Chaudhary, V. K., Jinno, Y., FitzGerald, D., and Pastan, I. (1990). *Pseudomonas* exotoxin contains a specific sequence at the carboxyl terminus that is required for cytotoxicity. *Proc. Natl. Acad. Sci. U.S.A. 87*:308–312.

Chopra, A. K., Houston, C. W., Peterson, J. W., Prasad, R., and Mekalanos, J. J. (1987). Cloning and expression of the *Salmonella* enterotoxin gene. *J. Bacteriol. 169*:5095–5100.

Clements, J. D., and Finkelstein, R. A. (1979). Isolation and characterization of homogeneous heat-labile enterotoxins with high specific activity from *Escherichia coli* cultures. *Infect. Immun. 24*:760–769.

Clements, J. D., Yancey, R. J., and Finkelstein, R. A. (1980). Properties of homogeneous heat-labile enterotoxin from *Escherichia coli*. *Infect. Immun. 29*:91–97.

Connell, T. D., and Holmes, R. K. (1992a). Molecular genetic analysis of ganglioside G_{D1b}-binding activity of *Escherichia coli* type IIa heat-labile enterotoxin by use of random and site-directed mutagenesis. *Infect. Immun. 60*:63–70.

Connell, T. D., and Holmes, R. K. (1992b). Characterization of hybrid toxins produced in *Escherichia coli* by assembly of A and B polypeptides from type I and type II heat-labile enterotoxins. *Infect. Immun. 60*:1653–1661.

Cuatrecasas, P. (1973). Gangliosides and membrane receptors for cholera toxin. *Biochemistry 12*:3558–3566.

Dallas, W. S. (1983). Conformity between heat-labile toxin genes from human and porcine enterotoxigenic *Escherichia coli*. *Infect. Immun. 40*:647–652.

Dallas, W. S., and Falkow, S. (1980). Amino acid sequence homology between cholera toxin and *Escherichia coli* heat-labile toxin. *Nature 288*:499–501.

Dallas, W. S., Gill, D. M., and Falkow, S. (1979). Cistrons encoding *Escherichia coli* heat-labile toxin. *J. Bacteriol. 139*:850–858.

Dams, E., De Wolf, M., and Dierick, W. (1991). Nucleotide sequence analysis of the CT operon of the *Vibrio cholerae* classical strain 569B. *Biochim. Biophys. Acta 1090*:139–141.

De, S. N. (1959). Enterotoxicity of bacteria-free culture-filtrate of *Vibrio cholerae*. *Nature 183*:1533–1534.

DeWolf, M. J. S., Fridkin, M., Epstein, M., and Kohn, L. D. (1981a). Structure–function studies of cholera toxin and its A and B protomers. *J. Biol. Chem. 256*:5481–5488.

DeWolf, M. J. S., Fridkin, M., and Kohn, L. D. (1981b). Tryptophan and residues of cholera toxin and its A and B protomers. Intrinsic fluorescence and solute quenching upon interacting with the ganglioside G_{M1}, oligo-G_{M1}, or dansylated oligo-G_{M1}. *J. Biol. Chem. 256*:5489–5496.

Donta, S. T., Moon, H. W., and Whipp, S. C. (1974). Detection of heat-labile *Escherichia coli* enterotoxin with the use of adrenal cells in tissue culture. *Science 183*:334–336.

Donta, S. T., Beristain, S., and Tomicic, T. K. (1993). Inhibition of heat-labile cholera and *Escherichia coli* enterotoxins by brefeldin A. *Infect. Immun. 61*:3282–3286.

Dubey, R. S., Lindblad, M. and Holmgren, J. (1990). Purification of El Tor cholera enterotoxins and comparisons with classical toxin. *J. Gen. Microbiol. 136*:1839–1847.

Duffy, L. K., and Lai, C. Y. (1979). Involvement of arginine residues in the binding site of cholera toxin B subunit. *Biochem. Biophys. Res. Commun. 91*:1005–1010.

Dutta, N. K., Panse, M. W., and Kulkarni, D. R. (1959). Role of cholera toxin in experimental cholera. *J. Bacteriol. 78*:594–595.

Eidels, L., Proia, R. L., and Hart, D. A. (1983). Membrane receptors for bacterial toxins. *Microbiol. Rev. 47*:596–620.

Etkin, S., and Gorbach, S. L. (1971). Studies on enterotoxin from *Escherichia coli* associated with acute diarrhea in man. *J. Lab. Clin. Med. 78*:81–87.

Fernandez, M., Sierra-Madero, J., de la Vega, H., Vazquez, M., Lopez-Vidal, Y., Ruiz-Palacios, G. M., and Calva, E. (1988). Molecular cloning of a *Salmonella typhi* LT-like enterotoxin gene. *Mol. Microbiol. 2*:821–825.

Field, M., Rao, M. C., and Chang, E. B. (1989a). Intestinal electrolyte transport and diarrheal disease (part 1). *N. Engl. J. Med. 321*:800–806.

Field, M., Rao, M. C., and Chang, E. B. (1989b). Intestinal electrolyte transport and diarrheal disease (part 2). *N. Engl. J. Med. 321*:879–883.

Finkelstein, R. A. (1973). Cholera. *CRC. Crit. Rev. Microbiol. 2*:553–623.

Finkelstein, R. A. (1983). Antigenic and structural variations in the cholera/coli family of enterotoxins. *Dev. Biol. Stand. 53*:93–95.

Finkelstein, R. A., and LoSpalluto, J. J. (1969). Pathogenesis of experimental cholera. Preparation and isolation of choleragen and choleragenoid. *J. Exp. Med. 130*:185–202.

Finkelstein, R. A., and LoSpalluto, J. J. (1972). Crystalline cholera toxin and toxoid. *Science 175*:529–530.

Finkelstein, R. A., Boesman, M., Neoh, S. H., LaRue, M. K., and Delaney, R. (1974a). Dissociation and recombination of the subunits of the cholera enterotoxin (choleragen). *J. Immunol. 113*:145–150.

Finkelstein, R. A., Vasil, M. L., and Holmes, R. K. (1974b). Studies on toxinogenesis in *Vibrio cholerae*. I. Isolation of mutants with altered toxinogenicity. *J. Infect. Dis. 129*:117–123.

Finkelstein, R. A., Marchlewicz, B. A., McDonald, R. J., and Boesman-Finkelstein, M. (1983). Isolation and characterization of a cholera-related enterotoxin from *Salmonella typhimurium*. *FEMS Microbiol. Lett. 17*:239–241.

Finkelstein, R. A., Burks, M. F., Zupan, A., Dallas, W. S., Jacob, C. O., and Ludwig, D. S. (1987a). Antigenic determinants of the cholera/coli family of enterotoxins. *Rev. Infect. Dis. 9*:S490–S502.

Finkelstein, R. A., Burks, M. F., Zupan, A., Dallas, W. S., Jacob, C. O., and Ludwig, D. S. (1987b). Epitopes of the cholera family of enterotoxins. *Rev. Infect. Dis. 9*:544–561.

Fishman, P. H. (1980). Mechanism of action of cholera toxin: studies on the lag period. *J. Membr. Biol. 54*:61–72.

Fishman, P. H. (1982a). Internalization and degradation of cholera toxin by cultured cells: relationship to toxin action. *J. Cell. Biol. 93*:860–865.

Fishman, P. H. (1982b). Role of membrane gangliosides in the binding and action of bacterial toxins. *J. Membr. Biol. 69*:85–97.

Fishman, P. H. (1986). Recent advances in identifying the functions of gangliosides. *Chem. Phys. Lipids 42*:137–151.

Fishman, P. H., and Atikkan, E. E. (1980). Mechanism of action of cholera toxin: effect of receptor density and multivaliant binding on activation of adenylate cyclase. *J. Membr. Biol. 54*:51–60.

Fishman, P. H., Moss, J., and Manganiello, V. C. (1977). Synthesis and uptake of gangliosides by choleragen-responsive human fibroblasts. *Biochemistry 16*:1871–1875.

Fishman, P. H., Moss, J., and Osborne, J. C., Jr. (1978). Interaction of choleragen with the oligosaccharide of ganglioside G_{M1}: evidence for multiple oligosaccharide binding sites. *Biochemistry 17*:711–716.

Fishman, P. H., Pacuszka, T., Hom, B., and Moss, J. (1980). Modification of ganglioside G_{M1}. Effect of lipid moiety on choleragen action. *J. Biol. Chem. 255*:7657–7664.

Fukuta, S., Magnani, J. L., Twiddy, E. M., Holmes, R. K., and Ginsburg, V. (1988). Comparison of the carbohydrate-binding specificities of cholera toxin and *Escherichia coli* heat-labile enterotoxins LTh-I, LT-IIa, and LT-IIb. *Infect. Immun. 56*:1748–1753.

Galloway, T. S., and van Heyningen, S. (1987). Binding of NAD$^+$ by cholera toxin. *Biochem. J. 244*:225–230.

Galloway, T. S., Tait, R. M., and van Heyningen, S. (1987). Photolabelling of cholera toxin by NAD$^+$. *Biochem. J. 242*:927–930.

Gardner, S. E., Fowlston, S. E., and George, W. L. (1987). In vitro production of cholera toxin-like activity by *Plesiomonas shigelloides*. *J. Infect. Dis. 156*:720–722.

Geary, S. J., Marchlewicz, B. A., and Finkelstein, R. A. (1982). Comparison of heat-labile enterotoxins from porcine and human strains of *Escherichia coli*. *Infect. Immun. 36*:215–220.

Gibbons, A. (1991). The shape of cholera. *Science 253*:382–383.

Gibert, I., Villegas, V., and Barbe, J. (1990). Expression of heat-labile enterotoxins genes is under cyclic AMP control in *Escherichia coli*. *Curr. Microbiol. 20*:83–90.

Gierschik, P. (1992). ADP-ribosylation of signal-transducing guanine nucleotide-binding proteins by pertussis toxin. *Curr. Top. Microbiol. Immunol. 175*:69–96.

Gill, D. M. (1976). The arrangement of subunits in cholera toxin. *Biochemistry 15*:1242–1248.

Gill, D. M., and Meren, R. (1978). ADP-ribosylation of membrane proteins catalyzed by cholera toxin: basis of the activation of adenylate cyclase. *Proc. Natl. Acad. Sci. U.S.A. 75*:3050–3054.

Gill, D. M., and Rappaport, R. S. (1979). Origin of the enzymatically active A1 fragment of cholera toxin. *J. Infect. Dis. 139*:674–680.

Gill, D. M., Clements, J. D., Robertson, D. C., and Finkelstein, R. A. (1981). Subunit number and arrangement in *Escherichia coli* heat-labile enterotoxin. *Infect. Immun. 33*:677–682.

Gilligan, P. H., Brown, J. C., and Robertson, D. C. (1983). Immunological relationships between cholera toxin and *Escherichia coli* heat-labile enterotoxin. *Infect. Immun. 42*:683–691.

Glass, R. I., Holmgren, J., Haley, C. E., Khan, M. R., Svennerholm, A. M. Stoll, B. J., Belayet Hossain, K. M., Black, R. E., Yunus, M., and Barua, D. (1985). Predisposition for cholera of individuals with O blood group. Possible evolutionary significance. *Am. J. Epidemiol. 121*:791–796.

Goins, B., and Freire, E. (1988). Thermal stability and intersubunit interactions of cholera toxin in solution and in association with its cell-surface receptor ganglioside G_{M1}. *Biochemistry 27*:2046–2052.

Green, B. A., Neill, R. J., Ruyechan, W. T., and Holmes, R. K. (1983). Evidence that a new enterotoxin of *Escherichia coli* which activates adenylate cyclase in eucaryotic target cells is not plasmid mediated. *Infect. Immun. 41*:383–390.

Griffiths, S. L., and Critchley, D. R. (1991). Characterization of the binding sites for *Escherichia coli* heat-labile toxin type I in intestinal brush borders. *Biochim. Biophys. Acta 1075*:154–161.

Griffiths, S. L., Finkelstein, R. A., and Critchley, D. R. (1986). Characterization of the receptor for cholera toxin and *Escherichia coli* heat-labile toxin in rabbit intestinal brush borders. *Biochem. J. 238*:313–322.

Guerrant, R. L., and Brunton, L. L. (1977). Characterization of the Chinese hamster ovary cell assay for the enterotoxins of *Vibrio cholerae* and *Escherichia coli* and for specific antisera, and toxoid. *J. Infect. Dis. 135*:720–728.

Guerrant, R. L., Brunton, L. L., Schnaitman, T. C., Rebhun, L. I., and Gilman, A. G. (1974). Cyclic adenosine monophosphate and alteration of Chinese hamster ovary cell morphology: a rapid, sensitive in vitro assay for the enterotoxins of *Vibrio cholerae* and *Escherichia coli*. *Infect. Immun. 10*:320–327.

Guth, B. E., Pickett, C. L., Twiddy, E. M., Holmes, R. K., Gomes, T. A., Lima, A. A., Guerrant, R. L., Franco, B. D., and Trabulsi, L. R. (1986a). Production of type II heat-labile enterotoxin by *Escherichia coli* isolated from food and human feces. *Infect. Immun. 54*: 587–589.

Guth, B. E., Twiddy, E. M., Trabulsi, L. R., and Holmes, R. K. (1986b). Variation in chemical properties and antigenic determinants among type II heat-labile enterotoxins of *Escherichia coli. Infect. Immun. 54*:529–536.

Gyles, C. L. (1992). *Escherichia coli* cytotoxins and enterotoxins. *Can. J. Microbiol. 38*:734–746.

Gyles, C. L., and Barnum, D. A. (1969). A heat-labile enterotoxin from strains of *Escherichia coli* enteropathogenic for pigs. *J. Infect. Dis. 120*:419–426.

Gyles, C., So, M., and Falkow, S. (1974). The enterotoxin plasmids of *Escherichia coli. J. Infect. Dis. 130*:40–48.

Hardy, S. J., Holmgren, J., Johansson, S., Sanchez, J., and Hirst, T. R. (1988). Coordinated assembly of multisubunit proteins: oligomerization of bacterial enterotoxins in vivo and in vitro, *Proc. Natl. Acad. Sci. U.S.A. 85*:7109–7113.

Harford, S., Dykes, C. W., Hobden, A. N., Read, M. J., and Halliday, I. J. (1989). Inactivation of the *Escherichia coli* heat-labile enterotoxin by in vitro mutagenesis of the A-subunit gene, *Eur. J. Biochem. 183*:311–316.

Hirst, T. R., and Holmgren, J. (1987). Conformation of protein secreted across bacterial outer membranes: a study of enterotoxin translocation from *Vibrio cholerae. Proc. Natl. Acad. Sci. U.S.A. 84*:7418–7422.

Hirst, T. R., and Welch, R. A. (1988). Mechanisms for secretion of extracellular proteins by gram-negative bacteria. *Trends. Biochem. Sci. 13*:265–269.

Hirst, T. R., Hardy, S. J., and Randall, L. L. (1983). Assembly in vivo of enterotoxin from *Escherichia coli*: formation of the B subunit oligomer. *J. Bacteriol. 153*:21–26.

Hirst, T. R., Randall, L. L., and Hardy, S. J. (1984a). Cellular location of enterotoxin in *Escherichia coli. Biochem. Soc. Trans. 12*:189–191.

Hirst, T. R., Randall, L. L., and Hardy, S. J. (1984b). Cellular location of heat-labile enterotoxin in *Escherichia coli, J. Bacteriol. 157*:637–642.

Hirst, T. R., Sanchez, J., Kaper, J. B., Hardy, S. J., and Holmgren, J. (1984c). Mechanism of toxin secretion by *Vibrio cholerae* investigated in strains harboring plasmids that encode heat-labile enterotoxins of *Escherichia coli. Proc. Natl. Acad. Sci. U.S.A. 81*:7752–7756.

Hofstra, H., and Witholt, B. (1984). Kinetics of synthesis, processing, and membrane transport of heat-labile enterotoxin, a periplasmic protein in *Escherichia coli. J. Biol. Chem. 259*:15182–15187.

Holmes, R. K., and Twiddy, E. M. (1983). Characterization of monoclonal antibodies that react with unique and cross-reacting determinants of cholera enterotoxin and its subunits. *Infect. Immun. 42*:914–923.

Holmes, R. K., Vasil, M. L., and Finkelstein, R. A. (1975). Studies on toxinogenesis in *Vibrio cholerae*. III. Characterization of nontoxinogenic mutants in vitro and in experimental animals. *J. Clin. Invest. 55*:551–560.

Holmes, R. K., Baine, W. B., and Vasil, M. L. (1978). Quantitative measurements of cholera enterotoxin in cultures of toxinogenic wild-type and nontoxinogenic mutant strains of *Vibrio cholerae* by using a sensitive and specific reversed passive hemagglutination assay for cholera enterotoxin. *Infect. Immun. 19*:101–106.

Holmes, R. K., Bramucci, M. G., and Twiddy, E. M. (1980). Genetic and biochemical studies of heat-labile enterotoxin of *Escherichia coli*. In *Proceedings of the Fifteenth Joint Conference on Cholera*, NIH Publ. 80-2003, National Institutes of Health, Bethesda, MD, pp. 187–201.

Holmes, R. K., Twiddy, E. M., and Bramucci, M. G. (1983). Antigenic heterogeneity among heat-labile enterotoxins from *Escherichia coli*. In *Advances in Research on Cholera and Related Diarrheas*, S. Kuwahara and N. F. Pierce (Eds.), KTK Scientific Publishers, Tokyo, pp. 293–300.

Holmes, R. K., Twiddy, E. M., and Neill, R. J. (1985) Recent advances in the study of heat-labile enterotoxins of *Escherichia coli*. In *Bacterial Diarrheal Diseases*, Y. Takeda and T. Miwatani (Eds.), Martinus Nijhoff, Boston, pp. 125–135.

Holmes, R. K., Twiddy, E. M., and Pickett, C. L. (1986). Purification and characterization of type II heat-labile enterotoxin of *Escherichia coli*. *Infect. Immun.* 53:464–473.

Holmes, R. K., Twiddy, E. M., Pickett, C. L., Marcus, H., Jobling, M. G., and Petitjean, F. M. J. (1991). The *Escherichia coli/Vibrio cholerae* family of enterotoxins. In *Molecular Toxins in Foods and Feeds—Cellular and Molecular Modes of Action*, E E. Pohland, R. D. Vulus, Jr., and J. L. Richard (Eds.), Plenum, New York, pp. 91–102.

Holmgren, J., and Svennerholm, A. M. (1977). Mechanisms of disease and immunity in cholera: a review. *J. Infect. Dis.* 136:S105–S112.

Holmgren, J., Fredman, P., Lindblad, M., Svennerholm, A. M., and Svennerholm, L. (1982). Rabbit intestinal glycoprotein receptor for *Escherichia coli* heat-labile enterotoxin lacking affinity for cholera toxin. *Infect. Immun.* 38:424–433.

Holmgren, J., Lindblad, M., Fredman, P., Svennerholm, L., and Myrvold, H. (1985). Comparison of receptors for cholera and *Escherichia coli* enterotoxins in human intestine. *Gastroenterology* 89:27–35.

Holmgren, J., Clemens, J., Sack, D. A., Sanchez, J., and Svennerholm, A. M. (1989). Oral immunization against cholera. *Curr. Top. Microbiol. Immunol.* 146:197–204.

Honda, T., Tsuji, T., Takeda, Y., and Miwatani, T. (1981). Immunological nonidentity of heat-labile enterotoxins from human and porcine enterotoxigenic *Escherichia coli*. *Infect. Immun.* 34:337–340.

Honda, T., Sato, M., Nishimura, T., Higashitsutsumi, M., Fukai, K., and Miwatani, T. (1985). Demonstration of cholera toxin-related factor in cultures of *Aeromonas* species by enzyme-linked immunosorbent assay. *Infect. Immun.* 50:322–323.

Huilan, S., Zhen, L. G., Mathan, M. M., Mathew, M. M., Olarte, J., Espejo, R., Khin Maung, U., Ghafoor, M. A., Khan, M. A., Sami, Z., et al. (1991). Etiology of acute diarrhoea among children in developing countries: a multicentre study in five countries. *Bull WHO* 69:549–555.

Ikemura, T. (1981). Correlation between the abundance of *Escherichia coli* transfer RNAs and the occurrence of the respective codons in its protein genes: a proposal for a synonymous codon choice that is optimal for the *E. coli* translational system. *J. Mol. Biol.* 151:389–409.

Jacob, C. O. (1990). Vaccination against the cholera/coli family of enterotoxins by synthetic peptides. *Adv. Biotechnol. Processes* 13:223–254.

Jacob, C. O., Arnon, R., and Finkelstein, R. A. (1986a). Immunity to heat-labile enterotoxins of porcine and human *Escherichia coli* strains achieved with synthetic cholera toxin peptides. *Infect. Immun.* 52:562–567.

Jacob, C. O., Arnon, R., and Sela, M. (1986b). Anti-cholera response elicited by a completely synthetic antigen with built-in adjuvanticity administered in aqueous solution. *Immunol. Lett.* 14:43–48.

Janicot, M., Fouque, F., and Desbuquois, B. (1991). Activation of rat liver adenylate cyclase by cholera toxin requires toxin internalization and processing in endosomes. *J. Biol. Chem.* 266:12858–12865.

Jobling, M. G., and Holmes, R. K. (1991). Analysis of structure and function of the B subunit of cholera toxin by the use of site-directed mutagenesis. *Mol. Microbiol.* 5:1755–1767.

Jobling, M. G., and Holmes, R. K. (1992). Fusion proteins containing the A2 domain of cholera toxin assemble with B polypeptides of cholera toxin to form immunoreactive and functional holotoxin-like chimeras [published erratum appears in *Infect Immun. 61*:1168 (1993)]. *Infect. Immun. 60*:4915–4924.

Johnson, W. M., and Lior, H. (1986). Cytotoxic and cytotonic factors produced by *Campylobacter jejuni, Campylobacter coli,* and *Campylobacter laridis. J. Clin. Microbiol.* 24:275–281.

Joseph, K. C., Stieber, A., and Gonatas, N. K. (1979). Endocytosis of cholera toxin in GERL-like structures of murine neuroblastoma cells pretreated with G_{M1} ganglioside. Cholera toxin internalization into neuroblastoma GERL. *J. Cell. Biol.* 81:543–554.

Kaper, J. B., and Srivastava, B. S. (1992). Genetics of cholera toxin. *Indian J. Med. Res.* 95:163–167.

Kaper, J. B., Bradford, H. B., Roberts, N. C., and Falkow, S. (1982). Molecular epidemiology of *Vibrio cholerae* in the U.S. Gulf Coast. *J. Clin. Microbiol. 16*:129–134.

Kassis, S., Hagmann, J., Fishman, P. H., Chang, P. P., and Moss, J. (1982). Mechanism of action of cholera toxin on intact cells. Generation of A1 peptide and activation of adenylate cyclase. *J. Biol. Chem. 257*:12148–12152.

Kazemi, M., and Finkelstein, R. A. (1990a). Study of epitopes of cholera enterotoxin-related enterotoxins by checkerboard immunoblotting. *Infect. Immun. 58*:2352–2360.

Kazemi, M., and Finkelstein, R. A. (1990b). Checkerboard immunoblotting (CBIB): an efficient, rapid, and sensitive method of assaying multiple antigen/antibody cross-reactivities. *J. Immunol. Methods 128*:143–146.

Klipstein, F. A., Guerrant, R. L., Wells, J. G., Short, H. B., and Engert, R. F. (1979). Comparison of assay of coliform enterotoxins by conventional techniques versus in vivo intestinal perfusion. *Infect. Immun. 25*:146–152.

Krasilnikov, O. V., Muratkhodjaev, J. N., Voronov, S. E., and Yezepchuk, Y. V. (1991). The ionic channels formed by cholera toxin in planar bilayer lipid membranes are entirely attributable to its B-subunit. *Biochim. Biophys. Acta 1067*:166–170.

Kunkel, S. L., and Robertson, D. C. (1979). Purification and chemical characterization of the heat-labile enterotoxin produced by enterotoxigenic *Escherichia coli*. *Infect. Immun. 25*:586–596.

Kurosky, A., Markel, D. E., Touchstone, B., and Peterson, J. W. (1976). Chemical characterization of the structure of cholera toxin and its natural toxoid. *J. Infect. Dis. 133*:14–22.

Lai, C. Y., Mendez, E., and Chang, D. (1975). Chemistry of cholera toxin: the subunit structure. *J. Infect. Dis. 133*:23–30.

Lee, C. M., Chang, P. P., Tsai, S. C., Adamik, R., Price, S. R., Kunz, B. C., Moss, J., Twiddy, E. M., and Holmes, R. K. (1991). Activation of *Escherichia coli* heat-labile enterotoxins by native and recombinant adenosine diphosphate-ribosylation factors. 20-kD guanine nucleotide-binding proteins, *J. Clin. Invest. 87*:1780–1786.

Lencer, W. I., Delp, C., Neutra, M. R., and Madara, J. L. (1992). Mechanism of cholera toxin action on a polarized human intestinal epithelial cell line: role of vesicular traffic. *J. Cell. Biol. 117*:1197–1209.

Leong, J., Vinal, A. C., and Dallas, W. S. (1985). Nucleotide sequence comparison between heat-labile toxin B-subunit cistrons from *Escherichia coli* of human and porcine origin. *Infect. Immun. 48*:73–77.

Levine, M. M. (1987). *Escherichia coli* that cause diarrhea: enterotoxigenic, enteropathogenic, enteroinvasive, enterohemorrhagic, and enteroadherent. *J. Infect. Dis. 155*:377–389.

Levner, M., Weiner, F. P., and Rubin, B. A. (1977). Induction of *Escherichia coli* and *Vibrio cholerae* enterotoxins by an inhibitor of protein synthesis. *Infect. Immun. 15*:132–137.

Levner, M. H., Urbano, C., and Rubin, B. A. (1980). Lincomycin increases synthetic rate and periplasmic pool size for cholera toxin. *J. Bacteriol. 143*:441–447.

Lindholm, L., Holmgren, J., Wikström, M., Karlsson, U., Andersson, K., and Lycke, N. (1983). Monoclonal antibodies to cholera toxin with special reference to cross-reactions with *Escherichia coli* heat-labile enterotoxin. *Infect. Immun. 40*:570–576.

Lobet, Y., Cluff, C. W., and Cieplak, W., Jr. (1991). Effect of site-directed mutagenic alterations on ADP-ribosyltransferase activity of the A subunit of *Escherichia coli* heat-labile enterotoxin. *Infect. Immun. 59*:2870–2879.

Lockman, H., and Kaper, J. B. (1983). Nucleotide sequence analysis of the A2 and B subunits of *Vibrio cholerae* enterotoxin. *J. Biol. Chem. 258*:13722–13726.

Lockman, H. A., Galen, J. E., and Kaper, J. B. (1984). *Vibrio cholerae* enterotoxin genes: nucleotide sequence analysis of DNA encoding ADP-ribosyltransferase, *J. Bacteriol. 159*: 1086–1089.

Longbottom, D., and van Heyningen, S. (1989). The activation of rabbit intestinal adenylate cyclase by cholera toxin. *Biochim. Biophys. Acta 1014*:289–297.

Ludwig, D. S., Holmes, R. K., and Schoolnik, G. K. (1985). Chemical and immunochemical studies on the receptor binding domain of cholera toxin B subunit. *J. Biol. Chem.* 260:12528–12534.

Maneval, D. R., Jr., Colwell, R. R. Joseph, S. W., Gray, R., and Donta, S. T. (1980). A tissue culture method for the detection of bacterial enterotoxins. *J. Tissue Cult. Methods* 6:85–90.

Marchlewicz, B. A., and Finkelstein, R. A. (1983). Immunological differences among the cholera/coli family of enterotoxins. *Diagn. Microbiol. Infect. Dis.* 1:129–138.

Masserini, M., Freire, E., Palestini, P., Calappi, E., and Tettamanti, G. (1992). Fuc-G$_{M1}$ ganglioside mimics the receptor function of G$_{M1}$ for cholera toxin. *Biochemistry 31*:2422–2426.

Mekalanos, J. J. (1983). Duplication and amplification of toxin genes in *Vibrio cholerae*. *Cell 35*:253–263.

Mekalanos, J. J. (1992). Environmental signals controlling expression of virulence determinants in bacteria. *J. Bacteriol. 174*:1–7.

Mekalanos, J. J., and Murphy, J. R. (1980). Regulation of cholera toxin production in *Vibrio cholerae*: genetic analysis of phenotypic instability in hypertoxinogenic mutants. *J. Bacteriol. 141*:570–576.

Mekalanos, J. J., Collier, R. J., and Romig, W. R. (1978a). Affinity filters, a new approach to the isolation of *tox* mutants of *Vibrio cholerae*. *Proc. Natl. Acad. Sci. U.S.A. 75*:941–945.

Mekalanos, J. J., Collier, R. J., and Romig, W. R. (1978b). Purification of cholera toxin and its subunits: new methods of preparation and the use of hypertoxinogenic mutants. *Infect. Immun. 20*:552–558.

Mekalanos, J. J., Collier, R. J., and Romig, W. R. (1979a). Enzymic activity of cholera toxin. I. New method of assay and the mechanism of ADP-ribosyl transfer. *J. Biol. Chem. 254*:5849–5854.

Mekalanos, J. J., Collier, R. J., and Romig, W. R. (1979b). Enzymic activity of cholera toxin. II. Relationships to proteolytic processing, disulfide bond reduction, and subunit composition. *J. Biol. Chem. 254*:5855–5861.

Mekalanos, J. J., Sublett, R. D., and Romig, W. R. (1979). Genetic mapping of toxin regulatory mutations in *Vibrio cholerae*. *J. Bacteriol. 139*:859–865.

Mekalanos, J. J., Swartz, D. J., Pearson, G. D., Harford, N., Groyne, F., and de Wilde, M. (1983). Cholera toxin genes: nucleotide sequence, deletion analysis and vaccine development. *Nature 306*:551–557.

Miller, V. L., and Mekalanos, J. J. (1984). Synthesis of cholera toxin is positively regulated at the transcriptional level by *toxR*. *Proc. Natl. Acad. Sci. U.S.A. 81*:3471–3475.

Miller, V. L., and Mekalanos, J. J. (1985). Genetic analysis of the cholera toxin-positive regulatory gene *toxR*. *J. Bacteriol. 163*:580–585.

Montesano, R., Roth, J., Robert, A., and Orci, L. (1982). Non-coated membrane invaginations are involved in binding and internalization of cholera and tetanus toxins. *Nature 296*:651–653.

Moon, H. W., and Bunn, T. O. (1993). Vaccines for preventing enterotoxigenic *Escherichia coli* infections in farm animals. *Vaccine 11*:213–200.

Morita, A., Tsao, D., and Kim, Y. S. (1980). Identification of cholera toxin binding glycoproteins in rat intestinal microvillus membranes. *J. Biol. Chem. 255*:2549–2553.

Moseley, S. L., and Falkow, S. (1980). Nucleotide sequence homology between the heat-labile enterotoxin gene of *Escherichia coli* and *Vibrio cholerae* deoxyribonucleic acid. *J. Bacteriol. 144*:444–446.

Moss, J., and Vaughan, M. (1977). Choleragen activation of solubilized adenylate cyclase: requirement for GTP and protein activator for demonstration of enzymatic activity. *Proc. Natl. Acad. Sci. U.S.A. 74*:4396–4400.

Moss, J., and Vaughan, M. (1979). Activation of adenylate cyclase by choleragen. *Annu. Rev. Biochem. 48*:581–600.

Moss, J., and Vaughan, M. (1993a). ADP-ribosylation factors: protein activators of cholera toxin. *Prog. Nucleic Acid Res. Mol. Biol. 45*:47–65.

Moss, J., and Vaughan, M. (1993b). ADP-ribosylation factors, 20,000 M_r guanine nucleotide-binding protein activators of cholera toxin and components of intracellular vesicular transport systems. *Cell. Signal.* 5:367–379.

Moss, J., and Vaughan, M. (1988) Cholera toxin and *E. coli* enterotoxins and their mechanisms of action. In *Handbook of Natural Toxins*, Vol. 4, *Bacterial Toxins*, M. Hardegree and A. T. Tu (Eds.), Marcel Dekker, New York, pp. 39–87.

Moss, J., Manganiello, V. C., and Fishman, P. H. (1977). Enzymatic and chemical oxidation of gangliosides in cultured cells: effects of choleragen. *Biochemistry* 16:1876–1881.

Moss, J., Garrison, S., Fishman, P. H., and Richardson, S. H. (1979). Gangliosides sensitize unresponsive fibroblasts to *Escherichia coli* heat-labile enterotoxin. *J. Clin. Invest.* 64:381–384.

Moss, J., Stanley, S. J., Watkins, P. A., and Vaughan, M. (1980). ADP-ribosyltransferase activity of mono- and multi-(ADP-ribosylated) choleragen. *J. Biol. Chem.* 255:7835–7837.

Moss, J., Tsuchiya, M., Tsai, S. C., Adamik, R., Bobak, D. A., Price, S. R., Nightingale, M. S., and Vaughan, M. (1990). Structural and functional characterization of ADP-ribosylation factors, 20 kDa guanine nucleotide-binding proteins that activate cholera toxin. *Adv. Second Messenger Phosphoprotein Res.* 24:83–88.

Mullin, B. R., Aloj, S. M., Fishman, P. H., Lee, G., Kohn, L. D., and Brady, R. O. (1976). Cholera toxin interactions with thyrotropin receptors on thyroid plasma membranes. *Proc. Natl. Acad. Sci. U.S.A.* 73:1679–1683.

Murer, H., Ammann, E., Biber, J., and Hopfer, U. (1976). The surface membrane of the small intestinal epithelial cell. I. Localization of adenyl cyclase. *Biochim. Biophys. Acta 433*:509–519.

Nakamura, K., Suzuki, M., Inagaki, F., Yamakawa, T., and Suzuki, A. (1987). A new ganglioside showing choleragenoid-binding activity in mouse spleen. *J. Biochem. (Tokyo)* 101:825–835.

Nambiar, M. P., Oda, T., Chen, C., Kuwazuru, Y., and Wu, H. C. (1993). Involvement of the Golgi region in the intracellular trafficking of cholera toxin. *J. Cell. Physiol.* 154:222–228.

Nashar, T. O., Amin, T., Marcello, A., and Hirst, T. R. (1993). Current progress in the development of the B subunits of cholera toxin and *Escherichia coli* heat-labile enterotoxin as carriers for the oral delivery of heterologous antigens and epitopes. *Vaccine 11*:235–240.

Neill, R. J., Ivins, B. E., and Holmes, R. K. (1983a). Synthesis and secretion of the plasmid-coded heat-labile enterotoxin of *Escherichia coli* in *Vibrio cholerae*. *Science 221*:289–291.

Neill, R. J., Twiddy, E. M., and Holmes, R. K. (1983b). Synthesis of plasmid-coded heat-labile enterotoxin in wild-type and hypertoxinogenic strains of *Escherichia coli* and in other genera of Enterobacteriaceae. *Infect. Immun.* 41:1056–1061.

Okamoto, K., Miyama, A., Tsuji, T., Honda, T., and Miwatani, T. (1988). Effect of substitution of glycine for arginine at position 146 of the A1 subunit on biological activity of *Escherichia coli* heat-labile enterotoxin. *J. Bacteriol.* 170:2208–2211.

Olsnes, S., Stenmark, H., Moskaug, J. O., McGill, S., Madshus, I. H., and Sandvig, K. (1990). Protein toxins with intracellular targets. *Microb. Pathogen.* 8:163–168.

Olsvik, O., Lund, A., Berdal, B. P., and Bergan, T. (1983). Differences in bindings to the G_{M1} receptor by heat-labile enterotoxin of human and porcine *Escherichia coli* strains. *NIPH Ann.* 6:5–15.

Orlandi, P. A., Curran, P. K., and Fishman, P. H. (1993). Brefeldin A blocks the response of cultured cells to cholera toxin. Implications for intracellular trafficking in toxin action. *J. Biol. Chem.* 268:12010–12016.

Ottemann, K. M., DiRita, V. J., and Mekalanos, J. J. (1992). ToxR proteins with substitutions in residues conserved with OmpR fail to activate transcription from the cholera toxin promoter. *J. Bacteriol.* 174:6807–6814.

Patzer, E. J., Wagner, R. R., and Dubovi, E. J. (1979). Viral membranes: model systems for studying biological membranes. *CRC. Crit. Rev. Biochem.* 6:165–217.

Pearson, G. D., and Mekalanos, J. J. (1982). Molecular cloning of *Vibrio cholerae* enterotoxin genes in *Escherichia coli* K-12. *Proc. Natl. Acad. Sci. U.S.A.* 79:2976–2980.

Pearson, G. D., Woods, A., Chiang, S. L., and Mekalanos, J. J. (1993). CTX genetic element encodes a site-specific recombination system and an intestinal colonization factor. *Proc. Natl. Acad. Sci. U.S.A. 90*:3750–3754.

Pelham, H. R. (1991a). Multiple targets for brefeldin A. *Cell 67*:449–451.

Pelham, H. R. (1991b). Recycling of proteins between the endoplasmic reticulum and Golgi complex. *Curr. Opin. Cell. Biol. 3*:585–591.

Peterson, J. W., and Ochoa, L. G. (1989). Role of prostaglandins and cAMP in the secretory effects of cholera toxin. *Science 245*:857–859.

Peterson, J. W., LoSpalluto, J. J., and Finkelstein, R. A. (1972). Localization of cholera toxin in vivo. *J. Infect. Dis. 126*:617–628.

Pickett, C. L., Twiddy, E. M., Belisle, B. W., and Holmes, R. K. (1986). Cloning of genes that encode a new heat-labile enterotoxin of *Escherichia coli. J. Bacteriol. 165*:348–352.

Pickett, C. L., Weinstein, D. L., and Holmes, R. K. (1987). Genetics of type IIa heat-labile enterotoxin of *Escherichia coli*: operon fusions, nucleotide sequence, and hybridization studies. *J. Bacteriol. 169*:5180–5187.

Pickett, C. L., Twiddy, E. M., Coker, C., and Holmes, R. K. (1989). Cloning, nucleotide sequence, and hybridization studies of the type IIb heat-labile enterotoxin gene of *Escherichia coli. J. Bacteriol. 171*:4945–4952.

Prasad, R., Chopra, A. K., Chary, P., and Peterson, J. W. (1992). Expression and characterization of the cloned *Salmonella typhimurium* enterotoxin. *Microb. Pathog. 13*:109–121.

Qu, Z. H., Boesman-Finkelstein, M., Kazemi, M., and Finkelstein, R. A. (1991). Heterogeneity of immunotypes of heat-labile enterotoxins of enterotoxigenic *Escherichia coli* of human origin. *J. Infect. Dis. 164*:796–799.

Rask-Madsen, J., Bukhave, K., and Beubler, E. (1990). Influence on intestinal secretion of eicosanoids. *J. Intern. Med. Supple. 732*:137–144.

Ribi, H. O., Ludwig, D. S., Mercer, K. L., Schoolnik, G. K., and Kornberg, R. D. (1988). Three-dimensional structure of cholera toxin penetrating a lipid membrane. *Science 239*:1272–1276.

Robb, M., Nichols, J. C., Whoriskey, S. K., and Murphy, J. R. (1982). Isolation of hybridoma cell lines and characterization of monoclonal antibodies against cholera enterotoxin and its subunits. *Infect. Immun. 38*:267–262.

Sack, R. B. (1990). Travelers' diarrhea: microbiologic bases for prevention and treatment. *Rev. Infect. Dis. 12* (Suppl. 1):S59–S63.

Sack, R. B., Gorbach, S. L., Banwell, J. G., Jacobs, B., Chatterjee, B. D., and Mitra, R. C. (1971). Enterotoxigenic *Escherichia coli* isolated from patients with severe cholera-like disease. *J. Infect. Dis. 123*:378–385.

Sandefur, P. D., and Peterson, J. W. (1976). Isolation of skin permeability factors from culture filtrates of *Salmonella typhimurium. Infect. Immun. 14*:671–679.

Sandkvist, M., Morales, V., and Bagdasarian, M. (1993). A protein required for secretion of cholera toxin through the outer membrane of *Vibrio cholerae. Gene 123*:81–86.

Sanyal, S. C. (1983). NAG vibrio toxin. *Pharmacol. Ther. 20*:183–201.

Schengrund, C. L., and Ringler, N. J. (1989). Binding of *Vibrio cholerae* toxin and the heat-labile enterotoxin of *Escherichia coli* to G_{M1}, derivatives of G_{M1}, and nonlipid oligosaccharide polyvalent ligands [published erratum appears in *J. Biol. Chem. 264*:18853 (1989)] *J. Biol. Chem. 264*:13233–13237.

Seriwatana, J., Echeverria, P., Taylor, D. N., Rasrinaul, L., Brown, J. E., Peiris, J. S., and Clayton, C. L. (1988). Type II heat-labile enterotoxin-producing *Escherichia coli* isolated from animals and humans. *Infect. Immun. 56*:1158–1161.

Sigler, P. B., Dryan, M. E., Kiuefer, H. C., and Finkelstein, R. A. (1977). Cholera toxin crystals suitable for X-ray diffraction. *Science 197*:1277–1279.

Sixma, T. K., Pronk, S. E., Kalk, K. H., Wartna, E. S., van Zanten, B. A., Witholt, B., and Hol, W. G. (1991). Crystal structure of a cholera toxin-related heat-labile enterotoxin from *E. coli Nature 351*:371–377.

Sixma, T. K., Aguirre, A., Terwisscha van Scheltinga, A. C., Wartna, E. S., Kalk, K. H., and

Hol, W. G. (1992a). Heat-labile enterotoxin crystal forms with variable A/B5 orientation. Analysis of conformational flexibility. *FEBS Lett.* *305*:81–85.

Sixma, T. K., Pronk, S. E., Kalk, K. H., van Zanten, B. A., Berghuis, A. M., and Hol, W. G. (1992b). Lactose binding to heat-labile enterotoxin revealed by X-ray crystallography. *Nature 355*:561–564.

Sixma, T. K., Kalk, K. H., van Zanten, B. A., Dauter, Z., Kingma, J., Witholt, B., and Hol, W. G. (1993). Refined structure of *Escherichia coli* heat-labile enterotoxin, a close relative of cholera toxin. *J. Mol. Biol.* *230*:890–918.

Smith, H. W., and Linggood, M. A. (1971). The transmissible nature of enterotoxin production in a human enteropathogenic strain of *Escherichia coli.* *J. Med. Microbiol.* *4*:301–305.

So, M., Dallas, W. S., and Falkow, S. (1978). Characterization of an *Escherichia coli* plasmid encoding for synthesis of heat-labile toxin: molecular cloning of the toxin determinant. *Infect. Immun. 21*:405–411.

Spangler, B. D. (1991). Binding to native proteins by antipeptide monoclonal antibodies. *J. Immunol. 146*:1591–1595.

Spangler, B. D. (1992). Structure and function of cholera toxin and the related *Escherichia coli* heat-labile enterotoxin. *Microbiol. Rev. 56*:622–647.

Spangler, B. D., and Westbrook, E. M. (1989). Crystallization of isoelectrically homogeneous cholera toxin. *Biochemistry 28*:1333–1340.

Spicer, E. K., Kavanaugh, W. M., Dallas, W. S., Falkow, S., Konigsberg, W. H., and Schafer, D. E. (1981). Sequence homologies between A subunits of *Escherichia coli* and *Vibrio cholerae* enterotoxins. *Proc. Natl. Acad. Sci. U.S.A. 78*:50–54.

Spiegel, S., Blumenthal, R., Fishman, P. H., and Handler, J. S. (1985). Gangliosides do not move from apical to basolateral plasma membrane in cultured epithelial cells. *Biochim. Biophys. Acta 821*:310–318.

Spira, W. M., and Fedorka-Cray, P. J. (1983). Production of cholera toxin-like toxin by *Vibrio mimicus* and non-O1 *Vibrio cholerae*: batch culture conditions for optimum yields and isolation of hypertoxigenic lincomycin-resistant mutants. *Infect. Immun. 42*:501–509.

Spira, W. M., and Fedorka-Cray, P. J. (1984). Purification of enterotoxins from *Vibrio mimicus* that appear to be identical to cholera toxin. *Infect. Immun. 45*:679–684.

Sporecke, I., Castro, D., and Mekalanos, J. J. (1984). Genetic mapping of *Vibrio cholerae* enterotoxin structural genes. *J. Bacteriol. 157*:253–261.

Surewicz, W. K., Leddy, J. J., and Mantsch, H. H. (1990). Structure, stability, and receptor interaction of cholera toxin as studied by Fourier-transform infrared spectroscopy. *Biochemistry 29*:8106–8111.

Takeda, Y., Honda, T., Taga, S., and Miwatani, T. (1981). In vitro formation of hybrid toxins between subunits of *Escherichia coli* heat-labile enterotoxin and those of cholera enterotoxin. *Infect. Immun. 34*:341–346.

Takeda, Y., Honda, T., Sima, H., Tsuji, T., and Miwatani, T. (1983). Analysis of antigenic determinants in cholera enterotoxin and heat-labile enterotoxins from human and porcine enterotoxigenic *Escherichia coli. Infect. Immun. 41*:50–53.

Tomasi, M., and Montecucco, C. (1981). Lipid insertion of cholera toxin after binding to G_{M1}-containing liposomes. *J. Biol. Chem. 256*:11177–11181.

Tran, D., Carpentier, J. L., Sawano, F., Gorden, P., and Orci, L. (1987). Ligands internalized through coated or noncoated invaginations follow a common intracellular pathway. *Proc. Natl. Acad. Sci. U.S.A. 84*:7957–7961.

Tsuji, T., Taga, S., Honda, T., Takeda, Y., and Miwatani, T. (1982). Molecular heterogeneity of heat-labile enterotoxins from human and porcine enterotoxigenic *Escherichia coli. Infect. Immun. 38*:444–448.

Tsuji, T., Honda, T., Miwatani, T., Wakabayashi, S., and Matsubara, H. (1985). Analysis of receptor-binding site in *Escherichia coli* enterotoxin. *J. Biol. Chem. 260*:8552–8558.

Tsuji, T., Iida, T., Honda, T., Miwatani, T., Nagahama, M., Sakurai, J., Wada, K., and Matsubara, H. (1987). A unique amino acid sequence of the B subunit of a heat-labile enterotoxin isolated from a human enterotoxigenic *Escherichia coli*. *Microb. Pathog.* 2:381–390.

Tsuji, T., Inoue, T., Miyama, A., Okamoto, K., Honda, T., and Miwatani, T. (1990). A single amino acid substitution in the A subunit of *Escherichia coli* enterotoxin results in a loss of its toxic activity. *J. Biol. Chem.* 265:22520–22525.

Tsuji, T., Inoue, T., Miyama, A., and Noda, M. (1991). Glutamic acid-112 of the A subunit of heat-labile enterotoxin from enterotoxigenic *Escherichia coli* is important for ADP-ribosyltransferase activity. *FEBS. Lett.* 291:319–321.

Vasil, M. L., Holmes, R. K., and Finkelstein, R. A. (1975). Conjugal transfer of a chromosomal gene determining production of enterotoxin in *Vibrio cholerae*. *Science 187*:849–850.

Wisnieski, B. J., and Bramhall, J. S. (1981). Photolabelling of cholera toxin subunits during membrane penetration. *Nature 289*:319–321.

Yamamoto, T., and Yokota, T. (1981). *Escherichia coli* heat-labile enterotoxin genes are flanked by repeated deoxyribonucleic acid sequences. *J. Bacteriol. 145*:850–860.

Yamamoto, T., and Yokota, T. (1983). Sequence of heat-labile enterotoxin of *Escherichia coli* pathogenic for humans. *J. Bacteriol. 155*:728–733.

Yamamoto, T., Nakazawa, T., Miyata, T., Kaji, A., and Yokota, T. (1984a). Evolution and structure of two ADP-ribosylation enterotoxins, *Escherichia coli* heat-labile toxin and cholera toxin. *FEBS Lett. 169*:241–246.

Yamamoto, T., Tamura, T., and Yokota, T. (1984b). Primary structure of heat-labile enterotoxin produced by *Escherichia coli* pathogenic for humans. *J. Biol. Chem. 259*:5037–5044.

Yamamoto, T., Suyama, A., Mori, N., Yokota, T., and Wada, A. (1985). Gene expression in the polycistronic operons of *Escherichia coli* heat-labile toxin and cholera toxin: a new model of translational control. *FEBS. Lett. 181*:377–380.

Yamamoto, T., Gojobori, T., and Yokota, T. (1987). Evolutionary origin of pathogenic determinants in enterotoxigenic *Escherichia coli* and *Vibrio cholerae* O1. *J. Bacteriol. 169*:1352–1357.

Yoh, M., Yamamoto, K., Honda, T., Takeda, Y., and Miwatani, T. (1983). Effects of lincomycin and tetracycline on production and properties of enterotoxins of enterotoxigenic *Escherichia coli*. *Infect. Immun. 42*:778–782.

Yoshida, T., Chen, C. C., Zhang, M. S., and Wu, H. C. (1991). Disruption of the Golgi apparatus by brefeldin A inhibits the cytotoxicity of ricin, modeccin, and *Pseudomonas* toxin. *Exp. Cell Res. 192*:389–395.

Young, D. B., and Broadbent, D. A. (1986). The effect of lincomycin on exoprotein production by *Vibrio cholerae*. *J. Med. Microbiol. 21*:13–17.

Zinnaka, Y., and Carpenter, C. C., Jr. (1972). An enterotoxin produced by noncholera vibrios. *Johns Hopkins Med. J. 131*:403–411.

11

ADP-Ribosylation Factors: A Family of Guanine Nucleotide-Binding Proteins That Activate Cholera Toxin and Regulate Vesicular Transport

Catherine F. Welsh, Joel Moss, and Martha Vaughan

National Heart, Lung, and Blood Institute, National Institutes of Health, Bethesda, Maryland

I. CHOLERA TOXIN ACTIVATION OF ADENYLYL CYCLASE

Cholera toxin, a heterooligomeric protein secreted by *Vibrio cholera*, causes the pathogenesis of cholera due to its ability to activate adenylyl cyclase of intestinal mucosal cells and increase intracellular cAMP levels (Finkelstein, 1973; Kelly, 1986). Cholera toxin is composed of one A and five B subunits, the latter forming the cell membrane binding domain (Gill, 1976, 1977; for review see Moss and Vaughan, 1990). The A subunit is cleaved into two polypeptides, A1 and A2 (Gill and Rappaport, 1979), and the active A1 species is generated after reduction of a disulfide bond that links the two (Mekalanos, 1979b; Moss and Vaughan, 1990). The A1 protein is responsible for the enzymatic activities of cholera toxin (Moss et al., 1979). These include (1) NAD hydrolysis to ADP-ribose and nicotinamide (Moss et al., 1976); (2) transfer of ADP-ribose to free arginine and simple guanidino compounds (Moss and Vaughan, 1977a; Mekalanos et al., 1979a); (3) ADP-ribosylation of nonspecific proteins, presumably due to available arginine residues that serve as toxin substrates (Moss and Vaughan, 1978); (4) auto-ADP-ribosylation of CTA1 (Moss et al., 1980; Trepel et al., 1977); and (5) ADP-ribosylation of $G_s\alpha$, the guanine nucleotide-binding protein responsible for stimulation of adenylyl cyclase (Cassel and Pfeuffer, 1978; Gill and Meren, 1978; Johnson et al., 1978; Northup et al., 1980).

The substrate for cholera toxin in intact cells is $G_s\alpha$. In general, the alpha subunits of G proteins are critical components of receptor-mediated signaling cascades, as they serve as sites for both regulation by guanine nucleotides and transmission of the activation signal from the cell surface receptor to the intracellular effector (Birnbaumer et al., 1987, 1990; Gilman, 1987). Under normal cellular conditions, G_s, which is composed of noncovalently associated α, β, and γ subunits in the basal state, is stimulated by specific cell surface receptors occupied by an agonist. Bound GDP is released from $G_s\alpha$, allowing GTP to bind and activate it. α-GTP dissociates from $\beta\gamma$ to interact with and stimulate its effector, adenylyl cyclase. The cascade is terminated by hydrolysis of the γ-phosphate of GTP by intrinsic GTPase activity of the α subunit. Finally, inactive α-GDP reassociates with $\beta\gamma$, completing the cycle. Cholera toxin-catalyzed ADP-ribosylation has several functional consequences, which result in persistent activation of $G_s\alpha$. Modification by toxin enhances release of bound GDP from the α subunit, promoting the subsequent binding of GTP (Burns et al., 1982, 1983). In addition, ADP-ribosylation reduces intrinsic GTPase activity of α (Cassel and Selinger, 1977) and promotes dissociation of α from $\beta\gamma$ (Kahn and Gilman, 1984b). Thus, the effects of cholera toxin mimic physiologic conditions that promote activation and inhibit normal regulatory mechanisms that terminate activation of $G_s\alpha$, producing an unregulated, GTP-bound state.

II. BIOCHEMICAL CHARACTERIZATION OF ARFs

A. Purification and Characterization of Native ARF Proteins

Activation of adenylyl cyclase by cholera toxin in broken cells was shown to require several soluble cofactors including NAD (Gill, 1975), GTP or analogues (Enomoto and Gill, 1979; Moss and Vaughan, 1977b; Lin et al., 1978; Nakaya et al., 1980), and a cytosolic component (Enomoto and Gill, 1979). In membrane preparations, cholera toxin-catalyzed ADP-ribosylation of $G_s\alpha$ also required GTP or Gpp(NH)p and a cytosolic factor and correlated well with activation of adenylyl cyclase (Enomoto and Gill, 1980). Subsequently, two distinct guanine nucleotide-binding sites were shown to be involved in this system: one responsible for activation of adenylyl cyclase and a functionally separate site that supported ADP-ribosylation but not cyclase activity (Gill and Meren, 1983).

Various cellular factors that augmented cholera toxin-catalyzed ADP-ribosylation of $G_s\alpha$ and/or activation of adenylyl cyclase were described (LeVine and Cuatrecasas, 1981; Schleifer et al., 1982; Enomoto and Gill, 1979, 1980; Pinkett and Anderson, 1982; Gill and Woolkalis, 1988; Schmidt et al., 1987; Gill and Coburn, 1987; Woolkalis et al., 1988; Kahn and Gilman, 1984a, 1986; Tsai et al., 1987, 1988). These factors, which were isolated from numerous tissues and species, varied in size, biochemical properties, and subcellular localization. A cytosolic factor partially purified from horse erythrocyte cytosol enhanced cholera toxin-catalyzed activation of adenylyl cyclase and ADP-ribosylation of a 43-kDa membrane protein (LeVine and Cuatrecasas, 1981). The factor was trypsin-sensitive and heat-stable and behaved as a protein of 13 kDa on gel filtration. Other soluble factors ranging in size from 16 to 20 kDa were described in wild-type or cyc⁻ S49 cells (Schleifer et al., 1982), pigeon erythrocytes (Enomoto and Gill, 1980), and bovine brain (Tsai et al., 1988). None of these required detergent for isolation, and the factor from cyc⁻ cells was inactivated by 1% cholate (Schleifer et al., 1982). Membrane-associated factors were described in preparations from rat kidney (Pinkett and Anderson, 1982), turkey erythrocytes, cyc⁻ cells (Schleifer et al., 1982), rabbit liver (Kahn and Gilman, 1984a), and bovine brain (Kahn and Gilman, 1986; Tsai et al., 1987). These ranged in size from 13 to 50 kDa and varied in their sensitivities to trypsin and heat.

A membrane-associated ADP-ribosylation factor (mARF) was purified 1800-fold from cholate extracts of rabbit liver membranes (Kahn and Gilman, 1984a). Detergent was required throughout the purification to maintain protein stability. The protein was trypsin- and chymotrypsin-sensitive, thermolabile, and N-ethylmaleimide-insensitive (Kahn and Gilman, 1984a). ARF migrated as a doublet on SDS polyacrylamide gel electrophoresis and exhibited a Stokes radius of 2.38 nm, $S_{20,w}$ of 2.1, partial specific volume of 0.74 mL/g, and M_r of 12,500 (Kahn and Gilman, 1984a). GTP was required for ARF to stimulate cholera toxin-catalyzed ADP-ribosylation of $G_s\alpha$, and ARF appeared to act catalytically in the assay. A delay in the activation of $G_s\alpha$ was eliminated by incubating ARF with G_s, GTP, and lipid or by increasing the concentration of ARF (Kahn and Gilman, 1984a).

ARF purified from bovine brain membranes was stabilized by Mg^{2+}, which also altered its chromatographic behavior (Kahn and Gilman, 1986). ARF itself bound guanine nucleotides (GTP, GDP, and GTPγS), but not adenine nucleotides, with high affinity in the presence of dimyristoylphosphatidylcholine (DMPC), $MgCl_2$, and high ionic strength (0.8 M NaCl) (Kahn and Gilman, 1986). The apparent dissociation constants for GTP and GDP were 90 and 40 nM, respectively. No GTPase activity was detected in vitro despite the fact that purified ARF contained equimolar quantities of bound GDP (Kahn and Gilman, 1986). The ARF-GTP species was active in supporting cholera toxin-catalyzed ADP-ribosylation of $G_s\alpha$, whereas ARF-GDP was inactive (Kahn and Gilman, 1986).

The mechanism by which ARF enhances the ability of cholera toxin to activate $G_s\alpha$ was further clarified using membrane and soluble ARF preparations (mARF, sARFs I and II) from bovine brain (Tsai et al., 1987, 1988). In addition to enhancing the ADP-ribosylation of $G_s\alpha$, ARF enhanced $G_s\alpha$-independent reactions catalyzed by the toxin A1 peptide (CTA1) and the holotoxin (Tsai et al., 1987, 1988). These included the auto-ADP-ribosylation of CTA1, ADP-ribosylation of simple guanidino compounds such as agmatine, and the hydrolysis of NAD to nicotinamide and ADP-ribose. ARF also stimulated the ADP-ribosylation by cholera toxin of numerous proteins unrelated to $G_s\alpha$, including ARF itself (Tsai et al., 1987, 1988). In all cases, ARF activity required GTP or a nonhydrolyzable analogue such as Gpp(NH)p or GTPγS; GDP, GDPβS, and the ATP analogue App(NH)p were not effective (Tsai et al., 1987, 1988). These data support a model for cholera toxin activation of adenylyl cyclase through a guanine nucleotide-binding protein cascade (Fig. 1). The central feature of this model is that ARF, activated by GTP, interacts with and directly activates the toxin A1 peptide, which in turn catalyzes the ADP-ribosylation of $G_s\alpha$. In the presence of GTP, $G_s\alpha$ stimulates the adenylyl cyclase catalytic subunit. In this model, no direct interaction between ARF and the toxin substrate is required for activation.

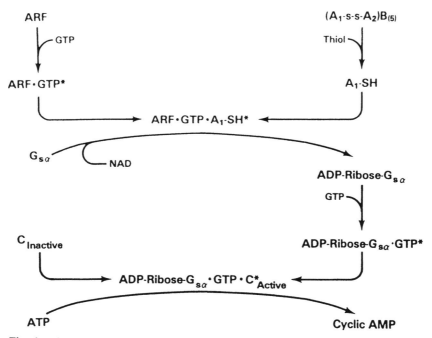

Fig. 1 Guanine nucleotide-binding protein cascade resulting in activation of adenylyl cyclase by cholera toxin. GTP-activated ARF interacts with and stimulates thiol-activated cholera toxin A1 subunit (A$_1$-SH), enhancing the transfer of ADP-ribose from NAD to G$_s\alpha$. After binding GTP, modified G$_s\alpha$ persistently stimulates the catalytic subunit of adenylyl cyclase (C), generating cyclic AMP from ATP. Holotoxin is represented by (A1-ss-A2)B$_{(5)}$. (From Tsai et al., 1988.)

Kinetic analysis of ARF-stimulated NAD:agmatine ADP-ribosyltransferase activity of CTA1 provided additional evidence in support of the model (Noda et al., 1990). ARF, in the presence of GTP, lowered the apparent K_m for NAD and agmatine when the reaction was performed in the absence of detergents; V_{max} was unaffected. In the presence of 0.003% SDS, the concentration at which activity was maximal, the apparent K_m values for both substrates were further lowered and V_{max} was increased (Noda et al., 1990). Stimulation of CTA1 ADP-ribosyltransferase activity by ARF was enhanced by cholate as well as by 0.003% SDS but was decreased by Triton X-100 and CHAPS (Noda et al., 1990). These effects are similar to those of certain detergents on cholera toxin activation of adenylyl cyclase (Neer et al., 1987). All observations are consistent with ARF being a direct allosteric activator of the toxin A1 subunit.

Detergents and phospholipids influenced the ability of ARF to stimulate different cholera toxin-catalyzed reactions, both G$_s\alpha$-dependent and G$_s\alpha$-independent (Tsai et al., 1988; Noda et al., 1990; Bobak et al., 1990). DMPC was required for sARF plus GTP to maximally stimulate ADP-ribosylation of G$_s\alpha$ but not auto-ADP-ribosylation of CT-A1, which was somewhat inhibited (Fig. 2) (Tsai et al., 1988; Bobak et al., 1990). Maximal ARF stimulation of NAD:agmatine ADP-ribosyltransferase activity (Noda et al., 1990; Bobak et al., 1990) and CTA1 auto-ADP-ribosylation (Noda et al., 1990; Tsai et al., 1991a) was observed in the presence of nondenaturing concentrations of SDS.

Since the ability of ARF to stimulate each of the cholera toxin-catalyzed reactions is

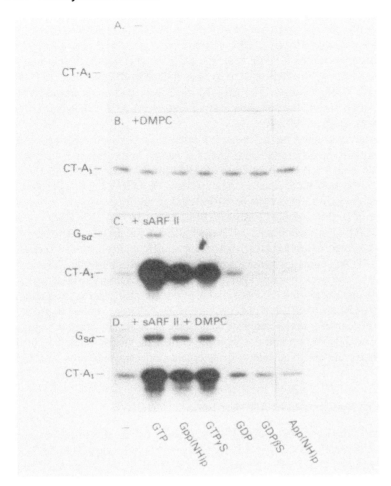

Fig. 2 Effect of sARF II, dimyristoylphosphatidylcholine (DMPC), and nucleotides on ADP-ribosylation of $G_{s\alpha}$ and auto-ADP-ribosylation of CTA1. Shown are autoradiograms of SDS-PAGE after incubations with CTA1, G_s, [^{32}P]NAD, the indicated nucleotide, and (A) no addition, (B) 1 mM DMPC, (C) 1.5 μg sARF II, (D) sARF II and DMPC. (From Tsai et al., 1988.)

dependent on the presence of GTP or its analogues, the effects of detergents and phospholipids on GTP binding by ARF were determined (Bobak et al., 1990). Maximal binding of GTP by sARF II was observed in the presence of 3 mM DMPC and 0.2% cholate; it required Mg^{2+} but not NaCl (Bobak et al., 1990). Using these conditions, the apparent K_d for GTP was 70 nM and binding was unaffected by components of the transferase reaction (Bobak et al., 1990). In the presence of low concentrations of SDS, conditions optimal for ARF stimulation of cholera toxin ADP-ribosyltransferase activity, however, minimal high affinity GTP binding by sARF II was observed (Bobak et al., 1990). Thus, ARF stimulated cholera toxin-catalyzed reactions under conditions that support both high-affinity (DMPC/cholate) and low-affinity (SDS) interactions with GTP.

There was a delay in sARF II plus GTP stimulation of cholera toxin ADP-ribosyltransferase activity in the presence of DMPC/cholate but not SDS (Bobak et al.,

1990). This was abolished by incubation of sARF II with GTP in DMPC/cholate under conditions that promoted high-affinity GTP binding prior to the transferase assay (Bobak et al., 1990). Consistent with the binding data, the EC_{50} for GTP in sARF II stimulation of cholera toxin ADP-ribosyltransferase activity was 5 μM in the presence of 0.003% SDS and 50 nM with DMPC/cholate. The EC_{50} for GTP was also lower in the presence of DMPC/cholate when stimulation of cholera toxin-catalyzed ADP-ribosylation of $G_s\alpha$ was measured (Bobak et al., 1990).

Additional studies evaluated the nature of the interaction between ARF and cholera toxin and helped explain the difference in activities observed in the presence of SDS and DMPC/cholate (Tsai et al., 1991a). A fraction of ARF formed a complex with CTA1 in the presence of GTPγS and SDS (Tsai et al., 1991a). The ARF–CTA1 complex was separated from the monomeric, uncomplexed ARF by gel filtration. Separated, monomeric ARF was incapable of further complex formation (Tsai et al., 1991a). The ability of ARF complexed with CTA1 to stimulate auto-ADP-ribosylation was greater than that of the monomeric species, which exhibited clearly different substrate specificities (Tsai et al., 1991a). No ARF–CTA1 complex was formed during incubation with GTPγS and DMPC/cholate. ARF, however, became aggregated under these conditions and was more active in stimulating cholera toxin ADP-ribosylation of added substrates, such as albumin, than auto-ADP-ribosylation of CT-A1 (Tsai et al., 1991a). These studies confirm that ARF stimulation of cholera toxin-catalyzed reactions is a result of direct physical interaction of ARF with the toxin and not with the G protein. In addition, they provide a biochemical basis for the observed effects of detergents and phospholipids on substrate specificity.

B. Expression of Recombinant ARF Proteins

Six mammalian ARF family members have been identified through molecular cloning (see Section IV for complete discussion). Each of these has been expressed in bacteria (Weiss et al., 1989; Kahn et al., 1991; Price et al., 1992), allowing the purified protein to be characterized without the ambiguities inherent in preparations from tissues, which contain multiple ARFs. Recombinant ARF 1 (rARF 1) bound GTPγS with high affinity in the presence of Mg^{2+} and DMPC/cholate (Weiss et al., 1989), similar to ARF purified from bovine brain (Kahn and Gilman, 1986; Bobak et al., 1990). Although the purified protein contained at least 0.91 mol GDP bound per mole of protein, no GTPase activity was detected (Weiss et al., 1989). In the presence of low Mg^{2+} concentrations and 50 mM NaCl, the K_d for GDP was 2 nM, whereas the K_d for GTPγS was 50 nM. rARF 1 bound GDP even in the virtual absence of free Mg^{2+}, whereas GTPγS binding required Mg^{2+} (Weiss et al., 1989). High concentrations of NaCl inhibited both GTPγS binding and ARF activity, as measured by stimulation of cholera toxin-catalyzed ADP-ribosylation of $G_s\alpha$ (Weiss et al., 1989).

Bacterially synthesized rARF 4 (rARF 4) also required lipid and cholate for high-affinity binding of GTPγS and had no detectable GTPase activity (Kahn et al., 1991). rARF 4 stimulated cholera toxin-catalyzed ADP-ribosylation of $G_s\alpha$ in the presence of DMPC/cholate (Kahn et al., 1991). Expression of either ARF 1 or ARF 4 in *Saccharomyces cerevisiae*, in which expression of one of the two yeast ARF genes is essential, conferred viability to the otherwise lethal *arf 1⁻arf2⁻* double mutant (Kahn et al., 1991). Thus, the recombinant mammalian proteins were apparently able to replace endogenous yeast ARFs in in vitro and in vivo functions despite the lack of posttranslational modifications.

TABLE 1 ADP-Ribosyltransferase Activity of Recombinant ARFs (rARFs) 2, 3, 5, and 6 and the Effect of Dimyristoylphosphatidylcholine (DMPC) and Cholate

Expt.	Additions	ADP-ribosylagmatine formed (pmol/min)	
		−DMPC/cholate	+DMPC/cholate
A	None	8.9	12.5
	rARF 3	7.1	34.4
	rARF 5	7.4	34.2
	rARF 6	32.9	59.3
B	None	14.0	17.9
	rARF 2	15.7	114
	rARF 6	63.9	137

Source: Price et al. (1992).

Bacterially expressed recombinant ARFs 2, 3, 5, and 6 all stimulated the NAD:agmatine ADP-ribosyltransferase activity of cholera toxin (Price et al., 1992). Table 1 shows the activity of 1 μg of rARF (except for 0.5 μg of rARF 5 in A) in the presence or absence of DMPC/cholate. rARFs 2, 3, and 5 required DMPC/cholate for activity, whereas rARF 6 was active in the presence of SDS, DMPC/cholate, or cholate alone or in the total absence of lipid or detergent (Price et al., 1992). rARF 2 and rARF 6 stimulated transferase activity with EC_{50} values for GTP of 4 μM and 50 nM, respectively. This value for rARF 6 was similar with or without DMPC/cholate (Price et al., 1992). rARF 2 activity was much more sensitive to high ionic strength than was that of rARF 6. Maximal GTPγS binding to both proteins occurred in the presence of DMPC/cholate (Price et al., 1992). These differences among the recombinant ARFs may be related to the differences in biochemical properties noted during early characterization and purification of factors required for cholera toxin activation.

III. DISTRIBUTION OF ARF PROTEINS AND DEVELOPMENTAL EXPRESSION

ARF protein expression among tissues and species has been assessed using polyclonal antibodies and antipeptide antibodies. Immunoreactive material was detected in every eukaryote tested, including human, cow, mouse, rat, frog, chicken, yeast, and slime mold (Kahn et al., 1988; Tsai et al., 1991b). It was found to predominate in neural tissue and was 50–90% cytosolic (Kahn et al., 1988; Tsai et al., 1991b). Polyclonal antibodies against bovine sARF II reacted with two soluble forms of ARF that migrate as a closely spaced doublet on SDS polyacrylamide gel electrophoresis. The upper band (sARF II) predominated in neural tissue, whereas the lower band (sARF I) was more intense in rat and bovine peripheral tissues (Tsai et al., 1991b).

In postnatal rat brain cytosol (Tsai et al., 1991b), the level of immunoreactive sARF II increased relative to sARF I between postnatal days 2 and 10. This age-related increase paralleled a rise in ARF activity of partially purified brain cytosol. ARF 3 mRNA also increased with age as determined by hybridization of mRNA blots with ARF-specific probes. There were simultaneous decreases in levels of ARF 2 and 4 mRNAs while ARFs 1, 5, and 6 mRNAs remained unchanged (Tsai et al., 1991b).

In general, mRNAs corresponding to ARFs 1–6 appear to be widespread and have been detected in a variety of species and tissues (Tsuchiya et al., 1989, 1991; Bobak et al., 1989; Monaco et al., 1990; Tsai et al., 1991b). The exception is ARF 2 mRNA, which was observed in bovine, rat, and mouse tissue but not detected in human or monkey tissues (Tsuchiya et al., 1989, 1991; Tsai et al., 1991b). Differences in ARF 4 mRNA during development have been observed in testis (Mishima et al., 1992). A short form of ARF 4 mRNA (~1.1 kb) was detected only in the testis of several species including human, rat, cow, dog, mouse, and rabbit, along with the ubiquitous 1.8-kb mRNA. The 1.1-kb species resulted from use of the first of three potential polyadenylation signals in the 3'-untranslated region (Mishima et al., 1992). Studies of developmental expression in rat testis showed that the 1.1-kb mRNA did not appear until the 35th postnatal day. In situ hybridization also suggested that appearance of the shorter message may be associated with a late stage of spermatogenesis (Mishima et al., 1992).

IV. MOLECULAR CHARACTERIZATION OF ARF

There are at least six mammalian ARFs based on identification and sequencing of cDNA clones (Sewell and Kahn, 1988; Bobak et al., 1989; Price et al., 1988; Monaco et al., 1990; Kahn et al., 1991; Tsuchiya et al., 1991; Peng et al., 1989). ARF genes have also been cloned from yeast (yARFs 1 and 2, Sewell and Kahn, 1988; Stearns et al., 1990b), *Drosophila melanogaster* (dARF 1, Murtagh et al., 1993), and the protozoan *Giardia lamblia* (gARF, Murtagh et al., 1992). Most of the mammalian ARFs were identified by screening human cDNA libraries (Bobak et al., 1989; Tsuchiya et al., 1991; Peng et al., 1989) or by amplifying human DNA using the polymerase chain reaction (PCR) (Monaco et al., 1990). ARF 2 was cloned from a bovine retinal cDNA library (Price et al., 1988) and has not been identified in human tissues by low stringency screening of an HL-60 cDNA library (Tsuchiya et al., 1991) or amplification of human placental DNA by PCR (Monaco et al., 1990). Likewise, ARF 2 mRNA was found in poly(A) $^+$ RNA from bovine, rat, and mouse species but not human or monkey tissues (Tsuchiya et al., 1989, 1991; Tsai et al., 1991b). Whether ARF 2 expression is truly species-specific or limited to certain developmental stages and/or tissues remains to be determined. With the exception of ARF 2, the individual ARFs appear to be highly conserved across species. For example, human and bovine deduced amino acid and coding region nucleotide sequences for ARF 1 are 100% and 91% identical, respectively (Bobak et al. 1989). Rat ARF 1 is also identical to bovine and human deduced amino acid sequences (Price et al., unpublished observations). Similarly, human ARF 6 is 99% identical in deduced amino acid sequence to a processed chicken pseudogene CPS 1, the only difference being a conservative serine/threonine substitution at position 158 (Tsuchiya et al., 1991). Table 2 shows the percentage identity of ARF sequences from human (hARFs 1, 3–6), bovine (bARF 1, 2), yeast (yARF 1), and *Giardia* (gARF). Clearly, there is considerable conservation in nucleotide and amino acid sequence among the lowest and highest eukaryotes.

The individual ARF family members exhibit a high degree of conservation in nucleotide and deduced amino acid sequences (Table 2, Fig. 3). Based on these comparisons, the mammalian ARFs have been assigned to three classes. Class I ARFs (ARFs 1, 2, and 3) are 95–96% identical to each other in deduced amino acid sequence, 79–80% identical with class II, and 64–69% identical with class III ARFs. Class I ARFs

TABLE 2 Percentage Identity of ARF Deduced Amino Acid Sequences and Coding Region Nucleotide Sequences[a]

	hARF1	bARF1	bARF2	hARF3	hARF4	hARF5	hARF6	yARF1	gARF
hARF1	—	100	96	96	80	80	68	77	70
bARF1	91	—	96	96	80	80	68	77	70
bARF2	79	80	—	95	80	80	69	77	70
hARF3	84	84	80	—	79	79	68	76	69
hARF4	67	68	68	71	—	90	64	72	69
hARF5	75	73	71	73	77	—	64	69	69
hARF6	68	69	64	66	60	65	—	65	63
yARF1	64	66	66	65	67	64	60	—	62
gARF	65	67	62	64	61	66	62	67	—

[a]Deduced amino acid sequence comparisons are above the dashes whereas coding region nucleotide sequence comparisons are below. Data are from references in legend to Figure 3.
Source: Tsuchiya et al. (1991); Murtagh et al. (1992).

are 181 amino acids in length. Class II (ARFs 4 and 5) are 90% identical and contain 180 amino acids. The sole known member of class III is ARF 6, which has only 175 amino acids and a deduced sequence that is <70% identical to those of class I and II ARFs. Sequence differences among classes I, II, and III are located primarily in the amino terminal region and carboxyl half of the proteins (Fig. 3).

An evaluation of evolutionary relationships among ARF family members underscored and confirmed the classification of ARF proteins (Tsuchiya et al., 1991). Phylogenetic analysis was performed using coding region nucleotide sequences of the individual ARFs and is shown in tree form in Figure 4. Branch points, indicating evolutionary divergence, correspond to each class of mammalian ARF and yARF 1.

In addition to mammalian and yeast ARFs, one ARF gene has been cloned from each of two different isolates of *Giardia* (Murtagh et al., 1992), a protozoan with characteristics of both eukaryotes and prokaryotes. The *Giardia* ARF gene encodes a protein of 191 amino acids 63–70% identical to mammalian ARFs—remarkable conservation considering that this protozoan and other eukaryotes are estimated to have diverged about 1.5 billion years ago (Murtagh et al., 1992) (Table 2, Fig. 3).

ARF-specific nucleotide probes were used to detect and identify individual mRNA species. The size of mRNA corresponding to individual ARFs was relatively constant across species on high-stringency probing of poly(A)$^+$ RNA. A single mRNA species of 1.7–2.1 kb represents ARF 1 in human, bovine, rat, mouse, and rabbit tissues (Bobak et al., 1989; Tsuchiya et al., 1989; Tsai et al., 1991b). ARF 2 mRNA is 2.1–2.6 kb in bovine and rat tissues (Tsuchiya et al., 1989, 1991; Tsai et al., 1991b). ARF 3 mRNA corresponds to bands of 3.7–3.8 and 1.2–1.3 kb (Bobak et al., 1989), which appear to result from the use of alternative polyadenylation signals (Tsai et al., 1991c). ARF 4 mRNA is a single 1.8-kb species, except in testis, where a 1.1-kb species is also found (Mishima et al., 1992). ARF 5 has a 1.3-kb mRNA and ARF 6 two mRNAs of 1.8 and 4.2 kb (Tsuchiya et al., 1991; Tsai et al., 1991b).

From conservation of mRNA size and sequence across species, a similar conservation of gene structure is inferred. Each of the three mammalian class I ARF genes contains five exons and four introns. The first exon is untranslated; translation is initiated in exon 2 (Fig.

```
                 10        20        30        40        50        60        70
hARF1    1   MGnifanLFk gLFGKKEMRI LMVGLDAAGK TTILYKLKLG EiVTTIPTIG FNVETVEYKN IsFTVWDVGG
bARF2    1   MGnvfekLFk sLFGKKEMRI LMVGLDAAGK TTILYKLKLG EiVTTIPTIG FNVETVEYKN IsFTVWDVGG
hARF3    1   MGnifgnLlk sLiGKKEMRI LMVGLDAAGK TTILYKLKLG EiVTTIPTIG FNVETVEYKN IsFTVWDVGG
hARF4    1   MGltiSsLFs rLFGKKqMRI LMVGLDAAGK TTILYKLKLG EiVTTIPTIG FNVETVEYKN IcFTVWDVGG
hARF5    1   MGltvSaLFs riFGKKqMRI LMVGLDAAGK TTILYKLKLG EiVTTIPTIG FNVETVEYKN IcFTVWDVGG
hARF6    1   MGkvlSk    iFGnKEMRI LMlGLDAAGK TTILYKLKLG qsVTTIPTvG FNVETVtYKN vkFnVWDVGG
yarf1    1   MGlfaSkLFs nLFGnKEMRI LMVGLDgAGK TTvLYKLKLG EviTTIPTIG FNVETVqYKN IsFTVWDVGG
yarf2    1   MGlyaSkLFs nLFGnKEMRI LMVGLDgAGK TTvLYKLKLG EviTTIPTIG FNVETVqYKN IsFTVWDVGG
gARF     1   MGqgaSkiFg kLFsKKEvRI LMVGLDAAGK TTILYKLmLG EvVTTvPTIG FNVETVEYKN InFTVWDVGG
CONSENSUS    MG          K   RI LM GLD AGK TT LYKL LG    TT PT G FNVETV YKN   F VWDVGG


                 80        90        100       110       120       130       140
hARF1    71  QDkIRPLWRH YfQNTQGLIF VVDSNDRE   Rv nEAReELmRM LaEDELRDAV LLVFANKQDL PnAMnaaEIT
bARF2    71  QDkIRPLWRH YfQNTQGLIF VVDSNDRE   Rv nEAReELtRM LaEDELRDAV LLVFvNKQDL PnAMnaaEIT
hARF3    71  QDkIRPLWRH YfQNTQGLIF VVDSNDRE   Rv nEAReELmRM LaEDELRDAV LLVFANKQDL PnAMnaaEIT
hARF4    71  QDrIRPLWkH YfQNTQGLIF VVDSNDRE   Ri qEvadELqkM LlvDELRDAV LLlFANKQDL PnAMaisEmT
hARF5    71  QDkIRPLWRH YfQNTQGLIF VVDSNDRE   Rv qEsadELqkM LqEDELRDAV LLVFANKQDm PnAMpvsElT
hARF6    67  QDkIRPLWRH YytgTQGLIF VVDcaDRd   Ri dEARqELhRi indrEmRDAi iLiFANKQDL PdAMkphEIq
yarf1    71  QDrIRsLWRH YyrNTeGvIF VVDSNDRs   Ri gEARevmqRM LnEDELRnAa wLVFANKQDL PeAMsaaEIT
yarf2    71  QDrIRsLWRH YyrNTeGvIF ViDSNDRs   Ri gEARevmqRM LnEDELRnAV wLVFANKQDL PeAMsaaEIT
gARF     71  QDsIRPLWRH YyQNTdaLIy ViDSaDlEpkRi edARnELhtl LgEDELRDAa LLVFANKQDL PkAMsttdlT
CONSENSUS    QD IR LW H Y   T   I V D  D    R            E R A   L F NKQD  P AM


                 150       160       170          180
hARF1   141  dKLGLhSLRh RnWYIQATCA TSGdGLYEGL DWL  SNqL rNqk       181
bARF2   141  dKLGLhSLRq RnWYIQATCA TSGdGLYEGL DWL  SNqL KNqk       181
hARF3   141  dKLGLhSLRn RnWYIQATCA TSGdGLYEGL DWL  aNqL KNkk       181
hARF4   141  dKLGLqSLRn RtWYvQATCA TqGtGLYEGL DWL  SNeLs Kr        180
hARF5   141  dKLGLqhLRs RtWYvQATCA TqGtGLYdGL DWL  SheLs Kr        180
hARF6   137  eKLGLtriRd RnWYSvQpsCA TSGdGLYEGL twL  tsNy Ks        175
yarf1   141  eKLGLhSiRn RpWfIQATCA TSGeGLYEGL eWL  SNsL KNst       181
yarf2   141  eKLGLhSiRn RpWfIQsTCA TSGeGLYEGL eWL  SNnL KNqs       181
gARF    143  erLGLqeLkk RdWYIQpTCA rSGdGLYqGL DWL  SdyifdkKNkkkgkkr 191
CONSENSUS    LGL        R W Q  CA  G GLY GL  WL
```

Fig. 3 Comparison of ARF amino acid sequences. Deduced amino acid sequences of human ARFs (hARFs 1, 3–6) (Bobak et al., 1989; Monaco et al., 1990; Tsuchiya et al., 1991), bovine ARF (bARF 2) (Price et al., 1988), *Drosophila* ARF (dARF 1) (Murtagh et al., 1993), yeast ARFs (yARF 1,2) (Sewell and Kahn, 1988; Stearns et al., 1990b), and *Giardia* ARF (gARF) (Murtagh et al., 1992) were aligned using Gene Works program. Capital letters within the alignment indicate amino acids identical in at least six ARFs. CONSENSUS indicates amino acids identical in all the ARFs. GAPs were introduced where needed for optimal alignment. (From Murtagh et al., 1993.)

5) (Tsai et al., 1991c; Lee et al., 1992; Serventi et al., 1993). Consensus sequences for GTP binding and hydrolysis, which are identical in all mammalian ARFs, are encoded in separate exons in each of the three genes. The sequence GLDAAGK, the putative phosphate binding loop, is encoded in exon 2, and DVGG, which is likely involved in Mg^{2+} coordination, in exon 3. The sequence encoding NKQD (important in guanine ring-binding specificity) is separated by intron 4 into exons 4 and 5 between the codons for glutamine and aspartate. Exon/intron boundaries occur at identical sites within the coding regions of all three genes (Fig. 5). Each of the promoter regions contains multiple transcription initiation sites, multiple potential Sp1 binding sites, and no TATA box, similar to promoters of housekeeping genes (Tsai et al., 1991c; Lee et al., 1992; Serventi et al., 1993).

An ARF gene from *Drosophila*, dARF 1, was cloned and assigned to class I based on 93–95% identity of deduced amino acid sequence to those of mammalian ARFs 1–3 (Murtagh et al., 1993). Splice sites for the two introns in the *Drosophila* gene are very similar to the exon/intron boundaries in the mammalian class I genes, differing by only one nucleotide in the first intron (Murtagh et al., 1993). Conservation of gene structure among class I ARFs thus appears to be maintained across rather divergent species.

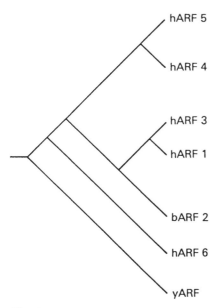

Fig. 4 Phylogenetic analysis of ARF cDNA sequences. Coding region nucleotide sequences of hARFs 1, 3–6, bARF 2, and yARF (yARF 1) were analyzed using DNADIST and KITSCH programs to determine evolutionary relationships. Branches indicate points of evolutionary divergence. (From Tsuchiya et al., 1991.)

Although the promoter regions of the ARF 1 and 3 genes lack a CAAT box (Tsai et al., 1991c; Lee et al., 1992), the ARF 2 promoter contains six inverted CCAAT motifs (Serventi et al., 1993). Studies using deletion constructs from the ARF 2 promoter region fused to a luciferase reporter gene suggest that the functional regulatory region of the promoter is within 400 bp of the initiation of transcription, a region that contains a CCAAT box (Serventi et al., 1993). Whether this region is important in the species-specific and developmentally regulated expression patterns noted previously is an intriguing speculation that requires further investigation.

Analysis of the promoter region of the ARF 3 gene was performed using deletion constructs linked to the gene encoding chloramphenicol acetyltransferase and transient expression in IMR-32, a neuroblastoma cell line (Haun et al., 1993b). Sequences located between −58 and −17 of the initiation of transcription were critical for promoter activity. This region, which contains a palindromic sequence, bound a protein from nuclear extracts (TLTF, TATA-less transcription factor) as evidenced by mobility shift assays (Haun et al., 1993b). A mutated palindromic sequence, which failed to compete for binding of TLTF, reduced transcription activity when linked to the reporter gene (Haun et al., 1993b). The identify of TLTF and its role in regulating ARF expression, or other TATA-less promoters, remains to be determined.

V. ARF PROTEIN STRUCTURE

Although ARF protein structure has not been determined crystallographically, certain functional domains have been inferred from sequence comparisons with other GTP-binding

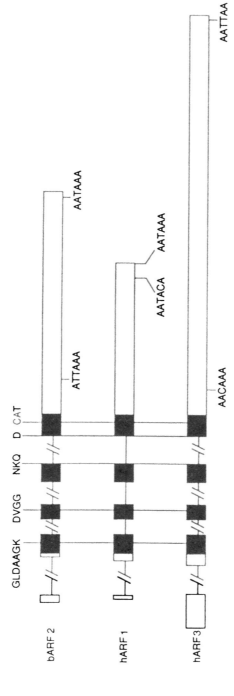

Fig. 5 Structures of mammalian class I ARF genes. Shown are bovine ARF 2 (bARF 2) (Serventi et al., 1993) and human ARFs 1 and 3 (hARF 1, 3) (Lee et al., 1992; Tsai et al., 1991c) genes, with exons represented by boxes and introns by horizontal lines. Shaded and open regions indicate coding and untranslated regions, respectively. Consensus sequences for GTP binding and hydrolysis are shown, and their locations within the gene indicated by vertical lines. Potential polyadenylation signal sequences and locations are indicated below each gene. (From Serventi et al., 1993.)

proteins, such as ras. All of the ARF gene products contain the consensus sequences for GTP binding and hydrolysis (Price et al., 1990). These are 100% identical among mammalian ARFs with minimal and usually conservative replacements in yeast and *Giardia* ARFs (Fig. 3). The consensus sequence for the first GTP-binding region $(GX_1X_2X_3X_4GK$, where X is any amino acid) represents the phosphate binding loop in the *ras* gene product. This sequence in the mammalian ARFs is GLDAAGK, spanning amino acids 24–29 in hARF1. Aspartate 26 in ARF corresponds to glycine 12 in c-Ha-ras, the X_2 position of the consensus sequence (Price et al., 1990), which is essential for normal GTP hydrolysis. In ras, replacement by any other amino acid except proline markedly reduces GTPase activity and promotes oncogenic transformation (Gibbs et al., 1984; McGrath et al., 1984; Sweet et al., 1984). The presence of aspartate may explain the absence of detectable GTPase activity in the ARF proteins. It should be noted, however, that *arl*, product of an ARF-like gene identified in *Drosophila* (Tamkun et al., 1991), hydrolyzed GTP with a derived rate of 0.05 min^{-1} despite the presence of an aspartate in the X_2 position of the consensus sequence. This observation may indicate the importance of amino acids other than X_2 in GTP hydrolysis.

The second GTP-binding region, with the consensus sequence $DX_1X_2G(Q)$, is involved in Mg^{2+} ion coordination. This sequence in all known ARFs is DVGGQ (amino acids 67–71 in hARF1), which differs in positions X_1 and X_2 from those of other small GTP-binding proteins, including products of the *ras, rho, rab, rap,* and *ral* genes, in which the sequence is DTAGQ. It is identical, however, to the cognate sequence in the alpha subunits of the heterotrimeric G proteins (Price et al., 1990). Similarly, the TCAT sequence (amino acids 158–161 in hARF 1) is identical to the sequence in most of the heterotrimeric G protein alpha subunits. The corresponding region in ras and other small GTP-binding proteins is SAK (amino acids 145–147 in c-Ha-ras) (Price et al., 1990). In ras, this region is thought to be involved in interaction with the guanine ring. The significance of such marked differences in this region among related guanine nucleotide-binding proteins is unclear.

The ARF family of proteins thus appears to have more in common with the G protein alpha subunits than with the ras-like proteins. This is true also of the posttranslational protein modifications. Like G$_o$ and G$_i$, many of the ARFs undergo N-terminal myristoylation (Kahn et al., 1988; Haun et al., 1993a; Kunz et al., 1993). Each of the ARFs contains a glycine in the position following the initiating methionine, which is the required acceptor site for myristoyl CoA:protein *N*-myristoyltransferase (NMT) (Towler et al., 1987). Neighboring amino acids are also important in determining substrate specificity (Towler et al., 1987). Of the mammalian ARFs, 1, 4, 5, and 6 contain an alanine or serine at position 5, consistent with the relatively restricted group of residues found in known substrates of NMT. ARF 2 has a glutamate at position 5 but did undergo myristoylation in a baculovirus expression system (Kunz et al., 1993). Amino-terminal myristoylation is necessary for ARF association with Golgi membranes (Haun et al., 1993a), where it functions in protein trafficking via vesicular transport. In contrast, ras proteins undergo carboxy-terminal processing including farnesylation and carboxymethylation, which is similarly required for membrane association.

Because it is the site for covalent lipid modification, the influence of the amino terminus on ARF structure and function has been investigated. An oligopeptide representing the first 16 amino acids of ARF 1 (excluding the initial methionine) was shown by CD spectra to likely form an alpha helix in a hydrophobic but not an aqueous environment (Kahn et al., 1992). Myristoylation of the amino-terminal glycine of the peptide stabilized the alpha helical structure even in the absence of DMPC/cholate. A mutant of ARF 1 lacking the amino-terminal 17 residues ([Δ1–17]mARF1p) could bind GTPγS but no longer required DMPC/cholate to do so (Kahn et al., 1992). Whether specific structural features, such as amino-terminal α-helix formation, are involved in the ability of ARF to bind or release guanine nucleotides and/or to associate with Golgi membranes is unknown. To what extent the myristoylation requirement in membrane targeting (Haun et al., 1993a) may also be a requirement for amino-terminal α-helix formation is also unknown.

The ARF 1 deletion mutant [Δ1–17]mARF1p failed to stimulate cholera toxin-catalyzed ADP-ribosylation of $G_s\alpha$ in the presence of DMPC/cholate (Kahn et al., 1992). Likewise, oligopeptides representing the amino-terminal sequences in very high concentrations inhibited ARF activity. It was concluded that the amino terminus is critical for ARF function (Kahn et al., 1992). Recent data appear inconsistent with this conclusion, however. ARD 1 is a 64-kDa ARF but lacks the 15 amino acids corresponding to the amino-terminal region (Mishima et al., 1993). ARD 1 stimulated cholera toxin ADP-ribosyltransferase activity in a GTP-dependent fashion in the presence of Tween 20 but not other lipids and detergents, including DMPC/cholate. Although these studies clearly indicate that the amino terminus is not responsible for ARF activity, it may be involved with phospholipid interaction, as suggested earlier.

VI. CELLULAR FUNCTIONS OF ARFs

ARF was initially discovered because of its ability to enhance cholera toxin activation of adenylyl cyclase via ADP-ribosylation of $G_s\alpha$. Whether such action occurs in cells, however, is unknown. NAD:arginine ADP-ribosyl transferases have been purified from turkey erythrocytes, but despite the similarity of reactions catalyzed by cholera toxin and these enzymes, ARFs did not affect their activities (Moss and Stanley, unpublished observations).

Accumulating evidence has implicated ARF as one of a growing number of proteins for which a role in intracellular protein transport has been recognized. Several ~20-kDa GTP-binding proteins regulate vesicular protein trafficking through the Golgi. The best characterized are Sec4 and Ypt1, which participate in the yeast secretory pathway at different stages, Ypt1 in early transport from the endoplasmic reticulum to the Golgi (Segev et al., 1988) and Sec4 in late post-Golgi transport (Goud et al., 1988). Mammalian homologues of these proteins are members of the rab family of GTP-binding proteins, which are also localized to intracellular organelles of the secretory pathway (Goud et al., 1990; for review see Rothman and Orci, 1992).

Two ARF proteins have been identified in *Saccharomyces cerevisiae*: yARF 1 (Sewell and Kahn, 1988) and yARF 2 (Stearns et al., 1990b). The encoded proteins are 96% identical and functionally homologous. *yARF 1* accounts for 90% of the ARF protein, and the *arf1* null mutant displayed a slowed rate of growth at 30°C, further sensitivity to cold, and inhibition of growth by otherwise nontoxic concentrations of fluoride ion (Stearns et al., 1990b). These phenotypes were suppressed by increasing the expression of yARF 2 protein. The *arf2* null mutant displayed a normal phenotype, but the *arf1arf2* double mutant was lethal; that is, expression of one of the two genes is essential for viability (Stearns et al., 1990b).

Disruption of *yARF 1* caused a defect in invertase secretion with intracellular accumulation of incompletely glycosylated invertase (Stearns et al., 1990a). The yARF

1-dependent step of secretion was located after a SEC18-dependent step as determined by phenotypic analysis of temperature-sensitive double mutants (Stearns et al., 1990a). Immunocytochemical and immunoelectron microscopy using antipeptide antibodies against a highly conserved region of the mammalian ARFs localized ARF to the cytoplasmic surface of the Golgi in NIH 3T3 cells, consistent with the yeast genetic studies (Stearns et al., 1990a).

The immunolocalization to Golgi membranes did not explain previous findings that ARF was predominantly cytosolic when purified from certain tissues (Kahn et al., 1988; Tsai et al., 1988) but raised the possibility that ARF might cycle between membranes and cytosol in its role in the secretory process. In PC-12 cells, ARF shifted from a predominantly cytosolic distribution to membrane fractions upon incubation of cellular homogenates with GTPγS, Gpp(NH)p, and Gpp(CH_2)p but not GTP, GDP, GMP, or ATP. ATPγS was moderately effective, probably resulting from thiophosphorylation of GDP (Walker et al., 1992). Cytosolic ARF also associated with anionic phospholipid vesicles (phosphatidylserine, phosphatidylinositol, and cardiolipin but not phosphatidylcholine) in the presence of nonhydrolyzable GTP analogues. ARF binding to phosphatidylserine vesicles was reversible, particularly in the presence of GDP, and ARF retained its activity throughout the cycle of binding and release (Walker et al., 1992).

In permeabilized RINm5F cells, a 20-kDa GTP-binding protein, p20, moved from cytosolic to microsomal fractions on incubation with GTPγS but not GTP, GDP, or GDPβS (Regazzi et al., 1991). The target microsomal fraction was enriched with Golgi and plasma membranes, not secretory granules. p20 was felt to be ARF because of the coincidence of ARF immunoreactivity and [^{32}P]GTP binding on blots after two-dimensional electrophoresis (Regazzi et al., 1991).

In NRK cells, ARF immunofluorescence was predominantly associated with the Golgi complex but moved to the cytoplasm after lowering cellular energy levels (Donaldson et al., 1991). Treatment with the fungal metabolite brefeldin A, which disrupts Golgi architecture and blocks secretion (Lippincott-Schwartz et al., 1989, 1990), also caused ARF to redistribute throughout the cytoplasm (Donaldson et al., 1991). In vitro, ARF bound to Golgi membranes in the presence of GTPγS. Brefeldin A treatment before, but not after, addition of GTPγS inhibited ARF association with membranes.

Intra-Golgi protein transport proceeds vectorially through the budding off and subsequent fusion of non-clathrin-coated vesicles from one Golgi compartment to another (for review see Rothman and Orci, 1992). The major constituents of the non-clathrin coat are four coat proteins, termed α-, β-, γ-, and δ-COP ranging in size from 61 to 160 kDa (Serafini et al., 1991a), which exist as a cytoplasmic complex, the coatomer (Waters et al., 1991), and associate reversibly with Golgi-derived transport vesicles. After binding to their target membranes, but before fusing with them, coated vesicles release the coat proteins. Dissection of stages of the secretory pathway has been possible through treatment with agents that inhibit specific steps. For example, GTPγS causes accumulation of coated vesicles (Melançon et al., 1987), and *N*-ethylmaleimide inhibits transport at a subsequent step, causing the accumulation of uncoated vesicles (Malhotra et al., 1988). Thus, uncoated vesicles are derived from coated vesicles (Orci et al., 1989).

Serafini et al. (1991b) identified an ~20-kDa GTP-binding protein that copurified with α–δ COPs in COP-coated vesicle preparations. The protein was felt to be ARF because of comigration of bands on ARF-specific immunoblots with GTP-binding blots. ARF, which was initially cytosolic, bound to Golgi membranes in the presence of GTPγS. ARF associated with membranes only under conditions that promoted accumulation of coated

(GTPγS) but not uncoated vesicles (*N*-ethylmaleimide). There were approximately three ARF molecules for each molecule of α-COP associated with coated vesicles, identifying ARF as a coat protein (Serafini et al., 1991b). Golgi membranes also contained a heat- and trypsin-sensitive component that was required for ARF binding. Thus, ARF association with Golgi membranes involves protein–protein interactions as well as interactions with the lipid bilayer (Serafini et al., 1991b).

Taylor et al. (1992) purified two 20-kDa soluble GTP-binding proteins that could support GTPγS inhibition of intra-Golgi transport. These GTP-dependent Golgi binding factors, termed GGBFs, were identified as ARFs on the basis of activity and immunoreactivity. Each stimulated cholera toxin-catalyzed ADP-ribosylation of $G_s\alpha$ and myristoylated recombinant ARF 1 was active in the GGBF assay. In addition, each reacted with antibodies specific for ARFs 1 and 3 but not ARF 4. The two proteins differed in activity by a factor of 3 and had different chromatographic properties. Depletion of GGBF from the cytosol abolished GTPγS inhibition of transport. The authors concluded that ARFs are solely responsible for GTPγS inhibition of transport despite the presence of other GTP-binding proteins, particularly those of the rab family, known to be associated with intra-Golgi transport (Taylor et al., 1992).

To define the function of ARF in intra-Golgi transport during its guanine nucleotide-dependent cycle of membrane association and dissociation, Donaldson et al. (1992a) showed that ARF was required for the GTPγS-dependent binding of β-COP and therefore, presumably, the coatomer. Recombinant ARF 1 restored β-COP binding in the presence of ARF-depleted cytosol. A sequential reaction occurred in which ARF bound to Golgi membranes in the presence of GTPγS and allowed subsequent binding of β-COP after the free components were removed by washing the membranes. Brefeldin A inhibition of β-COP binding occurred during the initial activation of membranes by ARF and GTPγS (Donaldson et al., 1992a). Golgi membranes enhanced the exchange of ARF-bound nucleotide for GTP, GDP, or GTPγS but not that of adenine nucleotides (Donaldson et al., 1992b; Helms and Rothman, 1992). The Golgi-associated exchange activity was trypsin- and proteinase K-sensitive and resistant to extraction with 1 M KCl. Brefeldin A inhibited both the nucleotide exchange activity and ARF binding to Golgi membranes. Membrane-independent guanine nucleotide exchange on ARF in the presence of DMPC/cholate was not inhibited by brefeldin A, indicative of the drug's specificity presumably for a nucleotide exchange protein associated with Golgi membranes (Donaldson et al., 1992b). Thus, a Golgi-bound protein appears to catalyze guanine nucleotide exchange on ARF and is the site of brefeldin A inhibition of transport. In additional experiments with ARF and Golgi membranes, after incubation with $[\alpha\text{-}^{32}P]$GTP or GTPγS, protein-bound nucleotide was recovered in the supernatant or membrane pellet, respectively, indicative of rapid hydrolysis following the exchange reaction (Donaldson et al., 1992b).

There is considerable experimental evidence to support the model of ARF function in vesicular transport proposed by Serafini et al. (1991b). In this model, cytoplasmic ARF-GDP interacts with a specific Golgi-bound nucleotide exchange protein, generating ARF-GTP, which is competent to associate with membranes. The sequence of coat assembly and vesicle budding ensues. Destabilization of the vesicle coat is triggered by the formation of ARF-GDP resulting from the action of a specific GTPase-activating protein (GAP) at the target membrane. ARF-GDP dissociates from the membrane followed by release of other coat proteins into the cytosol, where they may be recycled. As GTP hydrolysis must occur for vesicles to become uncoated, this model explains the inhibition of transport and accumulation of coated vesicles caused by GTPγS. In addition, it provides a mechanism for vectorial transport by employing an ARF-specific exchange protein and GAP on the budding and target membranes, respectively.

Specificity in targeting of coated vesicles may also be achieved if each individual ARF interacts exclusively with its own specifically localized exchange protein and GAP. Tsai et al. (1992) observed differences in membrane localization among ARFs 1, 3, and 5 (Fig. 6). Subcellular localization of individual ARFs was determined after incubation of rat brain homogenate with ATP, an ATP-regenerating system, and GTPγS. ARF 1 localized to the Golgi fraction as well as to a fraction containing Golgi plus microsomes; ARF 5 localized to the Golgi fraction preferentially; ARF 3 was rather diffusely distributed among the fractions. Similar results were obtained using purified ARF proteins. Under the same conditions, Golgi membranes bound almost twice as much ARF 1 as ARF 3 (Tsai et al., 1992). GTPγS-dependent binding of ARFs 1 and 3 to Golgi membranes was enhanced by a fraction of relatively large (>45 kDa) soluble proteins referred to as SAP or soluble accessory proteins (Tsai et al., 1993). The SAP-dependent binding of ARFs 1 and 3 was inhibited by brefeldin A. Binding of ARF 5 was neither SAP-dependent nor inhibited by brefeldin A (Tsai et al., 1993). Thus, individual ARFs clearly differ in their subcellular localization and in the conditions that promote membrane association.

ARF has been implicated in other events involving vesicular fusion including endoplasmic reticulum to cis-Golgi transport (Balch et al., 1992), endosome fusion during endocytosis (Lenhard et al., 1992), and nuclear vesicle fusion following mitosis (Boman et al., 1992). These processes also presumably involve the interaction of specific ARF proteins with specific exchange proteins and GAPs.

VII. SUMMARY

Understanding the mechanisms of cholera toxin action aided in defining the role of $G_s\alpha$ in stimulating adenylyl cyclase and the importance of specific structural and functional features

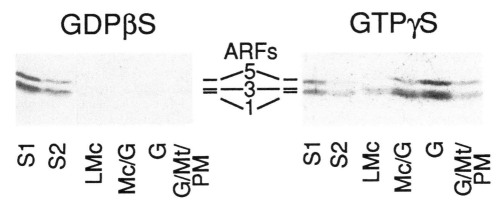

Fig. 6 Guanine nucleotide-dependent subcellular localization of mammalian ARFs. Rat brain homogenates were incubated with either GDPβS (left) or GTPγ (right), and after sucrose gradient centrifugation, localization of ARFs 1, 3, and 5 was determined by SDS-PAGE and immunoreactivity. The position of each ARF is indicated. S1 and S2, supernatants 1 and 2; LMc, light microsomes; Mc/G, microsomes and Golgi; G/Mt/PM, Golgi/mitochondria/plasma membrane. (From Tsai et al., 1992.)

of G proteins in the signaling cascade (Section I). During the course of these studies, ARF was discovered as a cofactor for cholera toxin-catalyzed ADP-ribosylation of $G_s\alpha$ and subsequently shown to enhance all of cholera toxin's catalytic activities (Section II). ARF itself was found to be a GTP-binding protein and became the object of more intense investigation. It was found that ARFs have properties distinct from those of other GTP-binding proteins, including little or no intrinsic GTPase activity, a relative dependence on phospholipids for GTP binding and activity, and amino-terminal modification by *N*-myristoyltransferase (Sections II and V). Molecular cloning studies revealed that the ARFs are a family of proteins that are highly homologous and are conserved structurally and functionally among eukaryotes (Section IV). After initial studies implicated GTP-binding proteins in the yeast secretory pathway, Bourne (1988) proposed a model for vectorial transport of Golgi-derived vesicles that involved regulation by unidentified GTPases. Efforts to ascribe such functions to ARF proved fruitful, and a clearer picture of its role in vectorial transport is emerging (Section VI). Most recently, Golgi-associated regulatory proteins that catalyze ARF guanine nucleotide exchange and GTPase activity have been detected. ARF also appears to be the stimulus for coatomer binding to Golgi membranes (Section VI). Thus ARF has quickly emerged as a critical component in vesicular transport. Future studies will define the roles of individual ARFs and regulatory proteins in the formation, targeting, and fusion of intracellular membrane vesicles.

Acknowledgment

We thank Mrs. Carol Kosh for expert secretarial assistance.

ABBREVIATIONS

ARF, ADP-ribosylation factor; sARF I and sARF II, soluble ADP-ribosylation factors purified from bovine brain; mARF, purified membrane-associated ARF; hARF, human ARF; bARF, bovine ARF; yARF, yeast ARF; rARF, bacterially expressed recombinant ARF; gARF *Giardia* ARF; dARF, *Drosophila* ARF; G protein, guanine nucleotide-binding protein; G_s, stimulatory G protein responsible for stimulation of adenylyl cyclase; Gpp(NH)p, guanylyl-($\beta\gamma$-imido)-diphosphate; GTPγS, guanosine-5'-*O*-(3-thiotriphosphate); GDPβS, guanosine-5'-*O*-(2-thiodiphosphate); App(NH)p, adenylyl-($\beta\gamma$-imido)-diphosphate; Gpp(CH$_2$)p, guanylyl-($\beta\gamma$-methylene)-diphosphate; ATPγS, adenosine-5'-*O*-(3-thiotriphosphate); CTA1, cholera toxin A1 subunit; DMPC, dimyristoylphosphatidylcholine; SDS, sodium dodecyl sulfate; CHAPS, 3-[3-cholamidopropyl) dimethyl-ammonio]-1-propane sulfonate; COP, coat protein.

REFERENCES

Balch, W. E., Kahn, R. A., and Schwaninger, R. (1992). ADP-ribosylation factor is required for vesicular trafficking between the endoplasmic reticulum and the *cis*-Golgi compartments. *J. Biol. Chem. 267*: 13053.

Birnbaumer, L., Codina, J., Mattera, R., Yatani, A., Scherer, N., Toro, M.-J., and Brown, A. M. (1987). Signal transduction by G proteins. *Kidney Int. 32* (Suppl. 23):S14.

Birnbaumer, L., Mattera, R., Yatani, A., Codina, J., Van Dongen, A. M. J., and Brown, A. M. (1990). Recent advances in the understanding of multiple roles of G proteins in coupling of receptors to ionic channels and other effectors. In *ADP-Ribosylating Toxins and G Proteins:*

Insights into Signal Transduction, J. Moss and M. Vaughan (Eds.), American Society for Microbiology, Washington, DC, p. 225.

Bobak, D. A., Nightingale, M. S., Murtagh, J. J., Price, S. R., Moss, J., and Vaughan, M. (1989). Molecular cloning, characterization, and expression of human ADP-ribosylation factors: two guanine nucleotide-dependent activators of cholera toxin. *Proc. Natl. Acad. Sci. U.S.A. 86*: 6101.

Bobak, D. A., Bliziotes, M. M., Noda, M., Tsai, S.-C., Adamik, R., and Moss, J. (1990). Mechanism of activation of cholera toxin by ADP-ribosylation factor (ARF): both low- and high-affinity interactions of ARF with guanine nucleotides promote toxin activation. *Biochemistry 29*: 855.

Boman, A. L., Taylor, T. C., Melançon, P., and Wilson, K. L. (1992). A role for ADP-ribosylation factor in nuclear vesicle dynamics. *Nature 358*: 512.

Bourne, H. R. (1988). Do GTPases direct membrane traffic in secretion? *Cell 53*: 669.

Burns, D. L., Moss, J., and Vaughan, M. (1982). Choleragen-stimulated release of guanyl nucleotides from turkey erythrocyte membranes. *J. Biol. Chem. 257*: 32.

Burns, D. L., Moss, J., and Vaughan, M. (1983). Release of guanyl nucleotides from the regulatory subunit of adenylate cyclase. *J. Biol. Chem. 258*: 1116.

Cassel, D., and Pfeuffer, T. (1978). Mechanism of cholera toxin action: covalent modification of the guanyl nucleotide-binding protein of the adenylate cyclase system. *Proc. Natl. Acad. Sci. U.S.A. 75*: 2669.

Cassel, D., and Selinger, Z. (1977). Mechanism of adenylate cyclase activation by cholera toxin: inhibition of GTP hydrolysis at the regulatory site. *Proc. Natl. Acad. Sci. U.S.A. 74*: 3307.

Donaldson, J. G., Kahn, R. A., Lippincott-Schwartz, J., and Klausner, R. D. (1991). Binding of ARF and β-COP to Golgi membranes: possible regulation by a trimeric G protein. *Science 254*: 1197.

Donaldson, J. G., Cassel, D., Kahn, R. A., and Klausner, R. D. (1992a). ADP-ribosylation factor, a small GTP-binding protein, is required for binding of the coatomer protein β-COP to Golgi membranes. *Proc. Natl. Acad. Sci. U.S.A. 89*: 6408.

Donaldson, J. G., Finazzi, D., and Klausner, R. D. (1992b). Brefeldin A inhibits Golgi membrane-catalysed exchange of guanine nucleotide onto ARF protein. *Nature 360*: 350.

Enomoto, K., and Gill, D. M. (1979). Requirement for guanosine triphosphate in the activation of adenylate cyclase by cholera toxin. *J. Supramol. Struct. 10*: 51.

Enomoto, K., and Gill, D. M. (1980). Cholera toxin activation of adenylate cyclase. Roles of nucleoside triphosphates and a macromolecular factor in the ADP ribosylation of the GTP-dependent regulatory component. *J. Biol. Chem. 255*: 1252.

Finkelstein, R. A. (1973). Cholera. *CRC Crit. Rev. Microbiol. 2*: 553.

Gibbs, J. B., Sigal, I. S., Poe, M., and Scolnick, E. M. (1984). Intrinsic GTPase activity distinguishes normal and oncogenic *ras* p21 molecules. *Proc. Natl. Acad. Sci. U.S.A. 81*: 5704.

Gill, D. M. (1975). Involvement of nicotinamide adenine dinucleotide in the action of cholera toxin in vitro. *Proc. Natl. Acad. Sci. U.S.A. 72*: 2064.

Gill, D. M. (1976). The arrangement of subunits in cholera toxin. *Biochemistry 15*: 1242.

Gill, D. M. (1977). Mechanism of action of cholera toxin. *Adv. Cyclic Nucleotide Res. 8*: 85.

Gill, D. M., and Coburn, J. (1987). ADP-ribosylation by cholera toxin: functional analysis of a cellular system that stimulates the enzymatic activity of cholera toxin fragment A_1. *Biochemistry 26*: 6364.

Gill, D. M., and Meren, R. (1978). ADP-ribosylation of membrane proteins catalyzed by cholera toxin: basis of the activation of adenylate cyclase. *Proc. Natl. Acad. Sci. U.S.A. 75*: 3050.

Gill, D. M., and Meren, R. (1983). A second guanyl nucleotide-binding site associated with adenylate cyclase. Distinct nucleotides activate adenylate cyclase and permit ADP-ribosylation by cholera toxin. *J. Biol. Chem. 258*: 11908.

Gill, D. M., and Rappaport, R. S. (1979). Origin of the enzymatically active A_1 fragment of cholera toxin. *J. Infect. Dis. 139*: 674.

Gill, D. M., and Woolkalis, M. J. (1988). [^{32}P]ADP-ribosylation of proteins catalyzed by cholera toxin and related heat-labile enterotoxins. *Methods Enzymol. 165*: 235.

Gilman, A. G. (1987). G proteins: transducers of receptor-generated signals. *Annu. Rev. Biochem. 56*: 615.

Goud, B., Salminen, A. Walworth, N. C., and Novick, P. J. (1988). A GTP-binding protein required for secretion rapidly associates with secretory vesicles and the plasma membrane in yeast. *Cell 53*: 753.

Goud, B., Zahraoui, A., Tavitian, A., and Saraste, J. (1990). Small GTP-binding protein associated with Golgi cisternae. *Nature 345*: 553.

Haun, R. S., Tsai, S.-C., Adamik, R., Moss, J., and Vaughan, M. (1993a). Effect of myristoylation on GTP-dependent binding of ADP-ribosylation factor to Golgi. *J. Biol. Chem. 268*: 7064.

Haun, R. S., Moss, J., and Vaughan, M. (1993b). Characterization of the human ADP-ribosylation factor 3 promoter. Transcriptional regulation of a TATA-less promoter. *J. Biol. Chem. 268*: 8793.

Helms, J. B., and Rothman, J. E. (1992). Inhibition by brefeldin A of a Golgi membrane enzyme that catalyses exchange of guanine nucleotide bound to ARF. *Nature 360*: 352.

Johnson, G. L., Kaslow, H. R., and Bourne, H. R. (1978). Genetic evidence that cholera toxin substrates are regulatory components of adenylate cyclase. *J. Biol. Chem. 253*: 7120.

Kahn, R. A., and Gilman, A. G. (1984a). Purification of a protein cofactor required for ADP-ribosylation of the stimulatory regulatory component of adenylate cyclase by cholera toxin. *J. Biol. Chem. 259*: 6228.

Kahn, R. A., and Gilman, A. G. (1984b). ADP-ribosylation of G_s promotes the dissociation of its α and β subunits. *J. Biol. Chem. 259*: 6235.

Kahn, R. A., and Gilman, A. G. (1986). The protein cofactor necessary for ADP-ribosylation of G_s by cholera toxin is itself a GTP binding protein. *J. Biol. Chem. 261*: 7906.

Kahn, R. A., Goddard, C., and Newkirk, M. (1988). Chemical and immunological characterization of the 21-kDa ADP-ribosylation factor of adenylate cyclase. *J. Biol. Chem. 263*: 8282.

Kahn, R. A., Kern, F. G., Clark, J., Gelmann, E. P., and Rulka, C. (1991). Human ADP-ribosylation factors. A functionally conserved family of GTP-binding proteins. *J. Biol. Chem. 266*: 2606.

Kahn, R. A., Randazzo, P., Serafini, T., Weiss, O., Rulka, C., Clark, J., Amherdt, M., Roller, P., Orci, L., and Rothman, J. E. (1992). The amino terminus of ADP-ribosylation factor (ARF) is a critical determinant of ARF activities and is a potent and specific inhibitor of protein transport. *J. Biol. Chem. 267*: 13039.

Kelly, M. T. (1986). Cholera: a worldwide perspective. *Pediatr. Infect. Dis. 5*: S101.

Kunz, B. C., Muczynski, K. A., Welsh, C. F., Stanley, S. J., Tsai, S.-C., Adamik, R., Chang, P. P., Moss, J., and Vaughan, M. (1993). Characterization of recombinant and endogenous ADP-ribosylation factors synthesized in Sf9 insect cells. *Biochemistry, 32*:6643.

Lee, C.-M., Haun, R. S., Tsai, S.-C., Moss, J., and Vaughan, M. (1992). Characterization of the human gene encoding ADP-ribosylation factor 1, a guanine nucleotide-binding activator of cholera toxin. *J. Biol. Chem. 267*: 9028.

Lenhard, J. M., Kahn, R. A., and Stahl, P. D. (1992). Evidence for ADP-ribosylation factor (ARF) as a regulator of in vitro endosome–endosome fusion. *J. Biol. Chem. 267*: 13047.

LeVine, H., III and Cuatrecasas, P. (1981). Activation of pigeon erythrocyte adenylate cyclase by cholera toxin. Partial purification of an essential macromolecular factor from horse erythrocyte cytosol. *Biochim. Biophys. Acta 672*: 248.

Lin, M. C., Welton, A. F., and Berman, M. F. (1978). Essential role of GTP in the expression of adenylate cyclase activity after cholera toxin treatment. *J. Cyclic Nucleotide Res. 4*: 159.

Lippincott-Schwartz, J., Yuan, L. C., Bonifacino, J. S., and Klausner, R. D. (1989). Rapid redistribution of Golgi proteins into the ER in cells treated with brefeldin A: evidence for membrane cycling from Golgi to ER. *Cell 56*: 801.

Lippincott-Schwartz, J., Donaldson, J. G., Schweizer, A., Berger, E. G., Hauri, H.-P., Yuan, L. C., and Klausner, R. D. (1990). Microtubule-dependent retrograde transport of proteins into the ER in the presence of brefeldin A suggests an ER recycling pathway. *Cell 60*: 821.

McGrath, J. P., Capon, D. J., Goeddel, D. V., and Levinson, A. D. (1984). Comparative biochemical properties of normal and activated human *ras* p21 protein. *Nature 310*: 644.

Malhotra, V., Orci, L., Glick, B. S., Block, M. R., and Rothman, J. E. (1988). Role of an *N*-ethylmaleimide-sensitive transport component in promoting fusion of transport vesicles with cisternae of the Golgi stack. *Cell 54*: 221.

Mekalanos, J. J., Collier, R. J., and Romig, W. R. (1979a). Enzyme activity of cholera toxin. I. New methods of assay and the mechanism of ADP-ribosyl transfer. *J. Biol. Chem. 254*: 5849.

Mekalanos, J. J., Collier, R. J., and Romig, W. R. (1979b). Enzymic activity of cholera toxin. II. Relationships to proteolytic processing, disulfide bond reduction, and subunit composition. *J. Biol. Chem. 254*: 5855.

Melançon, P., Glick, B. S., Malhotra, V., Weidman, P. J., Serafini, T., Gleason, M. L., Orci, L., and Rothman, J. E. (1987). Involvement of GTP-binding "G" proteins in transport through the Golgi stack. *Cell 51*: 1053.

Mishima, K., Price, S. R., Nightingale, M. S., Kousvelari, E., Moss, J., and Vaughan, M. (1992). Regulation of ADP-ribosylation factor (ARF) expression. Cross-species conservation of the developmental and tissue-specific alternative polyadenylation of ARF 4 mRNA. *J. Biol. Chem. 267*: 24109.

Mishima, K., Tsuchiya, M., Nightingale, M. S., Moss, J., and Vaughan, M. (1993). ARD 1, a 64-kDa guanine nucleotide-binding protein with a carboxyl-terminal ADP-ribosylation factor domain. *J. Biol. Chem. 268*: 8801.

Monaco, L., Murtagh, J. J., Newman, K. B., Tsai, S.-C., Moss, J., and Vaughan, M. (1990). Selective amplification of an mRNA and related pseudogene for a human ADP-ribosylation factor, a guanine nucleotide-dependent protein activator of cholera toxin. *Proc. Natl. Acad. Sci. U.S.A. 87*: 2206.

Moss, J., and Vaughan, M. (1977a). Mechanism of action of choleragen. Evidence for ADP-ribosyltransferase activity with arginine as an acceptor. *J. Biol. Chem. 252*: 2455.

Moss, J., and Vaughan, M. (1977b). Choleragen activation of solubilized adenylate cyclase: requirement for GTP and protein activator for demonstration of enzymatic activity. *Proc. Natl. Acad. Sci. U.S.A. 74*: 4396.

Moss, J., and Vaughan, M. (1978). Isolation of an avian erythrocyte protein processing ADP-ribosyltransferase activity and capable of activating adenylate cyclase. *Proc. Natl. Acad. Sci. U.S.A. 75*: 3621.

Moss, J., and Vaughan, M. (1990). Participation of guanine nucleotide-binding protein cascade in activation of adenylyl cyclase by cholera toxin (choleragen). In *G Proteins*, (R. Iyengar and L. Birnbaumer (Eds.), Academic, San Diego, p. 179.

Moss, J., Manganiello, V. C., and Vaughan, M. (1976). Hydrolysis of nicotinamide adenine dinucleotide by choleragen and its A protomer: possible role in the activation of adenylate cyclase. *Proc. Natl. Acad. Sci. U.S.A. 73*: 4424.

Moss, J., Stanley, S. J., and Lin, M. C. (1979). NAD glycohydrolase and ADP-ribosyltransferase activities are intrinsic to the A$_1$ peptide of choleragen. *J. Biol. Chem. 254*: 11993.

Moss, J., Stanley, S. J., Watkins, P. A., and Vaughan, M. (1980). ADP-ribosyltransferase activity of mono- and multi-(ADP-ribosylated) choleragen. *J. Biol. Chem. 255*: 7835.

Murtagh, J. J., Jr., Mowatt, M. R., Lee, C.-M., Lee, F.-J. S., Mishima, K., Nash, T. E., Moss, J., and Vaughan, M. (1992). Guanine nucleotide-binding proteins in the intestinal parasite *Giardia lamblia*. Isolation of a gene encoding an ~20 kDa ADP-ribosylation factor. *J. Biol. Chem. 267*: 9654.

Murtagh, J. J., Jr., Lee, F.-J. S., Deak, P., Hall, L. M., Monaco, L., Lee, C.-M., Stevens, L. A., Moss, J., and Vaughan, M. (1993). Molecular characterization of a conserved, guanine nucleotide-dependent ADP-ribosylation factor in *Drosophila melanogaster*. *Biochemistry, 32*:6011.

Nakaya, S., Moss, J., and Vaughan, M. (1980). Effects of nucleoside triphosphates on choleragen-activated brain adenylate cyclase. *Biochemistry 19*: 4871.

Neer, E. J., Wolf, L. G., and Gill, D. M. (1987). The stimulatory guanine-nucleotide regulatory unit of adenylate cyclase from bovine cerebral cortex. ADP-ribosylation and purification. *Biochem. J. 241*: 325.

Noda, M., Tsai, S.-C., Adamik, R., Moss, J., and Vaughan, M. (1990). Mechanism of cholera toxin activation by a guanine nucleotide-dependent 19 kDa protein. *Biochim. Biophys. Acta 1034*: 195.

Northup, J. K., Sternweis, P. C., Smigel, M. D., Schleifer, L. S., Ross, E. M., and Gilman, A. G. (1980). Purification of the regulatory component of adenylate cyclase. *Proc. Natl. Acad. Sci. U.S.A. 77*: 6516.

Orci, L., Malhota, V., Amherdt, M., Serafini, T., and Rothman, J. E. (1989). Dissection of a single round of vesicular transport: sequential intermediates for intercisternal movement in the Golgi stack. *Cell 56*: 357.

Peng, Z., Calvert, I., Clark, J., Helman, L., Kahn, R., and Kung, H.-F. (1989). Molecular cloning, sequence analysis and mRNA expression of human ADP-ribosylation factor. *Biofactors 2*: 45.

Pinkett, M. O., and Anderson, W. B. (1982). Plasma membrane-associated component(s) that confer(s) cholera toxin sensitivity to adenylate cyclase. *Biochim. Biophys. Acta 714*: 337.

Price, S. R., Nightingale, M. S., Tsai, S.-C., Williamson, K. C., Adamik, R., Chen, H.-C., Moss, J., and Vaughan, M. (1988). Guanine nucleotide-binding proteins that enhance choleragen ADP-ribosyltransferase activity: nucleotide and deduced amino acid sequence of an ADP-ribosylation factor cDNA. *Proc. Natl. Acad. Sci. U.S.A. 85*: 5488.

Price, S. R., Barber, A., and Moss, J. (1990). Structure–function relationships of guanine nucleotide-binding proteins. In *ADP-Ribosylating Toxins and G Proteins: Insights Into Signal Transduction*, J. Moss and M. Vaughan (Eds.), American Society for Microbiology, Washington, DC, p. 397.

Price, S. R., Welsh, C. F., Haun, R. S., Stanley, S. J., Moss, J., and Vaughan, M. (1992). Effects of phospholipid and GTP on recombinant ADP-ribosylation factors (ARFs). Molecular basis for differences in requirements for activity of mammalian ARFs. *J. Biol. Chem. 267*: 17766.

Regazzi, R., Ullrich, S., Kahn, R. A., and Wollheim, C. B. (1991). Redistribution of ADP-ribosylation factor during stimulation of permeabilized cells with GTP analogues. *Biochem. J. 275*: 639.

Rothman, J. E., and Orci, L. (1992). Molecular dissection of the secretory pathway. *Nature 355*: 409.

Schleifer, L. S., Kahn, R. A., Hanski, E., Northup, J. K., Sternweis, P. C., and Gilman, A. G. (1982). Requirements for cholera toxin-dependent ADP-ribosylation of the purified regulatory component of adenylate cyclase *J. Biol. Chem. 257*: 20.

Schmidt, G. J., Huber, L. J., and Weiter, J. J. (1987). A-protein catalyzes the ADP-ribosylation of G-protein from cow rod outer segments. *J. Biol. Chem. 262*: 14333.

Segev, N. J., Mulholland, J., and Botstein, D. (1988). The yeast GTP-binding YPT1 protein and a mammalian counterpart are associated with the secretion machinery. *Cell 52*: 915.

Serafini, T., Stenbeck, G., Brecht, A., Lottspeich, F., Orci, L., Rothman, J. E., and Wieland, F. T. (1991a). A coat subunit of Golgi-derived non-clathrin-coated vesicles with homology to the clathrin-coated vesicles coat protein β-adaptin. *Nature 349*: 215.

Serafini, T., Orci, L., Amherdt, M., Brunner, M., Kahn, R. A. and Rothman, J. E. (1991b). ADP-ribosylation factor is a subunit of the coat of Golgi-derived COP-coated vesicles: a novel role for a GTP-binding protein. *Cell 67*: 239.

Serventi, I. M., Cavanaugh, E., Moss, J., and Vaughan, M. (1993). Characterization of the gene for ADP-ribosylation factor (ARF) 2, a developmentally regulated, selectively expressed member of the ARF family of ~20kDa guanine nucleotide-binding proteins. *J. Biol. Chem. 268*: 4863.

Sewell, J. L., and Kahn, R. A. (1988). Sequences of the bovine and yeast ADP-ribosylation factor and comparison to other GTP-binding proteins. *Proc. Natl. Acad. Aci. U.S.A. 85*: 4620.

Stearns, T., Willingham, M. C., Botstein, D., and Kahn, R. A. (1990a). ADP-ribosylation factor is functionally and physically associated with the Golgi complex. *Proc. Natl. Acad. Sci. U.S.A.* 87: 1238.

Stearns, T., Kahn, R. A., Botstein, D., and Hoyt, M. A. (1990b). ADP ribosylation factor is an essential protein in *Saccharomyces cerevisiae* and is encoded by two genes. *Mol. Cell. Biol.* 10: 6690.

Sweet, R. W., Yokoyama, S., Kamata, T., Feramisco, J. R., Rosenberg, M., and Gross, M. (1984). The product of *ras* is a GTPase and the T24 oncogenic mutant is deficient in this activity. *Nature 311*: 273.

Tamkun, J. W., Kahn, R. A., Kissinger, M., Brizuela, B. J., Rulka, C., Scott, M. P., and Kennison, J. A. (1991). The arflike gene encodes an essential GTP-binding protein in *Drosophila. Proc. Natl. Acad. Sci. U.S.A.* 88: 3120.

Taylor, T. C., Kahn, R. A., and Melançon, P. (1992). Two distinct members of the ADP-ribosylation factor family of GTP-binding proteins regulate cell-free intra-Golgi transport. *Cell 70*: 69.

Towler, D. A., Adams, S. P., Eubanks, S. R., Towery, D. S., Jackson-Machelski, E., Glaser, L., and Gordon, J. I. (1987). Purification and characterization of yeast myristoyl CoA:protein *N*-myristoyltransferase. *Proc. Natl. Acad. Sci. U.S.A.* 84: 2708.

Trepel, J. B., Chuang, D.-M., and Neff, N. H. (1977). Transfer of ADP-ribose from NAD to choleragen: A subunit acts as a catalyst and acceptor protein. *Proc. Natl. Acad. Sci. U.S.A.* 74: 5440.

Tsai, S.-C., Noda, M., Adamik, R., Moss, J., and Vaughan, M. (1987). Enhancement of choleragen ADP-ribosyltransferase activities by guanyl nucleotides and a 19-kDa membrane protein. *Proc. Natl. Acad. Sci. U.S.A.* 84: 5139.

Tsai, S.-C., Noda, M., Adamik, R., Chang, P. P., Chen, H.-C., Moss, J., and Vaughan, M. (1988). Stimulation of choleragen enzymatic activities by GTP and two soluble proteins purified from bovine brain. *J. Biol. Chem.* 263: 1768.

Tsai, S.-C., Adamik, R., Moss, J., and Vaughan, M. (1991a). Guanine nucleotide dependent formation of a complex between choleragen (cholera toxin) A subunit and bovine brain ADP-ribosylation factor. *Biochemistry 30*: 3697.

Tsai, S.-C., Adamik, R., Tsuchiya, M., Chang, P. P., Moss, J., and Vaughan, M. (1991b). Differential expression during development of ADP-ribosylation factors, 20-kDa guanine nucleotide-binding protein activators of cholera toxin. *J. Biol. Chem.* 266: 8213.

Tsai, S.-C., Haun, R. S., Tsuchiya, M., Moss, J., and Vaughan, M. (1991c). Isolation and characterization of the human gene for ADP-ribosylation factor 3, a 20-kDa guanine nucleotide-binding protein activator of cholera toxin. *J. Biol. Chem.* 266: 23053.

Tsai, S.-C., Adamik, R., Haun, R. S., Moss, J., and Vaughan, M. (1992). Differential interaction of ADP-ribosylation factors 1, 3, and 5 with rat brain Golgi membranes. *Proc. Natl. Acad. Sci. U.S.A.* 89: 9272.

Tsai, S.-C., Adamik, R., Haun, R. S., Moss, J., and Vaughan, M. (1993). Effects of brefeldin A and accessory proteins on association of ADP-ribosylation factors 1, 3, and 5 with Golgi. *J. Biol. Chem.* 268: 10820.

Tsuchiya, M., Price, S. R., Nightingale, M. S., Moss, J., and Vaughan, M. (1989). Tissues and species distribution of mRNA encoding two ADP-ribosylation factors, 20-kDa guanine nucleotide binding proteins. *Biochemistry 28*: 9668.

Tsuchiya, M., Price, S. R., Tsai, S.-C., Moss, J., and Vaughan, M. (1991). Molecular identification of ADP-ribosylation factor mRNAs and their expression in mammalian cells. *J. Biol. Chem.* 266: 2772.

Walker, M. W., Bobak, D. A., Tsai, S.-C., Moss, J., and Vaughan, M. (1992). GTP but not GDP analogues promote association of ADP-ribosylation factors, 20-kDa protein activators of cholera toxin, with phospholipids and PC-12 cell membranes. *J. Biol. Chem.* 267: 3230.

Waters, M. G., Serafini, T., and Rothman, J. E. (1991). "Coatomer": a cytosolic protein complex containing subunits of non-clathrin-coated Golgi transport vesicles. *Nature 349*: 248.

Weiss, O., Holden, J., Rulka, C., and Kahn, R. A. (1989). Nucleotide binding and cofactor activities of purified bovine brain and bacterially expressed ADP-ribosylation factor. *J. Biol. Chem. 264*: 21066.

Woolkalis, M., Gill, D. M., and Coburn, J. (1988). Assay and purification of cytosolic factor required for cholera toxin activity. *Methods Enzymol. 165*: 246.

12

Heat-Stable Enterotoxin of *Escherichia coli*

Toshiya Hirayama

Institute of Tropical Medicine, Nagasaki University, Sakamoto, Nagasaki, Japan

I. INTRODUCTION

Bacterial diarrheal diseases are a very important problem for human health, causing the death of many people, especially infants and children, each year, particularly in developing countries. It has long been known that certain strains of *Escherichia coli* can cause acute diarrhea in humans and animals. At least five kinds of *E. coli* [enteropathogenic *E. coli* (EPEC), enterotoxigenic *E. coli* (ETEC), enteroinvasive *E. coli* (EIEC), enterohemorrhagic *E. coli* (EHEC), and enteroaggregative *E. coli* (EAggEC)] have been clearly identified as diarrheal pathogens. Recently, it became possible to understand the pathogenicity of these *E. coli* at a molecular level.

Enterotoxigenic strains of *E. coli*, when injected into ligated intestinal loops, act like *Vibrio cholerae* in causing fluid secretion, and this enterotoxic effect can be reproduced with cell-free preparations of culture fluids of ETEC. Most ETEC produce a heat-labile enterotoxin (LT) and sometimes also a heat-stable enterotoxin (ST). LT and cholera toxin (CT) are superficially similar in that both bind to monosialoganglioside (G_{M1}) on the cell membrane of intestinal cells and exert their effects at least in part through the ADP-ribosylation of a guanine nucleotide-binding protein involved in the regulation of adenylate cyclase. This results in prolonged accumulation of intracellular cyclic AMP (cAMP) (Field, 1980; Moss and Vaughan, 1988).

adenylate cyclase. This results in prolonged accumulation of intracellular cyclic AMP (cAMP) (Field, 1980; Moss and Vaughan, 1988).

ST is distinguished from LT by its resistance to boiling at 100°C for 30 min. Two different kinds of ST have been described, STa, or ST-I, which is methanol-soluble and active in suckling mice and piglets, and STb, or ST-II, which is methanol-insoluble and assumed until recently to be specifically active in weaned pigs. STa, characterized as a low molecular weight enterotoxin, activates the membrane-bound form of intestinal guanylate cyclase, thereby causing a secretory response to the increasing concentration of cyclic GMP (cGMP) within intestinal epithelial cells (Field et al., 1978; Guerrant et al., 1980; Rao et al., 1980). Since purification of STb was accomplished only recently, little is known about its structure and function.

II. HEAT-STABLE ENTEROTOXIN STa (ST-I)

A. Structure and Biological Properties of STa

Genetically distinct ST-Ia and ST-Ib have been named STp and STh, respectively, because they were originally found in *E. coli* isolated from pigs (and cows) and humans, respectively. STh and STp purified from culture supernatants of human and porcine strains have molecular weights of 2048 and 1978 with 19 and 18 amino acids, respectively (Fig. 1) (Aimoto et al., 1982; Takao et al., 1983). Molecular cloning also helped in the characterization of STa (So and McCarthy, 1980; Moseley et al., 1983). The incidence of ETEC producing STa alone is low but significant. STa is encoded in a transmissible plasmid. Similar heat-stable enterotoxins are produced by several gram-negative enteric bacteria such as *Yersinia enterocolitica* (Takao et al., 1985b), *V. cholerae* non-O1 (Takao et al., 1985a), *Vibrio mimicus* (Shimonishi, 1988), and *Citrobacter freundii* (Guarino et al., 1989). Most information is available, however, on *E. coli* STh and STp, which are considered here.

All heat-stable enterotoxins contain very similar sequences of 13 amino acids that correspond to those boxed in Figure 1. Six cysteines are also found in all heat-stable enterotoxins. The three disulfide bonds are arranged in the same way in STh and STp as

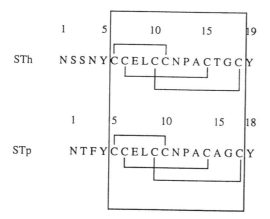

Fig. 1   Amino acid sequences and disulfide bond structures of heat-stable enterotoxins STh and STp of enterotoxigenic *Escherichia coli*. Box encloses sequences of 13 amino acids that are very similar in all heat-stabile enterotoxins.

well as in heat-stable enterotoxins from *Yersinia enterocolitica* and *Vibrio cholerae* non-O1 (Hidaka et al., 1990). Disulfide bonds may play a critical role in the tertiary structure of these heat-stable enterotoxins.

Shimonishi and coworkers chemically synthesized STh, STp, and their analogues and examined their biological and immunological properties (Aimoto et al., 1983; Ikemura et al., 1984a,b; Yoshimura et al., 1984a,b, 1985). Analysis of NMR spectra of synthesized STa and native STa showed that the synthetic STa had the same three-dimensional structure as native toxin, including the positions of disulfide bonds. Activity of synthetic STa and its analogues in the fluid accumulation assay in suckling mice showed that the core sequences (STh[6-18] and STp[5-17]) have the same toxicity as the holotoxins and are more stable to heat (Aimoto et al., 1983; Ikemura et al., 1984a,b; Yoshimura et al., 1984a,b, 1985). The minimum effective doses (MED) of the synthetic peptides were 0.4 pmol for STh[6-18] and 0.5 pmol for STp[5-17], which corresponded to those of native STh and STp. It may be inferred that the core sequences of the 13 amino acids linked intramolecularly by three disulfide bonds from the cysteine near the amino terminus to the cysteine near the carboxy terminus are responsible for the toxicity of STa and form spatially compact, very stable structures. Moreover, when these synthetic peptides were treated with antiserum raised against purified native STh, their biological activities were neutralized in a manner similar to that of native STh and STp.

To identify the amino acid(s) that are important for the enterotoxic activity of STa, Yamasaki et al. (1988b, 1990) synthesized analogues of STh[6-19] with single amino acid replacements at positions 12, 13, and 14 and examined their activities. Among eight analogues of STh[6-19] with amino acid replacements at position 12 (asparagine in native STh), the one with the valine replacement had the lowest MED, which was very similar to that of native STh (Table 1). Replacement by neutral amino acids, such as alanine, phenylalanine, serine, and glycine, influenced enterotoxic activity relatively little compared to replacement by basic amino acids such as arginine and lysine. Notably, activity was reduced by a factor of about 400 by the glutamate replacements, although the sizes of the side chains of asparagine and glutamic acid are not significantly different. It was concluded that the amino acid at position 12 should not be charged, and in particular should not be basic, to maintain the enterotoxicity.

TABLE 1 Biological Activities of Analogues of STh[6-19] with Amino Acid Replacements at Position 12

Amino acid sequence													MED (pmol)
6						12						19	
C	C	E	L	C	C	N	P	A	C	T	G	C	T	0.4
–	–	–	–	–	–	V	–	–	–	–	–	–	–	0.5
–	–	–	–	–	–	A	–	–	–	–	–	–	–	4.5
–	–	–	–	–	–	F	–	–	–	–	–	–	–	7.8
–	–	–	–	–	–	S	–	–	–	–	–	–	–	11
–	–	–	–	–	–	G	–	–	–	–	–	–	–	32
–	–	–	–	–	–	E	–	–	–	–	–	–	–	150
–	–	–	–	–	–	R	–	–	–	–	–	–	–	340
–	–	–	–	–	–	K	–	–	–	–	–	–	–	1300

MED=minimum effective dose.

TABLE 2 Biological Activities of Analogues of STh[6-19] with
Amino Acid Replacements at Position 13

Amino acid sequence														MED (pmol)
6							13					19		
C	C	E	L	C	C	N	P	A	C	T	G	C	T	0.4
–	–	–	–	–	–	–	V	–	–	–	–	–	–	1.4
–	–	–	–	–	–	–	A	–	–	–	–	–	–	6.9
–	–	–	–	–	–	–	Q	–	–	–	–	–	–	13
–	–	–	–	–	–	–	S	–	–	–	–	–	–	15
–	–	–	–	–	–	–	K	–	–	–	–	–	–	20
–	–	–	–	–	–	–	R	–	–	–	–	–	–	20
–	–	–	–	–	–	–	E	–	–	–	–	–	–	93
–	–	–	–	–	–	–	F	–	–	–	–	–	–	1200

MED=minimum effective dose.

The results obtained with toxin analogues in which proline at position 13 was replaced
are summarized in Table 2. Replacement by valine had a minimal effect. Replacement
by neutral amino acids such as alanine, glutamine, and serine and by basic amino acids
such as lysine and arginine reduced activity only a little. The analogue containing
glutamine, with an amide group at the terminal carboxylic acid of glutamic acid, was
seven times more toxic than the glutamic acid analogue. It is assumed, therefore, that the
amino acid at position 13 should not be acidic. Nor should it have a bulky side chain, as
substitution of phenylalanine at position 13 decreased potency approximately 3000-fold.

Table 3 summarizes the results obtained with the analogues containing substitutions
for alanine in position 14. Replacement by glycine or serine had minimal effects.
Replacement by aspartic acid, glutamic acid, or glutamine reduced activity considerably,
and replacement by valine, leucine, phenylalanine, lysine, or arginine totally abolished

TABLE 3 Biological Activities of Analogues of STh[6-19] with
Amino Acid Replacements at Position 14

Amino acid sequence														MED (pmol)
6								14					19	
C	C	E	L	C	C	N	P	A	C	T	G	C	T	0.4
–	–	–	–	–	–	–	–	G	–	–	–	–	–	10
–	–	–	–	–	–	–	–	S	–	–	–	–	–	14
–	–	–	–	–	–	–	–	D	–	–	–	–	–	506
–	–	–	–	–	–	–	–	E	–	–	–	–	–	540
–	–	–	–	–	–	–	–	Q	–	–	–	–	–	2090
–	–	–	–	–	–	–	–	V	–	–	–	–	–	>6600
–	–	–	–	–	–	–	–	L	–	–	–	–	–	>6600
–	–	–	–	–	–	–	–	F	–	–	–	–	–	>6400
–	–	–	–	–	–	–	–	K	–	–	–	–	–	>6400
–	–	–	–	–	–	–	–	R	–	–	–	–	–	>6500

MED=minimum effective dose.

of native STh. Thus, the size of the side chain of the amino acid in position 14 also has a great effect upon the enterotoxic activity of the toxin. All three amino acids in positions 12, 13, and 14 play an important role in the enterotoxicity of STa.

The importance of the three disulfide bonds for enterotoxic activity was proved by using synthetic peptides of STh with various combinations of two disulfide bonds (Yamasaki et al., 1988a; Hidaka et al., 1990). The disulfide bond between Cys-7 and Cys-15 is closely related to formation of the spatial structure that is necessary for expression of STh enterotoxic activity. This was tested using synthetic peptides 1–10 (Table 4) with two disulfide bonds at specific positions between four cysteines. As shown in Table 4, peptides 1 and 3, which had disulfide bonds between Cys-7 and Cys-15 and between Cys-6 and Cys-11 or Cys-10 and Cys-18 were toxic, whereas peptide 2 with no disulfide bond between Cys-7 and Cys-15 was inactive. Digestion of peptide 1 with aminopeptidase M, and of peptide 3 with carboxypeptidase A, gave peptides 4 and 5, which lack the original N-terminal amino acid and the three C-terminal amino acids, respectively. These products were about twice as toxic as the original peptides. These results indicate that a disulfide bond between Cys-7 and Cys-15 is necessary for the enterotoxicity of STh and that the active center of STh is located in a peptide from Cys-7 to Cys-15. A peptide with a disulfide bond between Cys-6 and Cys-11 and between Cys-7 and Cys-15 but without the peptide bond between Cys-6 and Cys-7 (peptide 6) and three peptides with only one disulfide bond between Cys-6 and Cys-11, Cys-7 and Cys-15, or Cys-10 and Cys-18 (peptides 7–10) were inactive. Therefore, the spatial structure of STh with a disulfide bond between Cys-7 and Cys-15 and between Cys-6 and Cys-11 or Cys-10 and Cys-18 is essential for the enterotoxigenic activity of STh.

A CPK model and schematic ribbon drawing (Fig. 2) of a short analogue of STh with 13 amino acids was constructed by Yamasaki et al. (1988a) from its three disulfide bonds (Hidaka et al., 1988) and nuclear Overhauser effects (NOE) data obtained by two-dimensional ^{1}H-NMR spectroscopy (Ohkubo et al., 1986). The main chain of the peptide is seemingly folded in a twisted-8 shape, and there are two loop structures consisting of two turns between amino acids 9–11 and 12–15. The three disulfide bonds are located in the hinge region binding these two turns between residues 9–11 and 12–15. These two loops presumably stabilize the spatial structure of STh. From the results shown in Tables 1–4, it is assumed that cleavage of the disulfide bond between Cys-6 and Cys-11or Cys-10 and Cys-18 does not appreciably disrupt the spatial structure necessary for expression of enterotoxic activity, and the twisted-8 structure shown in Figure 2 is still maintained, at least partially. However, cleavage of the disulfide bond between Cys-7 and Cys-15 may completely loosen the spatial conformation of STh, because peptide 2 in Table 4 was completely inactive. Cleavage of any two of the disulfide bonds shown in Figure 2 may completely disrupt the structure fixed by the hinge region. Therefore, a twisted-8 structure of the main chain seems important for expression of the enterotoxicity of STh or for stabilization of the spatial structure of the peptide chain from Cys-7 to Cys-15.

Ozaki et al. (1991) succeeded in the crystallization of a synthetic fully toxic analogue of STp, Mpr^5-STp[5-17], where Mpr was β-mercaptopropionic acid. It consists of 13 amino acids from Cys-5 to Cys-17 in STp deaminated at its N-terminus (Kubota et al., 1989). By X-ray diffraction analysis, STa had a right-hand spiral structure consisting of three β turns fixed by three disulfide bonds. Part of the surface of the

TABLE 4 Synthetic Analogues of STh with One or Two
Disulfide Bonds and Their Enterotoxic Activity

No.	Peptide[a]	MED (pmol)
	6 18 C C E L C C N P A C T G C	0.4
1	A C E L C A N P A C T G C	380
2	C A E L C C N P A A T G C	Inactive
3	C C E L A C N P A C T G A	290
4	C E L C A N P A C T G C	150
5	C C E L A C N P A C	110
6	C (above) C E L A C N P A C T G A	Inactive
7	* C C E L A C N P A C T G A *	Inactive
8	* * C C E L A C N P A C T G A	Inactive
9	* * C A E L C C N P A A T G C	Inactive
10	* C E L A C N P A C T G A	Inactive

MED-minimum effective dose
[a]C denotes a cysteine residue in which the thiol group is blocked by an
acetamidomethyl group.

toxin was amphiphilic, important for its interaction with its receptor on intestinal
epithelial cells.

B. Receptor for STa

STa has only recently been purified sufficiently or synthesized for the study of its mode
of action in vitro. The initial step in the biological action of STa is its interaction with
specific high-affinity receptors. The binding of STa to the plasma membrane of intestinal

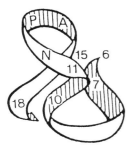

Fig. 2   Schematic ribbon drawing of the polypeptide backbone of an STh[6-18].

epithelial cells of rabbits and rats through these receptors activates membrane-bound guanylate cyclase, leading to an increase in the intracellular concentration of cGMP, followed by activation of cGMP-dependent protein kinase, culminating in the inhibition of absorption of Na^+ and Cl^- in the villi and stimulation of Cl^- secretion in the crypts.

Scatchard analysis (Frantz et al., 1984; Choen et al., 1988) indicated a single class of receptor for STh on rat intestinal epithelial cells and membranes. On the other hand, Kuno et al. (1986) demonstrated a 200-kDa protein that consisted of three specific proteins with molecular weights of 60,000, 68,000, and 80,000 as a high-affinity receptor for STh on rat intestinal cell membranes. Gariepy and Schoolnik (1987) also reported two different proteins with molecular weights of 57,000 and 75,000 as putative ST receptors in rat intestinal cell membranes. Robertson and Jaso-Friedmann (1988) attempted to purify the receptor of STh by lectin affinity chromatography and found that the affinity of the receptor for a ConA-Sepharose column is effective for its purification.

Two distinct glycoprotein receptors have also been identified and characterized (Hirayama et al., 1989b, 1992). Incubation of rat intestinal cell membranes with radioiodinated *N*-5-azido-nitrobenzoyl-STh[5-19] (^{125}I-ANB-STh[5-19]) followed by photolysis resulted in specific radiolabeling of two distinct proteins, STR-200A and STR-200B. Both bound to a ConA-Sepharose column and were separable by elution with α-methylglucoside and α-methylmannoside, respectively. STR-200A was composed of two molecules of a 70-kDa protein, whereas STR-200B was composed of two different protein molecules, one 53 kDa and the other 77 kDa. The 70-kDa protein was labeled most with ^{125}I-ANB-STh[5-19], suggesting that STR-200A is the main receptor protein for STa on the rat intestinal cell membrane. Binding of various synthetic STh analogues with a single amino acid replacement at position 14 was studied by competition of the analogues with ^{125}I-ANB-STh[5-19] for binding to the 70-kDa protein. Binding of ^{125}I-ANB-STh[5-19] was completely inhibited by STh[6-19] and partially by the STh analogue with a glycine replacement at position 14, which had about one-twentieth of the activity of the original toxin (Table 3). Other analogues with glutamic acid, glutamine, phenylalanine, arginine, valine, lysine, or leucine replacements did not inhibit the binding of the ^{125}I-ANB-STh[5-19] to the 70-kDa protein. These results indicate that the abilities of synthetic peptides to compete for binding to 70-kDa protein correlated well with their enterotoxic activities.

The carbohydrate moieties of STR-200A and STR-200B were examined by enzymatic deglycosylation. The 70-kDa protein of STR-200A contains N-linked high-mannose-type and/or hybrid-type oligosaccharides and possesses at least three N-glycosylation sites. The 53-kDa protein of STR-200B has an N-linked complex-type oligosaccharide side chain. The effect of the carbohydrate moiety of the 70-kDa STR-200A protein on its binding to STh was examined. The product of the 70-kDa protein, after digestion with *endo*-β-glycosidase H was still capable of binding STh and was labeled with ^{125}I-ANB-STh[5-19], suggesting that the N-linked oligosaccharides at three sites of the 70-kDa protein are not essential for ST binding. The sizes of three ST-binding proteins (subunits of STa receptor) reported by Kuno et al. (1986) are similar to those of subunits of STR-200A and STR-200B. A high-affinity ST receptor that has a 74-kDa subunit without guanylate cyclase activity was recently found (Hugues et al., 1991, 1992).

Recently, a guanylate cyclase-coupled STa receptor was reported by several investigators (Schulz et al., 1990; de Sauvage et al., 1991; Singh et al., 1991). Schulz et al. (1990) determined a novel genetic sequence encoding the STa receptor protein, proving it to be a new form of guanylate cyclase (GC-C), with an extracellular region completely different from those of two guanylate cyclase (GC-A and GC-B) atrial natriuretic peptide (ANP) receptors. Two other groups (de Sauvage et al., 1991; Singh et al., 1991) described a cDNA encoding the human GC-C. We also have confirmed and characterized the binding capability of GC-C for STa and the marked activation of GC-C by STa using a cloned cDNA of pig GC-C expressed in mammalian cells (Wada et al., 1992). Photoaffinity labeling with ^{125}I-ANB-STh[5-18] resulted in a radiolabeled protein with an apparent molecular weight of 140,000, suggesting that this protein is a glycoprotein.

As shown in Figure 3, the general features of GC-C from rat (Schulz et al., 1990), human (de Sauvage et al., 1991; Singh et al., 1991), and pig (Wada et al., 1992) are the same as those of GC-A and GC-B. GC-C has a single transmembrane region that divides it in about half. The intracellular region of GC-C contains a guanylate cyclase catalytic domain and a protein kinase-like domain that lacks the GXGXXG motif of tyrosine kinases. Apparent similarity between GC-C and other membrane-bound guanylate cyclases GC-A and GC-B was identified in intracellular regions, whereas no similarity was detected in the extracellular regions. The structure of the extracelluar regions of GC-C and the ANP-receptor guanylate cyclases might reflect the difference between their ligands. A deletion mutant of GC-A that completely lacked the protein kinase-like domain constitutively produced cGMP, independent of the presence of ANP (Chinkers and Garbers, 1989). The protein kinase-like domain of GC-A normally represses the guanylate cyclase activity of GC-A. Binding of ANP to the extracellular region of GC-A releases the inhibition of its guanylate cyclase activity. ATP and its nonhydrolyzable analogues stimulated the guanylate cyclase activity of GC-C (Gazzano et al., 1991) just as ATP enhanced stimulation of guanylate cyclase by ANP in membranes from various tissues (Kurose et al., 1987).

The site of action of ATP in GC-C has not been defined. For GC-A and GC-B, ATP

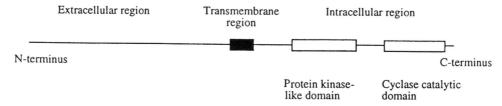

Fig. 3 Schematic model for GC-C.

appears to bind and act allosterically on activation of guanylate cyclase. Weikel et al. (1990) showed that a protein kinase C activator, phorbol dibutyrate, acted synergistically with STa to elevate cGMP in T84 human colon carcinoma cell lines. Crane et al. (1992) recently postulated that ATP has dual effects on activity of GC-C in T84 cells. ATP and its nonhydrolyzable analogues enhance stimulation of guanylate cyclase by STa and decrease the number of binding sites for STa, as has been observed for ANP receptors (Cole et al., 1989). Additionally, ATP (not the nonhydrolyzable analogues) serves as a substrate for protein kinase C, resulting in an additional increase in guanylate cyclase activity. In the predicted amino acid sequence of GC-C, there are consensus phosphorylation sites for protein kinase C but no direct proof of the phosphorylation of GC-C by protein kinase C.

There is an additional recent report of considerable interest with respect to the function of protein kinase-like domain in activation of the guanylate cyclase catalytic domain in the ANP receptor (Koller et al., 1992). Although comparison of amino acid sequences of the protein kinase-like domains of GC-A and GC-C reveals that they are only 30% identical (de Sauvage et al., 1991), most of the characteristic amino acids of the protein kinase-like domain conserved in GC-A and GC-B are present in GC-C, with a few exceptions (Koller et al., 1992). With a chimeric guanylate cyclase made by replacing the protein kinase-like domain of GC-A with the protein kinase-like domain of GC-C, the latter was not able to regulate guanylate cyclase activity as those of GC-A and GC-B did. The protein kinase-like domain of the ANP receptor guanylate cyclase requires strict sequence conservation to maintain proper regulation of the enzyme activity of the guanylate cyclase catalytic domain that occurs by direct interaction of the kinase-like domain with the cyclase domain (Koller et al., 1992). Furthermore, Koller et al. (1992) pointed out that GC-C completely lacks the glycine-rich nucleotide-binding consensus sequence, suggesting no interaction of ATP with GC-C. It is very likely, therefore, that the regulation of guanylate cyclase catalytic domain of GC-C is different from those of GC-A and GC-B (Koller et al., 1992).

To characterize the function of extracellular regions of GC-C, the binding properties of cells transfected with the truncated cDNA encoding rat GC-C without its intracellular region were studied (Hirayama et al., 1991). An approximately 1.7-kb fragment corresponding to amino acids 1–553 spanning both the extracellular and transmembrane regions and a portion of the intracellular regions was amplified using template cDNA prepared from rat intestinal cells by the PCR method, based on the nucleotide sequence of GC-C (Schulz et al., 1990). After the cloned 1.7-kb fragment was inserted into the mammalian expression vector pCGUT, the expressed protein on the surface of COS-7 cells was identified as the site of binding of GC-C to STa by photoaffinity labeling with ^{125}I-ANB-STh[5-19]. The truncated GC-C protein expressed in COS-7 cells binds ST with an affinity similar to that of native GC-C and a dissociation constant of 5 nM. A radioactive band with a molecular weight of approximately 180,000 was detected after SDS-PAGE, whereas two radioactive bands with molecular weights of 84,000 and 94,000 were detected under denaturing conditions, suggesting that the expressed protein is dimeric. The molecular weight of the recombinant protein in *E. coli* with pET-3 vector was calculated to be 66,000, an expected value. It seems that the expressed proteins on the cell surface of COS-7 cells are glycosylated. Cells transfected with another cloned cDNA (pCGUT-GCCN1.4) coding the extracellular region of GC-C with transmembrane anchor had intrinsic binding activity for STa, and the proteins labeled with ^{125}I-ANB-STh[5-19] formed dimers. These results suggest the possibility that GC-C forms a dimer and that its intracellular region does not play an important role in binding STa or in dimerization. It is not clear whether STa induces dimerization of the truncated GC-C or whether two molecules of truncated GC-C form a dimer on the cell surface of COS-7 cells regardless of STa binding. It is difficult to rule out the possibility of association of the truncated GC-C with another protein similar to the interleukin-6 receptor (Taga et al., 1989). More recently, Thorpe et al. (1991) reported that GC-A might be composed of homodimers interacting via their catalytic regions. A consideration of dimeric structure must account for cooperativity of membrane-bound guanylate cyclases (Garbers et al., 1974; Ramarao and Garbers, 1988). Dimerization of GC-C can be assigned as significant in terms of the mechanism of activation of GC-C by STa. It is not likely that STa induces dissociation of a dimeric form of GC-C to a monomeric state and subsequent activation of GC-C. Soluble guanylate cyclase is known to be a heterodimer. Its α and β subunits of 7.7 and 70 kDa are required for expression of catalytic activity and for its regulation by nitric oxide (Harteneck et al., 1990). Chinkers and Wilson (1992) showed that the oligomeric state of GC-A is important for the transduction of a conformational change across the plasma membrane. Activation of GC-C by STa may occur by means of conformational change of its extracellular domain and transmission of the activation signal to the cyclase catalytic domain across the plasma membrane.

These findings have demonstrated that GC-C is a cell surface receptor for STa in intestinal cells. Since STR-200A and STR-200B did not have guanylate cyclase activity, it is assumed that they are different STa receptor molecules from GC-C. The low molecular weight binding proteins for STa that lack guanylate cyclase appear to be degradation products (Vaandrager et al., 1993). GC-C fragments of 65 and 85 kDa were observed on immunoblots of rat GC-C expressed in 293 cells or rat intestinal cell membranes with anti-GC-C antiserum. However, further studies on the analysis of amino acid sequence of the purified STR-200A and STR-200B are necessary to clarify the relationships of STR-200A and STR-200B to GC-C and to establish the existence of multiple ST receptors in the intestinal cell membrane.

It is very noteworthy that guanylin was recently discovered as an endogenous activator of GC-C in extract of rat jejunum (Currie et al., 1992). Guanylin consists of 15 amino acids with the sequence PNTCEICAYAACTGC and two intramolecular disulfide bonds. Guanylin increased cGMP in T84 cells in a time- and concentration-dependent manner. Guanylin displaced the specific binding of ^{125}I-STp from T84 cells. This biological activity of guanylin was abolished by treatment with the reducing agent DTT. The sequence of guanylin is encoded in a 0.9-kb mRNA that is abundant in rat small intestine and is detected in the kidney, adrenal gland, and oviduct/uterus of rat (Schulz et al., 1992). Therefore, guanylin seems to be involved in the regulation of fluid and electrolyte transport in many tissues, although about 10-times higher concentrations of guanylin than of STa are required for an activation of guanylate cyclase of GC-C in T84 cells. The structure and function of membrane-bound guanylate cyclase have recently been reviewed by Garbers (Garbers, 1992; Yuen and Garbers, 1992). Details of biological function and molecular structure are not yet known.

C. Fluid Secretion Induced After cGMP Accumulation in Intestinal Cells by STa

The bacterial enterotoxins such as STa, LT, and CT induce watery diarrhea without causing histological damage. However, there are obvious differences between the mode of action of STa and that of LT or CT. After binding to the receptor on intestinal epithelial cells, STa rapidly transmits its signal across the membrane, bringing about a change in the activity of guanylate cyclase. This results in accumulation of cGMP in enterocytes. Initially, there was considerable doubt as to whether activation of guanylate cyclase could cause changes in electrolyte transport and water secretion leading to diarrhea, since hormones known to activate guanylate cyclase did not bring about fluid accumulation in ligated ileal loops. However, a stable analogue of cGMP such as 8-bromo-cGMP (Hughes et al., 1978; Giannella and Drake, 1979) can induce fluid accumulation under similar conditions. Thus, it is possible that STa causes diarrhea by activating guanylate cyclase, followed by accumulation of cGMP in the intestinal cells. Absorption and secretion of electrolytes in the intestine take place separately in different cells. Absorption of Na^+ and Cl^- occurs through mature intestinal epithelial cells at the tip of the villus, whereas Cl^- secretion occurs through crypt cells. CT and STa are known to cause diarrhea by stimulating Cl^- secretion in the crypts and by inhibiting absorption of Na^+ and Cl^- in the villi. However, the biochemical reactions that follow the increase of intracellular cGMP and cause secretory diarrhea are still not well understood. It is assumed that both cGMP and cAMP cause intestinal secretion by activation of specific protein kinases, which are thought to phosphorylate membrane proteins, thereby stimulating secretion of electrolytes and water. de Jonge and coworkers reported the presence of intestinal cGMP-dependent protein kinase that was autophosphorylated by addition of cGMP and cAMP, suggesting the possible importance of this enzyme in diarrhea induced by STa (de Jonge, 1976; de Jonge and Lohmann, 1985; van Dommelen and de Jonge, 1986). Moreover, using stripped rat proximal colon mounted in an Ussing chamber, de Jonge (1987) showed that the protein kinase inhibitor *N*-[2-methylamino)ethyl]-5-isoquinoline-sulfonamide (H-8) inhibited Cl^- secretion stimulated by STa or 8-bromo-cGMP, whereas 8-bromo-cAMP-induced secretion remained unaffected.

We also examined the effects of H-8 and another isoquinolinesulfonamide, *N*-(2-aminoethyl)-5-isoquinolinesulfonamide (H-9) on fluid accumulation in suckling mice induced by STa, 8-bromo-cGMP, and 8-bromo-cAMP (Hirayama et al., 1989a). It was found that H-8 and H-9 completely inhibited the fluid accumulation induced by STa and 8-bromo-cGMP whereas 8-bromo-cAMP-induced fluid accumulation was partially inhibited by H-8 and H-9. These results suggest the direct involvement of cGMP-dependent protein kinase and possibly of cAMP-dependent protein kinase in diarrhea caused by STa, CT, and LT. They also suggest that the brush border proteins phosphorylated in the presence of STa and CT through the action of cGMP- and cAMP-dependent protein kinases play roles in the production of watery diarrhea induced by these enterotoxins (Hirayama et al., 1990). Characterization of these phosphorylated proteins is necessary to elucidate the biochemical reaction cascade after their phosphorylation that may result in diarrhea. However, it is not known whether electrolyte and water secretion takes place through or parallel with the phosphorylation of the membrane proteins in the villi or in the crypts. A report by Peterson and Ochoa (1989) suggested a significant role of prostaglandins E1 and E2 synthesized from arachidonic acid in CT-mediated fluid accumulation in rabbit intestinal loop. Dreyfus et al. (1984) showed that STa did not alter the production of prostaglandin E2 or thromboxane in rat jejunum cells. Other arachidonic acid metabolites, leukotrienes such as 5-hydroxyeicosatetraenoic acid and 5-hydroperoxy-eicosatetraenoic acid, stimulated secretion in rabbit colon (Musch et al., 1992). It is quite possible that prostaglandins and leukotrienes also function as another second messenger in the induction of secretory diarrhea by STa. The importance of various links between cyclic nucleotides and arachidonic acid metabolites and the role of arachidonic acid metabolites in STa-induced diarrhea remains unclear. Thus, although structural details of

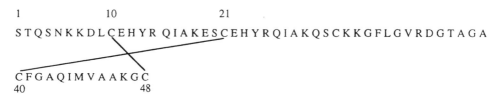

Fig. 4 Amino acid sequence of STb.

STa have been elucidated by protein chemical studies and DNA sequencing, the molecular mechanism by which STa induces secretory diarrhea is still not well understood.

III. HEAT-STABLE ENTEROTOXIN STb (ST-II)

The gene encoding STb was cloned, and its nucleotide sequence was determined (Lee et al., 1983; Picken et al., 1983). Fujii et al. (1991) purified STb and examined its chemical and biological properties. STb consists of 48 amino acids with a molecular weight of 5104 (Fig. 4). The amino acid sequence of STb corresponds to the inferred C-terminal amino acid sequence of STb predicted from the DNA sequence (Lee et al., 1983). STb has four cysteines at positions 10, 21, 36, and 48, which form intramolecular disulfide bonds between Cys-10 and Cys-48 and between Cys-21 and Cys-36. A crude preparation of STb induced fluid secretion within 30 min, with a maximal effect between 3 and 6 h in ligated pig intestinal loop assay (Kennedy et al., 1984). The only animal known to be susceptible to STb was the pig, until a recent report that STb can induce intestinal fluid secretion in an intestinal loop assay in rats and infant mice in the presence of a protease inhibitor (Whipp, 1987, 1990), indicating that host response specificity of STb is not dependent on the special properties of pig intestinal cells. Weikel and Guerrant (1985) indicated that STb induced no change in Cl^- or Na^+ fluxes as judged by the use of labeled ^{22}Na and ^{36}Cl. The removal of Ca^{2+} and the addition of La^{2+} did not inhibit fluid secretion due to STb, suggesting that this secretion is calcium-independent. Purified STb induced contraction of the mouse ileum and increased the amount of prostaglandin E2 in the ileum without affecting the intracellular concentration of cyclic nucleotides (Hitotsubashi et al., 1992).

REFERENCES

Aimoto, S., Takao, T., Shimonishi, Y., Hara, S., Takeda, T., Takeda, Y., and Miwatani, T. (1982). Amino-acid sequence of a heat-stable enterotoxin produced by human enterotoxigenic *Escherichia coli*. *Eur. J. Biochem. 129*: 257–263.

Aimoto, S., Watanabe, H., Ikemura, H., Shimonishi, Y., Takeda, T., Takeda, Y., and Miwatani, T. (1983) Chemical synthesis of a highly potent and heat-stable analog of an enterotoxin produced by a human strain of enterotoxigenic *Escherichia coli*. *Biochem. Biophys. Res. Commun. 112*: 320–326.

Chinkers, M., and Garbers, D. L. (1989). The protein kinase domain of the ANP receptor is required for signaling. *Science 245*: 1392–1394.

Chinkers, M., and Wilson, E. M. (1992). Ligand-independent oligomerization of natriuretic peptide receptors: identification of heterotrimeric receptors and a dominant negative mutant. *J. Biol. Chem. 267*: 18589–18597.

Choen, M. B., Guarino, A., Shukla, R., and Giannella (1988). Age-related differences in receptors for *Escherichia coli* heat-stable enterotoxin in the small and large intestine of children. *Gastroenterology 94*: 367–373.

Cole, F. E., Rondon, I., Iwata, T., Hardee, E., and Frohlich, E. D. (1989). Effect of ATP and amiloride on ANF binding and stimulation of cyclic GMP accumulation in rat glomerular membranes. *Life Sci. 45*: 477–484.

Crane, J. K., Wehner, M. S., Bolen, E. J., Sando, J. J., Linden, J., Guerrant, R. L., and Sears, C. L. (1992). Regulation of intestinal guanylate cyclase by the heat-stable enterotoxin of *Escherichia coli* (STa) and protein kinase C. *Infect. Immun. 60*: 5004–5012.

Currie, M. G., Fok, K. F., Kato, J., Moore, R. J., Hamra, F. K., Duffin, K. L., and Smith, C. E. (1992). Guanylin: an endogenous activator of intestinal guanylate cyclase. *Proc. Natl. Acad. Sci. U.S.A. 89*: 947–951.

de Jonge, H. R. (1976). Cyclic nucleotide-dependent phosphorylation of intestinal epithelium proteins. *Nature 262*:590–593.

de Jonge, H. R. (1987). Potential targets for phenothiazines and isoquinolinesulfonamides in intestinal epithelium. *Dig. Dis. Sci. 32*: 797.

de Jonge, H. R., and Lohmann, S. M. (1985). Mechanism by which cyclic nucleotides and other intracellular mediators regulate secretion. *Ciba Found. Symp. 112*: 116–138.

de Sauvage, F. J., Camerats, T. R., and Goeddel, D. V. (1991). Primary structure and functional expression of the human receptor for *Escherichia coli* heat-stable enterotoxin. *J. Biol. Chem. 266*: 17912–17918.

de Sauvage, F. J., Horuk, R., Bennett, G., Quan, C., Burnier, J. P., and Goeddel, D. V. (1992). Characterization of the recombinant human receptor for *Escherichia coli* heat-stable enterotoxin. *J. Biol. Chem. 267*: 6479–6482.

Dreyfus, L. A., Jaso-Friedman, L., and Robertson, D. C. (1984). Characterization of the mechanism of action of *Escherichia coli* heat-stable enterotoxin. *Infect. Immun. 44*: 493–501.

Field, M. (1980). Role of cyclic nucleotides in enterotoxic diarrhea. In *Advances in Cyclic Nucleotide Research*, Vol. 12, P. Hamet and H. Sand (Eds.), Raven, New York, pp. 267–277.

Field, M., Graf, L. H., Jr., Laird, W. J., and Smith, P. L. (1978). Heat-stable enterotoxin of *Escherichia coli*: in vitro effects of guanylate cyclase activity, cyclic GMP concentration, and ion transport in small intestine. *Proc. Natl. Acad. Sci. U.S.A. 75*: 2800–2804.

Frantz, J. C., Jaso-Friedman, L., and Robertson, D. C. (1984). Binding of *Escherichia coli* heat-stable enterotoxin to rat intestinal cells and brush border membranes. *Infect. Immun. 43*: 622–630.

Fujii, Y., Hayashi, M., Hitotsubashi, S., Fuke, Y., Yamanaka, H., and Okamoto, K. (1991). Purification and characterization of *Escherichia coli* heat-stable enterotoxin II. *J. Bacteriol. 173*: 5516–5522.

Garbers, D. L. (1992) Guanylyl cyclase receptors and their endocrine, pancrine, and autocrine ligands. *Cell 71*: 1–4.

Garbers, D. L., Hardaman, J. G., and Rudolph, F. G. (1974). Kinetic analysis of sea urchin sperm guanylate cyclase. *Biochemistry 13*: 4166–4171.

Gariepy, J., and Schoolnik, G. K. (1987). Design of a photoreactive analogue of the *Escherichia coli* heat-stable enterotoxin STIb: use in identifying its receptor on rat brush border membranes. *Proc. Natl. Acad. Sci. U.S.A. 83*: 483–487.

Gazzano, H., Wu, H. I., and Waldman, S. A. (1991). Activation of particulate guanylate cyclase by *Escherichia coli* heat-stable enterotoxin is regulated by adenine nucleotides. *Infect. Immun. 59*: 1552–1557.

Giannella, R. A., and Drake, K. W. (1979). Effect of purified *Escherichia coli* heat-stable enterotoxin on intestinal cyclic nucleotide metabolism and fluid secretion. *Infect. Immun. 24*: 19–23.

Guarino, A., Giannella, R. A., and Thompson, M. R. (1989). *Citrobacter freundii* produces an 18-amino-acid heat-stable enterotoxin identical to the 18-amino-acid *Escherichia coli* heat-stable enterotoxin (STIa). *Infect. Immun. 57*: 649–652.

Guerrant, R. L., Hughes, J. M., Chang, B., Robertson, D. C., and Murad, F. (1980). Activation of intestinal guanylate cyclase by heat-stable enterotoxin of *Escherichia coli*: studies of tissue specificity, potential receptors and intermediates. *J. Infect. Dis. 142*: 220–228.

Harteneck, C., Koesling, D., Soling, A., Schultz, G., and Bohne, E. (1990). Expression of soluble guanylate cyclase: catalytic activity requires two enzyme subunits. *FEBS Lett. 272*: 221–223.

Hidaka, Y., Kubota, H., Yamasaki, S., Ito, H., Takeda, Y., and Shimonishi, Y. (1988). Disulfide linkages in a heat-stable enterotoxin (STp) produced by a porcine strain of enterotoxigenic *Escherichia coli. Bull. Chem. Soc. Jpn. 61*: 1265–1271.

Hidaka, Y., Kubota, H., Yamasaki, S., Ito, H., Takeda, Y., and Shimonishi, Y. (1990). Structural requirements for expression of the toxicity of heat-stable enterotoxins produced by enteric bacteria. In *Advances in Research Cholera Related Diarrhea*, Vol. 7, R. B. Sack and Y. Zinnaka (Eds.), KTK Scientific Publishers, Tokyo, Japan, pp. 113–123.

Hirayama, T., Ito, H., and Takeda, Y. (1989a). Inhibition by the protein kinase inhibitors, isoquinolinesulfonamides, of fluid accumulation induced by *Escherichia coli* heat-stable enterotoxin, 8-bromo-cGMP and 8-bromo-cAMP in suckling mice. *Microb. Pathog. 7*: 255–261.

Hirayama, T., Shimonishi, Y., and Takeda, Y. (1989b). Glycoprotein receptors for heat-stable enterotoxin produced by *Escherichia coli*. Abstr. 7th Int Conf. on Cyclic Nucleotides, Calcium and Protein Phosphorylation, Kobe, Japan, p. 51.

Hirayama, T., Noda, M., Ito, H., and Takeda, Y. (1990). Stimulation of phosphorylation of rat brush-border membrane proteins by *Escherichia coli* heat-stable enterotoxin, cholera enterotoxin and cyclic nucleotides, and its inhibition by protein kinase inhibitors, isoquinoline sulfonamides. *Microb. Pathog. 8*: 421–431.

Hirayama, T., Wada, A., Hidaka, Y., and Shimonishi, Y. (1991). Photoaffinity labeling of ST-binding domain of guanylate cyclase (GC-C) expressed in COS-7 cells. Abstr. 27th U.S.-Japan Joint Conf. on Cholera and Related Diarrheal Diseases, Charlottesville, VA, pp. 95–96.

Hirayama, T., Wada, A., Iwata, N., Takasaki, S., Shimonishi, Y., and Takeda, Y. (1992). Glycoprotein receptors for a heat-stable enterotoxin (STh) produced by enterotoxigenic *Escherichia coli. Infect. Immun. 60*:4213–4220.

Hitotsubashi, S., Fujii, Y., Yamanaka, H., and Okamoto, K. (1992). Some properties of purified *Escherichia coli* heat-stable enterotoxin II. *Infect. Immun. 60*: 4468–4474.

Hughes, J. M., Murad, F., Chang, B., and Guerrant, R. L. (1978). Role of cyclic GMP in action of heat-stable enterotoxin of *Escherichia coli. Nature 271*: 755–756.

Hugues, M., Crane, M. R., Hakki, S., O'Hanley, P., and Waldman, S. A. (1991). Identification and characterization of a new family of high-affinity receptors for *Escherichia coli* heat-stable enterotoxin in rat intestinal membranes. *Biochemistry 30*: 1038–1045.

Hugues, M., Crane, M. R., Thomas, B. R., Robertson, D., Gazzano, H., O'Hanley, P., and Waldman, S. A. (1992). Affinity purification of functional receptors for *Escherichia coli* heat-stable enterotoxin from rat intestine. *Biochemistry 31*: 12–16.

Ikemura, H., Yoshimura, S., Aimoto, S., Shimonishi, Y., Hara, S., Takeda, T., Takeda, Y., and Miwatani, T. (1984a). Synthesis of a heat-stable enterotoxin (STh) produced by human strain SK-1 of enterotoxigenic *Escherichia coli. Bull. Chem. Soc. Jpn. 57*: 1381–1387.

Ikemura, H., Watanabe, H., Aimoto, S., Shiomonishi, Y., Hara, S., Takeda, T., Takeda, Y., and Miwatani, T. (1984b). Heat-stable enterotoxin (STh) of human enterotoxigenic *Escherichia coli* (strain SK-1): structure–activity relationship. *Bull. Chem. Soc. Jpn. 57*: 2550–2556.

Kennedy, D. J., Greenberg, R. N., Dunn, J. A., Abernathy, R., Ryerse, J. J., and Guerrant, R. L. (1984). Effects of *Escherichia coli* heat-stable enterotoxin STb on intestine of mice, rats, rabbits, and piglets. *Infect. Immun. 46*: 639–645.

Koller, K. J., de Sauvage, F. J., Lowe, D. G., and Goeddel, D. V. (1992). Conservation of the kinase like regulatory domain is essential for activation of the natriuretic peptide receptor guanylyl cyclases. *Mol. Cell. Biol. 12*: 2581–2590.

Kubota, H., Hidaka, Y., Ozaki, H., Ito, H., Hirayama, T., Takeda, Y., and Shimonishi, Y. (1989). A long-acting heat-stable enterotoxin analog of *Escherichia coli* with a single D-amino acid. *Biochem. Biophys. Res. Commun. 161*: 229–235.

Kuno, T., Kamisaki, Y., Waldman, S. A., Gariepy, J., Schoolnik, G. K., and Murad, F. (1986). Characterization of the receptor for heat-stable enterotoxin from *Escherichia coli* in rat intestine. *J. Biol. Chem. 261*: 1470–1476.

Kurose, H., Inagami, T., and Ui, M. (1987). Participation of adenosine 5′-triphosphate in the activation of membrane-bound guanylate cyclase by the atrial natriuretic factor. *FEBS Lett. 219*: 375–379.

Lee, C. H., Moseley, S. L., Moon, H. W., Whipp, S. C., Gyles, C. L., and So, M. (1983). Characterization of the gene encoding heat-stable enterotoxin II and preliminary molecular epidemiological studies of enterotoxigenic *Escherichia coli* heat-stable toxin II producers. *Infect. Immun. 42*: 264–268.

Moseley, S. L., Hardy, J. W., Imdadul, Hug, M. I., Escheverria, P., and Falkow, S. (1983). Isolation and nucleotide sequence determination of a gene encoding a heat-stable enterotoxin of *Escherichia coli. Infect. Immun. 39*: 1167–1174.

Moss, J., and Vaughan, M. (1988). ADP-ribosylation of guanyl nucleotide-binding regulatory proteins by bacterial toxins. *Adv. Enzymol. 61*: 303–379.

Musch, M. W., Miller, R. J., Field, M., and Siegel, M. I. (1982). Stimulation of colonic secretion by lipooxygenase metabolites of arachidonic acid. *Science 217*: 1255–1256.

Ohkubo, T., Kobayashi, Y., Shimonishi, Y., and Kyogoku, Y. (1986). A conformational study of polypeptides in solution by ^{1}H-NMR and distance geometry. *Biopolymers 25*: 123–134.

Ozaki, H., Sato, T., Kubota, H., Hata, Y., Katsube, Y., and Shimonishi, Y. (1991). Molecular structure of the toxic domain of heat-stable enterotoxin produced by a pathogenic strain of *Escherichia coli*: a putative binding site for a binding protein on rat intestinal epithelial cell membranes. *J. Biol. Chem. 266*: 5934–5941.

Peterson, J. W., and Ochoa, L. G. (1989). Role of prostaglandins and cAMP in the secretory effects of cholera toxin. *Science 245*: 857–859.

Picken, R. N., Mazaitis, A. J., Maas, W. K., and Heyneker, H. (1983). Nucleotide sequence of the gene for heat-stable enterotoxin II of *Escherichia coli. Infect. Immun. 42*: 269–275.

Ramarao, C. S., and Garbers, D. L. (1988). Purification and properties of the phosphorylated form of guanylate cyclase. *J. Biol. Chem. 263*: 1524–1529.

Rao, M. C., Guandalini, S., Smith, P. L., and Field, M. (1980). Mode of action of heat-stable *Escherichia coli* enterotoxin: tissue and subcellular specificities and role of cyclic GMP. *Biochem. Biophys. Acta 632*: 35–46.

Robertson, D. C., and Jaso-Friedmann, L. (1988). Partial purification and characterization of the intestinal receptor for *Escherichia coli* enterotoxin (STa). In *Advances in Cholera Related Diarrhea*, Vol. 4, S. Kuwahara and N. F. Pierce (Eds.), KTK Scientific Publishers, Tokyo, Japan, pp. 167–179.

Schulz, S., Green, C. K., Yuen, P. S. Y., and Garbers, D. L. (1990). Guanylyl cyclase is a heat-stable enterotoxin receptor. *Cell 63*: 941–948.

Schulz, S., Chrisman, T. D., and Garbers, D. L. (1992). Cloning of expression of guanylin: its existence in various mammalian tissues. *J. Biol. Chem. 267*: 16019–16021.

Schimonishi, Y. (1988). Molecular aspects of heat-stable enterotoxins of enteric bacteria. In *Advances in Cholera Related Diarrhea*, Vol. 6, N. Ohtomo and R. B. Sack (Eds.), KTK Scientific Publishers, Tokyo, Japan, pp. 9–21.

Singh, S., Singh, G., Heim, J.-M., and Gerzer, R. (1991). Isolation and expression of a guanylate cyclase-coupled heat-stable enterotoxin receptor cDNA from a human colonic cell line. *Biochem. Biophys. Res. Commun. 179*: 1455–1463.

So, M., and McCarthy, B. J. (1980). Nucleotide sequence of transposon Tn 1681 encoding a heat-stable toxin (ST) and its identification in enterotoxigenic *Escherichia coli* strains. *Proc. Natl. Acad. Sci. U.S.A. 77*: 4011–4015.

Taga, T., Hibi, M., Hirata, Y., Yamasaki, K., Yasukawa, K., Matsuda, T., Hirano, T., and Kishimoto, T. (1989). Interleukin-6 triggers the association of its receptor with a possible signal transducer gp 130. *Cell 58*: 573–581.

Takao, T., Hitouji, T., Aimoto, S., Shimonishi, Y., Hara, S., Takeda, T., Takeda, Y., and

Miwatani, T. (1983). Amino acid sequence of a heat-stable enterotoxin isolated from enterotoxigenic *Escherichia coli* strain 18D. *FEBS Lett.* *152*: 1–5.

Takao, T., Shimonishi, Y., Kobayashi, M., Nishimura, O., Arita, M., Takeda, T., Honda, T., and Miwatani, T. (1985a). Amino acid sequence of heat-stable enterotoxin produced by *V. cholerae* non-O1. *FEBS Lett.* *193*: 250–254.

Takao, T., Tominaga, N., Yoshimura, Y., Shimonishi, Y., Hara, S., Inoue, T., and Miyama, A. (1985b). Isolation, primary structure and synthesis of heat-stable enterotoxin produced by *Yersinia enterocolitica*. *Eur. J. Biochem.* *152*: 199–206.

Thorpe, D. S., Niu, S., and Morkin, E. (1991). Overexpression of dimeric guanylate cyclase cores of a natriuretic peptide receptor. *Biochem. Biophys. Res. Commun.* *180*: 538–544.

Vaandrager, A. B., Schulz, S., de Jonge, H. R., and Garbers, D. L. (1993). Guanylyl cyclase C is a N-linked glycoprotein receptor that accounts for multiple heat-stable enterotoxin-binding proteins in the intestine. *J. Biol. Chem.* *268*: 2174–2179.

van Dommelen, F. S., and de Jonge, H. R. (1986). Local change in fractional saturation of cGMP- and cAMP-receptor in intestinal microvilli in response to cholera toxin and heat-stable *Escherichia coli*. *Biochim. Biophys. Acta 886*: 135–142.

Wada, A., Hirayama, T., Fujisawa, J., Hidaka, Y., and Shimonishi, Y. (1992). Bindings of various analogs of *Escherichia coli* heat-stable enterotoxins (STa) to its recombinant pig receptor. Meeting Abstr. JASPEC '92, Shizuoka, Japan, p. 146.

Weikel, C. S., and Guerrant, R. L. (1985). STb enterotoxin of *Escherichia coli*: cyclic nucleotide-independent secretion. *Ciba Found. Symp.* *112*: 94–115.

Weikel, C. S., Spann, C. L., Chambers, C. P., Crane, J. K., Linden, J., and Hewlett, E. L. (1990). Phorbol esters enhance the cyclic GMP response of T84 cells to the heat-stable enterotoxin of *Escherichia coli* (STa). *Infect. Immun.* *58*: 1402–1407.

Whipp, S. C. (1987). Protease degradation of *Escherichia coli* heat-stable, mouse-negative, pig-positive enterotoxin. *Infect. Immun.* *55*: 2057–2060.

Whipp, S. C. (1990). Assay for enterotoxigenic *Escherichia coli* heat-stable toxin b in rats and mice. *Infect. Immun.* *58*: 930–934.

Yamasaki, S., Hidaka, Y., Ito, H., Takeda, Y., and Shimonishi, Y. (1988a). Structure requirements for the spatial structure and toxicity of heat-stable enterotoxin (STh) of enterotoxigenic *Escherichia coli*. *Bull. Chem. Soc. Jpn.* *61*: 1701–1706.

Yamasaki, S., Ito, H., Hirayama, T., Takeda, Y., and Shimonishi, Y. (1988b). Effects on the activity of amino acids replacement at positions 12, 13, and 14 heat-stable enterotoxin (STh) by chemical synthesis. Abstr. 24th Joint Conf. U.S.–Japan Cooperative Med. Sci. Program on Cholera and Related Diarrheal Disease Panel, Tokyo, Japan, p. 42.

Yamasaki, S., Sato, T., Hidaka, Y., Ozaki, H., Ito, H., Hirayama, T., Takeda, Y., Sugimura, T., Tai, A., and Shimonishi, Y. (1990). Structure–activity relationship of *Escherichia coli* heat-stable enterotoxin: role of Ala residue at position 14 in toxin-receptor interaction. *Bull. Chem. Soc. Jpn.* *63*: 2063–2070.

Yoshimura, S., Miki, M., Ikemura, H., Aimoto, S., Shimonishi, Y., Takeda, T., Takeda, Y., and Miwatani, T. (1984a). Chemical synthesis of a heat-stable enterotoxin produced by enterotoxigenic *Escherichia coli* strain 18B. *Bull. Chem. Soc. Jpn.* *57*: 125–133.

Yoshimura, S., Takao, T., Ikemura, H., Aimoto, S., Shimonishi, Y., Hara, S., Takeda, T., Takeda, Y., and Miwatani, T. (1984b). Chemical synthesis of fully active and heat-stable enterotoxin of enterotoxigenic *Escherichia coli* strain 18D. *Bull. Chem. Soc. Jpn.* *57*: 2543–2549.

Yoshimura, S., Ikemura, H., Watanabe, H., Aimoto, S., Shimonishi, Y., Hara, S., Takeda, T., Miwatani, T., and Takeda, Y. (1985). Essential structure for full enterotoxigenic activity of heat-stable enterotoxin produced by enterotoxigenic *Escherichia coli*. *Fed Eur. Biochem Soc. Lett.* *181*: 138–142.

Yuen, P. T. S., and Garbers, D. L. (1992). Guanylyl cyclase-linked receptors. *Annu. Rev. Neurosci.* *15*: 193–225.

13

Zot and Ace Toxins of *Vibrio cholerae*

Laurie E. Comstock, Michele Trucksis, and James B. Kaper

University of Maryland at Baltimore, Baltimore, Maryland

I. INTRODUCTION

Vibrio cholerae causes severe and life-threatening diarrhea in humans. The onset of the disease is often rapid, and as much as a liter of fluid may be purged each hour, leading to massive dehydration and death. Many of the aspects of the pathogenesis of *V. cholerae* have been elucidated. The bacteria adhere to epithelial cells of the proximal small intestine and elaborate a potent enterotoxin, cholera toxin (CT). CT is composed of five B subunits, which bind the holotoxin to the G_{M1} ganglioside receptor of intestinal mucosa, and one A subunit, which stimulates the production of intracellular cAMP, ultimately leading to copious secretion of fluid into the intestinal lumen (reviewed by Holmes et al., Chapter 10). Although CT is the major toxin elicited by *V. cholerae* and is responsible for the massive purging in cholera patients, *V. cholerae* strains specifically deleted for genes encoding CT (Dctx), are still able to cause mild to moderate diarrhea and additional clinical manifestations. Two additional toxins elaborated by *V. cholerae* are the zonula occludens toxin (Zot) and the accessory cholera enterotoxin (Ace). Zot increases the permeability of rabbit ileal tissue by altering the intercellular tight junctions. Ace acts as a classical enterotoxin, causing fluid secretion in the rabbit ligated ileal loop model and net ion secretion in the Ussing chamber (Trucksis et al., 1993). Although the mechanism of action of Ace has not yet been detailed, the predicted secondary structure and amino acid homology with other proteins suggests that it may act on the eukaryotic membrane as a pore-forming toxin.

II. DIARRHEA PRODUCTION FROM CHOLERA TOXIN NEGATIVE STRAINS

If all of the diarrhea from infection with *V. cholerae* was attributable to the action of CT, then removal of the *ctx* genes from a wild-type strain would result in a strain that is no longer able to cause diarrhea. Indeed this was the premise of initial efforts to construct a live oral attenuated *V. cholerae* vaccine. Since the A subunit alone is responsible for the ADP-ribosyltransferase activity of CT (Gill and King, 1975) and antibodies against the B subunit neutralize this activity (Peterson et al., 1979; Svennerholm, 1980), only the A subunit need be mutated to obtain a nontoxigenic vaccine strain and ensure an antitoxic immune response. In 1979, the first strain of this type, Texas Star, was developed by chemical mutagenesis (Honda and Finkelstein, 1979b). This strain was selected to secrete normal amounts of CT-B but no functional holotoxin. When a live oral vaccine of this strain was administered to volunteers, 25% developed mild diarrhea (Levine et al., 1984).

Subsequent vaccine strains, such as the classical strain CVD101 (Kaper et al., 1984a) and the El Tor strain JBK70 (Kaper et al., 1984b), were constructed using recombinant DNA techniques. Unlike the point mutation of Texas Star, JBK70 contained deletions of both the *ctxA* and *ctxB* genes, and CVD101 contained deletions of the *ctxA* genes only. When administered to volunteers, these vaccines were found to be highly effective in stimulating vibriocidal responses equaling those of the wild-type parent strains (Levine et al., 1988). These strains, however, still caused diarrhea in ca. 50% of volunteers. The diarrhea observed with the attenuated strains was much less than that seen with the wild-type strains. For example, the largest volume of diarrhea observed with CVD101 was 2.1 liters, whereas the parent strain, 395, has produced more than 40 liters of diarrhea in volunteers (Levine, 1980). In addition, abdominal cramps, anorexia, malaise, and low-grade fever were also reported from volunteers receiving CVD101 (Levine et al., 1988). This led to the hypothesis that *V. cholerae* may elaborate other toxins that account for diarrhea and other clinical symptoms induced by the CT$^-$ strains.

After the initial trials with the reactogenic CT$^-$ strains, other possible toxigenic products that *V. cholerae* was known to produce were investigated as a possible source of the residual diarrhea. One of the factors that was examined was the hemolysin/cytolysin. Hemolysis of sheep red blood cells was traditionally used to distinguish between the El Tor and classical biotypes of *V. cholerae*. Many El Tor strains produce a hemolysin from the *hlyA* locus (Goldberg and Murphy, 1984; Manning et al., 1984) that is proteolytically cleaved to a final active form of 65 kDa (Hall and Drasar, 1990). This toxin exhibits hemolysis of many mammalian erythrocytes as well as cytolytic activities on Vero and other mammalian cells in culture (Honda and Finkelstein, 1979a). In addition, the hemolysin caused fluid accumulation in the rabbit ileal loop model (Ichinose et al., 1987). Sequences homologous to *hlyA* are also present in classical strains, but the hemolysin from this locus is able to lyse only chicken and rabbit and not sheep erythrocytes (Richardson et al., 1986). A second separate hemolysin gene was cloned from the classical Ogawa strain 395 and was shown to be genetically distinct from *hlyA* (Richardson et al., 1986). This hemolysin was also capable of lysing only chicken and rabbit erythrocytes.

The role of the hemolysin encoded by the *hlyA* locus in diarrheal disease was assessed by Kaper et al. The *hlyA* gene was mutated in vitro by creating an internal deletion. This mutated gene was then introduced into the Δ*ctxA* strains JBK70 and CVD101 via allelic exchange to create strains CVD104 and CVD105 (Kaper et al., 1988). When fed to volunteers, 33% developed mild to moderate diarrhea (Levine et al., 1988). The outcome of these volunteer trials led the investigators to conclude that the hemolysin/cytolysin encoded by the *hlyA* locus of *V. cholerae* was most likely not the source of the diarrhea observed in recipients of Δ*ctxA* strains. Alm et al. (1991) reported that the hemolysin encoded by the *hlyA* locus may contain additional enterotoxic/cytotoxic activity at the N-terminal portion, which they believe may be active in the internal deletions of strains CVD104 and CVD105. However, since Alm et al. used strains that contained *ace* and *zot*, the true origin of the enterotoxic/cytotoxic activity is in question.

A Shiga-like toxin has also been associated with many strains of *V. cholerae*. This factor was characterized as a cell-associated, heat-stable material that is cytotoxic for Hela cells and is completely neutralized by antibodies to Shiga toxin (O'Brien et al., 1984). The gene(s) for the Shiga-like toxin have not been cloned, and therefore deletion mutants have not been analyzed in volunteer studies. Morris et al. (1990), however, showed that a non-O1 strain that contained the Shiga-like toxin activity was able to colonize humans without causing diarrhea. These studies argue against the Shiga-like toxin contributing to diarrhea in Δ*ctxA* strains.

Since the previously described toxins of *V. cholerae* had been shown not to be responsible for the residual diarrhea seen with Δ*ctxA* strains, additional, previously unrecognized, toxins of *V. cholerae* were sought. A classic technique used to study CT, namely the Ussing chamber, was employed to search for additional toxins in CT⁻ strains. The Ussing chamber is a sensitive technique for measuring transepithelial transport of electrolytes across intestinal tissue (Field et al., 1971, 1972), and most known enterotoxins exhibit activity in Ussing chambers. Supernatants from Δ*ctxA* strains were added to each side of the rabbit ileal tissue mounted in the Ussing chambers and were found to induce increases in the short-circuit current (I_{sc}) and tissue conductance (G_t) across the tissue. These changes were found to be due to two previously undescribed toxins, Ace (Trucksis et al., 1993) and Zot (Fasano et al., 1991). The genes encoding *zot* and *ace* are located immediately upstream of *ctx* (Trucksis et al., 1993; Baudry et al., 1992) in an area of the *V. cholerae* chromosome called the "core region" (Goldberg and Mekalanos, 1986). This area is often flanked by repeated sequences called RS1 elements that have transposase activity and are capable of causing amplification or deletion of the region (Goldberg and Mekalanos, 1986).

III. ZOT

A. Zot Activity

The activity that was given the name Zot was first detected in Ussing chambers as causing an immediate increase in G_t across rabbit ileal tissue. This activity was detected with the supernatants of strain CVD101 (Δ*ctxA*) and was indistinguishable from the activity of the supernatants of the wild-type 395 strain (Fasano et al., 1991). Since conductance is the reciprocal of resistance, the effect of Zot is to decrease tissue resistance (R_t). The R_t continued to decrease during the 2-h length of the assay to a point where the decrease was 20–25%.

In order to appreciate the significance of this decrease in resistance, an understanding of the relationship between resistance and tight junctions is necessary. A decrease in resistance across ileal tissue is representative of modifications to the intercellular spaces rather than of a change in the transcellular pathway (Powell, 1981). Intestinal epithelial cells are surrounded at adjacent apical membranes by zonula occludens (ZO), also referred to as tight junctions. Tight junctions function as both a molecular "fence" and a paracellular "gate." They serve as a fence by maintaining the polarized distribution of apical and basolateral membrane molecules by preventing their movement across the ZO. Therefore, tight junctions maintain the asymmetric localization of ion pumps and channels in epithelial membranes (Diamond, 1977). Tight junctions function as a gate by selectively controlling the transport of ions and water-soluble molecules between cells, that is, via the paracellular pathway (Frizzell and Schultz, 1972; Phillips et al., 1987; Diamond, 1977). A decrease in resistance is the most sensitive measure of changes in the paracellular pathway, that is, a loosening of or alteration to the tight junctions (Diamond, 1977).

To determine if the decrease in resistance caused by CVD101 supernatants was actually altering the arrangement of the tight junctions, Fasano et al. (1991) added an electron-dense marker to treated and untreated ileal tissues. This marker, wheat germ agglutinin-horseradish peroxidase (WGA-HRP), is normally incapable of passing beyond the ZO. Figure 1a shows tissue that was exposed to medium alone, showing integrity of the tight junction. Figure 1b represents the passage of the marker beyond the mucosal surface of the cells into the intercellular space of tissue treated with supernatants from strain CVD101. These results indicate that the tight junctions of tissue treated with CVD101 supernatants were being altered.

The exact composition of the tight junctional complex is unknown. Four proteins have been localized to the tight junction area: ZO-1, cingulin, ZO-2, and a newly described protein that reacts with monoclonal antibody 7H6 (Stevenson et al., 1986; Citi et al., 1991; Gumbiner et al., 1991; Zhong et al., 1993). All of these proteins, however, are contained on the cytoplasmic side of the membrane rather than as structural proteins of the junction. Many studies have been conducted to determine if the junctions are composed of protein or lipid or a combination of the two. Some experimental evidence is consistent with the hypothesis that the tight junctions are composed of lipid (Pinto da Silva and Kachar, 1982; Verkleij, 1984; Chevalier and Pinto da Silva, 1987; Balda et al., 1991); however, other data are inconsistent with this notion (Stevenson and Goodenough, 1984; Van Meer, 1989; Lazaro et al., 1991). In freeze-fracture electron microscopy, the fusion sites between lateral membranes are observed as anastomosing strands or grooves. A relationship has been established between the number of strands within tight junctions and the resistance of various epithelia. In general, as the number of ZO strands increases, the resistance of the tight junctions to passive ion flow also increases (Claude, 1978; Claude and Goodenough, 1973; Madara and Dharmsathaphorn, 1985). When ileal tissue was treated with supernatants from CVD101 and then prepared for freeze-fracture EM, a decrease in the strand complexity (Fig. 2) and the density of strand intersections (Table 1) was observed. Not all ZO, however, were affected; rather there was a mixture of affected and unaffected ZO. A preferential loss of the strands lying perpendicular to the long axis of the ZO was noted. Based on the activity of this factor it was named zonula occludens toxin (Zot).

B. The *zot* Gene

Supernatants from two Δ*ctxA* strains, CVD101 and 395N1 (Mekalanos et al., 1983), both derived from wild-type 395 and containing no other intentional mutations, reacted differently when tested in Ussing chambers. Supernatants from CVD101 caused an immediate increase in G_t that was indistinguishable from that seen with 395. In contrast, 395N1 exhibited a lag period when the variation in G_t was similar to the negative broth control. After 40 min, however, the supernatants from 395N1 caused G_t to rise, and after 2 h ΔG_t was equal to that seen with CVD101 and wild-type 395. The differing activities of these two Δ*ctxA* strains in Ussing chambers could also be correlated with differing results in volunteer studies. CVD101 caused diarrhea in 54% of volunteers (Levine et al., 1988), whereas 395N1 caused little if any diarrhea (Herrington et al., 1988). However, 395N1 caused other side effects such as vomiting, cramps, and fever. Since the only known mutation to these strains was a deletion of the *ctxA* genes, Baudry and coworkers examined the region flanking the *ctx* operon for an open reading frame that might encode this activity and may have been additionally mutated in the 395N1 construction. A plasmid was constructed that contained 2.7 kb of DNA upstream of the *ctx* operon and 1.4 kb downstream with a tetracycline gene interrupting the *ctxA* gene. The introduction of this plasmid into strain 395N1 was able to restore Zot activity to the wild-type level (Baudry et al., 1992). Therefore, the *zot* gene was contained within the 4.1 kb of DNA flanking the *ctx* locus. The *zot* gene was further localized to the 2.7-kb region upstream of *ctxA*. DNA sequence analysis of this region revealed one small open reading frame of approximately 300 bp and a 1.3-kb open reading frame that was located immediately upstream of *ctxA*. Both open reading frames would be transcribed in the same direction as the *ctx* operon. Random Tn*phoA* insertions were introduced into this 2.7-kb region of DNA cloned into the pUC19 vector in *E. coli* HB101. The culture supernatants from several Tn*phoA* insertion mutants were screened in Ussing chambers, and insertions in the 1.3-kb open reading frame showed loss of the Zot activity. This region was further subcloned, and the Zot activity was confirmed to be the product of the 1.3-kb open reading frame.

Based on the DNA sequence, the predicted Zot protein would consist of 399 amino acids and have a molecular mass of 44.8 kDa. Although a classic signal peptide sequence is not present, Zot is presumed to be excreted from *V. cholerae* since Zot activity is found in culture supernatants and because alkaline phosphatase activity is detected in the supernatants from *V. cholerae* with *zot-phoA* fusions (L. E. Comstock, unpublished).

The proximity of *zot* to the *ctx* operon is intriguing. Only 127 basepairs separate the *zot* stop codon from the start codon of *ctxA*. Additionally, the termination codon of the *zot* gene overlaps the initial TTTTGAT ToxR binding repeat of the *ctx* promoter (Fig. 3). ToxR is a transcriptional activator, and binding of ToxR to the repeated sequence between the *zot* and *ctx* genes enhances transcription of the *ctx* operon (see chapter by Mekalanos in this volume). Three studies examining the distribution of the *zot* gene among *V. cholerae* O1 toxigenic (i.e., containing *ctx*), O1 nontoxigenic, and non-O1 strains have revealed an interesting correlation between *zot* and the *ctx* genes (Johnson et al., 1993; Karasawa et al., 1993; Faruque et al., 1993). These studies demonstrated that, with very few exceptions, strains that contain the *ctx* genes also contain *zot* and vice versa.

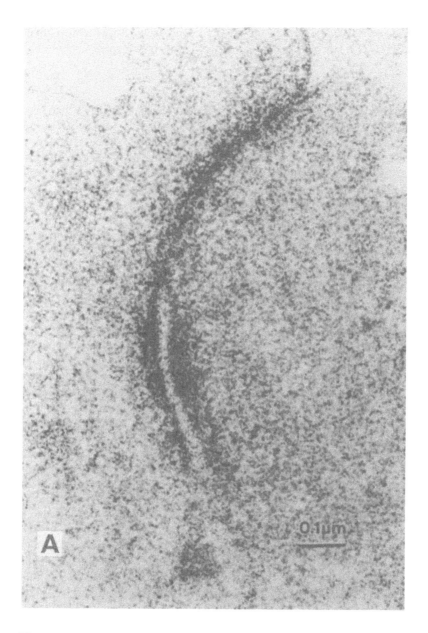

Fig. 1 WGA-HRP permeability assay on rabbit ileal tissues exposed for 60 min to (A) medium or (B) culture supernatants of *V. cholerae* CVD101. (From Fasano et al., 1991.)

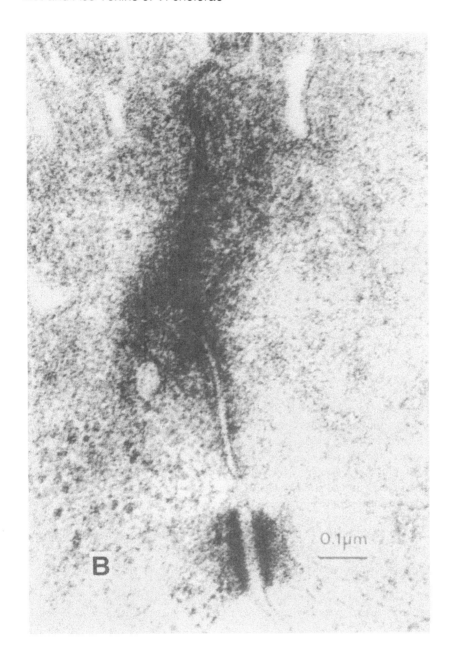

TABLE 1 Quantification of ZO Complexity[a] in Tissues Exposed to Culture Supernatants or Medium Control

Sample	n	Mean	SD	P
395	30	244.2	107.9	
				0.002
Medium	14	373.7	129.0	
CVD101	32	230.2	75.5	
				0.004
Medium	30	305.8	96.3	

[a]Complexity expressed as mean number of strand intersections per square micrometer. ZO complexity was quantified by counting the number of intersections between ridges or furrows in a representative area (0.1–0.25 μm^2) of each ZO and dividing by that area. To avoid bias in selecting ZO due to variation in crypts versus villus tips, the borders of replicated villi were followed and 16–48 consecutive ZO were photographed from a given replica. *n*, number of ZO examined.
Source: Fasano et al. (1991).

Therefore, *zot* and the *ctx* genes are very closely linked. The strong correlation of the *zot* gene with the *ctx* operon suggests that Zot plays a role in the pathogenesis of cholera.

Not only is the presence of *zot* conserved in toxigenic strains, but the DNA sequence of the *zot* gene is also highly conserved. Two strains within the classical biotype, 395 and 569B, shared 100% sequence homology in the *zot* gene (L. E. Comstock, unpublished). Strain E7946, of the El Tor biotype, contained only 14 nucleotide changes, 10 of which resulted in no amino acid change, with three coding for conserved substitutions, and only one encoding a nonconserved amino acid substitution. Therefore, even between biotypes, the sequence of *zot* is highly conserved. The apparent bias against mutational diversity additionally is consistent with Zot being important for *V. cholerae*.

C. Mechanism of Action

The ability of an infectious agent to alter the tight junctions of epithelial cells has not been described very often. Another pathogenic bacterium that produces a toxin capable of altering tight junctions is *Clostridium difficile*. *C. difficile* toxin A is a large 300-kDa protein that, like Zot, has been demonstrated to induce a substantial alteration in epithelial permeability by altering intercellular tight junctions (Hecht et al., 1988; Moore et al., 1990). Unlike Zot, however, toxin A induces cytopathic effects in cultured cells characterized by morphological changes such as retraction and cell rounding, nuclear polarization, and ultimately cell death (Fiorentini et al., 1990). These effects have been shown to be linked to cytoskeletal perturbations involving microfilaments (Fiorentini et al., 1990). No homology between Zot and toxin A (Dove et al., 1990) has been identified at the protein or nucleotide level. Unlike toxin A, Zot is a relatively small protein, approximately 44 kDa, and is not cytopathic for cultured cells. Additionally, unlike toxin

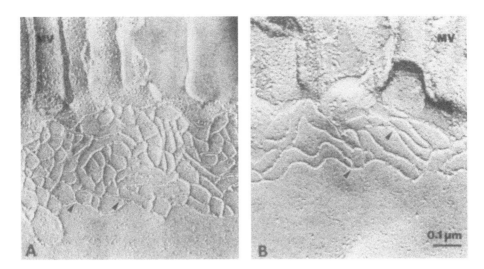

Fig. 2 Freeze-fracture studies of rabbit ileal tissue exposed to culture supernatants of *V. cholerae* for 60 min. (A) An intact ZO with numerous intersections between junctional strands. (B) An affected ZO from ileal tissue exposed to *V. cholerae* 395 supernatant; the reticulum appears simplified due to a greatly decreased incidence of strand intersections. (From Fasano et al., 1991.)

A, the activity of Zot is reversible once it is removed (Fasano et al., 1991). Therefore, the mechanisms by which these toxins exert their effects on target cells are most likely quite different.

Although the mechanism of action of Zot has not been elucidated, much is known about tight junctions and the ways in which these structures may be reversibly altered. There are extensive data implicating an important role of the cytoskeleton in the control of tight junctions (Bentzel et al., 1980; Madara et al., 1986; Mooseker, 1983; Meza et al., 1982). Condensation of actin microfilaments has been shown to diminish tight junctional resistance, thereby opening tight junctions (Madara et al., 1986, 1988; Madara and Pappenheimer, 1987). Intracellular mediators such as protein kinase C (Mullin and O'Brien, 1986), Ca^{2+} (Martinez-Palomo et al., 1980; Palant et al., 1983), and cAMP

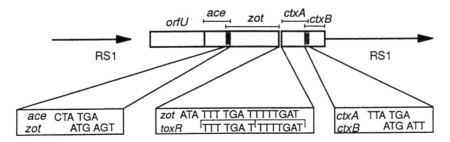

Fig. 3 *V. cholerae* E7946 "core region" with flanking RS1 elements. The enlarged regions depict the overlapping open reading frames of *ace* and *zot* and of *ctxA* and *ctxB*. In addition, the region where the stop codon of *zot* overlaps with the first ToxR binding repeat upstream of the *ctx* operon is also enlarged. (From Trucksis et al., 1993; Lockman and Kaper, 1983; and Miller et al., 1987.)

(Duffey et al., 1981) have all been shown to affect the permeability of the tight junction, and this effect is believed to be mediated through modifications to microfilaments.

With these data in mind, a possible mechanism of action for Zot emerges. A likely hypothesis is that Zot binds a membrane receptor that transmits an intracellular signal. There is no evidence for what type of receptor Zot may bind or what the intracellular signal is; however, an increase in PKC, Ca^{2+}, or cAMP may be involved. This second messenger may cause a transient contraction in the dynamic structure of the microfilament network, thereby loosening the tight junctions. A role for Ca^{2+} in the modulation of microfilament structure has been demonstrated (Janmey et al., 1985). This scheme would be consistent with a reversible toxin such as Zot, because once the toxin is removed the level of second messengers would return to preexposure levels, thereby allowing the structurally "tight" ZO to re-form.

The initial search of GenBank sequence databases did not yield significant similarity between *zot* and previously described genes. However, a more recent search revealed an interesting homology, the significance of which is unknown. The *zot* gene product is related to the predicted products of two open reading frames, one of the plasmid pKB740 and one of a bacteriophage Pf1, both of *Pseudomonas* (Koonin, 1992). Zot was further found to share three conserved regions with the gene II product of filamentous bacteriophages, which is believed to be an ATPase. The gene I product contains the classical NTP-binding motif and is thought to be involved in phage assembly and exit from the bacterial cell (Koonin, 1992). However, the putative NTP-binding domain of Zot was found to contain a drastic substitution of tyrosine for glycine, and the possibility that Zot contains ATPase activity has not yet been shown experimentally. If Zot is shown to contain this activity, its role in the modification of the tight junctions of intestinal epithelial cells will need to be tested. Elucidation of the actual mode of action of Zot awaits purification of the toxin, which is currently in progress.

IV. ACE

Ace activity was identified when the cloned 2.9-kb fragment of the "core region" (pCVD630) (Fig. 4) was introduced into strain CVD110 (Trucksis et al., 1993). *V. cholerae* strain CVD110 was constructed with a deletion of the entire core region, including all of the then known toxin genes of *V. cholerae*, *zot* and *ctx* (Michalski et al., 1993), from a virulent *V. cholerae* El Tor Ogawa strain E7946. The resulting strain, CVD110 (pCVD630), was tested in rabbit ligated ileal loops and the Ussing chamber model.

In the rabbit ligated ileal loops, CVD110 (pCVD630) caused significant fluid accumulation as compared to CVD110 (pCVD315), the negative control (0.73 ± 0.12 vs. 0.35 ± 0.1 mL/cm, $p < 0.05$). Parallel results were obtained in the Ussing chamber. CVD110 (pCVD630) showed a significant increase in short-circuit current (I_{sc}) relative to the negative control (37.7 ± 4.5 vs. 7.6 ± 5.9 μA/cm^2, $p < 0.01$). The increase in I_{sc} was secondary to an increase in PD with tissue resistance remaining stable.

Analysis of pCVD630 showed two open reading frames in this area of the core region (Fig. 4). The larger of the two open reading frames (*orfU*) was mutated by restriction enzyme digestion, resulting in an in-frame deletion mutation in this open reading frame and plasmid construct pCVD630A. When this plasmid was introduced into CVD110 and then tested in the rabbit ligated ileal loop model, significant fluid accumulation remained, equivalent to that seen with the intact plasmid pCVD630 (0.79 ± 0.18 vs. 0.73 ± 0.12

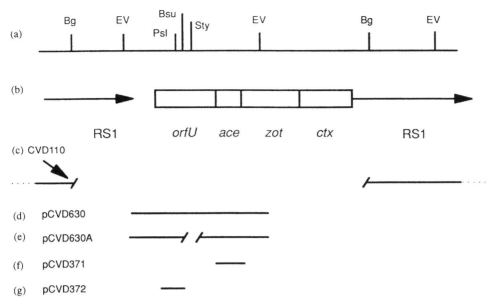

Fig. 4 Bacterial strains and plasmids used in this study. (a) Restriction map of the "core region" of *V. cholerae* E7946. Bg, *Bgl*II; EV, *Eco*RV; PsI, *Pst*I; Bsu, *Bsu*36I; Sty, *Sty*I. (b) Open reading frames identified by DNA sequence analysis in the "core region": *orfU*, open reading frame of unknown function; *ace*, accessory cholera enterotoxin; *zot*, zonula occludens toxin; *ctx*, cholera toxin operon including *ctxA* and *ctxB*. (c) *V. cholerae* strain CVD110, showing the sequences deleted from E7946. (d) pCVD630 containing the 2.9-kb *Eco*RV fragment immediately upstream of *ctx*. (e) pCVD630A, which was constructed from pCVD630 by deletion of a 309-bp *Bsu*361–*Sty*I fragment yielding *ace* Δ*orfU*. (f) Plasmid pCVD371 containing the Ace gene generated by PCR. (g) Plasmid pCVD372 containing the 5′ end of *orfU* generated by PCR. (From Trucksis et al., 1993.)

mL/cm). This result suggested that the smaller of the open reading frames, intact on both pCVD630 and pCVD630A, was responsible for the enterotoxic activity. This open reading frame was called *ace* for accessory cholera enterotoxin.

To confirm that *ace* was responsible for the activity detected in the rabbit ileal loop model and in the Ussing chamber model, two subclones were constructed using the polymerase chain reaction (PCR) technique that contained only the individual open reading frames. Plasmid pCVD372 contained a truncated *orfU* gene, equivalent to the fragment present in pCVD630A (Fig. 4). Plasmid pCVD371 contained the open reading frame *ace* (Fig. 3). These plasmids were introduced into CVD110 and tested in the Ussing chamber model. CVD110 (pCVD372), containing the truncated *orfU* fragment, showed no increase in I_{sc} and was comparable to the negative control CVD110 (pTTQ181K) (6.43 ± 4.7 vs. 7.6 ± 5.9 μA/cm^2). In contrast, CVD110 (pCVD371), the *ace* construct, showed an increase in I_{sc} equal in magnitude to that of the positive control, CVD110 (pCVD630) (32.5 ± 4.7 vs. 37.7 ± 4.5 μA/cm^2).

ace is located immediately upstream of *zot* and *ctx* on the *V. cholerae* chromosome (Fig. 3) (Trucksis et al., 1993). It consists of a 289-bp open reading frame counting from the first ATG, and this would predict a 96-residue peptide with a predicted M_r of 11.3 kDa. An interesting finding in gene structure is that the last codon of the *ace* open reading frame overlaps with the start codon of *zot* (Fig. 3). This same structure is seen with the two subunit genes of *ctx*; the last codon of *ctxA* overlaps with the initiation codon of *ctxB* (Lockman and Kaper, 1983) (Fig. 3). This pattern suggests translational coupling between *ace* and *zot*.

How *ace* acts to cause fluid secretion and ion transport is not yet clear; however, analysis of the amino acid sequence of Ace offers some insight. A search of the Swiss protein database using the program WORDSEARCH shows striking similarity between Ace and a family of ion-transporting ATPases including the product of the cystic fibrosis transmembrane conductance regulator (CFTR) (Trucksis et al., 1993). The similarity of Ace and the ion-transporting ATPases lies in the hydrophobic transmembrane domains of the latter. Analysis of the structure of Ace using the Chou-Fasman algorithm predicts that the C-terminal end of Ace may exist in an alpha-helix formation. This region of Ace has a high hydrophobic moment (Eisenberg et al., 1984), which is indicative of amphipathic structures and would be consistent with a model of Ace multimers inserting into the eukaryotic membrane with hydrophobic surfaces facing the lipid bilayer and the hydrophilic sides facing the interior of a transmembrane pore.

The predicted structure of Ace is similar to that of *Staphylococcus aureus* delta toxin (Freer and Birkbeck, 1982). Delta toxin is a 26-residue peptide that forms aggregates in aqueous solution but transforms into an alpha-helix conformation in the presence of phospholipid micelles (Mellor et al., 1988). Planar lipid bilayer analysis reveals that delta toxin forms ion channels that are cation-selective (Mellor et al., 1988). The proposed structure of the delta toxin channel is that of a hexameric cluster (Mellor et al., 1988). The C-terminal region of Ace shows 47% amino acid similarity with residues 2–20 of delta toxin (Trucksis et al., 1993), lending some support to the hypothesis that Ace acts by aggregating and inserting into the eukaryotic membrane, forming an ion channel.

The purification of Ace is currently under way. Studies on crude toxin preparations have established the proteinaceous nature of Ace. Ace is sensitive to treatment with proteinase K and is destroyed by boiling but is stable to freezing (Trucksis et al., 1993). A time course comparing the change in I_{sc} with Ace$^+$ and Ace$^-$ strains is presented in Figure 5.

V. ROLE OF ZOT AND ACE IN PATHOGENESIS

Although Zot and Ace have been shown to have toxic properties, it now appears that these toxins were not responsible for the residual diarrhea caused by Δ*ctxA* strains. *V. cholerae* CVD110 (Δ*ctxA*, Δ*zot*, Δ*ace*, and Δ*orfU*) was recently fed to volunteers (Tacket et al., 1993). This strain still caused mild to moderate diarrhea in 7 of 10 volunteers.

Although removal of *ace* and *zot* did not bring about a nonreactogenic strain, the results do not rule out a role for these toxins in disease due to *V. cholerae*. Indeed, there are several compelling observations supporting such a role. Mathan and coworkers, working in Villore, India, have examined biopsy specimens from cholera patients and have seen a Zot-like activity in vivo. Ileal biopsies taken from cholera patients revealed alterations to the intercellular tight junctions of the epithelium (V. I. Mathan, personal communication). The patients were treated with intravenous rehydration only, so that opening of the tight junctions due to glucose in oral rehydration solution is not responsible. Therefore, Zot may add to the transcellular CT-induced diarrhea by allowing additional fluid secretion via the paracellular pathway.

Unlike Zot, which primarily affects tissue resistance, Ace has a classic enterotoxic

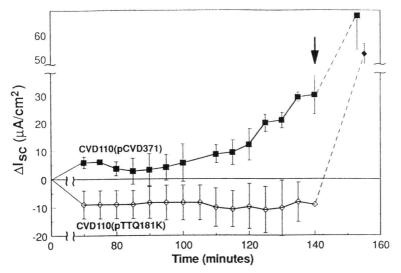

Fig. 5 Time course of ΔI_{sc} and ΔPD of supernatants of *V. cholerae* strains CVD110(pCVD371) (Ace[+] construct) versus CVD110(pTTQ181K) (negative control) (from Trucksis et al., 1993). Curves represent means and standard deviations of nine experiments for CVD110(pCVD371) and six experiments for CVD110(pTTQ181K). Time 0 is the time when the supernatant is added. There is an average lag time of 69 min (range 55–95 min) before toxin activity is apparent. Arrow indicates time of addition of 10 mM glucose (final concentration) to mucosal side of each chamber. (■), ΔI_{sc} CVD110(pCVD371) supernatants; (◇) ΔI_{sc} CVD110(pTTQ181K) supernatant. (From Trucksis et al., 1993.)

activity causing an increase in I_{sc} and PD in Ussing chambers and fluid secretion in the rabbit ligated ileal loop. Additionally, the sequence homology that Ace shares with channel-forming toxins further supports a role in pathogenesis. The actual role of each of these toxins in contributing to the pathogenesis of cholera awaits specific deletion of each of these genes separately from a wild-type strain.

REFERENCES

Alm, R. A., Mayrhofer, G., Kotlarski, I., and Manning, P. A. (1991). Amino-terminal domain of the El Tor haemolysin of *Vibrio cholerae* O1 is expressed in classical strains and is cytotoxic. *Vaccine 9*: 588.

Balda, M. S., Gonzalez-Mariscal, L., Contreras, R. G., Macias-Silva, M., Torres-Marquez, M. E., Garcia Sainz, J. A., and Cerijido, M. (1991). Assembly and sealing of tight junctions: possible participation of G-proteins, phospholipase C, protein kinase C and calmodulin. *J. Memb. Biol. 122*: 193.

Baudry, B., Fasano, A., Ketley, J., and Kaper, J. B. (1992). Cloning of a gene (*zot*) encoding a new toxin produced by *Vibrio cholerae. Infect. Immun. 60*: 428.

Bentzel, C. J., Hainau, B., Ho, S., Hui, S. W., Edelman, A., Anagnostopoulous, T., and Benedetti, E. L. (1980). Cytoplasmic regulation of tight junction permeability: effect of plant cytokinins. *Am. J. Physiol. 239*: C75.

Chevalier, J., and Pinto da Silva, P. (1987). Osmotic reversal induces assembly of tight junction strands at the basal pole of toad bladder epithelial cells but does not reverse cell polarity. *J. Memb. Biol. 95*: 199.

Citi, S., Amorosi, A., Franconi, F., Giotti, A., and Zampi, G. (1991). Cingulin, a specific protein component of tight junctions, is expressed in normal and neoplastic human epithelial tissues. *Am. J. Pathol. 138*: 781.

Claude, P. (1978). Morphologic factors influencing transepithelial permeability: a model for the resistance of the zonula occludens. *J. Memb. Biol. 39*: 219.

Claude, P., and Goodenough, D. A. (1973). Fracture faces of zonulae occludentes from "tight" and "leaky" epithelia. *J. Cell. Biol. 58*: 390.

Diamond, J. M. (1977). The epithelial junction: bridge, gate and fence. *Physiologist 20*: 10.

Dove, C. H., Wang, S. Z., Price, S. B., Phelps, C. J., Lyerly, D. M., Wilkens, T. D., and Johnson, J. L. (1990). Molecular characterization of the *Clostridium difficile* toxin A gene. *Infect. Immun. 58*: 480.

Duffey, M. E., Hainou, B., Ho, S., and Bentzel, C. J. (1981). Regulation of epithelial tight junction permeability by cyclic AMP. *Nature 294*: 451.

Eisenberg, D., Schwarz, E., Komaromy, M., and Wall, R. (1984). Analysis of membrane and surface protein sequences with the hydrophobic moment plot. *J. Mol. Biol. 179*: 125.

Faruque, S. M., Kaper, J. B., and Albert, M. J. (1993). Prevalence of zonula occludens toxin (*zot*) gene among clinical isolates of *Vibrio cholerae* O1 in Bangladesh. *J. Diarrhoeal Diseases Research*.

Fasano, A., Baudry, B., Pumplin, D. W., Wasserman, S. S., Tall, B. D., Ketley, J. M., and Kaper, J. B. (1991). *Vibrio cholerae* produces a second enterotoxin, which affects intestinal tight junctions. *Proc. Natl. Acad. Sci. U.S.A. 88*: 5242.

Field, M., Fromm, D., and McColl, I. (1971). Ion transport in rabbit ileal mucosa. I. Na and Cl fluxes and short-circuit current. *Am. J. Physiol. 220*: 1388.

Field, M., Fromm, D., Al-Awqati, Q., and Greenenough, W. B. (1972). Effect of cholera enterotoxin on ion transport across isolated ileal mucosa. *J. Clin. Invest. 51*: 796.

Fiorentini, C., Malorni, W., Paradisi, S., Giuliano, M., Mastrantonio, P., and Donelli, G. (1990). Interaction of *Clostridium difficile* toxin A with cultured cells: cytoskeletal changes and nuclear polarization. *Infect. Immun. 58*: 2329.

Freer, J. H., and Birkbeck, T. H. (1982). Possible conformation of delta-lysin, a membrane-damaging peptide of *Staphylococcus aureus*. *J. Theor. Biol. 94*: 535.

Frizzell, R. A., and Schultz, S. G. (1972). Ionic conductance of extracellular shunt pathway in rabbit ileum. *J. Gen. Physiol. 59*: 318.

Gill, D. M., and King, C. A. (1975). The mechanism of action of cholera toxin in pigeon erythrocyte lysates. *J. Biol. Chem. 250*: 6424.

Goldberg, I., and Mekalanos, J. J. (1986). Effect of a recA mutation on cholera toxin gene amplification and deletion events. *J. Bacteriol. 165*: 723.

Goldberg, S. L., and Murphy, J. R. (1984). Molecular cloning of the hemolysin determinant from *Vibrio cholerae* El Tor. *J. Bacteriol. 160*: 239.

Gumbiner, B., Lowenkopf, T., and Apatira, D. (1991). Identification of a 160-kDa polypeptide that binds to the tight junction protein ZO-1. *Proc. Natl. Acad. Sci. U.S.A. 88*: 3460.

Hall, R. H., and Drasar, B. S. (1990). *Vibrio cholerae hlyA* hemolysin is processed by proteolysis. *Infect. Immun. 58*: 3375.

Hecht, G., Pothoulakis, C., LaMont, J. T., and Madara, J. L. (1988). *Clostridium difficile* toxin A perturbs cytoskeletal structure and tight junction permeability of the cultured human intestine epithelium monolayers. *J. Clin. Invest. 82*: 1516.

Herrington, D. A., Hall, R. H., Losonsky, G., Mekalanos, J. J., Taylor, R. K., and Levine, M. M. (1988). Toxin, toxin-coregulated pili, and the *toxR* regulon are essential for *Vibrio cholerae* pathogenesis in humans. *J. Exp. Med. 168*: 1487.

Honda, T., and Finkelstein, R. A. (1979a). Purification and characterization of a hemolysin produced by *Vibrio cholerae* biotype El Tor: another toxic substance produced by cholera vibrios. *Infect. Immun. 26*: 1020.

Honda, T., and Finkelstein, R. A. (1979b). Selection and characteristics of a *Vibrio cholerae* mutant lacking the A (ADP-ribosylating) portion of the cholera enterotoxin. *Proc. Natl. Acad. Sci. U.S.A.* 76: 2052.

Ichinose, Y., Yamamoto, K., Nakasone, N., Tanabe, M. J., Takeda, T., Miwatani, T., and Iwanaga, M. (1987). Enterotoxicity of El Tor hemolysin of non-O1 *Vibrio cholerae*. *Infect. Immun.* 55: 1090.

Janmey, P. A., Chaponnier, C., Lind, S. E., Zaner, K. S., Stossel, T. P., and Yin, H. L. (1985). Interactions of gelsolin and gelsolin-actin complexes with actin. Effects of calcium on actin nucleation, filament severing and end blocking. *Biochemistry* 24: 3714.

Johnson, J. A., Morris, J. G., Jr., and Kaper, J. B. (1993). Gene encoding zonula occludens toxin (*zot*) does not occur independently from cholera enterotoxin genes (*ctx*) in *Vibrio cholerae*. *J. Clin. Microbiol.* 31(3): 732.

Kaper, J. B., Lockman, H., Baldini, M. M., and Levine, M. M. (1984a). A recombinant live oral cholera vaccine. *Bio/Technology* 2: 345.

Kaper, J. B., Lockman, H., Baldini, M. M., and Levine, M. M. (1984b). Recombinant nontoxinogenic *Vibrio cholerae* strains as attenuated cholera vaccine candidates. *Nature 308*: 655.

Kaper, J. B., Mobley, H. L. T., and Michalski, J. M. (1988). Recent advances in developing a safe and effective live oral attenuated *Vibrio cholerae* vaccine. In *Advances in Research on Cholera and Related Diarrheas*, N. Ohtomo and R. B. Sack (Eds.), KTK Scientific Publishers, Tokyo, pp. 161–167.

Karasawa, T., Mihara, T., Kurazono, H., Nair, G. B., Garg, S., Ramamurthy, T., and Takeda, Y. (1993). Distribution of the *zot* (zonula occludens toxin) gene among strains of *Vibrio cholerae* O1 and non-O1. *FEMS Microbiol. Lett.* 106(2): 143.

Koonin, E. V. (1992). The second cholera toxin, Zot, and its plasmid-encoded and phage-encoded homologues constitute a group of putative ATPases with an altered purine NTP-binding motif. *FEBS Lett.* 312: 3.

Lazaro, A., Calderon, V., Gonzalez-Mariscal, L., Contreras, R. G., Valdez, J., Zampighi, G., and Cereijido, M. (1991). Tight junctions, lipids, and transepithelial electrical resistance (TER). *J. Cell Biol.* 115: 481a.

Levine, M. M. (1980). Immunity to cholera as evaluated in volunteers. In *Cholera and Related Diarrheas*, O. Ouchterlony and J. Holmgren (Eds.), Karger, Basel, pp. 195–203.

Levine, M. M., Black, R. E., Clements, M. L., Lanata, C., Sears, S., Honda, T., Young, C. R., and Finkelstein, R. A. (1984). Evaluation in humans of attenuated *Vibrio cholerae* El Tor Ogawa strain Texas Star-SR as a live oral vaccine. *Infect. Immun.* 43: 515.

Levine, M. M., Kaper, J. B., Herrington, D., Losonsky, G., Morris, J. G., Clements, M. L., Black, R. E., Tall, B., and Hall, R. (1988). Volunteer studies of deletion mutants of *Vibrio cholerae* O1 prepared by recombinant techniques. *Infect. Immun.* 56: 161.

Lockman, H., and Kaper, J. B. (1983). Nucleotide sequence analysis of the A2 and B subunits of *Vibrio cholerae* enterotoxin. *J. Biol. Chem.* 258: 13722.

Madara, J. L., and Dharmsathaphorn, K. (1985). Occluding junction structure–function relationships in cultured epithelial monolayer. *J. Cell Biol.* 101: 2124.

Madara, J. L., and Pappenheimer, J. R. (1987). The structural basis for physiological regulation of paracellular pathways in intestinal epithelia. *J. Memb. Biol.* 100: 149.

Madara, J. L., Barenberg, D., and Carlson, S. (1986). Effects of cytochalasin D on occluding junctions of intestinal absorptive cells: further evidence that the cytoskeleton may influence paracellular permeability and junctional charge selectivity. *J. Cell Biol.* 102: 2125.

Madara, J. L., Stafford, J., Barenberg, D., and Carlson, S. (1988). Functional coupling of tight junctions and microfilaments in T84 monolayers. *Am. J. Physiol.* 254: G416.

Manning, P. A., Brown, M. H., and Heuzenroeder, M. W. (1984). Cloning of the structural gene (*hly*) for the haemolysin of *Vibrio cholerae* El Tor strain O17. *Gene 31*: 225.

Martinez-Palomo, W., Meza, I., Beaty, G., and Cereijido, M. (1980). Experimental modulation of occluding junctions in a cultured transporting epithelium. *J. Cell Biol.* 87: 736.

Mekalanos, J. J., Swartz, D. J., Pearson, G. D., Harford, N., Groyne, F., and de Wilde, M. (1983). Cholera toxin genes: nucleotide sequence, deletion analysis and vaccine development. *Nature 306*: 551.

Mellor, I. R., Thomas, D. H., and Sansom, M. S. P. (1988). Properties of ion channels formed by *Staphylococcus aureus* delta-toxin. *Biochim. Biophys. Acta 942*: 280.

Meza, I., Sabanero, M., Stefani, E., and Cereijido, M. (1982). Occluding junctions in MDCK cells: modulation of transepithelial permeability by the cytoskeleton. *J. Cell. Biochem. 18*: 407.

Michalski, J., Galen, J. E., Fasano, A., and Kaper, J. B. (1993). CVD110: an attenuated *Vibrio cholerae* O1 El Tor live oral vaccine strain. *Infect. Immun. 61*: 4462.

Miller, V. L., Taylor, R. K., and Mekalanos, J. J. (1987). Cholera toxin transcriptional activator ToxR is a transmembrane DNA binding protein. *Cell 48*: 271.

Moore, R., Pothoulakis, C., LaMont, J. T., Carlson, S., and Madara, J. (1990). *C. difficile* toxin A increases intestinal permeability and induces Cl$^-$ secretion. *Am. J. Physiol. 259*: G165.

Mooseker, M. S. (1983). Actin binding proteins of the brush order. *Cell 35*: 11.

Morris, J. G., Jr., Takeda, T., Tall, B. D., Losonsky, G. A., Bhattacharya, S. K., Forrest, B. D., Kay, B. A., and Nishibuchi, M. (1990). Experimental non-O1 group 1 *Vibrio cholerae* gastroenteritis in humans. *J. Clin. Invest. 85*: 697.

Mullin, J. M., and O'Brien, T. (1986). Effects of tumor promoters on LLC-pk, renal epithelial tight junctions and transepithelial fluxes. *Am. J. Physiol. 251*: C597.

O'Brien, A. D., Chen, M. E., Holmes, R. K., Kaper, J., and Levine, M. M. (1984). Environmental and human isolates of *Vibrio cholerae* and *Vibrio parahaemolyticus* produce a *Shigella dysenteriae* 1 (Shiga)-like cytotoxin. *Lancet 1*: 77.

Palant, C. E., Duffey, M. E., Mookerjee, B. K., Ho, S., and Bentzel, C. J. (1983). Ca regulation of tight junction permeability and structure in *Necturus* gall bladder. *Am. J. Physiol. 245*: C203.

Peterson, J. W., Hejmancik, K. E., Markel, D. E., Craig, J. P., and Kurosky, A. (1979). Antigenic specificity of neutralizing antibody to cholera toxin. *Infect. Immun. 24*: 774.

Phillips, T. E., Phillips, T. L., and Neutra, M. R. (1987). Macromolecules can pass through occluding junctions of rat ileal epithelium during cholinergic stimulation. *Cell Tissue Res. 247*: 547.

Pinto da Silva, P., and Kachar, B. (1982). On tight junction structure. *Cell 28*: 441.

Powell, D. W. (1981). Barrier function of epithelia. *Am. J. Physiol. 241*: G275–G288.

Richardson, K., Michalski, J., and Kaper, J. B. (1986). Hemolysin production and cloning of two hemolysin determinants from classical *Vibrio cholerae*. *Infect. Immun. 54*: 415.

Stevenson, B. R., and Goodenough, D. A. (1984). Zonulae occludentes in junctional complex-enriched fractions from mouse liver: preliminary morphological and biochemical characterization. *J. Cell Biol. 98*: 1209.

Stevenson, B. R., Siliciano, J. D., Mooseker, M. S., and Goodenough, D. A. (1986). Identification of ZO-1: a high molecular weight polypeptide associated with tight junction (zonula occludens) in a variety of epithelia. *J. Cell Biol. 103*: 755.

Svennerholm, A.-M. (1980). The nature of protective immunity in cholera. In *Cholera and Related Diseases, 43rd Nobel Symposium*, O. Ouchterlony and J. Holmgren (Eds.), Karger, Basel, pp. 171–184.

Tacket, C. O., Losonsky, G., Nataro, J. P., Cryz, S. J., Edelman, R., Fasano, A., Michalski, J., Kaper, J. B., and Levine, M. M. (1993). Safety and Immunogenicity of live oral cholera vaccine candidate CVD110, a ΔctxA Δzot Δace derivative of El Tor Ogawa *Vibrio cholerae*. *J. Infect. Dis. 168*: 1536.

Trucksis, M., Galen, J. E., Michalski, J., Fasano, A., and Kaper, J. B. (1993). Accessory cholera enterotoxin (Ace), the third toxin of a *Vibrio cholerae* virulence cassette. *Proc. Natl. Acad. Sci. U.S.A. 90*: 5267.

Van Meer, G. (1989). Polarity and polarized transport of membrane lipids in a cultured epithelium. In *Functional Epithelial Cells in Culture*, K. S. Matlin and J. D. Valentich (Eds.), Liss, New York, pp. 43–69.

Verkleij, A. J. (1984). Lipidic intramembranous particles. *Biochim. Biophys. Acta 779*: 43.

Zhong, Y., Saitoh, T., Minase, T., Sawada, N., Enomoto, K., and Mori, M. (1993). Monoclonal antibody 7H6 reacts with a novel tight junction-associated protein distinct from ZO-1, cingulin and ZO-2. *J. Cell Biol. 120*(2): 477.

14

Shiga and Shiga-like (Vero) Toxins

Yoshifumi Takeda*

Kyoto University, Yoshida, Kyoto, Japan

I. INTRODUCTION

The discovery of *Shigella dysenteriae* was reported by Kiyoshi Shiga in Japanese (Shiga, 1897) and then in German (Shiga, 1898). Soon after this report, several investigators attempted to isolate the toxin from the organism (Flexner, 1900; Neisser and Shiga, 1903; Conradi, 1903). Although extensive progress was achieved in characterizing the toxin produced by *Shigella dysenteriae* (Olitsky and Kligler, 1920; van Heyningen and Gladstone, 1953; Keusch et al., 1972), highly purified Shiga toxin was obtained only about 10 years ago (Olsnes and Eiklid, 1980; O'Brien et al., 1980).

O'Brien et al. (1982) discovered that the ability of a cytotoxin produced by a strain of *Escherichia coli* O26 to kill a line of HeLa cells was neutralized by anti-Shiga toxin antiserum. Around the same time, in 1982, *E. coli* O157:H7, named enterohemorrhagic *E. coli* (Levine, 1987), was isolated as a causative bacterium of food poisoning due to

---

Current affiliation: Research Institute, International Medical Center of Japan, Toyama, Shinjuku-ku, Tokyo, Japan

the consumption of contaminated hamburgers in the United States (Riley et al., 1983). O'Brien et al. (1983) found that the cytotoxin produced by enterohemorrhagic *E. coli* O157:H7 reported by Johnson et al. (1983) was also neutralized by anti-Shiga toxin antiserum. Thus, O'Brien et al. (1983) proposed the name Shiga-like toxin for the cytotoxin produced by enterohemorrhagic *E. coli*.

Independent of these studies, Konowalchuk et al. (1977) reported that *E. coli* strains of serotypes O18, O26, O68, O111, and O138 produced a cytotoxin against Vero cells and named it Verocytotoxin. This finding was confirmed by Wade et al. (1979) and Scotland et al. (1980), but these findings remained virtually unnoticed until the unique outbreak of *E. coli* food poisoning mentioned above occurred (Riley et al., 1983). Johnson et al. (1983) found that enterohemorrhagic *E. coli* O157:H7 isolated from a patient produced the Verotoxin previously reported by Konowalchuk et al. (1977).

Thus, two names, Shiga-like toxin and Verocytotoxin (or Verotoxin), have been used for the same toxin, one that is produced by some *E. coli* strains and is cytotoxic to HeLa and Vero cells.

II. NOMENCLATURE OF THE ORGANISM AND THE TOXIN

The *E. coli* that produces the toxin discussed in this chapter was named enterohemorrhagic *E. coli* by Levine (1987). This was based on the assumption that this organism causes the disease in humans and thus followed the names of the previously described diarrheagenic *E. coli*, enteropathogenic *E. coli*, enteroinvasive *E. coli*, and enterotoxigenic *E. coli*.

However, later it was found that a toxin belonging to the same family is a causative agent of the edema disease of pigs. Thus the name Verocytoxin (Verotoxin)-producing *E. coli* proposed by Karmali (1989) is now widely accepted.

As mentioned above, the toxin has two names, Shiga-like toxin (SLT) and Verocytotoxin (or Verotoxin, VT). This is unusual and confusing. Although O'Brien et al. (1982, 1983) were the first to report that the toxin produced by *E. coli* is immunologically related to Shiga toxin, as far as the nomenclature of the toxin is concerned, Verocytotoxin (or Verotoxin) (Konowalchuk et al., 1977) has priority.

SLT-I (VT1) was first purified and characterized by O'Brien and LaVeck (1982). SLT-II (VT2) was first described by Scotland et al. (1985) and purified by Yutsudo et al. (1987). Consequently, several variants of SLT-II (VT2) were reported, and confusing names were proposed for each variant. The first variant reported was SLT-IIv by Marques et al. (1987). As this toxin was from the pig strain causing edema disease, it was later named VTe by Johnson et al. (1990) and VT2vpl by Lin et al. (1993a). Subsequently, from *E. coli* strains isolated from a patient with hemolytic uremic syndrome, two variants of VT2 were isolated and named VT2vha and VT2vhb, respectively (Ito et al., 1990). Later, new names, VT2v-a and VT2v-b, respectively, were proposed for these variants (Tyler et al., 1991). Another variant of SLT-II was described by Gannon et al. (1990). This was reported to be most closely related to SLT-IIv (VTe, VT2vpl), thus the name SLT-IIva was proposed. As Lin et al. (1993a) proposed the name VT2vpl for SLT-IIv, they designated SLT-IIva as VT2vp2. As will be discussed below, several more variants of SLT-II (VT2) have been reported.

A systematic nomenclature for SLT-II (VT2) variants is urgently needed. Numbering in order of the finding would be appropriate. A question would be whether the variants could be named according to their source of isolation. If we accept the nomenclature

VT2vh and VT2vp, VT2vha and VT2vhb can be renamed VT2vh1 and VT2vh2, respectively. The SLT-IIc reported by Schmitt et al. (1991) would be VT2vh3. VT2vp1 and VT2vp2 are as discussed above, but VT2vp2 originated from humans, not from pigs.

III. MOLECULAR STRUCTURE

Extensive efforts have been made to purify Shiga toxin, following the early work on Shiga toxin by several investigators (Olitsky and Kligler, 1920; van Heyningen and Gladstone, 1953; Keusch et al., 1972), without significant success. Highly purified toxin was obtained in 1980 by Olsnes and Eiklid (1980) and by O'Brien et al. (1980). Subsequently a variety of purification procedures were applied to obtain significantly large quantities of purified Shiga toxin that include chromatofocusing column chromatography (Yutsudo et al., 1986), receptor molecule mediated affinity chromatography (Donohue-Rolfe et al., 1989a), and immunoaffinity chromatography (Donohue-Rolfe et al., 1984; Kongmuaug et al., 1987).

The subunit structure of Shiga toxin was first noted by Olsnes et al. (1981). One molecule of the A subunit (about 31 kDa) and five molecules of B subunit (about 7 kDa) form a holotoxin of about 66 kDa (Donohue-Rolfe et al., 1989b). The A subunit is enzymatically active and inhibits protein synthesis in eukaryotic cells (Reisbig et al., 1981), whereas the B subunits bind to receptor molecules of the target cells (Mobassaleh et al., 1988).

There are two types of Verotoxin, VT1 (SLT-I) (O'Brien et al., 1982; Scotland et al., 1985; Strockbine et al., 1986; Calderwood et al., 1987; Jackson et al., 1987a,b; DeGrandis et al., 1987; Noda et al., 1987; Takao et al., 1988) and VT2 (or SLT-II) (Scotland et al., 1985; Strockbine et al., 1988; Jackson et al., 1987a; Newland et al., 1987; Yutsudo et al., 1987; Nishibuchi et al., 1994), and several variants of VT2, such as SLT-IIv (VTe, VT2vpl) (Marques et al., 1987; Gyles et al., 1988; Weinstein et al., 1988; MacLeod and Gyles, 1990; Lin et al., 1993a; Cao et al., 1994b), VT2vha (VT2v-a) (Oku et al., 1989; Ito et al., 1990), VT2vhb (VT2v-b) (Oku et al., 1989; Ito et al., 1990; Paton et al., 1993b), SLT-IIva (VT2vp2) (Gannon et al., 1990; Lin et al., 1993a; Ohmura et al., 1993a), SLT-IIc (Schmitt et al., 1991), and others (Meyer et al., 1992; Paton et al., 1992; Lin et al., 1993b), have been reported.

Among these Verotoxins, VT1 (SLT-I) (O'Brien and LaVeck, 1983; Noda et al., 1987; Donohue-Rolfe et al., 1989a), VT2 (SLT-II) (Yutsudo et al., 1987; Head et al., 1988; Donohue-Rolfe et al., 1989a), VTvh (Oku et al., 1989), SLT-IIv (VT2vpl) (MacLeod and Gyles, 1990; Cao et al., 1994b), and SLT-IIva (VT2vp2) (Ohmura et al., 1993a) have been purified. Like Shiga toxin, the purified Verotoxins consist of one enzymatically active A subunit and five B subunits that bind to a receptor. The molecular masses of the A and B subunits are about 33,000 and 7500 Da, respectively, but differ in each VT as summarized in Table 1.

E. coli clones harboring the gene for the B subunit of SLT-I were independently constructed by Calderwood et al. (1990) and Ramotar et al. (1990). Using a preparation purified from the cloned strain as described by Donohue-Rolfe et al. (1989b), Hart et al. (1991) crystallized the SLT-I B subunit and structurally analyzed it by X-ray crystallography. Stein et al. (1992) also crystallized SLT-IB, and Sixma et al. (1993) compared it with crystals of the B subunit of heat-labile enterotoxin produced by enterotoxigeic *E. coli*. Further crystallization of the holotoxins may elucidate the stoichiometric structures of these toxins.

TABLE 1 Number of Amino Acid Residues and Molecular Weight of Various Verotoxins

	A subunit		B subunit	
	Number of amino acids	Mol wt	Number of amino acids	Mol wt
VT1	293	32226	69	7690
VT2	297	33193	70	7817
VT2vha	297	33147	70	7772
VT2vhb	297	33165	70	7772
VT2vp1	297	33068	68	7583
VT2vp2	297	33090	68	7555

Source: Data from Takao et al. (1988) for VT1, from Nishibuchi et al. (1994) for VT2, from Ito et al. (1990) for VT2vha and VT2vhb, from Weinstein et al. (1988) for VT2vp1, and from Gannon et al. (1990) for VT2vp2.

The structures of the genes for Shiga and Shiga-like (Vero) toxins have also been intensively studied. Strockbine et al. (1988) cloned and sequenced the structural gene for Shiga toxin. It was found that, like the cholera toxin gene, the gene for the A subunit was located upstream of that for the B subunit. The nucleotide sequence of the VT1 gene was first reported by Calderwood et al. (1987), and the sequence was confirmed by Jackson et al. (1987b) and DeGrandis et al. (1987). These investigators reported that there were three nucleotide differences in three different codons of the A subunit between the Shiga toxin and VT1 genes that result in a one-amino-acid difference between Shiga toxin and VT1 (the amino acid at position 45 from the N terminus was serine in Shiga toxin and threonine in VT1). However, Takao et al. (1988) demonstrated that the nucleotide sequence of VT1 was identical to that of Shiga toxin. Furthermore, Takao et al. (1988) confirmed this by analyzing the amino acid sequence of the A subunit of VT1. Thus, it is assumed that the structure of the VT1 prototype is exactly the same as that of Shiga toxin and that there are some strains that have a one-amino-acid difference. Paton et al. (1993a) reported a variant of SLT-1 that differed in three nucleotides, resulting in a one-amino-acid difference in the A subunit. The amino acid at position 45 from the N terminus of the A subunit was the same as that of the Shiga toxin.

The nucleotide sequences of VT2 and various VT2 variants were also determined (Jackson et al., 1987a; Weinstein et al., 1988; Gyles et al., 1988; Ito et al., 1990; Lin et al., 1993b; Nishibuchi et al., 1994). The overall nucleotide sequence homology of the VT2 family of genes with that of the VT1 gene is about 55–60%.

The genes encoding VT1 and VT2 are carried by phages. Williams-Smith et al. (1983) were the first to recognize this in *E. coli* O26:H19. They found two distinct phages, H19A and H19B. O'Brien et al. (1984) reported a similar finding with *E. coli* O157:H7 and designated two phages as 933W and 933J. No evidence is yet available that genes for the various VT2 variants are phage-mediated.

IV. BIOLOGICAL ACTIVITIES

The biological activities of Shiga toxin have been well characterized (Keusch et al., 1986). It has cytotoxicity to Vero cells and a certain line of HeLa cells, lethal toxicity to mice and other small experimental animals, and enterotoxicity that causes fluid accumulation in the rabbit ileal loop. These activities are expressed by a single protein, Shiga toxin (Keusch and Jacewiez, 1975; Eiklid and Olsnes, 1983). Shiga and Verotoxins share the same biological properties, and VT1, VT2, and VT2 variants exert essentially the same activities as Shiga toxin. The toxicity of the purified Shiga and Verotoxins reported by several investigators differed considerably (Yutsudo et al., 1986, 1987; Noda et al., 1987; Oku et al., 1989; MacLeod and Gyles, 1990; Samuel et al., 1990; Ohmura et al., 1993a; Cao et al., 1994a), probably because different batches were studied and the animals and assays also differed. Table 2 summarizes the biological activities of the purified Shiga and Vero toxins studied in our laboratory (Yutsudo et al., 1986, 1987; Noda et al., 1987; Oku et al., 1989; Ohmura et al., 1993a; Cao et al., 1994a). Although there are some differences, it can be concluded that these toxins all have similar biological activities.

TABLE 2 Biological Activities of Various Verotoxins

Toxin	Vero cell cytotoxicity CD_{50} (pg)	Mouse lethality LD_{50} (ng)
Shiga toxin	1.0	27
VT1	1.0	30
VT2	0.75	1
VT2vh	6	2.7
VT2vp1	6	22
VT2vp2	20	55

Source: Data are from Yutsudo et al. (1986) for Shiga toxin, from Noda et al. (1987) for VT1, from Yutsudo et al. (1987) for VT2, from Oku et al. (1989) for VT2vh, from Ohmura et al. (1993a) for VT2vp1, and from Cao et al. (1994a) for VT2vp2.

V. MOLECULAR MECHANISM OF ACTION

Inhibition of protein synthesis in eukaryotic cells by Shiga toxin was first reported by Thompson et al. (1976) and confirmed by Brown and coworkers (1980a,b). Further studies revealed that Shiga toxin inactivates 60 S ribosomal subunits (Reisbig et al., 1981), which results in the inhibition of elongation factor 1 (EF-1)-dependent aminoacyl-tRNA binding to the 60 S ribosomal subunits (Brown et al., 1986). The same mode of action was subsequently demonstrated for VT1 and VT2 (Igarashi et al., 1987; Ogasawara et al., 1988); both toxins inactivate 60 S ribosomal subunits and inhibit EF-1-dependent aminoacyl-tRNA binding.

In 1987, Endo and coworkers (Endo and Tsurugi, 1987; Endo et al., 1987) demonstrated that a plant lectin, ricin, had RNA *N*-glycosidase activity and cleaved the *N*-glycosidic bond of an adenosine at position 4 324 from the 5′ terminal of the 28 S ribosomal RNA of the 60 S ribosomal subunit of eukaryotic cells, which resulted in the inhibition of protein synthesis. Since the similarity in the mode of action of Shiga toxin to that of ricin, a ribosome-inactivating protein, has long been recognized, Endo et al. (1988) studied the molecular mode of action of Shiga toxin and VT2. They found that both toxins exhibited RNA N-glycosidase activity and cleaved the N-glycosidic bond of the adenosine at position 4 324 from the 5′ terminal of the 28 S ribosomal RNA of 60 S ribosomal subunit in rabbit reticulocytes (Endo et al., 1988). It was subsequently demonstrated that SLT-IIv (VT2vpl) (Saxena et al., 1989) and VT2vh (Furutani et al., 1990) also showed the RNA N-glycosidase activity. Although it has not been demonstrated in all Verotoxin variants, it can be concluded that Shiga and Shiga-like (Vero) toxins have RNA-N-glycosidase activity and act enzymatically on the target cells. This is consistent with the finding that a few molecules of Shiga toxin are sufficient to kill target cells (Olsnes et al., 1981).

The demonstration of exactly the same molecular mode of action of Shiga toxin, Verotoxins, and ricin suggested that a putative active center for the enzymatic activity in the A subunit of Shiga and Verotoxins could be identified by comparing their amino acid sequences. Thus the amino acid sequence analyzed either directly or deduced from the nucleotide sequence of the A subunits of Shiga toxin (Strockbine et al., 1988), VT1 (SLT-I) (Calderwood et al., 1987; Jackson et al., 1987a,b; DeGrandis et al., 1987; Takao et al., 1988), VT2 (SLT-II) (Jackson et al., 1987a; Nishibuchi et al., 1994), VT2vpl (SLT-IIv) (Weinstein et al., 1988; Gyles et al., 1988), VT2vha (Ito et al., 1990), and VT2vhb (Ito et al., 1990), were compared (Hovde et al., 1988; Yamasaki et al., 1991) with that of ricin (Halling et al., 1985; Lamb and Roberts, 1985). Although the overall homology of the A subunits between Shiga toxin or Verotoxins and ricin is only about 20%, there are three regions that have more than 80% homology containing conserved (identical or chemically similar) amino acid residues (Yamasaki et al., 1991). These regions were located in the catalytic site of the A chain of ricin by X-ray crystal diffraction (Montfort et al., 1987; Ready et al., 1988), suggesting that they form an active center for the enzymatic activity. Thus, several investigators attempted to substitute amino acid residues in these regions by site-directed mutagenesis of the gene. Hovde et al. (1988) replaced glutamic acid at position 167 from the N terminus of the A subunit of VT1 (Glu-167) by aspartic acid and demonstrated the importance of Glu-167 for its toxicity. Jackson et al. (1990a) also demonstrated the similar importance of Glu-167 of VT2.

Yamasaki et al. (1991) constructed the 22 mutant genes listed in Table 3 that encoded mutant VT1's with a single amino acid replacement of 14 amino acid residues in the three regions described above. Among these, two mutants that had Glu-167 replaced by glutamine (E167Q) and Arg-170 by leucine (R170L) had significantly reduced activity both in cytotoxicity to Vero cells and in inhibiting protein synthesis in the rabbit reticulocyte lysate system (Table 3) (Yamasaki et al., 1991). Ohmura et al. (1993b) further constructed a mutant gene of VT1 that encoded a mutant toxin in which Glu-167 and Arg-170 were replaced by glutamine and leucine, respectively (E167Q-R170L), and demonstrated reduced biological activities. Similar mutants of SLT-IIv (VT2vpl), E167Q, E167D, and E167D-R170K with reduced toxic activities were prepared by Gordon et al. (1992). Cao et al. (1994a) also manipulated the VT2vpl gene to construct mutant genes for E167Q, R170L, and E167Q-R170L and showed that these mutant VT2vpl's were less toxic. Furthermore, another relatively nontoxic mutant VT1, Y77S, has recently been reported (Deresiewicz et al., 1992).

Several mutant VT1's (E167Q, R170L, and E167Q-R170L) and mutant VT2vpl's (E167Q, R170L, and E167Q-R170L) were purified, and their activities were examined (Ohmura et al., 1993b; Cao et al., 1994a). As shown in Table 4, the inhibition of protein synthesis, cytotoxicity to Vero cells, and lethal toxicity in the mouse were all markedly decreased in the purified E167Q and E167Q-R170L. However, the decrease in these

TABLE 3 Activities of Mutant VT1 Prepared by
Site-Directed Mutagenesis[a]

Mutant[b]	Vero cell cytotoxicity	Inhibition of protein synthesis
VT1	1	1
Asp-53→Ala	2	10
Arg-55→Ile	1	1
Arg-55→Lys	1/2	1/2
Glu-167→Leu	1/10	1/40
Glu-167→Gln	1/2000	1/400
Ala-168→Gly	1/3	1/2
Leu-169→Ile	1/3	1/3
Arg-170→His	1/100	1/70
Arg-170→Lys	1/200	1/100
Arg-170→Leu	1/1000	1/200
Phe-171→Tyr	1	1/3
Arg-172→Leu	1/100	1/70
Arg-172→Lys	1/5	1/15
Asn-202→Asp	1/8	1/30
Trp-203→Phe	1/10	1/60
Trp-203→Leu	1/50	1/25
Trp-203→His	1/50	1/65
Gly-204→Ala	1	1
Arg-205→Lys	1	1
Arg-205→Thr	1/10	1/10
Leu-206→Val	1	1/6
Ser-207→Ala	1/4	1/3

[a]The activity of VT1 is represented as 1.
[b]Sites and mutated amino acids are indicated.
Source: Data from Yamasaki et al. (1991).

activities of purified R170L was less than those of the other two mutants. Since mutant
VT1's and VT2vpl's possess antigenicities similar to those of the respective wild types
(Ohmura et al., 1993b; Cao et al., 1994a), they may be used as potential toxoids to protect
VT-mediated diseases as suggested by Gordon et al. (1992).

VI. RECEPTORS

Jacewicz et al. (1986) were the first to report that the B subunit of Shiga toxin binds to
globotriosylceramide Gb3 (galactose α 1→4 galactose β1→4 glucose-ceramide). Lindberg
et al. (1987) also reported the binding of Shiga toxin to a P1 blood group glycolipid that
has a terminal galactose α 1→4 galactose residue. The evidence that the age-dependent
susceptibility of the rabbit to Shiga toxin correlated with the Gb3 content in rabbit jejunal
microvillus membranes (Mobassaleh et al., 1988) and that cultured cell lines that are
sensitive to Shiga toxin contain Gb3 whereas those that are insensitive do not (Jacewicz
et al., 1986, 1989) confirmed that the receptor of the Shiga toxin is globotriosylceramide
Gb3.

VT1 binds to globotriosylceramide Gb3 (Cohen et al., 1987; Lingwood et al., 1987).

TABLE 4 Biological Activities of Purified VT1 and VT2vp1 Mutants and Their Respective Wild Types

Toxin	Cytotoxic activity CD_{50} (mg)	Mouse lethality LD_{50} (μg)	Inhibition of protein synthesis ED_{50} (μg)
VT1			
Wild type	0.0001	0.04	0.008
E167Q-R170L	30	60	22
E167Q	30	80	>100
R170L	9	10	2
VT2vp1			
Wild type	0.006	0.025	0.009
E167Q-R170L	750	50	17
E167Q	75	12	9
R170L	30	3	7

Source: Data are from Ohmura et al. (1993b) for VT1 and its mutants, and from Cao et al. (1994a) for VT2vp1 and its mutants.

This is obvious as the primary structure of the B subunits of Shiga toxin and VT1 are essentially the same (Calderwood et al., 1987; Jackson et al., 1987a,b; DeGrandis et al., 1987; Takao et al., 1988). The receptor for VT2 was also demonstrated to be Gb3 (Cohen et al., 1987; Lingwood et al., 1987; Waddell et al., 1988; DeGrandis et al., 1989). On the other hand, VT2vpl was found to bind to globotetraosylceramide Gb4 in addition to Gb3 (DeGrandis et al., 1989; Samuel et al., 1990).

The binding of VT2 and VT2vpl to different receptors is due to the different primary structures of their B subunits. Thus, site-directed mutagenesis was applied to the VT2 gene to identify the amino acid(s) responsible for the differential binding to Gb3 and Gb4. Jackson et al. (1990b) demonstrated that aspartic acid at position 17 and lysine at position 53 are important for toxin binding to Gb3. Tyrrell et al. (1992) studied VT1 and found that replacing aspartic acid at position 18 with asparagine resulted in binding to both Gb3 and Gb4. They also showed that a double mutant of VT2vpl with replacements consisting of glutamine at position 65 and lysine at position 67 with glutamic acid and glutamine, respectively, prevented binding to Gb4. These currently available data do not provide a definite conclusion as to which epitope(s) or amino acid(s) are important for the binding of the B subunits to the receptors.

REFERENCES

Brown, J. E., Rothman, S. W., and Doctor, B. P. (1980a). Inhibition of protein synthesis in intact HeLa cells by *Shigella dysenteriae* 1 toxin. *Infect. Immun. 29*: 98–107.

Brown, J. E., Ussery, M. A., Leppla, S. H., and Rothman, S. W. (1980b). Inhibition of protein synthesis by Shiga toxin. Activation of the toxin and inhibition of peptide elongation. *FEBS Lett. 177*: 84–88.

Brown, J. E., Obrig, T. G., Ussery, M. A., and Moran, T. P. (1986). Shiga toxin from *Shigella dysenteriae* 1 inhibits protein synthesis in reticulocyte lysate by inactivation of aminoacyl-tRNA binding. *Microb. Pathog. 1*: 325–334.

Calderwood, S. B., Auclair F., Donohue-Rolfe, A., Keusch, G. T., and Mekalanos, J. J. (1987).

Nucleotide sequence of the Shiga-like toxin genes of *Escherichia coli*. *Proc. Natl. Acad. Sci. U.S.A. 84*: 4364–4368.

Calderwood, S. B., Acheson, D. W. K., Goldberg, M. B., Boyko, S. A., and Donohue-Rolfe, A. (1990). A system for production and rapid purification of large amounts of the Shiga toxin/Shiga-like toxin I B subunit. *Infect. Immun. 58*: 2977–2982.

Cao, S., Kurazono, H., Yamasaki, S., and Takeda, Y. (1994a). Construction of mutant genes for a Verotoxin 2 variant (VT2vpl) of *Escherichia coli* and characterization of purified mutant toxins. *Microbiol. Immunol., 38*: 441–447.

Cao, S., Yamasaki, S., Lin, Z., Kurazono, H., and Takeda, Y. (1994b). Specific detection of a Verotoxin 2 variant, VT2vpl, by a bead-enzyme-linked immunosorbent assay. *Microbiol. Immunol., 38*: 435–440.

Cohen, A., Hannigan, G. E., Williams, B. R. G., and Lingwood, C. A. (1987). Roles of globotriosyl- and galabiosylceramide in Verotoxin binding and high affinity interferon receptor. *J. Biol. Chem. 262*: 17088–17091.

Conradi, H. (1903). Über lösliche, durch aseptische Autolyse erhaltene Giftstoffe von Ruhr- und typhus-bazillen. *Deut. Med. Wochschr. 20*: 26–28.

De Grandis, S., Ginsberg, J., Toone, M., Climie, S., Friesen, J., and Brunton, J. (1987). Nucleotide sequence and promoter mapping of the *Escherichia coli* Shiga-like toxin operon of bacteriophage H-19B. *J. Bacteriol. 169*: 4313–4319.

De Grandis, S., Law, H., Brunton, J., Gyles, C., and Lingwood, C. A. (1989). Globotetraosylceramide is recognized by the pig edema disease toxin. *J. Biol. Chem. 264*: 12520–12525.

Deresiewicz, R. L., Calderwood, S. B., Robertus, J. D., and Collier, R. J. (1992). Mutations affecting the activity of the Shiga-like toxin I A-chain. *Biochemistry 31*: 3272–3280.

Donohue-Rolfe, A., Keusch G. T., Edson, C., Thorley-Lawson, D., and Jacewicz, M. (1984). Pathogenesis of *Shigella* diarrhea. IX. Simplified high yield purification of *Shigella* toxin and characterization of subunit composition and function by the use of subunit-specific monoclonal and polyclonal antibodies. *J. Exp. Med. 160*: 1767–1781.

Donohue-Rolfe, A., Acheson, D. W. K., Kane, A. V., and Keusch, G. T. (1989a). Purification of Shiga toxin and Shiga-like toxins I and II by receptor analog affinity chromatography with immobilized P1 glycoprotein and production of cross-reactive monoclonal antibodies. *Infect. Immun. 57*: 3888–3893.

Donohue-Rolfe, A., Jacewicz, M., and Keusch, G. T. (1989b). Isolation and characterization of functional Shiga toxin subunits and renatured holotoxin. *Mol. Microbiol. 3*: 1231–1236.

Eiklid, K., and Olsnes, S. (1983). Animal toxicity of *Shigella dysenteriae* cytotoxin: evidence that the neurotoxic, enterotoxic, and cytotoxic activities are due to one toxin. *J. Immunol. 130*: 380–384.

Endo, Y., and Tsurugi, K. (1987). RNA *N*-glycosidase activity of ricin A-chain. *J. Biol. Chem. 262*: 8128–8130.

Endo, Y., Mitsui, K., Motizuki, M., and Tsurugi, K. (1987). The mechanism of action of ricin and related toxic lectins on eukaryotic ribosomes. *J. Biol. Chem. 262*: 5908–5912.

Endo, Y., Tsurugi, K., Yutsudo, T., Takeda, Y., Ogasawara, T., and Igarashi, K. (1988). Site of action of a Vero toxin (VT2) from *Escherichia coli* O157:H7 and of Shiga toxin on eukaryotic ribosomes. *Eur. J. Biochem. 171*: 45–50.

Flexner, S. (1900). On the etiology of tropical dysentery. *Bull. Johns Hopkins Hosp. 11*: 231–242.

Furutani, M., Ito, K., Oku, Y., Takeda, Y., and Igarashi, K. (1990). Demonstration of RNA *N*-glycosidase activity of a Verotoxin (VT2 variant) produced by *Escherichia coli* O91:H21 from a patient with the hemolytic uremic syndrome. *Microbiol. Immunol. 34*: 387–392.

Gannon, V. P. J., Teerling, C., Masri, S. A., and Gyles, C. L. (1990). Molecular cloning and nucleotide sequence of another variant of the *Escherichia coli* Shiga-like toxin II family. *J. Gen. Microbiol. 36*: 1125–1135.

Gordon, V. M., Whipp, S. C., Moon, H. W., O'Brien, A. D., and Samuel, J. E. (1992). An enzymatic mutant of Shiga-like toxin II variant is a vaccine candidate for edema disease of swine. *Infect. Immun. 60*: 485–490.

Gyles, C. L., De Grandis, S. A., MacKenzie, C., and Brunton, J. L. (1988). Cloning and nucleotide sequence analysis of the genes determining Verocytotoxin production in a porcine edema disease isolate of *Escherichia coli*. *Microb. Pathog. 5*: 419–426.

Halling, K. C., Halling, A. C., Murray, E. E., Ladin, B. F., Houston, L. L., and Weaver, R. F. (1985). Genomic cloning and characterization of a ricin gene from *Ricinus communis*. *Nucleic Acids Res. 13*: 8019–8033.

Hart, P. J., Monzingo, A. F., Donohue-Rolfe, A., Keusch, G. T., Calderwood, S. B., and Robertus, J. D. (1991). Crystallization of the B chain of Shiga-like toxin I from *Escherichia coli*. *J. Mol. Biol. 218*: 691–694.

Head, S. C., Petric, M., Richardson, S., Roscoe, M., and Karmali, M. A. (1988). Purification and characterization of Verocytotoxin 2. *FEMS Microbiol. Lett. 51*: 211–216.

Hovde, C. J., Calderwood, S. B., Mekalanos, J. J., and Collier, R. J. (1988). Evidence that glutamic acid 167 is an active-site residue of Shiga-like toxin I. *Proc. Natl. Acad. Sci. U.S.A. 85*: 2568–2572.

Igarashi, K., Ogasawara, T., Ito, K., Yutsudo, T., and Takeda, Y. (1987). Inhibition of elongation factor 1-dependent aminoacyl-tRNA binding to ribosomes by Shiga-like toxin I (VT1) from *Escherichia coli* O157:H7 and by Shiga toxin. *FEMS Microbiol. Lett. 44*: 91–94.

Ito, H., Terai, A., Kurazono, H., Takeda, Y., and Nishibuchi, M. (1990). Cloning and nucleotide sequencing of Verotoxin 2 variant genes from *Escherichia coli* O91:H21 isolated from a patient with hemolytic uremic syndrome. *Microb. Pathog. 8*: 47–60.

Jacewicz, M., Clausen, H., Nudelman, E., Donohue-Rolfe, A., and Keusch, G. T. (1986). Pathogenesis of *Shigella* diarrhea. XI. Isolation of a shigella toxin-binding glycolipid from rabbit jejunum and HeLa cells and its identification as globotriosylceramide. *J. Exp. Med. 163*: 1391–1404.

Jacewicz, M., Feldman, H. A., Donohue-Rolfe, A., Balasubramanian, K. A., and Keusch, G. T. (1989). Pathogenesis of *Shigella* diarrhea. XIV. Analysis of Shiga toxin receptors on cloned HeLa cells. *J. Infect. Dis. 159*: 881–889.

Jackson, M. P., Neill, R. J., O'Brien, A. D., Holmes, R. K., and Newland, J. W. (1987a). Nucleotide sequence analysis and comparison of the structural genes for Shiga-like toxin I and Shiga-like toxin II encoded by bacteriophages from *Escherichia coli* 933. *FEMS Microbiol. Lett. 44*: 109–114.

Jackson, M. P., Newland, J. W., Holmes, R. K., and O'Brien, A. D. (1987b). Nucleotide sequence analysis of the structural genes for Shiga-like toxin I encoded by bacteriophage 933J from *Escherichia coli*. *Microb. Pathog. 2*: 147–153.

Jackson, M. P., Deresiewicz, R. L., and Calderwood, S. B. (1990a). Mutational analysis of the Shiga toxin and Shiga-like toxin II enzymatic subunits. *J. Bateriol. 172*: 3346–3350.

Jackson, M. P., Wadolkowski, E. A., Weinstein, D. L., Holmes, R. K., and O'Brien, A. D. (1990b). Functional analysis of the Shiga toxin and Shiga-like toxin type II variant binding subunits by using site-directed mutagenesis. *J. Bacteriol. 172*: 653–658.

Johnson, W. M., Lior, H., and Bezanson, G. S. (1983). Cytotoxic *Escherichia coli* O157:H7 associated with hemorrhagic colitis in Canada. *Lancet i*: 76.

Johnson, W. M., Pollard, D. R., Lior, H., Tyler, S. D., and Rozee, K. R. (1990). Differentiation of genes coding for *Escherichia coli* Verotoxin 2 and the Verotoxin associated with porcine edema disease (VTe) by the polymerase chain reaction. *J. Clin. Microbiol. 28*: 2351–2353.

Karmali, M. A. (1989). Infection by Verocytotoxin-producing *Escherichia coli*. *Clin. Microbiol. Rev. 2*: 15–38.

Keusch, G. T., and Jacewicz, M. (1975). The pathogenesis of *Shigella* diarrhea. V. Relationship of Shiga enterotoxin, neurotoxin and cytotoxin. *J. Infect. Dis. 131*: S33–S39.

Keusch, G. T., Grady, G. F., Mata, L. J., and McIver, J. (1972). The pathogenesis of *Shigella* diarrhea. I. Enterotoxin production by *Shigella dysenteriae* 1. *J. Clin. Invest. 51*: 1212–1218.

Keusch, G. T., Donohue-Rolfe, A., and Jacewicz, M. (1986). *Shigella* toxin(s): description and role in diarrhea and dysentery. In *Pharmacology of Bacterial Toxins*, F. Dorner and J. Drews (Eds.), Pergamon, Oxford, pp. 235–270.

Kongmuang, U., Honda, T., and Miwatani, T. (1987). A simple method for purification of Shiga and Shiga-like toxin from *Shigella dysenteriae* and *Escherichia coli* O157:H7 by immunoaffinity column chromatography. *FEMS Microbiol. Lett. 48*: 379–383.

Konowalchuk, J., Speirs, J. I., and Stavric, S. (1977). Vero response to a cytotoxin of *Escherichia coli. Infect. Immun. 18*: 775–779.

Lamb, F. I., Roberts, L. M., and Lord, J. M. (1985). Nucleotide sequence of cloned cDNA coding for preproricin. *Eur. J. Biochem. 148*: 265–270.

Levine, M. M. (1987). *Escherichia coli* that cause diarrhea: enterotoxigenic, enteropathogenic, enteroinvasive, enterohemorrhagic, and enteroadherent. *J. Infect. Dis. 155*: 377–389.

Lin, Z., Kurazono, H., Yamasaki, S., and Takeda, Y. (1993a). Detection of various variant Verotoxin genes in *Escherichia coli* by polymerase chain reaction. *Microbiol. Immunol. 37*: 543–548.

Lin, Z., Yamasaki, S., Kurazono, H., Ohmura, M., Karasawa, T., Inoue, T., Sakamoto, S., Suganami, T., Takeoka, T., Taniguchi, Y., and Takeda, Y. (1993b). Cloning and sequencing of two new Verotoxin 2 variant genes of *Escherichia coli* isolated from cases of human and bovine diarrhea. *Microbiol. Immunol. 37*: 451–459.

Lindberg, A. A., Brown, J. E., Strömberg, N., Westling-Ryd, M., Schultz, J. E., and Karlsson, K.-A. (1987). Identification of the carbohydrate receptor for Shiga toxin produced by *Shigella dysenteriae* type 1. *J. Biol. Chem. 262*: 1779–1785.

Lingwood, C. A., Law, H., Richardson, S., Petric, M., Bruton, J. L., De Grandis, S., and Karmali, M. (1987). Glycolipid binding of purified and recombinant *Escherichia coli* produced Verotoxin in vitro. *J. Biol. Chem. 262*: 8834–8839.

MacLeod, D. L., and Gyles, C. L. (1990). Purification and characterization of an *Escherichia coli* Shiga-like toxin II variant. *Infect. Immun. 58*: 1232–1239.

Marques, L. R. M., Peiris, J. S. M., Cryz, S. J., and O'Brien, A. D. (1987). *Escherichia coli* strains isolated from pigs with edema disease produce a variant of Shiga-like toxin II. *FEMS Microbiol. Lett. 44*: 33–38.

Meyer, T., Karch, H., Hacker, J., Bocklage, H., and Heesemann, J. (1992). Cloning and sequencing of a Shiga-like toxin II-related gene from *Escherichia coli* O157:H7 strain 7279. *Zbl. Bakt. 276*: 176–188.

Mobassaleh, M., Donohue-Rolfe, A., Jacewicz, M., Grand, R. J., and Keusch, G. T. (1988). Pathogenesis of *Shigella* diarrhea: evidence for a developmentally regulated glycolipid receptor for *Shigella* toxin involved in the fluid secretory response of rabbit small intestine. *J. Infect. Dis. 157*: 1023–1031.

Montfort, W., Villafranca, J. E., Monzingo, A. F., Ernst, S. R., Katzin, B., Ruteuber, E., Xuong, N. H., Hamlin, R., and Robertus, J. D. (1987). The three-dimensional structure of ricin at 2.8 Å. *J. Biol. Chem. 262*: 5398–5403.

Neisser, M., and Shiga, K. (1903). Ueber freire Receptoren von typhus- und dysenterie Bazillen und über das dysenterie Toxin. *Deut. Med. Wochschr. 29*: 61–62.

Newland, J. W., Strockbine, N. A., and Neill, R. J. (1987). Cloning of genes for production of *Escherichia coli* Shiga-like toxin type II. *Infect. Immun. 55*: 2675–2680.

Nishibuchi, M., Yamasaki, S., Ito, H., Terai, A., Kurazono, H., Furatani, M., Ito, K., Igarashi, K., and Takeda, Y. (1994). Comparative analysis of structure–function relationship of Verotoxins produced by enterohemorrhagic *Escherichia coli*. Proceedings of the 25th Joint Conf. US-Japan Cooperative Medical Science Program, Cholera Panel, in press.

Noda, M., Yutsudo, T., Nakabayashi, N., Hirayama, T., and Takeda, Y. (1987). Purification and some properties of Shiga-like toxin from *Escherichia coli* O157:H7 that is immunologically identical to Shiga toxin. *Microb. Pathog. 2*: 339–349.

O'Brien, A. D., and LaVeck, G. D. (1983). Purification and characterization of a *Shigella dysenteriae* 1-like toxin produced by *Escherichia coli. Infect. Immun. 40*: 675–683.

O'Brien, A. D., LaVeck, G. D., Griffin, D. E., and Thompson, M. R. (1980). Characterization of *Shigella dysenteriae* 1 (Shiga) toxin purified by anti-Shiga toxin affinity chromatography. *Infect. Immun. 30*: 170–179.

O'Brien, A. D., LaVeck, G. D., Thompson, M. R., and Formal, S. B. (1982). Production of *Shigella dysenteriae* type 1-like cytotoxin by *Escherichia coli*. *J. Infect. Dis. 146*: 763–769.

O'Brien, A. D., Lively, T. A., Chen, M. E., Rothman, S. W., and Formal, S. B. (1983). *Escherichia coli* O157:H7 strains associated with haemorrhagic colitis in the United States produce a *Shigella dysenteriae* 1 (Shiga) like cytotoxin. *Lancet i*: 702.

O'Brien, A. D., Newland, J. W., Miller, S. F., Holmes, R. K., Smith, H. W., and Formal, S. B. (1984). Shiga-like toxin-converting phages from *Escherichia coli* strains that cause hemorrhagic colitis or infantile diarrhea. *Science 226*: 694–696.

Ogasawara, T., Ito, K., Igarashi, K., Yutsudo, T., Nakabayashi, N., and Takeda, Y. (1988). Inhibition of protein synthesis by a Verotoxin (VT2 or Shiga-like toxin II) produced by *Escherichia coli* O157:H7 at the level of elongation factor 1-dependent aminoacyl-tRNA binding to ribosomes. *Microb. Pathog. 4*: 127–135.

Ohmura, M., Cao, C., Karasawa, T., Okuda, J., Kurazono, H., Gannon, V. P. J., Gyles, C. L., and Takeda, Y. (1993a). Purification and some properties of a Verotoxin 2 (Shiga-like toxin II) variant (SLT-IIva) of *Escherichia coli* isolated from infantile diarrhea. *Microb. Pathog. 15*: 399–405.

Ohmura, M., Yamasaki, S., Kurazono, H., Kashiwagi, K., Igarashi, K., and Takeda, Y. (1993b). Characterization of non-toxic mutant toxins of Verotoxin 1 that were constructed by replacing amino acids in the A subunit. *Microb. Pathog. 15*:169–176.

Oku, Y., Yutsudo, T., Hirayama, T., O'Brien, A. D., and Takeda, Y. (1989). Purification and some properties of a Verotoxin from a human strain of *Escherichia coli* that is immunologically related to Shiga-like toxin II (VT2). *Microb. Pathog. 6*: 113–122.

Olitsky, P. K., and Kligler, I. J. (1920). Toxins and antitoxins of *Bacillus dysenteriae* Shiga. *J. Exp. Med. 31*: 19–33.

Olsnes, S., and Eiklid, K. (1980). Isolation and characterization of *Shigella shigae* cytotoxin. *J. Biol. Chem. 255*: 284–289.

Olsnes, S., Reisbig, R., and Eiklid, K. (1981). Subunit structure of *Shigella* cytotoxin. *J. Biol. Chem. 256*: 8732–8738.

Paton, A. W., Paton, J. C., Heuzenroeder, M. W., Goldwater, P. N., and Manning, P. A. (1992). Cloning and nucleotide sequence of a variant Shiga-like toxin II gene from *Escherichia coli* OX3:H21 isolated from a case of sudden infant death syndrome. *Microb. Pathog. 13*: 225–236.

Paton, A. W., Paton, J. C., Goldwater, P. N., Heuzenroeder, M. W., and Manning, P. A. (1993a). Sequence of a variant Shiga-like toxin type-I operon of *Escherichia coli* O111:H⁻. *Gene 129*: 87–92.

Paton, A. W., Paton, J. C., and Manning, P. A. (1993b). Polymerase chain reaction amplification, cloning and sequencing of variant *Escherichia coli* Shiga-like toxin type II operons. *Microb. Pathog. 15*: 77–82.

Ramotar, K., Boyd, B., Tyrrell, G., Gariepy, J., Lingwood, C., and Brunton, J. (1990). Characterization of Shiga-like toxin I B subunit purified from overproducing clones of the SLT-I B cistron. *Biochem. J. 272*: 805–811.

Ready, M. P., Katzin, B. J., and Robertus, J. D. (1988). Ribosome-inhibiting proteins, retroviral reverse transcriptases, and RNase H share common structural elements. *Proteins 3*: 53–59.

Reisbig, R., Olsnes, S., and Eiklid, K. (1981). The cytotoxic activity of *Shigella* toxin. Evidence for catalytic inactivation of the 60 S ribosomal subunit. *J. Biol. Chem. 256*: 8739–8744.

Riley, L. W., Remis, R. S., Helgerson, S. D., McGee, H. B., Wells, J. G., Davis, B. R., Hebert, R. J., Olcott, E. S., Johnson, L. M., Hargrett, N. T., Blake, P. A., and Cohen, M. L. (1983). Hemorrhagic colitis associated with a rare *Escherichia coli* serotype. *N. Engl. J. Med. 308*: 681–685.

Samuel, J. E., Perera, L. P., Ward, S., O'Brien, A. D., Ginsburg, V., and Krivan, H. C. (1990). Comparison of the glycolipid receptor specificities of Shiga-like toxin type II and Shiga-like toxin type II variants. *Infect. Immun. 58*: 611–618.

Saxena, S. K., O'Brien, A. D., and Ackerman, E. J. (1989). Shiga toxin, Shiga-like toxin II variant, and ricin are all single-site RNA *N*-glycosidases of 28 S RNA when microinjected into *Xenopus* oocytes. *J. Biol. Chem. 264*: 596–601.

Schmitt, C. K., McKee, M. L., and O'Brien, A. D. (1991). Two copies of Shiga-like toxin II-related genes common in enterohemorrhagic *Escherichia coli* strains are responsible for the antigenic heterogeneity of the O157:H⁻ strain E32511. *Infect. Immun. 59*: 1065–1073.

Scotland, S. M., Day, N. P., and Rowe, B. (1980). Production of a cytotoxin affecting Vero cells by strains of *Escherichia coli* belonging to traditional enteropathogenic serogroups. *FEMS Microbiol. Lett. 7*: 15–17.

Scotland, S. M., Smith, H. R., and Rowe, B. (1985). Two distinct toxins active on Vero cells from *Escherichia coli* O157. *Lancet 2*: 885–886.

Shiga, K. (1897). Study on the cause of shigellosis. I (in Japanese). *Saikingaku-Zasshi 25*: 790–810.

Shiga, K. (1898). Ueber den Dysenteriebacillus (*Bacillus dysenteriae*). *Zbl. Bakt. Parasit. Abt. 1, Orig. 24*: 817–824.

Sixma, T. K., Stein, P. E., Hol, W. G. J., and Read, R. J. (1993). Comparison of the B-pentamers of heat-labile enterotoxin and Verotoxin-1: two structures with remarkable similarity and dissimilarity. *Biochemistry 32*: 191–198.

Stein, P. E., Boodhoo, A., Tyrrell, G. J., Brunton, J. L., and Read, R. J. (1992). Crystal structure of the cell-binding B oligomer of Verotoxin-1 from *E. coli. Nature 355*: 748–750.

Strockbine. N. A., Marques, L. R. M., Newland, J. W., Smith, H. W., Holmes, R. K., and O'Brien, A. D. (1986). Two toxin-converting phages from *Escherichia coli* O157:H7 strain 933 encode antigenically distinct toxins with similar biologic activities. *Infect. Immun. 53*: 135–140.

Strockbine, N. A., Jackson, M. P., Sung, L. M., Holmes, R. K., and O'Brien, A. D. (1988). Cloning and sequencing of the genes for Shiga toxin from *Shigella dysenteriae* type 1. *J. Bacteriol. 170*: 1116–1122.

Takao, T., Tanabe, T., Hong, Y.-M., Shimonishi, Y., Kurazono, H., Yutsudo, T., Sasakawa, C., Yoshikawa, M., and Takeda, Y. (1988). Identity of molecular structure of Shiga-like toxin I (VT1) from *Escherichia coli* O157:H7 with that of Shiga toxin. *Microb. Pathog. 5*: 357–369.

Thompson, M. R., Steinberg, M. S., Gemski, P., Formal, S. B., and Doctor, B. P. (1976). Inhibition of in vitro protein synthesis by *Shigella dysenteriae* 1 toxin. *Biochem. Biophys. Res. Commun. 71*: 783–788.

Tyler, S. D., Johnson, W. M., Lior, H., Wang, G., and Rozee, K. R. (1991). Identification of Verotoxin type 2 variant B subunit genes in *Escherichia coli* by the polymerase chain reaction and restriction fragment length polymorphism analysis. *J. Clin. Microbiol. 29*: 1339–1343.

Tyrrell, G. J., Ramotar, K., Toye, B., Boyd, B., Lingwood, C. A., and Brunton, J. L. (1992). Alteration of the carbohydrate binding specificity of Verotoxins from Galα1-4Gal to GalNAcβ1-3Galα1-4Gal and vice versa by site-directed mutagenesis of the binding subunit. *Proc. Natl. Acad. Sci. U.S.A. 89*: 524–528.

van Heyningen, W. E., and Gladstone, G. P. (1953). The neurotoxin of *Shigella shigae* 1. Production, purification and properties of the toxin. *Br. J. Exp. Pathol. 34*: 202–216.

Waddell, T., Head, S., Petric, M., Cohen, A., and Lingwood, C. (1988). Globotriosyl ceramide is specifically recognized by the *Escherichia coli* Verocytotoxin 2. *Biochem. Biophys. Res. Commun. 152*: 674–679.

Wade, W. G., Thom, B. T., and Evans, N. (1979). Cytotoxic enteropathogenic *Escherichia coli. Lancet ii*: 1235–1236.

Weinstein, D. L., Jackson, M. P., Samuel, J. E., Holmes, R. K., and O'Brien, A. D. (1988). Cloning and sequencing of a Shiga-like toxin type II variant from an *Escherichia coli* strain responsible for edema disease of swine. *J. Bacteriol. 170*: 4223–4230.

Williams-Smith, H., Green, P., and Parsell, Z. (1983). Vero cell toxins in *Escherichia coli* and related bacteria: transfer by phage and conjugation and toxic action in laboratory animals, chickens and pigs. *J. Gen. Microbiol. 129*: 3121–3137.

Yamasaki, S., Furutani, M., Ito, K., Igarashi, K., Nishibuchi, M., and Takeda, Y. (1991). Importance of arginine at position 170 of the A subunit of Verotoxin 1 produced by enterohemorrhagic *Escherichia coli* for toxin activity. *Microb. Pathog. 11*: 1–9.

Yutsudo, T., Honda, T., Miwatani, T., and Takeda, Y. (1986). Characterization of purified Shiga toxin from *Shigella dysenteriae* 1. *Microbiol. Immunol. 30*: 1115–1127.

Yutsudo, T., Nakabayashi, N., Hirayama, T., and Takeda, Y. (1987). Purification and some properties of a Verotoxin from *Escherichia coli* O157:H7 that is immunologically unrelated to Shiga toxin. *Microb. Pathog. 3*: 21–30.

15

Clostridium difficile and Its Toxins

Nathan M. Thielman and Richard L. Guerrant

University of Virginia School of Medicine, Charlottesville, Virginia

I. INTRODUCTION

Clostridium difficile is the etiologic agent of pseudomembranous colitis and the most frequent cause of antibiotic-associated colitis (Bartlett, 1990; Fekety, 1990). In hospitals, it is the most common identifiable cause of antibiotic-associated diarrhea (O'Neill et al., 1991). As a growing body of literature implicates this pathogen in hospital and nursing home epidemics (Treloar and Kalra, 1987; Bender et al., 1986), an understanding of its nosocomial acquisition is emerging (McFarland et al., 1990).

In the 15 years since *C. difficile* was linked with pseudomembranous colitis, two potent lethal toxins have been characterized; it is their production that ultimately leads to disease manifestations. In the context of a clinical understanding of *C. difficile*-mediated disease, this chapter reviews the current wealth of information available, describing physicochemical and biological characteristics of these unusual toxins.

II. HISTORICAL PERSPECTIVES

The sentinel works of Bartlett et al. (1977a,b, 1978a), Rifkin et al. (1977, 1978), Larson et al. (1977), and Larson and Price (1977) in the late 1970s established toxigenic *C. difficile* as the cause of pseudomembranous colitis. These studies provided the critical link between the first anatomic description of pseudomembranous colitis one century ago and the discovery of *C. difficile* in the stools of neonates in 1935.

Despite the close association with antibiotics, the first anatomic description of pseudomembranous colitis predates the antibiotic era. Finney (1893) reported the case of a young woman who, 10 days after gastric surgery, developed hemorrhagic diarrhea and eventually died. An autopsy revealed "diphtheritic colitis." Also from the prepenicillin era, data from the Mayo Clinic record three cases per year of pseudomembranous lesions in the intestinal tract (Pettet et al., 1954). After the introduction of antibiotics, pseudomembranous colitis became more commonly recognized; it was frequently associated with the broad-spectrum antibiotics chloramphenicol and tetracycline (Bartlett, 1988). In one surgical series, an enterocolitis was reported in up to 30% of patients receiving these antibiotics (Hummel et al., 1964). Based on some intriguing microbiological and experimental data, the major etiologic agent then was presumed to be *Staphylococcus aureus*. However, this was never firmly established in the majority of cases, and today it is considered only a rare cause of pseudomembranous colitis (Bartlett, 1990).

In a study of intestinal flora of newborn infants, Hall and O'Toole (1935) were the first to isolate *Clostridium difficile*. They described an obligate anaerobic, spore-producing, gram-positive rod that was toxigenic, and "because of the unusual difficulty which was encountered in its isolation and study" it was named *Bacillus difficilis*. These investigators, and Snyder (1937), in a follow-up study, further characterized the toxins as thermolabile; lethal for guinea pigs, rabbits, cats, dogs, and pigeons; and immunogenic.

Despite the thorough work of Hall and O'Toole and Snyder, the clinical relevance of *C. difficile* and its toxin(s) escaped recognition until the 1970s. Though Hamre et al. (1943) found that penicillin was lethal for guinea pigs, inducing hemorrhage and edema in their ceca, and Small (1968) reported a fatal enterocolitis associated with lincomycin in the Syrian hamster, a link with *C. difficile* was not suspected. In 1974, Tedesco et al.

(1974) described "clindamycin colitis." In this prospective study of 200 patients treated with clindamycin, 42 (21%) developed diarrhea, and 20 (10%) had pseudomembranous colitis by proctoscopic examination; there was little to suggest an underlying bacterial cause for this disease (Tedesco et al., 1974).

The first to hypothesize a role for bacterial toxins in pseudomembranous colitis were Larson et al. (1977), who attempted to isolate viruses from the feces of a 12-year-old girl with a prior history of penicillin exposure, diarrhea, negative stool cultures, and endoscopically proven pseudomembranous colitis. They found that fecal suspensions from this patient (and subsequently those from four out of five others with pseudomembranous colitis) had cytopathic effects on HeLa, rhesus monkey kidney cells, and human embryonic lung fibroblast cells (Larson et al., 1977). A flurry of scientific investigation followed. Bartlett et al. (1977b), using the hamster model of clindamycin-associated enterocolitis, found that intracecal material transferred the disease from affected animals to healthy ones, that broth cultures of a *Clostridium* species and their cell-free supernatants produced disease, and that this activity was neutralized by gas gangrene antiserum. Rifkin et al. (1977) simultaneously reported that stool filtrates from humans with pseudomembranous colitis were lethal for hamsters; produced edema, hemorrhage, and increased vascular permeability in rabbit skin; and possessed cytotoxic activity that was neutralized by *Clostridium sordelli* antitoxin. Finally, Bartlett et al. (1978a) and George et al. (1978) implicated the species *Clostridium difficile*, demonstrating that toxigenic strains were readily isolated from the stool of patients with pseudomembranous colitis.

Until 1980, it was widely held that the cytotoxic factor was the only toxin produced by *C. difficile*. However, Bartlett et al. (1980) reported that a second protein could be separated from the cytotoxin with ion-exchange chromatography and that this substance was enterotoxic whereas the previously described toxin was not. Other investigators (Taylor et al., 1981; Sullivan et al., 1982; Banno et al., 1984) confirmed these findings; thus it was concluded that *C. difficile* produced two distinct toxins: the previously recognized cytotoxin, designated toxin B, and an enterotoxin, toxin A.

III. CLINICAL CONSIDERATIONS

A. Overview

Clostridium difficile is now recognized as the major pathogen of antibiotic-associated diarrhea and colitis (Bartlett, 1990). It is implicated in 20–30% of patients with antibiotic-associated diarrhea (George et al., 1982a; Bartlett et al., 1980; Viscidi et al., 1981), in 50–75% of those with antibiotic-associated colitis (Bartlett, 1990), and in more than 90% of those with antibiotic-associated pseudomembranous colitis (Bartlett et al., 1980; George et al., 1982b). The incidence of *C. difficile* colitis in outpatients being treated with oral antibiotics is approximately 1–3 cases per 100,000, while in hospitalized patients the frequency may be as high as 1 in 1000 to 1 in 100 (Andrejak et al., 1991; Hirschhorn et al., 1992).

Estimates of total health care dollars spent in diagnosing and treating this condition are unknown, but in one community hospital, charges for an admitting diagnosis of *C. difficile* averaged $5000 per admission, and cases acquired in the hospital resulted in costs of nearly $2000 per patient (Kofsky et al., 1991).

B. Acquisition of *Clostridium difficile*

1. Inciting Agents

The intimate association of antibiotics with *C. difficile* colitis is unprecedented. It is now apparent that this is a result of antibiotic-induced perturbation of the normal intestinal flora, leading to overgrowth of toxigenic strains of *C. difficile*. Oral challenge with *C. difficile* alone will not produce enteric disease in any animal model studied to date unless the bowel flora is modified with an antibiotic (Bartlett, 1990) or unless gnotobiotic animals are used (Corthier et al., 1991; Czuprynski et al., 1983). Although almost all antibiotic classes have been associated with the disease (including drugs such as vancomycin and metronidazole that are usually effective in treating it)—reports of large clinical experiences most commonly implicate clindamycin, ampicillin or amoxicillin, and cephalosporins (Andrejak et al., 1991; Bartlett, 1981; Silva et al., 1984). In the hamster model, ampicillin, carbenicillin, cefamandole, cefaclor, cefazolin, cefoxitin, cephalexin, cephaloridine, cephalothin, cephradine, clindamycin, oral gentamicin, imipenem, metronidazole, nafcillin, penicillin, and ticarcillin regularly produce lethal hemorrhagic cecitis due to *C. difficile* (Bartlett, 1992; Small, 1968). Rarely or inconsistently do tetracyclines, chloramphenicol, sulfonamides, or trimethoprim-sulfamethoxazole produce disease. To date there are no published data on the quinolones in the animal model, though there are case reports of *C. difficile*-associated diarrhea in humans (Dan and Samra, 1989; Andrejak et al., 1991).

In addition, *C. difficile* disease has been associated with several antineoplastic agents including adriamycin, cyclophosphamide, 5-fluorouracil, and methotrexate, all of which have antibacterial activity (George, 1988; Cudmore et al., 1982; Silva et al., 1984; Bartlett, 1981), and possibly with the neuroleptics cyamemazine and chlorpromazine (Barc et al., 1992). There is one report of a patient with AIDS developing *C. difficile* colitis following antiviral therapy with zidovudine and acyclovir with no exposure to antibacterial agents (Colarian, 1988). Rarely the disease occurs without apparent exposure to agents that alter the microecology of the gut (Ellis et al., 1983; Kim et al., 1983; Moskovitz and Bartlett, 1981; Peikin et al., 1980; Wald et al., 1980).

2. Risk Factors

In a comprehensive prospective study of risk factors for *C. difficile* disease in hospitalized patients, McFarland et al. (1990) found that age and severity of underlying illness were associated with increased risk of *C. difficile* carriage and diarrhea. In addition, exposure to factors that alter normal intestinal flora or motility, specifically antibiotics, enemas, and intestinal stimulants, all increased the risk of developing *C. difficile*-associated diarrhea (McFarland et al., 1990). Others report that critically ill burn patients (Grube et al., 1987), uremic patients (Leung et al., 1985; Aronsson et al., 1987), patients with hematologic malignancies (Rampling et al., 1985; Heard et al., 1988), and those undergoing gastrointestinal surgery (Keighley et al., 1978) are at increased risk for *C. difficile* diarrhea and colitis.

The purported role of *C. difficile* in exacerbations of inflammatory bowel disease in adults (LaMont and Trnka, 1980; Bolton et al., 1980) was recently thoroughly reviewed by Bartlett et al. (1989), who tallied the experience of 11 investigators involving 550 patients with inflammatory bowel disease. Two percent had evidence of toxin during relapse, and the majority of these patients had recently been exposed to antimicrobial agents. It is thus argued that *C. difficile* does not play an important role in the relapse of inflammatory bowel disease in adults. However, Gryboski (1991) found *C. difficile* toxin in 16% of 65 children with exacerbations of inflammatory bowel disease.

Independent of exposure to more antimicrobial agents, AIDS patients do not appear to be at greater risk for *C. difficile* colitis than immunocompetent patients (McFarland et al., 1990).

3. Reservoirs

The source of *C. difficile* may be either endogenous or environmental. As discussed above, the organism was actually first recovered by Hall and O'Toole (1935) from the stools of healthy neonates. Studies since report the prevalence of *C. difficile* (or its cytopathic toxin) in the stools of healthy neonates ranging from 15% to 70% (Snyder, 1940; Viscidi et al., 1981; Bolton et al., 1984; Donta and Myers, 1982; Holst et al., 1981; Al-Jumaili et al., 1984). Despite evidence for toxigenic organisms in this population, the prevalence of *C. difficile* colitis is relatively low (Viscidi et al., 1981; Donta and Myers, 1982). Eglow et al. (1992) recently demonstrated age-related susceptibility to toxin A, correlating increased numbers of toxin A receptors in rabbit ileum with aging, and suggested that a similar pattern may be operative in humans to explain this observation.

In the healthy adult population, intestinal carriage of *C. difficile* has been reported to range from less than 3% to more than 11% (Larson et al., 1978; George et al., 1979; Marrie et al., 1982; Aronsson et al., 1985; Nakamura et al., 1981). As many as 21–46% of adults given antibiotics who remain without gastrointestinal complications harbor the organism (Viscidi et al., 1981; George et al., 1982b).

While the intestinal tract is considered the most important endogenous reservoir of *C. difficile*, it was reportedly isolated in urethral and vaginal cultures, respectively, from 100% of men and 75% of women attending a sexually transmitted diseases clinic (Hafiz and Oakley, 1976). Tabaqchali et al. (1984) detected vaginal carriage of the organism in 11–18% of pregnant women cultured. Larson et al. (1982), however, were unable to detect urogenital carriage of *C. difficile*. Thus the frequency of extraintestinal carriage of this organism is not yet clearly defined.

It has been convincingly shown that an existing reservoir of *C. difficile* is not a prerequisite for symptomatic infection and that the disease-causing organism may be acquired from exogenous sources. Arguing for the importance of environmental acquisition of *C. difficile* are the number of outbreaks that have been reported in hospitals (Greenfield et al., 1981; Testore et al., 1988; Delmee et al., 1986; Poxton et al., 1984; Han et al., 1983), long-term care facilities (Cartmill et al., 1992; Bender et al., 1986; Bennett et al., 1989), and day care centers (Kim et al., 1983). In fact, the classic "clindamycin colitis" study (Tedesco et al., 1974) is now retrospectively recognized as the first confirmed hospital epidemic—owing to the unusually high rates of *C. difficile* toxin in subsequent stool analysis (Snook et al., 1989). In such settings, *C. difficile* is readily cultured from multiple inanimate environmental sources and from the hands and stools of hospital personnel (Fekety et al., 1981; McFarland et al., 1989). McFarland et al. (1989), using immunoblot typing, demonstrated time-space clustering in nosocomially acquired *C. difficile* infection in 399 patients followed prospectively. Twenty-one percent of patients admitted to the hospital acquired the organism prior to discharge, and of that number, 37% developed diarrhea.

In addition, *C. difficile* is found in many sources outside the hospital. Hafiz and Oakley (1976) cultured the organism from soil, mud, and sand and several animals. Carriage rates in household pets such as dogs and cats ranges from 20 to 40%, and some propose that it may thus be zoonotically acquired (Riley et al., 1991; Borriello et al., 1983). Foodborne transmission has also been suggested but not definitively documented (Gurian et al., 1982).

C. Disease Manifestations

The symptoms of C. *difficile* disease usually appear after 5–10 days of antibacterial treatment, but they may develop as early as the first day of therapy or as late as 10 weeks after cessation of therapy (Tedesco, 1982). Most frequently, C. *difficile* disease is manifested by diarrhea, which may be brief and self-limited or cholera-like, resulting in more than 20 stools per day (Bartlett, 1992). Findings suggesting progression to colitis include crampy abdominal pain, fever, hypoalbuminemia, leukocytosis, and watery, green, foul-smelling stools (Andrejak et al., 1991; Bartlett, 1992; Fekety and Shah, 1993). Fecal leukocytes are found in approximately 50% of cases (Bartlett, 1992), and occult blood may be detected in one fourth of diarrheal stools; frank blood is seen in 5–10% of those with pseudomembranous colitis (George, 1984).

Rarely C. *difficile* colitis presents without diarrhea as an acute abdominal syndrome (Triadafilopoulos and Hallstone, 1991) or toxic megacolon (Burke et al., 1988). Other reported complications include reactive arthritis (Cope et al., 1992; Hannonen et al., 1989; Hayward et al., 1990; Mermel and Osborn, 1989; Atkinson and McLeod, 1988), protein-losing enteropathy (Rybolt et al., 1989), colonic perforation (Morris et al., 1990; Snooks et al., 1984), bacteremia from bowel flora or rarely C. *difficile* (George, 1988), and persistent colitis or diarrhea (Tedesco, 1982; Sutphen et al., 1983). Unrecognized and untreated, C. *difficile* colitis may be fatal; death rates of 10–20% have been reported in elderly and debilitated untreated patients (Fekety, 1990). With specific therapy the mortality rate is less than 2% (Morris et al., 1990).

D. Diagnosis

Direct visualization of pseudomembranes is considered the definitive means of diagnosing pseudomembranous colitis (Fekety and Shah, 1993). Gebhard et al. (1985) found C. *difficile* in 95% of stool cultures from patients with endoscopically confirmed pseudo-membranous colitis. Flexible sigmoidoscopy alone will not detect up to 10% of cases without colonoscopy (Tedesco et al., 1982). When C. *difficile* colitis is not accompanied by pseudomembrane formation, endoscopic findings are relatively nonspecific, and biopsy may reveal changes typical of pseudomembranous colitis (Fekety and Shah, 1993).

Stool cultures using selective media containing cycloserine, cefoxitin, and fructose readily grow C. *difficile* and are considered highly sensitive tests for C. *difficile* colitis. However, because of the high carriage rate of C. *difficile* in patients without disease, the specificity is low, and its utility is limited to epidemiologic surveys, screening, and back-up to other diagnostic methods (Bartlett, 1992; Fekety and Shah, 1993).

The most widely used means to establish C. *difficile* diarrhea and colitis clinically is the tissue culture assay to detect the presence of toxin B. Used in the appropriate clinical setting, this test is both sensitive and specific, and it is considered the "gold standard" laboratory test at present. More than 90% of patients with pseudomembranous colitis have cytotoxic activity in their stools detected by this assay (Lyerly et al., 1988b). Sensitivity may vary, however, with the use of different cell lines and potential inactivation of toxin B during storage and handling of samples (Fekety and Shah, 1993). Because many stool specimens contain other factors that cause rounding of cultured cells, its specificity must be confirmed by neutralizing cytotoxic activity with *C. difficile* or *C. sordelli* antitoxin (Lyerly et al., 1988b; Fekety, 1990).

Several alternative diagnostic tests have been introduced to detect *C. difficile* and its toxins. The latex particle agglutination test, introduced commercially in 1986, is rapid and simple but relatively insensitive and nonspecific, as it actually detects a protein that has no definable role in enteric disease and is not specific for *C. difficile* (Lyerly and Wilkins, 1986; Lyerly et al., 1988a). Rapid enzyme-linked immunosorbent assays (ELISAs) that detect toxins A or B or both have recently been introduced, and four are FDA-approved (Bartlett, 1992). These tests are rapid, relatively inexpensive, and specific but do not appear to be as sensitive as cytotoxicity assays (Doern et al., 1992). A dot immunobinding assay is now commercially available as well, and in one report its results agreed with those of cytotoxicity assays in 92% of specimen tests (Woods and Iwen, 1990). Although the ELISA tests and the dot immunobinding assays look promising, neither is recommended as the sole means for diagnosing *C. difficile* disease (Bartlett, 1992; Fekety and Shah, 1993).

The polymerase chain reaction has been used to detect the *C. difficile* 16S rRNA gene (Gumerlock et al., 1991) and the toxin A gene (Kato et al., 1993) in human feces, but the clinical utility of this method has yet to be established.

E. Pathology

Pseudomembranous colitis has a distinct macroscopic appearance that may be appreciated by endoscopic examination. The colonic mucosa is studded with adherent raised white to yellowish plaques. Initially, they are small and discrete and are easily dislodged; with progression of disease, they enlarge and coalesce (Gerding et al., 1988). The intervening mucosa may be inflamed and covered with mucus, but often it is entirely normal. Pseudomembranes typically exist throughout the entire colon, but they are usually most pronounced in the rectosigmoid colon; rarely does the disease progress proximal to the ileocecal valve (Fekety, 1990).

Histologic criteria for pseudomembranous colitis and a method for grading lesions have been described (Price and Davies, 1977). The principal features are inflamed mucosa with a neutrophilic predominance and mucin-distended glands. A loose network of mucin, polymorphonuclear leukocytes, nuclear debris, and fibrin comprise the attached pseudomembranes (Price and Davies, 1977; Gerding et al., 1988).

F. Treatment

Initial therapy of patients with pseudomembranous colitis includes discontinuing the offending antibiotic regimen and replacing fluid and electrolyte losses; up to 25% of patients may respond to these conservative measures, but most patients with severe disease require specific antibacterial therapy (Fekety and Shah, 1993). The majority of *C. difficile* isolates are susceptible to vancomycin, metronidazole, rifampin, and bacitracin, though generally only vancomycin and metronidazole are used in current clinical practice. Because the *C. difficile* rarely invades the mucosa, oral therapy is most effective, and intestinal absorption of the drug is disadvantageous.

The greatest amount of clinical experience in specific anticlostridial therapy for this disease is with oral vancomycin (Batts et al., 1980; Teasley et al., 1983; Young et al., 1985; Dudley et al., 1986; Fekety et al., 1989), and it is the drug of choice for severe disease. Fever generally resolves within 24 h, and diarrhea diminishes over the next 3–4 days. Oral metronidazole, though not approved by the U.S. Food and Drug Administration for the treatment of *C. difficile* colitis, is much less expensive and was as effective as vancomycin in a randomized controlled trial (Teasley et al., 1983).

Bacitracin (Young et al., 1985; Dudley et al., 1986) and teicoplanin (de Lalla et al., 1992), given orally, were comparable to vancomycin in resolving the symptoms of *C. difficile*-associated diarrhea in randomized clinical trials. Bacitracin, however, cleared the pathogen and its cytotoxin significantly less effectively than vancomycin (Young et al., 1985). Cholestyramine binds *C. difficile* toxins (Taylor and Bartlett, 1980) and is considered a treatment option in mild or moderately severe disease by at least one authority (Fekety and Shah, 1993); however, because it may also bind vancomycin, it is not recommended for use with low-dose vancomycin therapy.

The use of antiperistaltic agents is contraindicated, as intestinal stasis may lead to the retention of toxins and further complications (Church and Fazio, 1986; Walley and Milson, 1990; Bartlett, 1992; Fekety and Shah, 1993; Novak et al., 1976).

G. Recurrent Disease

Despite initial adequate antimicrobial treatment, recurrent *C. difficile*-associated diarrhea has been reported in 5–55% of patients in several small studies (Young et al., 1985, 1986; Walters et al., 1983; Teasley et al., 1983; Dudley et al., 1986). Bartlett (1990) records the greatest clinical experience with this problem, observing relapses in 46 of 189 patients treated with vancomycin (24%). A "repeated relapse pattern" characterized by five or more relapses with each sequential course of treatment has been described and may be seen in 5–8% of patients treated (Bartlett, 1985, 1990). Elderly patients and those who have recently undergone abdominal surgery are more prone to relapse (Young et al., 1986).

The leading explanation for relapsing *C. difficile* disease has been the persistence of *C. difficile* spores, which may survive even the high intraluminal levels of vancomycin seen during standard oral therapy (Wilcox and Spencer, 1992; Baird, 1989). However, two independent groups (Johnson et al., 1989; O'Neill et al., 1991), using DNA restriction endonuclease analysis, recently found that up to half of patients with apparent relapse were actually infected with a new strain of *C. difficile*. It is also suggested that patients may be reinfected with the same strain from environmental sources that they have contaminated (O'Neill et al., 1991).

Several novel regimens have been described for the treatment of recurrent infection, including prolonged therapy with vancomycin for several weeks followed by a slow tapering, alternating periods of antimicrobial therapy with "drug holidays," and the use of toxin-binding agents after cessation of antibiotics to ameliorate symptoms without continued disruption of the normal flora by antimicrobial agents (Sterne and Wentzel, 1950).

Introduction of competing, nonpathogenic organisms into the intestinal tract may also prove useful in treating recurrent disease. *Lactobacillus* preparations were effective in helping to prevent recurrence in four of the five patients reported by Gorbach et al. (1987), but there remains a paucity of clinical experience with this treatment. *Saccharomyces boulardii*, a nonpathogenic yeast, has been shown to suppress *C. difficile* overgrowth and significantly decrease the mortality of clindamycin-induced enterocolitis in hamsters (Elmer and McFarland, 1987); it has been used extensively in Europe for the treatment and prevention of antibiotic-associated diarrhea and colitis. In a controlled double-blind study of 180 patients, *S. boulardii*, given concomitantly with antibiotics, significantly reduced the overall incidence of antibiotic-associated diarrhea and had a favorable (but not statistically significant) effect in reducing *C. difficile*-mediated disease (Surawicz et al., 1989a). When *S. boulardii* was given with vancomycin in an uncontrolled study for recurrent *C. difficile* colitis, 11 of 13 patients were cured (Surawicz et al., 1989b). Further study of *S. boulardii* in the prevention of recurrent disease is ongoing (Fekety and Shah, 1993).

Nontoxigenic *C. difficile* administered orally was reported to be curative for two patients with refractory relapsing *C. difficile* colitis (Seal et al., 1987). Others report cures of *C. difficile* colitis by rectal infusions of a mixture of aerobic and anaerobic bacteria cultured from feces (Tvede and Rask-Madsen, 1989) and by rectal instillation of feces from healthy donors (Tvede and Rask-Madsen, 1989; Schwan et al., 1984).

H. Prevention

As it has become increasingly evident that much of *C. difficile*-mediated disease is a function of nosocomial spread of the organism, there is growing interest in developing ways to interrupt transmission within hospitals and nursing homes. Johnson et al. (1992b) recently compared vancomycin with metronidazole for eradicating *C. difficile* excretion from asymptomatic hospitalized carriers as a potential means of controlling nosocomial outbreaks. Though treatment with vancomycin was temporarily effective at suppression of organism excretion, patients soon began to excrete *C. difficile* again after cessation of therapy. The authors thus concluded that treating asymptomatic excretors of *C. difficile* would likely be ineffective in controlling nosocomial epidemics (Johnson et al., 1992b). On the other hand, rigorous adherence to simple infection control measures, such as the use of vinyl gloves (Johnson et al., 1990) and disposable thermometers (Brooks et al., 1992), has significantly reduced the incidence of *C. difficile*-associated diarrhea in hospitals and extended care facilities. In addition, some authorities justifiably recommend isolation and formal enteric precautions in patients with known *C. difficile* diarrhea to prevent its spread (Wilcox and Spencer, 1992).

IV. *CLOSTRIDIUM DIFFICILE* TOXINS

A. Overview

Since the late 1970s, a number of toxic products from *C. difficile* have been characterized. Toxin A, an enterotoxin that causes extensive mucosal damage and hemorrhagic fluid accumulation in animal models, is thought by many to be the toxin primarily responsible for the colitis and diarrhea associated with *Clostridium difficile*. Toxin B is a potent cytotoxin with little, if any, enteropathic effect. It likely has a synergistic role with toxin A in producing disease, though the exact mechanism of this synergy has yet to be worked out. Both toxins A and B rank among the most lethal bacterial toxins studied (Gill, 1982).

In addition, a high molecular weight myoelectric factor that alters intestinal motility has been reported (Justus et al., 1982), but the exact role of this toxin in disease is uncertain. It causes neither secretion nor mucosal damage and appears to be distinct from toxins A and B.

Fraction C denotes a family of three enterotoxic proteins described by Torres and Lonnroth (1989). They differ from toxin A in that they produce intestinal secretion that is nonhemorrhagic; unlike toxin B they are not cytotoxic but produce cytotonic effects on Chinese hamster ovary cells. The three proteins contain similar but not identical antigenic determinants to toxin A, and cytotonic activity was neutralized with antisera to toxin A (Torres and Lonnroth, 1989). Nonhemorrhagic enterotoxic activity from partially purified toxins of smaller size were previously described by others (Banno et al., 1984; Lonnroth and Lange, 1983; Giuliano et al., 1988), and there is some speculation that these products might be fragments of fraction C (Torres and Lonnroth, 1989).

Finally, an actin-specific ADP-ribosyltransferase was detected in one of 15 strains of *C. difficile* examined. This product, which is not cytotoxic, is not related to either toxin A or B (Popoff et al., 1988). ADP-ribosylating activity has not been identified by other investigators in toxins A (von Eichel-Streiber et al., 1991) or B (von Eichel-Streiber et al., 1991; Florin and Thelestam, 1991).

The remainder of this section focuses on toxins A and B.

B. Factors Affecting Toxin Production

Clostridium difficile must initially colonize the relevant area of the bowel (Wilson, 1988; Viscidi et al., 1981). A relatively fastidious anaerobe, *C. difficile* causes diarrhea and colitis almost exclusively when the intestinal flora is modified—usually from prior antibiotic exposure. (See the subsection Inciting Agents in Section III.B). The exact manner in which toxins are liberated from the *C. difficile* bacillus is not known. Lyerly et al. (1988b) found that toxin concentration in supernatant fluid parallels the growth of the organism and that most of the toxin is released in the late logarithmic or early stationary phase, suggesting that toxin release may be an active process. Alternatively, Ketley et al. (1984) reported that toxins A and B are released after the stationary phase of growth, arguing for their release after cell lysis.

There are some data to suggest that spore formation may be linked to toxin production in *C. difficile*. Although Ketley et al. (1986) and Onderdonk et al. (1979) found that the kinetics of spore production by *C. difficile* were not paralleled by toxin release in vitro, Kamiya et al. (1992) recently demonstrated that the organism releases cytotoxin during sporulation.

Several investigators (Barc et al., 1992; Nakamura et al., 1982; Honda et al., 1983; Onderdonk et al., 1979) have sought to establish the effect of antibiotics and other agents on toxin production in vitro, independent of their effects on the complex bacterial milieu in animals and humans. In the first study of this sort, Onderdonk et al. (1979) found that subinhibitory concentrations of vancomycin and penicillin, but not clindamycin, increased toxin levels in continuous culture. In contrast, Honda et al. (1983), assaying a different clinical strain, found that clindamycin increased enterotoxic and cytotoxic activity 16-fold and fourfold, respectively, compared with controls. Not unexpectedly, it was later demonstrated with some 80 clinical isolates of *C. difficile* that undifferentiated cytotoxin production in response to various antibiotics tended to vary depending on the particular toxigenic strain tested (Nakamura et al., 1982). Barc et al. (1992), in studies with the toxigenic strain *C. difficile* VPI 10463, found that neither clindamycin, chlorpromazine, nor methotrexate increased the production of toxins A or B in either in vitro or in vivo systems.

Diet may affect toxin production in vivo. In gnotobiotic mice monoassociated with toxigenic strains of *C. difficile* and fed various semisynthetic diets, fecal cytotoxin and enterotoxin production were highly reduced, in the absence of significant changes in the population of the organism (Mahe et al., 1987). Moreover, these diets also tended to prevent mortality. It remains unknown exactly how such semisynthetic diets altered toxin production: the different diets may have possessed inhibitors or may have been devoid of nutrients essential for toxin manufacture and release.

C. Experimental Production and Purification of *C. difficile* Toxins

1. Production

Toxins A and B are found in the supernatant of cultures and can be purified from filtrates. To maximize the recovery of toxins, most investigators culture the organism in a brain-heart infusion broth. The use of a dialysis flask procedure, similar to that used by Sterne and Wentzel (1950) for recovering botulinum toxin, results in both slower growth of organisms (and hence increased production of toxin) and less contamination with the high molecular weight components of the brain-heart infusion, allowing for relatively pure starting material for purification (Lyerly et al., 1988b). This method has been compared to cultivation directly with brain-heart infusion in an anaerobic chamber and was found superior (Torres and Lonnroth, 1988b).

2. Purification Processes

Various purification processes have been described for the preparation of toxins A and B (Sullivan et al., 1982; Torres and Lonnroth, 1988a; Meador and Tweten, 1988; Rihn et al., 1988; Bisseret et al., 1989; Kamiya et al., 1989; Katoh et al., 1988; Lonnroth and Lange, 1983; Rihn et al., 1984; Krivan and Wilkins, 1987; Pothoulakis et al., 1986; von Eichel-Streiber et al., 1987; Taylor et al., 1981). To purify toxin A, the method described by Sullivan et al. (1982) is most commonly employed (Shahrabadi et al., 1984; Fiorentini et al., 1989, 1992; Hecht et al., 1988; Kushnaryov et al., 1992; Kushnaryov and Sedmak, 1989; Pothoulakis et al., 1991; Clark et al., 1987). It involves a concentration step using ultrafiltration or ammonium sulfate precipitation followed by anion-exchange chromatography to separate toxins A and B. Toxin A can then be purified to homogeneity by acetic acid precipitation at pH 5.5–5.6. Modifications to the method by Sullivan et al. have employed gel filtration (Banno et al., 1984), preparative electrophoresis (Lyerly et al., 1986b), hydrophobic interaction chromatography (Rothman et al., 1984), and fast protein liquid chromatography (Rihn et al., 1984; von Eichel-Streiber et al., 1987). Thermal affinity chromatography on fixed bovine thyroglobulin (which bears the carbohydrate sequence Galα 1-3 Galβ 1-4 1-4GlcNAc) binds toxin A at 4°C but not at 37°C, thus offering a relatively simple and gentle alternative method to purify toxin A (Krivan and Wilkins, 1987). However, small amounts of toxin B and other proteins have been detected in the thermal eluent (Kamiya et al., 1988), and a modified method of purification adding two sequential anion-exchange chromatographic steps has been proposed (Kamiya et al., 1989).

After ultrafiltration, toxin B may be separated from toxin A by anion-exchange chromatography (Sullivan et al., 1982; Bisseret et al., 1989) and immunoadsorption (Sullivan et al., 1982). Others have used gel filtration and high-resolution ion-exchange chromatography (Torres and Lonnroth, 1988a; Meador and Tweten, 1988), sequential ammonium sulfate precipitation, DEAE-Sepharose chromatography (Taylor et al., 1981), DEAE-Sepharose chromatography with high-performance liquid chromatography (Pothoulakis et al., 1986), and fast protein liquid chromatography (von Eichel-Streiber et al., 1987) to further purify this toxin.

D. Physical and Chemical Characteristics of Toxins A and B

1. Molecular Weight and Structure

Toxin A appears to be the largest bacterial protein toxin identified to date (Fiorentini and Thelestam, 1991). In its native form, it bands at positions corresponding to 400–600 kDa on polyacrylamide gel electrophoresis (PAGE) as measured by numerous investigators (Sullivan et al., 1982; Lyerly et al., 1986a; Rothman et al., 1988; Torres and Lonnroth, 1988b; Kamiya et al., 1988; Krivan and Wilkins, 1987). Using gel filtration, Banno et al. (1984) similarly reported the molecular weight to be 550–600 kDa. Even under denaturing conditions the toxin still has an M_r in the 220–320 kDa range (Banno et al., 1984; Lyerly et al., 1986a; von Eichel-Streiber et al., 1987; Torres and Lonnroth, 1988b; Kamiya et al., 1988; Rothman et al., 1988).

Torres et al. (1992) have argued that toxin A exists as an aggregate of many subunits. The finding by Kamiya et al. (1989) of multiple lower molecular weight bands under reducing conditions and that of Rautenberg and Stender (1986), who reported one or more copies of a 35-kDa protein under immunoblotting conditions, lend support to this hypothesis. Borriello et al. (1990) presented indirect evidence of subunit structure activity. Noting that peaks of activity of toxin A for hemagglutination and cytotoxicity are not coincidental as the toxin is eluted by anion-exchange chromatography, it was suggested that pure toxin A may consist of cytotoxin-rich, hemagglutination-rich and cytotoxin-rich, hemagglutination-poor forms of subunit structure.

The hypothesis of subunit structure seems less tenable since the recent construction and expression of the complete toxin A gene in *E. coli* (Phelps et al., 1991). The product of this 8.1-kb structure is a single polypeptide that is neutralized by antiserum to *C. difficile* toxin A and exerts all the biological activities of native toxin—hemagglutination, cytotoxicity, enterotoxicity, and lethality.

Like toxin A, toxin B in its native form is extremely large, with molecular mass estimates by most researchers of 360–500 kDa (Banno et al., 1984; Sullivan et al., 1982; Rothman et al., 1984, 1988; Meador and Tweten, 1988). Meador and Tweten (1988) reported a molecular mass of 250 kDa by SDS PAGE and 500 kDa by gel filtration and hypothesize that toxin B normally forms a dimer in solution. In contrast, Katoh et al. (1988) found their purified cytotoxin to weigh 260,000 in native form and 50,000 in its reduced form. Some controversy has arisen concerning this 50-kDa protein identified under denaturing conditions (Pothoulakis et al., 1986; Rihn et al., 1988; Bisseret et al., 1989; Katoh et al., 1988; Rolfe and Finegold, 1979); this is described by some as a subunit of toxin B (Pothoulakis et al., 1986; Rihn et al., 1988; Bisseret et al., 1989; Katoh et al., 1988), while others such as Meador and Tweten (1988) suggest that it is a contaminant. The latter group showed that a 50-kDa protein that copurified with toxin B was a contaminant, but that under nondenaturing conditions it has physical characteristics similar to those of toxin B. Further controversy has arisen from the findings by Torres and Lonnroth (1988a), who suggest that two forms of toxin B can be purified: one of high molecular weight and similar to the toxin B described by most investigators, and another that is 43 kDa. Many of the above discrepancies may be resolved once further work is done with the recently cloned and sequenced gene of toxin B (Barroso et al., 1990) (see below).

2. Molecular Composition

Both toxins A and B are digested into more than 30 peptides by chymotrypsin (Sullivan et al., 1982; Lyerly et al., 1986b), pronase (Banno et al., 1984; Katoh et al., 1988),

Staphylococcus aureus V-8 protease, papain, and proteinase K (Lyerly et al., 1986b). Unlike toxin A, toxin B is sensitive to trypsin inactivation (Torres and Lonnroth, 1988b; Banno et al., 1984; Sullivan et al., 1982; Taylor et al., 1981; Katoh et al., 1988). In some early reports, toxin A was described as sensitive to digestion with trypsin (Sullivan et al., 1982; Banno et al., 1984), while others reported resistance (Taylor et al., 1981). It was subsequently shown that highly purified trypsin did not inactivate toxin A but that crude trypsin contaminated with chymotrypsin did (Torres and Lonnroth, 1988b; Lyerly et al., 1989).

The peptide digest findings corroborate formal amino acid analyses performed by several groups. Both toxins A and B possess a high percentage of hydrophobic amino acids (Rothman et al., 1988; Lyerly et al., 1986b) and low amounts of histidine (Rothman et al., 1988; Lyerly et al., 1986b; Banno et al., 1984), methionine (Rothman et al., 1988; Lyerly et al., 1986b; Banno et al., 1984), and cysteine (Lyerly et al., 1986b).

Though Rolfe and Finegold (1979) suggested early on that toxin B was a glycoprotein because of inactivation by large amounts of bacterial amylase, others found toxin B active even at high concentrations of amylase (Sullivan et al., 1982; Katoh et al., 1988). Banno et al. (1984) observed neutralization of toxin B by bacterial amylase but attributed its inactivation to possible proteolytic contaminants of the bacterial amylase preparation. Additionally, only small amounts of hexose and pentose have been detected in toxin B, and it is not adsorbed when passed through a concanavalin A column (Banno et al., 1984; Sullivan et al., 1982).

3. *Inhibitors*

Toxin A is inactivated by the arginine-specific reagent 1,2-cyclohexanedione as evidenced by drastically decreased cytotoxicity and mouse lethality (Balfanz and Rautenberg, 1989). Oxidizing agents inactivate the toxin; reducing agents such as 2-mercaptoethanol and dithiothreitol and sulfhydryl-inactivating agents (*p*-hydroxymercuribenzoate, *N*-ethylmaleimide have no effect.

Like toxin A, toxin B is sensitive to a number of oxidizing agents but resistant to sulfhydryl-inactivating agents such as *p*-hydroxymercuribenzoate, *N*-ethymaleimide, and iodoacetate, suggesting that sulfhydryl groups are not involved in toxic or binding activities of the toxin (Lyerly et al., 1986b). Ooi et al. (1984) reported that toxin B loses its cytotoxic activity when exposed to myeloperoxidase and demonstrated that myeloperoxidase is important in neutrophil-mediated detoxification.

4. *Physical Characteristics*

Both toxins A and B can be rapidly inactivated by freezing and thawing (Fiorentini and Thelestam, 1991; Lyerly et al., 1986b), and they both are stable at 37°C and inactivated at 56–60°C (Taylor et al., 1981; Lyerly et al., 1986b; Sullivan et al., 1982; Banno et al., 1984; Katoh et al., 1988). Unlike toxin A, which is stable at pH 4 and 10, toxin B is inactivated in more acidic and more alkaline environments (Sullivan et al., 1982; Taylor et al., 1981).

Both toxins are acidic molecules. Reported isoelectric points range from 3.5 to 5.7 for toxin A (Rothman et al., 1988; Lyerly et al., 1986b, 1988b; Rihn et al., 1984) and from 4.1 to 4.8 for toxin B (von Eichel-Streiber et al., 1987; Lyerly and Wilkins, 1988).

E. Immunocharacterization of Toxins A and B

Toxins A and B are largely immunologically distinct. Specific antisera raised against toxin A that neutralize its biological activity do not affect toxin B (Taylor et al., 1981;

Libby and Wilkins, 1982; Sullivan et al., 1982; Banno et al., 1984). Similarly, antibodies that specifically block the activity of toxin B have no effect on toxin A (Katoh et al., 1988; Libby and Wilkins, 1982). In addition, initial studies detected no cross-reactivity by immunoassay (Lyerly et al., 1983).

Subsequently, von Eichel-Streiber et al. (1987) and Rothman et al. (1988) produced monoclonal antibodies to both toxins A and B and found cross-reactivity with pure toxin A and toxin B. Thus it appears that the toxins share conformational epitopes, but as previous immunoassays did not detect significant cross-reactivity, such homology likely is not vast. Kamiya et al. (1991) identified a monoclonal antibody (37B5) that neutralized the enterotoxicity but not the hemagglutination activity of toxin A and concluded that the epitopes associated with the enterotoxicity and hemagglutinating activity of toxin A are different entities.

Another monoclonal antibody, PCG-4, actually precipitates the toxin; it has no effect on cytotoxicity but neutralizes its enterotoxicity (Lyerly et al., 1986a). By subcloning portions of the recently cloned and sequenced gene for toxin A, Frey and Wilkins (1992) identified two epitopes on the toxin recognized by the monoclonal antibody PCG-4. Because PCG-4 neutralizes this activity, these regions appear to be important for the expression of its enterotoxicity (Lyerly et al., 1986a; Lima et al., 1988). Further, since these sites alone are not toxic, this portion of the toxin A molecule may mediate enterotoxicity by binding to appropriate target receptors (Price et al., 1987; Lyerly et al., 1990; Frey and Wilkins, 1992).

F. Effects of Toxins on Cultured Cells

1. *General Characteristics of Toxin Interaction with Cells*

Clostridium difficile toxins A and B are active against the more than 20 different cell lines from different mammalian species and tissues tested to date (Lyerly et al., 1988b; Florin, 1991). The primary morphological effect induced by toxins A and B in cell monolayers is similar to that caused by cytochalasin B, namely, retraction and rounding up of the cell body caused by a rearrangement of actin and α-actinin microfilaments (Fiorentini and Thelestam, 1991). These changes may not be readily distinguished from one cell type to another or from the rounding effects exerted by other unrelated toxins in some cell lines such as cholera toxin, *E. coli* LT, ACTH, and trypsin (Donta and Shaffer, 1980), and in most cell lines studied to date, the morphological changes elicited by toxin A and B exposure are identical, a notable exception being the HEp-2 cell line (Fiorentini et al., 1989). In contrast, however, the rounding effects of *C. difficile* toxins on CHO cells are quite different from the elongation effects of cholera toxin or *E. coli* LT on CHO cells (Guerrant et al., 1974).

Compared to toxin B, toxin A is much less cytotoxic, with the reported activity 1000–10,000-fold less than that of toxin B in HeLa cells (Chang et al., 1978), Chinese hamster ovary (CHO) cells (Sullivan et al., 1982; Lyerly et al., 1986b), and human lung fibroblasts (Taylor et al., 1981). In CHO cells, toxins A and B appear to have additive cytopathic effects (Lima et al., 1988). A human colon carcinoma cell line (HT-29) and rhesus monkey kidney cells (MA-104, epithelial) have been shown to be exquisitely sensitive to toxin A, demonstrating cytopathic changes to picogram and subpicogram quantities, respectively. In contrast to all other cell lines investigated to date, HT-29 and MA-104 were much more sensitive to toxin A and less responsive to toxin B, which may make them suitable models for testing clinical samples for toxin A (Torres et al., 1992).

There is contradictory evidence regarding the reversibility of toxin A-induced cytopathogenic changes. Henriques et al. (1987) found that once internalization occurs, the toxic action in human lung fibroblasts is irreversible; in contrast, Kushnaryov and Sedmak (1989) observed that CHO cells returned to normal when toxin A was removed from them within 2 h of its introduction, and Lima et al. (1988) reported a 100-fold decrease in toxicity when toxin A was rinsed off CHO monolayers after 5 min of incubation. Cytopathic effects in this study were unchanged by rinsing at 24 h. More recent evidence from our laboratory suggests that the effects of toxin A on T84 cell monolayer resistance become progressively less reversible over 0.5–2 h (Yotseff et al., 1993). In macrophages, the action of either toxin was found to be reversible, taking 7–10 days for cells to regain their original morphology (von Eichel-Streiber et al., 1991). Fiorentini and Thelestam (1991) have found that the morphological changes induced by toxin A are generally irreversible.

2. Effects on Plasma Membrane

The mechanism by which these toxins exert their effects on cells has yet to be clearly defined. Though Kushnaryov and Sedmak (1989) detected fenestration of the plasma membrane in CHO cells within 1 h of toxin A administration, this did not portend cell lysis, and its relationship to other ultrastructural events observed was largely unexplained. In contrast, von Eichel-Streiber and coworkers (1991) found no evidence of pore formation as assessed by monitoring for hemoglobin release from erythrocytes. Furthermore, CHO and T-84 cells (Lima et al., 1988) and L929 cells (von Eichel-Streiber et al., 1991) remain viable (as determined by trypan blue exclusion testing) after 48–96 h of treatment with toxin A or B (Lima et al., 1988; von Eichel-Streiber et al., 1991). In accordance with this, T. J. Mitchell et al. (1987) found that toxins A and B did not damage the membranes of McCoy cells or isolated intestinal cells as judged by the leakage of labeled uridine. Others (Fiorentini and Thelestam, 1991) have observed leakage of ^{3}H-labeled uridine nucleotides from toxin A-treated cells but noted significant cytopathogenic effects prior to release of the radiolabeled marker, suggesting that altered membrane permeability is an effect rather than a cause of morphology changes. While the exact nature of the *Clostridium difficile* toxin activity on the plasma membrane is not clearly defined, there is an accumulating body of literature to suggest that the toxin primarily targets cytosolic structures rather than the plasma membrane per se.

3. Ultrastructural Changes

Toxins A and B cause some changes in cell organelles, but this observation is inconsistent between different cell lines. In CHO cells, Kushnaryov and Sedmak (1989) observed extensive alteration of the Golgi apparatus and of the smooth endoplasmic reticulum and lysosomal changes within 3 h of the initial toxic effect, suggesting disturbances of the synthetic and secretory functions of these cells. The nucleus was also irregularly shaped, with cytoplasmic intermediate filaments collapsed toward it; nucleoplasmic filamentous changes were also noted (see below). In three types of macrophages studied, rounding early after exposure to either toxin was followed by ultrastructural changes in mitochondria and the rough endoplasmic reticulum (von Eichel-Streiber et al., 1991). Similarly, Triadafilopoulos et al. (1987) found widespread nonspecific dilatation of the endoplasmic reticulum and mitochondrial swelling in rabbit ileal epithelial cells exposed to toxin A. In HeLa cells exposed to toxin A, during the final stages of intoxication, after retraction of the cell body, the nucleus becomes kidney-shaped and polarizes to one pole; Golgi apparatus stalks are markedly increased, and numerous vesicles are visible (Fiorentini et al., 1990; Fiorentini and Thelestam, 1991). In contrast to these findings, Hecht et al. (1988) observed no ultrastructural differences between control and toxin A-treated human intestinal epithelial T84 cells. The exact significance of these findings is uncertain, but it is likely that they are ultimately related to changes in the cytoskeleton, as positioning and shape of intracellular organelles is strongly dependent on cytoskeletal components.

4. Cytoskeletal Changes

Evidence is accumulating that *C. difficile* toxin A (Fiorentini et al., 1989, 1990; Fiorentini and Thelestam, 1991) and toxin B (T. J. Mitchell et al., 1987; Donta and Shaffer, 1980; Ottlinger and Lin, 1988; Pothoulakis et al., 1986; Wedel et al., 1983) mediate morphological changes in cells primarily by targeting elements of the cytoskeleton. The action of both toxins of *C. difficile* is comparable to that of cytochalasins B and C—all of these agents bind to actin, lead to an increase of the monomeric fraction of actin, and yield nearly identical morphological changes in multiple cell lines (Ottlinger and Lin, 1988; M. J. Mitchell et al., 1987; von Eichel-Streiber et al., 1991). Thelestam and Bronnegard (1980), using antibodies against indirect immunofluorescence and antiactin antisera, provided the first direct evidence that the microfilament bundles are disrupted by crude *C. difficile* toxin preparations. Ottlinger and Lin (1988) demonstrated that the morphological effects of toxin B are directed specifically against actin and its related components, noting that, along with F-actin, the adhesion plaque proteins vinculin and talin are displaced. Fiorentini et al. (1989) similarly showed that toxins A and B modify the cytoskeleton by disaggregation of actin filaments in HEp-2 cells. In the human leukemic T-cell line JURKAT, toxin A arrested cell division and induced multinucleation, suggesting that the cytoskeletal disturbance can inhibit the formation of the contractile ring and cytokinesis (Fiorentini et al., 1992).

M. J. Mitchell et al. (1987) used direct fluorescence staining of actin filaments to study the activity of toxin B on HeLa and 3T3 cells and demonstrated that morphological changes were preceded by a decrease in the number and length of stress fibers, followed by their disappearance, with condensation of actin around the nucleus. Thus it was suggested that the action of toxin B on cells is related to its effect(s) on cellular proteins involved in the regulation of actin assembly. This argument is supported by von Eichel-Streiber et al. (1991) and Ottlinger and Lin (1988), who reason that because toxin B can exert its effect at ratios as low as 1–10 molecules per cell, its toxicity is targeted at some regulatory mechanism of actin assembly rather than actin itself, and likely via the liberation of an enzymatic activity.

Some have suggested that the cytoskeletal changes induced by toxin A may be responsible for the profound secretory fluid response observed in vivo in intestinal loop studies and ultimately for the diarrhea it produces in animals and humans. Hecht et al. (1988) showed that toxin A produces a marked permeability defect in human intestinal epithelial (T84) monolayers that localizes to the intercellular tight junction. Further, over the same time course, F-actin staining is markedly diminished. Hence, the investigators hypothesized that toxin A alters epithelial barrier function by altering the cytoskeleton (Hecht et al., 1988). Similarly, in experiments with toxin B, actin rearrangement was shown to correlate with increased tight junction permeability and altered transepithelial resistance as measured in modified Ussing chambers (Hecht et al., 1992). The effect of toxin B on transepithelial resistance was not altered by the inhibition of protein synthesis with cycloheximide, suggesting that the interruption of actin assembly by toxin B is not secondary to induction of a new protein.

In addition, in CHO cells, toxin A appears to disrupt bundles of 11-nm filaments within the nucleoplasm (Kushnaryov and Sedmak, 1989). These intermediate nuclear filaments are likely related to the lamina proteins, which are important for maintaining nuclear shape.

5. Toxin Binding

Specific glycoconjugate receptors for toxin A have been identified on rabbit erythrocytes (Krivan et al., 1986; Clark et al., 1987), hamster brush border membrane (Krivan et al., 1986), human intestinal epithelial cells (Tucker and Wilkins, 1991), and rabbit ileal brush border (Tucker et al., 1990) and are likely present on human granulocytes (Tucker and Wilkins, 1991). In pivotal investigations of *C. difficile* toxin receptors, Krivan et al. (1986) found that toxin A bound to rabbit erythrocytes with a high degree of specificity. It was further deduced that the receptor on rabbit erythrocytes and brush border membranes of the hamster (a mammal known to be extremely sensitive to toxin A) included the trisaccharide Galα 1-3 Galβ1-4 GlcNAc (Krivan et al., 1986), the full sequence of which is necessary for recognition by the toxin (Clark et al., 1987). This toxin's avid binding to Galα 1-3 Galβ1-4 GlcNAc at 4°C was since employed in a thermal affinity purification process using bovine thyroglobulin, which expresses this glycoprotein (Krivan and Wilkins, 1987; Kamiya et al., 1989). That three mouse teratocarcinoma cell lines that express the trisaccharide receptors are more sensitive to toxin A suggests that this receptor may be involved in the mediation of toxin A-induced cytotoxicity (Tucker et al., 1990). Evidence that this receptor actually mediates enterotoxicity is unclear to date and based largely on indirect findings, namely that antisera raised against the carbohydrate-binding portion of toxin A protects against enterotoxicity (Lyerly et al., 1990; Tucker and Wilkins, 1991). Sialic acid residues do not appear to be required for binding toxin A (Shahrabadi et al., 1984). That tunicamycin treatment of cells reduced the binding of toxin A suggests that the toxin receptor may be glycosylated (Shahrabadi et al., 1984).

Pothoulakis et al. (1991) reported that the toxin A receptor on rabbit ileal brush border is a protease-sensitive glycoprotein containing α-D-galactose. Further, they found that ^{3}H-labeled toxin A binding was modulated by guanine nucleotides, suggesting that the toxin A receptor may be coupled directly to a G protein (Pothoulakis et al., 1991).

In a separate study, newborn rabbit ileum was relatively devoid of toxin A brush border binding sites—a finding that correlates with a diminished biological response to the toxin in this species and lends credence to an analogous theory explaining human infant resistance to *C. difficile* diarrhea and colitis (Eglow et al., 1992). This is also consistent with the finding of Chang et al. (1986) that fetal intestinal mucosal cells were more resistant to the effects of both toxins A and B than were adult intestinal mucosal cells.

The carbohydrate receptor Galα1-3 Galβ1-4 GlcNAc identified by Krivan et al. (1986) is not present on normal human cells (Galili, 1989; Galili et al., 1988) and therefore likely has no role in the pathogenesis of *C. difficile* toxin A-mediated disease in humans (Tucker and Wilkins, 1991). Three carbohydrate antigens, I, X, and Y, that do exist on human intestinal epithelium were found to bind toxin A (Tucker and Wilkins, 1991). These all share the disaccharide Galβ1-4 GlcNAc, the minimum carbohydrate structure bound by the toxin A. Furthermore, it has been noted that human granulocytes express large amounts of the X antigen (Gooi et al., 1983; Huang et al., 1983; Spooncer et al., 1984). As granulocytes likely participate in the pathogenesis of *C. difficile*-associated diarrhea and colitis (Triadafilopoulos et al., 1987, 1991; Pothoulakis et al., 1988), there is increasing speculation that toxin A produces colonic damage not only through its direct effects on epithelial cells but also via inflammatory mediators released from granulocytes (Triada-filopoulos et al., 1987, 1989; Fang et al., 1993).

No receptor has been identified for toxin B to date.

6. Toxin Uptake

It has been convincingly demonstrated that cellular internalization is necessary for intoxication by both toxin A (Henriques et al., 1987) and toxin B (Florin and Thelestam, 1983; Ciesielski-Treska et al., 1989). Ferritin-labeled toxin A has been shown to enter mouse L cells and localize within cytoplasmic vacuoles as visualized by electron microscopy (Shahrabadi et al., 1984). Further studies used indirect biochemical evidence to demonstrate that internalization and processing is necessary for toxin A to be cytopathogenic (Henriques et al., 1987). Specifically, antitoxin, trypsin, monensin, lysotropic amines, and inhibitors of energy metabolism all rendered the toxin ineffectual in producing morphological changes (Henriques et al., 1987). In accordance with these findings, Kushnaryov and Sedmak (1989) demonstrated that toxin A internalization in CHO cells occurs at least in part via receptor-mediated endocytosis.

There is similar indirect evidence to suggest that in order for toxin B to exert its effects, it must be internalized and exposed to an acidic environment (Florin and Thelestam, 1983). Von Eichel-Streiber et al. (1991) reported that, like toxin A, gold-labeled toxin B enters macrophages via endocytosis, localizes to lysosome, and later associates with membrane structures when altered cell morphology is detected. Further direct evidence of its internalization was provided by Ciesielski-Treska et al. (1989), who visualized penetration of the toxin into cultured astroglial cells using specific immunoaffinity purified antitoxin B immunoglobulin. Internalization of the toxin required extracellular calcium and was noted to precede characteristic morpho-logical changes. Katoh et al. (1988) showed that cytotoxic activity of toxin B was completely inhibited by adding anticytotoxin serum at 2 min after challenge to CHO cells. This effect was not seen when anticytotoxin was added 5 min after toxin challenge, suggesting that binding and internalization steps necessary for toxic effects are likely completed within 5 min. Caspar et al. (1987), finding that calcium channel blockers and calmodulin antagonists protected cells against intoxication with toxin B when added within 30 min of toxin B binding, suggest that calcium and active calmodulin are required for internalization of the toxin but not for its cytosolic action. Further, because entry of toxin B into cells is temperature-sensitive and may involve endocytosis, it is likely an energy-dependent process (Florin and Thelestam, 1983; Ciesielski-Treska et al., 1989).

7. Intracellular Processing

Latency. While it is fairly well established that the cytoskeleton is the primary target of both toxin A and B intoxication, the intracellular mechanisms by which they exert their effects are largely unknown. Toxins A and B are not immediately active, and latency periods of between 30 min and 4 h have been reported (von Eichel-Streiber et al., 1991; M. J. Mitchell et al., 1987; Kushnaryov and Sedmak, 1989; Henriques et al., 1987; Ciesielski-Treska et al., 1989; Fiorentini et al., 1989). In general, this finding is suggestive of intracellular processing, activation, or both (von Eichel-Streiber et al., 1991).

Energy Dependency. Given the energy requirements necessary for internalization, it is not surprising that starvation of energy supplied by mitochondria (using 2,4-dinitrophenol or sodium fluoride) blocks the cytopathogenic effects of both toxin A (Henriques et al., 1987; von Eichel-Streiber et al., 1991) and toxin B (von Eichel-Streiber et al., 1991; Florin and Thelestam, 1983). It still remains unclear what energy requirements exist for later stages of the intoxication process (Fiorentini and Thelestam, 1991).

Effects on Protein Synthesis. Rothman et al. (1984) reported that HeLa cells exposed to either toxin at standard experimental doses demonstrated a loss of intracellular potassium and inhibition of protein synthesis—after the appearance of altered cell morphology. On the other hand, in McCoy cells, protein synthesis was inhibited by both toxins A and B before the onset of altered cell morphology, but in companion in vivo experiments no decrement in protein was detected, and the significance of the cell culture assay was questioned (T. J. Mitchell et al., 1987). In human lung fibroblasts, the cytopathic effects were not diminished by inhibitors of protein synthesis, RNA, or DNA (Florin and Thelestam, 1983). Furthermore, Fiorentini and Thelestam (1991) report that in IEC-6 cells treated with toxin A, inhibition of [^{3}H]thymidine into DNA occurred only after cytopathogenic effects were observed.

Thus, while there are some conflicting data concerning the effects of intoxication on protein synthesis, the time course suggests that any inhibition in protein synthesis is a consequence of the toxins' cytopathic effects rather than a cause (Fiorentini and Thelestam, 1991).

Lysosomal Processing. After toxins A and B are internalized, they are processed differently. Florin and Thelestam (1983, 1986) have shown that toxin B-induced cytopathogenic changes were inhibited by lysosomal proteases; further, they found that CHO cell mutants that were defective in their ability to acidify lysosomes were resistant to the effects of toxin B. Toxin B does not depolymerize filamentous actin in vitro but does decrease F-actin in intact cells (M. J. Mitchell et al., 1987). Together, these findings provide further indirect evidence to support the notion that toxin B requires activation in lysosomes to effect cytoskeletal disruption (Cuzzolin et al., 1992).

In contrast, toxin A probably does not have to undergo lysosomal processing. Henriques et al. (1987) showed that a number of lysosomal proteases had no effect on the cytopathogenicity of toxin A for cultured human lung fibroblasts. They further hypothesized that in order to be delivered to the cytosol, toxin A needs to be transported from endosomes by vesicular fusion to some other compartment, perhaps the Golgi complex, where it may undergo some enzymatic activation (Henriques et al., 1987). Work by Fiorentini et al. (1990) supported this model by showing that toxin A cytopathogenicity was dependent on a serine protease. The serine protease inhibitor chymostatin, added long after the end of the binding step, was protective against the cytopathogenicity of toxin A in five human and one murine cell line.

The Role of Calcium. Although calcium is necessary for the binding of toxin A to IEC-6 cells (Fiorentini and Thelestam, 1991) and for the internalization of toxin B into astroglial cells (Ciesielski-Treska et al., 1989), it is not required for the cytosolic action of these toxins (Fiorentini et al., 1990). In addition, Caspar et al. (1987) demonstrated that calmodulin and extracellular calcium are needed for internalization of toxin B as calcium channel blockers and calmodulin antagonists protected human lung fibroblasts against intoxication. However, when added after internalization, they were not protective.

Lima et al. (1988) reported that toxin A-induced cytopathic changes in CHO cells were not affected by pretreatment of cells with diltiazem or the calmodulin inhibitor trifluoperazine. These findings suggest that, while calcium-dependent pathways may be used for binding and internalization in some cell lines, they do not appear to be necessary for toxins A and B to mediate cytosolic activities.

Interactions with Cellular Proteins.   It is also possible that *C. difficile* toxins induce cytopathogenicity by altering cellular proteins. Ciesielski-Treska et al. (1991) found that toxin B (and *Clostridium sordelli* toxin L) provokes alteration in the phosphorylation of several cellular proteins, including tropomyosin, which stabilizes microfilaments. It does not, however, appear to change protein kinase C activity, nor does it result in increased levels of cAMP, suggesting that protein kinase A is not activated by this toxin (Ciesielski-Treska et al., 1991).

The Role of Cyclic Nucleotides.   The toxins of *C. difficile* do not appear to mediate their cellular effects via alteration in cyclic nucleotide activity. Vesely et al. (1981), in an early study, showed that partially purified cytotoxin from four strains of *C. difficile* stimulated hamster colonic guanylate cyclase activity while inhibiting adenylate cyclase and noted that toxins from the strains that produced more marked cytopathic changes were the same toxins associated with increased guanylate cyclase activity. Other workers, however, have been unable to identify any changes in cyclic nucleotide activity in cells exposed to crude extracts of *C. difficile* toxins (Hughes et al., 1983; Donta and Shaffer, 1980), purified toxin A (Lonnroth and Lange, 1983), or purified toxin B (Ciesielski-Treska et al., 1991). The findings of Vesely et al. (1981) likely reflected guanylate cyclase activity from sites other than epithelial cells, since crude homogenized whole gut tissue was used in their assays (Hughes et al., 1983).

G. Biological Activity of *C. difficile* Toxins

1. Lethality

Both toxins A and B are lethal. Hamsters (Taylor et al., 1981), mice (Taylor et al., 1981; Banno et al., 1984; Lyerly et al., 1986b; Sullivan et al., 1982), and infant rhesus monkeys (Arnon et al., 1984) receiving small quantities of either toxin through a variety of routes die. Physiologic monitoring and necroscopy after intraperitoneal and intravenous administration of the toxins to infant rhesus monkeys did not help to clarify the manner in which they caused death (Arnon et al., 1984). When injected intraperitoneally into mice, toxin A was found by some (Banno et al., 1984; Taylor et al., 1981) to be more lethal than toxin B; others found them roughly equipotent (Lyerly et al., 1986b; von Eichel-Streiber et al., 1987). The mechanism by which intraperitoneal toxin administration causes death is not defined, though Ehrich (1982) showed that with either toxin there is evidence of liver damage.

2. Vascular Permeability

Intradermal injections of small quantities of toxins A and B into rabbits (Banno et al., 1984; Lyerly et al., 1982) and guinea pigs (Taylor et al., 1981) produce erythematous and hemorrhagic lesions; slightly high doses increase vascular permeability as assessed by Evans blue dye studies (Lyerly et al., 1982; Taylor et al., 1981). Only Taylor et al. (1981) note a qualitative difference between responses to toxins A and B, reporting that the lesion produced by toxin B had well-defined boundaries and little or no edema whereas that of toxin A was diffuse and edematous.

3. Enterotoxicity

Overview. Using cecal filtrates from hamsters and clostridial culture filtrates, Bartlett (1985) demonstrated the enterotoxicity of *Clostridium difficile* toxins in whole animals. Later, Taylor et al. (1981), Banno et al. (1984), and Lima et al. (1988) demonstrated that this enterotoxicity was attributable to purified toxin A and that toxin B had no enterotoxic effect.

Infant Mouse Studies. In the infant or suckling mouse assay, intestinal fluid response is measured following intragastric administration of the toxin. Lyerly et al. (1982) reported a positive fluid response with submicrogram quantities of toxin A. Toxin B was much less active; at the high doses (1–5 μg) necessary to produce a response, more than half of the mice died. In contrast, Taylor et al. (1981) found that neither toxin actively produced a fluid response, but mice given toxin A were moribund at 3 h.

Hamster Models. The Golden Syrian hamster has been used extensively to study the activity of *Clostridium difficile* toxins in vivo in the context of active infection with the bacillus (Small, 1968; Lusk et al., 1978; Chang et al., 1978; Bartlett et al., 1978b; Allo et al., 1979; Abrams et al., 1980; Libby et al., 1982; Rolfe and Iaconis, 1983; Kim et al., 1987; Lyerly et al., 1991; Chang and Rohwer, 1991). This model is useful for several reasons: It mimics the usual clinical situation in which prior antibiotic administration precipitates the disease (Small, 1968; Lusk et al., 1978), hamsters are sensitive to human-derived isolates of *C. difficile* (Gemmell, 1984), and disease activity appears to be mediated by the toxins of *C. difficile* (Chang et al., 1978; Bartlett et al., 1978b; Allo et al., 1979; Abrams et al., 1980; Lyerly and Wilkins, 1988). This model does have limitations, however, particularly with regard to localization of pathology. Whereas *C. difficile* disease in humans is limited to the colon, in hamsters the disease causes an enterocolitis with intense cecal involvement (Small, 1968; Bartlett et al., 1977b).

The most striking gross pathologic finding is distention in the cecum and ileum and to a lesser extent in the jejunum, duodenum, and large intestine. Microscopically, numerous polymorphonuclear leukocytes crowd the lamina propria and extrude into the lumen, and epithelial cells are disrupted over the tips of villi (Small, 1968).

Purified toxin A instilled directly into the cecum of hamsters induces severe hemorrhage, disrupted villi, mucosal edema, fluid accumulation, and death (Taylor et al., 1981; Libby et al., 1982). Taylor et al. (1981) found that toxin B produced only small areas of focal hemorrhage; it did not result in death. In contrast, Libby et al. (1982) observed extensive hemorrhage and inflammation with toxin B, but little or no fluid accumulation.

When administered intragastrically to hamsters, toxin A produces hemorrhage and fluid accumulation in the cecum and small intestine with diarrhea and eventual death. Alone, toxin B has no effects, but when given with small amounts of toxin A or given to hamsters with bruised ceca, toxin B administration results in death without antecedent symptoms (Lyerly et al., 1985). These findings led Lyerly et al. (1985) to hypothesize that toxins A and B act synergistically and that the tissue damage caused by toxin A may facilitate the activity of toxin B.

Organ Models and in vitro *Systems.*

Histopathology: Mitchell et al. (1986), Lyerly et al. (1982), and Triadafilopoulos et al. (1987) all demonstrated that toxin A, but not toxin B, caused tissue damage and permeability changes with fluid accumulation in rabbit ileal and colonic loops. A sequential study of early histopathological changes in rabbit small intestine by Lima et al. (1989) revealed an initial diffuse lymphocytic infiltrate in the lamina propria followed by edema and bulging of the lamina propria in the apical portions of villi, and eventual cytolysis and separation of the vasal portions of the apical epithelial cells. At this early stage, not associated with significant intestinal fluid accumulation, the epithelial cell brush border and the basement membranes remained intact—lending some credence to the hypothesis that fluid accumulation results primarily from cell disruption at the basement membrane (Lima et al., 1989). An acute neutrophilic inflammatory response was notably absent from these early changes (Lima et al., 1989). Triadafilopoulos et al. (1987), however, in examining histologic changes later after toxin A exposure, reported an acute inflammatory response in which polymorphonuclear leukocytes and mononuclear cells infiltrate the lamina propria. This was followed by epithelial cell necrosis and ulceration with hemorrhagic edema. In addition, the toxin induced secretion of a protein-rich hemorrhagic fluid that contained significant numbers of neutrophils after a 16-h incubation. In contrast, toxin B had no demonstrable effect on permeability, neutrophil migration, or changes in intestinal morphology (Triadafilopoulos et al., 1987).

Similarly, Mitchell et al. (1986), investigating the effects of toxins A and B on both rabbit ileum and colon, found that only toxin A was active, that the ileum was more vulnerable morphologically and more prone to secrete fluid than the colon, and that damage progressed from localized destruction of the villus tips to entire destruction of villi with hemorrhage in the ileum (Mitchell et al., 1986).

Secretion studies and permeability: When administered in equimolar amounts to ligated loops of rabbit small intestine, toxin A promotes fluid secretion as efficiently as cholera toxin, though the mechanisms of action for the two are completely different (Lima et al., 1988). There has been limited evidence that *C. difficile* toxins alone are secretagogues. Using rabbit ileal loops, Lima et al. (1988) demonstrated that toxin A causes a significant net accumulation of sodium, chloride, potassium, and total protein at 6 h; but such fluid secretion was observed only in the presence of significant mucosal damage. This is in contrast to a previous report by Hughes et al. (1983), who suggested that a crude extract of *C. difficile* toxins could abolish net sodium absorption and induce net chloride secretion in the absence of visible histological damage in Ussing chambers. These ion fluxes appeared to be calcium-dependent.

In guinea pig ileum in vitro, toxin A appears to assault intestinal epithelial barrier structure and function by increasing tight junction permeability. Moore et al. (1990) correlated perturbations in intercellular tight junctions of these epithelial cells with increased chloride secretion and diminished sodium absorption in Ussing chambers. It is notable that permeability was increased in systems that included intact lamina propria that were relatively devoid of neutrophils.

Similarly, in a human intestinal epithelial cell line (T84), perturbations of barrier function and increased tight junction permeability were associated with alteration of the cytoskeletal structure (see fourth subsection of Section IV.F) (Hecht et al., 1988).

Thus it appears that toxin A may alter intestinal barrier function directly through its effects on tight junction permeability in vitro; it may also elicit a secretory response indirectly activating an inflammatory response.

Effects on smooth muscle: Justus et al. (1982) reported that unpurified supernatants from *C. difficile* cultures, but not toxins A and B, induced specific smooth muscle myoelectrical responses in rabbit ileal loops and hypothesized that this may in part explain the diarrhea associated with disease states. Although this myoelectric factor has not been further characterized, Smith et al. (1993) have recently shown that *C. difficile* toxin A elicits bursts of spontaneous contractile activity, which suggests hyperactivity of intrinsic sensory neurons in the enteric nervous system.

Others (Gilbert et al., 1989b) have found that in vivo administration of toxin A into a rabbit intestinal loop causes significant alterations of smooth muscle excitation–contraction coupling. That this phenomenon is not observed with in vitro preparations suggests that toxin A may exert its effect on smooth muscle via products of local inflammatory cells (Gilbert et al., 1989b). In contrast, toxin B has no effects on smooth muscle strips when administered in vivo, but in vitro it causes membrane depolarization in association with inhibition of electromechanical activity (Gilbert et al., 1989a). It is uncertain how the effects of these toxins on smooth muscle in these studies contribute to altered intestinal motility accompanied by *C. difficile*-associated diarrhea.

H. Interactions with the Immune System

1. Introduction: The Inflammatory Response

The final pathogenic mechanisms in *C. difficile*-mediated colitis and diarrhea likely involve some interplay between the direct effects of its toxins on intestinal epithelium and their indirect effects via activation of the host inflammatory response. As toxin A has no demonstrable effect on ileal explants compared with the in vivo ileum, Triadafilopoulos et al. (1987) argue that much of the pathology of toxin A tissue damage in vivo is likely mediated by its activation of inflammatory cells rather than the toxin itself. This is further substantiated by their observation that, in vivo, this toxin not only produces secretion of a protein-rich hemorrhagic fluid in ligated rabbit ileal loops but also elicits an acute inflammatory response in which polymorphonuclear leukocytes and mononuclear cell infiltration of the lamina propria is followed by epithelial cell necrosis and ulceration and hemorrhagic edema (Triadafilopoulos et al., 1987). These experimental findings are certainly consistent with prior histopathological descriptions of inflammation with polymorphonuclear neutrophils (PMNs) prominent in rectal biopsies of patients with pseudomembranous colitis (Price and Davies, 1977). Furthermore, we have recently shown that pertussis toxin, which blocks the PMN effect but has no effect on the epithelial effect of toxin A, completely blocks the secretory response in vivo, further implicating the PMNs in the intestinal secretion seen (Fang et al., 1993).

In contrast, toxin B had no demonstrable effect on permeability, neutrophil migration, or changes in intestinal morphology (Triadafilopoulos et al., 1987).

2. Cellular Immunity

Polymorphonuclear Leukocytes. Polymorphonuclear leukocytes likely play a significant role in pathophysiology of this disease as they can be found located within the pseudomembrane and in the intestinal mucosal layer underneath the pseudomembrane (Czuprynski et al., 1983; Sumner and Tedesco, 1975; Onderdonk et al., 1980; Dailey et al., 1987). In contrast to Gemmell (1984), who reported that purified cytotoxin inhibits phagocytic activity of PMNs, Dailey et al. (1987), using crude preparations with both enterotoxic and cytotoxic activity, found no effect on either phagocytic acitivity or viability of human polymorphonuclear leukocytes. Multiple isolates of *C. difficile* (especially those that were toxigenic), however, proved to be very resistant to phagocytosis. Thus, it appears that the relative resistance of *C. difficile* to polymorphonuclear phagocytosis is not mediated by its toxins but is attributable to other virulence factors of the bacillus (Dailey et al., 1987).

More recent studies suggest a stimulatory effect for toxin A on granulocytes. Toxin A directly activates human granulocytes in vitro, inducing a transient rise in unbound cytosolic calcium that is rapid and dose-dependent while promoting chemotaxis and chemokinesis (Pothoulakis et al., 1988). That this effect was abolished by pretreating granulocytes with pertussis toxin (which catalyzes ADP-ribosylation of several members of the guanine nucleotide regulatory protein family, including G_i, G_o, and G_n) suggests that toxin A-induced intracellular calcium fluxes are mediated through a pertusis-toxin-inhibitable G protein (Pothoulakis et al., 1988). Triadafilopoulos et al. (1991) demonstrated that toxin A is more chemoattractive in granulocytes of elderly subjects than in those of younger subjects, arguing that this may explain in part the epidemiological observation that *C. difficile* colitis occurs more frequently and with greater severity in the elderly.

The effects of highly purified toxin B on granulocytes is unknown.

Monocytes and Lymphocytes.   Toxin A activates mouse peritoneal macrophages to release IL-1 (Miller et al., 1990). This (and possibly other soluble factors) is comitogenic with the calcium ionophore ionomycin for murine B lymphocytes as assessed by [^{3}H]thymidine incorporation (Miller et al., 1990). Flegel et al. (1991) showed that *C. difficile* toxins (toxin B more so than A) are potent activators of human monocytes, inducing production of IL-1, TNF, and IL-6. Further, noting that *C. difficile* toxins are presented to the intestinal epithelium in the presence of various other enteric bacteria, predominantly gram-negative, it was shown that lipopolysaccharide with toxin A or B produced synergistic release of cytokines (Flegel et al., 1991). Malorni et al. (1991) showed that treatment of human peripheral blood mononuclear cells with toxin A enhances their cytotoxic efficiency.

Interestingly, toxin B was also shown to be extremely toxic to monocytes (Flegel et al., 1991). Hence, in humans, toxin B may first activate monocytes, thereby promoting inflammation, and then kill these cells, rendering them incapable of further damaging the *C. difficile* bacillus (Flegel et al., 1991). It is notable that a similar pattern of both activation and toxicity has been observed in several other bacterial toxins including *Pseudomonas aeruginosa* toxin A (Misfeldt et al., 1990) and staphylococcal toxin A (Bhakdi et al., 1989) as well as *E. coli* hemolysin (Bhakdi et al., 1990; Flegel et al., 1991).

Daubener et al. (1988) showed that purified human monocytes exposed to toxin A or B had a markedly reduced capacity to stimulate T-cell proliferation. The toxins had no effect on the proliferation of T cells, however, independent of monocytes (Daubener et al., 1988).

3.   Humoral Immunity

Although *Clostridium difficile* is not invasive, its toxins are capable of inducing an antibody response by the host, suggesting an ability to cross the mucosal barrier (Viscidi et al., 1983; Flegel et al., 1991). Individual serum response to toxin A varied widely in a recent clinical study characterizing toxin A-specific serum and secretory antibody responses in patients with *C. difficile* colitis (Johnson et al., 1992a). Whereas some patients with *C. difficile*-associated diarrhea had dramatic increases in convalescent titers, especially of IgA antibodies, others had no change. Though the statistical analysis was hampered by small numbers, it appears that the serum antibody response to natural infection did not effectively protect against relapse (Johnson et al., 1992a).

Libby et al. (1982) found that to protect hamsters colonized with highly toxigenic strains of *C. difficile*, vaccination with both toxins was necessary. Others (Kim et al., 1987) have shown that vaccination against toxin A alone is protective in hamsters. Corthier et al. (1991) reported that in gnotobiotic mice passive immunity conferred by monoclonal antibodies to toxin A was protective against pseudomembranous cecitis. Antibody-mediated protection is transferrable from mothers to infants through milk (Kim et al., 1987).

Recently, Kelly et al. (1992) demonstrated the presence of IgA antibody to *C. difficile* toxin A in colonic (but not duodenal) aspirates in 57% of human subjects undergoing diagnostic endoscopy and found that these aspirates inhibited the binding of ^{3}H-labeled toxin A to rabbit ileal brush border membrane. It is thus suggested that binding of toxin A by specific secretory antibody in the colon hinders toxin A binding to its intestinal epithelial receptor, thus protecting against *C. difficile* colitis. The investigators propose that this finding may account for the presence of toxin B in feces of asymptomatic patients and the transmission of toxin-producing *C. difficile* by asymptomatic carriers (Kelly et al., 1992).

I. The Role of Inflammatory Mediators

There is a growing body of evidence to suggest that arachidonic acid metabolites ultimately participate in both the inflammatory and secretory effects of *C. difficile*-mediated colitis and diarrhea. Like cholera toxin, toxin A mediates the release of prostaglandin E_2 in rabbit ileal loops in association with increases in fluid secretion and intestinal permeability (Triadafilopoulos et al., 1989). But unlike cholera toxin, the clostridial toxin increases intraluminal leukotriene B_4 and elicits an inflammatory response characterized by a neutrophil-rich exudate (Triadafilopoulos et al., 1987, 1989). Triadafilopoulos et al. (1991) suggest that toxin A recruits and activates neutrophils that release leukotrienes and other inflammatory mediators that ultimately contribute to intestinal secretion (Triadafilopoulos et al., 1989). It is also possible that toxin A stimulates the release of eicosanoids directly from the subepithelium, where the majority of intestinal prostaglandins are synthesized (Gaginella, 1990), to produce its secretory effects. Eicosanoids have been measured in vivo using equilibrium dialysis of the rectum in patients with ulcerative colitis, Crohn's colitis, and *Clostridium difficile* colitis. Concentrations of prostaglandin E_2, prostaglandin $F_{2\alpha}$, and thromboxane B_2 were statistically significantly elevated in seven patients with *C. difficile* colitis compared with 10 healthy controls (Lauritsen et al., 1988). Levels of leukotriene B_4 and 6-keto-PGF$_{1\alpha}$ were not different from controls. While this study was small and demonstrated more eicosanoids in patients with ulcerative colitis and Crohn's disease than in those with *C. difficile* colitis, it nevertheless lends some support to the hypothesis that arachidonic acid metabolites participate in the underlying pathophysiology of this disease. Furthermore, our recent findings implicate platelet activating factor as playing a key role in the pathogenesis of *C. difficile* toxin A-induced inflammation and secretion (Guerrant et al., 1992; Fonteles et al., 1994).

J. Genetics of *C. difficile* Toxins

Genes for both toxins A and B have been cloned and sequenced. Muldrow et al. (1987) cloned a 0.3 kb fragment of the toxin A gene in a lambda bacteriophage expression vector that expressed peptides reactive with toxin A polyvalent antisera. Subsequently, von Eichel-Streiber et al. (1989) cloned portions of a 4.7-kb fragment into a plasmid expression vector to obtain a product that reacted with toxin A antisera. When Wren et al. (1987) cloned toxin A in a lambda phage, the resultant protein had a molecular weight of 235,000, exhibited hemagglutinating activities, and caused elongation of Chinese hamster ovary cells. The gene for toxin A has 8133 base pairs that encode a protein with a deduced molecular weight of 308,000 that consists of 2710 amino acids (Dove et al., 1990). Phelps et al. (1991) reconstructed cloned fragments into *E. coli.* The resultant protein expressed was cytotoxic, enterotoxic, and lethal. The toxic lysate caused rabbit erythrocytes to hemagglutinate. Furthermore, the above biological activities of this protein were all neutralized with specific toxin A antibody. Hence, all of the biological activities of native *C. difficile* toxin A were reproduced by the single polypeptide encoded by the 8.1-kb toxin A gene (Phelps et al., 1991).

Toxin B has also been cloned and sequenced. The open reading frame is 7.1 kb long, coding for 2366 amino acids and yielding a deduced polypeptide molecular weight of 270,000 (Barroso et al., 1990).

V. ADDITIONAL VIRULENCE FACTORS

It is clear that toxin A is the single most important virulence factor of *C. difficile* and that toxin B likely contributes to the pathogenicity of this organism in ways yet to be defined. But, because not all toxigenic strains of *Clostridium difficile* are equally virulent, several nontoxin virulence factors have been suggested (Borriello et al., 1990).

Borriello et al. (1988b) demonstrated that highly virulent strains of *C. difficile* adhere to hamster intestinal mucosa. When culture filtrates were coadministered with nontoxigenic, avirulent, poorly adhering strains, the level of mucosal association rose to that of highly virulent strains. It remains unknown whether this effect was mediated by toxins A or B, the host inflammatory response, or other soluble factors not identified to date (Borriello, 1990). It is possible that adherence is mediated at least in part by fimbriae as some *C. difficile* isolates possess multiple polar fimbriae (Borriello et al., 1988a).

Another feature of this organism contributing to its pathogenicity is its known resistance to polymorphonuclear leukocytes. While purified cytotoxin was shown by Gemmell (1984) to inhibit the phagocytic activity of polymorphonuclear cells, Dailey et al. (1987) demonstrated antiphagocytic properties in toxigenic strains that appeared to be independent of the toxins themselves and hypothesized that this antiphagocytic factor could be attributable to a polysaccharide capsule. Davies and Borriello (1990) recently identified such a capsule in both toxigenic and nontoxigenic strains of *C. difficile.* They found no correlation between capsule size in vitro with virulence in vivo. No studies have correlated capsule size with antiphagocytic properties to date.

Hydrolytic enzyme production (Seddon et al., 1990) and cell-associated proteolytic activity (Seddon and Borriello, 1992) have also been described in multiple strains of *C. difficile.* With the former, no correlation between toxigenicity or virulence was noted (Seddon et al., 1990); the latter was correlated more with strains highly virulent in the hamster model (Seddon and Borriello, 1992).

In addition, Borriello et al. (1992) recently described the highly unusual occurrence of a strain of *C. difficile* (*C. difficile* 8864) that produces toxin B without toxin A and yet still, in hamsters, caused diarrhea with cecal damage characteristic of fully toxigenic strains. Though toxin B in this strain was slightly different from that of other strains, it failed to produce hemorrhage or fluid accumulation in ligated rabbit ileal loops in one investigation (Borriello et al., 1992). Others found this modified toxin B weakly enterotoxic (Lyerly et al., 1992). Thus it is suggested that yet additional enterotoxic factors are produced by some strains of *C. difficile* that are devoid of toxin A (Borriello et al., 1992). Work is ongoing to further characterize these putative agents, their relationship to toxins A and B, and their role in the pathogenesis of *C. difficile* disease.

VI. CONCLUSION

In the 15 years since *Clostridium difficile* was established as the cause of pseudomem-branous colitis, thorough scientific investigation has characterized the major toxins, defined their cytotoxic and biological activity, and yielded the genes that lead to their production.

Clinical research established the importance of nosocomial acquisition and suggested measures for interrupting outbreaks. Though antimicrobial treatment regimens are generally initially effective, recurrent *C. difficile* disease poses a significant clinical dilemma. The aging population, intense antibiotic regimens required for increasing numbers of immunocompromised hosts, and the continued use and development of broad-spectrum antimicrobial agents all suggest that *C. difficile* diarrhea and colitis will remain a serious clinical problem for some time.

It is to be hoped that future efforts will define the mechanisms of secretion evoked by toxin A, further characterize toxin interactions with inflammatory cells and their mediators, and establish the role of other putative virulence factors. In so doing, more effective means to diagnose and treat *C. difficile* diarrhea and colitis may be established.

REFERENCES

Abrams, G. D., Allo, M., Rifkin, G. D., Fekety, R., and Silva, J., Jr. (1980). Mucosal damage mediated by clostridial toxin in experimental clindamycin-associated colitis. *Gut 21*: 493–499.

Al-Jumaili, I. J., Shibley, M., Lishman, A. H., and Record, C. O. (1984). Incidence and origin of *Clostridium difficile* in neonates. *J. Clin. Microbiol. 19*: 77–78.

Allo, M., Silva, J., Jr., Fekety, R., Rifkin, G. D., and Waskin, H. (1979). Prevention of clindamycin-induced colitis in hamsters by *Clostridium sordellii* antitoxin. *Gastroenterology 76*: 351–355.

Andrejak, M., Schmit, J. L., and Tondriaux, A. (1991). The clinical significance of antibiotic-associated pseudomembranous colitis in the 1990s. *Drug Safety 6*: 339–349.

Arnon, S. S., Mills, D. C., Day, P. A., Henrickson, R. V., Sullivan, N. M., and Wilkins, T. D. (1984). Rapid death of infant rhesus monkeys injected with *Clostridium difficile* toxins A and B: physiologic and pathologic basis. *J. Pediatr. 104*: 34–40.

Aronsson, B., Mollby, R., and Nord, C. E. (1985). Antimicrobial agents and *Clostridium difficile* in acute enteric disease: epidemiological data from Sweden, 1980–1982. *J. Infect. Dis. 151*: 476–481.

Aronsson, B., Barany, P., Nord, C. E., Nystrom, B., and Stenvinkel, P. (1987). *Clostridium difficile*-associated diarrhoea in uremic patients. *Eur. J. Clin. Microbiol. 6*: 352–356.

Atkinson, M. H., and McLeod, B. D. (1988). Reactive arthritis associated with *Clostridium difficile* enteritis. *J. Rheumatol. 15*: 520–522.

Baird, D. R. (1989). Comparison of two oral formulations of vancomycin for treatment of diarrhoea associated with *Clostridium difficile* [Letter]. *J. Antimicrob. Chemother. 23*: 167–169.

Balfanz, J., and Rautenberg, P. (1989). Inhibition of *Clostridium difficile* toxin A and B by 1,2-cyclohexanedione modification of an arginine residue. *Biochem. Biophys. Res. Commun. 165*: 1364–1370.

Banno, Y., Kobayashi, T., Kono, H., Watanabe, K., Ueno, K., and Nozawa, Y. (1984). Biochemical characterization and biologic actions of two toxins (D-1 and D-2) from *Clostridium difficile. Rev. Infect. Dis. 6*(Suppl. 1): S11–S20.

Barc, M. C., Depitre, C., Corthier, G., Collignon, A., Su, W. J., and Bourlioux, P. (1992). Effects of antibiotics and other drugs on toxin production in *Clostridium difficile* in vitro and in vivo. *Antimicrob. Agents Chemother. 36*: 1332–1335.

Barroso, L. A., Wang, S. Z., Phelps, C. J., Johnson, J. L., and Wilkins, T. D. (1990). Nucleotide sequence of *Clostridium difficile* toxin B gene. *Nucleic Acids Res. 18*: 4004.

Bartlett, J. G. (1981). Antimicrobial agents implicated in *Clostridium difficile* toxin-associated diarrhea of colitis. *Johns Hopkins Med. J. 149*: 6–9.

Bartlett, J. G. (1985). Treatment of *Clostridium difficile* colitis [Editorial]. *Gastroenterology 89*: 1192–1195.

Bartlett, J. (1988). Introduction. In *Clostridium difficile: Its Role in Intestinal Disease*, R. D. Rolfe and S. M. Finegold (Eds.), Academic, New York, pp. 1–13.

Bartlett, J. G. (1990). *Clostridium difficile*: clinical considerations. *Rev. Infect. Dis. 12*(Suppl. 2): S243–S251.

Bartlett, J. G. (1992). Antibiotic-associated diarrhea. *Clin. Infect. Dis. 15*: 573–581.

Bartlett, J. G., Onderdonk, A. B., and Cisneros, R. L. (1977a). Clindamycin-associated colitis in hamsters: protection with vancomycin. *Gastroenterology 73*: 772–776.

Bartlett, J. G., Onderdonk, A. B., Cisneros, R. L., and Kasper, D. L. (1977b). Clindamycin-associated colitis due to a toxin-producing species of *Clostridium* in hamsters. *J. Infect. Dis. 136*: 701–705.

Bartlett, J. G., Chang, T. W., Gurwith, M., Gorbach, S. L., and Onderdonk, A. B. (1978a). Antibiotic-associated pseudomembranous colitis due to toxin-producing clostridia. *N. Engl. J. Med. 298*: 531–534.

Bartlett, J. G., Chang, T. W., Moon, N., and Onderdonk, A. B. (1978b). Antibiotic-induced lethal enterocolitis in hamsters: studies with eleven agents and evidence to support the pathogenic role of toxin-producing clostridia. *Am. J. Vet. Res. 39*: 1525–1530.

Bartlett, J. G., Taylor, N. S., Chang, T., and Dzink, J. (1980). Clinical and laboratory observations in *Clostridium difficile* colitis. *Am. J. Clin. Nutr. 33*: 2521–2526.

Bartlett, J. G., Laughon, B. E., and Bayless, T. M. (1989). Role of microbial agents in relapses of idiopathic inflammatory bowel disease. In *Current Management of Inflammatory Bowel Disease*, T. M. Bayless and B. C. Decker (Eds.), C. V. Mosby, St. Louis.

Batts, D. H., Martin, D., Holmes, R., Silva, J., and Fekety, F. R. (1980). Treatment of antibiotic-associated *Clostridium difficile* diarrhea with oral vancomycin. *J. Pediatr. 97*: 151–153.

Bender, B. S., Bennett, R., Laughon, B. E., Greenough, W. B., Gaydos, C., Sears, S. D., Forman, M. S., and Bartlett, J. G. (1986). Is *Clostridium difficile* endemic in chronic-care facilities? *Lancet 2*: 11–13.

Bennett, R. G., Laughon, B. E., Mundy, L. M., Bobo, L. D., Gaydos, C. A., Greenough, W. B., and Bartlett, J. G. (1989). Evaluation of a latex agglutination test for *Clostridium difficile* in two nursing home outbreaks. *J. Clin. Microbiol. 27*: 889–893.

Bhakdi, S., Muhly, M., Korom, S., and Hugo, F. (1989). Release of interleukin-1 beta associated with potent cytocidal action of staphylococcal alpha-toxin on human monocytes. *Infect. Immun. 57*: 3512–3519.

Bhakdi, S., Muhly, M., Korom, S., and Schmidt, G. (1990). Effects of *Escherichia coli* hemolysin on human monocytes. Cytocidal action and stimulation of interleukin 1 release. *J. Clin. Invest. 85*: 1746–1753.

Bisseret, F., Keith, G., Rihn, B., Amiri, I., Werneburg, B., Girardot, R., Baldacini, O., Green, G., Nguyen, V. K., and Monteil, H. (1989). *Clostridium difficile* toxin B: characterization and sequence of three peptides. *J. Chromatogr. 490*: 91–100.

Bolton, R. P., Sherriff, R. J., and Read, A. E. (1980). *Clostridium difficile* associated diarrhoea: a role in inflammatory bowel disease? *Lancet 1*: 383–384.

Bolton, R. P., Tait, S. K., Dear, P. R., and Losowsky, M. S. (1984). Asymptomatic neonatal colonisation by *Clostridium difficile. Arch. Dis. Child. 59*: 466–472.

Borriello, S. P. (1990). 12th C. L. Oakley Lecture. Pathogenesis of *Clostridium difficile* infection of the gut. *J. Med. Microbiol. 33*: 207–215.

Borriello, S. P., Honour, P., Turner, T., and Barclay, F. (1983). Household pets as a potential reservoir for *Clostridium difficile* infection. *J. Clin. Pathol. 36*: 84–87.

Borriello, S. P., Davies, H. A., and Barclay, F. E. (1988a). Detection of fimbriae amongst strains of *Clostridium difficile. FEMS Microbiol. Lett. 49*: 65–67.

Borriello, S. P., Welch, A. R., Barclay, F. E., and Davies, H. A. (1988b). Mucosal association by *Clostridium difficile* in the hamster gastrointestinal tract. *J. Med. Micribiol. 25*: 191–196.

Borriello, S. P., Davies, H. A., Kamiya, S., Reed, P. J., and Seddon, S. (1990). Virulence factors of *Clostridium difficile. Rev. Infect. Dis. 12*(Suppl. 2): S185–S191.

Borriello, S. P., Wren, B. W., Hyde, S., Seddon, S. V., Sibbons, P., Krishna, M. M., Tabaqchali, S., Manek, S., and Price, A. B. (1992). Molecular, immunological, and biological characterization of a toxin A-negative, toxin B-positive strain of *Clostridium difficile. Infect. Immun. 60*: 4192–4199.

Brooks, S. E., Veal, R. O., Kramer, M., Dore, L., Schupf, N., and Adachi, M. (1992). Reduction in the incidence of *Clostridium difficile*-associated diarrhea in an acute care hospital and a skilled nursing facility following replacement of electronic thermometers with single-use disposables. *Infect. Control Hosp. Epidemiol. 13*: 98–103.

Burke, G. W., Wilson, M. E., and Mehrez, I. O. (1988). Absence of diarrhea in toxic megacolon complicating *Clostridium difficile* pseudomembranous colitis. *Am. J. Gastroenterol. 83*: 304–307.

Cartmill, T. D., Shrimpton, S. B., Panigrahi, H., Khanna, V., Brown, R., and Poxton, I. R. (1992). Nosocomial diarrhoea due to a single strain of *Clostridium difficile*: a prolonged outbreak in elderly patients. *Age Ageing 21*: 245–249.

Caspar, M., Florin, I., and Thelestam, M. (1987). Calcium and calmodulin in cellular intoxication with *Clostridium difficile* toxin B. *J. Cell. Physiol. 132*: 168–172.

Chang, J., and Rohwer, R. G. (1991). *Clostridium difficile* infection in adult hamsters. *Lab. Anim. Sci. 41*: 548–552.

Chang, T. W., Bartlett, J. G., Gorbach, S. L., and Onderdonk, A. B. (1978). Clindamycin-induced enterocolitis in hamsters as a model of pseudomembranous colitis in patients. *Infect. Immun. 20*: 526–529.

Chang, T. W., Sullivan, N. M., and Wilkins, T. D. (1986). Insusceptibility of fetal intestinal mucosa and fetal cells to *Clostridium difficile* toxins. *Chung-Kuo Yao Li Hsueh Pao—Acta Pharm. Sin. 7*: 448–453.

Church, J. M., and Fazio, V. W. (1986). A role for colonic stasis in the pathogenesis of disease related to *Clostridium difficile. Dis. Colon Rectum 29*: 804–809.

Ciesielski-Treska, J., Ulrich, G., Rihn, B., and Aunis, D. (1989). Mechanism of action of *Clostridium difficile* toxin B: role of external medium and cytoskeletal organization in intoxicated cells. *Eur. J. Cell. Biol. 48*: 191–202.

Ciesielski-Treska, J., Ulrich, G., Baldacini, O., Monteil, H., and Aunis, D. (1991). Phosphorylation of cellular proteins in response to treatment with *Clostridium difficile* toxin B and *Clostridium sordellii* toxin L. *Eur. J. Cell. Biol. 56*: 68–78.

Clark, G. F., Krivan, H. C., Wilkins, T. D., and Smith, D. F. (1987). Toxin A from *Clostridium difficile* binds to rabbit erythrocyte glycolipids with terminal Gal alpha 1-3Gal beta 1-4GlcNAc sequences. *Arch. Biochem. Biophys. 257*: 217–229.

Colarian, J. (1988). *Clostridium difficile* colitis following antiviral therapy in the acquired immunodeficiency syndrome. *Am. J. Med. 84*: 1081.

Cope, A., Anderson, J., and Wilkins, E. (1992). *Clostridium difficile* toxin-induced reactive arthritis in a patient with chronic Reiter's syndrome. *Eur. J. Clin. Microbiol. Infect. Dis.* *11*: 40–43.

Corthier, G., Muller, M. C., Wilkins, T. D., Lyerly, D., and L'Haridon, R. (1991). Protection against experimental pseudomembranous colitis in gnotobiotic mice by use of monoclonal antibodies against *Clostridium difficile* toxin A. *Infect. Immun. 59*: 1192–1195.

Cudmore, M. A., Silva, J., Jr., Fekety, R., Liepman, M. K., and Kim, K. H. (1982). *Clostridium difficile* colitis associated with cancer chemotherapy. *Arch. Intern. Med. 142*: 333–335.

Cuzzolin, L., Zambreri, D., Donini, M., Griso, C., and Benoni, G. (1992). Influence of radiotherapy on intestinal microflora in cancer patients. *J. Chemother. 4*: 176–179.

Czuprynski, C. J., Johnson, W. J., Balish, E., and Wilkins, T. (1983). Pseudomembranous colitis in *Clostridium difficile*-monoassociated rats. *Infect. Immun. 39*: 1368–1376.

Dailey, D. C., Kaiser, A., and Schloemer, R. H. (1987). Factors influencing the phagocytosis of *Clostridium difficile* by human polymorphonuclear leukocytes. *Infect. Immun. 55*: 1541–1546.

Dan, M., and Samra, Z. (1989). *Clostridium difficile* colitis associated with ofloxacin therapy. *Am. J. Med. 87*: 479.

Daubener, W., Leiser, E., von Eichel-Streiber, C., and Hadding, U. (1988). *Clostridium difficile* toxins A and B inhibit human immune response in vitro. *Infect. Immun. 56*: 1107–1112.

Davies, H. A., and Borriello, S. P. (1990). Detection of capsule in strains of *Clostridium difficile* of varying virulence and toxigenicity. *Microb. Pathogen. 9*: 141–146.

de Lalla, F., Nicolin, R., Rinaldi, E., Scarpellini, P., Rigoli, R., Manfrin, V., and Tramarin, A. (1992). Prospective study of oral teicoplanin versus oral vancomycin for therapy of pseudomembranous colitis and *Clostridium difficile*-associated diarrhea. *Antimicrob. Agents Chemother. 36*: 2192–2196.

Delmee, M., Bulliard, G., and Simon, G. (1986). Application of a technique for serogrouping *Clostridium difficile* in an outbreak of antibiotic-associated diarrhoea. *J. Infect. 13*: 5–9.

Doern, G. V., Coughlin, R. T., and Wu, L. (1992). Laboratory diagnosis of *Clostridium difficile*-associated gastrointestinal disease: comparison of a monoclonal antibody enzyme immunoassay for toxins A and B with a monoclonal antibody enzyme immunoassay for toxin A only and two cytotoxicity assays. *J. Clin. Microbiol. 30*: 2042–2046.

Donta, S. T., and Myers, M. G. (1982). *Clostridium difficile* toxin in asymptomatic neonates. *J. Pediatr. 100*: 431–434.

Donta, S. T., and Shaffer, S. J. (1980). Effects of *Clostridium difficile* toxin on tissue-cultured cells. *J. Infect. Dis. 141*: 218–222.

Dove, C. H., Wang, S. Z., Price, S. B., Phelps, C. J., Lyerly, D. M., Wilkins, T. D., and Johnson, J. L. (1990). Molecular characterization of the *Clostridium difficile* toxin A gene. *Infect. Immun. 58*: 480–488.

Dudley, M. N., McLaughlin, J. C., Carrington, G., Frick, J., Nightingale, C. H., and Quintiliani, R. (1986). Oral bacitracin vs vancomycin therapy for *Clostridium difficile*-induced diarrhea. A randomized double-blind trial. *Arch. Intern. Med. 146*: 1101–1104.

Eglow, R., Pothoulakis, C., Itzkowitz, S., Israel, E. J., O'Keane, C. J., Gong, D., Gao, N., Xu, Y. L., Walker, W. A., and LaMont, J. T. (1992). Diminished *Clostridium difficile* toxin A sensitivity in newborn rabbit ileum is associated with decreased toxin A receptor. *J. Clin. Invest. 90*: 822–829.

Ehrich, M. (1982). Biochemical and pathological effects of *Clostridium difficile* toxins in mice. *Toxicon 20*: 983–989.

Ellis, M. E., Watson, B. M., Milewski, P. J., and Jones, G. (1983). *Clostridium difficile* colitis unassociated with antibiotic therapy. *Br. J. Surg. 70*: 242–243.

Elmer, G. W., and McFarland, L. V. (1987). Suppression by *Saccharomyces boulardii* of toxigenic *Clostridium difficile* overgrowth after vancomycin treatment in hamsters. *Antimicrob. Agents Chemother. 31*: 129–131.

Fang, G. D., Fonteles, M. C., Barrett, L. J., and Guerrant, R. L. (1993). Inhibition by platelet activating factor (PAF) antagonists of the effects of cholera toxin on intestinal secretion and cytoskeleton of Chinese hamster ovary (CHO) cells. *Clin. Res.* (Abstract)

Fekety, R. (1990). Antibiotic-associated colitis. In *Principles and Practice of Infectious Diseases*, G. L. Mandell, R. G. Douglas, and J. E. Bennett (Eds.), Churchill Livingstone, New York, pp. 863–869.

Fekety, R., and Shah, A. B. (1993). Diagnosis and treatment of *Clostridium difficile* colitis. *J. Am. Med. Assoc. 269*: 71–75.

Fekety, R., Kim, K. H., Brown, D., Batts, D. H., Cudmore, M., and Silva, J., Jr. (1981). Epidemiology of antibiotic-associated colitis; isolation of *Clostridium difficile* from the hospital environment. *Am. J. Med. 70*: 906–908.

Fekety, R., Silva, J., Kauffman, C., Buggy, B., and Deery, H. G. (1989). Treatment of antibiotic-associated *Clostridium difficile* colitis with oral vancomycin: comparison of two dosage regimens. *Am. J. Med. 86*: 15–19.

Finney, J. M. T. (1893). Gastro-enterostomy for cicatrizing ulcer of the pylorus. *Johns Hopkins Hosp. Bull. 4*: 53–55.

Fiorentini, C., and Thelestam, M. (1991). *Clostridium difficile* toxin A and its effects on cells. *Toxicon 29*: 543–567.

Fiorentini, C., Arancia, G., Paradisi, S., Donelli, G., Giuliano, M., Piemonte, F., and Mastrantonio, P. (1989). Effects of *Clostridium difficile* toxins A and B on cytoskeleton organization in HEp-2 cells: a comparative morphological study. *Toxicon 27*: 1209–1218.

Fiorentini, C., Malorni, W., Paradisi, S., Giuliano, M., Mastrantonio, P., and Donelli, G. (1990). Interaction of *Clostridium difficile* toxin A with cultured cells: cytoskeletal changes and nucelar polarization. *Infect. Immun. 58*: 2329–2336.

Fiorentini, C., Chow, S. C., Mastrantonio, P., Jeddi-Tehrani, M., and Thelestam, M. (1992). *Clostridium difficile* toxin A induces multinucleation in the human leukemic T cell line JURKAT. *Eur. J. Cell. Biol. 57*: 292–297.

Flegel, W. A., Muller, F., Daubener, W., Fischer, H. G., Hadding, U., and Northoff, H. (1991). Cytokine response by human monocytes to *Clostridium difficile* toxin A and toxin B. *Infect. Immun. 59*: 3659–3666.

Florin, I. (1991). Isolation of a fibroblast mutant resistant to *Clostridium difficile* toxins A and B. *Microb. Pathogen. 11*: 337–346.

Florin, I., and Thelestam, M. (1983). Internalization of *Clostridium difficile* cytotoxin into cultured human lung fibroblasts. *Biochim. Biophys. Acta 763*: 383–392.

Florin, I., and Thelestam, M. (1986). Lysosomal involvement in cellular intoxication with *Clostridium difficile* toxin B. *Microb. Pathogen. 1*: 373–385.

Florin, I., and Thelestam, M. (1991). ADP-ribosylation in *Clostridium difficile* toxin-treated cells is not related to cytopathogenicity of toxin B. *Biochim. Biophys. Acta 1091*: 51–54.

Fonteles, M., Fang, G. D., Thielman, N. M., Yotseff, P. S., and Guerrant, R. L. (1994). Role of platelet activatory factor in the inflammatory and secretory effects of *Clostridium difficile* toxin A. *J. Lipid Med. Cell Sig.*, in press.

Frey, S. M., and Wilkins, T. D. (1992). Localization of two epitopes recognized by monoclonal antibody PCG-4 on *Clostridium difficile* toxin A. *Infect. Immun. 60*: 2488–2492.

Gaginella, T. S. (1990). Eicosanoid-mediated intestinal secretion. In *Textbook of Secretory Diarrhea*, E. Lebenthal and M. Duffey (Eds.), Raven Press, New York, pp. 15–30.

Galili, U. (1989). Abnormal expression of alpha-galactosyl epitopes in man. *Lancet 2*: 358–361.

Galili, U., Shohet, S. B., Kobrin, E., Stults, C. L. M., and Macher, B. A. (1988). Man, apes, and old world monkeys differ from other animals in the expression of alpha-galactosyl epitopes on nucleated cells. *J. Biol. Chem. 263*: 17755–17762.

Gebhard, R. L., Gerding, D. N., Olson, M. M., Peterson, L. R., McClain, C. J., Ansel, H. J., Shaw, M. J., and Schwartz, M. L. (1985). Clinical and endoscopic findings in patients early in the course of *Clostridium difficile*-associated pseudomembranous colitis. *Am. J. Med. 78*: 45–48.

Gemmell, C. G. (1984). Interaction of *Clostridium difficile* toxin with human polymorphonuclear leukocyte function. In *Bacterial Protein Toxins*, J. E. Alouf, F. J. Fehrenbach, J. H. Freer, and J. Jeljaszewicz (Eds.), Academic, New York, pp. 287–295.

George, R. H., Symonds, J. M., Dimock, F., Brown, J. D., Arabi, Y., Shinagawa, N., Keighley, M. R., Alexander-Williams, J., and Burdon, D. W. (1978). Identification of *Clostridium difficile* as a cause of pseudomembranous colitis. *Br. Med. J. 1*: 695.

George, W. L. (1984). Antimicrobial agent-associated colitis and diarrhea: historical background and clinical aspects. *Rev. Infect Dis. 6*(Suppl. 1): S208–S213.

George, W. L. (1988). Antimicrobial agent-associated diarrhea in adult humans. In *Clostridium difficile: Its Role in Intestinal Disease*, R. D. Rolfe and S. M. Finegold (Eds.), Academic, New York, pp. 31–44.

George, W. L., Rolfe, R. D., Sutter, V. L., and Finegold, S. M. (1979). Diarrhea and colitis associated with antimicrobial therapy in man and animals. *Am. J. Clin. Nutr. 32*: 251–257.

George, W. L., Rolfe, R. D., and Finegold, S. M. (1982a). *Clostridium difficile* and its cytotoxin in feces of patients with antimicrobial agent-associated diarrhea and miscellaneous conditions. *J. Clin. Microbiol. 15*: 1049–1053.

George, W. L., Rolfe, R. D., Harding, G. K., Klein, R., Putnam, C. W., and Finegold, S. M. (1982b). *Clostridium difficile* and cytotoxin in feces of patients with antimicrobial agent-associated pseudomembranous colitis. *Infection 10*: 205–208.

Gerding, D. N., Gebhard, R. L., Sumner, H. W., and Peterson, L. R. (1988). Pathology and diagnosis of *Clostridium difficile* disease. In *Clostridium difficile: Its Role in Intestinal Disease*, R. D. Rolfe and S. M. Finegold (Eds.), Academic, New York, pp. 260–286.

Gilbert, R. J., Pothoulakis, C., and LaMont, J. T. (1989a). Effect of purified *Clostridium difficile* toxins on intestinal smooth muscle. II. Toxin B. *Am. J. Physiol. 256*: G767–G772.

Gilbert, R. J., Triadafilopoulos, G., Pothoulakis, C., Giampaolo, C., and LaMont, J. T. (1989b). Effect of purified *Clostridium difficile* toxins on intestinal smooth muscle. I. Toxin A. *Am. J. Physiol. 256*: G759–G766.

Gill, D. M. (1982). Bacterial toxins: a table of lethal amounts. *Microbiol. Rev. 46*: 86–94.

Giuliano, M., Piemonte, F., and Gianfrilli, P. M. (1988). Production of an enterotoxin different from toxin A by *Clostridium difficile*. *FEMS Microbiol. Lett. 50*: 191–194.

Gooi, H. C., Thorpe, S. J., Hounsell, E. F., Rumpold, H., Kraft, D., Forester, O., and Feizi, T. (1983). Marker of peripheral blood granulocytes and monocytes of man recognized ty two monoclonal antibodies VEO8 and VEP9 involves the trisaccharide 3-fucosyl-*N*-acetyl-lactosamine. *Eur. J. Immunol. 13*: 306–312.

Gorbach, S. L., Chang, T. W., and Goldin, B. (1987). Successful treatment of relapsing *Clostridium difficile* colitis with *Lactobacillus* GG [Letter]. *Lancet 2*: 1519.

Greenfield, C., Burroughs, A., Szawathowski, M., Bass, N., Noone, P., and Pounder, R. (1981). Is pseudomembranous colitis infectious? *Lancet 1*: 371–372.

Grube, B. J., Heimbach, D. M., and Marvin, J. A. (1987). *Clostridium difficile* diarrhea in critically ill burned patients. *Arch. Surg. 122*: 655–661.

Gryboski, J. D. (1991). *Clostridium difficile* in inflammatory bowel disease relapse. *J. Pediatr. Gastroenterol. Nutr. 13*: 39–41.

Guerrant, R. L., Brunton, L. L., Schnaitman, R. C., Rebhun, L. I., and Gilman, A. G. (1974). Cyclic adenosine monophosphate and alteration of Chinese hamster ovary cell morphology: a rapid, sensitive in vitro assay for the enterotoxins of *Vibrio cholerae* and *Escherichia coli*. *Infect. Immun. 10*: 320–327.

Guerrant, R. L., Fang, G., Lima, A. A. M., Lyerly, D. M., and Fonteles, M. C. (1992). PAF antagonists, inhibitors of phospholipase A_2 and cyclooxygenase block the secretory and cytoskeletal effects of *C. difficile* toxin A. Presented at Fourth Int. Congr. on PAF and Related Lipid Mediators, Utah.

Gumerlock, P. H., Tang, Y. J., Meyers, F. J., and Silva, J., Jr. (1991). Use of the polymerase chain reaction for the specific and direct detection of *Clostridium difficile* in human feces. *Rev. Infect. Dis. 13*: 1053–1060.

Gurian, L., Ward, T. T., and Katon, R. M. (1982). Possible foodborne transmission in a case of pseudomembranous colitis due to *Clostridium difficile*: influence of gastrointestinal secretions on *Clostridium difficile* infection. *Gastroenterology 83*: 465–469.

Hafiz, S., and Oakley, C. L. (1976). *Clostridium difficile*: isolation and characteristics. *J. Med. Microbiol. 9*: 129–136.

Hall, I. C., and O'Toole, E. (1935). Intestinal flora in new-born infants. *Am. J. Dis. Child. 49*: 390–402.

Hamre, D. M., Rake, G., McKee, C. M., and MacPhillamy, H. B. (1943). The toxicity of penicillin as prepared for clinical use. *Am. J. Med. Sci. 206*: 642–652.

Han, V. K., Sayed, H., Chance, G. W., Brabyn, D. G., and Shaheed, W. A. (1983). An outbreak of *Clostridium difficile* necrotizing enterocolitis: a case for oral vancomycin therapy? *Pediatrics 71*: 935–941.

Hannonen, P., Hakola, M., Mottonen, T., and Oka, M. (1989). Reactive oligoarthritis associated with *Clostridium difficile* colitis. *Scand. J. Rheumatol. 18*: 57–60.

Hayward, R. S., Wensel, R. H., and Kibsey, P. (1990). Relapsing *Clostridium difficile* colitis and Reiter's syndrome. *Am. J. Gastroenterol. 85*: 752–756.

Heard, S. R., Wren, B., Barnett, M. J., Thomas, J. M., and Tabaqchali, S. (1988). *Clostridium difficile* infection in patients with haematological malignant disease. Risk factors, faecal toxins and pathogenic strains. *Epidemiol. Infect. 100*: 63–72.

Hecht, G., Pothoulakis, C., LaMont, J. T., and Madara, J. L. (1988). *Clostridium difficile* toxin A perturbs cytoskeletal structure and tight junction permeability of cultured human intestinal epithelial monolayers. *J. Clin. Invest. 82*: 1516–1524.

Hecht, G., Koutsouris, A., Pothoulakis, C., LaMont, J. T., and Madara, J. L. (1992). *Clostridium difficile* toxin B disrupts the barrier junction of T84 monolayers. *Gastroenterology 102*: 416–423.

Henriques, B., Florin, I., and Thelestam, M. (1987). Cellular internalization of *Clostridium difficile* toxin A. *Microb. Pathogen. 2*: 455–463.

Hirschhorn, L. R., Trnka, Y., Onderdonk, A., and Platt, R. (1992). Epidemiology of community-acquired *Clostridium difficile* associated diarrhea. *Clin. Res. 40*: 394A (Abstract).

Holst, E., Helin, I., and Mardh, P. A. (1981). Recovery of *Clostridium difficile* from children. *Scand. J. Infect. Dis. 13*: 41–45.

Honda, T., Hernandez, I., Katoh, T., and Miwatani, T. (1983). Stimulation of enterotoxin production of *Clostridium difficile* by antibiotics [Letter]. *Lancet 1*: 655.

Huang, L. C., Civin, C. I., Magnani, J. L., Shaper, J. H., and Ginsburg, V. (1983). My-1, the human myeloid-specific antigen detected by mouse monoclonal antibodies, is a sugar sequence found in lacto-*N*-fucopentaose III. *Blood 61*: 1020–1023.

Hughes, S., Warhurst, G., Turnberg, L. A., Higgs, N. B., Giugliano, L. G., and Drasar, B. S. (1983). *Clostridium difficile* toxin-induced intestinal secretion in rabbit ileum in vitro. *Gut 24*: 94–98.

Hummel, R. P., Altemeier, W. A., and Hill, E. O. (1964). Iatrogenic staphylococcal enterocolitis. *Ann. Surg. 160*: 551–559.

Johnson, S., Adelmann, A., Clabots, C. R., Peterson, L. R., and Gerding, D. N. (1989). Recurrences of *Clostridium difficile* diarrhea not caused by the original infecting organism. *J. Infect Dis. 159*: 340–343.

Johnson, S., Gerding, D. N., Olson, M. M., Weiler, M. D., Hughes, R. A., Clabots, C. R., and Peterson, L. R. (1990). Prospective, controlled study of vinyl glove use to interrupt *Clostridium difficile* nosocomial transmission. *Am. J. Med. 88*: 137–140.

Johnson, S., Gerding, D. N., and Janoff, E. N. (1992a). Systemic and mucosal antibody responses to toxin A in patients infected with *Clostridium difficile*. *J. Infect. Dis. 166*: 1287–1294.

Johnson, S., Homann, S. R., Bettin, K. M., Quick, J. N., Clabots, C. R., Peterson, L. R., and Gerding, D. N. (1992b). Treatment of asymptomatic *Clostridium difficile* carriers (fecal excretors) with vancomycin or metronidazole. A randomized, placebo-controlled trial. *Ann. Intern. Med. 117*: 297–302.

Justus, P. G., Martin, J. L., Goldberg, D. A., Taylor, N. S., Bartlett, J. G., Alexander, R. W., and Mathias, J. R. (1982). Myoelectric effects of *Clostridium difficile*: motility-altering factors distinct from its cytotoxin and enterotoxin in rabbits. *Gastroenterology 83*: 836–843.

Kamiya, S., Reed, P. J., and Borriello, S. P. (1988). Analysis of purity of *Clostridium difficile* toxin A derived by affinity chromatography on immobilized bovine thyroglobulin. *FEMS Microbiol. Lett. 56*: 331–336.

Kamiya, S., Reed, P. J., and Borriello, S. P. (1989). Purification and characterisation of *Clostridium difficile* toxin A by bovine thyroglobulin affinity chromatography and dissociation in denaturing conditions with or without reduction. *J. Med. Microbiol. 30*: 69–77.

Kamiya, S., Yamakawa, K., Meng, X. Q., Ogura, H., and Nakamura, S. (1991). Production of monoclonal antibody to *Clostridium difficile* toxin A which neutralizes enterotoxicity but not haemagglutination activity. *FEMS Microbiol. Lett. 65*: 311–315.

Kamiya, S., Ogura, H., Meng, X. Q., and Nakamura, S. (1992). Correlation between cytotoxin production and sporulation in *Clostridium difficile*. *J. Med. Microbiol. 37*: 206–210.

Kato, N., Ou, C.-Y., Kato, H., Bartley, S. L., Luo, C.-C., Killgore, G. E., and Ueno, K. (1993). Detection of toxigenic *Clostridium difficile* in stool specimens by the polymerase chain reaction. *J. Infect. Dis. 167*: 455–458.

Katoh, T., Honda, T., and Miwatani, T. (1988). Purification and some properties of cytotoxin produced by *Clostridium difficile*. *Microbiol. Immunol. 32*: 551–564.

Keighley, M. R., Burdon, D. W., Alexander-Williams, J., Shinagawa, N., Arabi, Y., Thompson, H., Youngs, D., Bentley, S., and George, R. H. (1978). Diarrhoea and pseudomembranous colitis after gastrointestinal operations. A prospective study. *Lancet 2*: 1165–1167.

Kelly, C. P., Pothoulakis, C., Orellana, J., and LaMont, J. T. (1992). Human colonic aspirates containing immunoglobulin A antibody to *Clostridium difficile* toxin A inhibit toxin A-receptor binding. *Gastroenterology 102*: 35–40.

Ketley, J. M., Haslam, S. C., Mitchell, T. J., Stephen, J., Candy, D. C., and Burdon, D. W. (1984). Production and release of toxins A and B by *Clostridium difficile*. *J. Med. Microbiol. 18*: 385–391.

Ketley, J. M., Mitchell, T. J., Haslam, S. C., Stephen, J., Candy, D. C., and Burdon, D. W. (1986). Sporogenesis and toxin A production by *Clostridium difficile*. *J. Med. Microbiol. 22*: 33–38.

Kim, K., DuPont, H. L., and Pickering, L. K. (1983). Outbreaks of diarrhea associated with *Clostridium difficile* and its toxin in day-care centers: evidence of person-to-person spread. *J. Pediatr. 102*: 376–382.

Kim, P. H., Iaconis, J. P., and Rolfe, R. D. (1987). Immunization of adult hamsters against *Clostridium difficile*-associated ileocecitis and transfer of protection to infant hamsters. *Infect. Immun. 55*: 2984–2992.

Kofsky, P., Rosen, L., Reed, J., Tolmie, M., and Ufberg, D. (1991). *Clostridium difficile*—a common and costly colitis. *Dis. Colon Rectum 34*: 244–248.

Krivan, H. C., and Wilkins, T. D. (1987). Purification of *Clostridium difficile* toxin A by affinity chromatography on immobilized thyroglobulin. *Infect. Immun. 55*: 1873–1877.

Krivan, H. C., Clark, G. F., Smith, D. F., and Wilkins, T. D. (1986). Cell surface binding site for *Clostridium difficile* enterotoxin: evidence for a glycoconjugate containing the sequence Gal alpha 1-3Gal beta 1-4GlcNAc. *Infect. Immun. 53*: 573–581.

Kushnaryov, V. M., and Sedmak, J. J. (1989). Effect of *Clostridium difficile* enterotoxin A on ultrastructure of Chinese hamster ovary cells. *Infect. Immun. 57*: 3914–3921.

Kushnaryov, V. M., Redlich, P. N., Sedmak, J. J., Lyerly, D. M., Wilkins, T. D., and Grossberg, S. E. (1992). Cytotoxicity of *Clostridium difficile* toxin A for human colonic and pancreatic carcinoma cell lines. *Cancer Res. 52*: 5096–5099.

LaMont, J. T., and Trnka, Y. M. (1980). Therapeutic implications of *Clostridium difficile* toxin during relapse of chronic inflammatory bowel disease. *Lancet 1*: 381–383.

Larson, H. E., and Price, A. B. (1977). Pseudomembranous colitis: presence of clostridial toxin. *Lancet 2*: 1312–1314.

Larson, H. E., Parry, J. V., Price, A. B., Davies, D. R., Dolby, J., and Tyrrell, D. A. J. (1977). Undescribed toxin in pseudomembranous colitis. *Br. Med. J. 1*: 1246–1248.

Larson, H. E., Price, A. B., Honour, P., and Borriello, S. P. (1978). *Clostridium difficile* and the aetiology of pseudomembranous colitis. *Lancet 1*: 1063–1066.

Larson, H. E., Barclay, F. E., Honour, P., and Hill, I. D. (1982). Epidemiology of *Clostridium difficile* in infants. *J. Infect. Dis. 146*: 727–733.

Lauritsen, K., Laursen, L. S., Bukhave, K., and Rask-Madsen, J. (1988). In vivo profiles of eicosanoids in ulcerative colitis, Crohn's colitis, and *Clostridium difficile* colitis. *Gastroenterology 95*: 11–17.

Leung, A. C., Orange, G., McLay, A., and Henderson, I. S. (1985). *Clostridium difficile*-associated colitis in uremic patients. *Clin. Nephrol. 24*: 242–248.

Libby, J. M., and Wilkins, T. D. (1982). Production of antitoxins to two toxins of *Clostridium difficile* and immunological comparison of the toxins by cross-neutralization studies. *Infect. Immun. 35*: 374–376.

Libby, J. M., Jortner, B. S., and Wilkins, T. D. (1982). Effects of the two toxins of *Clostridium difficile* in antibiotic-associated cecitis in hamsters. *Infect. Immun. 36*: 822–829.

Lima, A. A., Lyerly, D. M., Wilkins, T. D., Innes, D. J., and Guerrant, R. L. (1988). Effects of *Clostridium difficile* toxins A and B in rabbit small and large intestine in vivo and on cultured cells in vitro. *Infect. Immun. 56*: 582–588.

Lima, A. A., Innes, D. J., Jr., Chadee, K., Lyerly, D. M., Wilkins, T. D., and Guerrant, R. L. (1989). *Clostridium difficile* toxin A. Interactions with mucus and early sequential histopathologic effects in rabbit small intestine. *Lab. Invest. 61*: 419–425.

Lonnroth, I., and Lange, S. (1983). Toxin A of *Clostridium difficile*: production, purification and effect in mouse intestine. *Acta Pathol. Microbiol. Immunol. Scand. 91*: 395–400.

Lusk, R. H., Fekety, R., Silva, J., Browne, R. A., Ringler, D. H., and Abrams, G. D. (1978). Clindamycin-induced enterocolitis in hamsters. *J. Infect. Dis. 137*: 464–475.

Lyerly, D. M., and Wilkins, T. D. (1986). Commercial latex test for *Clostridium difficile* toxin A does not detect toxin A. *J. Clin. Microbiol. 23*: 622–623.

Lyerly, D. M., and Wilkins, T. D. (1988). Purification and properties of toxins A and B of *Clostridium difficile*. In *Clostridium difficile: Its Role in Intestinal Disease*, R. D. Rolfe and S. M. Finegold (Eds.), Academic, New York, pp. 145–168.

Lyerly, D. M., Lockwood, D. E., Richardson, S. H., and Wilkins, T. D. (1982). Biological activities of toxins A and B of *Clostridium difficile*. *Infect. Immun. 35*: 1147–1150.

Lyerly, D. M., Sullivan, N. M., and Wilkins, T. D. (1983). Enzyme-linked immunosorbent assay for *Clostridium difficile* toxin A. *J. Clin. Microbiol. 17*: 72–78.

Lyerly, D. M., Saum, K. E., MacDonald, D. K., and Wilkins, T. D. (1985). Effects of *Clostridium difficile* toxins given intragastrically to animals. *Infect. Immun. 47*: 349–352.

Lyerly, D. M., Phelps, C. J., Toth, J., and Wilkins, T. D. (1986a). Characterization of toxins A and B of *Clostridium difficile* with monoclonal antibodies. *Infect. Immun. 54*: 70–76.

Lyerly, D. M., Roberts, M. D., Phelps, C. J., and Wilkins, T. D. (1986b). Purification and properties of toxins A and B of *Clostridium difficile*. *FEMS Microbiol. Lett. 33*: 31–35.

Lyerly, D. M., Ball, D. W., Toth, J., and Wilkins, T. D. (1988a). Characterization of cross-reactive proteins detected by Culturette brand rapid latex test for *Clostridium difficile*. *J. Clin. Microbiol. 26*: 397–400.

Lyerly, D. M., Krivan, H. C., and Wilkins, T. D. (1988b). *Clostridium difficile*: its disease and toxins. *Clin. Microbiol. Rev. 1*: 1–18.

Lyerly, D. M., Carrig, P. E., and Wilkins, T. D. (1989). Susceptibility of *Clostridium difficile* toxins A and B to trypsin and chymotrypsin. *Microb. Ecol. Health Dis. 2*: 219–221.

Lyerly, D. M., Johnson, J. L., Frey, S. M., and Wilkins, T. D. (1990). Vaccination against lethal *Clostridium difficile* enterocolitis with a nontoxic recombinant peptide of toxin A. *Curr. Microbiol. 21*: 29–32.

Lyerly, D. M., Bostwick, E. F., Binion, S. B., and Wilkins, T. D. (1991). Passive immunization of hamsters against disease caused by *Clostridium difficile* by use of bovine immunoglobulin G concentrate. *Infect. Immun. 59*: 2215–2218.

Lyerly, D. M., Barroso, L. A., Wilkins, T. D., Depitre, C., and Corthier, G. (1992). Characterization of a toxin A-negative, toxin B-positive strain of *Clostridium difficile*. *Infect. Immun. 60*: 4633–4639.

McFarland, L. V., Mulligan, M. E., Kwok, R. Y., and Stamm, W. E. (1989). Nosocomial acquisition of *Clostridium difficile* infection. *N. Engl. J. Med. 320*: 204–210.

McFarland, L. V., Surawicz, C. M., and Stamm, W. E. (1990). Risk factors for *Clostridium difficile* carriage and *C. difficile*-associated diarrhea in a cohort of hospitalized patients. *J. Infect. Dis. 162*: 678–684.

Mahe, S., Corthier, G., and Dubos, F. (1987). Effect of various diets on toxin production by two strains of *Clostridium difficile* in gnotobiotic mice. *Infect. Immun. 55*: 1801–1805.

Malorni, W., Paradisi, S., Dupuis, M. L., Fiorentini, C., and Ramoni, C. (1991). Enhancement of cell-mediated cytotoxicity by *Clostridium difficile* toxin A: an in vitro study. *Toxicon 29*: 417–428.

Marrie, T. J., Furlong, M., Faulkner, R. S., Sidorov, J., Haldane, E. V., and Kerr, E. A. (1982). *Clostridium difficile*: epidemiology and clinical features. *Can. J. Surg. 25*: 438–442.

Meador, J., and Tweten, R. K. (1988). Purification and characterization of toxin B from *Clostridium difficile*. *Infect. Immun. 56*: 1708–1714.

Mermel, L. A., and Osborn, T. G. (1989). *Clostridium difficile* associated reactive arthritis in an HLA-B27 positive female: report and literature review. *J. Rheumatol. 16*: 133–135.

Miller, P. D., Pothoulakis, C., Baeker, T. R., LaMont, J. T., and Rothstein, T. L. (1990). Macrophage-dependent stimulation of T cell-depleted spleen cells by *Clostridium difficile* toxin A and calcium ionophore. *Cell. Immunol. 126*: 155–163.

Misfeldt, M. L., Legaard, P. K., Howell, S. E., Fornella, M. H., and LeGrand, R. D. (1990). Induction of interleukin-1 from murine peritoneal macrophages by *Pseudomonas aeruginosa* exotoxin A. *Infect. Immun. 58*: 978–982.

Mitchell, M. J., Laughon, B. E., and Lin, S. (1987). Biochemical studies on the effect of *Clostridium difficile* toxin B on actin in vivo and in vitro. *Infect. Immun. 55*: 1610–1615.

Mitchell, T. J., Ketley, J. M., Haslam, S. C., Stephen, J., Burdon, D. W., Candy, D. C., and Daniel, R. (1986). Effect of toxin A and B of *Clostridium difficile* on rabbit ileum and colon. *Gut 27*: 78–85.

Mitchell, T. J., Ketley, J. M., Burdon, D. W., Candy, D. C., and Stephen, J. (1987). The effects of *Clostridium difficile* toxins A and B on membrane integrity and protein synthesis in intestinal cells in vivo and in vitro and in McCoy cells in vitro. *J. Med. Microbiol. 23*: 205–210.

Moore, R., Pothoulakis, C., LaMont, J. T., Carlson, S., and Madara, J. L. (1990). *C. difficile* toxin A increases intestinal permeability and induces Cl⁻ secretion. *Am. J. Physiol. 259*: G165–G172.

Morris, J. B., Zollinger, R. M., Jr., and Stellato, T. A. (1990). Role of surgery in antibiotic-induced pseudomembranous enterocolitis. *Am. J. Surg. 160*: 535–539.

Moskovitz, M., and Bartlett, J. G. (1981). Recurrent pseudomembranous colitis unassociated with prior antibiotic therapy. *Arch. Intern. Med. 141*: 663–664.

Muldrow, L. L., Ibeanu, G. C., Lee, N. I., Bose, N. K., and Johnson, J. (1987). Molecular cloning of *Clostridium difficile* toxin A gene fragment in lambda gt11. *FEBS Lett. 213*: 249–253.

Nakamura, S., Mikawa, M., Nakashio, S., Takabatake, M., Okado, I., Yamakawa, K., Serikawa, T., Okumura, S., and Nishida, S. (1981). Isolation of *Clostridium difficile* from the feces and the antibody in sera of young and elderly adults. *Microbiol. Immunol. 25*: 345–351.

Nakamura, S., Mikawa, M., Tanabe, N., Yamakawa, K., and Nishida, S. (1982). Effect of clindamycin on cytotoxin production by *Clostridium difficile*. *Microbiol. Immunol. 26*: 985–992.

Novak, E., Lee, J. G., Seckman, C. E., Phillips, J. P., and DiSanto, A. R. (1976). Unfavorable effect of atropine-diphenoxylate (lomotil) therapy in lincomycin-caused diarrhea. *J. Am. Med. Assoc. 235*: 1451–1454.

O'Neill, G. L., Beaman, M. H., and Riley, T. V. (1991). Relapse versus reinfection with *Clostridium difficile. Epidemiol. Infect. 107*: 627–635.

Onderdonk, A. B., Lowe, B. R., and Bartlett, J. G. (1979). Effect of environmental stress on *Clostridium difficile* toxin levels during continuous cultivation. *Appl. Environ. Microbiol. 38*: 637–641.

Onderdonk, A. B., Cisneros, R. L., and Bartlett, J. G. (1980). *Clostridium difficile* in gnotobiotic mice. *Infect. Immun. 28*: 277–282.

Ooi, W., Levine, H. G., LaMont, J. T., and Clark, R. A. (1984). Inactivation of *Clostridium difficile* cytotoxin by the neutrophil myeloperoxidase system. *J. Infect. Dis. 149*: 215–219.

Ottlinger, M. E., and Lin, S. (1988). *Clostridium difficile* toxin B induces reorganization of actin, vinculin, and talin in cultured cells. *Exp. Cell. Res. 174*: 215–229.

Peikin, S. R., Galdibini, J., and Bartlett, J. G. (1980). Role of *Clostridium difficile* in a case of nonantibiotic-associated pseudomembranous colitis. *Gastroenterology 79*: 948–951.

Pettet, J. D., Baggenstoss, A. H., Dearing, W. H., and Judd, E. S. (1954). Postoperative pseudomembranous enterocolitis. *Surg. Gynecol. Obstet. 8*: 546–552.

Phelps, C. J., Lyerly, D. L., Johnson, J. L., and Wilkins, T. D. (1991). Construction and expression of the complete *Clostridium difficile* toxin A gene in *Escherichia coli. Infect. Immun. 59*: 150–153.

Popoff, M. R., Rubin, E. J., Gill, D. M., and Boquet, P. (1988). Actin-specific ADP-ribosyltransferase produced by a *Clostridium difficile* strain. *Infect. Immun. 56*: 2299–2306.

Pothoulakis, C., Barone, L. M., Ely, R., Faris, B., Clark, M. E., Franzblau, C., and LaMont, J. T. (1986). Purification and properties of *Clostridium difficile* cytotoxin B. *J. Biol. Chem. 261*: 1316–1321.

Pothoulakis, C., Sullivan, R., Melnick, D. A., Triadafilopoulos, G., Gadenne, A. S., Meshulam, T., and LaMont, J. T. (1988). *Clostridium difficile* toxin A stimulates intracellular calcium release and chemotactic response in human granulocytes. *J. Clin. Invest. 81*: 1741–1745.

Pothoulakis, C., LaMont, J. T., Eglow, R., Gao, N., Rubins, J. B., Theoharides, T. C., and Dickey, B. F. (1991). Characterization of rabbit ileal receptors for *Clostridium difficile* toxin A. Evidence for a receptor-coupled G protein. *J. Clin. Invest. 88*: 119–125.

Poxton, I. R., Aronsson, B., Mollby, R., Nord, C. E., and Collee, J. G. (1984). Immunochemical fingerprinting of *Clostridium difficile* strains isolated from an outbreak of antibiotic-associated colitis and diarrhoea. *J. Med. Microbiol. 17*: 317–324.

Price, A. B., and Davies, D. R. (1977). Pseudomembranous colitis. *J. Clin. Pathol. 30*: 1–12.

Price, S. B., Phelps, C. J., Wilkins, T. D., and Johnson, J. L. (1987). Cloning of the carbohydrate-binding portion of the toxin A gene of *Clostridium difficile. Curr. Microbiol. 16*: 55–60.

Rampling, A., Warren, R. E., Bevan, P. C., Hoggarth, C. E., Swirsky, D., and Hayhoe, F. G. (1985). *Clostridium difficile* in haematological malignancy. *J. Clin. Pathol. 38*: 445–451.

Rautenberg, P. and Stender, F. (1986). Characterization of immunogenic p230 as the toxin A of *Clostridium difficile. FEMS Microbiol. Lett. 37*: 1–7.

Rifkin, G. D., Fekety, F. R., Silva, J., and Sack, R. B. (1977). Antibiotic-induced colitis implication of a toxin neutralised by *Clostridium sordelli* antitoxin. *Lancet 2*: 1103–1106.

Rifkin, G. D., Silva, J., Jr., and Fekety, R. (1978). Gastrointestinal and systemic toxicity of fecal extracts from hamsters with clindamycin-induced colitis. *Gastroenterology 74*: 52–57.

Rihn, B., Scheftel, J. M., Girardot, R., and Monteil, H. (1984). A new purification procedure for *Clostridium difficile* enterotoxin. *Biochem. Biophys. Res. Commun. 124*: 690–695.

Rihn, B., Bisseret, F., Girardot, R., Scheftel, J. M., Nguyen, V. K., and Monteil, H. (1988). Fast protein purification of *Clostridium difficile* cytotoxin. *J. Chromatogr. 428*: 408–414.

Riley, T. V., Adams, J. E., O'Neill, G. L., and Bowman, R. A. (1991). Gastrointestinal carriage of *Clostridium difficile* in cats and dogs attending veterinary clinics. *Epidemiol. Infect. 107*: 659–665.

Rolfe, R. D., and Finegold, S. M. (1979). Purification and characterization of *Clostridium difficile* toxin. *Infect. Immun. 25*: 191–201.

Rolfe, R. D., and Iaconis, J. P. (1983). Intestinal colonization of infant hamsters with *Clostridium difficile*. *Infect. Immun. 42*: 480–486.

Rothman, S. W., Brown, J. E., Diecidue, A., and Foret, D. A. (1984). Differential cytotoxic effects of toxins A and B isolated from *Clostridium difficile*. *Infect. Immun. 46*: 324–331.

Rothman, S. W., Gentry, M. K., Brown, J. E., Foret, D. A., Stone, M. J., and Strickler, M. P. (1988). Immunochemical and structural similarities in toxin A and toxin B of *Clostridium difficile* shown by binding to monoclonal antibodies. *Toxicon 26*: 583–597.

Rybolt, A. H., Bennett, R. G., Laughon, B. E., Thomas, D. R., Greenough, W. B., and Bartlett, J. G. (1989). Protein-losing enteropathy associated with *Clostridium difficile* infection. *Lancet 1*: 1353–1355.

Schwan, A., Sjolin, S., Trottestam, U., and Aronsson, B. (1984). Relapsing *Clostridium difficile* enterocolitis cured by rectal infusion of normal faeces. *Scand. J. Infect. Dis. 16*: 211–215.

Seal, D., Borriello, S. P., Barclay, F., Welch, A., Piper, M., and Bonnycastle, M. (1987). Treatment of relapsing *Clostridium difficile* diarrhoea by administration of non-toxigenic strain. *Eur. J. Clin. Microbiol. 6*: 51–53.

Seddon, S. V., and Borriello, S. P. (1992). Proteolytic activity of *Clostridium difficile*. *J. Med. Microbiol. 36*: 307–311.

Seddon, S. V., Hemingway, I., and Borriello, S. P. (1990). Hydrolytic enzyme production by *Clostridium difficile* and its relationship to toxin production and virulence in the hamster model. *J. Med. Microbiol. 31*: 169–174.

Shahrabadi, M. S., Bryan, L. E., and Lee, P. W. (1984). Interaction of *Clostridium difficile* toxin A with L cells in culture. *Can. J. Microbiol. 30*: 874–883.

Silva, J., Fekety, R., Werk, C., Ebright, J., Cudmore, M., Batts, D., Syrjamaki, C., and Lukens, J. (1984). Inciting and etiologic agents of colitis. *Rev. Infect. Dis. 6*(Suppl. 1): S214–S221.

Small, J. D. (1968). Fatal enterocolitis in hamsters given lincomycin hydrochloride. *Lab. Anim. Care 18*: 411–420.

Smith, T. K., Feng, T., Guerrant, R. L., Lyerly, D. M., McCallum, R. W., and Roche, J. K. (1993). Effect of *Clostridium difficile* toxin A upon the ascending excitatory response to distension of the isolated guinea pig ileum. *Gastroenterology 104*: A783.

Snook, S. S., Canfield, D. R., Sehgal, P. K., and King, N. W., Jr. (1989). Focal ulcerative ileocolitis with terminal thrombocytopenic purpura in juvenile cotton top tamarins (*Saguinus oedipus*). *Lab. Anim. Sci. 39*: 109–114.

Snooks, S. J., Hughes, A., and Horsburgh, A. G. (1984). Perforated colon complicating pseudomembranous colitis. *Br. J. Surg. 71*: 291–292.

Snyder, M. L. (1937). Further studies on *Bacillus difficilis*. *J. Infect. Dis. 60*: 223–231.

Snyder, M. L. (1940). The normal fecal flora of infants between two weeks and one year of age. *J. Infect. Dis. 66*: 1–16.

Spooncer, E., Fukuda, M., Klock, J., Oates, J., and Dell, A. (1984). Isolation and characterization of polyfucosylated lactosaminoglycan from human granulocytes. *J. Biol. Chem. 259*: 4792–4801.

Sterne, M., and Wentzel, L. M. (1950). A new method for the large-scale production of high-titre botulinum formol-toxoid types C and D. *J. Immunol. 65*: 175–183.

Sullivan, N. M., Pellett, S., and Wilkins, T. D. (1982). Purification and characterization of toxins A and B of *Clostridium difficile*. *Infect. Immun. 35*: 1032–1040.

Sumner, H. W., and Tedesco, F. J. (1975). Rectal biopsy in clindamycin-associated colitis. An analysis of 23 cases. *Arch. Pathol. 99*: 237–241.

Surawicz, C. M., Elmer, G. W., Speelman, P., McFarland, L. V., Chinn, J., and van Belle,

G. (1989a). Prevention of antibiotic-associated diarrhea by *Saccharomyces boulardii*: a prospective study. *Gastroenterology 96*: 981–988.

Surawicz, C. M., McFarland, L. V., Elmer, G., and Chinn, J. (1989b). Treatment of recurrent *Clostridium difficile* colitis with vancomycin and *Saccharomyces boulardii*. *Am. J. Gastroenterol. 84*: 1285–1287.

Sutphen, J. L., Grand, R. J., Flores, A., Chang, T. W., and Bartlett, J. G. (1983). Chronic diarrhea associated with *Clostridium difficile* in children. *Am. J. Dis. Child. 137*: 275–278.

Tabaqchali, S., O'Farrell, S., Nash, J. Q., and Wilks, M. (1984). Vaginal carriage and neonatal acquisition of *Clostridium difficile*. *J. Med. Microbiol. 18*: 47–53.

Taylor, N. S., and Bartlett, J. G. (1980). Binding of *Clostridium difficile* cytotoxin and vancomycin by anion-exchange resins. *J. Infect. Dis. 141*: 92–97.

Taylor, N. S., Thorne, G. M., and Bartlett, J. G. (1981). Comparison of two toxins produced by *Clostridium difficile*. *Infect. Immun. 34*: 1036–1043.

Teasley, D. G., Gerding, D. N., Olson, M. M., Peterson, L. R., Gebhard, R. L., Schwartz, M. J., and Lee, J. T., Jr. (1983). Prospective randomized trial of metronidazole versus vancomycin for *Clostridium difficile*-associated diarrhoea and colitis. *Lancet 2*: 1043–1046.

Tedesco, F. J. (1982). Pseudomembranous colitis. *Med. Clin. North Am. 66*: 655–665.

Tedesco, F. J., Barton, R. W., and Alpers, D. H. (1974). Clindamycin-associated colitis. *Ann. Intern. Med. 81*: 429–433.

Tedesco, F. J., Corless, J. K., and Brownstein, R. E. (1982). Rectal sparing in antibiotic-associated pseudomembranous colitis: a prospective study. *Gastroenterology 83*: 1259–1260.

Testore, G. P., Pantosti, A., Cerquetti, M., Babudieri, S., Panichi, G., and Gianfrilli, P. M. (1988). Evidence for cross-infection in an outbreak of *Clostridium difficile*-associated diarrhoea in a surgical unit. *J. Med. Microbiol. 26*: 125–128.

Thelestam, M., and Bronnegard, M. (1980). Interaction of cytopathogenic toxin from *Clostridium difficile* with cells in tissue culture. *Scand. J. Infect. Dis. (Suppl. 22)* 16–29.

Torres, J. F., and Lonnroth, I. (1988a). Purification and characterization of two forms of toxin B produced by *Clostridium difficile*. *FEBS Lett. 233*: 417–420.

Torres, J. F., and Lonnroth, I. (1988b). Comparison of methods for the production and purification of toxin A from *Clostridium difficile*. *FEMS Microbiol. Lett. 52*: 41–46.

Torres, J. F., and Lonnroth, I. (1989). Production, purification and characterisation of *Clostridium difficile* toxic proteins different from toxin A and from toxin B. *Biochim. Biophys. Acta 998*: 151–157.

Torres, J., Camorlinga-Ponce, M., and Munoz, O. (1992). Sensitivity in culture of epithelial cells from rhesus monkey kidney and human colon carcinoma to toxins A and B from *Clostridium difficile*. *Toxicon 30*: 419–426.

Treloar, A. J., and Kalra, L. (1987). Mortality and *Clostridium difficile* diarrhoea in the elderly [Letter]. *Lancet 2*: 1279.

Triadafilopoulos, G., and Hallstone, A. E. (1991). Acute abdomen as the first presentation of pseudomembranous colitis. *Gastroenterology 101*: 685–691.

Triadafilopoulos, G., Pothoulakis, C., O'Brien, M. J., and LaMont, J. T. (1987). Differential effects of *Clostridium difficile* toxins A and B on rabbit ileum. *Gastroenterology 93*: 273–279.

Triadafilopoulos, G., Pothoulakis, C., Weiss, R., Giampaolo, C., and LaMont, J. T. (1989). Comparative study of *Clostridium difficile* toxin A and cholera toxin in rabbit ileum. *Gastroenterology 97*: 1186–1192.

Triadafilopoulos, G., Shah, M. H., and Pothoulakis, C. (1991). The chemotactic response of human granulocytes to *Clostridium difficile* toxin A is age dependent. *Am. J. Gastroenterol. 86*: 1461–1465.

Tucker, K. D., and Wilkins, T. D. (1991). Toxin A of *Clostridium difficile* binds to the human carbohydrate antigens I, X, and Y. *Infect. Immun. 59*: 73–78.

Tucker, K. D., Carrig, P. E., and Wilkins, T. D. (1990). Toxin A of *Clostridium difficile* is a potent cytotoxin. *J. Clin. Microbiol. 28*: 869–871.

Tvede, M., and Rask-Madsen, J. (1989). Bacteriotherapy for chronic relapsing *Clostridium difficile* diarrhoea in six patients [see comments]. *Lancet 1*: 1156–1160.

Vesely, D. L., Straub, K. D., Nolan, C. M., Rolfe, R. D., Finegold, S. M., and Monson, T. P. (1981). Purified *Clostridium difficile* cytotoxin stimulates guanylate cyclase activity and inhibits adenylate cyclase activity. *Infect. Immun. 33*: 285–291.

Viscidi, R., Willey, S., and Bartlett, J. G. (1981). Isolation rates and toxigenic potential of *Clostridium difficile* isolates from various patient populations. *Gastroenterology 81*: 5–9.

Viscidi, R., Laughon, B. E., Yolken, R., Bo-Linn, P., Moench, T., Ryder, R. W., and Bartlett, J. G. (1983). Serum antibody response to toxins A and B of *Clostridium difficile*. *J. Infect Dis. 148*: 93–100.

von Eichel-Streiber, C., Harperath, U., Bosse, D., and Hadding, U. (1987). Purification of two high molecular weight toxins of *Clostridium difficile* which are antigenically related. *Microb. Pathogen. 2*: 307–318.

von Eichel-Streiber, C., Suckau, D., Wachter, M., and Haddig, U. (1989). Cloning and characterization of overlapping DNA fragments of the toxin A gene of *Clostridium difficile*. *J. Gen. Microbiol. 135*: 55–64.

von Eichel-Streiber, C., Warfolomeow, I., Knautz, D., Sauerborn, M., and Hadding, U. (1991). Morphological changes in adherent cells induced by *Clostridium difficile* toxins. *Biochem. Soc. Trans. 19*: 1154–1160.

Wald, A., Mendelow, H., and Bartlett, J. B. (1980). Nonantibiotic associated pseudomembranous colitis due to toxin producing clostridia. *Ann. Intern. Med. 92*: 798–799.

Walley, T., and Milson, D. (1990). Loperamide related toxic megacolon in *Clostridium difficile* colitis [Letter]. *Postgrad. Med. J. 66*: 582.

Walters, B. A., Roberts, R., Stafford, R., and Seneviratne, E. (1983). Relapse of antibiotic associated colitis: endogenous persistence of *Clostridium difficile* during vancomycin therapy. *Gut 24*: 206–212.

Wedel, N., Toselli, P., Pothoulakis, C., Faris, B., Oliver, P., Franzblau, C., and LaMont, T. (1983). Ultrastructural effects of *Clostridium difficile* toxin B on smooth muscle cells and fibroblasts. *Exp. Cell. Res. 148*: 413–422.

Wilcox, M. H., and Spencer, R. C. (1992). *Clostridium difficile* infection: responses, relapses and re-infections. *J. Hosp. Infect. 22*: 85–92.

Wilson, K. H. (1988). Microbial ecology of *Clostridium difficile*. In *Clostridium difficile: Its Role in Intestinal Disease*, R. D. Rolfe and S. M. Finegold (Eds.), Academic, New York, pp. 183–200.

Woods, G. L., and Iwen, P. C. (1990). Comparison of a dot immunobinding assay, latex agglutination, and cytotoxin assay for laboratory diagnosis of *Clostridium difficile*-associated diarrhea. *J. Clin. Microbiol. 28*: 855–857.

Wren, B. W., Clayton, C. L., Mullany, P. P., and Tabaqchali, S. (1987). Molecular cloning and expression of *Clostridium difficile* toxin A in *Escherichia coli* K12. *FEBS Lett. 225*: 82–86.

Yotseff, P. S., Roche, J. K., and Guerrant, R. L. (1993). In vitro studies on the reversibility of *Clostridium difficile* toxin A effects on intestinal epithelial cells and on the efficacy of PAF antagonists in preserving barrier function. *Gastroenterolgy 104*: A805.

Young, G. P., Ward, P. B., Bayley, N., Gordon, D., Higgins, G., Trapani, J. A., McDonald, M. I., Labrooy, J., and Hecker, R. (1985). Antibiotic-associated colitis due to *Clostridium difficile*: double-blind comparison of vancomycin with bacitracin. *Gastroenterology 89*: 1038–1045.

Young, G. P., Bayley, N., Ward, P., St. John, D. J., and McDonald, M. I. (1986). Antibiotic-associated colitis caused by *Clostridium difficile*: relapse and risk factors. *Med. J. Austral. 144*: 303–306.

16

Bordetella pertussis: Determinants of Virulence

David A. Relman

*Stanford University School of Medicine, Stanford, and
Palo Alto Department of Veterans Affairs Medical Center,
Palo Alto, California*

I. INTRODUCTION

Far from becoming a disease of the past, pertussis (or whooping cough) continues to cause substantial morbidity and mortality around the world. Interest in the etiologic agent, *Bordetella pertussis*, reflects this practical problem as well as the increasingly apparent sophistication of this organism as a host-adapted pathogen.

II. PERTUSSIS AS A DISEASE: SOCIAL AND BIOLOGICAL FEATURES

It is only from the epidemiologic, pathologic, and clinical features of an infectious disease that one begins to understand and appreciate microbial pathogenetic mechanisms and determinants of virulence. From these features arise the relevant and critical questions that guide scientific inquiry. As with many other infectious diseases, these principles apply to pertussis and to our current understanding of *Bordetella pertussis* virulence determinants.

A. Epidemiology

Pertussis was first described over 350 years ago, and the causative agent was isolated almost 100 years ago (Hewlett, 1990). Pertussis is a respiratory disease and is spread by aerosol droplets. It is highly transmissible among persons with close contact, especially in household and classroom situations. In the former situation, attack rates vary from 70 to 100%. The disease occurs in all regions of the world and is not particularly seasonal. Epidemic peaks of disease activity tend to occur every 3–5 years.

Over the past decade, 1.0–1.8 pertussis cases have been reported per 100,000 persons annually in the United States (Davis et al., 1992). With a total of 6132 cases reported in 1993, pertussis would appear to be an uncommon disease in this country. However, these numbers represent a dramatic change since the early 1940s when there were approximately 150 cases and six deaths annually per 100,000 population in the United States. The major reason for this decrease was the development and implementation of whole-cell pertussis vaccines and a widespread vaccination program. Theoretical considerations predict that vaccination rates of greater than 80% lead to substantial reductions in pertussis incidence (Thomas, 1989). In the United States the annual case incidence rate reached a nadir in 1976. The same pattern was observed in other developed nations with similar vaccination programs. In underdeveloped areas of the world without organized vaccination programs, pertussis has continued to be a cause of substantial morbidity (approximately 60 million cases per year worldwide) and mortality (approximately 600,000 deaths per year) (Muller et al., 1986). Vaccination has led to a shift in the peak age incidence of disease from children of ages 1–5 years to children under the age of 1 year. [The latter group now accounts for approximately one-half of all reported cases (Farizo et al., 1992).] And because of waning vaccine-induced immunity, the population of pertussis-susceptible adults in the United States has grown rapidly to greater than 50 million people (Bass and Stephenson, 1987; Herwaldt, 1991).

Waning immunity has not been the only problem to plague pertussis vaccines and thus indirectly impact on pertussis epidemiology. A number of side effects have been associated with current whole-cell vaccines, ranging from local reactions to fever and a variety of purported neurological problems (Howson and Fineberg, 1992). Despite the lack of supporting data for an association with chronic neurologic damage (Cherry, 1990; Howson and Fineberg, 1992), vaccine acceptance has fallen in many communities and developed nations. The result has been a resurgence of disease in these regions as well as an imperative to develop a less reactogenic, well-defined acellular pertussis vaccine. Since 1976, case incidence rates in the United States have slowly increased (Davis et al., 1992). Because of concerns over increased vaccine reactogenicity in older persons, adults have not been offered vaccine booster doses, and disease incidence rates have increased disproportionately among persons 15 years of age and older. Although 12% of pertussis cases are reported among persons 15 years of age or older (Davis et al., 1992), in outbreak settings a much higher proportion of cases are documented among adolescents and adults (Centers for Disease Control, 1993). In addition, most of the clinical disease attributable to *B. pertussis* in adults may go unrecognized. Studies specifically designed to address this issue reveal that approximately 20–26% of young adults with persistent cough have serologic evidence of recent *B. pertussis* infection (Robertson et al., 1987; Mink et al., 1992).

Adults may play an even larger role in pertussis epidemiology than was previously appreciated. *B. pertussis* is a highly host-adapted human pathogen; the organism has never been isolated from other species or from the environment. Studies of nosocomial disease and disease in institutional settings suggest that adults may in fact serve as the primary reservoir for *B. pertussis* and facilitate its spread to susceptible children (Nelson, 1978; Biellik et al., 1988; Steketee et al., 1988; Cherry et al., 1989; Long et al., 1990). Adults with impaired immune mechanisms (e.g., secondary to human immunodeficiency virus infection) may be more susceptible to overt disease (Doebbeling et al., 1990). However, the manner and degree to which adults serve this purpose cannot be accurately assessed at the present time. In part, this is due to the insensitivity of current detection methods for *B. pertussis*. Clinical case definitions, culture, direct fluorescent antibody staining, and serology all suffer from poor sensitivity and specificity (Davis et al., 1992). In addition, current clinical samples may not come from the most appropriate anatomic sites. Therefore, it is not surprising that in one study nasopharyngeal cultures for *B. pertussis* were negative in HIV seropositive individuals with cough (Cohn et al., 1993). Assays based on the polymerase chain reaction (PCR) may offer improvements in both sensitivity and specificity (He et al., 1993). Nonetheless, adults are now viewed as important targets for refined prophylactic interventions (Edwards et al., 1993). In addition, a goal of newer vaccines should be prevention of colonization as well as disease.

These epidemiologic features raise a number of questions about *B. pertussis* pathogenesis strategies and mechanisms for virulence. What are the primary determinants of protective immunity? What is the duration of host colonization following initial infection or disease? Where does the bacterium reside in the host? And how might anatomic location change with duration of disease? How does the bacterium adapt to various host anatomic niches? Are some *B. pertussis* strains relatively more competent at either establishing infection, causing disease, or persisting within the host?

B. Pathology

Early descriptions of the pathology associated with whooping cough were frequently confused by concurrent secondary superinfection with "influenza bacilli" (*H. influenzae*) or other bacteria that were known to provoke a purulent bronchopneumonia (Rich, 1932). And the picture was biased by the overrepresentation of cases that ended in death (usually as a result of pneumonia), since only postmortem examinations were available. Nevertheless, a number of consistent histologic findings emerged and were linked to a characteristic clinical syndrome (a prolonged illness with coughing paroxysms followed by vomiting, and peripheral lymphocytosis) and to the isolation of the Bordet-Gengou bacillus (*B. pertussis*). These findings almost exclusively concerned the upper and lower respiratory tract.

In studies by Mallory and coworkers, the key histologic features of pertussis were first described in human cases of naturally occurring disease (Mallory and Hornor, 1912; Mallory et al., 1913). In these cases, clumps of bacteria were attached to and wedged between cilia of the epithelial cells lining the large airways. Many cilia were reduced to short stubs. Mallory hypothesized that the bacteria interfered mechanically with normal ciliary function and impaired clearance of secretions. Small to moderate numbers of migratory polymorphonuclear leukocytes were found adjacent to infected ciliated epithelial cells. The additional possibility that a secreted toxin might damage ciliary action and lead to the accumulation of lumenal debris, resulting in cough, could not be ruled out. These observations provide the foundation of our current understanding of pertussis pathogenesis. Other important pathologic observations from later studies have included peribronchial and peribronchiolar mononuclear cell infiltrates within the airway walls, consisting of lymphocytes, plasma cells, and macrophages (Lapin, 1943). Pathologists have also described peribronchial lymphoid hyperplasia and tracheobronchial lymph node hyperplasia. In a minority of cases, airway inflammation, epithelial desquamation, and mucus-containing exudates extend into the most peripheral portions of the respiratory tree, causing alveolar impairment and true airspace pneumonia. *B. pertussis* has been only rarely isolated from the lung in humans. Upper respiratory tract histologic findings are limited to mucosal erythema, edema, and hemorrhage. The extent and frequency of histologic changes in the human lower respiratory tract during mild to moderate cases of pertussis are still unclear. Also missing is a clear sense of *B. pertussis* tissue localization at different times during the course of the natural disease.

Pathology outside of the respiratory tract has always been attributed to "a circulating endo- or exotoxin" (Lapin, 1943) since *B. pertussis* is never isolated from the bloodstream or from distant tissues. More severe cases of pertussis often include signs of pathophysiology involving the central nervous system. Examination of the brain in terminal cases reveals multiple diffuse punctate parenchymal hemorrhages, capillary dilatation, and occasional neuronal cell damage with glial proliferation. However, three clinical features of severe, complicated pertussis—hypoxia, hypoglycemia, and convulsions—could themselves account for some or all of these neuropathologic findings.

Despite the lack of an animal model that adequately duplicates the natural disease, a few experimental systems simulate certain features of tissue colonization, cell tropism, and cytopathology. Some of these systems have led to important insights into mechanisms of pertussis pathogenesis. *B. pertussis* colonization of the ciliated respiratory epithelia and adherence to ciliary tufts are consistently demonstrated in respiratory infections of monkeys, dogs, rabbits, and mice. At the same time, tracheal organ culture can provide more immediate access to viable, differentiated respiratory epithelium and has been particularly useful in studying the effects of *B. pertussis* attachment on the ciliated epithelial cell (Collier et al., 1977; Muse et al., 1977). Selective damage to hamster ciliated tracheal epithelial cells begins within the first 24 h of attachment with alterations in metabolic function. Within the next 48 h, one observes ciliostasis, cytoplasmic vacuolization, organelle destruction, cell swelling, and finally cell detachment from the underlying basement membrane and expulsion (see Fig. 1). As the ciliated columnar cells

Fig. 1   *B. pertussis* adherent to the cilia of a respiratory columnar epithelial cell that is detaching from the underlying basement membrane. Scanning electron micrograph of a hamster tracheal explant, ×6000. (From Muse et al.; © 1977 by The University of Chicago.)

are selectively lost from the lumenal surface of the respiratory mucosa, nonciliated columnar cells and basal cells are left, neither of which seem to be as attractive targets for the remaining bacteria in this model system. Studies of *B. bronchiseptica* with mouse tracheal epithelia demonstrate some of the same features, as well as selective rupture of ciliary membranes, loss of cilia, and some damage to nonciliated cells (Sekiya et al., 1988, 1989). Additional data from these sorts of studies suggest a role for a variety of bacterial factors at different steps in this process and are described in more detail in later sections of this review.

C.   Clinical Manifestations

After a mean incubation period of approximately 14 days, the clinical manifestations of pertussis begin in a subacute, mild, and nonspecific fashion (Lapin, 1943). The disease is often divided into three clinical stages: catarrhal, paroxysmal, and convalescent. During the initial stage, mild nasal and pharyngeal inflammation predominate along with sneezing, rhinorrhea, and occasional cough. There is usually slight fever but not higher than 38°C. The protean features most resemble those of a common cold and do not provide a clue to the true diagnosis.

   As the frequency and severity of the cough increase, the disease enters the classic, paroxysmal stage approximately 7–14 days following the initial onset of cough. During this stage, periods of violent coughing occur, followed in some cases by a sudden and rapid inspiration of air through the swollen upper airways, resulting in a characteristic whoop. In 1578, DeBaillou described the coughing episode in the following manner (Lapin, 1943):

> The lung is so irritated that . . . it can neither admit breath nor easily give it forth again. The sick person seems to swell up, and, as if about to strangle, holds his breath clinging in the midst of his jaws—for they are free from this annoyance of coughing sometimes for the space of four or five hours, then the paroxysm of coughing returns. Very frequently the belly happens to be upset.

The severity of the disease at this stage is age-dependent and is most dramatic in children less than 1 year of age. These children are more likely to be hospitalized (46–69% vs. 4% for older patients) and to die of complications (0.6% vs. <0.1%) (Gan and Murphy, 1990; Davis et al., 1992; Farizo et al., 1992; Sutter and Cochi, 1992). Pneumonia (often secondary to *H. influenzae*, *S. pneumoniae*, and *S. pyogenes*), seizures, and encephalitis are the most significant of these complications. They occur in approximately 20–30%, 2–4%, and 0.2–1.2%, respectively, of hospitalized pertussis patients. However, because most cases of pertussis are mild, as many as 92–97% of all cases may not be reported (Sutter and Cochi, 1992). The duration of the paroxysmal stage varies from 1 to 6 weeks. As the coughing episodes become less severe and less frequent, the patient enters the convalescent stage, which may last for weeks to months.

Pertussis may be particularly mild in adolescents and in adults and therefore unrecognized. Persistent dry cough may be the principal clinical finding among adults with serologic evidence of recent *B. pertussis* infection. In studies of adults of a wide range of ages and with persistent cough, approximately 26% have serologic evidence of recent infection (Robertson et al., 1987; Mink et al., 1992). The duration of cough often exceeds 3 weeks.

A number of biochemical pathophysiological abnormalities have been described in individuals with pertussis. These include peripheral leuko- and lymphocytosis (B and T cells), hyperinsulinemia, and hypoglycemia, although the last is only infrequently documented. Many of these responses are reproduced in an intranasally-infected mouse model of *B. pertussis* respiratory infection (Pittman et al., 1980). Diagnosis of pertussis still leaves much to be desired. The current reference method remains isolation of the organism from nasopharyngeal swab cultures (Onorato and Wassilak, 1987). One of the perplexing features of pertussis is the early loss of culture positivity (about 4 weeks following onset of cough) relative to the duration of the disease. However, throat cultures are less sensitive than nasopharyngeal ones (Marcon et al., 1987), and direct immunofluorescent staining of nasopharyngeal specimens is less specific. Serologic methods may be more sensitive than any of the preceding techniques but are not timely, since they require acute and convalescent samples (Halperin et al., 1989; Hallander et al., 1991). PCR-based methods may offer improved sensitivity and specificity but are not yet widely available (Glare et al., 1990; He et al., 1993).

The treatment of pertussis is largely supportive. Erythromycin may diminish the severity of the illness when given early in the course (Farizo et al., 1992) and also reduce person-to-person transmission (Steketee et al., 1988). The usefulness of whole-cell and acellular pertussis vaccines in protecting children from clinically significant disease was mentioned earlier in this chapter and is discussed elsewhere (Cherry, 1992, 1993; Edwards, 1993). Current acellular vaccines are most often composed of inactivated pertussis toxin (PT), filamentous hemagglutinin (FHA), and sometimes pertactin and fimbrial protein; these vaccines are associated with fewer and less frequent systemic adverse effects than are whole-cell pertussis vaccines (Centers for Disease Control, 1992a,b; Howson and Fineberg, 1992). There are no data bearing on the efficacy of any pertussis vaccine in reducing *B. pertussis* colonization of susceptible hosts.

B. parapertussis has been only occasionally associated with disease, always in humans and usually of a milder form than that caused by *B. pertussis*. It may be coisolated with the latter in some cases (Linnemann and Perry, 1977). *B. bronchiseptica* has been isolated from a small number of humans, including HIV-seropositive persons, with a range of respiratory disease (Woolfrey and Moody, 1991; Ng et al., 1992). *B. bronchiseptica* and *B. avium* are adapted to the respiratory tract of animals, where they cause disease in probably only a small proportion of those they colonize.

III. PATHOGENESIS OF PERTUSSIS: AN OVERVIEW

The events that characterize the process of bacterial pathogenesis, in general, define several common microbial strategies for multiplication, survival, and transmission (Finlay and Falkow, 1989; Relman and Falkow, 1994). These events tend to follow a temporal sequence and consist of entry into the host, acquisition of a suitable niche, local multiplication, subversion or evasion of host defenses, occasionally dissemination and/or persistence within the host, and then transmission to a new host. The key features of pertussis pathogenesis follow from this scheme (see Fig. 2). Our conceptualization of pertussis pathogenesis has undergone some revision during the past decade (Weiss and Hewlett, 1986). In this period, the notion of pertussis as a monofactorial process (pertussis toxin-mediated) (Pittman, 1984) has yielded to the recognition of multiple virulence factors, many interacting in a cooperative or synergistic fashion. In the following discussions, the term *virulence factor* is used to refer to any microbial component that is so important to microbial survival within the host that its loss negatively impacts upon the usual subsequent disease pathology and pathophysiology. This is a more inclusive definition than that which depends upon death as an end point, but it is meant to exclude

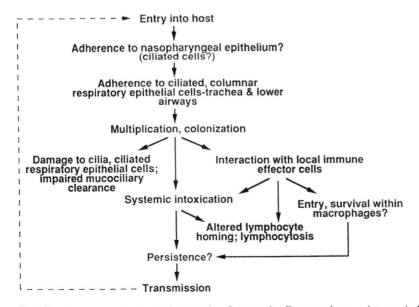

Fig. 2 Key events in the pathogenesis of pertussis. Events whose existence is based primarily upon speculation are indicated with a question mark.

gene products that play housekeeping roles in microbial growth or multiplication outside of the host.

Bordetella pertussis is a small coccobacillary organism that causes disease only in humans. It is piliated, aerobic, nonmotile, and nonfermentative. The population structure of the *Bordetella* species is clonal; in addition, most disease is caused by a few clones, or electrophoretic types (Musser et al., 1986, 1987). Despite the fact that the different species (1) can cause similar diseases, (2) have highly homologous chromosomes by DNA-DNA hybridization, and (3) appear to be highly related by multilocus enzyme electrophoresis methods, there are important differences among them concerning various growth, biochemical, and morphologic properties as well as some virulence determinants. These phenotypic characteristics form the basis for the continued recognition of several distinct species. Where particularly relevant, differences between species will be mentioned in the following discussions.

Following transmission to a susceptible host by aerosol droplet, *B. pertussis* probably makes initial contact with the host nasopharyngeal mucosa. Although ciliated cells in this portion of the upper respiratory tract are confined solely to the posterior nares, it is clear that pertussis as well as *B. bronchiseptica*-associated diseases in animals involve a variety of nasal clinical findings and pathology [e.g., sneezing and nasal discharge in humans, atrophic rhinitis in pigs (Duncan et al., 1966; Switzer et al., 1966)]. *B. bronchiseptica* can be reproducibly cultivated from the nasal passage following experimental intranasal inoculation of animals. Therefore, it seems reasonable to assume that during natural infection *Bordetella* species colonize these tissues prior to, or concurrent with, colonization of the ciliated epithelia of the trachea and lower airways.

Adherence is a critical process for an organism that remains essentially restricted to a mucosal surface. The multiplicity and complexity of *B. pertussis* adherence factors reflect its dependence upon this process and are discussed in some detail in later sections of this chapter. Based upon the previously described pathology, the primary adhesion events probably take place at the ciliary tufts of the respiratory epithelium and within the overlying mucus layer. It remains unclear whether *B. pertussis* adheres to any significant degree to other epithelial cell types during the course of natural human infection. Both in vivo and in vitro observations suggest that *B. pertussis* may adhere specifically to host macrophages (Cheers and Gray, 1969; Relman et al., 1990; Bromberg et al., 1991; Saukkonen et al., 1991; Friedman et al., 1992). Among the bacterial factors involved in adherence are FHA, PT, fimbriae, and probably pertactin as well as other components. Survival at the surface of the respiratory tract mucosa may also require resistance to the antimicrobial effects of a family of small, cationic, membrane-active peptides (Diamond et al., 1993).

The acquisition of an initial site for multiplication brings about a rapid series of events involving passive leakage and active secretion of toxic substances by *B. pertussis*. Tracheal cytotoxin (TCT), a fragment of the *Bordetella* cell wall peptidoglycan, is released from replicating organisms and exerts a potent toxic effect on ciliated epithelial cell metabolism. Dermonecrotic toxin (DNT) and a secreted bifunctional adenylate cyclase-hemolysin (AC) may also cause structural and metabolic damage to these epithelial cells. Within relatively short periods of time, the local environment has been modified and a means established for later evading components of the host immune system. According to data from in vitro experiments, PT and AC inhibit a number of phagocyte functions, including chemotaxis, phagocytosis, and bacterial killing. It has not been established if all of these effects occur in vivo or what role they might play in determining the outcome of infection. On the other hand, there is convincing evidence that PT is crucial for many of the systemic manifestations of pertussis. This oligomeric protein encompasses six subunits with a variety of biological activities including that of an ADP-ribosyltransferase, B-cell mitogen, and lectin. Cellular intoxication and the consequent disruption of intracellular signalling pathways are irreversible and hence long-lasting. For years, these facts have been invoked to explain the persistence of dramatic clinical manifestations (i.e., paroxysmal phase) despite the apparent absence of the causative organism. As an alternative explanation, *B. pertussis* might become sequestered in an intracellular or other anatomic niche, perhaps continuing to release biologically active products for a limited time and/or remaining within the host for an even longer period of time.

One assumes that *B. pertussis* quickly senses and adapts to the initial and subsequently evolving environment (Miller et al., 1989a; Mekalanos, 1992), although the exact environmental cues recognized by this pathogen in the host have not been defined. The apparent complexity of a *Bordetella* regulatory system, the BvgAS regulon, suggests that environmental sensing plays a central role in pertussis pathogenesis, that signal amplification may be possible, as well as limited or partial responses, and that negatively regulated gene products as well as the better-known positively regulated factors mentioned above might be important virulence factors. By means of this system, *B. pertussis* can alter its characteristics significantly. Thus, there is no reason to assume that the organism that leaves a recently infected host resembles the organism that might have been isolated during the earlier stages of that infection.

IV. REGULATION OF VIRULENCE

Bordetella pertussis exists in at least two antigenically distinct phases; these phases have been given various names (phase I or domed/hemolytic; and phase IV or flat/nonhemolytic) ever since the first descriptions of the organism (Bordet and Sleeswyck, 1910; Leslie and Gardner, 1931; Peppler, 1982). The two phases were renamed virulent and avirulent when it became clear that phase I organisms express a variety of factors associated with pathogenicity such as PT, FHA, and AC and phase IV organisms do not (Weiss et al., 1983; Weiss and Falkow, 1984). Two forms of regulation effect a switch between these two phases: a spontaneous shift that can be inherited, known as phase variation, and a readily reversible transition (between "X mode" and "C mode") induced by a variety of environmental conditions, known as antigenic or phenotypic modulation (Lacey, 1960). In the laboratory, the most common signals for modulating *B. pertussis* from virulent to avirulent mode include elevated Mg^{2+}, SO_4^{2-}, and nicotinic acid concentrations, and growth at 25°C. Melton and Weiss (1993) have proposed that the in vivo environmental signal for *B. pertussis* is likely to be an anionic molecule small enough to diffuse through outer membrane pores into the periplasm and similar in structure to various nicotinic acid derivatives.

Weiss et al. (1983) identified a *B. pertussis* chromosomal locus, using transposon mutagenesis, that appeared to be responsible for trans-activation of the virulent phase-associated gene products. The latter included FHA, PT, AC, and DNT. The transposon insertion mutant strain had a phenotype (Vir⁻) similar to that of spontaneously occurring avirulent phase mutants, and a cloned copy of the locus could complement the defect in the latter (Stibitz et al., 1988); hence, the locus was named *vir*. Sequence analysis of this chromosomal region, later renamed *bvg* (*Bordetella* virulence gene), reveals two open reading frames, designated *bvgA* and *bvgS* (Arico et al., 1989; Stibitz and Yang, 1991).

These genes are cotranscribed; in turn, the two gene products are sufficient for transcription of the FHA structural gene *fhaB* in *E. coli*, although overproduction of BvgA can overcome the absence of BvgS in this system (Miller et al., 1989b; Roy et al., 1989). While the *bvgAS* locus was not sufficient to activate transcription of a PT operon fusion in *E. coli*, an in-frame deletion of BvgA in *B. pertussis* exhibited the typical Vir⁻ phenotype. Early investigations demonstrated that environmental modulation is mediated by the two Bvg proteins.

An understanding of the role of the BvgAS regulatory system in facilitating *B. pertussis* survival in the host will require an appreciation of the "wiring" and "timing" of this system. Besides a lack of information about relevant environmental cues, the dynamics of the *B. pertussis* phase transition have not been clearly defined. Early work by Lacey suggests that there may be intermediate states between the virulent and avirulent phases (Lacey, 1960); however, it is unclear whether these intermediate states might be important during natural infection. It does seem likely that multiple, simultaneous environmental stimuli create an integrated response and that different Bvg-regulated genes respond in different fashions (Melton and Weiss, 1993). In addition, it is clear that the *BvgAS* operon is subject to autogenous regulation. BvgA expression is controlled by at least two promoters. One of these promoters, bvgP2, allows low levels of *bvgAS* transcription in the absence of BvgA. The other, bvgP$_1$, is activated by BvgA and is responsive to environmental modulating stimuli. BvgA synthesis is then significantly augmented, to a greater degree than is BvgS production. Each promoter is silenced when the other is active. Thus, this system is designed to maintain levels of BvgS and BvgA under modulating conditions, but then to be able to respond rapidly with a dramatic increase in the amount of BvgA (Scarlato et al., 1991a; Stibitz and Yang, 1991) upon removal of these conditions. Scarlato and coworkers have examined the temporal sequence of promoter activation following a temperature shift from 25°C to 37°C, based on the notion that this manuever may simulate some of the environmental changes encountered by *B. pertussis* as it enters the human host. Within 10 min after relief of modulating conditions, transcription of *bvgAS*, *fhaB*, and fimbrial genes is up-regulated; however, transcriptional activity at the PT and AC promoters does not begin for 2 h and does not reach maximum levels until 6 h after the temperature shift (Scarlato et al., 1991a; Rappuoli et al., 1992). Of course, these events take place on a rapid time scale in comparison to the more slowly developing gross pathologic events of the natural infection.

Database homology searches yielded some immediate insights into the molecular basis of BvgAS function (Arico et al., 1989). BvgA and BvgS share significant degrees of sequence similarity with members of a family of bacterial regulatory proteins that respond to environmental signals (Stock 1990; Mekalanos, 1992; Parkinson, 1993). This "two-component" family consists of one group of proteins, known as sensors, that usually span the cytoplasmic membrane and contain a C-terminal "transmitter" communication module. The family also includes a group of cytoplasmic proteins, known as response regulators, that contain an N-terminal "receiver" communication module. Sensor proteins often have the ability to phosphorylate themselves at a conserved histidine residue. This high-energy phosphate group is then transferred to a conserved aspartate residue within the response regulator receiver module. The overall phosphorylation state of the response regulators determines their level of activity, either as DNA binding proteins or as effectors of protein–protein signalling. *bvgA* encodes a 23-kDa protein with an N-terminal domain that contains sequences highly similar to known receiver communication modules. *bvgS* encodes a 135-kDa protein with a more complex arrangement of predicted functional domains (Arico et al., 1989; Stibitz and Yang, 1991). BvgS contains both transmitter and receiver module homologies, as do *Myxococcus xanthus* FrzE and *Agrobacterium tumefaciens* VirA. In addition, BvgS contains a C-terminal regulatory domain.

Experimental data confirm, so far, the biochemical properties of BvgA and BvgS predicted from the initial sequence analysis. BvgS localizes to the inner membrane of fractionated whole *B. pertussis* (Stibitz and Yang, 1991). Genetic complementation studies suggest that BvgS may depend upon multimerization (e.g., as a dimer) for its activity. Mutations within either a 541-amino-acid periplasmic region or within a 161-amino-acid linker region that connects the transmembrane segment with the transmitter module abolish sensitivity to environmental modulating signals, resulting in either a constitutive Bvg^+ or a Bvg^- phenotype (Miller et al., 1992). Presumably, these signals are "sensed" by the BvgS periplasmic domain and the information is then transmitted across the cytoplasmic membrane through the linker. One hypothesis holds that negatively charged modulators interact with positively charged residues within the periplasmic domain, induce conformational changes, and thereby prevent signal transmission required for subsequent phosphorylation (Melton and Weiss, 1993). BvgS is thought to autophosphorylate at a conserved transmitter domain histidine residue and then transfer the phosphate to a conserved aspartate residue within its own receiver domain, prior to phosphorylating BvgA (Uhl and Miller, 1994). The BvgS cytoplasmic domain is required for this process. (See Fig. 3.)

Phosphorylation enables most regulator proteins to implement their specified response. In the case of BvgA, the response is DNA binding and transcriptional activation. Gel retardation and DNase protection analyses have demonstrated that protein–DNA interactions occur at sites upstream of both the *fhaB* promoter and the positively regulated $bvgAP_1$ promoter, and that this binding protein is most probably BvgA (Roy and Falkow, 1991). Mutations within a specific seven-base repeat, found at both of these two sites, abolish *fhaB* and *bvgAS* transcription, respectively. A similar recognition sequence has also been found upstream of *fimB*, at the 5' end of a Bvg positively regulated gene cluster (Willems et al., 1992). It also appears to be conserved in *B. parapertussis* (Scarlato et al., 1991b). However, the binding recognition sequence is not found upstream of the *ptx* or *cyaA* promoters. In addition, BvgA extracts do not retard or protect *ptx* promoter-containing DNA fragments (Roy and Falkow, 1991). Thus, BvgA, although required for transcription of these and other positively regulated genes, may not mediate this effect directly. If this is true, then other regulatory factors must be involved. In this regard, one study offers evidence from southwestern analysis that a Bvg positively regulated 23-kDa protein, distinct from BvgA, binds to a 20-bp tandem repeat found in the upstream regions of the PT and AC promoters (Huh and Weiss, 1991). A competing theory suggests that higher concentrations of BvgA than those needed for *bvgAS* and *fhaB* activation may be sufficient for activation of *ptxABDEC* and *cyaABDE* (Scarlato et al., 1991a).

The Bvg regulon consists of more than 20 chromosomal loci whose expression is subject to reversible trans-activation and trans-repression. A physical map of the *B. pertussis* chromosome reveals that none of the well-characterized bvg-regulated genes are linked to each other, with the exception of *bvgAS*, *fhaB*, and a fimbrial gene cluster *fimABCDfhaC*, which are located within a 27-kb region (Stibitz and Garletts, 1992). The proximity of *bvg*, *fha*, and *fim* genes and their possible coevolution makes sense in view of the direct role of BvgA in *fhaB* (and perhaps *fimBCDfhaC*) gene expression and the relative importance of FHA-mediated adherence during early stages of *B. pertussis* infection (see below). Relatively little information is available concerning genes and gene products that are expressed maximally in the presence of modulators, for example, *bvg*

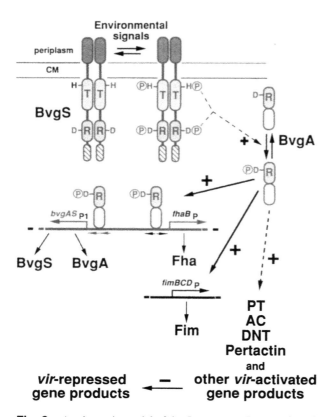

Fig. 3 A schematic model of the *B. pertussis* Bvg regulon. CM, bacterial cytoplasmic membrane; T, transmitter protein domain; R, receiver protein domain; H, histidine; D, aspartic acid; P phosphorylated residue; subscript P, transcriptional promoter. Thick black horizontal lines indicate regions of chromosome; small arrows underneath indicate BvgA DNA binding sites; + and − indicate positive and negative regulatory signals, respectively. See text for further explanations. (Modified from Relman and Falkow, 1994.)

(*vir*)-repressed genes, or *vrg* genes. Beattie and colleagues have identified a group of five *vrg* genes (Beattie et al., 1990) and have found that at least one of them, *vrg-6*, is important for trachea and lung colonization in the mouse aerosol infection model (Beattie et al., 1992). Repression of these genes occurs at the level of transcription and may involve binding of a 34-kDa *bvg*-activated gene product to a consensus DNA sequence within the upstream coding region of these genes (Beattie et al., 1993). Akerley and Miller have shown that the structural gene for flagellin from *B. bronchiseptica* is also a *vrg* gene (Akerley et al., 1992; Akerley and Miller, 1993). As other *bvg*-repressed gene products are identified, a more complete understanding of the role of these genes in pathogenesis should evolve. It is interesting that a serologic response to C-mode (*bvg*-repressed) antigens has been detected primarily in adults following mild disease (Lacey, 1960). Constitutive sensory transduction *B. pertussis* mutants, similar to *B. bronchiseptica* mutants characterized by Miller et al. (1992), might be useful in determining the importance of phenotypic modulation during the course of *B. pertussis* infection in various animal models.

Phase variation in *B. pertussis*, in contrast to phenotypic modulation, is due to frameshift mutations within a C-rich region of the BvgS-coding sequence (Stibitz et al., 1989). Some spontaneous *B. bronchiseptica* avirulent phase variants contain deletions within the BvgS coding sequence (Arico et al., 1991). Although *B. pertussis* avirulent phase variants have been isolated from patients convalescing from whooping cough (Kasuga et al., 1954), it is not known whether phase variation plays any significant role in natural infection.

V. VIRULENCE DETERMINANTS

A. Pertussis Toxin (PT)

Pertussis toxin (PT) (islet-activating protein, lymphocytosis-promoting factor) is a hexameric protein secreted by *B. pertussis* that has a variety of biological properties. The importance of these properties in the pathophysiology of pertussis is reflected by the number of studies of this molecule (Weiss and Hewlett, 1986) and in the prominence of PT in earlier formulations of pertussis pathogenesis (Pittman, 1984). Many of these properties are discussed in separate chapters by Burns, Rappuoli, and Katada elsewhere in this volume. The following discussion will emphasize the role of PT as a virulence factor.

PT is best known for its enzymatic activity as an NAD-dependent ADP-ribosyltransferase (Katada and Ui, 1982). As such, it belongs to a family of bacterial exotoxins with similar enzymatic activity, including diphtheria toxin, cholera toxin, *E. coli* heat-labile enterotoxin, and *Pseudomonas* exotoxin A. The substrates for ADP-ribosylation by these exotoxins differ. In the case of PT the enzyme substrate is the regulatory GTP-binding protein, G_i, which exerts inhibitory activity on membrane-associated eukaryotic adenylate cyclase. PT causes irreversible inactivation of G_i α subunits, thereby blocking homeostatic inhibition of adenylate cyclase activity and leading to unchecked cAMP production. Other GTP-binding regulatory membrane proteins are also inactivated by PT, such that a variety of receptors become uncoupled from inhibitory input. The result is an extensive disruption of intracellular signaling pathways (e.g., some phosphatidylinositol-specific phospholipase C-mediated signals) and a sensitivity to the actions of a variety of stimulatory hormones and other molecules such as histamine. PT is an invaluable reagent in the identification of regulatory G_i proteins and in the study of these signaling pathways (Chaffin et al., 1990).

ADP-ribosyltransferase activity resides within the S1 subunit (the A monomer); the other five subunits form a B oligomer comprising S2S4 and S3S4 dimers that are connected by the S5 subunit. The B oligomer interacts with glycoproteins and glycolipids on eukaryotic cell surfaces and facilitates translocation of the S2 subunit into the target cell membrane. PT is encoded by an operon consisting of genes for the five subunits in the order S1 (*ptxA*), S2 (*ptxB*), S4 (*ptxD*), S5 (*ptxE*), S3 (*ptxC*) (Locht and Keith, 1986; Nicosia et al., 1986). *B. parapertussis* and *B. bronchiseptica* contain copies of the *ptx* operon that are transcriptionally silent by virtue of numerous point mutations within the operon promoter (Arico and Rappuoli, 1987). *B. pertussis* S2 and S3 are 67% homologous at the amino acid level. Each of the five predicted amino acid sequences contains an N-terminal signal sequence, suggesting secretion of the individual subunits into the periplasm. Recently, an additional operon (*ptl* for "pertussis toxin liberation") was identified immediately downstream of *ptx* that appears to play a role in PT secretion into culture supernatants (Covacci and Rappuoli, 1993; Weiss et al., 1993). An insertion mutation in this operon causes intracellular PT accumulation. Predicted translated products from the seven *ptl* open reading frames bear significant similarity to *Agrobacterium tumefaciens* VirB proteins that are involved in transfer of T-DNA out of the bacterium and into plant cells.

Considerable structure–function correlations are available for some of the PT subunits, and for S1 in particular. Specific amino acids within the S1 subunit have been identified that are essential for ADP-ribosyltransferase enzymatic activity (Pizza et al., 1988). Some of these residues, for example, Glu-129, are shared at similar regions within diphtheria toxin and *Pseudomonas* exotoxin A and are thought to play key roles in the catalytic site. Similarly, common regions thought to bind NAD have also been mapped (Domenighini et al., 1991). By means of site-directed mutational analysis, certain biological properties can be attributed to the ADP-ribosyltransferase activity of PT (Black et al., 1988; Pizza et al., 1989). These properties include induction of lymphocytosis, potentiation of anaphylaxis, and enhanced histamine sensitivity in mice. S1 enzyme activity is also responsible for aberrant lymphocyte trafficking since PT-treated bloodborne lymphocytes are unable to home to lymph nodes and Peyer's patches (Spangrude et al., 1984), and transgenic mice expressing the PT S1 subunit under the control of the lymphocyte-specific *lck* promoter demonstrate faulty thymocyte homing (Chaffin et al., 1990). The mechanism of faulty homing may be related to the finding of a PT-sensitive activation step in lymphocyte attachment to high endothelial venules (Bargatze and Butcher, 1993).

It has long been recognized that patients convalescing from pertussis develop a prominent humoral immune response to PT, including secretory IgA, and that this response correlates to some extent with protection against disease (Zackrisson et al., 1990a,b). Linear and conformational protective epitopes for humoral and cellular immunity have been mapped within the S1 subunit (DeMagistris et al., 1989) at regions that are distinct from the catalytic site (Pizza et al., 1989). For this reason, it has been proposed that newer generation acellular pertussis vaccines should include genetically engineered PT containing those site-specific mutations that eliminate enzymatic activity and yet preserve native conformational folding (Pizza et al., 1989).

The comparison of isogenic mutants with a parental wild-type strain allows the most accurate assessment of the contribution of putative virulence factors to bacterial patho-genesis. This type of approach with *B. pertussis* first became possible after the development of transposon mutagenesis (Weiss and Falkow, 1983) and later with techniques for allelic exchange (Stibitz et al., 1986). Isogenic mutant analysis confirms a central role for PT in pertussis pathogenesis but by no means suggests that other virulence factors are not critical at various stages in this process. In 1983, Weiss and coworkers described the isolation of Tn*5* insertion mutations in a variety of virulence-associated loci, including the one encoding PT (Weiss and Falkow, 1983). In comparison with wild-type organisms, more than 1000-fold greater numbers of a PT-deficient (*ptx* insertional inactivation) mutant strain were required for death of 50% of intranasally infected newborn mice (Weiss et al., 1984; Weiss and Goodwin, 1989). A *ptl* mutant strain also exhibits reduced lethality for these mice but to a much smaller degree. Interestingly, an AC⁻ mutant that secretes normal amounts of PT does not complement at wild-type lethal doses this virulence defect of the *ptx*-insertion PT⁻ mutant (Weiss and Goodwin, 1989; Goodwin and Weiss, 1990). This finding indicates that the lethality of PT, a secreted product, may require local association with or presentation by "competent" organisms.

Analysis of mutant strains in a mouse aerosol infection model indicates that PT plays a role in the persistence of *B. pertussis* within the lung, although it is not a crucial factor in tracheal colonization. PT mutant strains are recovered in lower numbers than wild-type strains from the lungs of intranasally infected mice and are cleared somewhat more quickly (Weiss and Goodwin, 1989; Goodwin and Weiss, 1990; Finn et al., 1991). In addition, mice infected with PT⁻ strains do not develop lymphocytosis. Lung colonization may reflect as much the role of host immune cells and the ability of PT to modify their bactericidal activity as it does the true adherence capabilities of the bacterium. This finding raises the question of PT as a classical adherence factor for host cells.

A number of studies have suggested that PT may mediate bacterial adherence to a number of different host cell types. Tuomanen and Weiss (1985) reported a role for PT in *B. pertussis* binding to human ciliated respiratory epithelial cells. Incubation with exogenous PT restored the binding of PT mutant strains and led to the proposal that both secreted PT and FHA might promote the binding of not only *B. pertussis* but also other secondary bacterial pathogens in the natural host (Tuomanen, 1986). These data were based on the use of PT mutant strains that were in some cases also defective in expression of pili. They also relied upon anti-FHA and anti-PT antibodies to block all FHA and PT adherence activities. Isolated PT B oligomer has been shown to bind eukaryotic glycoproteins and glycolipids, with two different types of specificity: one for sialylated glycoconjugates, and one for lactose-containing glycoconjugates such as lactosylceramide (Brennan et al., 1988; Tuomanen et al., 1988). These two binding activities have been mapped to regions within S3 and S2, respectively, each of which shares sequence similarity with a galactose-specific eukaryotic carbohydrate recognition domain of the C type. Some data suggest that the S2 lectin-like domain may mediate binding to glycoconjugates on ciliated respiratory epithelial cells and macrophages, while the S3 lectin-like domain may mediate binding to macrophage sialylated glycoconjugates (Saukkonen et al., 1992; van't Wout et al., 1992).

PT, however, seems to be less important than FHA for *B. pertussis* binding to ciliated respiratory epithelial cells. Binding of a PT-deletion strain to these cells is approximately 50% that of the wild-type organism, whereas an FHA-deletion strain binds to these cells at levels approximately equivalent to 6% of wild type (Relman et al., 1989). With respect to *B. pertussis* adherence to human macrophages, PT and FHA seem to each play an independent role. An FHA-PT double mutant is more than 90% impaired in binding these cells, relative to the wild-type strain, whereas single mutants demonstrate no apparent adherence defect (Relman et al., 1990). To evaluate the possible impact of S2 and S3 carbohydrate recognition and S1 enzymatic activity on host colonization and persistence, it would preferable to study the relevant isogenic site-directed mutant strains in a suitable animal model of *B. pertussis* respiratory tract colonization. These data are not currently available.

B. Filamentous Hemagglutinin (FHA)

The name filamentous hemagglutinin refers to both structural and biological features of a large surface-associated and secreted *B. pertussis* protein. Early observations revealed two distinct secreted *B. pertussis* proteins that were capable of agglutinating erythrocytes from a number of different animal species (Arai and Sato, 1976). One was named lymphocytosis-promoting factor and is now known as pertussis toxin (see above). The other formed fine filaments with dimensions of 2 × 40–100 nm under electron microscopy and therefore was named filamentous hemagglutinin, or FHA (Sato et al., 1983). Contrary to earlier assumptions, FHA plays no part in the composition of *Bordetella* fimbriae. Much of the initial interest in this protein stemmed from its ability to protect animals by passive and active immunization against respiratory aerosol challenge with *B. pertussis* (Oda et al., 1984; Sato and Sato, 1984) as well as from its prominent role in the serologic response of patients convalescing from pertussis (Burstyn et al., 1983; Thomas et al., 1989a,b). We now recognize that FHA is the dominant factor in *B. pertussis* adherence to a wide variety of host cells and is a crucial determinant of upper respiratory tract colonization. Recent work suggests that many FHA functional domains have yet to be identified.

Physical characterization of purified FHA protein with polyacrylamide gel electrophoresis reveals a collection of heterogenous polypeptides with major species of 220, 140, 125, and 95 kDa (Irons et al., 1983; der Lan et al., 1986). The multiple polypeptides are all derivatives of a large precursor protein that is highly susceptible to proteolytic cleavage and mechanical degradation. Most biological activities have been associated with the 200-kDa species. A more precise understanding of the origin of the various FHA polypeptides and their biological activities was slow to evolve prior to the isolation and characterization of the FHA structural gene.

The complete FHA structural gene, *fhaB*, was initially isolated using chromosomal probes that were generated from the original FHA-deficient Tn5 insertion mutants created by Weiss and coworkers (Weiss et al., 1983; Stibitz et al., 1988). This work followed a number of preliminary reports that described partial clones of this locus based upon antibody screening of expression libraries (Reiser et al., 1985; Mattei et al., 1986; Brown and Parker, 1987). The cosmid clones isolated by Stibitz and coworkers containing these Tn5 elements also hybridized with probes from the *vir* locus. Subsequent genetic analysis demonstrated that *fhaB* is located adjacent to *bvgAS* and is transcribed in the opposite orientation (Stibitz et al., 1988). Complete sequence determination of this gene reveals a single open reading frame sufficient to encode a 367-kDa precursor polypeptide (Domenighini et al., 1990) (see Fig. 4). The 220 kDa mature FHA product derives from the N terminus of this precursor, although the exact proteolytic cleavage site is not known. The other FHA polypeptide species derive from co- or posttranslational proteolytic processing of the 220-kDa moiety. One processing site has been mapped to residues 1073–1074 immediately following an arginine-rich sequence.

FHA is found both in culture supernatants and associated with the bacterial surface. Export of FHA protein by *B. pertussis* is not well understood. The N-terminal amino acid sequence does not closely fit the classical consensus for a signal peptide. Furthermore, the genetic mutations *fha-1*::Tn5(BP353) and *fha-2*::Tn5(BP354) isolated by Weiss et al. (1983) that affected FHA expression were subsequently mapped to regions downstream of the FHA structural gene and were used to define accessory loci, originally named *fhaA* and *fhaC*, that seemed to be involved in FHA secretion (Stibitz et al., 1988). More recent sequence analysis of this region reveals a cluster of open reading frames, now designated *fimABCD*, that share strong sequence similarity to well-characterized genes involved in fimbrial subunit export and assembly (Locht et al., 1992; Willems et al., 1992) (see Fig. 5). Overlapping the 3′ end of *fimD* is an additional open reading frame that retains the designation *fhaC*. The predicted 61-kDa protein product FhaC has sequence similarity to ShlB and HpmB, outer membrane proteins of *S. marcescens* and *P. mirabilis* that are involved in activation and export of hemolysin (Willems, 1993). A site-specific mutation in *fhaC* abolishes detectable cell-associated and secreted FHA protein, and it is now believed that the BP353 and BP354 FHA phenotypes reflected the polar effects of *fimC*

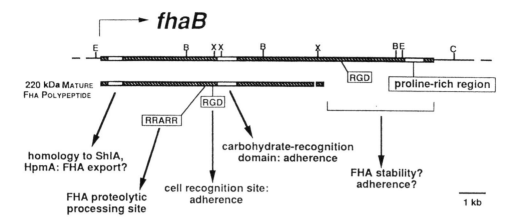

Fig. 4 Schematic diagram of the FHA structural gene *fhaB*. The gene is transcribed from left to right. Regions of the translated product with probable functional significance are indicated. RGD, Arg-Gly-Asp peptide sequence; RRARR, Arg-Arg-Ala-Arg-Arg peptide sequence; E, *Eco*RI restriction endonuclease recognition site; B, *Bam*HI restriction endonuclease recognition site; X, *Xho*I restriction endonuclease recognition site; C, *Cla*I restriction endonuclease recognition site (not all *Cla*I sites are shown).

and *fimD* insertion mutations, respectively, on *fhaC* expression (Willems et al., 1992; Willems, 1993). A putative role for FhaC in FHA export is strengthened by the finding of N-terminal FHA sequence similarity to the N-terminal sequences of the hemolysins ShlA and HpmA that are believed to be required for ShlB- and HpmB-mediated activation, respectively (Delisse-Gathoye et al., 1990). Two surface-associated proteins from nontypable *H. influenzae*, HMW1 and HMW2, also contain similar sequences in the N

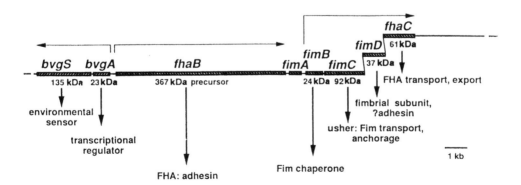

Fig. 5 The *bvgAS-fhaB-fimABCDfhaC* region of the *B. pertussis* chromosome. Open reading frames (ORFs) are indicated by hatched bars. The direction of transcription is indicated by the thin arrows above the ORFs; *fim BCDfhaC* may be cotranscribed. The 5' end of *fimD* overlaps the 3' end of *fimC*, and the 5' end of *fhaC* overlaps the 3' end of *fimD*. These three genes may be translationally coupled. Molecular weights of the predicted polypeptides are listed below each ORF, as well as its postulated function.

terminus (Barenkamp and Leininger, 1992). Like FHA, these two proteins are believed to serve as adherence factors (St. Geme et al., 1993).

Until isogenic FHA mutant strains were available, assessment of FHA as an adherence factor relied on the use of anti-FHA antisera to block function and the incubation of strains with FHA protein to restore function. Studies of this type suggested that FHA played a major role in mediating *B. pertussis* binding to a variety of cultured eukaryotic cell types, including erythrocytes, and HeLa, Vero, WiDr, and 3T3 cells (Lenin et al., 1986; Urisu et al., 1986). In addition, incubation of *H. influenzae* with FHA protein enhanced bacterial binding to ciliated respiratory epithelial cells (Tuomanen, 1986). Tn*5* mutants BP353 and BP354 were used in a number of studies to implicate FHA in bacterial adherence to ciliated respiratory epithelial cells (Tuomanen and Weiss, 1985; Urisu et al., 1986), although these strains were not ideal since they synthesized small amounts of FHA and also had defects in fimbrial assembly. With the FHA structural gene and downstream regions cloned and sequenced, analysis of FHA function has become more straightforward. Studies with in-frame *fhaB* deletion mutants confirm that FHA is the dominant adherence factor in *B. pertussis* binding to ciliated respiratory epithelial cells (Relman et al., 1989) and is a critical factor for tracheal colonization at early time points during respiratory infection in mice (Kimura et al., 1990). The reappearance of BP101, an *fhaB* partial deletion strain, in the trachea at later time points suggests that the tracheal mucosa may become modified during the infection process, revealing new adherence ligands for *Bordetella*. FHA does not appear to be important for colonization of the lung in these animals (Goodwin and Weiss, 1990; Kimura et al., 1990; Finn et al., 1991); however, as discussed earlier, it is not clear that lung colonization is a regular feature of pertussis in humans. Loss of FHA does not alter *B. pertussis* lethality for infant mice infected by the respiratory route (Weiss et al., 1984; Weiss and Goodwin, 1989), although, again, the relevance of this clinical end point for human infection is not clear. FHA has also been shown to be a dominant adhesin of *B. pertussis* for macrophages (Relman et al., 1990) and by virtue of that fact may play an important role in bacterial entry into these host cells (Friedman et al., 1992). This issue of *Bordetella* entry into host cells is discussed in greater detail in a later section.

Several regions of the FHA polypeptide contain predicted sequence motifs that are found in various prokaryotic and eukaryotic proteins. These regions may constitute functional domains and suggest possible biological activities for this complex protein. There are two predicted Arg-Gly-Asp (RGD) sequences, one of which is located within the mature 220-kDa FHA product at a site (residues 1097–1099) that corresponds to the N terminus of the cleaved 125- and 14-kDa breakdown products (Relman et al., 1989; Domenighini et al., 1990). Arg-Gly-Asp sequences serve as cell recognition sites for some eukaryotic extracellular matrix proteins (Ruoslahti and Pierschbacher, 1987; D'Souza et al. 1991) and mediate binding to members of the integrin receptor superfamily (Ruoslahti, 1991). Although a conservative alteration of this 1097–1099 FHA RGD site causes no gross defect in *B. pertussis* binding to isolated ciliated respiratory epithelial cells, this conservative mutation does contribute to a defect in FHA-mediated binding to human macrophages (Relman et al., 1990) and may affect intracellular uptake of bacteria by these cells (Saukkonen et al., 1991). In a study of PT⁻ *B. pertussis* strains in the infant mouse aerosol infection model, the addition of a conservative substitution in the 1097–1099 FHA RGD site significantly reduces tracheal colonization (D. A. Relman et al., unpublished data). The second RGD site is located in the C terminus of the FHA precursor polypeptide, a portion of the precursor that probably is not contained within the mature 220-kDa molecule. This C-terminal polypeptide includes a region of repeated sequences with high proline content. The function of the C terminus is unclear. Preliminary data suggest that it may be necessary for secretion of mature FHA (Locht et al., 1993); sequence features suggest that it may serve a structural role in the export or presentation of mature FHA at the cell surface (Domenighini et al., 1990).

Two other types of binding activities have been attributed to FHA. The first is an affinity for lactosamines, lactosylceramide, and other lactose-containing glycolipids that may be found in the ciliary membranes of the respiratory epithelium and on alveolar macrophages (Tuomanen et al., 1988). This binding affinity has been associated with an FHA region located between residues 1141 and 1279 ("carbohydrate recognition domain") (Prasad et al., 1993). Within this region is a short sequence that is similar to that of a proposed PT S2 subunit lectin-like domain (Saukkonen et al., 1992). The second type of binding affinity is for sulfatides and glycosphingolipids (Brennan et al., 1991; Menozzi et al., 1991b). For example, sulfated sugars such as heparin and dextran sulfate inhibit binding of *B. pertussis* to hamster tracheal cells. It has been postulated that glycos-aminoglycans in respiratory mucus or on the respiratory epithelial surface may facilitate *B. pertussis* colonization of this site by this kind of binding interaction. Heparin-binding activity is currently being mapped within FHA regions of the mature protein (Locht et al., 1993).

Interest in FHA as an immunogen reflects our appreciation of this protein as an adherence factor, as well as convincing evidence of humoral and cellular immune responses to FHA following natural infection. FHA and PT are dominant antigens in the serologic responses of patients following pertussis (Burstyn et al., 1983). A study of antibody class-specific responses suggests that serum and respiratory secretion anti-FHA IgA levels increase significantly after infection (Thomas et al., 1989a; Zackrisson et al., 1990a). Although the data are conflicting with regard to the protective value of anti-FHA antibodies against disease in humans, multiple studies demonstrate that FHA generates a long-lived immune response in animals (Amsbaugh et al., 1993) and can protect them against respiratory mucosal colonization (Kimura et al., 1990; Shahin et al., 1991), especially when given by intranasal route (Cahill et al., 1993). FHA-responsive human T-cell clones have also been identified in hosts with a history of pertussis (DeMagistris et al., 1988). Multiple T-cell epitopes occur within the mature FHA protein, but they have not yet been precisely mapped (DiTommaso et al., 1991). Given the known immunomodulatory activities of various secreted *B. pertussis* proteins, it would not be surprising that B- and T-cell responses to this organism are significantly modified during natural infection. However, little is known about this issue.

C. Fimbriae

Fimbriae, or pili, are important adherence factors for many bacterial pathogens that colonize mucosal surfaces (Jones and Isaacson, 1983). *B. pertussis* is fimbriated, and it might then be expected that these *Bordetella* structures play a role in attachment to and colonization of the host. Data to support these assumptions are available but are less than overwhelming. Based upon current information, it appears that the *Bordetella* fimbriae, while highly immunogenic, function as respiratory mucosal adhesins of secondary importance.

In the 1980s, *Bordetella pertussis* fimbriae ("agglutinogens") were purified and distinguished from FHA (Zhang et al., 1985) and were classified into serologic types 2 and 3 (Ashworth et al., 1985) or 2 and 6 (Zhang et al., 1985). The structural genes for the major fimbrial subunit protein for each of these fimbrial types (Fim2 and Fim3) have been cloned and sequenced (Livey et al., 1987; Mooi et al., 1990). These two proteins share significant sequence similarity. In addition to these two genes, *B. pertussis* contains other fimbrial structural subunit genes [*fimX* (Pedroni et al., 1988), *fimY* (Mooi et al., 1992), *fimA* (Willems et al., 1992)], some of which are silent. *fimA* appears to be a truncated remnant of a fimbrial subunit gene and is located immediately downstream of *fhaB* (Willems et al., 1992). Despite the fact that all of the expressed fimbrial subunit genes are located at physically unlinked locations within the *Bordetella* chromosome (Stibitz and Garletts, 1992), they are subject to *bvg*-mediated, positive coordinate regulation at the level of transcription. In addition, individual organisms exhibit phase variation; that is, they switch between high and low levels of fimbrial subunit expression. Thus, an individual Bvg^+ organism can be isolated in either a $Fim2^+Fim3^+$, a $Fim2^+Fim3^-$, a $Fim2^-Fim3^+$, or a $Fim2^-Fim3^-$ state. Phase variation may be more common in vivo than in vitro (Willems et al., 1990), especially in persistent, nonlethal infections (Weiss and Goodwin, 1989). Small DNA insertions and deletions mediate this event within a Fim gene promoter region of consecutive deoxycytodine nucleosides (Willems et al., 1990). Phase variation has been postulated as a means by which an organism avoids host immune recognition.

The first hint that fimbriae-associated genes might be located immediately downstream of *fhaB* arose from observations of a Fim⁻ phenotype in the FHA expression mutants BP353 and BP354 (Weiss et al., 1983). Recent sequence analysis reveals at least four fimbrial genes in this region, *fimABCD* (Locht et al., 1992; Willems et al., 1992) (see Fig. 5). As previously mentioned, *fimA* is a cryptic subunit structural gene. *fimB* appears to encode a 24-kDa homologue of the fimbrial chaperone-like proteins, *K. pneumoniae* MrkB and *E. coli* PapD. *fimC* appears to encode a 92-kDa fimbrial anchorage protein, or usher, with similarity to MrkC and PapC. The Tn5 insertions in BP353 and BP354 have been mapped to locations within *fimC*. Insertion mutations within *fimB* abolish detection of Fim2 and Fim3 by immunoblotting of whole-cell lysates. Downstream of the latter gene is *fimD*, which encodes a minor fimbrial subunit protein with similarity to known fimbrial adhesins such as MrkD and *E. coli* type 1 fimbrial FimH (Willems et al., 1993). MrkD mediates binding of *K. pneumoniae* to the basolateral surface of respiratory epithelial cells (Hornick et al., 1992), suggesting a similar role for *B. pertussis* FimD.

With isolation of the fimbrial structural and accessory genes and the ability to construct isogenic mutant strains, studies have begun to address definitively the role of fimbriae in pertussis pathogenesis. A *B. pertussis* strain with insertional inactivation mutations of the Fim2 and Fim3 genes demonstrated no apparent defect in binding to CHO cells or ciliated respiratory epithelial cells (Relman et al., 1989). However, this strain did demonstrate a defect in colonization of the mouse trachea (Mooi et al., 1992). Lung and nasopharyngeal colonization were unaffected. Since deletions within *fimD* eliminate assembly of fimbrial structures (Willems et al., 1993), the precise role of FimD has not yet been determined. In contrast, a partial FHA-deletion strain colonized the trachea for shorter periods of time and to a lesser degree than the Fim mutant; in addition, the FHA mutant was defective in nasopharyngeal colonization (Mooi et al., 1992). Fimbriae may play a relatively greater adherence role later in the course of infection when altered target surfaces are available. Certain fimbrial serotypes may be characteristic of *B. bronchiseptica* isolates from specific animal host species, suggesting a role for fimbriae in host species specificity (Burns et al., 1993). It is also interesting that the fimbrial accessory genes and the FHA structural gene are physically linked on the chromosome and that they both seem to be transcribed at early time points following relief of phenotypic modulation. One might speculate that fimbriae have evolved along with FHA for the purpose of cooperative or complementary adherence interactions with the host.

D. Adenylate Cyclase Toxin (AC)

Bordetella pertussis adenylate cyclase toxin [AC, cyclolysin (Glaser et al., 1988b), adenylate cyclase-hemolysin] was one of the first defined virulence factors for this organism. Its relationship to the *B. pertussis* hemolysin phenotype was established by Weiss et al. (1983) in the analysis of single Tn*5* mutants that had lost both AC and hemolytic activities. However, not all nonhemolytic mutants had reduced AC activity. In a subsequent study, Weiss et al. (1984) demonstrated that AC/hemolysin-defective mutants were severely impaired in their ability to kill infant mice. We now know that the two activities are encoded within a single polypeptide gene product (Glaser et al., 1988a,b). A great deal of subsequent work has focused on the biochemical activities of this molecule and the mechanisms by which it finds its target (see chapter by Hewlett in this volume). The following discussion emphasizes the molecular basis for AC expression and its possible roles in pathogenesis.

B. *pertussis* AC is both cell surface-associated and secreted as a 200-kDa protein (apparent mass), which then enters host eukaryotic cells, where it is activated by host calmodulin (Bellalou et al. 1990a). It then catalyzes the production of large amounts of cAMP, thereby disrupting host cell functions (Confer and Eaton, 1982; Farfel et al., 1987; Friedman et al., 1987). This 200-kDa protein has cytotoxic as well as membrane pore-forming activities. Analysis of the AC structural gene *cyaA* reveals a 1706-residue, 177-kDa predicted product with multiple domains (Glaser et al., 1988a). Within a region of approximately 400 residues at the N terminus reside adenylate cyclase catalytic and calmodulin-binding activities corresponding to an enzymatically active proteolytic break-down product of 45 kDa (Glaser et al., 1989). Site-directed mutations in *B. pertussis* corresponding to residues 58 and 188–189 or a deletion of 373 residues at the N terminus reduce or abolish adenylate cyclase activity of the 200-kDa AC molecule but do not affect hemolytic activity (Sakamoto et al., 1992). Conversely, the C-terminal 1250 residues bear sequence similarity to *E. coli* alpha-hemolysin HlyA and are responsible for the *B. pertussis* hemolytic phenotype. The 45-kDa product and molecules with C-terminal in-frame deletions retain catalytic activity in vitro but are no longer hemolytic, nor are they able to enter and intoxicate eukaryotic cells (Bellalou et al., 1990b). Thus, the 200-kDa AC molecule comprises domains with independent catalytic and hemolytic activities. The intact molecule seems to be required for entry into and intoxication of host cells.

Secretion of AC, like *E. coli* HlyA, requires *B. pertussis* accessory gene products that are encoded by loci downstream of *cyaA* (*cyaBDE*) (Glaser et al., 1988b) (see Fig. 6). CyaB and CyaD are similar to *E. coli* HlyB and HlyD, which mediate HlyA transport and secretion from the bacterium. In addition, AC requires "activation" by the product of the *cyaC* gene before it acquires hemolytic and toxin activity (Barry et al., 1991). CyaC is believed to modify AC so that the latter becomes capable of host cell membrane insertion and translocation of its catalytic domain (Hewlett et al., 1993).

AC has profound physiologic effects on a wide variety of eukaryotic cells. It remains unclear, however, which cell types are the most relevant targets during natural infection

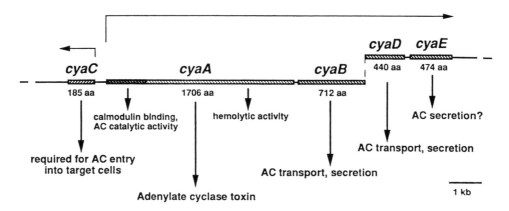

Fig. 6 The *B. pertussis* adenylate cyclase toxin chromosomal locus. Open reading frames (ORFs) are indicated by hatched bars. The direction of transcription is indicated by the thin arrows above the ORFs; *cyaABDE* may be cotranscribed. The 5' end of the *cyaD* ORF overlaps the 3' end of the *cyaB* ORF. These two genes may be translationally coupled. Sizes (total amino acids) of the predicted polypeptides are listed below each ORF, as well as its postulated functions.

and whether it is primarily secreted AC, AC delivered by adherent bacteria, or AC delivered by bacteria following host cell internalization (Mouallem et al., 1990) that is the most relevant form of this molecule. Perhaps some of the most important target cells are host leukocytes. Cell-free AC impairs monocyte and neutrophil chemotaxis, superoxide generation, and microbial killing without affecting leukocyte viability (Pearson et al., 1987). Studies in mice infected by the intranasal route demonstrate a crucial role for AC in lethality and in persistence of *B. pertussis* in the lungs of these animals (Weiss et al., 1984; Weiss and Goodwin, 1989; Goodwin and Weiss, 1990). Virulence defined in this manner has also been associated specifically with adenylate cyclase and hemolysin activities, independently, using site-specific *cyaA B. pertussis* mutants (Gross et al., 1992; Khelef et al., 1992). In addition, AC may facilitate greater degrees of bacterial multiplication during the early phase of mouse lung infection (Khelef et al., 1992).

E. Pertactin

B. pertussis 69-kDa outer membrane protein (P.69), or pertactin, first came to attention after its homologue in *B. bronchiseptica*, P.68 protein, was found to be a protective antigen in piglets and in mice against homologous challenge (Montaraz et al., 1985; Novotny et al., 1985). Monoclonal antibodies to P.68 cross-reacted with 69- and 70-kDa proteins in *B. pertussis* and *B. parapertussis*, respectively. The *B. pertussis* 69-kDa protein was discovered to be a surface-exposed component of the outer membrane and a prominent antigen among individuals convalescing from pertussis as well as those immunized with the whole-cell pertussis vaccine (Novotny et al., 1991). P.69, like P.68, confers protection to aerosol-challenged mice when used as either a passive or an active immunogen. However, claims of a possible role in pathogenesis were not addressed directly until the characterization of the pertactin structural gene (Charles et al., 1989).

Pertactin is synthesized as a 93.5-kDa precursor, which is then processed by proteolytic cleavage. The structural gene, *prn*, is positively regulated by the *bvg* locus

(Charles et al., 1989) (i.e., a *vag* gene) and is unlinked to other known virulence-associated *Bordetella* loci (Stibitz and Garletts, 1992). *B. pertussis* mutants with insertional inactivation of *prn* do not adhere as well as wild-type to CHO and HeLa cells in vitro (Leininger et al., 1991) but have no apparent defect in binding to or entering HEp2 cells (Roberts et al., 1991). Because of the known involvement of some Arg-Gly-Asp protein sequences in eukaryotic cell recognition (see above discussion regarding FHA), each of two Arg-Gly-Asp pertactin sites was postulated to play a role in the adherence activity of this protein. In support of this theory are experiments in which a pertactin Arg-Gly-Asp-containing peptide inhibited CHO cell adhesion to purified pertactin and also inhibited *B. pertussis* entry into HeLa cells in monolayers (Leininger et al., 1991, 1992). The relevance of *B. pertussis* entry into epithelial cells is unclear.

Our understanding of the role played by pertactin in pertussis pathogenesis is incomplete. Data currently available would suggest that pertactin acts as an accessory adherence factor, perhaps in conjunction with other surface proteins such as FHA. The notion of a bacterial multicomponent "attack complex" proposes that a specific and intimate assembly of various proteins takes place on the cell surface (Relman and Falkow, 1994) (see below). The precise configuration of this complex allows the bacterium to ligate different host cell receptors simultaneously at one contact point, leading to heterologous receptor cross-linking and signaling. In this scheme, a prominent surface protein such as pertactin may be most important as a scaffolding or support component of an adhesin complex. The additive effect in an FHA⁻ genetic background of an otherwise silent *prn* mutation on bacterial adherence to HEp-2 cells supports this notion (Roberts et al., 1991).

F. Tracheal Cytotoxin (TCT)

One of the most prominent features of pertussis pathology in both humans and animal models is the dysfunction, destruction, and extrusion of ciliated epithelial cells from the respiratory mucosa (see preceding sections). Given the abundance of biologically active macromolecular protein products produced by *B. pertussis* (and reviewed above), it was assumed for many years that one or more of these products were responsible for these effects. In 1982, Goldman and colleagues reported that a low molecular weight fraction of *B. pertussis* culture supernatant could duplicate this pathology when added to hamster tracheal organ cultures (Goldman et al., 1982). The active component in the culture supernatant was a 921-Da muramyl peptide named tracheal cytotoxin (TCT), a natural disaccharide tetrapeptide breakdown product of *Bordetella* cell wall peptidoglycan that is released during bacterial growth (Cookson et al., 1989a,b). (see chapter by Goldman in this volume). All *Bordetella* species produce TCT, including *B. avium*, a cause of respiratory disease in turkeys (Gentry et al., 1988). TCT production is not *bvg*-regulated, but it is released only by actively replicating cells.

Physiologically relevant concentrations of TCT cause ciliostasis, ciliated cell destruction within cultured hamster respiratory epithelia, and also inhibit DNA synthesis in isolated cultured hamster tracheal epithelial cells (Cookson et al., 1989a). Examination of nasal ciliated mucosa from children with pertussis corroborated earlier observations of reduced ciliated cell number and ciliated cell death and extrusion (Wilson et al., 1991). Purified TCT caused the same pathology in human nasal tissue organ cultures. No other known *B. pertussis* virulence factor or *B. pertussis* endotoxin demonstrates these biological effects. Structure–function studies of TCT indicate that the diaminopimelic acid moiety is critical for its activity against ciliated epithelial cells (Luker et al., 1993).

On the basis of these in vitro studies it has been proposed that the in vivo consequences of TCT action are impaired mucus clearance, accumulation of cellular and inflammatory debris within the small airways, and hence the characteristic cough of pertussis. The inhibitory activity of TCT on host cell DNA synthesis might perpetuate this situation by blocking the division and differentiation of underlying basal cells, which would normally replace the extruded ciliated epithelial cells (Cookson et al., 1989a). One mechanism by which TCT may exert this effect is by stimulation of intracellular IL-1α production in these epithelial cells (Heiss et al., 1993). It is interesting that *Neisseria gonorrhoeae*, which also damages human ciliated cells during the course of natural infection (within the fallopian tube), also produces significant amounts of TCT (Luker et al., 1993).

G. Dermonecrotic Toxin (DNT)

Dermonecrotic toxin (DNT) (or 56°C heat-labile toxin) was first identified by Bordet and Gengou in 1909 and defined on the basis of its biological activity upon intradermal injection into animals. Both local skin necrosis and animal lethality have been observed after intradermal administration (Zhang and Sekura, 1991). DNT and TCT have been identified in all four *Bordetella* species (Gentry et al., 1988). This fact has been used to argue that it may be a virulence factor. In addition, it is found only in Bvg$^+$ organisms, suggesting that its expression is Bvg-activated. The structural gene, *dnt*, is not linked to known virulence-associated loci on the *B. pertussis* chromosome (Stibitz and Garletts, 1992). However, the role of this molecule in disease pathogenesis remains unclear (Weiss and Goodwin, 1989).

DNT is a 140-kDa cytoplasmic protein in *B. pertussis*. Given that it is not secreted in appreciable amounts, the tissues in which DNT biological activity has been defined may in reality never be exposed to this molecule. Weiss and Goodwin (1989) have argued that DNT may not play an important role in disease pathogenesis, on the basis of their observation that a DNT$^-$ *B. pertussis* strain was as lethal as the wild-type strain for infant mice.

H. Other Proposed Virulence Determinants

Most of the virulence-associated factors described so far are positively regulated by the *bvg* locus. Yet it is not known to what extent *B. pertussis* may exist in the Bvg$^-$ phase state or at some intermediary transition state between the two phases during the course of natural infection. Lacey and others have reported a humoral response among pertussis patients directed against avirulent phase antigens (Lacey, 1960). But the timing of phase transition and the role of Bvg$^-$ phase antigens in natural infection are poorly understood. Flagellin expression is a *bvg* negatively regulated trait in *B. bronchiseptica* (Akerley et al., 1992; Akerley and Miller, 1993). In *B. pertussis*, Beattie et al. (1992) identified a *vir*-repressed gene (*vrg*) that is required for virulence. Vrg-6 is an 11-kDa predicted protein product that appears to play a role in tracheal and lung colonization in the mouse aerosol model. It has been postulated that the Vrg proteins are expressed either early or late in the course of infection and that Vrg-6 may be important for *B. pertussis* intracellular survival (Beattie et al., 1992).

The generation of random transposon insertion mutations in the *B. pertussis* chromosome led to the preliminary identification of new putative *vir*-activated virulence factors (Weiss et al., 1989; Finn et al., 1991). BPM2041 and SK8 contain insertion mutations in *vir*-activated genes that lead to either reduced infant mouse lethality

(BPM2041) (Weiss and Goodwin, 1989) or reduced lung colonization and lymphocytosis (SK8) (Finn et al., 1991). SK8 is defective for production of a 95-kDa outer membrane protein. In each of these cases, the novel factor could conceivably contribute to virulence indirectly by assisting in the presentation of other known virulence factors in a bacterial surface complex (see above discussion concerning pertactin).

In some cases, directed investigations of previously proposed *Bordetella* virulence factors have revealed novel putative virulence-associated factors. For example, previous claims that *B. avium* and *B. bronchiseptica* DNT were inhibitory for osteoblast cells and might therefore contribute to atrophic rhinitis led to a study of osteotoxin activity. Unexpectedly, in *B. avium*, toxicity for osteogenic cells and tracheal cells reflects the biological activities of a bacterial 41-kDa β-cystathionase (Gentry et al., 1993). This enzyme is capable of creating unstable and toxic sulfane sulfur derivatives in the presence of L-cystine. An immunoreactive protein of similar size is found in extracts of *B. bronchiseptica* and *B. pertussis*. However, the role of this factor in disease pathogenesis has yet to be defined. Iron acquisition by bacterial pathogens is critical for survival in the animal host. *B. pertussis* may acquire iron by means of a siderophore as well as with transferrin- and lactoferrin-binding proteins (Menozzi et al., 1991a; Agiato and Dyer, 1992). Each of these factors may be an important virulence determinant under circumstances in the human host where iron is a limiting nutrient factor.

VI. COOPERATION AMONG VIRULENCE FACTORS

The foregoing discussion of individual *B. pertussis* virulence-associated factors carries the implicit assumption that each factor is presented by the bacterium as an independent entity and that biological activity in vivo reflects the isolated action of each factor. In fact, there is no reason to think that this is the case. There are a number of examples where two or more factors may provide duplication of function or cooperate in effecting a particular virulence-associated phenotype. A knock-out mutation in just one factor may or may not cause a discernible effect on such a phenotype. In addition, as suggested earlier, surface-associated virulence factors may form heterologous complexes that then dictate a precise pattern of host receptor recognition and bacterium–host cell interactions (Relman and Falkow, 1994). The role of a particular factor may simply be to assist in the proper surface presentation of another virulence factor.

A. Adherence and Colonization

For an organism that remains on or near a mucosal surface during the course of infection, adherence is a critical aspect of pathogenesis. It is not surprising, therefore, that *B. pertussis* has multiple adhesins and mechanisms for host cell adherence. The expression of an unusually large protein, FHA, with multiple putative adherence domains is an example of the biosynthetic energy commitment by this microorganism directed toward the process of adherence (Locht et al., 1993). Although FHA is probably the dominant adherence factor for this organism, PT, fimbriae, and pertactin all probably play important roles in bacterial binding to either the respiratory tract mucosa, respiratory mucus, or the immune cells associated with this organ.

The relative importance of each *B. pertussis* adhesin may correlate with different host cell types, the timing of the adherence event with respect to the course of infection, and/or the specific outcome of the adherence event (i.e., bacterial multiplication at the cell surface, delivery of toxins, or bacterial entry into the host cell). For example, some adhesins may be critical for early colonization of the nasal mucosa or upper airway epithelial surface, whereas other adhesins may play more important roles in attachment to damaged respiratory mucosal surfaces that become revealed only later in the course of the disease. It has been proposed that in at least one situation, one adhesin (PT-S2 and S3 may enhance binding of another adhesin (FHA) by altering the host cell (macrophage) surface (Hoepelman and Tuomanen, 1992; van't Wout et al., 1992). The same kind of cooperative effect may also apply to the action of AC and PT enzymatic activities in facilitating FHA or other adhesin binding, following modification of host cell surfaces. Finally, *Bordetella* adhesins may act cooperatively to determine host species specificity of infection. For example, minor variations in FHA and fimbrial structure together may explain why *B. pertussis* preferentially infects humans and why *B. bronchiseptica* preferentially infects animal species (Tuomanen et al., 1983; Burns et al., 1993).

Not all *B. pertussis* adherence events are direct bacterium–host receptor interactions; some of the adhesins described above can act at a distance. FHA as a secreted product may facilitate *B. pertussis* adherence to host cell surfaces as well as the adherence of secondary superinfecting bacterial pathogens (Tuomanen, 1986). Secreted FHA may also serve as a blocking factor for host cell receptors, including epithelial cell glycolipid receptors (Tuomanen et al., 1988) as well as the leukocyte complement receptor CR3 (Relman et al., 1990).

B. Local Damage and Systemic Intoxication

The anatomic and physiologic disruptions that can result from *B. pertussis* infection may be viewed as advantageous to the organism, to some degree. With disruption of the original respiratory tract surface there are new adherence receptors revealed on basal cells, on the basement membrane, and perhaps on damaged epithelial cells. And with intoxication of host immune cells there is impaired direct bacterial killing, modified humoral and cellular immune responses, and perhaps the possibility of intracellular bacterial persistence (see below). Coughing serves to disseminate *B. pertussis* to new susceptible hosts. These disruptions are probably the result of coordinated action between a number of *B. pertussis* factors; TCT, PT, and AC dominate among these factors. In addition, a number of other factors may play important secondary roles. These may include DNT, β-cystathionase, and a variety of uncharacterized outer membrane and cell wall components.

It is important to remember that significant degrees of tissue pathology and pathophysiology probably do not occur with most *B. pertussis* infections. The majority of infected humans, especially adults, may display only mild clinical manifestations or perhaps none at all. This form of bacterium–host interaction may bias the outcome toward bacterial persistence in the host, although there are few data currently available to support this speculation.

C. *B. pertussis* Interactions with the Host Immune System

Most bacterial pathogens that survive on a host mucosal surface have evolved a specific array of mechanisms for avoiding immune cell-mediated microbial death. These mechanisms are often separate from those used by microorganisms that routinely pass through multiple cells and tissues, become bloodborne, and ultimately multiply in privileged anatomic sites. *B. pertussis* clearly contains a variety of secreted and cell-associated products with the potential for interacting with or altering host immune cell functions. It remains unclear, however, to what degree and for what specific purpose these interactions take place (see Fig. 7).

Specific mechanisms exist for *B. pertussis* adherence to human leukocytes. These mechanisms involve at least dual involvement of FHA and PT as has been described above. Coating of the bacterium with host factors probably does not constitute a significant additional mechanism. Whether other *B. pertussis* factors are involved is unknown. Elimination of FHA and PT results in the substantial reduction of bacterial binding to human macrophages from the levels of a Bvg$^+$ wild-type strain to nearly those of a Bvg$^-$ phase variant (Relman et al., 1990). Again, this observation does not rule out a role for pertactin, fimbriae, or other surface factors. Careful analysis of single and multiple mutations in an isogenic background is needed. Available data favor the concept of FHA direct binding to CR3 (Relman et al., 1990). Concerning macrophage receptors, there are almost certainly more than CR3 and various glycoconjugates. Receptor activity and receptor–receptor signaling are issues that may bear on *B. pertussis* leukocyte adherence and warrant further investigation.

Of more global relevance are questions about the importance of *B. pertussis*–leukocyte interactions for the course of the natural infection. One aspect of this issue involves understanding the outcome of *B. pertussis* binding to leukocytes. The possibilities include delivery of potent toxins, blockade of leukocyte receptors with secreted bacterial adhesins, and bacterial entry into certain leukocytes. The last possibility might result in either bacterial death or survival, multiplication, and perhaps persistence. A few observations in humans and in animals suggest that *B. pertussis* does enter alveolar macrophages and may survive for at least brief periods of time (Cheers and Gray, 1969; Bromberg et al., 1991; Saukkonen et al., 1991). In vitro studies suggest that this bacterium can enter and survive for short periods of time within human peripheral monocyte-derived macrophages and polymorphonuclear leukocytes (Steed et al., 1991; Friedman et al., 1992). One of the difficulties in addressing these questions is the absence of an attractive animal model.

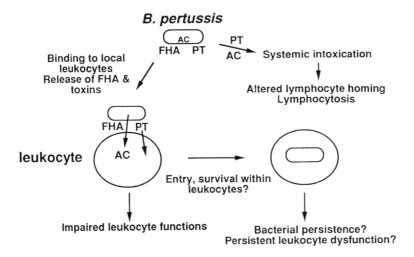

Fig. 7   *B. pertussis* interactions with host leukocytes and their potential outcomes. Events whose existence is based primarily upon speculation are indicated with a question mark.

The type and location of leukocytes that are most relevant to *B. pertussis*–host interactions may not be alveolar macrophages but rather tracheal mucosal leukocytes or leukocytes found within Waldeyer's ring or the bronchus-associated lymphoid tissue (Sminia et al., 1989).

While it is clear that purified PT and AC, in particular, induce profound changes in leukocyte function, the role of bacterial toxin delivery by direct leukocyte contact is unknown. For example, when studied in isolation, PT is capable of blocking receptor-mediated signaling events in a variety of leukocytes (Chaffin et al., 1990; Hiemstra et al., 1992). However, it is not clear whether these signaling events are representative of those generated by a leukocyte that bears a number of adherent *B. pertussis* organisms. And to what extent does one leukotoxin or adhesin render a leukocyte more or less susceptible to the actions of another?

VII. UNRESOLVED ISSUES

Among a large group of unanswered questions, two issues are especially crucial to a more complete understanding of pertussis pathogenesis.

A.   The *bvg* Regulon and In Vivo Regulatory Signals

Patients recovering from pertussis display immune responses to antigens that are expressed by bacteria in either of the two Bvg phases. However, there is little or no available information concerning the timing of phase switching during natural infection. Some data lead one to believe that *B. pertussis* is avirulent (Bvg⁻) when inside macrophages (Masure, 1992), but this has not been convincingly demonstrated. To the degree that they may be important in natural infection, we know very little about *vrg* products and the role they play in pathogenesis. Finally, we do not appreciate the relevant physiologic signals that modulate *B. pertussis* and control Bvg phase switching within the human host. In all likelihood, there is no one factor that is sensed by the organism in vivo, but rather a combination of signals that are integrated.

B.   Persistence and a *B. pertussis* Intracellular State

The lack of a known environmental or animal reservoir for *B. pertussis* suggests that humans may serve in this capacity. It has been suggested that adults with waning postvaccination immunity may be the primary source of *B. pertussis* transmission to susceptible children (Nelson, 1978; Biellik et al., 1988; Steketee et al., 1988; Cherry et al., 1989; Long et al., 1990). These hypotheses imply that asymptomatic adults harbor *B. pertussis* in a relatively sequestered site. A number of findings suggest that this site is an intracellular one. Studies of *B. pertussis* interaction with human epithelial cells and human leukocytes demonstrate *B. pertussis* entry, albeit at relatively low levels (Ewanowich et al., 1989; Ewanowich and Peppler, 1990; Lee et al., 1990; Friedman et al., 1992; Leininger et al., 1992). Any organism with high affinity for a wide variety of eukaryotic cells is likely to become internalized to some degree. However, the degree to which *B. pertussis* internalization takes place in vivo and the degree to which it survives are unknown. In the mouse aerosol model of pertussis, *B. pertussis* can be recovered by culture from the lungs until 5 weeks after inoculation (Mills et al., 1993), CD4⁺ T cells play a role in clearance of the organisms. However, when more sensitive methods are employed, *B. pertussis* can be detected in murine lungs even as late as 26 weeks postinfection (Amsbaugh et al., 1993). It would be of great interest if these observations of *B. pertussis* persistence could be extended to humans.

VIII. CONCLUSION

A proper understanding of bacterial virulence and pathogenesis must rely heavily upon clinical and histologic observations. The foregoing discussion of *B. pertussis* determinants of virulence has attempted to follow this dictum. This is especially difficult, however, since the disease does not occur in animals and human infection is relatively infrequent and poorly accessible to study. Nonetheless, the following conclusions can be made. *B. pertussis* is highly evolved in its ability to sense environmental conditions and to regulate expression of virulence-associated factors accordingly. Pertussis pathogenesis should no longer be viewed simply as a reflection of PT expression. *B. pertussis* pathogenesis instead involves a wide array of adherence factors and toxins that may act in a cooperative fashion. Some virulence factors such as TCT are not highly regulated but are expressed simply by virtue of bacterial growth and cell turnover. Many important questions about pertussis pathogenesis cannot be answered at the present time, including questions concerning *B. pertussis* environmental sensing in vivo and the possible significance of a bacterial intracellular state. With rapidly improving techniques for organ and cell culture and for bacterial and eukaryotic genetic manipulations, it should not be long before answers are available to these and other questions.

Acknowledgement

This work was supported in part by a grant from the Lucille P. Markey Charitable Trust (D.A.R. is a Lucille P. Markey Scholar) and by the American Federation for Clinical Research Foundation.

REFERENCES

Agiato, L. A., and Dyer, D. W. (1992). Siderophore production and membrane alterations by *Bordetella pertussis* in response to iron starvation. *Infect. Immun. 60*: 117–213.

Akerley, B. J., and Miller, J. F. (1993). Flagellin gene transcription in *Bordetella bronchiseptica* is regulated by the BvgAS virulence control system. *J. Bacteriol. 175*: 3468–3479.

Akerley, B. J., Monack, D. M., Falkow, S., and Miller, J. F. (1992). The *bvgAS* locus negatively controls motility and synthesis of flagella in *Bordetella bronchiseptica*. *J. Bacteriol. 174*: 980–990.

Amsbaugh, D. F., Li, Z. M., and Shahin, R. D. (1993). Long-lived respiratory immune response to filamentous hemagglutinin following *Bordetella pertussis* infection. *Infect. Immun. 61*: 1447–1452.

Arai, H., and Sato, Y. (1976). Separation and characterization of two distinct hemagglutinins contained in purified leukocytosis-promoting factor from *Bordetella pertussis. Biochim. Biophys. Acta 444*: 765–782.

Arico, B., and Rappuoli, R. (1987). *Bordetella parapertussis* and *Bordetella bronchiseptica* contain transcriptionally silent pertussis toxin genes. *J. Bacteriol. 169*: 2847–2853.

Arico, B., Miller, J. F., Roy, C., Stibitz, S., Monack, D., Falkow, S., Gross, R., and Rappuoli, R. (1989). Sequences required for expression of *Bordetella pertussis* virulence factors share homology with prokaryotic signal transduction proteins. *Proc. Natl. Acad. Sci. U.S.A. 86*: 6671–6675.

Arico, B., Scariato, V., Monack, D. M., Falkow, S., and Rappuoli, R. (1991). Structural and genetic analysis of the *bvg* locus in *Bordetella* species. *Mol. Microbiol. 5*: 2481–2491.

Ashworth, L. A., Dowsett, A. B., Irons, L. I., and Robinson, A. (1985). The location of surface antigens of *Bordetella pertussis* by immuno-electron microscopy. *Dev. Biol. Stand. 61*: 143–152.

Barenkamp, S. J., and Leininger, E. (1992). Cloning, expression, and DNA sequence analysis of genes encoding nontypeable *Haemophilus influenzae* high-molecular-weight surface-exposed proteins related to filamentous hemagglutinin of *Bordetella pertussis. Infect. Immun. 60*: 1302–1313.

Bargatze, R. F., and Butcher, E. C. (1993). Rapid G protein-regulated activation event involved in lymphocyte binding to high endothelial venules. *J. Exp. Med. 178*: 367–372.

Barry, E. M., Weiss, A. A., Ehrmann, I. E., Gray, M. C., Hewlett, E. L., and Goodwin, M. S. (1991). *Bordetella pertussis* adenylate cyclase toxin and hemolytic activities require a second gene, *cyaC*, for activation. *J. Bacteriol. 173*: 720–726.

Bass, J. W., and Stephenson, S. R. (1987). The return of pertussis. *Pediatr. Infect. Dis. J. 6*: 141–144.

Beattie, D. T., Knapp, S., and Mekalanos, J. J. (1990). Evidence that modulation requires sequences downstream of the promoters of two *vir*-repressed genes of *Bordetella pertussis. J. Bacteriol. 172*: 6997–7004.

Beattie, D. T., Shahin, R., and Mekalanos, J. J. (1992). A *vir*-repressed gene of *Bordetella pertussis* is required for virulence. *Infect. Immun. 60*: 571–577.

Beattie, D. T., Mahan, M. J., and Mekalanos, J. J. (1993). Repressor binding to a regulatory site in the DNA coding sequence is sufficient to confer transcriptional regulation of the *vir*-repressed genes (*vrg* genes) in *Bordetella pertussis. J. Bacteriol. 175*: 519–527.

Bellalou, J., Ladant, D., and Sakamoto, H. (1990a). Synthesis and secretion of *Bordetella pertussis* adenylate cyclase as a 200-kilodalton protein. *Infect. Immun. 58*: 1195–1200.

Bellalou, J., Sakamoto, H., Ladant, D., Geoffroy, C., and Ullmann, A. (1990b). Deletions affecting hemolytic and toxin activities of *Bordetella pertussis* adenylate cyclase. *Infect. Immun. 58*: 3242–3247.

Biellik, R. J., Patriarca, P. A., Mullen, J. R., Rovira, E. Z., Brink, E. W., Mitchell, P., Hamilton, G. H., Sullivan, B. J., and Davis, J. P. (1988). Risk factors for community- and household-acquired pertussis during a large-scale outbreak in central Wisconsin. *J. Infect. Dis. 157*: 1134–1141.

Black, W. J., Munoz, J. J., Peacock, M. G., Schad, P. A., Cowell, J. L., Burchall, J. J., Lim, M., Kent, A., Steinman, L., and Falkow, S. (1988). ADP-ribosyltransferase activity of pertussis toxin and immunomodulation by *Bordetella pertussis. Science 240*: 656–659.

Bordet, J., and Sleeswyck (1910). Serodiagnostic et variabilite des microbes suivant le milieu de culture. *Ann. Inst. Pasteur (Paris) 24*: 476–494.

Brennan, M. J., David, J. L., Kenimer, J. G., and Manclark, C. R. (1988). Lectin-like binding of pertussis toxin to a 165-kilodalton Chinese hamster ovary cell glycoprotein. *J. Biol. Chem. 263*: 4895–4899.

Brennan, M. J., Hannah, J. H., and Leininger, E. (1991). Adhesion of *Bordetella pertussis* to sulfatides and to the GalNAc beta 4Gal sequence found in glycosphingolipids. *J. Biol. Chem. 266*: 18827–18831.

Bromberg, K., Tannis, G., and Steiner, P. (1991). Detection of *Bordetella pertussis* associated with alveolar macrophages of children with human immunodeficiency virus infection. *Infect. Immun. 59*: 4715–4719.

Brown, D. R., and Parker, C. D. (1987). Cloning of the filamentous hemagglutinin of *Bordetella pertussis* and its expression in *Escherichia coli. Infect. Immun. 55*: 154–161.

Burns, E. H., Norman, J. M., Hatcher, M. D., and Bemis, D. A. (1993). Fimbriae and determination of host species specificity of *Bordetella bronchiseptica. J. Clin. Microbiol. 31*: 1838–1844.

Burstyn, D. G., Baraff, L. J., Peppler, M. S., Leake, R. D., St. Geme, J. J., and Manclark, C. R. (1983). Serological response to filamentous hemagglutinin and lymphocytosis-promoting toxin of *Bordetella pertussis. Infect. Immun. 41*: 1150–1156.

Cahill, E. S., O'Hagan, D. T., Illum, L., and Redhead, K. (1993). Mice are protected against *Bordetella pertussis* infection by intra-nasal immunization with filamentous haemagglutinin. *FEMS Microbiol. Lett. 107*: 211–216.

Centers for Disease Control (1992a). Pertussis vaccination: acellular pertussis vaccine for reinforcing and booster use—supplementary ACIP statement. Recommendations of the Immunization Practices Advisory Committee (ACIP). *MMWR 41* (RR-1): 1–10.

Centers for Disease Control (1992b). Pertussis vaccination: acellular pertussis vaccine for the fourth and fifth doses of the DTP series update to supplementary ACIP statement. Recommendations of the Advisory Committee on Immunization Practices (ACIP). *MMWR 41* (RR-15):1–5.

Centers for Disease Control (1993). Pertussis outbreaks—Massachusetts and Maryland, 1992. *MMWR 42*: 197–200.

Chaffin, K. E., Beals, C. R., Wilkie, T. M., Forbush, K. A., Simon, M. I., and Perlmutter, R. M. (1990). Dissection of thymocyte signaling pathways by in vivo expression of pertussis toxin ADP-ribosyltransferase. *EMBO J. 9*: 3821–3829.

Charles, I. G., Dougan, G., Pickard, D., Chatfield, S., Smith, M., Novotny, P., Morrissey, P., and Fairweather, N. F. (1989). Molecular cloning and characterization of protective outer membrane protein P. 69 from *Bordetella pertussis*. *Proc. Natl. Acad. Sci. U.S.A. 86*: 3554–3558.

Cheers, C., and Gray, D. F. (1969). Macrophage behaviour during the complaisant phase of murine pertussis. *Immunology 17*: 875–887.

Cherry, J. D. (1990). "Pertussis vaccine encephalopathy": It is time to recognize it as the myth that it is [Editorial] [published erratum appears in *JAMA 263*(16): 2182 (1990)] [see comments]. *J. Am. Med. Assoc. 263*: 1679–1680.

Cherry, J. D. (1992). Pertussis: the trials and tribulations of old and new pertussis vaccines. *Vaccine 10*: 1033–1038.

Cherry, J. D. (1993). Acellular pertussis vaccines—a solution to the pertussis problem. *J. Infect. Dis. 168*: 21–24.

Cherry, J. D., Baraff, L. J., and Hewlett, E. (1989). The past, present, and future of pertussis. The role of adults in epidemiology and future control [see comments]. *West. J. Med. 150*: 319–328.

Cohn, S. E., Knorr, K. L., Gilligan, P. H., Smiley, M. L., and Weber, D. J. (1993). Pertussis is rare in human immunodeficiency virus disease. *Am. Rev. Respir. Dis. 147*: 411–413.

Collier, A. M., Peterson, L. P., and Baseman, J. B. (1977). Pathogenesis of infection with *Bordetella pertussis* in hamster tracheal organ culture. *J. Infect. Dis. 136*: 5196–5203.

Confer, D. L., and Eaton, J. W. (1982). Phagocyte impotence caused by an invasive bacterial adenylate cyclase. *Science 217*: 948–950.

Cookson, B. T., Cho, H. L., Herwaldt, L. A., and Goldman, W. E. (1989a). Biological activities and chemical composition of purified tracheal cytotoxin of *Bordetella pertussis*. *Infect. Immun. 57*: 2223–2229.

Cookson, B. T., Tyler, A. N., and Goldman, W. E. (1989b). Primary structure of the peptidoglycan-derived tracheal cytotoxin of *Bordetella pertussis*. *Biochemistry 28*: 1744–1749.

Covacci, A., and Rappuoli, R. (1993). Pertussis toxin export requires accessory genes located downstream from the pertussis toxin operon. *Mol. Microbiol. 8*: 429–434.

D'Souza, S. E., Ginsberg, M. H., and Plow, E. F. (1991). Arginyl-glycyl-aspartic acid (RGD): a cell adhesion motif. *Trends Biochem. Sci. 16*: 246–250.

Davis, S. F., Strebel, P. M., Cochi, S. L., Zell, E. R., and Hadler, S. C. (1992). *Pertussis Surveillance—United States, 1989–1991*, In CDC Surveillance Summaries (Dec. 11), *MMWR 41*:11–19.

Delisse-Gathoye, A. M., Locht, C., Jacob, F., Raaschou, N. M., Heron, I., Ruelle, J. L., De Wilde, M., and Cabezon, T. (1990). Cloning, partial sequence, expression, and antigenic analysis of the filamentous hemagglutinin gene of *Bordetella pertussis*. *Infect. Immun. 58*: 2895–2905.

DeMagistris, M., Romano, M., Nuti, S., Rappuoli, R., and Tagliabue, A. (1988). Dissecting human T cell responses against *Bordetella* species. *J. Exp. Med. 168*: 1351–1362.

DeMagistris, M. T., Romano, M., Bartoloni, A., Rappuoli, R., and Tagliabue, A., (1989). Human T cell clones define S1 subunit as the most immunogenic moiety of pertussis toxin and determine its epitope map. *J. Exp. Med. 169*: 1519–1532.

der Lan, B. A., Cowell, J. L., Burstyn, D. G., Manclark, C. R., and Chrambach, A. (1986). Characterization of the filamentous hemagglutinin from *Bordetella pertussis* by gel electrophoresis. *Mol. Cell. Biochem. 70*: 31–55.

Diamond, G., Jones, D. E., and Bevins, C. L. (1993). Airway epithelial cells are the site of expression of a mammalian antimicrobial peptide gene. *Proc. Natl. Acad. Sci. U.S.A. 90*: 4596–4600.

DiTommaso, A., Domenighini, M., Bugnoli, M., Tagliabue, A., Rappuoli, R., and DeMagistris, M. T. (1991). Identification of subregions of *Bordetella pertussis* filamentous hemagglutinin that stimulate human T-cell responses. *Infect. Immun. 59*: 3313–3315.

Doebbeling, B. N., Feilmeier, M. L., and Herwaldt, L. A. (1990). Pertussis in an adult man infected with the human immunodeficiency virus. *J. Infect. Dis. 161*: 1296–1298.

Domenighini, M., Relman, D., Capiau, C., Falkow, S., Prugnola, A., Scarlato, V., and Rappuoli, R. (1990). Genetic characterization of *Bordetella pertussis* filamentous haemagglutinin: a protein processed from an unusually large precursor. *Mol. Microbiol. 4*: 787–800.

Domenighini, M., Montecucco, C., Ripka, W. C., Rappuoli, R. (1991). Computer modelling of the NAD binding site of ADP-ribosylating toxins: active-site structure and mechanism of NAD binding. *Mol. Microbiol. 5*: 23–31.

Duncan, J. R., Ross, R. F., Switzer, W. P., and Ramsey, F. K. (1966). Pathology of experimental *Bordetella bronchiseptica* infection in swine: atrophic rhinitis. *Am. J. Vet. Res. 27*: 457–466.

Edwards, K. M. (1993). Acellular pertussis vaccines—a solution to the pertussis problem? *J. Infect. Dis. 168*: 15–20.

Edwards, K. M., Decker, M. D., Graham, B. S., Mezzatesta, J., Scott, J., and Hackell, J. (1993). Adult immunization with acellular pertussis vaccine *J. Am. Med. Assoc. 269*: 53–56.

Ewanowich, C. A., and Peppler, M. S. (1990). Phorbol myristate acetate inhibits HeLa 229 invasion by *Bordetella pertussis* and other invasive bacterial pathogens. *Infect. Immun. 58*: 3187–3193.

Ewanowich, C. A., Melton, A. R., Weiss, A. A., Sherburne, R. K., and Peppler, M. S. (1989). Invasion of HeLa 229 cells by virulent *Bordetella pertussis*. *Infect. Immun. 57*: 2698–2704.

Farfel, Z., Friedman, E., and Hanski, E., (1987). The invasive adenylate cyclase of *Bordetella pertussis*. Intracellular localization and kinetics of penetration into various cells. *Biochem. J. 243*: 153–158.

Farizo, K. M., Colchi, S. L., Zell, E. R., Brink, E. W., Wassilak, S. G., and Patriarca, P. A. (1992). Epidemiological features of pertussis in the United States, 1980–1989. *Clin. Infect. Dis. 14*: 708–719.

Finlay, B. B., and Falkow, S. (1989). Common themes in microbial pathogenicity. *Microbiol. Rev. 53*: 210–230.

Finn, T. M., Shahin, R., and Mekalanos, J. J. (1991). Characterization of *vir*-activated TnphoA gene fusions in *Bordetella pertussis*. *Infect. Immun. 59*: 3273–3279.

Friedman, E., Farfel, Z., and Hanki, E. (1987). The invasive adenylate cyclase of *Bordetella pertussis*. Properties and penetration kinetics. *Biochem. J. 243*: 145–151.

Friedman, R. L., Nordensson, K., Wilson, L., Akporiaye, E. T., and Yocum, D. E. (1992). Uptake and intracellular survival of *Bordetella pertussis* in human macrophages. *Infect. Immun. 60*: 4578–4585.

Gan, V. N., and Murphy, T. V., (1990). Pertussis in hospitalized children. *Am. J. Dis. Child. 144*: 1130–1134.

Gentry-Weeks, C. R., Cookson, B. T., Goldman, W. E., Rimler, R. B., Porter, S. B., and Curtiss, R. III (1988). Dermonecrotic toxin and tracheal cytotoxin, putative virulence factors of *Bordetella avium*. *Infect. Immun. 56*: 1698–1707.

Gentry-Weeks, C. R., Keith, J. M., and Thompson, J. (1993). Toxicity of *Bordetella avium* beta-cystathionase toward MC3T3-E1 osteogenic cells. *J. Biol. Chem. 268*: 7298–7314.

Glare, E. M., Paton, J. C., Premier, R. R., Lawrence, A. J., and Nisbet, I. T. (1990). Analysis of a repetitive DNA sequence from *Bordetella pertussis* and its application to the diagnosis of pertussis using the polymerase chain reaction. *J. Clin. Microbiol. 28*: 1982–1987.

Glaser, P., Ladant, D., Sezer, O., Pichot, F., Ullmann, A., and Danchin, A. (1988a). The calmodulin-sensitive adenylate cyclase of *Bordetella pertussis*: cloning and expression in *Escherichia coli*. *Mol. Microbiol. 2*: 19–30.

Glaser, P., Sakamoto, H., Bellalou, J., Ullmann, A., and Danchin, A. (1988b). Secretion of cyclolysin, the calmodulin-sensitive adenylate cyclase-haemolysin bifunctional protein of *Bordetella pertussis*. *EMBO J. 7*: 3997–4004.

Glaser, P., Elmaoglou, L. A., Krin, E., Ladant, D., Barzu, O., and Danchin, A. (1989). Identification of residues essential for catalysis and binding of calmodulin in *Bordetella pertussis* adenylate cyclase by site-directed mutagenesis. *EMBO J. 8*: 967–972.

Goldman, W. E., Klapper, D. G., and Baseman, J. B. (1982). Detection, isolation, and analysis of a released *Bordetella pertussis* product toxic to cultured tracheal cells. *Infect. Immun. 36*: 782–794.

Goodwin, M. S., and Weiss, A. A. (1990). Adenylate cyclase toxin is critical for colonization and pertussis toxin is critical for lethal infection by *Bordetella pertussis* in infant mice. *Infect. Immun. 58*: 3445–3447.

Gross, M. K., Au, D. C., Smith, A. L., and Storm, D. R. (1992). Targeted mutations that ablate either the adenylate cyclase or hemolysin function of the bifunctional *cyaA* toxin of *Bordetella pertussis* abolish virulence. *Proc. Natl. Acad. Sci. U.S.A. 89*: 4898–4902.

Hallander, H. O., Storsaeter, J., and Molby, R. (1991). Evaluation of serology and nasopharyngeal cultures for diagnosis of pertussis in a vaccine efficacy trial. *J. Infect. Dis. 163*: 1046–1054.

Halperin, S. A., Bortolussi, R., and Wort, A. J. (1989). Evaluation of culture, immunofluorescence, and serology for the diagnosis of pertussis. *J. Clin. Microbiol. 27*: 752–757.

He, Q., Mertsola, J., Soini, H., Skurnik, M., Ruuskanen, O., and Viljanen, M. K. (1993). Comparison of polymerase chain reaction with culture and enzyme immunoassay for diagnosis of pertussis. *J. Clin. Microbiol. 31*: 642–645.

Heiss, L. N., Moser, S. A., Unanue, E. R., and Goldman, W. E. (1993). Interleukin-1 is linked to the respiratory epithelial cytopathology of pertussis. *Infect. Immun. 61*: 3123–3128.

Herwaldt, L. A. (1991). Pertussis in adults: what physicians need to know. *Arch. Intern. Med. 151*: 1510–1512.

Hewlett, E. L. (1990). *Bordetella* species. In *Principles and Practice of Infectious Diseases* (Mandell, G. L., Douglas, R. G., Bennett, J. E., eds.), 3rd ed., Churchill Livingstone, New York, pp. 1756–1762.

Hewlett, E. L., Gray, M. C., Ehrmann, I. E., Maloney, N. J., Otero, A. S., Gray, L., Allietta, M., Szabo, G., Weiss, A. A., and Barry, E. M. (1993). Characterization of adenylate cyclase toxin from a mutant of *Bordetella pertussis* defective in the activator gene, *cyaC*. *J. Biol. Chem. 268*: 7842–7848.

Hiemstra, P. S., Annema, A., Schippers, E. F., and van Furth, R. (1992). Pertussis toxin partially inhibits phagocytosis of immunoglobulin G-opsonized *Staphylococcus aureus* by human granulocytes but does not affect intracellular killing. *Infect. Immun. 60*: 202–205.

Hoepelman, A. I. M., and Tuomanen, E. I. (1992). Consequences of microbial attachment: directing host cell functions with adhesins. *Infect. Immun. 60*: 1729–1733.

Hornick, D. B., Allen, B. L., Horn, M. A., and Clegg, S. (1992). Adherence to respiratory epithelia by recombinant *Escherichia coli* expressing *Klebsiella pneumoniae* type 3 fimbrial gene products. *Infect. Immun. 60*: 1577–1588.

Howson, C. P., and Fineberg, H. V. (1992). Adverse events following pertussis and rubella vaccines. Summary of a report of the Institute of Medicine. *J. am. Med. Assoc. 267*: 392–396.

Huh, Y. J., and Weiss, A. A. (1991). A 23-kilodalton protein, distinct from BvgA, expressed by virulent *Bordetella pertussis* binds to the promoter region of *vir*-regulated toxin genes. *Infect. Immun.* 59: 2389–2395.

Irons, L. I., Ashworth, L. A., and Wilton-Smith, P. (1983). Heterogeneity of the filamentous haemagglutinin of *Bordetella pertussis* studied with monoclonal antibodies. *J. Gen. Microbiol.* 129: 2769–2778.

Jones, G. W., and Isaacson, R. E. (1983). Proteinaceous bacterial adhesins and their receptors. *Crit. Rev. Microbiol.* 10: 229–260.

Kasuga, T., Nakase, Y., Ukishima, K., and Takatsu, K. (1954). Studies on *Haemophilus pertussis*. Pt. V. Relation between the phase of bacilli and the progress of the whooping cough. *Kitasato Arch. Exp. Med.* 27: 57–62.

Katada, T., and Ui, M. (1982). Direct modification of the membrane adenylate cyclase system by islet-activating protein due to ADP-ribosylation of a membrane protein. *Proc. Natl. Acad. Sci. U.S.A.* 79: 3129–3133.

Khelef, N., Sakamoto, H., and Guiso, N. (1992). Both adenylate cyclase and hemolytic activities are required by *Bordetella pertussis* to initiate infection. *Microb. Pathog.* 12: 227–235.

Kimura, A., Mountzouros, K. T., Relman, D. A., Falkow, S., and Cowell, J. L. (1990). *Bordetella pertussis* filamentous hemagglutinin: evaluation as a protective antigen and colonization factor in a mouse respiratory infection model. *Infect. Immun.* 58: 7–16.

Lacey, B. W. (1960). Antigenic modulation of *Bordetella pertussis*. *J. Hyg. (Camb)* 58: 57–93.

Lapin, J. H. (1943). *Whooping Cough*, Charles C. Thomas, Springfield, IL.

Lee, C. K., Roberts, A. L., Finn, T. M., Knapp, S., and Mekalanos, J. J. (1990). A new assay for invasion of HeLa 229 cells by *Bordetella pertussis*: effects of inhibitors, phenotypic modulation, and genetic alterations. *Infect. Immun.* 58: 2516–2522.

Leininger, E., Roberts, M., Kenimer, J. G., Charles, I. G., Fairweather, N., Novotny, P., and Brennan, M. J. (1991). Pertactin, an Arg-Gly-Asp-containing *Bordetella pertussis* surface protein that promotes adherence of mammalian cells. *Proc. Natl. Acad. Sci. U.S.A.* 88: 345–349.

Leininger, E., Ewanowich, C. A., Bhargava, A., Peppler, M. S., Kenimer, J. G., and Brennan, M. J. (1992). Comparative roles of the Arg-Gly-Asp sequence present in the *Bordetella pertussis* adhesins pertactin and filamentous hemagglutinin. *Infect. Immun.* 60: 2380–2385.

Lenin, S., Alonso, J. M., Brezin, C., Rocancourt, M., and Poupel, O. (1986). Effects of antibodies to the filamentous hemagglutinin and to the pertussis toxin of *Bordetella pertussis* on adherence and toxic effects to 3T3 cells. *FEMS Microbiol. Lett.* 37: 89–94.

Leslie, P. H., and Gardner, A. D. (1931). The phases of *Haemophilus pertussis*. *J. Hyg.*(Camb.) 31: 423–434.

Linnemann, C. C., and Perry, E. B. (1977). *Bordetella parapertussis*. Recent experience and a review of the literature. *Am. J. Dis. Child.* 131: 560–563.

Livey, I., Duggleby, C. J., and Robinson, A. (1987). Cloning and nucleotide sequence analysis of the serotype 2 fimbrial subunit gene of *Bordetella pertussis*. *Mol. Microbiol.* 1: 203–209.

Locht, C., and Keith, J. M. (1986). Pertussis toxin gene: nucleotide sequence and genetic organization. *Science* 232: 1258–1264.

Locht, C., Geoffroy, M. C., and Renauld, G. (1992). Common accessory genes for the *Bordetella pertussis* filamentous hemagglutinin and fimbriae share sequence similarities with the *papC* and *papD* gene families. *EMBO J.* 11: 3175–3183.

Locht, C., Bertin, P., Menozzi, F., and Renauld, G. (1993). The filamentous haemagglutinin, a multifaceted adhesin produced by virulent *Bordetella* spp. *Mol. Microbiol.* 9: 653–660.

Long, S. S., Welkon, C. J., and Clark, J. L. (1990). Widespread silent transmission of pertussis in families: antibody correlates of infection and symptomatology [see comments]. *J. Infect. Dis.* 161: 480–486.

Luker, K. E., Collier, J. L., Kolodziej, E. W., Marshall, G. R., and Goldman, W. E. (1993). *Bordetella pertussis* tracheal cytotoxin and other muramyl peptides: distinct structure–activity relationships for respiratory epithelial cytopathology. *Proc. Natl. Acad. Sci. U.S.A.* 90: 2365–2369.

Mallory, F. B., and Hornor, A. A. (1912). Pertussis: the histological lesion in the respiratory tract. *J. Med. Res. 27*: 115–123.

Mallory, F. B., Hornor, A. A. and Henderson, F. F. (1913). The relation of the Bordet-Gengou bacillus to the lesion of pertussis. *J. Med. Res. 27*: 391–397.

Marcon, M. J., Hamoudi, A. C., Cannon, H. J., and Hribar, M. M. (1987). Comparison of throat and nasopharyngeal swab specimens for culture diagnosis of *Bordetella pertussis* infection. *J. Clin. Microbiol. 25*: 1109–1110.

Masure, H. R. (1992). Modulation of adenylate cyclase toxin production as *Bordetella pertussis* enters human macrophages. *Proc. Natl. Acad. Scu. U.S.A. 89*: 6521–6525.

Mattei, D., Pichot, F., Bellalou, J., Mercereau, P. O., and Ullmann, A. (1986). Cloning of a coding sequence of *"Bordetella pertussis"* filamentous hemagglutinin gene. *Ann. Sclavo Collana Monogr. 3*: 307–311.

Mekalanos, J. J. (1992). Environmental signals controlling expression of virulence determinants in bacteria. *J. Bacteriol. 174*: 1–7.

Melton, A. R., and Weiss, A. A. (1993). Characterization of environmental regulators of *Bordetella pertussis*. *Infect. Immun. 61*: 807–815.

Menozzi, F. D., Gantiez, C., and Locht, C. (1991a). Identification and purification of transferrin- and lactoferrin-binding proteins of *Bordetella pertussis* and *Bordetella bronchiseptica*. *Infect. Immun. 59*: 3982–3988.

Menozzi, F. D., Gantiez, C., and Locht, C. (1991b). Interaction of the *Bordetella pertussis* filamentous hemagglutinin with heparin. *FEMS Microbiol. Lett. 62*: 59–64.

Miller, J. F., Mekalanos, J. J., and Falkow, S. (1989a). Coordinate regulation and sensory transduction in the control of bacterial virulence. *Science 243*: 916–922.

Miller, J. F., Roy, C. R., and Falkow, S. (1989b). Analysis of *Bordetella pertussis* virulence gene regulation by use of transcriptional fusions in *Escherichia coli*. *J. Bacteriol. 171*: 6345–6348.

Miller, J. F., Johnson, S. A., Black, W. J., Beattie, D. T., Mekalanos, J. J., and Falkow, S. (1992). Constitutive sensory transduction mutations in the *Bordetella pertussis bvgS* gene. *J. Bacteriol. 174*: 970–979.

Mills, K. H., Barnard, A., Watkins, J., and Redhead, K. (1993). Cell-mediated immunity to *Bordetella pertussis*: role of Th1 cells in bacterial clearance in a murine respiratory infection model. *Infect. Immun. 61*: 399–410.

Mink, C. M., Cherry, J. D., Christenson, P., Lewis, K., Pineda, E., Shlian, D., Dawson, J. A., and Blumberg, D. A. (1992). A search for *Bordetella pertussis* infection in university students. *Clin. Infect. Dis. 14*: 464–471.

Montaraz, J. A., Novotny, P., and Ivanyi, J. (1985). Identification of a 68-kilodalton protective protein antigen from *Bordetella bronchiseptica*. *Infect. Immun. 47*: 744–751.

Mooi, F. R., ter Avest, A., and van der Heide, H. G. (1990). Structure of the *Bordetella pertussis* gene coding for the serotype 3 fimbrial subunit. *FEMS Microbiol. Lett. 54*: 327–331.

Mooi, F. R., Jansen, W. H., Brunings, H., Gielen, H., van der Heide, H. G. J., Walvoort, H. C., and Guinee, P. A. (1992). Construction and analysis of *Bordetella pertussis* mutants defective in the production of fimbriae. *Microb. Pathog. 12*: 127–135.

Mouallem, M., Farfel, Z., and Hanski, E. (1990). *Bordetella pertussis* adenylate cyclase toxin: intoxication of host cells by bacterial invasion. *Infect. Immun. 58*: 3759–3764.

Muller, A. S., Leeuwenberg, J., and Pratt, D. S. (1986). Pertussis: epidemiology and control. *Bull. WHO 64*: 321–331.

Muse, K. E., Collier, A. M., and Baseman, J. B. (1977). Scanning electron microscopic study of hamster tracheal organ cultures infected with *Bordetella pertussis*. *J. Infect. Dis.* (University of Chcago Press) *136*: 768–777.

Musser, J. M., Hewlett, E. L., Peppler, M. S., and Selander, R. K. (1986). Genetic diversity and relationships in populations of *Bordetella* spp. *J. Bacteriol. 166*: 230–237.

Musser, J. M., Bemis, D. A., Ishikawa, H., and Selander, R. K. (1987). Clonal diversity and host distribution in *Bordetella bronchiseptica*. *J. Bacteriol. 169*: 2793–2803.

Nelson, J. D. (1978). The changing epidemiology of pertussis in young infants: the role of adults as reservoirs of infection. *Am. J. Dis. Child. 132*: 371–373.

Ng, V. L., Boggs, J. M., York, M. K., Golden, J. A., Hollander, H., and Hadley, W. K. (1992). Recovery of *Bordetella bronchiseptica* from patients with AIDS. *Clin. Infect. Dis. 15*: 376–377.

Nicosia, A., Perugini, M., Franzini, C., Casagli, C., Borri, M. G., Antoni, G., Almoni, M., Neri, P., Ratti, G., and Rappuoli, R. (1986). Cloning and sequencing of the pertussis toxin genes: operon structure and gene duplication. *Proc. Natl. Acad. Sci. U.S.A. 83*: 4631–4635.

Novotny, P., Kobisch, M., Cownley, K., Chubb, A. P., and Montaraz, J. A. (1985). Evaluation of *Bordetella bronchiseptica* vaccines in specific-pathogen-free piglets with bacterial cell surface antigens in enzyme-linked immunosorbent assay. *Infect. Immun. 50*: 190–198.

Novotny, P., Chubb, A. P., Cownley, K., and Charles, I. G. (1991). Biologic and protective properties of the 69-kDa outer membrane protein of *Bordetella pertussis*: a novel formulation for an acellular pertussis vaccine. *J. Infect. Dis. 164*: 114–122.

Oda, M., Cowell, J. L., Burstyn, D. G., and Manclark, C. R. (1984). Protective activities of the filamentous hemagglutinin and the lymphocytosis-promoting factor of *Bordetella pertussis* in mice. *J. Infect. Dis. 150*: 823–833.

Onorato, I. M., and Wassilak, S. G. (1987). Laboratory diagnosis of pertussis: the state of the art. *Pediatr. Infect. Dis. J. 6*: 145–151.

Parkinson, J. S. (1993). Signal transduction schemes of bacteria. *Cell 73*: 857–71.

Pearson, R. D., Symes, P., Conboy, M., Weiss, A. A., and Hewlett, E. L. (1987). Inhibition of monocyte oxidative responses by *Bordetella pertussis* adenylate cyclase toxin. *J. Immunol. 139*: 2749–2754.

Pedroni, P., Riboli, B., de Ferra, F., Grandi, G., Toma, S., Arico, B., and Rappuoli, R. (1988). Cloning of a novel pilin-like gene from *Bordetella pertussis*: homology to the *fim2* gene. *Mol. Microbiol. 2*: 539–543.

Peppler, M. S. (1982). Isolation and characterization of isogenic pairs of domed hemolytic and flat nonhemolytic colony types of *Bordetella pertussis*. *Infect. Immun. 35*: 840–851.

Pittman, M. (1984). The concept of pertussis as a toxin-mediated disease. *Pediatr. Infect. Dis. 3*: 467–486.

Pittman, M., Furman, B. L., and Wardlaw, A. C. (1980). *Bordetella pertussis* respiratory tract infection in the mouse: pathophysiological responses. *J. Infect. Dis. 142*: 56–66.

Pizza, M., Bartoloni, A., Prugnola, A., Silvestri, S., and Rappuoli, R. (1988). Subunit S1 of pertussis toxin: mapping of the regions essential for ADP-ribosyltransferase activity. *Proc. Natl. Acad. Sci. U.S.A. 85*: 7521–7525.

Pizza, M., Covacci, A., Bartoloni, A., Perugini, M., Nencioni, L., DeMagistris, M. T., Villa, L., Nucci, D., Manetti, R., Bugnoli, M., et al. (1989). Mutants of pertussis toxin suitable for vaccine development. *Science, 246*: 497–500.

Prasad, S. M., Yin, Y., Rodzinski, E., Tuomanen, E. I., and Masure, H. R. (1993). Identification of a carbohydrate recognition domain in filamentous hemagglutinin from *Bordetella pertussis*. *Infect. Immun. 61*: 2780–2785.

Rappuoli, R., Arico, b., and Scarlato, V. (1992). Thermoregulation and reversible differentiation in *Bordetella*: a model for pathogenic bacteria. *Mol. Microbiol. 6*: 2209–2211.

Reiser, J., Friedman, R. L., and Germanier, R. (1985). *Bordetella pertussis* filamentous hemagglutinin gene: molecular cloning of a potential coding sequence. *Dev. Biol. Stand. 61*: 265–271.

Relman, D. A., Domenighini, M., Tuomanen, E., Rappuoli, R., and Falkow, S. (1989). Filamentous hemagglutinin of *Bordetella pertussis*: nucleotide sequence and crucial role in adherence. *Proc. Natl. Acad. Sci. U.S.A. 86*: 2637–2641.

Relman, D., Tuomanen, E., Falkow, S., Golenbock, D. T., Saukkonen, K., and Wright, S. D. (1990). Recognition of a bacterial adhesion by an integrin: macrophage CR3 (alpha M beta 2, CD11b/CD18) binds filamentous hemagglutinin of *Bordetella pertussis*. *Cell 61*: 1375–1382.

Relman, D. A., and Falkow, S. (1994). A molecular perspective of microbial pathogenicity. In

Principles and Practice of Infectious Diseases (Mandell, G. L. Douglas, R. G., Bennett, J. E., eds.) 4th ed., Churchill Livingstone, New York, pp 19–29.

Rich, A. R. (1932). On the etiology and pathogenesis of whooping cough. *Johns Hopkins Hosp. Bull. 51*: 346–363.

Roberts, M., Fairweather, N. F., Leininger, E., Pickard, D., Hewlett, E. L., Robinson, A., Hayward, C., Dougan, G., and Charles, I. G. (1991). Construction and characterization of *Bordetella pertussis* mutants lacking the *vir*-regulated P.69 outer membrane protein. *Mol. Microbiol. 5*: 1393–1404.

Robertson, P. W., Goldberg, H., Jarvie, B. H., Smith, D. D., and Whybin, L. R. (1987). *Bordetella pertussis* infection: a cause of persistent cough in adults. *Med. J. Aust. 146*: 522–525.

Roy, C. R., and Falkow, S. (1991). Identification of *Bordetella pertussis* regulatory sequences required for transcriptional activation of the *fhaB* gene and autoregulation of the *bvgAS* operon. *J. Bacteriol. 173*: 2385–2392.

Roy, C. R., Miller, J. F., and Falkow, S. (1989). The *bvgA* gene of *Bordetella pertussis* encodes a transcriptional activator required for coordinate regulation of several virulence genes. *J. Bacteriol. 171*: 6338–6344.

Ruoslahti, E. (1991). Integrins. *J. Clin. Invest. 87*: 1–5.

Ruoslahti, E., and Pierschbacher, M. D. (1987). New perspectives in cell adhesion: RGD and integrins. *Science 238*: 491–497.

Sakamoto, H., Bellalou, J., Sebo, P., and Ladant, D. (1992). *Bordetella pertussis* adenylate cyclase toxin. Structural and functional independence of the catalytic and hemolytic activities. *J. Biol. Chem. 267*: 13598–13602.

Sato, H., and Sato, Y. (1984). *Bordetella pertussis* infection in mice: correlation of specific antibodies against two antigens, pertussis toxin, and filamentous hemagglutinin with mouse protectivity in an intracerebral or aerosol challenge system. *Infect. Immun. 46*: 415–421.

Sato, Y., Cowell, J. L., Sato, H., Burstyn, D. G., and Manclark, C. R. (1983). Separation and purification of the hemagglutinins from *Bordetella pertussis*. *Infect. Immun. 41*: 313–320.

Saukkonen, K., Cabellos, C., Burroughs, M., Prasad, S., and Tuomanen, E. (1991). Integrin-mediated localization of *Bordetella pertussis* within macrophages: role in pulmonary colonization. *J. Exp. Med. 173*: 1143–1149.

Saukkonen, K., Burnette, W. N., Mar, V. L., Masure, H. R., and Tuomanen, E. I. (1992). Pertussis toxin has eukaryotic-like carbohydrate recognition domains. *Proc. Natl. Acad. Sci. U.S.A. 89*: 118–122.

Scarlato, V., Arico, B., Prugnola, A., and Rappuoli, R. (1991a). Sequential activation and environmental regulation of virulence genes in *Bordetella pertussis*. *EMBO J. 10*: 3971–3975.

Scarlato, V., Prugnola, A., Arico, B., and Rappuoli, R. (1991b). The *bvg*- dependent promoters show similar behavior in different *Bordetella* species and share sequence homologies. *Mol. Microbiol. 5*: 2493–2498.

Sekiya, K., Futaesaku, Y., and Nakase (1988). Electron microscopic observations on tracheal epithelia of mice infected with *Bordetella bronchiseptica*. *Microbiol. Immunol. 32*: 461–472.

Sekiya, K., Futaesaku, Y., and Nakase Y. (1989). Electron microscopic observations on ciliated epithelium of tracheal organ cultures infected with *Bordetella bronchiseptica*. *Microbiol. Immunol. 33*: 111–121.

Shahin, R. D., Amsbaugh, D. F., and Leef, M. F. (1992). Mucosal immunization with filamentous hemagglutinin protects against *Bordetella pertussis* respiratory infection. *Infect. Immun. 60*: 1482–1488.

Siminia, T., van der Brugge-Gamelkoorn, G. J., and Jeurissen, S. H. (1989). Structure and function of bronchus-associated lymphoid tissue (BALT). *Crit. Rev. Immunol. 9*: 119–150.

Spangrude, G. J., Braaten, B. A., and Daynes, R. A. (1984). Molecular mechanisms of lymphocyte extravasation. I. Studies of two selective inhibitors of lymphocyte recirculation. *J. Immunol. 132*: 354–362.

Steed, L. L., Setareh, M., and Friedman, R. L. (1991). Intracellular survival of virulent *Bordetella pertussis* in human polymorphonuclear leukocytes. *J. Leukocyte Biol. 50*: 321–330.

Steketee, R. W., Wassilak, S. G., Adkins, W. J., Burstyn, D. G., Manclark, C. R., Berg, J., Hopfensperger, D., Schell, W. L., and Davis, J. P. (1988). Evidence for a high attack rate and efficacy of erythromycin prophylaxis in a pertussis outbreak in a facility for the developmentally disabled. *J. Infect. Dis. 157*: 434–440.

St. Geme, J. W. III., Falkow, S., and Barenkamp, S. J. (1993). High-molecular-weight proteins of nontypable *Haemophilus influenzae* mediate attachment to human epithelial cells. *Proc. Natl. Acad. Sci. U.S.A. 90*: 2875–2879.

Stibitz, S., and Garletts, T. L. (1992). Derivation of a physical map of the chromosome of *Bordetella pertussis* Tohama I. *J. Bacteriol. 174*: 7770–7777.

Stibitz, S., and Yang, M. S. (1991). Subcellular localization and immunological detection of proteins encoded by the *vir* locus of *Bordetella pertussis*. *J. Bacteriol. 173*: 4288–4296.

Stibitz, S., Weiss, A. A., and Falkow, S. (1986). The construction of a cloning vector designed for gene replacement in *Bordetella pertussis*. *Gene 50*: 133–140.

Stibitz, S., Weiss, A. A., and Falkow, S. (1988). Genetic analysis of a region of the *Bordetella pertussis* chromosome encoding filamentous hemagglutinin and the pleiotropic regulatory locus *vir*. *J. Bacteriol. 170*: 2904–2913.

Stibitz, S., Aaronson, W., Monack, D., and Falkow, S. (1989). Phase variation in *Bordetella pertussis* by frameshift mutation in a gene for a novel two-component system. *Nature 338*: 266–269.

Stock, J. B., Stock, A. M., and Mottonen, J. M. (1990). Signal transduction in bacteria. *Nature 344*: 395–400.

Sutter, R. W., and Cochi, S. L. (1992). Pertussis hospitalizations and mortality in the United States, 1985–1988. Evaluation of the completeness of national reporting. *J. Am. Med. Assoc. 267*: 386–391.

Switzer, W. P., Mare, C. J., and Hubbard, E. D. (1966). Incidence of *Bordetella bronchiseptica* in wildlife and man in Iowa. *Am. J. Vet. Res. 27*: 1134–1136.

Thomas, M. G. (1989). Epidemiology of pertussis. *Rev. Infect. Dis. 11*: 255–262.

Thomas, M. G., Ashworth, L. A., Miller, E., and Lambert, H. P. (1989a). Serum IgG, IgA, and IgM responses to pertussis toxin, fialmentous hemagglutinin, and agglutinogens 2 and 3 after infection with *Bordetella pertussis* and immunization with whole-cell pertussis vaccine. *J. Infect. Dis. 160*: 838–845.

Thomas, M. G., Redhead, K., and Lambert, H. P. (1989b). Human serum antibody responses to *Bordetella pertussis* infection and pertussis vaccination. *J. Infect. Dis. 159*: 211–218.

Tuomanen, E. (1986). Piracy of adhesins: attachment of superinfecting pathogens to respiratory cilia by secreted adhesins of *Bordetella pertussis*. *Infect. Immun. 54*: 905–908.

Tuomanen, E., and Weiss, A. (1985). Characterization of two adhesins of *Bordetella pertussis* for human ciliated respiratory-epithelial cells. *J. Infect. Dis. 152*: 118–125.

Tuomanen, E. I., Nedelman, J., Hendley, J. O., and Hewlett, E. L. (1983). Species specificity of *Bordetella* adherence to human and animal ciliated respiratory epithelial cells. *Infect. Immun. 42*: 692–695.

Tuomanen, E., Towbin, H., Rosenfelder, G., Braun, D., Larson, G., Hansson, G. C., and Hill, R. (1988). Receptor analogs and monoclonal antibodies that inhibit adherence of *Bordetella pertussis* to human ciliated respiratory epithelial cells. *J. Exp. Med. 168*: 267–277.

Uhl, M. A., and Miller, J. F. (1994). Autophosphorylation and phosphotransfer in the *Bordetella pertussis* BvgAS signal transduction cascade. *Proc. Natl. Acad. Sci. USA 91*:1163–1167.

Urisu, A., Cowell, J. L., and Manclark, C. R. (1986). Filamentous hemagglutinin has a major role in mediating adherence of *Bordetella pertussis* to human WiDr cells. *Infect. Immun. 52*: 695–701.

van't Wout, J., Burnette, W. N., Mar, V. L., Rozdzinski, E., Wright, S. D., and Tuomanen, E. I. (1992). Role of carbohydrate recognition domains of pertussis toxin in adherence of *Bordetella pertussis* to human macrophages. *Infect. Immun. 60*: 3303–3308.

Weiss, A. A., and Falkow, S. (1983). Transposon insertion and subsequent donor formation promoted by Tn501 in *Bordetella pertussis*. *J. Bacteriol. 153*: 304–309.

Weiss, A. A., and Falkow, S. (1984). Genetic analysis of phase change in *Bordetella pertussis*. *Infect. Immun. 43*: 263–269.

Weiss, A. A., and Goodwin, M. S. (1989). Lethal infection by *Bordetella pertussis* mutants in the infant mouse model. *Infect. Immun. 57*: 3757–3764.

Weiss, A. A., and Hewlett, E. L. (1986). Virulence factors of *Bordetella pertussis*. *Annu. Rev. Microbiol. 40*: 661–686.

Weiss, A. A., Hewlett, E. L., Myers, G. A., and Falkow, S. (1983). Tn5-induced mutations affecting virulence factors of *Bordetella pertussis*. *Infect. Immun. 42*: 33–41.

Weiss, A. A., Hewlett, E. L., Myers, G. A., and Falkow, S. (1984). Pertussis toxin and extracytoplasmic adenylate cyclase as virulence factors of *Bordetella pertussis*. *J. Infect. Dis. 150*: 219–222.

Weiss, A. A., Melton, A. R., Walker, K. E., Andraos, S. C., and Meidl, J. J. (1989). Use of the promoter fusion transposon Tn5 *lac* to identify mutations in *Bordetella pertussis* vir-regulated genes. *Infect. Immun. 57*: 2674–2682.

Weiss, A. A., Johnson, F. D., and Burns, D. L. (1993). Molecular characterization of an operon required for pertussis toxin secretion. *Proc. Natl. Acad. Sci. U.S.A. 90*: 2970–2974.

Willems, R. J. L. (1993). Genetic and functional studies on *Bordetella pertussis* fimbriae. Thesis, Rijksuniversiteit, Utrecht.

Willems, R., Paul, A., van der Heide, H. G., ter Avest, A. R., and Mooi, F. R. (1990). Fimbrial phase variation in *Bordetella pertussis*: a novel mechanism for transcriptional regulation. *EMBO J. 9*: 2803–2809.

Willems, R. J., van der Heide, H. G. J., and Mooi, F. R. (1992). Characterization of a *Bordetella pertussis* fimbrial gene cluster which is located directly downstream of the filamentous haemagglutinin gene. *Mol. Microbiol. 6*: 2661–2671.

Willems, R. J. L., Geuijen, C., van der Heide, H. G. J., Matheson, M., Robinson, A., Versluis, L. F., Ebberink, R., Theelen, J., and Mooi, F. R. (1993). Isolation of a putative fimbrial adhesin from *Bordetella pertussis* and the identification of its gene. *Mol. Microbiol. 9*: 623–634.

Wilson, R., Read, R., Thomas, M., Rutman, A., Harrison, K., Lund, V., Cookson, B., Goldman, W., Lambert, H., and Cole, P. (1991). Effects of *Bordetella pertussis* infection on human respiratory epithelium in vivo and in vitro. *Infect. Immun. 59*: 337–345.

Woolfrey, B. F., and Moody, J. A. (1991). Human infections associated with *Bordetella bronchiseptica*. *Clin. Microbiol. Rev. 4*: 243–255.

Zackrisson, G., Lagergard, T., Trollfors, B., and Krantz, I. (1990a). Immunoglobulin A antibodies to pertussis toxin and filamentous hemagglutinin in salvia from patients with pertussis. *J. Clin. Microbiol. 28*: 1502–1505.

Zackrisson, G., Taranger, J., and Trollfors, B. (1990b). History of whooping cough in non-vaccinated Swedish children, related to serum antibodies to pertussis toxin and filamentous hemagglutinin. *J. Pediatr. 116*: 190–194.

Zhang, J. M., Cowell, J. L., Steven, A. C., Carter, P. H., McGrath, P. P., and Manclark, C. R. (1985). Purification and characterization of fimbriae isolated from *Bordetella pertussis*. *Infect. Immun. 48*: 422–427.

Zhang, Y. L., and Sekura, R. D. (1991). Purification and characterization of the heat-labile toxin of *Bordetella pertussis*. *Infect. Immun. 59*: 3754–3759.

17

Muramyl Peptides as Exotoxins: *Bordetella* Tracheal Cytotoxin

Linda Nixon Heiss, Kathryn E. Luker, and William E. Goldman

Washington University School of Medicine, St. Louis, Missouri

I. INTRODUCTION

As seen in the chapters of this book, the diversity among bacterial toxins reflects an impressive variety of strategies by which pathogens interfere with host cellular processes. *Bordetella pertussis*, the gram-negative bacterium that causes whooping cough (pertussis), produces at least five toxins, all of which have distinctly different structures and functions. The three best-studied are released during logarithmic phase growth and are therefore functionally defined as exotoxins, but the similarity ends there. The first is pertussis toxin, a classical bipartite protein toxin with ADP-ribosylating activity and a wide range of target cells (see chapters by Burns and Katada). The second exotoxin is adenylate cyclase toxin, a chimeric molecule with an N-terminal portion that functionally resembles anthrax toxin edema factor and a C-terminal portion with homology to the RTX family of pore-forming toxins (see Hewlett's chapter). These two vastly different *B. pertussis* toxins have pronounced effects on cells involved in the immune and inflammatory responses to infection.

    However, it is the third exotoxin that accounts for the primary cellular pathology of pertussis. In the respiratory tract, *B. pertussis* causes selective and widespread destruction of ciliated cells, essential for normal clearance of mucus in the large airways. Tracheal cytotoxin (TCT) was functionally identified as the exotoxin that reproduces this cellular pathology (Goldman et al., 1982), which leads to the debilitating coughing episodes that characterize pertussis (see Olson, 1975). However, TCT defies all conventional categorization of bacterial exotoxins: it is a "muramyl peptide," a low molecular weight fragment of cell wall peptidoglycan. This chapter describes the unique structure of TCT and its mechanism of action, which explains how a seemingly harmless molecule can participate in the central pathology and transmission of infectious disease.

II. TRACHEAL CYTOTOXIN EFFECTS ON THE RESPIRATORY EPITHELIUM

Since the first human pertussis autopsy studies were documented (Mallory and Hornor, 1912), the specific pattern of *B. pertussis* tropism and pathology has been clear. The organisms selectively adhere to ciliated epithelial cells in the large airways, and colonization is confined to the surface of these cells. Similar colonization can be seen in the airways of mice infected intranasally with *B. pertussis*, but mice do not develop the tracheobronchial pneumonia and coughing that are characteristic of the human pertussis syndrome (Burnet and Timmins, 1937). In the absence of a nonprimate animal model that accurately reflects the human disease, in vitro model systems proved to be of most use in understanding the basis for the respiratory epithelial pathology caused by *B. pertussis*.

A. Hamster Tracheal Organ Culture

The first pathologically accurate in vitro model of pertussis was *B. pertussis* infection of sectioned hamster tracheal tissue (Collier et al., 1977). These tracheal rings maintain ciliary activity and epithelial integrity for weeks in culture, unless inoculated with *B. pertussis*. As in human infection, the organisms adhere exclusively to the apical surface of ciliated cells, eventually causing ciliostasis and extrusion of these cells. Only the ciliated cells are destroyed, and because other cell types migrate to fill the gaps created by lost cells, the result is an intact but nonciliated epithelium.

    The same ciliated cell-specific damage can result from exposure of tracheal rings to supernatants from log-phase *B. pertussis* cultures, providing the first clue that an exotoxin could be responsible for this pathology. TCT is the only molecule from *B. pertussis* culture supernatant that has been demonstrated to duplicate this activity (Goldman et al., 1982). When monitored by light microscopy, hamster tracheal rings treated with TCT for 72 h show a significant decline in ciliary activity; by 96 h, ciliary activity is totally absent and nearly all ciliated cells have been extruded from the epithelium. The time course of ciliostasis can be accelerated or slowed, depending upon the concentration of TCT used. Electron microscopy reveals that whereas nonciliated cells remain ultrastructurally normal after exposure to TCT, ciliated cells exhibit a reproducible pattern of pathology (Goldman and Herwaldt, 1985). Changes visible in ciliated cells include their rounding and constriction at the apical end, and reduced numbers of cilia and basal bodies are associated with each cell. Mitochondria appear swollen and have fewer cristae, and intercellular junctions are lost (see Fig. 1).

B. Cultured Hamster Trachea Epithelial Cells

Although tracheal organ cultures are an architecturally correct model of the respiratory epithelium, the requirement for freshly excised hamster tissue imposes severe limitations on sample number and quantitation. Hamster tracheal epithelial (HTE) cells, a homogenous epithelial cell population isolated from tracheal tissue, provide a much simpler model of the respiratory epithelium (Goldman and Baseman, 1980). These are primary cultures that proliferate for 35–40 generations before entering senescence. *B. pertussis* culture supernatant inhibits DNA synthesis by serum-stimulated HTE cells in a dose-dependent manner, though there is little effect on overall RNA or protein synthesis (Goldman et al., 1982). This became the basis for a quantitative assay to monitor the purification of TCT from *B. pertussis* culture supernatant. In fact, since TCT has no demonstrated effect on established cell lines, the development of the HTE cell model was an essential step in the identification of this toxin.

The precise cellular identity of HTE cells is speculative, since they lack morphological features that are typically used to define airway epithelial cell types. HTE cells, though epithelial in origin, differ greatly from ciliated cells: HTE cells lack cilia and are capable of cell division, while ciliated cells are considered to be terminally differentiated. One speculation is that HTE cells may represent the basal cell population, responsible for division and differentiation into the specialized cell types of the pseudostratified airway mucosa (see Goldman and Baseman, 1980). TCT inhibition of HTE cell proliferation may therefore reflect a secondary effect of TCT on the capacity of respiratory epithelium to regenerate the lost ciliated cell population. This could be at least as important as the initial TCT damage to ciliated cells, since the failure to replace them would result in long-term effects on respiratory tract function. Clinical observations are certainly consistent with this: the paroxysmal coughing phase often continues for many weeks after *B. pertussis* can no longer be cultured from patients, and the clinical course of disease is unaltered by antibiotic therapy (see Olson, 1975).

C. Explants of Human Nasal Epithelium

More recently, the effects of *B. pertussis* culture supernatant and purified TCT on human nasal epithelial biopsies have also been assessed (Wilson et al., 1991). Pathological changes included loss of ciliated cells, an increased frequency of sparsely ciliated cells, cell blebbing, and mitochondrial damage. These abnormalities were consistent with the pathology seen in nasal epithelial biopsies from children with pertussis, most of whom showed a dramatic reduction in the percentage of epithelium that was ciliated. Interest-

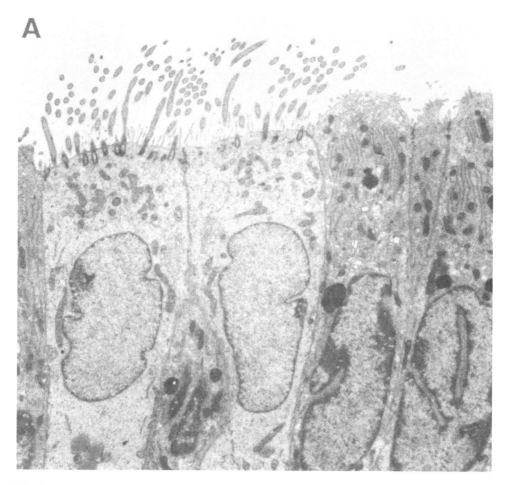

Fig. 1   Electron photomicrographs representative of the destruction of ciliated cells in hamster tracheal rings treated with TCT. (A) Control ring after 102 h of incubation; (B) ring treated with 3 μM TCT for 102 h. Magnification, ×6900. (Reproduced by permission of the American Society for Microbiology; see Heiss et al., 1993.)

ingly, the cellular damage was evident despite the absence of bacteria in nearly all of the samples examined, further evidence of a toxin at work. This study represents the only detailed and quantitative analysis of respiratory pathology in human pertussis; the parallels with the hamster trachea model are clear, and TCT is again directly implicated in causing the relevant damage.

III.   STRUCTURE AND BIOLOGICAL ORIGIN OF TCT

A.   Purification and Structural Analysis

By monitoring toxicity with the HTE cell model described in Section II.B, the biologically active fraction in culture supernatant was sequentially purified by liquid chromatography

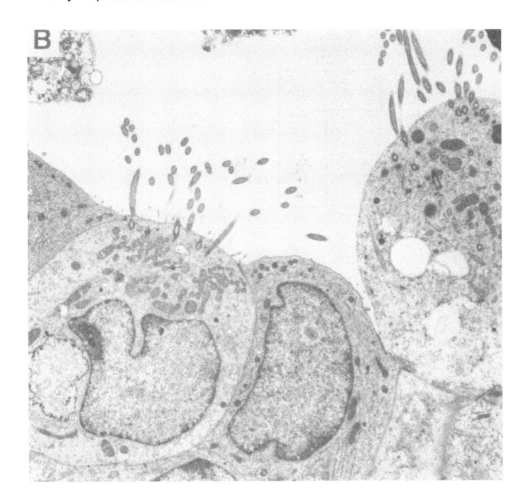

and high-voltage paper electrophoresis (Goldman and Herwaldt, 1985; Goldman et al., 1982). However, unequivocal determination of TCT structure was possible only after development of a more rigorous purification procedure involving solid-phase extraction and reversed-phase high-pressure liquid chromatography (HPLC) (Cookson et al., 1989a). Using fast atom bombardment mass spectrometry, TCT was determined to be a single molecular species with a mass of 921 Da (Cookson et al., 1989b).

The small size of TCT was an early indication that the factor responsible for respiratory epithelial damage was substantially different from all other known *B. pertussis* toxins. Furthermore, testing of purified preparations of the other toxins (pertussis toxin, adenylate cyclase toxin, heat-labile toxin, endotoxin) revealed that they had no effect on respiratory epithelium in vitro, even in concentrations greatly exceeding those that would be found in *B. pertussis* culture supernatant (Endoh et al., 1986; Goldman et al., 1982; Heiss et al., 1993).

The most surprising information, however, came from determination of the composition of TCT. Data from amino acid and amino sugar analysis, when combined

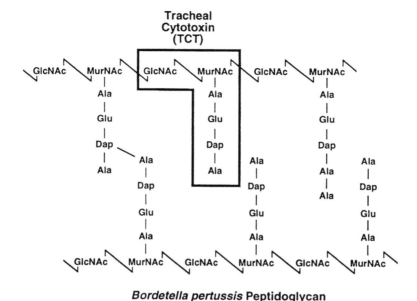

Fig. 2 TCT (outlined) as a subset of gram-negative peptidoglycan. Mur, muramic acid; Dap, diaminopimelic acid.

with the mass determination mentioned above, defined the primary structure of TCT as *N*-acetylglucosaminyl-1,6-anhydro-*N*-acetylmuramylalanyl-γ-glutamyldiaminopimelylalanine (Cookson et al., 1989b). The incorporation of muramic acid and diaminopimelic acid residues occurs in only one molecule in nature: peptidoglycan, the macromolecular network that provides structural rigidity to the bacterial cell wall (Schleifer and Kandler, 1972). TCT's structure indicated that it was a fragment of peptidoglycan, equivalent to one monomeric subunit of the polymer (see Fig. 2).

B. Peptidoglycan Processing by *Bordetella*

TCT release is strictly correlated with growth of *B. pertussis*, and the onset of stationary phase marks the end of accumulation of TCT in the culture supernatant (Goldman et al., 1990). Therefore, TCT release is not the result of bacterial lysis, but of peptidoglycan turnover during normal *B. pertussis* growth. This processing of peptidoglycan results in a remarkably homogeneous product, since nearly all of the peptidoglycan released by *B. pertussis* is in the form of TCT (Rosenthal et al., 1987). Concentrations exceeding 1 μM TCT can be found in late log-phase culture supernatants (Cookson et al., 1989a), and this is consistent with the concentrations required for respiratory epithelial cell damage.

TCT production has also been evaluated in other *Bordetella* species, and chemically identical TCT is produced by *B. parapertussis*, *B. bronchiseptica*, and *B. avium* (Cookson and Goldman, 1987). All of these organisms show the same ciliotropism as *B. pertussis* but colonize other animal hosts; in each case, the resulting disease syndrome features coughing. *Alcaligenes faecalis*, the organism most closely related to the Bordetellae, does not produce detectable levels of TCT, nor does *E. coli* (Cookson and Goldman, unpublished data). Thus, release of TCT is not a general property of gram-negative organisms, despite the fact that they share a common peptidoglycan composition.

IV. BIOLOGICAL ACTIVITIES OF MURAMYL PEPTIDES

The determination of TCT structure places it in the family of "muramyl peptides," low molecular weight fragments and derivatives of peptidoglycan that are naturally or synthetically produced. Studies of muramyl peptides have demonstrated that TCT and its relatives have a remarkably broad range of biological activites, including cytotoxicity, potentiation of immunological responses, and effects on sleep and body temperature. The activity of TCT in each of these systems is summarized below.

A. Cytotoxicity in Fallopian Tube Organ Culture

In experiments analogous to the studies of *B. pertussis* effects on respiratory epithelium, Melly et al. (1981) reported that *Neisseria gonorrhoeae* infection of human fallopian tube tissue in vitro features ciliated cell-specific damage and extrusion as the first step of pathogenesis. Log-phase gonococcal culture supernatant could duplicate this cytopathology. Earlier work had already demonstrated that *N. gonorrhoeae* releases a variety of muramyl peptides into its growth medium, including a 921-Da fragment that is chemically identical to TCT (Sinha and Rosenthal, 1980). Purified gonococcal peptidoglycan fragments, including the TCT homologue, were able to duplicate the ciliated cell damage observed with culture supernatant (Melly et al., 1984). Although *N. gonorrhoeae* and *B. pertussis* are not closely related, they have apparently evolved similar mechanisms to aid in successful infection of a ciliated mucosal surface.

B. Inhibition of Neutrophil Function

Recent work by Cundell et al. (1994) demonstrated that TCT is toxic for human neutrophils, with concentrations as low as 10 nM causing significant loss of viability after only 90 min of exposure. Neutrophil responses to the chemotactic peptide *N*-formyl-L-methionyl-L-leucyl-L-phenylalanine (FMLP) were affected by even lower concentrations of TCT: just 10 pM TCT inhibited neutrophil migration toward FMLP, and as little as 10 fM TCT inhibited neutrophil chemiluminescence stimulated by FMLP. Considering that culture supernatants of *B. pertussis* contain at least 1 μM TCT (see Section III.B), the potency of the effects on neutrophils suggests that these may be long-range activities of TCT.

C. Immunological Effects

The examples cited above have focused on toxic and inhibitory actions of TCT, but TCT and its muramyl peptide relatives have stimulatory effects on the humoral and cell-mediated immune response. For example, the well-known immunopotentiation by Freund's complete adjuvant is due to the presence of mycobacterial cell wall in this preparation. Studies to define the minimal active structure responsible for this adjuvant effect led to the synthesis of muramyl dipeptide (MDP, *N*-acetylmuramyl-L-analyl-D-isoglutamine) (Ellouz et al., 1974), now the most extensively studied peptidoglycan fragment. MDP, like other muramyl peptides, stimulates macrophages to produce the inflammatory cytokine interleukin-1 (IL-1), which is thought to be the common mediator of a wide range of muramyl peptide activities (see Section VI).

D. Central Nervous System Effects

Muramyl peptides also have been implicated as important modulators of sleep and fever, responses that may also involve IL-1 as a central signal (Walter et al., 1986). Krueger et al. (1982a) isolated a slow-wave sleep-promoting factor (FS_u) that accumulates in cerebrospinal fluid during sleep deprivation (Pappenheimer et al., 1975) and is present in all mammalian brain tissue examined (Krueger et al., 1982b). When administered intracranially to rabbits, FS_u increases the amount of time that the animals spend in slow-wave (non-REM) sleep; a concomitant rise in brain temperature is also observed (Krueger et al., 1984a). Structural analysis of FS_u unambiguously defined it as a peptidoglycan monomer chemically identical to TCT (Martin et al., 1984). Because animal cells do not have the enzymatic capability to synthesize peptidoglycan, the origin of FS_u must be bacterial sources, presumably normal flora. However, it is also not unlikely that these host responses to muramyl peptides reflect a general alarm to the presence of a bacterial infectious illness, where fever and sleep could each play a part as a beneficial host defense mechanism.

V. STRUCTURE–ACTIVITY RELATIONSHIPS FOR TCT IN RESPIRATORY EPITHELIUM

The remarkably broad spectrum of cellular responses to muramyl peptides has prompted a number of laboratories to search for a specific receptor or family of receptors. Ligand-binding studies have shown that MDP competes with serotonin for binding to macrophages and brain tissue (Silverman et al., 1985; 1989), although it is not yet clear whether this binding site corresponds to a functional muramyl peptide receptor. In other studies, muramyl peptide interaction sites have only been characterized inferentially by their ligand specificity as described by structure–activity correlations (see Adam and Lederer, 1984; Krueger et al., 1984b). However, these data have not yet been used to establish a classification of muramyl peptide receptor types; in fact, these studies have generally contained insufficient structural overlap to make solid comparisons.

In the case of TCT and the cytopathology of pertussis, it has been possible to probe the putative receptor–ligand interaction in much greater detail. This is because of an early observation that the sugar portion of the molecule has no apparent role in the toxicity of TCT for respiratory epithelial cells (Goldman et al., 1990). Without the disaccharide moiety, synthesis of a logical series of TCT peptide analogues is greatly simplified, although preserving the unusual stereochemistry of TCT still complicates the process. These analogues have made it possible to evaluate quantitatively the relative contribution of amino acid residues, functional groups, and chiral centers to the toxicity of TCT (Luker et al., 1993a,b).

A. Key Structural Features for Respiratory Epithelial Toxicity

The HTE cell model system described earlier has revealed a number of biologically important structural features of TCT, all of which are contained within the peptide portion of the molecule (see Fig. 3). Compounds tested have included truncated versions of TCT, analogues with missing or blocked functional groups, and analogues with stereochemical

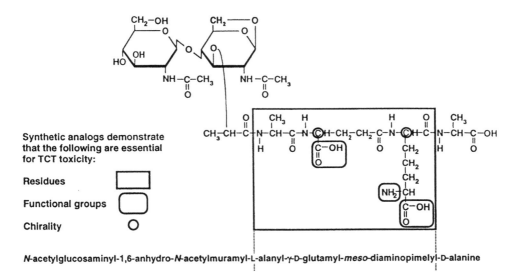

Fig. 3 Structural features of TCT important for respiratory epithelial toxicity. This diagram summarizes toxicity data derived from testing a library of TCT analogues on respiratory epithelial cells and tissue (Luker et al., 1993a,b).

changes at one of the six chiral centers. The irrelevance of the disaccharide moiety, as mentioned above, was demonstrated by the preservation of toxicity among several compounds lacking the disaccharide portion. The other end of the TCT molecule also appears to be relatively unimportant, since an analogue lacking the terminal alanine shows partial toxicity. However, analogues with further amino acid truncation lose more than 100-fold potency, indicating the importance of a core tripeptide, alanyl-γ-glutamyl-diaminopimelic acid (see Fig. 3).

Within this central core, altering a number of key structural features has dramatic effects on TCT activity. Elimination of any one of the three internal functional groups results in abrogation of measurable toxicity. Peptides with chirality variations have also demonstrated that the main-chain carbons of glutamic acid and diaminopimelic acid must retain their D- and L-chirality, respectively (see Fig. 3).

In summary, toxicity of muramyl peptides for the respiratory epithelium appears to rely primarily on a tripeptide core, within which the charged groups and the atomic arrangement near main-chain chiral centers appear to be critical. These results strongly implicate the existence of a specific muramyl peptide binding site that recognizes these features. The HTE cell system has allowed quantitative comparisons to assign all of these structure–function relationships, but tracheal organ cultures have been also been used to confirm the importance of some the major structural motifs.

B. Comparison with Other Muramyl Peptide Structure–Activity Correlations

Can one receptor or receptor family account for the range of biological activities of muramyl peptides? Previous work with MDP would have supported that concept, since MDP is capable of immune potentiation, pyrogenicity, and somnogenicity. However,

MDP is not toxic for respiratory epithelial cells (Luker et al., 1993a), consistent with the structure–function data discussed above. Furthermore, the muramic acid moiety has been considered essential for adjuvanticity of MDP and other muramyl peptides, though it is dispensable for TCT-mediated pertussis pathology. Other immune system responses to muramyl peptides may be elicited with the central dipeptide, γ-Glu-*meso*-Dap (Kitaura et al., 1982), which has no detectable activity in respiratory epithelial systems.

Similar inconsistencies are seen in comparisons with the requirements for somnogenicity. The internal dehydration featured in the muramic acid of FS_u (equivalent to TCT) is essential for maximum somnogenic potential of muramyl peptides (Krueger et al., 1987); diaminopimelic acid, however, is not required for somnogenicity (Krueger and Johannsen, 1989). Both of these findings are contrary to the structure–activity relationships for respiratory epithelial pathology, where muramic acid is a dispensable residue and diaminopimelic acid is an absolutely essential one.

The only structure–function analysis that is in full agreement with the pertussis models is the study of gonococcal muramyl peptide effects on fallopian tube organ cultures (see Section IV.A). In this system, MDP is also inactive, and the presence of the anhydro linkage on muramic acid is not important for toxicity (Melly et al., 1984). Although this study tested only a few analogues, the available evidence suggests that these ciliated mucosal tissues may share a common muramyl peptide interaction site or receptor, while other muramyl peptide activities employ one or more different receptors.

VI.   MOLECULAR MECHANISM OF TCT ACTION

The muramyl peptide structure of TCT gave an important clue to the mechanism by which TCT may damage respiratory epithelial cells. As mentioned in Section IV, many of the biological effects of muramyl peptides appear to be mediated by the cytokine IL-1. IL-1 belongs to a family of polypeptide mediators that have a broad range of biological activities, including central roles in stimulating immune and inflammtory responses to microbial infection and tissue injury (see Dinarello, 1992). Many of the biological effects of IL-1 are shared by muramyl peptides, consistent with the observation that muramyl peptides (including TCT) trigger macrophages to synthesize and release IL-1 (Dinarello and Krueger, 1986; Oppenheim et al., 1980).

Correlations between muramyl peptide effects and IL-1 have been demonstrated in a number of experimental systems. The adjuvant activity of muramyl peptides corresponds to their ability to trigger macrophage membrane-associated IL-1 (Bahr et al., 1987). In rabbits, the kinetics of the sleep response to recombinant IL-1 or muramyl peptides are consistent with the notion that muramyl peptides modulate sleep effects by triggering production of IL-1 (see Krueger, 1990). Muramyl peptide pyrogenic activities have also been correlated with production of and response to IL-1 (Dinarello et al., 1978).

A.   Interleukin-1 and TCT Toxicity

Despite the linkage between muramyl peptide actions and IL-1, it was not obvious that TCT would function through a similar mechanism in generating pertussis cytopathology. The structure–activity relationships described in Section V suggest that TCT relies on a different pathway to cause respiratory epithelial damage, since the inferred TCT receptor/interaction site is distinct from that required for other muramyl peptide biological effects. Furthermore, macrophages are not necessary for the TCT effects on cultured respiratory epithelial cells.

Nevertheless, IL-1 does appear to be involved in TCT toxicity in pertussis and may even play a central role (Heiss et al., 1993). Consistent with its ability to reproduce other muramyl peptide actions, recombinant IL-1 causes TCT-like damage to the respiratory epithelium. IL-1 inhibits DNA synthesis by HTE cells and generates pertussis-like destruction of ciliated cells in hamster tracheal organ culture. Furthermore, TCT stimulates IL-1α production by respiratory epithelial cells. The IL-1 produced remains intracellular, consistent with observations that the effects of TCT cannot be blocked using either anti-IL-1α antibodies or the IL-1 receptor antagonist (IL-1ra).

Could IL-1 produced by respiratory epithelial cells act intracellularly to generate the damage triggered by TCT? Although IL-1 is typically considered to function intercellularly, there is growing evidence that intracellular IL-1α may have an important biological role. Intracellular IL-1α is thought to regulate human endothelial cell senescence (Maier et al., 1990). The intracellular form of IL-1α contains a putative nuclear translocation sequence similar to that characterized for the SV40 large T antigen, and nuclear binding sites for IL-1 have been demonstrated (Grenfell et al., 1989). Moreover, epithelial cells have been found to produce an intracellular form of IL-1ra that presumably acts as an antagonist of endogenous or internalized IL-1 (Haskill et al., 1991). The ability of exogenously added recombinant IL-1 to duplicate TCT effects is not inconsistent with an intracellular IL-1 function, since exogenous IL-1 has been shown to be translocated to the nucleus (Mizel et al., 1987).

B. Nitric Oxide as the Mediator of TCT Damage

Interleukin-1 can cause target cell damage by a variety of mechanisms (see Dinarello, 1991), and one of these is by inducing nitric oxide synthase. This enzyme produces the free radical nitric oxide from the guanidino nitrogen atom of L-arginine (Hibbs et al., 1987). Cytokines have been shown to activate an inducible isoform of nitric oxide synthase in many cell types; this is distinct from the constitutive nitric oxide synthases responsible for endothelium-dependent relaxation and neural transmission (see Nathan, 1992). The formation of nitrite, which is a stable oxidation product of nitric oxide, serves as an indicator of nitric oxide synthase activity (Stuehr and Marletta, 1987; Stuehr and Nathan, 1989).

When treated with TCT or recombinant IL-1, HTE cells and hamster tracheal rings produce large amounts of nitrite. In TCT-or IL-1-treated HTE cells, the degree of inhibition of DNA synthesis corresponds to levels of nitrite accumulating in the culture supernatants. Two inhibitors of nitric oxide synthase, N^G-monomethyl-L-arginine and aminoguanidine, have been shown to block TCT-induced nitrite production. These nitrite experiments indicate that both TCT and IL-1 trigger production of nitric oxide in respiratory epithelial cells (Heiss et al., 1994).

The binding of nitric oxide to iron-sulfur centers of enzymes has been implicated as a mechanism for the cytotoxic actions of nitric oxide (see Hibbs et al., 1990). Target enzymes include aconitase, complex I and complex II of the mitochondrial electron transport chain, and ribonucleotide reductase. TCT reduces aconitase activity in HTE cells by 80%, suggesting a role for nitric oxide in cellular damage by TCT. This is also supported by electron paramagnetic resonance spectroscopy experiments demonstrating the formation of iron-nitrosyl complexes in TCT-treated HTE cells (Heiss et al., 1994).

Can the production of nitric oxide account for the cytostatic and cytotoxic actions of TCT? When HTE cells are treated with TCT in the presence of inhibitors of nitric oxide synthase, there is a dramatic reduction in TCT toxic activity. Nitric oxide synthase inhibitors also block the destruction of ciliated cells in hamster tracheal rings treated with TCT. These experiments implicate nitric oxide-mediated inactivation of target enzymes as the mechanism of TCT toxicity in pertussis (Heiss et al., 1994).

C. Target Cell Sensitivity to TCT

The actions of a number of bacterial products, including muramyl peptides, endotoxin, and toxic shock syndrome-associated staphylococcal and streptococcal exotoxins, have been associated with their abilities to stimulate cytokine production by mononuclear phagocytes. In addition, these bacterial products have been shown to induce nitric oxide synthase in macrophages (see Hibbs et al., 1990; Palacios et al., 1992; Zembowicz and Vane, 1992). Endotoxin, in the presence of macrophage- or lymphocyte-derived cytokines, can also directly stimulate certain target cells to produce nitric oxide (Curran et al., 1990; Palmer et al., 1992; Roberts et al., 1992). In contrast to endotoxin, TCT does not require the involvement of macrophages or lymphocytes as the producers of cytokines or nitric oxide. Instead, TCT stimulates IL-1 and nitric oxide production by the epithelial target cells themselves.

As mentioned in Section II.B, HTE cells represent the only cultured cell line that is known to be responsive to TCT (Goldman et al., 1982). The basis for this specificity is not yet understood. It is also not known whether other cells in the respiratory tract produce IL-1 and/or nitric oxide in response to TCT. However, the cytotoxic activities of TCT on intact respiratory epithelium are most evident among the ciliated cell population, which show dramatic ultrastructural changes, lose ciliary activity, and are eventually extruded. The basis for this sensitivity to TCT could relate to the energy required for their primary function, ciliary activity. Nitric oxide's inhibitory effects on a number of enzymes involved in ATP synthesis (see above) would make ciliated cells particularly susceptible to TCT. Similarly, the primary activity of basal cells—epithelial regeneration—would be dependent on enzymes important for DNA synthesis, such as the nitric oxide target ribonucleotide reductase. Thus, TCT's devastating effect on ciliated cells and its speculated interference with basal cell function can both be linked to the nitric oxide mechanism, an appealing explanation that unifies the diverse biological effects of this toxin.

VII. CONCLUSION

The proposed model in Figure 4 illustrates current thinking on the interaction of *B. pertussis* with respiratory epithelial cells. The implied pathway of TCT toxicity, with IL-1 as the intermediary signal that induces nitric oxide synthase, is consistent with the available data but not yet proved. Furthermore, although it is known that nitric oxide is a necessary effector molecule, it is not yet clear whether nitric oxide alone is sufficient for toxicity. Work with the HTE cell system has unambiguously implicated epithelial cells as the source of both IL-1 and nitric oxide. In the complex respiratory epithelium, however, the precise cells that bind TCT and respond via this toxic pathway have not yet been identified. What is certain is that both TCT and IL-1 can reproduce the respiratory epithelial damage

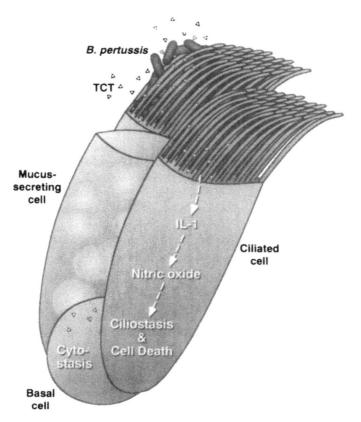

Fig. 4 Proposed mechanism of pertussis respiratory cytopathology. *B pertussis* colonizing the surface of ciliated cells releases TCT that triggers intracellular IL-1 production by respiratory epithelial cells. IL-1 induces nitric oxide synthase, and the production of nitric oxide affects two cell populations: ciliated cells, which show functional impairment and are eventually extruded from the epithelium, and basal cells, which should normally divide and differentiate to replace the lost ciliated cells. Since mucus-secreting cells are still active but ciliary clearance remains deficient, the static airways become blocked with mucus, bacteria, and inflammatory debris. Coughing is the expected physiological response to clear these blockages, and secondary infection (a common complication of pertussis) becomes more likely with the loss of this primary defense barrier.

of pertussis and that nitric oxide is a mediator of this destruction. It is the first time that nitric oxide has been causally linked to a bacterial exotoxin's role in disease.

The process of exotoxin production by bacteria is energetically costly, and the synthesis of many exotoxins is therefore regulated so that they are manufactured only in the presence of appropriate environmental signals. TCT violates this rule and is synthesized as long as the organisms are growing and expanding their cell wall (see Section III.B). Although this constant production may seem wasteful, TCT release accounts for less than 1% of the available peptidoglycan in the intact bacteria (Rosenthal et al., 1987). TCT production is therefore a rather economical alternative to the conventional synthesis and secretion of a classical protein exotoxin. The cost of generating TCT is particularly justified by the lifestyle of *Bordetella*, which is found exclusively in the respiratory tract and relies entirely on direct host-to-host transmission. Without an alternative environmental reservoir, these organisms must ensure their efficient delivery to another host before they are overwhelmed by inflammatory and immune responses. Coughing is a remarkably effective transmission technique, and pertussis is historically considered to be one of the most contagious bacterial diseases (see Olson, 1975).

The direct advantages of muramyl peptide production to *N. gonorrhoeae* are less obvious, though muramyl peptides are clearly implicated in the intense inflammatory reactions characteristic of gonococcal disease (Fleming et al., 1986). Both the *Bordetella* and *Neisseria* examples are compelling evidence that the release of muramyl peptides can play an important role in pathogenesis. Though quite different from conventional exotoxins in structure, synthesis, and mechanism of action, TCT nevertheless fulfills the requirements of an exotoxin at its most basic level of host cellular damage and specificity. Peptidoglycan, ubiquitous and seemingly innocuous, is an unlikely source material for such a toxic activity. However, the wide-ranging physiological responses to peptidoglycan fragments should make it less surprising that some pathogens would evolve to exploit muramyl peptides as a mechanism of virulence.

Acknowledgments

Our laboratory's work on pertussis is supported by Public Health Service grant AI22243 from the National Institutes of Health and by the Monsanto/Washington University Biomedical Research Contract.

REFERENCES

Adam, A., and Lederer, E. (1984). Muramyl peptides: immunomodulators, sleep factors, and vitamins. *Med. Res. Rev. 4*: 111–152.

Bahr, G. M., Chedid, L. A., and Behbehani, K. (1987). Induction, *in vivo* and *in vitro*, of macrophage membrane interleukin-1 by adjuvant-active synthetic muramyl peptides. *Cell. Immunol. 107*: 443–454.

Burnet, F. M. and Timmins, C. (1937). Experimental infection with *Haemophilus pertussis* in the mouse by intranasal inoculation. *Br. J. Exp. Pathol. 18*: 83–90.

Collier, A. M., Peterson, L. P., and Baseman, J. B. (1977). Pathogenesis of infection with *Bordetella pertussis* in hamster tracheal organ culture. *J. Infect. Dis. 136*: S196–S203.

Cookson, B. T., and Goldman, W. E. (1987). Tracheal cytotoxin: a conserved virulence determinant of all *Bordetella* species. *J. Cell. Biochem. (Suppl.) 11B*: 124 (Abstr. H208).

Cookson, B. T., Cho, H.-L., Herwaldt, L. A., and Goldman, W. E. (1989a). Biological activities and chemical composition of purified tracheal cytotoxin of *Bordetella pertussis*. *Infect. Immun. 57*: 2223–2229.

Cookson, B. T., Tyler, A. N., and Goldman, W. E. (1989b). Primary structure of the peptidoglycan-derived tracheal cytotoxin of *Bordetella pertussis*. *Biochemistry 28*: 1744–1749.

Cundell, D. R., Kanthakumar, K., Taylor, G. W., Goldman, W. E., Flak, T., Cole, P. J., and Wilson R. (1994). The effect of tracheal cytotoxin from *Bordetella pertussis* on human neutrophil function *in vitro*. *Infect. Immun. 62*: 639–643.

Curran, R. D., Billiar, T. R., Stuehr, D. J., Ochoa, J. B., Harbrecht, B. G., Flint, S. G., and Simmons, R. L. (1990). Multiple cytokines are required to induce hepatocyte nitric oxide production and inhibit total protein synthesis. *Ann. Surg. 212*: 462–469.

Dinarello, C. A. (1991). Inflammatory cytokines: interleukin-1 and tumor necrosis factor as effector molecules in autoimmune diseases. *Curr. Opin. Immunol. 3*: 941–948.

Dinarello, C. A. (1992). Role of interleukin-1 in infectious diseases. *Immunol. Rev. 127*: 119–146.

Dinarello, C. A., and Krueger, J. M. (1986). Induction of interleukin 1 by synthetic and naturally occurring muramyl peptides. *Fed. Proc. 45*: 2545–2548.

Dinarello, C. A., Elin, R. J., Chedid, L., and Wolff, S. M. (1978). The pyrogenicity of the synthetic adjuvant muramyl dipeptide and two structural analogues. *J. Infect. Dis. 138*: 760–767.

Ellouz, F., Adam, A., Ciorbaru, R., and Lederer, E. (1974). Minimal structural requirements for adjuvant activity of bacterial peptidoglycan derivatives. *Biochem. Biophys. Res. Commun. 59*: 1317–1325.

Endoh, M., Nagai, M., and Nakase, Y. (1986). Effect of *Bordetella* heat-labile toxin on perfused lung preparations of guinea pigs. *Microbiol. Immunol. 30*: 1239–1246.

Fleming, T. J., Wallsmith, D. E., and Rosenthal, R. E. (1986). Arthropathic properties of gonococcal peptidoglycan fragments: implications for the pathogenesis of disseminated gonococcal disease. *Infect. Immun. 52*: 600–608.

Goldman, W. E., and Baseman, J. B. (1980). Selective isolation and culture of a proliferating epithelial cell population from the hamster trachea. *In Vitro 16*: 313–319.

Goldman, W. E., and Herwaldt, L. A. (1985). *Bordetella pertussis* tracheal cytotoxin. *Dev. Biol. Stand. 61*: 103–111.

Goldman, W. E., Klapper, D. G., and Baseman, J. B. (1982). Detection, isolation, and analysis of a released *Bordetella pertussis* product toxic to cultured tracheal cells. *Infect. Immun. 36*: 782–794.

Goldman, W. E., Collier, J. L., Cookson, B. T., Marshall, G. R., and Erwin, K. M. (1990). Tracheal cytotoxin of *Bordetella pertussis*: biosynthesis, structure and specificity. In *Proceedings of the Sixth International Symposium on Pertussis*, (C. R. Manclark (Ed.), Dept. of Health and Human Services, U.S. Public Health Service, Bethesda, MD, pp. 5–12.

Grenfell, S., Smithers, N., Miller, K., and Solari, R. (1989). Receptor-mediated endocytosis and nuclear transport of human interleukin 1α. *Biochem. J. 264*: 813–822.

Haskill, S., Martin, G., Van Le, L., Morris, J., Peace, A., Bigler, C. F., Jaffe, G. J., Hammerberg, C., Sporn, S. A., Fong, S., Arend, W. P., and Ralph, P. (1991). cDNA cloning of an intracellular form of the human interleukin 1 receptor antagonist associated with epithelium. *Proc. Natl. Acad. Sci. U.S.A. 88*: 3681–3685.

Heiss, L. N., Moser, S. A., Unanue, E. R., and Goldman, W. E. (1993). Interleukin-1 is linked to the respiratory epithelial cytopathology of pertussis. *Infect. Immun. 61*: 3123–3128.

Heiss, L. N., Lancaster, J. R., Jr., Corbett, J. A., and Goldman, W. E. (1994). Epithelial autotoxicity of nitric oxide: role in the respiratory cytopathology of pertussis. *Proc. Natl. Acad. Sci. U.S.A. 91*: 267–270.

Hibbs, J. B., Jr., Taintor, R. R., and Vavrin, Z. (1987). Macrophage cytotoxicity: role for L-arginine deiminase and imino nitrogen oxidation to nitrite. *Science 235*: 473–476.

Hibbs, J. B., Jr., Taintor, R. R., Vavrin, Z., Granger, D. L., Drapier, J.-C., Amber, I. J., and Lancaster, J. R., Jr. (1990). Synthesis of nitric oxide from a terminal guanidino nitrogen atom of L-arginine: a molecular mechanism regulating cellular proliferation that targets intracellular iron. In *Nitric Oxide from L-Arginine: A Bioregulatory System*, S. Moncada and E. A. Higgs (Eds.), Elsevier, Amsterdam, pp. 189–223.

Kitaura, Y., Nakaguchi, O., Takeno, H., Okada, S., Yonishi, S., Hemmi, K., Mori, J., Senoh, H., Mine, Y., and Hashimoto, M. (1982). N^2-(γ-D-Glutamyl)-*meso*-2(L),2'(D)-diaminopimelic acid as the minimal prerequisite structure of FK-156: its acyl derivatives with potent immunostimulating activity. *J. Med. Chem. 25*: 335–337.

Krueger, J. M. (1990). Somnogenic activity of immune response modifiers. *Trends Pharmacol. Sci. 11*: 122–126.

Krueger, J. M., and Johannsen, L. (1989). Bacterial products, cytokines and sleep. *J. Rheumatol. (Suppl.) 16*: 52–57.

Krueger, J. M., Pappenheimer, J. R., and Karnovsky, M. L. (1982a). The composition of sleep-promoting factor isolated from human urine. *J. Biol. Chem. 257*: 1664–1699.

Krueger, J. M., Pappenheimer, J. R., and Karnovsky, M. L. (1982b). Sleep-promoting effects of muramyl peptides. *Proc. Natl. Acad. Sci. U.S.A. 79*: 6102–6106.

Krueger, J. M., Karnovsky, M. L., Martin, S. A., Pappenheimer, J. R., Walter, J., and Biemann, K. (1984a). Peptidoglycans as promoters of slow-wave sleep. II. Somnogenic and pyrogenic activities of some naturally occurring muramyl peptides; correlations with mass spectrometric structure determination. *J. Biol. Chem. 259*: 12659–12662.

Krueger, J. M., Walter, J., Karnovsky, M. L., Chedid, L., Choay, J. P., Lefrancier, P., and Lederer, E. (1984b). Muramyl peptides. Variation of somnogenic activity with structure. *J. Exp. Med. 159*: 68–76.

Krueger, J. M., Rosenthal, R. S., Martin, S. A., Walter, J., Davenne, D., Shoham, S., Kubillus, S. L., and Biemann, K. (1987). Bacterial peptidoglycans as modulators of sleep. I. Anhydro forms of muramyl peptides enhance somnogenic potency. *Brain Res. 403*: 249–257.

Luker, K. E., Collier, J. L. Kolodziej, E. W., Marshall, G. R., and Goldman, W. E. (1993a). *Bordetella pertussis* tracheal cytotoxin and other muramyl peptides: distinct structure–activity relationships for respiratory epithelial cytopathology. *Proc. Natl. Acad. Sci. U.S.A. 90*: 2365–2369.

Luker, K. E., Collier, J. L., Marshall, G. R., and Goldman, W. E. (1993b). Structural requirements for activity of *Bordetella pertussis* tracheal cytotoxin define a new class of muramyl peptide interactions. *Abstr. Annu. Meet. Am. Soc. Microbiol. 93*: 32 (Abstr. B-35).

Maier, J. A. M., Voulalas, P., Roeder, D., and Maciag, T. (1990). Extension of the life-span of human endothelial cells by an interleukin-1α antisense oligomer. *Science 249*: 1570–1574.

Mallory, R. B., and Hornor, A. A. (1912). Pertussis: the histological lesion in the respiratory tract. *J. Med. Res. 27*: 115–123.

Martin, S. A., Karnovsky, M. L., Krueger, J. M., Pappenheimer, J. R., and Biemann, K. (1984). Peptidoglycans as promoters of slow-wave sleep. I. Structure of the sleep-promoting factor isolated from human urine. *J. Biol. Chem. 259*: 12652–12658.

Melly, M. A., Gregg, C. R., and McGee, Z. A. (1981). Studies of toxicity of *Neisseria gonorrhoeae* for human fallopian tube mucosa. *J. Infect. Dis. 143*: 423–431.

Melly, M. A., McGee, Z. A., and Rosenthal, R. S. (1984). Ability of monomeric peptidoglycan fragments from *Neisseria gonorrhoeae* to damage human fallopian-tube mucosa. *J. Infect. Dis. 149*: 378–386.

Mizel, S. B., Kilian, P. L., Lewis, J. C., Paganelli, K. A., and Chizzonite, R. A. (1987). The interleukin 1 receptor. Dynamics of interleukin 1 binding and internalization in T cells and fibroblasts. *J. Immunol 138*: 2906–2912.

Nathan, C. (1992). Nitric oxide as a secretory product of mammalian cells. *FASEB J. 6*: 3051–3064.

Olson, L. C. (1975). Pertussis. *Medicine 54*: 427–469.

Oppenheim, J. J., Togawa, A., Chedid, L., and Mizel, S. (1980). Components of mycobacteria and muramyl dipeptide with adjuvant activity induce lymphocyte activating factor. *Cell. Immunol. 50*: 71–81.

Palacios, M., Knowles, R. G., and Moncada, S. (1992). Enhancers of nonspecific immunity induce nitric oxide synthase: induction does not correlate with toxicity or adjuvancy. *Eur. J. Immunol. 22*: 2303–2307.

Palmer, R. M., Bridge, L., Foxwell, N. A., and Moncada, S. (1992). The role of nitric oxide in endothelial cell damage and its inhibition by glucocorticoids. *Br. J. Pharm. 105*: 11–12.

Pappenheimer, J. R., Koski, G., Fencl, V., Karnovsky, M. L., and Krueger, J. (1975). Extraction of sleep-promoting factor S from cerebrospinal fluid and from brains of sleep-deprived animals. *J. Neurophysiol. 38*: 1299–1311.

Roberts, A. B., Vodovotz, Y., Roche, N. S., Sporn, M. B., and Nathan, C. F. (1992). Role of nitric-oxide in antagonistic effects of transforming growth-factor-β and interleukin-1-β on the beating rate of cultured cardiac myocytes. *Mol. Endocrinol. 6*: 1921–1930.

Rosenthal, R. S., Nogami, W., Cookson, B. T., Goldman, W. E., and Folkening, W. J. (1987).

Major fragment of soluble peptidoglycan released from growing *Bordetella pertussis* is tracheal cytotoxin. *Infect. Immun. 55:* 2117–2120.

Schleifer, K. H., and Kandler, O. (1972). Peptidoglycan types of bacterial cell walls and their taxonomic implications. *Bacteriol. Rev. 36:* 407–477.

Silverman, D. H. S., Wu, H., and Karnovsky, M. L. (1985). Muramyl peptides and serotonin interact at specific binding sites on macrophages and enhance superoxide release. *Biochem. Biophys. Res. Commun. 131:* 1160–1167.

Silverman, D. H. S., Imam, K., and Karnovsky, M. L. (1989). Muramylpeptide/serotonin receptors in brain-derived preparations. *Peptide Res. 2:* 338–344.

Sinha, R. K., and Rosenthal, R. S. (1980). Release of soluble peptidoglycan from growing gonococci: demonstration of anhydro-muramyl-containing fragments. *Infect. Immun. 29:* 914–925.

Stuehr, D. J., and Marletta, M. A. (1987). Induction of nitrite/nitrate synthesis in murine macrophages by BCG infection, lymphokines, or intereferon-γ. *J. Immunol. 139:* 518–525.

Stuehr, D. J., and Nathan, C. F. (1989). Nitric oxide. A macrophage product responsible for cytostasis and respiratory inhibition in tumor target cells. *J. Exp. Med. 169:* 1543–1555.

Walter, J., Davenne, D., Shoham, S., Dinarello, C. A., and Krueger, J. M. (1986). Brain temperature changes coupled to sleep states persist during interleukin 1-enhanced sleep. *Am. J. Physiol. 250:* R96–R103.

Wilson, R., Read, R., Thomas, M., Rutman, A., Harrison, K., Lund, V., Cookson, B., Goldman, W., Lambert, H., and Cole, P. (1991). Effects of *Bordetella pertussis* infection on human respiratory epithelium in vivo and in vitro. *Infect. Immun. 59:* 337–345.

Zembowicz, A., and Vane, J. R. (1992). Induction of nitric oxide synthase activity by toxic shock syndrome toxin 1 in a macrophage-monocyte cell line. *Proc. Natl. Acad. Sci. U.S.A. 89:* 2051–2055.

18

Adenylyl Cyclase Toxin from *Bordetella pertussis*

Erik L. Hewlett and Nancy J. Maloney*

University of Virginia School of Medicine, Charlottesville, Virginia

I. INTRODUCTION

Bordetella pertussis produces an adenylyl cyclase toxin (AC toxin) that is primarily extracytoplasmic in location. This novel toxin is an adenylyl cyclase, some or all of which enters target cells, is activated by endogenous target cell calmodulin, and increases intracellular cAMP to supraphysiologic levels. The activities of this AC toxin can be separated into its catalytic or enzyme activity (ability to convert ATP to cAMP), its

---

Current affiliation: Medical College of Virginia, Virginia Commonwealth University, Richmond, Virginia.

invasive or toxin activity (the ability to enter target cells and raise intracellular cAMP concentration), its hemolytic activity (ability to hemolyze red blood cells), and its pore-forming activity (ability to produce ion conductance in a lipid bilayer). In this chapter, the discovery of the various activities of this toxin is reviewed along with its operon organization, the present understanding of its structure and function, and its homology with other bacterial toxins. Finally, the role of AC toxin in the pathogenesis of whooping cough and its potential usefulness in an acellular vaccine are discussed.

II. DISCOVERY OF *B. PERTUSSIS* ADENYLYL CYCLASE AND RECOGNITION OF TOXIN ACTIVITY

The adenylyl cyclase activity associated with *Bordetella pertussis* was first observed by two groups of investigators in the 1970s. Fishel et al. (1970) reported adenylyl cyclase activity in a *B. pertussis* extract while studying the basis for histamine hypersensitivity in mice vaccinated against *B. pertussis*. That observation did not yield further information on AC toxin, but independently Wolff and Cook (1973) detected AC activity in a commercial pertussis vaccine. In the pursuit of the latter observation, Hewlett and Wolff (1976) purified soluble adenylyl cyclase with mass of ~70 kDa from the culture medium of *B. pertussis*. During characterization of that activity and investigation of its source, Hewlett et al. (1976) noted that the majority of the enzyme was cell-associated but extracytoplasmic and trypsin-sensitive. The extracytoplasmic location and the accumulation in the medium suggested possible toxin activity, but, for reasons that became clear subsequently, the 70-kDa purified molecule had no effect on cAMP levels of target cells (unpublished data). Later studies by Cowell et al. (1979) demonstrated that at least a portion of the cell-associated enzyme was located on the external side of the bacterial cytoplasmic membrane. The adenylyl cyclase activity was increased 100–1000-fold by a eukaryotic protein (Hewlett et al., 1978, 1979a,b) that was later identified as calmodulin (Berkowitz et al., 1980; Wolff et al., 1980; Goldhammer et al., 1981). Goldhammer and Wolff (1981) demonstrated that the activation of AC toxin was calmodulin-concentration-dependent and could, in fact, be used as an assay for calmodulin. AC toxin was also activated by a variety of detergents and phospholipids. In these experiments, phosphatidylglycerol was the most effective, eliciting a fivefold increase in AC toxin enzymatic activity alone and potentiating the activation by calmodulin (Wolff and Cook, 1982). The enzymatic activities of similar, but probably not identical, AC toxins from *B. parapertussis* and *B. bronchiseptica* have been described (Endoh et al., 1980). In the case of *B. bronchiseptica*, a large amount of the activity is present in the culture medium of exponentially growing organisms (Endoh et al., 1980; Maoury et al., unpublished data).

The toxin activity of *B. pertussis* AC toxin was first reported by Confer and Eaton (1982). This important discovery was based on two background observations in addition to the information described above. First, earlier work by Utsumi et al. (1978) had shown that urea extracts from *B. pertussis* could inhibit chemotaxis and oxygen consumption by neutrophils. Subsequently, Hoidal et al. (1978) reported impaired stimulation of oxygen consumption, chemiluminescence, and glucose oxidation in alveolar macrophages from rabbits infected with *B. bronchiseptica*, a related animal pathogen. These defects were reversed after clearance of the infection. In attempting to explain the suppression of these macrophage activities, Confer and Eaton demonstrated that extracts from *Bordetella pertussis* contained high levels of adenylyl cyclase activity and caused an increase in intracellular cAMP in a time- and concentration-dependent manner in human neutrophils and alveolar macrophages. The degree of inhibition of oxidative activity appeared to be correlated with cAMP accumulation, in keeping with earlier recognition of the inhibitory effects of cAMP on phagocytic cell functions (Bergman et al., 1978). Since this initial report of toxin activity, there have been many studies of intoxication by AC toxin in a wide variety of target cells. As expected, the biological effects of AC toxin vary depending on the target cell because of the wide assortment of responses elicited by increased intracellular cAMP concentrations. As will be discussed in Section V, it is believed that the role of AC toxin in the pathogenesis of whooping cough is principally to inhibit immune effector cells by increasing intracellular cAMP. This phenomenon is illustrated by the studies of Confer and Eaton (1982) described above. Similarly, Pearson et al. (1987) demonstrated that cAMP accumulation generated by partially purified AC toxin inhibits functions of human mononuclear phagocytes including ingestion of particulate stimuli and production of superoxide and other reactive molecules.

It should be noted that when used as a laboratory reagent, AC toxin is a promiscuous toxin that is able to intoxicate a wide range of target cells, most of which are not likely to be relevant to the pathogenesis of whooping cough (Hewlett and Gordon, 1988). For example, by increasing cAMP levels, AC toxin stimulates hormone release from rat pituitary cells (Cronin et al., 1986), increases the rate of contraction of chick heart cells (Selfe et al., 1987), and elicits a calcium current in frog heart cells (Hewlett et al., 1993). A wide range of cultured cell lines have been screened for sensitivity to AC toxin (Gordon, 1989). Several, including human erythrocytes and EL-4, a mouse lymphoma, have been found to be relatively resistant to intoxication; that is, there is little or no accumulation of cAMP (Hanski and Farfel, 1985; Maloney et al., 1992; Maloney, 1992). There is no known specific receptor for AC toxin, and the mechanism by which the toxin enters target cells is unknown.

III. ORGANIZATION AND REGULATION OF THE *cya* OPERON

Since the first AC isolated from *B. pertussis* was the 70-kDa molecule with enzyme activity only (Hewlett and Wolff, 1976), the size of the intact holotoxin was unknown until the mid-1980s. As will be discussed in Section IV, a number of forms of AC toxin with different apparent sizes have been isolated and studied (Hewlett and Gordon, 1988). With the cloning and sequencing of the AC toxin structural gene, *cyaA*, Glaser et al. (1988a) established that the toxin consisted of 1706 amino acids (177 kDa). Because of the smaller size associated with the catalytic activity in earlier reports (Hewlett and Wolff, 1976; Shattuck et al., 1985; Ladant et al., 1986; Kessin and Franke, 1986), the 177-kDa molecule was described as a "precursor form" by Glaser et al. (1988a). In a footnote to that publication, it was stated that a portion of the amino acid sequence of CyaA possesses significant sequence homology to the gene product of *hlyA*, *E. coli* hemolysin. A subsequent study investigating the secretion of AC toxin (referred to as "cyclolysin" by Glaser et al., 1988b) yielded the identification of additional genes (*cyaB*, *cyaD*, and *cyaE*), which are downstream from *cyaA* and are required for secretion of the toxin (Fig. 1). The gene products of *cyaB* and *cyaD* are highly similar to proteins HlyB and HlyD, which are required for *E. coli* hemolysin secretion. An upstream gene, *cyaC*, which encodes a protein apparently involved in posttranslational modification (activation) of AC toxin and is similar to an activator molecule for *E. coli* hemolysin, HlyC, was discovered by Barry et al. (1991). The gene (*cyaC*) is transcribed in the opposite direction from the remainder of the *cya* operon. Thus, the synthesis, activation, and release of AC toxin

Fig. 1 Schematic of the *cya* operon from *B. pertussis*. Numbers below the line indicate the calculated size of the protein encoded by each gene. Arrows indicate direction of transcription.

appears to be dependent on an operon consisting of five genes (Fig. 1). Transcription of the four genes *CyaA*, *B*, *D*, and *E* is regulated by a promoter upstream from *cyaA*, but there appears to be a concurrent constitutive, low-level expression of *cyaB*, *D*, and *E* from a separate promoter (Laoide and Ullmann, 1990).

The operon is registered by the *bvg* locus, which controls coordinate expression of a collection of virulence factors, including pertussis toxin, filamentous hemagglutinin (FHA), fimbriae, pertactin, and others (Arico et al., 1989). This system was first identified in *B. pertussis* by loss of multiple virulence factors in conjunction with a single transposon insertion into what was then termed the *"vir"* locus (Weiss et al., 1983). The *bvg* regulation means that when *B. pertussis* organisms are placed in a variety of culture conditions, they cease to produce AC toxin and other *bvg*-regulated factors (Idigbe et al., 1981; McPheat et al., 1983; Roy et al., 1990). Scarlato et al. (1992) demonstrated, however, that activation of *bvg* by a return to optimal temperature or by other manipulation of the environment to remove inhibitory agents does not result in simultaneous onset of expression of all genes. FHA and fimbriae are expressed within a few minutes, whereas production of the toxins such as AC toxin and pertussis toxin is delayed 4–6 h. These observations suggest that attachment of the organism is the first order of priority and generation of toxins to disarm the host defenses and cause disease is delayed.

IV. STRUCTURE AND FUNCTION OF ADENYLYL CYCLASE TOXIN

A. Structural Domains and Similarities with Other Toxins

Elucidation of the amino acid sequence of AC toxin from the cloned DNA by Glaser et al. (1988a) provided an important perspective for study of the toxin structure and function. Two major regions or domains were described: (1) a catalytic domain consisting of the first 400 amino acids and (2) the COOH-terminal 1300 amino acids, which include an alanine/glycine-rich hydrophobic region with the last 700 amino acids containing glycine aspartate-rich repeats, similar to elements present in a group of bacterial cytolytic exotoxins, which are provided by gram-negative organisms (termed RTX proteins for repeat in toxin) (Welch, 1991; Coote, 1992). This distal portion of AC toxin shows 25% similarity with *E. coli* hemolysin, the prototype of the RTX protein family (Glaser et al. 1988a,b; Welch, 1991). Other toxins in this family include *Pasteurella haemolytica* leukotoxin, *Actinobacillus pleuropneumonia* hemolysin, and *A. actinomycetemcomitans* leukotoxin. The toxins in the RTX family including AC toxin share several common elements; they (1) contain glycine-rich repeat regions (Leu-X-Gly-Gly-X-Gly-Asn-Asp), (2) have hemolytic or leukotoxic activity, (3) have genes necessary for secretion in the same operon as the structural gene; and (4) do not have an amino-terminal leader sequence for secretion targeting. These similarities are also reflected in immunological cross-reactivity between AC toxin and several other members of the RTX family (Ehrmann et al., 1991). Recently immunologically related molcules without clear toxin activity, such as an iron-regulated protein, FrpA, from *N. meningitidis* (Thompson et al., 1993), have been identified.

AC toxin from *B. pertussis* is one of only two known bacterial toxins with endogenous AC enzymatic activity. The other is produced by *Bacillus anthracis*, the gram-positive organism that is the etiologic agent of anthrax (Leppla, 1982). Although catalyzing the same reaction intracellularly, namely calmodulin-activated cyclic AMP accumulation, the AC toxin from *B. anthracis* is different structurally and in its entry mechanism (Leppla, 1982; Gordon et al., 1988, 1989). The catalytic domain resides in an 88.8-kDa protein known as edema factor (EF), which requires a separate proteolytically activated protein, protective antigen (PA), for its interaction with and entry into the target cell (Leppla, 1982; Klimpel et al., 1992).

B.   Enzyme Activity

The ability of *B. pertussis* organisms to catalyze cAMP formation from exogenous ATP provided the basis for discovery of AC toxin (Wolff and Cook, 1973; Hewlett and Wolff, 1976). The activity is, however, novel in that it is increased 100–1000-fold by the eukaryotic calcium-dependent regulatory protein calmodulin (Berkowitz et al., 1980; Goldhammer et al., 1981). Importantly, this activation can occur in the absence of calcium, unlike other recognized actions of calmodulin (Greenlee et al., 1982; Kilhoffer et al., 1983). The interaction between AC toxin (or smaller proteolytic fragments containing the calmodulin-binding site) and calmodulin has provided (1) a means for assaying calmodulin activity in vitro (Goldhammer and Wolff, 1982), (2) an approach to cloning the toxin in an *E. coli* strain that is expressing calmodulin (Glaser et al., 1988a), and (3) a basis for affinity purification of AC toxin (Ladant et al., 1986; Hewlett et al., 1988, 1991; Friedman, 1987; Rogel et al., 1989).

The location of the calmodulin-binding domains has been pursued using site-directed mutagenesis of the 43-kDa fragment of the holotoxin molecule expressed in *E. coli*. Glaser et al. (1989) demonstrated a gross reduction in calmodulin binding by substitutions for tryptophan 242. This work, confirmed by Gross et al. (1992), supported the concept that the calmodulin-binding site and the catalytic domain are located between amino acids 1 and 399 (Ladant et al., 1986). It appears, however, that calmodulin may interact with the 43-kDa fragment at more than one site. Ladant (1988) showed subsequently that in the presence of calmodulin, the 43-kDa peptide can be cleaved by trypsin into 25-kDa and 18-kDa peptides that remain associated in the continued presence of calmodulin.

Residues involved in the catalytic active site have been identified by Glaser et al. (1989) and Au et al. (1989). Both of these studies showed that replacement of lysine 58 resulted in loss of enzyme activity. Au et al. (1989) also showed reduced catalytic activity with substitution at lysine 65. Substitution of a lysine for methionine at position 58 in *B. pertussis* and evaluation of the activity of the resultant organism (Ehrmann et al., 1992) led to a better understanding of the role of the enzyme activity in other activities of the molecule, as described below.

C.   Toxin Activity

Prior to the cloning and sequencing of the AC toxin structural gene (*cyaA*) by Glaser et al. (1988a), there was uncertainty about the size of the holotoxin molecule (summarized in Hewlett and Gordon, 1988). It is now clear that the product of the *cyaA* gene is a protein of ~177 kDa (1706 amino acids) that possesses toxin activity (the ability to increase cAMP levels in intact target cells). The apparent mass of this molecule as determined by SDS-PAGE, however, is 200–216 kDs (Hewlett et al., 1991; Masure and Storm, 1989; Rogel et al., 1989). The difference from its calculated mass is apparently due to incomplete denaturation in SDS, since its sedimentation properties in sucrose are consistent with the calculated size of the molecule (Gentile et al., 1990).

Although the calmodulin activation of AC enzyme activity can occur in the absence of calcium, the toxin activity is clearly calcium-dependent (Confer et al., 1984; Hanski and Farfel, 1985; Gentile et al., 1990; Hewlett et al., 1991). When addressing this calcium dependence, Hanski and Farfel (1985) noted a reduction in the apparent mass of AC toxin on Ultragel ACA 34 chromatography from 340 kDa to 190 kDa with addition of calcium. In contrast, Masure et al. (1988) observed a greater molecular mass and Stokes radius of the 45-kDa catalytic fragment in the presence of calcium. Most recently, Hewlett et al. (1991) demonstrated that AC toxin is a calcium-binding protein that undergoes a conformational change in the presence of calcium at concentrations required for its toxin activity (Fig. 2). Although the carboxy-terminal domain of the molecule contains a series of glycine-rich repeats, which are analogous to those present in other proteins that bind calcium (Welch et al., 1986), the location, nature, and stoichiometry of the calcium interaction(s) are unknown.

The mechanisms by which AC toxin binds to and enters its target cell are complex, as illustrated by the sequence of steps illustrated in Figure 3. Although productive interaction of AC toxin with a target cell (which goes on to intoxication) requires calcium, there is substantial binding or adsorption in the absence of calcium. Binding that has occurred under these conditions cannot yield intoxication even when calcium is added subsequently (Maloney et al., 1992). This observation is in contrast to that of Rogel and Hanski (1992), who found that binding at low levels of calcium was followed by

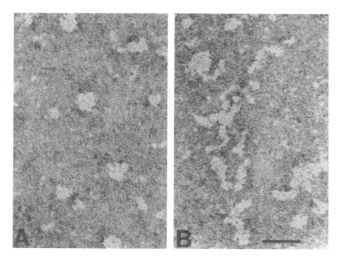

Fig. 2 Conformational change elicited in AC toxin by interaction with calcium. Electron micrographs show AC toxin from BP338 in the absence (A) or presence (B) of 2 mM calcium. Samples were prepared by Margaretta Allietta with negative staining; magnification is 480,000, and bar in panel B is 20.8 nm.

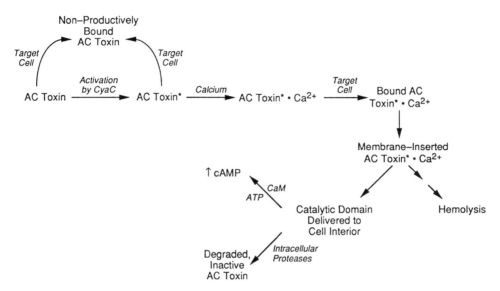

Fig. 3 Schematic representation of the process involved in AC toxin action. Activated toxin, as a result of the action of CyaC, is denoted by asterisk (*).

intoxication when additional calcium was provided. The difference between these results is probably explained by the absence of toxin-associated calcium in the studies of Maloney et al. (1992) due to toxin preparation in the presence of chelators (Hewlett et al., 1989, 1991). Productive binding also requires the posttranslational modification catalyzed by CyaC (Barry et al., 1991). The nature of this activation process has not been elucidated for AC toxin, but the activation of *E. coli* hemolysin by HlyC, a homologous protein in *E. coli*, appears to involve a covalent acylation (Hardie et al., 1991; Issartel et al., 1991).

There are no known "receptors" for AC toxin, although intoxication can be blocked by incubation of toxin with gangliosides (Gordon et al., 1988). Furthermore, protein-free multilamellar liposomes can be lysed by AC toxin, indicating that interaction with a lipid bilayer does not require a protein receptor. The current working model for productive binding involves attraction of the toxin to negatively charged lipids, perhaps via bound calcium ions, with guidance or orientation provided by the posttranslational modification. The toxin then inserts itself into the lipid bilayers with the catalytic domain still exposed to the medium. Without delay at 37°C, the catalytic domain is delivered to the cytoplasm, where it is activated by CaM to start producing cAMP from endogenous ATP. Interestingly, when a few cell types (such as human erythrocytes or the histiocytic lymphoma line, U-937) are used as targets, the addition of calmodulin to the AC toxin prior to addition to cells prevents intoxication, perhaps by blocking the "delivery" step (Shattuck and Storm, 1985; Gentile et al., 1988). It has been clearly demonstrated by several investigators that cell entry does not involve receptor-mediated endocytosis (Confer and Eaton, 1982; Hanski and Farfel, 1985; Gordon et al., 1988; Gentile et al., 1988; Donovan and Storm, 1990). This is of particular note because the adenylyl cyclase toxin from *Bacillus anthracis* does appear to be dependent on the receptor-mediated endocytosis process for intoxication (Gordon et al., 1988). Rogel and Hanski (1992) have demonstrated that although delivery of the catalytic domain to the target cell interior (translocation) is unaffected by treatment of the cells with trypsin, N-ethylmaleimide, or sodium carbonate, cleavage of the toxin to release a 45-kDa catalytic fragment into the cytosol is blocked by N-ethylmaleimide. It is likely that the enzymatically active fragment and the remainder of the toxin molecule that is known to be very proteolytically sensitive (Hewlett et al., 1976) are rapidly degraded by cellular proteases. In fact, Gilboa-Ron et al. (1989) showed that lysates of lymphocytes degrade AC toxin by an ATP-dependent mechanism, but the relationship of that process of intracellular degradation remains speculative.

A somewhat different process for cell entry and intoxication is proposed by Storm and coworkers, who consider the holotoxin (216 kDa) to be a cell-associated precursor that is cleaved by bacterial protease to the 45-kDa form during release into the medium (Masure and Storm, 1989). They postulate that additional peptides that can be separated by a wheat germ agglutinin-affinity column are responsible for interaction with the 45-kDa catalytic domain to facilitate its entry (Donovan et al., 1989).

Although it is clear that extracted, intact holotoxin is capable of intoxication, the nature of the process of intoxication in vivo during infection remains to be determined. The majority of AC toxin is cell-associated when organisms are cultured in synthetic medium in vitro (Hewlett et al., 1976). Since there is not a large accumulation of toxin in the culture medium, as there is with pertussis toxin, a role of bacterium–target cell interaction in toxin delivery has been suggested (Pearson et al., 1990; Mouallem et al., 1990). The addition of intact *B. pertussis* organisms to target cells (Chinese hamster ovary cells or Jurkat cells) can result in intoxication, suggesting that contact-mediated delivery of toxin can occur and may be pathophysiologically relevant.

D. Hemolysin Activity

Colonies of phase 1 *Bordetella pertussis* are hemolytic on blood agar, but the basis for that hemolysis was previously unknown. The first clue to a relationship between AC toxin and hemolysis came with the transposon mutagenesis studies of Weiss et al. (1983) in which concurrent loss of AC activity and hemolysis was observed. Subsequently, Gordon et al. (1989) demonstrated that purified AC toxin would elicit marker release from multilamellar liposomes, suggesting possible lytic activity of the molecule. After the discovery of sequence homology between AC toxin and *E. coli* hemolysin by Glaser et al. (1988a,b), work from a number of laboratories demonstrated hemolytic activity. Ehrmann et al. (1991) and Rogel et al. (1991) showed that purified AC toxin is hemolytic for sheep erythrocytes, but with a time delay of 45–60 min, compared to the immediate onset demonstrated by *E. coli* hemolysin. Since cAMP accumulation has immediate onset, it appears that additional steps or processing are required for hemolysis to occur (Fig. 3). In fact, intoxication and hemolysis can be dissociated with the use of calcium and nickel (Hewlett et al., 1992). Calcium concentrations in excess of 1 mM inhibit hemolysis without affecting intoxication, whereas nickel inhibits intoxication but not hemolysis. A clear understanding of the relationship between these activities requires additional studies, which are currently under way. The catalytic activity of the AC toxin is not, however, required for hemolysis since enzyme activity can be eliminated by site-directed mutagenesis (Ehrmann et al., 1992; Gross et al., 1992) or deletion of the catalytic domain (Sakamoto et el., 1992), without reducing hemolytic activity. There is general agreement that hemolysis per se is not a significant activity of the AC toxin but rather reflects a mechanism acquired, perhaps via gene fusion, and adapted for delivery of the catalytic domain to its site of action at the cell interior (Bellalou et al., 1990; Ehrmann et al., 1991; Rogel et al., 1991; Sakamoto et al., 1992; Gross et al., 1992).

E. Ion-Permeable Pore Formation

Hemolysis of sheep erythrocytes by AC toxin appears to occur by osmotic lysis, and protection against lysis by sugars of different sizes implicate a pore of <0.6 nm produced by the toxin in this process (Ehrmann et al., 1991). The hypothesis that toxin insertion may create an ion-permeable structure in membranes has been evaluated using a phospholipid bilayer system and measurement of transmembrane conductance (Szabo et al., 1994). Purified AC toxin elicits conductance across a bilayer composed of phosphatidylcholine/phosphatidylserine in a calcium- and polarity-dependent manner. The apparent pore formation does not occur with mutants of AC toxin that are not able to intoxicate cells by virtue of a defective CyaC or a deletion of the hydrophobic domain between amino acids 621 and 778. Since there is not generalized lysis of target cells (other than some erythrocytes) by AC toxin, it is hypothesized that pore formation may be only transient during the process of delivering the catalytic domain to the cell interior. Further studies of the relationship between toxin binding and insertion and pore formation are necessary to understand more clearly the distal sequence of events illustrated in Figure 3.

V. ROLE IN PATHOGENESIS AND AS A PROTECTIVE ANTIGEN

From the time of discovery of the toxin activity of *B. pertussis* AC, it appeared likely that its major role in pertussis might be inactivation of host phagocytic cell function (Weiss et al., 1986). The superoxide generation by neutrophils and alveolar macrophages in response to soluble and particulate stimuli is virtually abolished by intoxication (Confer and Eaton, 1982). In addition, exposure of human peripheral blood monocytes to AC toxin results in impaired phagocytic responses for several stimuli (Pearson et al., 1987). These data are consistent with the clinical observation that secondary infections are a common cause of morbidity and mortality in pertussis patients (Cherry et al., 1988) but do not establish that AC toxin alone is responsible. Pertussis toxin, for example, is also able to alter immune effector cell function (Meade et al., 1984a,b). Furthermore, there are no animal or human data to document that this activity of the toxin occurs in vivo.

As noted above, AC toxin is predominantly cell-associated and does not accumulate in culture medium as does pertussis toxin. This observation suggests that the primary effects of the toxin would be seen in host cells with which the *B. pertussis* organism has direct contact. In addition to inflammatory cells responding to the localized infection in the respiratory tract, there may be intoxication of other respiratory epithelial cells and goblet cells. Increased cAMP levels could enhance mucous secretion, as occurs in the GI tract, resulting in the exaggerated quantity of secretions well recognized in children with pertussis (Hewlett, 1994). There are, however, no data to support this latter hypothesis or any suggestion that AC toxin is disseminated systematically, as is likely the case for pertussis toxin.

Weiss et al. (1983, 1984) demonstrated that AC toxin is a virulence factor for *B. pertussis* using *Tn*-5 mutants defective in toxin production that were unable to elicit a lethal infection in suckling mice. Since that time, infections with AC toxin-deficient mutants have been characterized more extensively (Weiss and Goodwin, 1989). Goodwin and Weiss (1990) observed enhanced clearance of AC toxin-deficient *B. pertussis* from the lungs of infant mice, suggesting a role for the toxin in establishing infection. More recently, using site-directed mutagenesis, several groups have demonstrated that catalytic adenylate cyclase activity and the hemolytic activity of AC toxin are both required for *B. pertussis* virulence in murine respiratory infection models (Ehrmann et al., 1992; Gross et al., 1992; Khelef et al., 1992). Hemolytic activity alone was not sufficient to confer virulence, indicating that it is the ability of the hemolytic fragment to deliver the catalytic domain to the target cell interior that is critical to the contribution of this toxin to virulence.

With the recognition that *B. pertussis* can enter and survive within nonphagocytic cells (Ewanowich et al., 1989; Lee et al., 1990; Mouallem et al., 1990) and phagocytic cells (Friedman et al., 1992), a role for AC toxin in that process was proposed. The results from several laboratories suggest some interesting hypotheses. Ewanowich et al. (1989) found that FHA-deficient mutants were unable to enter HeLa cells, presumably because of impaired binding, but AC toxin-deficient mutants were enhanced (twofold relative to wild type) in cell entry. Lee et al. (1990) observed that an AC toxin mutant was comparable to wild-type organisms in invading HeLa cells. Friedman et al. (1992), on the other hand, found a marked decrease in invasion and intracellular survival of *B. pertussis* with defective AC toxin in human peripheral blood macrophages. Perhaps the increased cAMP levels that are known to result from intoxication even at the time of bacterium–cell contact (Pearson et al., 1987; Mouallem et al., 1990) impair the entry process into nonphagocytic cells, explaining the increased invasiveness of AC toxin-deficient organisms. On the other hand, the ability of oxidatively active phagocytes to kill ingested bacteria is clearly inhibited by AC toxin-induced cAMP accumulation (Confer and Eaton, 1982; Pearson et al., 1987), and AC toxin may well provide a survival advantage for interaction with those cells.

Patients in their convalescent phase following infection with *B. pertussis* and some recipients of whole-cell pertussis vaccine produce antibodies to AC toxin (Farfel et al., 1990; Arciniega et al., 1991). Although in humans the contribution of those antibodies to protection is yet to be demonstrated, several animal studies suggest that AC toxin may be a protective antigen. Passive or active immunization with the catalytic domain provides some protection against infection by *B. pertussis* in intranasal and intracerebral mouse models (Guiso et al., 1989). Most recently, Khelef et al. (1992) showed protective effects of holotoxin or the catalytic fragment from *B. pertussis* as well as *B. parapertussis* against challenge with the homologous species. They noted that the mouse virulent strain of *B. pertussis* (18323) expressed a higher level of AC enzymatic activity. That this antigen may be a protective antigen for inclusion in acellular pertussis vaccine for humans remains to be determined, limited at present by availability of a sufficient quantity of purified material for evaluation. It is likely, however, that such a possibility will be evaluated in the near future.

VI. SUMMARY AND CONCLUSIONS

Adenylyl cyclase toxin is an important virulence factor for *Bordetella pertussis*, probably by its ability to disable host immune effector cells. It has been demonstrated to be related by sequence homology and immunological cross-reactivity with other members of the RTX family of bacterial toxins, especially *E. coli* hemolysin. It is therefore serving as a prototype for the study of these and other bacterial toxins that insert themselves into the cytoplasmic membrane of target cells and deliver a domain or subunit to the cell interior to carry out a catalytic function. By virtue of its adenylyl cyclase enzymatic activity, it is also a novel reagent for the study of cAMP-dependent processes in eukaryotic cells. In light of the recent data suggesting that adenylyl cyclase toxin may be another protective antigen against *B. pertussis* infection, it will be evaluated in the future for possible inclusion in component or acellular pertussis vaccines.

NOTE ADDED IN PROOF

Hackett et al. have recently identified the post-translational modification which is dependent on CyaC, the protein product of *cyaC* (Hackett et al., 1994). Using mass spectrometry, wild type AC toxin was found to contain a palmitoylation on the epsilon amino group of lys^{983}, while toxin from the strain in which *cyaC* was disrupted by a 4bp insertion (BPDE386) was unmodified. These data suggest that this acylation is, in fact, required for toxin and hemolytic activities of this molecule, but leave unresolved the mechanism by which this occurs.

REFERENCES

Arciniega, J. L., Hewlett, E. L., Johnson, F. D., Deforest, A., Wassilak, S. S. F., Onorato, I. M., Manclark, C. R., and Burns, D. L. (1991). Human serologic response to envelope-associated proteins and adenylate cyclase toxin of *Bordetella pertussis*. *J. Infect. Dis. 163*: 135–142.

Arico, B., Miller, J. F., Roy, C., Stibitz, S., Monack, D., Falkow, S., Gross, R., and Rappuoli, R. (1989). Sequences required for expression of *Bordetella pertussis* virulence factors share homology with prokaryotic signal transduction proteins. *Proc. Natl. Acad. Sci. U.S.A. 86*: 6671–6675.

Au, D. C., Masure, H. R., and Storm, D. R. (1989). Site-directed mutagenesis of lysine 58 in a putative ATP-binding domain of the calmodulin-sensitive adenylate cyclase from *Bordetella pertussis* abolishes catalytic activity. *Biochemistry 28*:2772–2776.

Barry, E. M., Weiss, A. A., Ehrmann, S. E., Gray, M. C., Hewlett, E. L., and Goodwin, M. S. (1991). *Bordetella pertussis* adenylate cyclase toxin and hemolytic activities require a second gene, *cyaC*, for activation. *J. Bacteriol. 173*: 720–726.

Bellalou, J., Sakamoto, H., Ladant, D., Geoffroy, C., and Ullmann, A. (1990). Deletions affecting hemolytic activity and toxin activity of *Bordetella pertussis* adenylate cyclase. *Infect. Immun. 58*: 3242–3247.

Bergman, M. J., Guerrant, R. L., Murad, F., Richardson, S. H., and Weaver, D. (1978). Interaction of polymorphonuclear neutrophils with *Escherichia coli*: effect of enterotoxin on phagocytosis, killing, chemotaxis and cyclic AMP. *J. Clin. Invest. 61*: 227–234.

Berkowitz, S. A., Goldhammer, A. R., Hewlett, E. L., and Wolff, J. (1980). Activation of prokaryotic adenylate cyclase by calmodulin. *Ann. N.Y. Acad. 1*: 356–360.

Cherry, J. D., Brunell, P. A., Golden, G. S., and Karzon, D. (1988). Report on the Pertussis Task Force. *Pediatrics 81*: 939–984.

Confer, D., and Eaton, J. (1982). Phagocyte impotence caused by invasive bacterial adenylate cyclase. *Science 217*: 948–950.

Confer, D., Slungaard, A., Graf, E., Panter, S., and Eaton, J. (1984). *Bordetella* adenylate cyclase toxin: entry of bacterial adenylate cyclase into mammalian cells. *Adv. Cyclic Nucleotide, Protein Phos. Res. 17*: 183–187.

Coote, J. G. (1992). Structural and functional relationships among the RTX toxin determinants of gram-negative bacteria. *FEMS Microbiol. Rev. 88*: 137–162.

Cowell, J., Hewlett, E., and Manclark, C. (1979). Intracellular localization of dermonecrotic toxin of *Bordetella pertussis*. *Infect. Immun. 25*: 896–901.

Cronin, M. J., Evans, W. S., Rogol, A. D., Weiss, A. A., Thorner, M. O., Orth, D. N.,

Nicholson, W. E., Yasumoto, T., and Hewlett, E. L. (1986). Prokaryotic adenylate cyclase toxin stimulates anterior pituitary cells in culture. *Am. J. Physiol.* *251*: 164–171.

Donovan, M. G., and Storm, D. R. (1990). Evidence that the adenylate cyclase secreted from *Bordetella pertussis* does not enter animal cells by receptor-mediated endocytosis. *J. Cell. Physiol.* *145*: 444–449.

Donovan, M. G., Masure, H. R., and Storm, D. R. (1989). Isolation of a protein fraction from *Bordetella pertussis* that facilitates entry of the calmodulin-sensitive adenylate cyclase into animal cells. *Biochemistry* *28*: 8124–8129.

Ehrmann, I., Gray, M., Gordon, V., Gray, L., and Hewlett, E. L. (1991). Hemolytic activity of adenylate cyclase toxin from *Bordetella pertussis*. *FEBS* *278*: 79–83.

Ehrmann, I., Weiss, A., Goodwin, M. S., Barry, E., and Hewlett, E. L. (1992). Enzymatic activity of adenylate cyclase toxin from *Bordetella pertussis* is not required for hemolysis. *FEBS* *304*: 51–56.

Endoh, M., Takezawa, T., and Nakase, Y. (1980). Adenylate cyclase activity of *Bordetella* organisms. *Microbiol. Immunol.* *24*: 95–104.

Ewanowich, C., Melton, A., Weiss, A., Sherburne, R., and Peppler, M. (1989). Invasion of HeLa 229 cells by virulent *Bordetella pertussis*. *Infect. Immun.* *57*: 2698–2704.

Farfel, Z., Könen, S., Wiertz, E., Klapmuts, R., Addy, P., and Hanski, E. (1990). Antibodies to *Bordetella pertussis* adenylate cyclase are produced in man during pertussis infection and after vaccination. *J. Med. Microbiol.* *32*: 173–177.

Fishel, C. W., Cronholm, L. S., and Keller, K. F. (1970). The influence of *Bordetella pertussis* on the adenylate cyclase and phosphodiesterase enzymes of mouse tissue. In *International Symposium on Pertussis*, S. Karger, New York, 1970, pp. 190–197.

Friedman, R. (1987). *Bordetella pertussis* adenylate cyclase: isolation and purification by calmodulin-Sepharose 4B chromatography. *Infect. Immun.* *55*: 129–134.

Friedman, R. L., Nordensson, K., Wilson, L., Akporiaye, E. T., and Yocum, D. E. (1992). Uptake and intracellular survival of *Bordetella pertussis* in human macrophages. *Infect. Immun.* *60*: 4578–4585.

Gentile, F., Raptis, A., Knipling, L., and Wolff, J. (1988). *Bordetella pertussis* adenylate cyclase: penetration into host cells. *Eur. J. Biochem.* *175*: 447–453.

Gentile, F., Knipling, L. G., Sackett, L., and Wolff, J. (1990). Invasive adenylyl cyclase of *Bordetella pertussis*. *J. Biol. Chem.* *265*: 10686–10692.

Gilboa-Ron, A., Rogel, A., and Hanski, E. (1989). *Bordetella pertussis* adenylate cyclase inactivation by the host cells. *Biochem. J.* *262*: 25–31.

Glaser, P., Ladant, D., Sezer, O., Pichot, F., Ullmann, A., and Danchin, A. (1988a). The calmodulin-sensitive adenylate cyclase of *Bordetella pertussis*: cloning and expression in *Escherichia coli*. *Mol. Microbiol.* *2*: 19–30.

Glaser, P., Sakamoto, H., Bellalou, J., Ullmann, A., and Danchin, A. (1988b). Secretion of cyclolysin, the calmodulin-sensitive adenylate cyclase-haemolysin bifunctional protein of *Bordetella pertussis*. *EMBO J.* *7*: 3997–4004.

Glaser, P., Elmaoglou-Lazaridou, A., Krin, E., Ladant, D., Barzu, O., and Danchin, A. (1989). Identification of residues essential for catalysis and binding of calmodulin in *Bordetella pertussis* adenylate cyclase by site-directed mutagenesis. *EMBO J.* *8*: 967–972.

Goldhammer, A., and Wolff, J. (1981). Assay of calmodulin with *Bordetella pertussis* adenylate cyclase. *Anal. Biochem.* *124*: 45–52.

Goldhammer, A. R., Wolff, J., Hope Cook, G., Berkowitz, S. A., Klee, C. B., Manclark, C. R., and Hewlett, E. L. (1981). Spurious protein activators of *Bordetella pertussis* adenylate cyclase. *Eur. J. Biochem.* *115*: 605–609.

Goodwin, M., and Weiss, A. A. (1990). Adenylate cyclase toxin is critical for colonization and pertussis toxin is critical for lethal infection by *Bordetella pertussis* in infant mice. *Infect. Immun.* *58*: 3445–3447.

Gordon, V. (1989). A comparative study of the calmodulin-activated adenylate cyclase toxins from *Bordetella pertussis* and *Bacillus anthracis*: interactions with target cells. Doctoral dissertation, Dept. of Pharmacology, Univ. Virginia, Charlottesville, VA.

Gordon, V., Leppla, S., and Hewlett, E. (1988). Inhibitors of receptor-mediated endocytosis block entry of *Bacillus anthracis* adenylate cyclase toxin but not that of *Bordetella pertussis* adenylate cyclase toxin. *Infect. Immun. 56*: 1066–1069.

Gordon, V. M., Young, W. W., Lechler, S. M., Leppla, S. H., and Hewlett, E. L. (1989). Adenylate cyclase toxins from *Bacillus anthracis* and *Bordetella pertussis*: different processes for interaction with and entry into target cells. *J. Biol. Chem. 264*: 14792–14796.

Greenlee, D., Andreasen, T., and Storm, D. (1982). Calcium-independent stimulation of *Bordetella pertussis* adenylate cyclase by calmodulin. *Biochemistry 21*: 2759–2764.

Gross, M. K., Au, D. C., Smith, A. L., and Storm, D. R. (1992). Targeted mutations that ablate either the adenylate cyclase or hemolysin function of the bifunctional cyaA toxin of *Bordetella pertussis* abolish virulence. *Proc. Natl. Acad. Sci. U.S.A. 89*: 4898–4902.

Guiso, N., Rocancourt, M., Szatanik, S., and Alonso, J. (1989). *Bordetella* adenylate cyclase is a virulence associated factor and an immunoprotective antigen. *Microb. Pathog. 7*: 373–380.

Hackett, M., Guo, L., Shabanowitz, J., Hunt, D. F., and Hewlett, E. L. (1994). Internal lysine palmitoylation in adenylate cyclase toxin from *Bordetella pertussis*. *Science 266*: 433–435.

Hanski, E., and Farfel, Z. (1985). *Bordetella pertussis* invasive adenylate cyclase. *J. Biol. Chem. 290*: 5526–5532.

Hardie, K. R., Issartel, J. P., Koronakis, E., Hughes, C., and Koronakis, V. (1991). In vitro activation of *Escherichia coli* prohaemolysin to the mature membrane-targeted toxin requires HlyC and a low molecular weight cytosol polypeptide. *Mol. Microbiol. 5*: 1669–1679.

Hewlett, E. L. (1994). Bordetella species. In *Principles and Practice of Infectious Diseases*, G. L. Mandell, R. G. Douglas, and J. E. Bennett (Eds.), Churchill Livingstone, New York, pp. 2078–2083.

Hewlett, E. L., and Gordon, V. M. (1988). Adenylate cyclase toxin of *Bordetella pertussis*. In *Pathogenesis and Immunity in Pertussis*, A. C. Wardlaw and R. Parton (Eds.), Wiley, New York, pp. 193–209.

Hewlett, E., and Wolff, J. (1976). Soluble adenylate cyclase from the culture medium of *Bordetella pertussis*: purification and characterization. *J. Bacteriol. 127*: 890–898.

Hewlett, E., Urban, M., Manclark, C., and Wolff, J. (1976). Extracytoplasmic adenylate cyclase of *Bordetella pertussis*. *Proc. Natl. Acad. Sci. U.S.A. 73*: 1926–1930.

Hewlett, E. L., Wolff, J., and Manclark, C. (1978). Regulation of *Bordetella pertussis* extracytoplasmic adenylate cyclase. *Adv. Cycl. Nucleotide Res. 9*: 621–628.

Hewlett, E., Underhill, L., Cook, G., Manclark, C., and Wolff, J. (1979a). A protein activator for the adenylate cyclase of *Bordetella pertussis*. *J. Biol. Chem. 254*: 5602–5605.

Hewlett, E. L., Underhill, L. H., Vargo, S. A., Wolff, J., and Manclark, C. (1979b). *Bordetella pertussis* adenylate cyclase: regulation of activity and its loss in degraded strains. In *International Symposium on Pertussis*, C. Manclark and J. Hill (Eds.), U.S. DHEW, Washington, DC.

Hewlett, E. L., Gordon, V. M., McCaffery, J., Sutherland, W., and Gray, M. C. (1989). *Bordetella pertussis* adenylate cyclase toxin: identification and purification of the holotoxin molecule. *J. Biol. Chem. 264*: 19379–19384.

Hewlett, E. L., Gray, L., Allietta, M., Ehrmann, I., Gordon, V. M., and Gray, M. C. (1991). Adenylate cyclase toxin from *Bordetella pertussis*: conformational change associated with toxin activity. *J. Biol. Chem. 266*: 17503–17508.

Hewlett, E. L., Ehrmann, I. E., Maloney, N. J., Fremgen, P. R., Barry, E. M., Weiss, A. A., and Gray, M. C. (1992). Adenylate cyclase toxin from *Bordetella pertussis*: characterization of toxin-catalyzed intoxication and hemolysis: In *Bacterial Protein Toxins*, Witholt et al. (Eds.), Gustav Fischer, Stuttgart, pp. 241–248.

Hewlett, E. L., Gray, M. C., Ehrmann, I. E., Maloney, N. J., Otero, A. S., Gray, L., Allietta, M., Szabo, G., Weiss, A. A., and Barry, E. M. (1993). Characterization of adenylate cyclase toxin from a mutant of *Bordetella pertussis* defective in the activator gene *cyaC. J. Biol. Chem. 268*: 7842–7848.

Hoidal, J., Beale, G., Rasp, F., Holmes, B., White, J. G., and Repine, J. (1978). Comparison of the metabolism of alveolar macrophages from humans, rats, and rabbits: response to heat-killed bacteria on phorbol myristate acetate. *J. Lab. Clin. Med. 92*: 787–794.

Idigbe, E. O., Parton, R., and Wardlaw, A. C. (1981). Rapidity of antigenic modulation of *Bordetella pertussis* in modified Hornibrook medium. *J. Med. Microbiol. 14*: 409–418.

Issartel, J., Vassilis, K., and Hughes, C. (1991). Activation of *Escherichia coli* prohaemolysin to the mature toxin by acyl carrier protein-dependent fatty acylation. *Nature 351*: 759–761.

Kessin, R., and Franke, J. (1986). Secreted adenylate cyclase of *Bordetella pertussis*: calmodulin requirements and partial purification. *J. Bacteriol. 166*: 290–296.

Khelef, N., Sakamoto, H., and Guiso, N. (1992). Both adenylate cyclase and hemolytic activities are required by *Bordetella pertussis* to initiate infection. *Microb. Pathog. 12*: 227–235.

Kilhoffer, M., Cook, G., and Wolff, J. (1983). Calcium-independent activation of adenylate cyclase by calmodulin. *Eur. J. Biochem. 133*: 11–15.

Klimpel, K. R., Molloy, S. S., Thomas, G., and Leppla, S. H. (1992). Anthrax toxin protective antigen is activated by a cell surface protease with the sequence specificity and catalytic properties of furin. *Proc. Natl. Acad. Sci. U.S.A. 89*: 10277–10281.

Ladant, D. (1988). Interaction of *Bordetella pertussis* adenylate cyclase with calmodulin. *J. Biol. Chem. 88*: 2612–2618.

Ladant, D., Brezin, C., Alonso, J., Crenon, I., and Guiso, N. (1986). *Bordetella pertussis* adenylate cyclase: purification, characterization and radioimmunoassay. *J. Biol. Chem. 261*: 16264–16269.

Laoide, B. M., and Ullmann, A. (1990). Virulence dependent and independent regulation of *Bordetella pertussis* cya operon. *EMBO J. 9*: 999–1005.

Lee, C. K., Roberts, A. L., Finn, T. M., Knapp, S., and Mekalanos, J. J. (1990). A new assay for invasion of HeLa 229 cells by effects of inhibitors, phenotypic modulation, and genetic alterations. *Infect. Immun. 58*: 2516–2522.

Leppla, S. (1982). Anthrax toxin edema factor: a bacteria adenylate cyclase that increases cyclic AMP concentrations in eukaryotic cells. *Proc. Natl. Acad. Sci. U.S.A. 79*: 3162–3166.

McPheat, W. L., Wardlaw, A. C., and Novotny, P. (1983). Modulation of *Bordetella pertussis* by nicotinic acid. *Infect. Immun. 41*: 516–522.

Maloney, N. J. (1992). Interaction of the adenylate cyclase toxin from *Bordetella pertussis* with target cells. Master's thesis, Dept. of Pharmacology, Univ. Virginia, Charlottesville, VA.

Maloney, N. J., Gray, M. C., Fremgen, P., Gordon, V. M., and Hewlett, E. L. (1992). The identification and characterization of cell line resistant to intoxication by *B. pertussis* adenylate cyclase toxin. Program of the 92nd General Meeting of the American Society for Microbiology, Abstract B19.

Masure, H. R., and Storm, D. R. (1989). Characterization of the bacterial cell associated calmodulin-sensitive adenylate cyclase from *Bordetella pertussis*. *Biochemistry 28*: 438–442.

Masure, H. R., Oldenburg, D. J., Donovan, M. G., Shattuck, R. L., and Storm, D. R. (1988). The interaction of Ca^{2+} with the calmodulin-sensitive adenylate cyclase from *Bordetella pertussis*. *J. Biol. Chem. 263*: 6933–6940.

Meade, B. D., Kind, P. D., Ewell, J. B., McGrath, P. P., and Manclark, C. R. (1984a). In vitro inhibition of murine macrophage migration by *Bordetella pertussis* lymphocytosis promoting factor. *Infect. Immun. 45*: 718–725.

Meade, B. D., Kind, P. D., and Manclark, C. R. (1984b). Lymphocytosis promoting factor of *Bordetella pertussis* alters mononuclear phagocyte circulation in response to inflammation. *Infect. Immun. 46*: 733–739.

Mouallem, M., Farfel, Z., and Hanski, E. (1990). *Bordetella pertussis* adenylate cyclase toxin: intoxication of host cells by bacterial invasion. *Infect. Immun. 58*: 3759–3764.

Pearson, R., Symes, P., Conboy, M., Weiss, A., and Hewlett, E. (1987). Inhibition of monocyte oxidative responses by *Bordetella pertussis* adenylate cyclase. *J. Immunol. 139*: 2749–2754.

Pearson, R. D., Borland, M. K., Jeronimo, S. M., and Hewlett, E. L. (1990). Intoxication of human leukocytes by the adenylate cyclase toxin of viable *Bordetella pertussis*. *Clin. Res.* 38: 270A.

Rogel, A., and Hanski, E. (1992). Distinct steps in the penetration of adenylate cyclase toxin of *Bordetella pertussis* into sheep erythrocytes. *J. Biol. Chem.* 267: 22599–22605.

Rogel, A., Schultz, J., Brownlie, M., Coote, J., Parton, R., and Hanski, E. (1989). *Bordetella pertussis* adenylate cyclase: purification and characterization of the toxic form of the enzyme. *EMBO J.* 8: 2755–2760.

Rogel, A., Meller, R., and Hanski, E. (1991). Adenylate cyclase from *Bordetella pertussis*: the relationship between induction of cAMP and hemolysis. *J. Biol. Chem.* 266: 3154–3161.

Roy, C. R., Miller, J. F., and Falkow, S. (1990). Autogenous regulation of the *Bordetella pertussis* bvg ABC operon. *Proc. Natl. Acad. Sci. U.S.A.* 87: 3763–3767.

Sakamoto, H., Bellalou, J., Sebo, P., and Ladant, D. (1992). *Bordetella pertussis* adenylate cyclase toxin. *J. Biol. Chem.* 267: 13598–13602.

Scarlato, V., Prugnola, A., Arico, B., and Rappuoli, R. (1990). Positive transcriptional feedback at the *bvg* locus controls expression of virulence factors in *Bordetella pertussis*. *Proc. Natl. Acad. Sci. U.S.A.* 87: 6753–6757.

Selfe, S., Hunder, D. D., Shattuck, R. L., Nathanson, N. M., and Storm, D. R. (1987). Alteration of intracellular cAMP levels and beating rates of cultured chick cardiac cells by *Bordetella pertussis* adenylate cyclase. *Mol. Pharmacol.* 31: 529–534.

Shattuck, R. L., and Storm, D. R. (1985). Calmodulin inhibits entry of *Bordetella pertussis* adenylate cyclase into animal cells. *Biochemistry* 24: 6323–6328.

Shattuck, R. L., Oldenburg, D., and Storm, D. (1985). Purification and characterization of a calmodulin-sensitive adenylate cyclase from *Bordetella pertussis*. *Biochemistry* 24: 6356–6362.

Szabo, G., Gray, M. C., and Hewlett, E. L. (1994). Adenylate cyclase toxin from *Bordetella pertussis* produces ion conductance across artificial lipid bilayers in a calcium- and polarity-dependent manner. *J. Biol. Chem.* 269: 22496–22499.

Thompson, S. A., Wang, L. L., West, A., and Sparling, F. (1993). *Neisseria meningitidis* produces iron-regulated proteins related to the RTX family of exoproteins. *J. Bacteriol.* 175: 811–818.

Utsumi, S., Sonoda, S., Imagawa, T., and Kanoh, M. (1978). Polymorphonuclear leukocyte inhibitor factor of *Bordetella pertussis*. *Biochem. J.* 21: 121–135.

Weiss, A., and Goodwin, M. (1989). Lethal infection by *Bordetella pertussis* in the infant mouse model. *Infect. Immun.* 57: 3757–3764.

Weiss, A., Hewlett, E. L., Myers, S., and Falkow, S. (1983). Tn-5 induced mutations affecting virulence factors of *Bordetella pertussis*. *Infect. Immun.* 42: 33–41.

Weiss, A., Hewlett, E. L., Myers, G. A., and Falkow, S. (1984). Pertussis toxin and extracytoplasmic adenylate cyclase as virulence factors of *Bordetella pertussis*. *J. Infect. Dis.* 150: 219–222.

Weiss, A., Myers, G., Crane, J., and Hewlett, E. (1986). *Bordetella pertussis* adenylate cyclase toxin: structure and possible function in whooping cough and the pertussis vaccine. In *Microbiology—1986*, L. Leive (Ed.), American Society for Microbiology, Washington, DC, pp. 70–74.

Welch, R. (1991). Pore-forming cytolysins of gram-negative bacteria. *Mol. Microbiol.* 5: 521–528.

Welch, R. A., Felmlee, T., Pellett, S., and Chenoweth, D. (1986). In *Protein–Carbohydrate Interactions in Biological Systems*, D. L. Lark, S. Normack, B. E. Uhlin, and H. Wolf-Watz (Eds.), Academic, New York, pp. 433–438.

Wolff, J., and Cook, G. (1973). Activation of thyroid membrane adenylate cyclase by purine nucleotides. *J. Biol. Chem.* 248: 350–355.

Wolff, J., and Cook, G. (1982). Amphiphile-mediated activation of soluble adenylate cyclase of *Bordetella pertussis*. *Arch. Biochem. Biophys.* 215: 524–531.

Wolff, J., Cook, G., Goldhammer, A., and Berkowitz, S. (1980). Calmodulin activates prokaryotic adenylate cyclase. *Proc. Natl. Acad. Sci. U.S.A.* 77: 3841–3844.

19

Biosynthesis and Targeting of Pertussis Toxin

Drusilla L. Burns

Center for Biologics Evaluation and Research, Food and Drug Administration, Bethesda, Maryland

I. INTRODUCTION

Pertussis toxin (PT) is a protein secreted by the gram-negative bacterium *Bordetella pertussis*. PT interacts with many types of mammalian cells and has multiple biological activities, which include affecting the recirculation of leukocytes (Spangrude et al., 1984), altering the response of neutrophils to *N*-formylated peptides that are chemoattractants (Bokoch and Gilman, 1984), and inhibiting the migration of macrophages (Meade et al.,

1984). Thus, the toxin may contribute to the pathogenicity of the organism by impairing the immune system of the host.

B. pertussis is the causative agent of the disease pertussis (whooping cough), which kills approximately 500,000 people per year (Muller et al., 1986), mostly children in developing countries. A species closely related to *B. pertussis*, *B. parapertussis*, causes a disease in humans that is similar to pertussis but usually not as severe. Since both *B. pertussis* and *B. parapertussis* produce a number of the same virulence factors, but only *B. pertussis* produces PT, this toxin has been implicated in the pathogenicity of the organism.

A role for PT in the disease process was substantiated in part by the finding that inactivated pertussis toxin, when given as a vaccine to children, can protect the children against serious disease (Ad Hoc Group for the Study of Pertussis Vaccines, 1988). Moreover, antibodies to pertussis toxin can passively protect experimental animals from challenge with *B. pertussis* (Sato et al., 1984). Thus neutralization of the toxin by the immune system likely renders the infection less effective. Because of these findings, newly developed acellular pertussis vaccines contain inactivated pertussis toxin as one of their components.

The mechanism by which PT is synthesized, assembled, secreted, and introduced into eukaryotic cells has been studied in order to better understand this important virulence factor of *B. pertussis* and the role it plays in pathogenesis.

II. SUBUNIT STRUCTURE OF PERTUSSIS TOXIN

Pertussis toxin is a member of a family of bacterial protein toxins that have A-B structures in that they are composed of an active enzyme and a binding component (Tamura et al., 1982; Gill, 1978). The structure of PT is shown in Figure 1.

The A or S1 subunit catalyzes the ADP-ribosylation of GTP-binding regulatory proteins (G proteins) in the eukaryotic cell that are involved in signal transduction (Katada and Ui, 1982a,b; Katada et al., 1983) (Fig. 2). Normally, G proteins transmit to effector proteins the information generated upon binding of a hormone or neurotransmitter to its receptor. The effector protein then alters production of second messengers within the cell. When a G protein is ADP-ribosylated by PT, it becomes incapable of transducing these signals. The result is that the cell no longer responds to a variety of hormones or neurotransmitters (for reviews see Gilman, 1984, 1987; Casey and Gilman, 1988). Concentrations of PT as low as 3 pM can completely ADP-ribosylate G protein substrates in Chinese hamster ovary cells within 48 h (Burns et al., 1987a).

Fig. 1   Subunit structure of pertussis toxin.

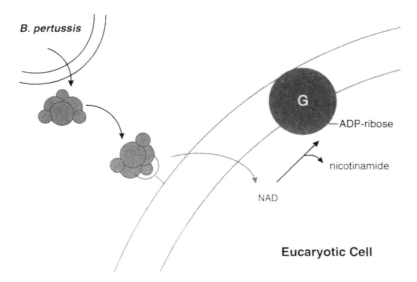

Fig. 2 Mechanism of action of pertussis toxin.

The B oligomer of pertussis toxin is responsible for binding the toxin to its receptor on the eukaryotic cell (Tamura et al., 1982, 1983). This moiety is made up of five subunits, one copy each of S2, S3, and S5 and two copies of S4. The B oligomer appears to be a complex of two dimers, S2-S4 and S3-S4, which are held together by the S5 subunit (Tamura et al., 1982).

The S1 subunit can be separated from the B oligomer by incubating the toxin with ATP and the detergent 3-[(3-cholamidopropyl)dimethylammonio]-1-propanesulfonate (CHAPS) followed by column chromatography (Arciniega et al., 1987; Kaslow and Lesikar, 1987). In this manner it has been possible to investigate the biological properties of each of the individual moieties of the toxin. The isolated S1 subunit exhibits both ADP-ribosyltransferase activity and NAD glycohydrolase activity (Katada et al., 1983; Mosset al., 1986). The latter reaction probably represents an abortive hydrolytic reaction that occurs when the G-protein substrate is not available, resulting in the hydrolysis of NAD to ADP-ribose and nicotinamide. The B oligomer is a stable moiety that is resistant to proteolysis (Burns et al., 1987b; Krueger et al., 1991), binds glycoproteins (Tamura et al., 1982; Irons and MacLennan, 1979), and at relatively high concentrations exhibits certain biological activities in vitro such as the ability to induce mitogenesis (Tamura et al., 1983), possibly because it is able to bind bivalently to cells, resulting in cross-linking or aggregation of proteins on the surface of cells.

The structure–function relationships of the S1 subunit have been examined in some detail. The S1 subunit is composed of 235 amino acids, has a molecular weight of about 26,000 (Nicosia et al., 1986), and contains two cysteine residues, located at positions 41 and 201 (Nicosia et al., 1986), which form a disulfide bond (Sekura et al., 1983). Reduction of this bond is essential for manifestation of enzymatic activity (Katada et al., 1983; Moss et al., 1983). Regions important for NAD glycohydrolase activity and ADP-ribosyltransferase activity have been identified. The N-terminal 180 amino acids appear to contain all the information needed for NAD glycohydrolase activity (Cortina and Barbieri, 1991); however, the ADP-ribosyltransferase activity of this truncated species was less than 1% of that of the native S1 subunit, suggesting that additional residues are required for efficient ADP-ribosyltransferase activity (Cortina and Barbieri, 1991). Cortina et al. (1991) found that amino acids 195–219 were needed to confer high-affinity binding to the G-protein transducin. The interaction of this portion of the S1 subunit with transducin was primarily with the $\beta\gamma$ subunits of transducin, whereas the first 180 amino acids of S1 interact with the α subunit of transducin.

III. ASSEMBLY OF PERTUSSIS TOXIN IN VITRO

Tamura et al. (1982) were the first to elucidate the pathway of assembly of PT subunits in vitro. Holotoxin was separated into individual subunits by incubating the toxin with urea followed by chromatography to isolate each subunit. Addition of either S2 or S3 to S4 resulted in the formation of dimers. No other pairs of subunits generated dimers. While these dimers showed no biological activity, they were capable of binding to haptoglobin-Sepharose. Addition of S1 to a mixture of S2-S4 and S3-S4 did not generate a biologically active molecule as measured by the ability of this species to enhance the insulin secretory response of rats to a glucose load. Addition of S5 to this mixture was necessary in order to observe islet activation, which suggested that S5 might stabilize the toxin structure.

B oligomer has been assembled from individual subunits produced in *Escherichia coli* using recombinant DNA techniques (Burnette et al., 1992). Each B-subunit polypeptide was synthesized individually in *E. coli* and was isolated as insoluble inclusion bodies. After individual subunits were solubilized with 6 M guanidine hydrochloride under reducing conditions, the subunits were reoxidized. Spontaneous association of the subunits occurred upon dialysis against 2 M urea. Recombinant B oligomer was then purified by affinity chromatography. This recombinant B oligomer possessed the mitogenic and hemagglutinating properties of the native B oligomer. At present, assembly of the complete PT molecule from polypeptide chains produced in *E. coli* has not been accomplished.

IV. GENETIC ANALYSIS OF THE PT STRUCTURAL GENES

The region of the *B. pertussis* chromosome encoding the pertussis toxin structural genes has been cloned and sequenced (Nicosia et al., 1986; Locht and Keith, 1986). As shown in Figure 3, the genes are clustered within 3.2 kb and appear in the order S1, S2, S4, S5, S3. These genes encode mature proteins having molecular weights of 26,220, 21,920, 12,060, 10,940, and 21,860, respectively. Coding regions for the mature protein subunits are preceded by sequences that have the features of bacterial signal sequences and appear to encode peptides that are 27–42 amino acids long each (Nicosia et al., 1986). The presence of signal sequences on these proteins may be expected because PT is a secreted protein.

Fig. 3   The *ptx* and *ptl* regions. Structure of the *ptx* region with the genes encoding subunits S1–S5 is shown. Also shown are the open reading frames (B–H) of the *ptl* region.

The genes encoding S2 and S3 are highly homologous (Nicosia et al., 1986; Locht and Keith, 1986; Capiau et al., 1986). They exhibit 75% homology at the nucleotide sequence level and 67% homology at the amino acid level. Others have postulated that this homology suggests a common evolutionary origin for these two subunits, possibly originating by gene duplication (Nicosia et al., 1986). Whereas the amino acid sequences of these subunits are similar, S2 cannot substitute for S3 when the toxin is assembled either in vitro (Tamura et al., 1982) or in vivo (Marchitto et al., 1987).

Analysis of the sequence of the S1 subunit yielded several interesting homologies. Two regions of the S1 subunit, amino acids 8–15 and amino acids 51–58, exhibit homology to the A subunit of cholera toxin (Nicosia et al., 1986; Locht and Keith, 1986; Capiau et al., 1986). Cholera toxin A subunit, like the S1 subunit, is an ADP-ribosyltransferase (Moss and Vaughan, 1979). The protein substrate for cholera toxin is a G protein that is similar to, but distinct from, the G proteins that are substrates for pertussis toxin. Alterations of the amino acid sequence of the S1 subunit in the region of amino acids 8–15 result in virtual inactivation of the molecule (Burnette et al., 1988). In contrast, changes in amino acids 51–58 for the most part have relatively small effects on the enzymatic activity of the molecule (Kaslow et al., 1992), suggesting that this region is not critical for enzymatic activity but may play other important roles such as aiding secretion of the PT molecule from *B. pertussis* or facilitating entry of the toxin into the eukaryotic cell.

V. REGULATION OF EXPRESSION OF PT

The PT structural genes are organized as shown in Fig. 3. A sequence homologous to *E. coli* promoters and a Shine-Dalgarno sequence precedes the first gene, and a possible rho-independent transcriptional terminator follows the last gene (Nicosia et al., 1986). The promoter region has been analyzed in some detail. The region directly upstream from the S1 structural gene contains –35 and –10 regions that are in good agreement with the consensus sequence of *E. coli* promoters except that the spacing is 21 base pairs (bp) instead of 17 bp (Nicosia and Rappuoli, 1987). Primer extension results indicated that transcription of the PT operon begins at an adenine nucleotide located 7 bp downstream from the –10 region (Nicosia and Rappuoli, 1987).

Pertussis toxin is expressed only by the virulent form of *B. pertussis* and not by the avirulent form. Attempts to express pertussis toxin in *E. coli* have met with little success. A possible explanation for this finding is that additional *B. pertussis* proteins are needed for proper regulation of the PT structural genes. Weiss and Falkow (1983) showed that a second chromosomal locus exists in *B. pertussis* that is required for expression of a number of virulence factors of the organism including PT. This region, termed the *bvg* locus, encodes two positive regulatory proteins, BvgA and BvgS, which share homology with a family of signal-transducing proteins found in a number of prokaryotic cells (Arico et al., 1989; Stibitz and Yang, 1991). Because of sequence homologies between the *bvg* products and the proteins of several two-component systems, BvgS is believed to be a sensor protein that responds to a number of environmental signals including low temperature, $MgSO_4$, and nicotinic acid (Arico et al., 1989; Roy et al., 1989; Stibitz and Yang, 1991). BvgA is believed to be a positive regulator of transcription. It has been proposed that in the absence of modulating signals, BvgS activates BvgA (Roy et al., 1989; Stibitz and Yang, 1991), which then turns on the transcription of certain virulence genes including the *ptx* genes that encode pertussis toxin structural subunits. Although the *bvg* genes are necessary for expression of pertussis toxin, they may not be sufficient (Miller et al., 1990). At least one additional regulatory gene is thought to also be needed for PT expression (Miller et al., 1989, 1990). Huh and Weiss (1991) have demonstrated that a 23-kDa protein produced by Bvg$^+$ but not Bvg$^-$ strains binds to the −157 to −117 region upstream from the PT promoter, which contains two tandemly repeated DNA sequences that have been shown to be necessary for transcription (Gross and Rappuoli, 1988). They have called this protein Act for activator of toxin expression and have proposed that Act is an additional protein necessary for the expression of PT.

B. *parapertussis* and B. *bronchiseptica* do not produce PT as determined by ELISA (Nicosia et al., 1987); however, regions homologous to the PT genes have been identified on the B. *parapertussis* and B. *bronchiseptica* chromosomes and have been found to be 98.5% and 96% homologous, respectively, to the PT genes of B. *pertussis* (Arico and Rappuoli, 1987). The fourth member of the *Bordetella* genus, B. *avium*, did not contain DNA homologous to the PT genes. When the S1 subunits of B. *pertussis*, B. *parapertussis*, and B. *bronchiseptica* are produced in E. *coli*, they possess ADP-ribosyltransferase activity (Arico and Rappuoli, 1987). Many of the mutations in the genes from B. *parapertussis* and B. *bronchiseptica* are found in the promoter region, and no PT mRNA could be detected by dot blot hybridization of the native organism, suggesting that although these species contain genes for PT, the genes are transcriptionally silent (Arico and Rappuoli, 1987).

VI. ASSEMBLY OF PERTUSSIS TOXIN IN VIVO

Once the toxin subunits are synthesized, they must fold, assemble, and pass through two bacterial membranes before the toxin can interact with the target eukaryotic cell. Although the order of these events has not been elucidated, some information is available on the pathway of assembly in vivo and the structural elements of the PT molecule required for secretion.

Interaction of the S1 subunit with the B oligomer appears to be required for stable expression of the S1 subunit in B. *pertussis*. Both an intact disulfide bond in the S1 subunit connecting Cys-41 to Cys-201 and an intact C terminus of the S1 subunit are necessary for interaction of the S1 subunit with the B oligomer.

In vitro experiments have demonstrated that when the disulfide bond of the S1 subunit is reduced, this subunit no longer associates efficiently with the B oligomer as detected either by native gel electrophoresis or by assay of the ability of the preparation to cluster Chinese hamster ovary cells (Burns and Manclark, 1989). In vivo experiments support the idea that the disulfide bond of the S1 subunit must form before this subunit can associate with the B oligomer. When Cys-41 was changed to either Ser or Gly using recombinant DNA techniques and this protein was expressed in B. *pertussis*, no S1 subunit could be detected either associated with the bacterial cells or in culture supernatants using immunoblot analysis (Antoine and Locht, 1990). However, when these mutant S1 subunits are synthesized in E. *coli*, they react with a panel of monoclonal antibodies that are specific for conformational epitopes (Francotte et al., 1989), suggesting that the S1 subunits are capable of folding. Since the S1 subunit in the absence of the B oligomer is much more susceptible to proteolysis than S1 that is part of the holotoxin molecule (Burns et al., 1987b), it is likely that these mutant S1 subunits do not assemble with the B oligomer and are therefore rapidly degraded (Antoine and Locht, 1990).

The extreme C-terminal end of the S1 subunit is also essential for assembly of S1 with the B oligomer. When the S1 subunit is cleaved by trypsin, probably at residues 181, 182, and/or 193, the resulting species does not bind efficiently to the B oligomer in vitro (Burns et al., 1987b). Truncated forms of the S1 subunit do not appear to assemble with the B oligomer in vivo to form a stable holotoxin species. Deletion of the C-terminal region of the S1 subunit after residue 187, which deletes Cys-201, results in unstable expression of the S1 subunit (Antoine and Locht, 1990) in *B. pertussis*, as would be expected since the disulfide bond cannot form. However, deletion of the C-terminal region after residue 207 also results in unstable expression of the S1 subunit (Antoine and Locht, 1990), suggesting that residues 208–235 either directly interact with the B oligomer or are essential for maintaining a conformation of the S1 subunit that is capable of assembly with the B oligomer. Truncated forms of the S1 subunit, however, can fold into species that have NAD glycohydrolase activity (Locht et al., 1987; Cortina and Barbieri, 1991), confirming that, at least when expressed in *E. coli*, they are capable of folding into native-like species.

In the absence of a stably expressed S1 subunit, the B oligomer assembles; however, this species is only poorly secreted from the bacterium (Pizza et al., 1990). These B-oligomer subunits remain associated with the bacterium and have the sizes expected for the native subunits, suggested that the signal sequences have likely been cleaved (Pizza et al., 1990). Thus association of the S1 subunit with the B oligomer is required for efficient release of the toxin from the bacterial cell.

Complete assembly of the B oligomer is required for PT secretion. A mutant of *B. pertussis*, BP356, has a Tn5 insertion within the structural gene for S3 at nucleotide position 3436 (Weiss et al., 1983; Nicosia and Rappuoli, 1987). This mutant should therefore produce S1, S2, S4, S5, and the first 136 residues of S3. Whereas this mutant is capable of producing active PT that remains associated with the bacterial cells as measured by the Chinese hamster ovary (CHO) cell assay (Nicosia and Rappuoli, 1987), secretion of the toxin from this mutant was not detected by either immunoblot analysis (Marchitto et al., 1987) or ELISA (Nicosia and Rappuoli, 1987). Since a stable S1 subunit that was associated with mutant cells was easily detected, and since toxin activity could be detected, the toxin likely assembles within the cell at least to the extent that the S1 subunit was associated with some of the B polypeptides. These results suggest that the C-terminal 63 amino acids of S3 are critical for secretion of the molecule.

VII. SECRETION OF PT FROM *B. PERTUSSIS*

The question of how the toxin subunits cross the inner and outer membranes of the bacterium has not been answered in entirety; however, clues to the mechanism have recently surfaced. Since each subunit has a signal sequence (Nicosia et al., 1986; Locht and Keith, 1986), the subunits are likely exported individually at least across the inner membrane into the periplasmic space, where the signal sequences would then be cleaved by signal peptidase. Transport of the individual chains across the inner membrane might use the general export pathway of the cell as exemplified by the *sec* pathway of *E. coli* (Wickner et al., 1991), or the subunits may use a more specialized pathway to cross the inner membrane. The mechanism by which the toxin might cross the outer membrane is unknown.

Proteins that are secreted from gram-negative organisms often require specialized transport apparatus (Lory, 1992). For example, the adenylate cyclase toxin of *B. pertussis* requires at least three accessory genes for its export (Glaser et al., 1988). Pullulanase, an enzyme exported to the outer membrane of *Klebsiella oxytoca*, requires at least 14 additional genes for its transport to the cell surface when it is expressed in *E. coli* (Pugsley et al., 1990, 1991).

Recently, data have suggested that specialized accessory proteins are involved in PT transport. At least seven genes that appear to encode proteins necessary for secretion of PT have been identified and were shown to belong to a family of genes necessary for transport of macromolecules across membranes (Weiss et al., 1993). Identification of these accessory genes was made possible by mutational analysis. A mutant of *B. pertussis* (BPM 3171) that produced PT but was less virulent in an animal model for disease (Weiss and Goodwin, 1989) was analyzed. It was found to secrete pertussis toxin at much lower levels than the parent strain (Weiss et al., 1993). Production of PT did not appear to be affected, since the mutant cells contained higher levels of cell-associated PT than did the parent strain. The mutant was originally generated by insertion of Tn*5lac* approximately 3.2 kb downstream from the PT structural genes (Weiss et al., 1989, 1993). More detailed mutational analysis of this region of the *B. pertussis* chromosome suggested that a region approximately 9.5 kb in length was necessary for secretion of the toxin, and this region was organized as an operon that was named the *ptl* operon (Weiss et al., 1993). Since Tn*5lac* is a transcriptional fusion, the direction of transcription could be determined and was found to be in the same direction as that for the PT structural genes (Weiss et al., 1993).

Nucleotide sequence analysis of the *ptl* region revealed seven open reading frames (orf's) named orf's B–H (Weiss et al., 1993) (Fig. 3). OrfH was homologous to several genes involved in secretion in other bacterial systems, including the *pulE* gene needed for secretion of pullulanase in *Klebsiella oxytoca* and the *pilB* gene necessary for export of pili from *Pseudomonas aeruginosa*. The best homology, however, was found to be to the *virB11* gene of *Agrobacterium tumefaciens*. Orf's B, C, D, E, F, and G were found to have sequences that would encode proteins homologous to the VirB3, VirB4, VirB6, VirB8, VirB9, and VirB10 proteins, respectively, from *A. tumefaciens*. The VirB proteins are thought to be involved in the transport of a DNA element, t-DNA, across the membranes of *A. tumefaciens* and into plant cells where the t-DNA eventually is transported into the nucleus and codes for proteins involved in the biosynthesis of plant hormones that stimulate growth of tumors in the plant (Zambryski, 1992; Ward et al., 1988; Kuldau et al., 1990). Thus the VirB proteins are necessary for the secretion of a macromolecule from this gram-negative organism. These findings support the idea that the *ptl* genes may code for proteins necessary for transport of PT across the bacterial membranes and suggest that these proteins may be members of a broader family of proteins involved in secretion of macromolecules from bacteria.

VIII. BINDING OF PT TO EUKARYOTIC CELL RECEPTORS

A. Characterization of the Eukaryotic Cell Receptor

After its release from *B. pertussis*, PT binds to the surface of mammalian cells, where the S1 subunit then gains access to its intracellular substrates by an unknown mechanism. Since most mammalian cells are susceptible to intoxication by PT, the receptor for PT must be prevalent, although its identity has not been established.

Early work on PT indicated that the toxin could bind to glycoproteins such as haptoglobin and fetuin (Irons and MacLennan, 1979; Tamura et al., 1982; Sekura et al.,

1985), suggesting that the toxin may bind to the oligosaccharide chains of glycoproteins. Removal of the carbohydrate residues of fetuin by treatment with a crude enzyme preparation from *Diplococcus pneumoniae* yielded an aglycofetuin derivative that was ineffective as an inhibitor of ^{125}I-labeled fetuin binding to PT (Sekura et al., 1985). Further analysis indicated that PT could bind to a number of proteins that contained N-linked oligosaccharide chains, but the toxin was not capable of binding to glycophorin, which contains only O-linked oligosaccharide chains (Witvliet et al., 1989). Moreover, when fetuin, which contains both O-linked and N-linked oligosaccharide chains, was treated with alkaline borohydride to selectively remove the O-linked chains, a derivative that inhibits binding of PT to ^{125}I-labeled fetuin with an activity indistinguishable from that of native fetuin was produced (Sekura et al., 1985). These results suggested that the binding of the toxin is specific for N-linked chains.

Armstrong et al. (1987) used exoglycosidases and glycosyltransferases to examine the binding of PT to the oligosaccharide chains of fetuin. The N-linked chains of fetuin end in a NeuAc→GalβGlcNAc residue linked to a mannose residue (Nilsson and Norden, 1979). Monosaccharides from the nonreducing ends of the chain were selectively removed using the appropriate exoglycosidases. Removal of NeuAc residues from the termini of the N-linked chains by treatment with neuraminidase resulted in a derivative that was less potent than fetuin in inhibiting PT-induced agglutination of red blood cells and in inhibiting binding of radiolabeled fetuin to PT-coated polystyrene tubes. Treatment with β-galactosidase resulted in the formation of the derivative asialoagalactofetuin, which showed no further reduction of inhibition in the hemagglutination system and showed partial recovery of inhibitory activity in the radiolabeled-fetuin-PT binding assay, suggesting that the presence of acetamido-containing sugar groups at the nonreducing terminus of the oligosaccharide chain may be recognized by PT. Further treatment of this derivative with β-N-acetylhexosaminidase resulted in formation of asialoagalacto-a[N-acetylhexosamino]fetuin, which exhibited a further decrease in the ability to compete with fetuin in both assays. The inhibitory activity of asialoagalactofetuin could be restored to that of fetuin by using glycotransferases to add back the missing sugar residues. This study suggested that the terminal NeuAc→GalβGlcNAc residues of oligosaccharide chains of fetuin are critical for PT binding and that optimal binding requires the presence of the terminal NeuAc residues.

The results of these studies using purified glycoproteins as models for PT cellular receptors were corroborated when binding of PT to cellular receptors was examined. PT is known to bind to and intoxicate CHO cells (Hewlett et al., 1983). A series of variant CHO cells that are defective in the biosynthesis of oligosaccharide chains were used to examine the binding of PT to cellular receptors (Brennan et al., 1988; Witvliet et al., 1989). The structures of oligosaccharide chains on the surface of wild-type and variant CHO cells are shown in Figure 4. LEC 2 cells lack terminal NeuAc residues, whereas LEC 8 cells lack terminal NeuAc-Gal residues on the oligosaccharide chains of both glycoproteins and glycolipids (Deutscher et al., 1984; Deutscher and Hirschberg, 1986). The defect in 15B cells is such that they lack the carbohydrate sequence NeuAc→Galβ4GlcNAc on complex-type glycoproteins (Brennan et al., 1988). All variant cells were very resistant to PT (Brennan et al., 1988; Witvliet et al., 1989). Since the defect of 15B cells is specific for glycoproteins, the functional receptor for PT on wild-type CHO cells is most likely a glycoprotein. Moreover, since LEC 2 cells, which lack only terminal NeuAc residues, are resistant to PT action, the functional receptor likely contains terminal sialic acid residues.

Fig. 4 Alterations in carbohydrate structures of variant CHO cells. Variations in carbohydrate structures are illustrated by the structures of biantennary Asn-linked complex-type oligosaccharide side chains of the indicated cells types. (From Li et al., 1988.)

Although the oligosaccharide structure of the cellular receptor for PT, at least on CHO cells, has been elucidated, much less is known about the identity of the receptor protein. PT binds to a 165 kDa protein of CHO cells that is absent from both LEC 2 and 15B cell extracts (Brennan et al., 1988; Witvliet et al., 1989). Both 70- and 43-kDa surface proteins of T lymphocytes have been observed to bind to PT (Clark and Armstrong, 1990; Rogers et al., 1990). Detergent extracts of goose erythrocytes contain a binding protein with M_r of 115,000 (Witvliet et al., 1989). Thus, proteins of varying sizes appear to be capable of binding to PT on the surface of different cell types. While a specific protein might serve as the receptor on many cell types, the possibility exists that several different glycoproteins might act as functional receptors.

The question remains as to whether other glycoconjugates might serve as functional receptors for PT. Certain studies have shown that PT bound to lactosylceramide, glucosylceramide, and mixed gangliosides that had been separated by thin-layer chromatography (Saukkonen et al., 1992); however, others have been unable to detect binding of PT to glycolipids using similar assays (Brennan et al., 1988). More recently, PT was shown to bind to glycolipids that had been incorporated into liposomes (Hausman and Burns, 1993). Both PT and the B oligomer bound well to the ganglioside G_{D1a}; however, little binding of PT or B oligomer was observed to other glycolipids, including G_{M1}, glucocerebrosides, and lactocerebrosides. G_{D1a} contains terminal NeuAc→GalβGlcNAc residues that are similar to those found on glycoprotein receptors for PT. As yet, no evidence exists that would suggest that G_{D1a} might act as a functional receptor for PT on any cell type.

B. Location of the Receptor Binding Site on PT

The receptor binding site on the PT molecule is located on the B oligomer. The simple act of binding the B oligomer to cell surfaces has it own biological effects that are quite distinct from the biological effects due to toxin-catalyzed ADP-ribosylation. The B oligomer stimulates mitogenesis of lymphocytes, increases phosphatidylinositol turnover, and induces a rise in cytosolic calcium levels in T lymphocytes (Tamura et al., 1983; Gray et al., 1989; MacIntyre et al., 1988; Stewart et al., 1989; Strnad and Carchman, 1987; Thom and Casnellie, 1989). These effects can occur quite rapidly (within 30 s after addition of the toxin to the cells), and most of them require 100–1000 times more PT than is required for biological activities caused by the ADP-ribosyltransferase activity of the toxin.

The B oligomer appears to contain more than one receptor binding site. Isolated S2-S4 and S3-S4 dimers are each capable of agglutinating red blood cells (Nogimori et al., 1984). Moreover, both S2-S4 and S3-S4 were capable of binding to a 115-kDa protein found in goose erythrocyte membrane extracts and were found to bind to a variety of glycoproteins that contained N-linked oligosaccharide chains (Witvliet et al., 1989). The specificities of binding of the two dimers to the various glycoproteins were similar, although some differences were observed. Lang et al. (1989) used immunofluorescence staining of cells with an anti-S4 monoclonal antibody to demonstrate that both S2-S4 and S3-S4 were capable of binding to the surface of CHO cells, consistent with the idea that both dimers may be involved in target cell recognition. They also found that antibody binding to one dimer was sufficient to block attachment of PT to sensitive cells, either by steric hindrance or by disruption of the tertiary structure necessary for target cell binding.

The location of the binding sites on the dimers has begun to be elucidated. Inclusion body preparations of either S2 or S3 that had been produced in *E. coli* using recombinant methods have been shown to bind to glycolipid preparations, suggesting that binding sites may be found on S2 and S3 (Saukkonen et al., 1992). Consistent with this hypothesis is the finding that antibodies raised to peptides corresponding to regions of either S2 or S3 could block PT-induced agglutination of red blood cells (Schmidt and Schmidt, 1989; Schmidt et al., 1991). The N-terminal regions of S2 and S3 may be involved in receptor recognition, for amino acid alterations in the regions corresponding to amino acids 31–61 altered the specificity of binding of each of the subunits (Saukkonen et al., 1992). Moreover, most of the antipeptide antibodies that inhibited the ability of PT to agglutinate red blood cells were raised to peptides corresponding to areas in the N-terminal region of each of the subunits. Antibodies raised to peptides corresponding to amino acids 1–12, 12–23, 14–29, and 36–51 of S3 inhibited PT-induced hemagglutination of red blood cells, whereas antibodies to peptides corresponding to amino acids 95–107, 134–150, and 164–178 did not inhibit PT-induced hemagglutination but did recognize PT in an ELISA (Schmidt et al., 1991). Antibodies directed against synthetic peptides corresponding to amino acids 1–7, 35–50, and 91–106 of S2 inhibited PT-induced hemagglutination of red blood cells (Schmidt and Schmidt, 1989).

IX. ENTRY OF PT INTO EUKARYOTIC CELLS

Once PT binds to receptors located on the outer surface of the mammalian cell, the S1 subunit must somehow gain access to the G protein, which is believed to be in the inner leaflet of the plasma membrane, and to NAD, which is thought to be in the cytoplasm.

In general, bacterial protein toxins are thought to enter eukaryotic cells by one of two mechanisms—endocytosis with subsequent entry into the cytoplasm of the cell from an intracellular vesicle or direct pentration of the plasm membrane. Either mechanism requires that the toxin or part of the toxin penetrate a biological membrane. At present, little information is available concerning the mechanism by which PT interacts with biological membranes.

Montecucco et al. (1986) examined the ability of PT subunits to interact with radioactive photoreactive phospholipid probes that were dispersed in detergent micelles. Two types of photoreactive probes were used—1-palmitoyl-2-(2-azido-4-nitro)benzoyl-sn-glycero-3-[^{3}H]phosphocholine (PC I), which has a photoreactive group at the polar head group level, and 1-myristoyl-2-[12-amino-(4-N-3-nitro-1-azidophenyl)]dodecanoyl-sn-glycero-3-[^{14}C]phosphocholine (PC II), which has a photoreactive group located at the fatty acid methyl terminus. When a mixture of PT and micelles of Triton X-100 containing the photoreactive lipid probes was illuminated, S2, S3, S4, and possibly S5 were labeled with PC I, but only S2 and S3 were labeled with PC II. Thus, S2 and S3 appear to penetrate deeper into the detergent micelle than S4 or S5. S1 was not labeled at all, indicating that it was not exposed to the lipid probes. When micelles of lysolecithin were used, slightly different results were obtained. In this case, S2, S3, S4, and possibly S5 were all labeled with PC I, but none of these subunits were labeled with PC II, indicating that different detergents interact with PT in different ways.

The interaction of PT, B oligomer, and the S1 subunit with lipids was also studied by examining the binding of these proteins to phospholipid vesicles composed of phosphatidylcholine (Hausman and Burns, 1992). In the absence of ATP or reducing agent, compounds that are required for the activation of PT, the toxin did not interact in a stable manner with the lipid vesicles. However, in the presence of both ATP (0.5 mM) and dithiothreitol (200 mM), the S1 subunit bound to the lipid vesicles, while none of the B polypeptides bound. The S1 subunit was also capable of releasing radiolabeled NAD that had been incorporated into the vesicles. The B oligomer released a small amount of trapped radiolabeled NAD, suggesting that it may be capable of transient interaction with the vesicles. These results suggested that once the toxin is activated, the S1 subunit may be capable of interacting directly with lipids. The possibility exists that if, after binding of PT to its cell surface receptor, the S1 subunit were freed from the constraints imposed by the B oligomer, the S1 subunit might directly interact with the membrane, perhaps gaining access to its cellular substrates.

The interaction of PT and its subunits with model membranes is strikingly similar to that of another ADP-ribosylating toxin, cholera toxin (CT), and its subunits with membranes. After binding of CT to its cellular receptor, the A subunit of CT is thought to insert into the membrane while the B oligomer remains bound to its receptor on the surface of the cell (Ribi et al., 1988; Wisnieski and Bramhall, 1981; Tomasi and Montecucco, 1981). Reduction of the single disulfide bond found in the A subunit of CT causes the A subunit to insert further into the membrane, and this subunit may then even separate from the rest of the toxin molecule (Ribi et al., 1988; Tomasi and Montecucco, 1981).

X. ACTIVATION OF PERTUSSIS TOXIN

Upon its release from the bacterial cell, PT is in an inactive state. Sometime after PT binds to eukaryotic cells but before it ADP-ribosylates G proteins in those cells, the toxin must be activated. Three agents are required for activation, all of which may be supplied by the eukaryotic cell. These are lipids, ATP, and a reducing agent.

A variety of phospholipids and detergents are able to stimulate both the ADP-ribosyltransferase and NAD glycohydrolase activities of PT (Moss et al., 1986). Phosphatidylcholine, lysophosphatidylcholine, and cholate stimulated the ATP-ribosylation of the G protein transducin by PT. Phosphatidylcholine and lysophosphatidylcholine were also able to stimulate the NAD glycohydrolase activity of the toxin. The detergent CHAPS greatly stimulated the NAD glycohydrolase activity of the toxin but appeared to inhibit the ability of the toxin to ADP-ribosylate transducin. Because activation of the toxin is not dependent on any specific lipid moiety, lipids found in eukaryotic cell membranes may be able to activate the toxin.

Katada and Ui (1982a) were the first to note that ATP enhanced the ADP-ribosyltransferase activity of the toxin. ATP enhances this activity by interacting directly with PT and not with the toxin substrates since ATP stimulates toxin-catalyzed ADP-ribosylation of heat-treated bovine serum albumin as well as the NAD glyco-hydrolase activity associated with the toxin (Lim et al., 1985). Other nucleotides such as ADP and GTP can also stimulate the NAD glycohydrolase activity of the toxin but not to the extent of ATP. Hydrolysis of ATP does not appear to be necessary for activation, since the nonhydrolyzable analogue App(NH)p can substitute for ATP (Moss et al., 1986). ATP acts by binding to the B oligomer of the toxin (Hausman et al., 1990) and weakening the intersubunit bonds between S1 and the B oligomer (Burns and Manclark, 1986). In the presence of the detergent CHAPS, addition of ATP to the holotoxin actually results in dissociation of about 50% of the S1 subunit from the B oligomer. Since free S1 subunit exhibits an NAD glycohydrolase activity that is almost 3–5 times higher than that of the holotoxin in the absence of ATP (Burns and Manclark, 1986), the B oligomer may constrain the S1 subunit in a less active form. ATP relieves this constraint either by promoting dissociation of subunits or by shifting the S1 subunit from a less active form that is tightly bound to the B oligomer to an active form that is only weakly bound to the B oligomer but free of the constraints imposed by the B oligomer.

Reducing reagents are an absolute requirement for PT activity (Katada et al., 1983). The single disulfide bond of the S1 subunit that connects Cys-41 to Cys-201 must be reduced for manifestation of enzyme activity. Concentrations of dithiothreitol greater than 100 mM are needed to efficiently activate the holotoxin in the absence of ATP or lipids (Moss et al., 1986). Much lower concentrations of reducing reagents are required for activation in the presence of ATP and lipids. For example, less than 1 mM dithiothreitol is needed to activate PT in the presence of 100 μM ATP and 1% CHAPS (Moss et al., 1986).

Thus, after binding of the toxin to the surface of the eukaryotic cell, exposure to membrane lipids, ATP, and a reducing environment would result in conversion of the toxin to its active form. Upon exposure to lipids and ATP, the S1 subunit may actually separate from the B oligomer and then carry out is task of ADP-ribosylating the G proteins of the cell.

XI. SUMMARY

Pertussis toxin is a complex oligomeric protein that plays a role in the pathogenesis of *B. pertussis*. After synthesis, this toxin must assemble and then cross two bacterial membranes and a eukaryotic cell membrane in order to reach its target substrates. The toxin apparently uses several accessory proteins to exit the bacterial cell. From the time the toxin is secreted from the bacterial cell until it reaches the eukaryotic cell surface, it is likely exposed to a hostile environment, which may contain proteases or other agents that could inactivate it. Once outside the bacterial cell, the toxin remains in an inactive state that is very resistant to proteolysis. Thus the toxin appears to have evolved in such a way that it has the necessary characteristics to ensure that it can reach its targets in the eukaryotic cell in an intact state.

Upon interaction with the eukaryotic cell, PT binds to its glycoconjugate receptor and then crosses the final barrier, the eukaryotic membrane, without the aid of other bacterial accessory proteins, although it is possible that the toxin receptor might play an active role in this transport. Only after the toxin reached the eukaryotic cell would it finally be exposed to the agents necessary for activation of the S1 subunit of the toxin.

REFERENCES

Ad Hoc Group for the Study of Pertussis Vaccines. (1988). Placebo-controlled trial of two acellular pertussis vaccines in Sweden—protective efficacy and adverse events. *Lancet i*: 955–960.

Antoine, R., and Locht, C. (1990). Roles of the disulfide bond and the carboxy-terminal region of the S1 subunit in the assembly and biosynthesis of pertussis toxin. *Infect. Immun. 58*: 1518–1526.

Arciniega, J. L., Burns, D. L., Garcia-Ortigoza, E., and Manclark, C. R. (1987). Immune response to the B oligomer of pertussis toxin. *Infect. Immun. 55*: 1132–1136.

Arico, B., and Rappuoli, R. (1987). *Bordetella parapertussis* and *Bordetella bronchiseptica* contain transcriptionally silent pertussis toxin genes. *J. Bacteriol. 169*: 2847–2853.

Arico, B., Miller, J. F., Roy, C., Stibitz, S., Monack, D., Falkow, S., Gross, R., and Rappuoli, R. (1989). Sequences required for expression of *Bordetella pertussis* virulence factors share homology with prokaryotic signal transduction proteins. *Proc. Natl. Acad. Sci. U.S.A. 86*: 6671–6675.

Armstrong, G. D., Howard, L. A., and Peppler, M. S. (1987). Use of glycosyltransferases to restore pertussis toxin receptor activity to asialoagalactofetuin. *J. Biol. Chem. 263*: 8677–8684.

Bokoch, G. M., and Gilman, A. G. (1984). Inhibition of receptor mediated release of arachidonic acid by pertussis toxin. *Cell 39*: 301–308.

Brennan, M. J., David, J. L., Kenimer, J. G., and Manclark, C. R. (1988). Lectin-like binding of pertussis toxin to a 165-kilodalton Chinese hamster cell glycoprotein. *J. Biol. Chem. 263*: 4895–4899.

Burnette, W. N., Cieplak, W., Mar, V. L., Kaljot, K. T., Sato, H., and Keith, J. M. (1988). Pertussis toxin S1 mutant with reduced enzyme activity and a conserved protective epitope. *Science 242*: 72–74.

Burnette, W. N., Arciniega, J. L., Mar, V. L., and Burns, D. L. (1992). Properties of pertussis toxin B oligomer assembled in vitro from recombinant polypeptides produced by *Escherichia coli. Infect. Immun. 60*: 2252–2256.

Burns, D. L., and Manclark, C. R. (1986). Adenine nucleotides promote dissociation of pertussis toxin subunits. *J. Biol. Chem. 261*: 4324–4327.

Burns, D. L., and Manclark, C. R. (1989). Role of cysteine 41 of the A subunit of pertussis toxin. *J. Biol. Chem. 264*: 564–568.

Burns, D. L., Kenimer, J. G., and Manclark, C. R. (1987a). Role of the A subunit of pertussis toxin in alteration of Chinese hamster ovary cell morphology. *Infect. Immun. 55*: 24–28.

Burns, D. L., Hausman, S. Z., Lindner, W., Robey, F. A., and Manclark, C. R. (1987b). Structural characterization of pertussis toxin A subunit. *J. Biol. Chem. 262*: 17677–17682.

Capiau, C., Petre, J., Van Damme, J., Puype, M., and Vandekerckhove, J. (1986). Protein-chemical analysis of pertussis toxin reveals homology between the subunits S2 and S3, between the S1 and A chains of enterotoxins of *Vibrio cholerae* and *Escherichia coli* and identifies S2 as the haptoglobin binding subunit. *FEBS Lett 204*: 336–340.

Casey, P. J., and Gilman, A. G. (1988). G protein involvement in receptor–effector coupling. *J. Biol. Chem. 263*: 2577–2580.

Clark, C. G., and Armstrong, G. D. (1990). Lymphocyte receptors for pertussis toxin. *Infect. Immun. 58*: 3840–3846.

Cortina, G., and Barbieri, J. T. (1991). Localization of a region of the S1 subunit of pertussis toxin required for efficient ADP-ribosyltransferase activity. *J. Biol. Chem. 266*: 3022–3030.

Cortina, G., Krueger, K. M., and Barbieri, J. T. (1991). The carboxyl terminus of the S1 subunit of pertussis toxin confers high affinity binding to transducin. *J. Biol. Chem. 266*: 23810–23814.

Deutscher, S. L., and Hirschberg, C. B. (1986). Mechanism of galactosylation in the Golgi apparatus. *J. Biol. Chem. 261*: 96–100.

Deutscher, S. L., Nuwayhid, N., Stanley, P., Briles, E., and Hirschberg, C. (1984). Translocation across Golgi vesicle membranes: a CHO glycosylation mutant deficient in CMP-sialic acid transport. *Cell 39*: 295–299.

Francotte, M., Locht, C., Feron, C., Capiau, C., and de Wilde, M. (1989). Monoclonal antibodies specific for pertussis toxin subunits and identification of the haptoglobin-binding site. In *Vaccines 89*, R. A. Lerner, H. Ginsberg, R. M. Chanock, and F. Brown (Eds.), Cold Spring Harbor Laboratory, Cold Spring Harbor, NY, pp. 243–247.

Gill, D. M. (1978). Seven toxic peptides that cross cell membranes. In *Bacterial Toxins and Cell Membranes*, J. Jeljaszewica and T. Wadstrom (Eds.), Academic, New York, pp. 291–332.

Gilman, A. G. (1984). G proteins and dual control of adenylate cyclase. *Cell 36*: 577–579.

Gilman, A. G. (1987). G proteins: transducers of receptor generated signals. *Annu. Rev. Biochem. 56*: 615–649.

Glaser, P., Sakamoto, H., Bellalou, J., Ullmann, A., and Danchin, A. (1988). Secretion of cyclolysin, the calmodulin-sensitive adenylate cyclase-hemolysin bifunctional protein of *Bordetella pertussis*. *EMBO J. 7*: 3997–4004.

Gray, L. S., Huber, K. S., Gray, M. C., Hewlett, E. L., and Engelhard, V. H. (1989). Pertussis toxin effects on T lymphocytes are mediated through CD3 and not by PT catalyzed modification of a G protein. *J. Immunol. 142*: 1631–1638.

Gross, R., and Rappuoli, R. (1988). Positive regulation of pertussis toxin expression. *Proc. Natl. Acad. Sci. U.S.A. 85*: 3913–3917.

Hausman, S. Z., and Burns, D. L. (1992). Interaction of pertussis toxin with cells and model membranes. *J. Biol. Chem. 267*: 13735–13739.

Hausman, S. Z., and Burns, D. L. (1993). Binding of pertussis toxin to lipid vesicles containing glycolipids. *Infect. Immun. 61*: 335–337.

Hausman, S. Z., Manclark, C. R., and Burns, D. L. (1990). Binding of ATP by pertussis toxin and isolated toxin subunits. *Biochemistry 29*: 6128–6131.

Hewlett, E. L., Sauer, K. T., Myers, G. A., Cowell, J. L., and Guerrant, R. L. (1983). Induction of a novel morphological response in Chinese hamster ovary cells by pertussis toxin. *Infect. Immun. 40*: 1198–1203.

Huh, Y. J., and Weiss, A. A. (1991). A 23-kilodalton protein, distinct from BvgA, expressed by virulent *Bordetella pertussis* binds to the promoter region of *vir*-regulated toxin genes. *Infect. Immun. 59*: 2389–2395.

Irons, L. I., and MacLennan, A. P. (1979). Isolation of the lymphocytosis promoting factor-haemagglutinin of *Bordetella pertussis* by affinity chromatography. *Biochim. Biophys. Acta 580*: 175–185.

Kaslow, H. R., and Lesikar, D. D. (1987). Sulfhydryl-alkylating reagents inactivate the NAD glycohydrolase activity of pertussis toxin. *Biochemistry 26*: 4397–4402.

Kaslow, H. R., Platler, B. W., Schlotterbeck, J. D., Mar, V. L., and Burnette, W. N. (1992).

Site-specific mutagenesis of the pertussis toxin S1 subunit gene: effects of amino acid substitutions involving residues 50–58. *Vaccine Res. 1:* 47–54.

Katada, T., and Ui, M. (1982a). ADP-ribosylation of the specific membrane protein of C6 cells by islet-activating protein associated with modification of adenylate cyclase activity. *J. Biol. Chem. 257:* 7210–7216.

Katada, T., and Ui, M. (1982b). Direct modification of the membrane adenylate cyclase system by islet-activating protein due to ADP-ribosylation of a membrane protein. *Proc. Natl. Acad. Sci. U.S.A. 79:* 3129–3133.

Katada, T., Tamura, M., and Ui, M. (1983). The A promoter of islet-activating protein, pertussis toxin, as an active peptide catalyzing ADP-ribosylation of a membrane protein. *Arch. Biochem. Biophys. 224:* 290–298.

Krueger, K. M., Mende-Mueller, L. M., and Barbieri, J. T. (1991). Protease treatment of pertussis toxin identifies the preferential cleavage of the S1 subunit. *J. Biol. Chem. 266:* 8122–8128.

Kuldau, G. A., de Vos, G., Owen, J., McCaffrey, G., and Zambryski, P. (1990). The *virB* operon of *Agrobacterium tumefaciens* pTiC58 encodes 11 open reading frames. *Mol. Gen. Genet. 221:* 256–266.

Lang, A., Ganss, M. T., and Cryz, S. J. (1989). Monoclonal antibodies that define neutralizing epitopes of pertussis toxin: conformational dependence and epitope mapping. *Infect. Immun. 57:* 2660–2665.

Li, E., Becker, A., and Stanley, S. L., Jr. (1988). Use of Chinese hamster ovary cells with altered glycosylation patterns to define the carbohydrate specificity of *Entamoeba histolytica* adhesion. *J. Exp. Med. 167:* 1725–1730.

Lim, L. K., Sekura, R. D., and Kaslow, H. R. (1985). Adenine nucleotides directly stimulate pertussis toxin. *J. Biol. Chem. 260:* 2585–2588.

Locht, C., and Keith, J. M. (1986). Pertussis toxin gene: nucleotide sequence and genetic organization. *Science 232:* 1258–1264.

Locht, C., Cieplak, W., Marchitto, K. S., Sato, H., and Keith, J. M. (1987). Activities of complete and truncated forms of pertussis toxin subunits S1 and S2 synthesized by *Escherichia coli. Infect. Immun. 55:* 2546–2553.

Lory, S. (1992). Determinants of extracellular protein secretion in gram-negative bacteria. *J. Bacteriol. 174:* 3423–3428.

MacIntyre, E. A., Tatham, P. E. R., Abdul-Gaffar, R., and Linch, D. C. (1988). The effects of pertussis toxin on human T lymphocytes. *Immunology 64:* 427–432.

Marchitto, K. S., Munoz, J. J., and Keith, J. M. (1987). Detection of subunits of pertussis toxin in Tn5-induced *Bordetella* mutants deficient in toxin biological activity. *Infect. Immun. 55:* 1309–1313.

Meade, B. D., Kind, P. D., and Manclark, C. R. (1984). Lymphocytosis-promoting factor of *Bordetella pertussis* alters mononuclear phagocyte circulation and response to inflammation. *Infect. Immun. 46:* 733–739.

Miller, J. F., Roy, C. R., and Falkow, S. (1989). Analysis of *Bordetella pertussis* virulence gene regulation by use of transcriptional fusions in *Escherichia coli. J. Bacteriol. 171:* 6345–6348.

Miller, J. F., Roy, C. R., and Falkow, S. (1990). Regulation of *fhaB, bvg,* and *ptx* transcription in *Escherichia coli:* a comparative analysis. In *Proceedings of the Sixth International Symposium on Pertussis,* C. R. Manclark (Ed.), Dept. of Health and Human Services Publ. 90-1164, U.S. Public Health Service, Bethesda, MD, pp. 217–224.

Montecucco, C., Tomasi, M., Schiavo, G., and Rappuoli, R. (1986). Hydrophobic photolabelling of pertussis toxin subunits interacting with lipids. *FEBS Lett. 194:* 301–304.

Moss, J., and Vaughan, M. (1979). Activation of adenylate cyclase by choleragen. *Annu. Rev. Biochem. 48:* 581–600.

Moss, J., Stanley, S. J., Burns, D. L., Hsia, J. A., Yost, D. A., Myers, G. A., and Hewlett, E. L. (1983). Activation by thiol of the latent NAD glycohydrolase and ADP-ribosyltransferase activities of *Bordetella pertussis* toxin (islet-activating protein). *J. Biol. Chem. 258*: 11879–11882.

Moss, J., Stanley, S. J., Watkins, P. A., Burns, D. L., Manclark, C. R., Kaslow, H. R., and Hewlett, E. L. (1986). Stimulation of the thiol-dependent ADP-ribosyltransferase and NAD glycohydrolase activities of *Bordetella pertussis* toxin by adenine nucleotides, phospholipids, and detergents. *Biochemistry 25*: 2720–2725.

Muller, A. S., Leeuwenburg, J., and Pratt, D. S. (1986). Pertussis: epidemiology and control. *Bull. WHO 64*: 321–331.

Nicosia, A., and Rappuoli, R. (1987). Promoter of the pertussis toxin operon and production of pertussis toxin. *J. Bacteriol. 169*: 2843–2846.

Nicosia, A., Perugini, M., Franzini, C., Casagli, M. C., Borri, M. G., Antoni, G., Almoni, M., Neri, P., Ratti, G., and Rappuoli, R. (1986). Cloning and sequencing of the pertussis toxin genes: operon structure and gene duplication. *Proc. Natl. Acad. Sci. U.S.A. 83*: 4631–4635.

Nilsson, B., Norden, N. E., and Svensson (1979). Structural studies in the carbohydrate portion of fetuin. *J. Biol. Chem. 254*: 4545–4553.

Nogimori, K., Tamura, M, Yajima, M., Ito, K., Nakamura, T., Kajikawa, N., Maruyama, Y., and Ui, M. (1984). Dual mechanisms involved in development of diverse biological activities of islet-activating protein, pertussis toxin, as revealed by chemical modification of lysine residues in the toxin molecule. *Biochim. Biophys. Acta 801*: 232–243.

Pizza, M., Bugnoli, M., Manetti, R., Covacci, A., and Rappuoli, R. (1990). The subunit S1 is important for pertussis toxin secretion. *J. Biol. Chem. 265*: 17759–17763.

Pugsley, A., d'Enfert, C., and Kornacker, M. G. (1990). Genetics of extracellular secretion of gram-negative bacteria. *Annu. Rev. Genet. 24*: 67–90.

Pugsley, A. P., Poquet, I., and Kornacker, M. G. (1991). Two distinct steps in pullulanase secretion by *Escherichia coli* K12. *Mol. Microbiol. 5*: 865–873.

Ribi, H. O., Ludwig, D. S., Mercer, K. L., Schoolnik, G. K., and Kornberg, R. D. (1988). Three dimensional structure of cholera toxin penetrating a lipid membrane. *Science 239*: 1272–1276.

Rogers, T. S., Corey, S. J., and Rosoff, P. M. (1990). Identification of a 43-kilodalton human T lymphocyte membrane protein as a receptor for pertussis toxin. *J. Immunol. 145*: 678–683.

Roy, C. R., Miller, J. F., and Falkow, S. (1989). The *bvgA* gene of *Bordetella pertussis* encodes a transcriptional activator required for coordinate regulation of several virulence genes. *J. Bacteriol. 171*: 6338–6344.

Sato, H., Ito, A., Chiba, J., and Sato, Y. (1984). Monoclonal antibodies against pertussis toxin: effect on toxin activity and pertussis infections. *Infect. Immun. 46*: 422–428.

Saukkonen, K., Burnette, W. N., Mar, V. L., Masure, H. R., and Tuomanen, E. I. (1992). Pertussis toxin has eucaryotic-like carbohydrate recognition domains. *Proc. Natl. Acad. Sci. U.S.A. 89*: 118–122.

Schmidt, M. A., and Schmidt, W. (1989). Inhibition of pertussis toxin binding to model receptors by antipeptide antibodies directed at an antigenic domain of the S2 subunit. *Infect. Immun. 57*: 3828–3833.

Schmidt, M. A., Raupach, B., Szulczynski, M., and Marzillier, J. (1991). Identification of linear B-cell determinants of pertussis toxin associated with the receptor recognition site of the S3 subunit. *Infect. Immun. 59*: 1402–1408.

Sekura, R. D., Fish, F., Manclark, C. R., Meade, B., and Zhang, Y. (1983). Pertussis toxin: affinity purification of a new ADP-ribosyltransferase. *J. Biol. Chem. 258*: 14647–14651.

Sekura, R. D., Zhang, Y., and Quentin-Millet, M.-J. (1985). Pertussis toxin: structural elements involved in the interaction with cells. In *Pertussis Toxin*, R. D. Sekura, J. Moss, and M. Vaughan (Eds.), Academic, New York, pp. 45–64.

Spangrude, G. J., Braaten, B. A., and Daynes, R. A. (1984). Molecular mechanisms of lymphocyte extravasation. I. Studies of two selective inhibitors of lymphocyte recirculation. *J. Immunol. 132*: 354–362.

Stewart, S. J., Prpic, V., Johns, J. A., Powers, F. S., Graber, S. E., Forbes, J. T., and Exton, J. H. (1989). Bacterial toxins affect early events of T lymphocyte activation. *J. Clin. Invest.* *83*: 234–242.

Stibitz, S., and Yang, M.-S. (1991). Subcellular localization and immunochemical detection of proteins encoded by the *vir* locus of *Bordetella pertussis*. *J. Bacteriol.* *173*: 4288–4296.

Strnad, C. F., and Carchman, R. A. (1987). Human T lymphocyte mitogenesis in response to the B oligomer of pertussis toxin is associated with an early elevation in cytosolic calcium concentrations. *FEBS Lett 225*: 16.

Tamura, M., Nogimori, K., Murai, S., Yajima, M., Ito, K., Katada, T., Ui, M., and Ishii, S. (1982). Subunit structure of islet-activating protein, pertussis toxin, in conformity with the A-B model. *Biochemistry 21*: 5516–5522.

Tamura, M., Nogimori, K., Yajima, M., Ase, K., and Ui, M. (1983). A role of the B oligomer moiety of islet-activating protein, pertussis toxin, in development of the biological effects on intact cells. *J. Biol. Chem. 258*: 6756–6761.

Thom, R. E., and Casnellie, J. E. (1989). Pertussis toxin activated protein kinase C and a tyrosine protein kinase in the human T cell line Jurkat. *FEBS Lett. 244*: 181–184.

Tomasi, M., and Montecucco, C. (1981). Lipid insertion of cholera toxin after binding to G_{M1}-containing liposomes. *J. Biol. Chem. 256*: 11177–11181.

Ward, J. E., Akiyoshi, D. E., Reiger, D., Datta, A., Gordon, M. P., and Nester, E. W. (1988). Characterization of the *virB* operon from an *Agrobacterium tumefaciens* Ti plasmid. *J. Biol. Chem. 263*: 5804–5814.

Weiss, A. A., and Falkow, S. (1983). Genetic analysis of phase change in *Bordetella pertussis*. *Infect. Immun. 43*: 263–269.

Weiss, A. A., and Goodwin, M. S. M. (1989). Lethal infection by *Bordetella pertussis* mutants in the infant mouse model. *Infect. Immun. 57*: 3757–3764.

Weiss, A. A., Melton, A. R., Walker, K. E., Andraos-Selim, C., and Meidl, J. J. (1989). Use of the promoter fusion transposon Tn*5 lac* to identify mutations in *Bordetella pertussis* *vir*-regulated genes. *Infect. Immun. 57*: 2674–2682.

Weiss, A. A., Johnson, F. D., and Burns, D. L. (1993). Molecular characterization of an operon required for pertussis toxin secretion. *Proc. Natl. Acad. Sci. U.S.A. 90*: 2970–2974.

Wickner, W., Driessen, A. J. M., and Hartl, F.-U. (1991). The enzymology of protein translocation across the *Escherichia coli* plasma membrane. *Annu. Rev. Biochem. 60*: 101–124.

Wisnieski, B. J., and Bramhall, J. S. (1981). Photolabelling of cholera toxin subunits during membrane penetration. *Nature 289*: 319–321.

Witvliet, M. H., Burns, D. L., Brennan, M. J., Poolman, J. T., and Manclark, C. R. (1989). Binding of pertussis toxin to eucaryotic cells and glycoproteins. *Infect. Immun. 57*: 3324–3330.

Zambryski, P. (1992). Chronicles from the *Agrobacterium*-plant cell DNA transfer story. *Annu. Rev. Plant Physiol. Plant Mol. Biol. 43*: 465–490.

20

GTP-Binding Proteins as the Substrates of Pertussis Toxin-Catalyzed ADP-Ribosylation

Toshiaki Katada

University of Tokyo, Tokyo, Japan

Taroh Iiri, Katsunobu Takahashi, Hiroshi Nishina, and Yasunori Kanaho

Tokyo Institute of Technology, Yokohama, Japan

I. INTRODUCTION

Pertussis toxin produced by *Bordetella pertussis* bacteria was first introduced by Ui and his colleagues into research on signal transduction under the name islet-activating protein (IAP) in 1979, when a mechanism of the toxin-induced modification of insulin secretory responses in rat pancreatic islets was reported (Katada and Ui, 1979a,b). The action of IAP was soon proved to be due to the ADP-ribosylation of a 41-kDa membrane protein (Katada and Ui, 1982a), which was later identified as the α subunit of the inhibitory guanine nucleotide-binding regulatory component of adenylyl cyclase, G_i (Katada et al., 1984a). Since demonstration of the molecular mechanism of IAP, the toxin has been widely used as the best probe for identifying and analyzing major $\alpha\beta\gamma$-trimeric G proteins that are involved in a variety of membrane-bound receptor-induced signal transductions. The purpose of this review is to summarize representative results of successful applications of IAP to research on cell signaling, together with recent advances in the studies of IAP-sensitive G proteins. The earlier use of IAP as a probe for signal transducers has been reviewed by Ui and his colleagues (Ui, 1984, 1986, 1990; Ui et al., 1984).

II. PERTUSSIS TOXIN AS AN ADP-RIBOSYLATION ENZYME

A. Lack of Receptor-Mediated Inhibition of Insulin Release and cAMP Formation in Pancreatic Islets Isolated from IAP-Pretreated Rats

Insulin secretion from pancreatic islet B cells is regulated by various hormones and nutrients. A typical example is illustrated in Figure 1. The rate of insulin release from the isolated pancreas was stimulated by a high concentration (16.7 mM) of glucose, and its stimulation was inhibited upon the occupation of α_2-adrenergic receptors in the cells by the agonist epinephrine. When pancreas isolated from rats that had been injected once with IAP 3 days before was subjected to the same perfusion experiments, the rate of insulin release in response to 16.7 mM glucose was greatly increased (Katada and Ui, 1979a). Moreover, the inhibitory action of the α_2-receptor agonist was completely abolished. The loss of the α_2-receptor function effected by IAP in vivo was observed from 1 day after its injection and was surprisingly persistent, lasting for at least a week.

   The molecular mechanism of the toxin action in mammalian cells was next studied in vitro by Katada and Ui (1980), who took advantage of the unique duration or irreversibility of its action in vivo. Pancreatic islet cells isolated from IAP-injected rats still retained the altered responses in vitro to glucose and the α_2-receptor agonist. At that time, cAMP appeared to be an important second messenger for receptor-coupled regulation of insulin release in rat pancreatic islets (Katada et al., 1981a,b; Yamazaki et al., 1982). Figure 2 shows insulin release from and cAMP content of rat pancreatic islet cells in

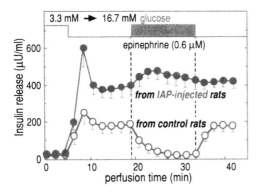

Fig. 1 Profiles of insulin secretion during perfusion of rat pancreas. Pancreases isolated from (○) nontreated control and (●) IAP-injected rats were perfused with a medium containing glucose (3.3 or 16.7 mM). Epinephrine (0.6 μM) was also added to the perfusion medium where indicated. (Data modified from Katada and Ui, 1979a.)

response to 16.7 mM glucose, together with the effect of epinephrine on the cellular response. In islet cells isolated from normal rats, the inhibition of insulin release by epinephrine was accompanied by a marked decrease in the cellular cAMP content. Not only α_2-adrenergic but also somatostatin receptor-mediated inhibition in insulin release appeared to result from the receptor-induced decrease in the cAMP content. When islet cells from IAP-injected rats were incubated with epinephrine, there was no decrease in the cellular cAMP content. In addition, the decrease in the cAMP content was not observed upon addition of somatostatin to islet cells from the toxin-injected rats. The same cellular effects of IAP were produced by prolonged exposure of the islet cells to the toxin in vitro as well as by prior injection into cell donor rats in vivo (Katada and Ui, 1981a).

Cells or tissues susceptible to IAP were not limited to pancreatic islet cells. Decreases in cellular cAMP content in response to muscarinic or α_2-adenosine agonists were also abolished in rat cardiac cells that had been incubated with IAP for 3 h (Hazeki and Ui, 1981). These changes or the lack of inhibitory effects on cellular cAMP content were still observed in the presence of a phosphodiesterase inhibitor, suggesting that the site of

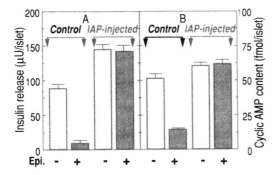

Fig. 2 Effects of epinephrine on insulin secretion and cAMP content in rat pancreatic islets. Pancreatic islets isolated from nontreated control and IAP-injected rats were incubated with 16.7 mM glucose in the presence or absence of 5 μM epinephrine, and insulin release and cAMP content in the islets were then determined. (Data modified from Katada and Ui, 1979b.)

interaction of IAP with cells is proximal to the site of cAMP formation or membrane adenylyl cyclase.

B. Catalytic Action of IAP in Cell Cultures

Since epinephrine-induced inhibition of insulin release from rat pancreatic islets was very effectively abolished by IAP, this characteristic was used for studies of the mode of IAP action in the primary culture of islet cells (Katada and Ui, 1980). The effect of IAP on insulin release in the presence of 16.7 mM glucose and 5 μM epinephrine was investigated kinetically in detail, and the results are illustrated in Figure 3. The presence of epinephrine in the culture medium caused a marked decrease in the rate of insulin release. The addition of IAP at the start of the primary culture gave rise to increases in the secretion rate. The reversal rate of the epinephrine-induced inhibition was increased as the concentration of IAP was increased from 1 to 100 ng/mL. Figure 3 also indicates clearly that the action of IAP took place with a definite delay of about 1 h, following which insulin release proceeded in a manner dependent on the concentration of IAP. The lag period before the onset of IAP action was highly dependent on the temperature of the culture. The continuous existence of IAP molecules in the culture medium was, however, not essential for duation of the IAP action. These results suggested that the uptake of IAP molecules by the islets was temperature-dependent and that the absolute lag corresponded to the time required for the penetration of the toxin molecules across the plasma membranes (Section II.D).

The effect of IAP on rat pancreatic islet cells was so persistent that receptor-coupled adenylyl cyclase displayed the uniquely modified behavior in membranes isolated from the islets that had been exposed to IAP (Katada and Ui, 1981b). The basal activity of membrane adenylyl cyclase measured without inclusion of any receptor agonist or GTP in the assay mixture as well as its elevated activity by forskolin were essentially the same in membranes from IAP-treated and nontreated islet cells. However, α_2-adrenergic receptor-mediated inhibition of the membrane cyclase was markedly attenuated by prior exposure of membrane donor islet cells to IAP.

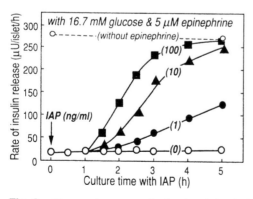

Fig. 3 Progressive reversal of epinephrine-induced inhibition of insulin release by increasing concentrations of IAP in rat pancreatic islets. Islets isolated from rat pancreas were cultured with 16.7 mM glucose plus 5 μM epinephrine in the presence of various concentrations of IAP. Insulin contents were determined at various times, and the rates of insulin release at the given times are plotted against the culture times. (Data modified from Katada and Ui, 1980.)

Not only receptor-coupled adenylyl cyclase but also GTP-stimulated activity was susceptible to IAP in several types of cells. In rat C6 glioma cells, the cyclase activity stimulated with GTP was strikingly augmented after IAP treatment (Katada et al., 1982). Figure 4 shows the effects of prior culture of the C6 cells with IAP for various times on the GTP-dependent cyclase activity in the membranes. As the time of IAP treatment was prolonged from 2 to 18 h, the concentration-dependence curve was shifted to the left; the concentrations of the toxin required for half-maximal effects were around 100, 1, and 0.001 ng/mL for 2, 6, and 18 h of treatment, respectively. Thus, the action of IAP on C6 cells (and rat pancreatic islets, see Fig. 3) appeared to be catalytic in the sense that the effect caused by its minute amounts was amplified by prolongation of the exposure time. Conceivably, only one molecule of the toxin that had entered the cells would be enough to alter the GTP-stimulated cyclase activity in the intact C6 cells just as had been reported for diphtheria toxin-induced cell toxicity.

C. Requirement of NAD and ATP for the Direct Action of IAP in Cell-Free Systems

Until 1982, it was not possible to observe the direct effect of IAP on cell-free preparations. The effect of the toxin even on intact cells was so slow in onset, the cells had to be exposed to IAP in the incubation medium for more than several hours before the toxin-induced specific modification of receptor-mediated responses became evident (see Figs. 3 and 4, for example). C6 cell membranes were expected to provide a cell-free system suitable for further studies on the action of IAP, since the index for toxin sensitivity (GTP-dependent adenylyl cyclase) was higher in this type of cell than in others. Direct addition of IAP to washed membrane preparations from various cells including C6 glioma cells, however, failed to affect the membrane adenylyl cyclase activated by GTP. Therefore, a concentrated homogenate of C6 cells, instead of the washed membranes, was used for the initial evaluation of IAP action. Thus, the direct action of IAP on the membrane cyclase was first observed in a cell-free system fortified with the cell cytosol (Katada and Ui, 1982a).

As shown in Figure 5, addition of the cell cytosol in increasing concentrations enhanced IAP-induced activation of GTP-stimulated cyclase activity progressively,

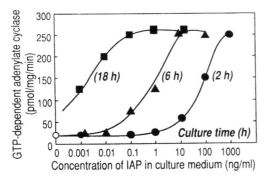

Fig. 4 Enhancement of GTP-dependent adenylate cyclase activity in C6 cell membranes by the culture of the donor cells with IAP. C6 cells were cultured for the indicated times in the presence of various concentrations of IAP, and membrane fractions prepared from the cells were assayed for adenylate cyclase activity with 10 μM GTP. (Data modified from Katada et al., 1982.)

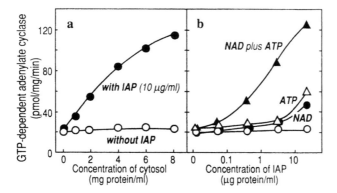

Fig. 5 Requirement of cytosol or NAD plus ATP for IAP-induced enhancement of GTP-dependent adenylate cyclase activity in C6 cell membranes. (a) C6 cell membranes and various concentrations of the cell cytosol were incubated (●) with or (○) without 10 μg/mL IAP. (b) The membranes were incubated with various concentrations of IAP alone (○) or in the presence (●), of 1 mM NAD, (△) 1 mM ATP, or (▲) NAD *plus* ATP. The membrane fractions were washed and assayed for adenylate cyclase activity with 10 μM GTP. (Data modified from Katada and Ui, 1982a.)

suggesting that factor(s) present in the cytosol was required for the IAP action. The action of the cytosol to support the IAP-induced activation of GTP-stimulated adenylyl cyclase was observed even with cytosol that had been heated but not with cytosol that had been depleted of nucleotides by charcoal adsorption. The heat-stable factors were then identified as NAD and ATP. When the washed membranes were incubated with IAP in the presence of both NAD and ATP, an increase in the concentration of IAP resulted in progressive enhancement of the GTP-stimulated cyclase activity (Fig. 5). It was soon proved that NAD served as the substrate of IAP-catalyzed ADP-ribosylation of a 41-kDa protein during incubation of C6 cell membranes. ADP-ribosylation of the 41-kDa membrane protein increased progressively as the concentration of IAP in the incubation medium was increased, in parallel with the concurrent graded increases in GTP-stimulated adenylyl cyclase activity of the same membranes. These results strongly suggested that the enhancement of GTP-stimulated adenylyl cyclase activity was caused by the ADP-ribosylation of the 41-kDa protein.

ADP-ribosylation of the 41-kDa protein also occurs in intact cells during exposure to IAP. Since radiolabeled NAD does not enter cells, evidence for the occurrence of ADP-ribosylation in intact cells is rather indirect but substantial. Intact C6 cells were exposed to increasing concentrations of IAP, and the membranes prepared from these cells were then subjected to ADP-ribosylation with [α-^{32}P]NAD and IAP. The amount of [^{32}P]ADP-ribose incorporated into the 41-kDa protein decreased progressively as the concentration of IAP in the cell-exposed medium was increased. The results are reasonably explained in terms of previous ADP-ribosylation of the 41-kDa protein by intracellular nonradioactive NAD during exposure of cells to IAP. The decreased incorporation of radioactivity into the membrane protein therefore reflects the ADP-ribosylation of the same protein in intact cells from which the membranes was prepared.

D. A-B Structure of IAP and the A Protomer as an ADP-Ribosylation Enzyme

The structure of the IAP molecule is characterized by a hexamer with five dissimilar subunits that are named in order of decreasing masses: S1 (M_r of 26,220), S2 (21,920), S3 (21,860), S4 (12,060), and S5 (10,940). Exposure of the toxin to 5 M urea gave four separate peaks upon subsequent column chromatography on carboxymethyl-Sepharose (see Fig. 6); two of these peaks were composed of S1 and S5, and the other two were of dimers, termed D1 and D2 (Tamura et al., 1982). These two dimers could be further resolved into their constituent subunits by exposure to 8 M urea, D1 to S2 and S4, and D2 to S3 and S4. S5 is also referred to as the C subunit because it connects two dimers of D1 and D2. Based on the relative intensities of the individual subunits stained after SDS-PAGE, the molecular ratio of these subunits in the toxin molecule were calculated as 1:1:1:2:1 (S1:S2:S3:S4:S5). The genes coding for these five subunits of IAP were later cloned and sequenced by Locht and Keith (1986) and Nicosia et al. (1986). S2 and S3 share approximately 70% homology at the amino acid level.

When the native toxin that had been exposed to urea under milder conditions was applied to a column of haptoglobin-Sepharose, a protein was passed through the column, and it was identified as S1. Proteins retained in the column were further eluted by 0.5 M NaCl plus 3 M potassium isothiocyanate, again as a sharp single peak of the pentamer consisting of S2–S5. Thus, IAP was readily dissociated into S1 and the rest of the pentamer. S1 was enzymatically active, as described below, and it was hence referred to as an A (active) protomer. The pentamer appeared to act as a B (binding) oligomer, since the A protomer is incapable of entering mammalian cells unless it forms a hexamer by association with the pentamer. Thus, IAP was characterized as an A-B structure as had been proposed for cholera and diphtheria toxins (Gill, 1978; Domenighini et al., 1991).

The A protomer as an active enzyme of NAD glycohydrolase was demonstrated under conditions lacking cellular components but containing dithiothreitol (Katada et al., 1983). This indicates that reductive cleavage of an intrachain disulfide bond in the A protomer is responsible for the induction of NAD glycohydrolase (and ADP-ribosyltransferase) activity. The A protomer and holotoxin in the presence of NAD and ATP were equipotent

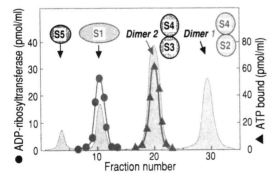

Fig. 6 Elution profile of urea-treated IAP from carboxymethyl-Sepharose column. IAP that had been treated with 5 M urea was resolved into the constituent subunits by carboxymethyl-Sepharose column chromatography (Tamura et al., 1982). (●) ADP-ribosyltransferase and (▲) ATP-binding activities were then measured.

in enhancing GTP-dependent adenylyl cyclase activity and in causing ADP-ribosylation of the 41-kDa protein when added directly to the cell-free membrane preparation from C6 cells. However, similar actions of holotoxin IAP observed upon its addition to the intact cells were not mimicked by its A protomer, indicating that the A protomer had to be associated with the B oligomer to become accessible to its site (G protein) on the inner surface of the cell membranes. As described above, the essential role of ATP was evident in 1982, when IAP was shown for the first time to be effective directly on C6 cell membranes only if the incubation medium was fortified with NAD and ATP (see Fig. 5b). Studies were later done on the mechanism by which ATP supports the action of IAP in cell-free systems. Burns and Manclark (1986) reported that ATP (or ADP) promotes detergent-induced dissociation of the A protomer from the B oligomer. The binding of ATP to the B oligomer has indeed been demonstrated with the isolated toxin subunits (Hausman et al., 1990). We have also found (Katada et al., unpublished observation) that ATP or its analogue binds to S3 in dimer 2 (see Fig. 6).

Figure 7 illustrates a proposed model for the action of IAP in mammalian cells. The first step of interaction of IAP with intact cells is the binding of its B-oligomer moiety to certain sites on the cell surface. The toxin-binding sites have not yet been fully identified, though a glycoprotein has been suggested to serve as the toxin receptor (Brennan et al., 1988, 1991; Clark and Armstrong, 1990; Heerze et al., 1992; Kaslow and Burns, 1992; Saukkonen et al., 1992). The next step is the entry of the toxin molecule into the cells. The holotoxin could enter the cells together with the putative binding proteins at a slow but steady rate under physiological conditions. After the entry, the A-protomer moiety is released from the holotoxin molecule as a result of an allosteric effect of intracellular ATP that binds to the S3 subunit of the B oligomer (Hausman and Burns, 1992). The active center of the ADP-ribosyltransferase (Cortina and Barbieri, 1991; Cortina et al., 1991) in the released A-protomer molecule is now unmasked to interact with intracellular reduced glutathione, which causes cleavage of disulfide bonds essential for the enzymatic activity. It has been reported that the cysteine at position 41 in the A protomer is important for ADP-ribosyltransferase and NADase activities (Burns and Manclark, 1989; Kaslow et al., 1989). However, recombinant S1 analogues with alanine or serine substituted for cysteine 41 retain considerable ADP-ribosyltransferase activity (Locht et al., 1990). Thus, Cys-41 appears to be located within the NAD-binding site. Tryptophan and glutamic acid residues in the S1 subunit are also essential for the enzymatic activities of IAP (Locht et al., 1989).

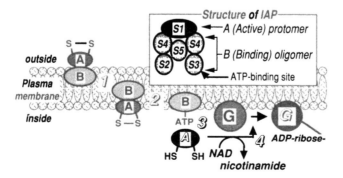

Fig. 7   Structure of IAP and its action on intact cells. See text for explanation.

III. G PROTEINS AS THE TARGET OF PERTUSSIS TOXIN-CATALYZED ADP-RIBOSYLATION

A. Attenuation by IAP of Membrane Receptor-Mediated Inhibition of Adenylyl Cyclase in a Variety of Mammalian Cells

In addition to C6 rat glioma cells, a variety of mammalian tissues and cells are susceptible to IAP-induced modification of the adenylyl cyclase system. Membranes from adipocytes were uniquely suitable for the studies of IAP action. The adipocyte adenylyl cyclase was stimulated and inhibited at lower and higher GTP concentrations, respectively, by isoproterenol, a β-adrenergic agonist. No inhibition was observed, but stimulation was still observed in membrances prepared from adipocytes that had been exposed to IAP (Murayama and Ui, 1983, 1984). In cholera toxin-treated cell membranes, inhibition of adenylyl cyclase was still observed at higher GTP concentrations in which GTP-stimulated cyclase activity was generally very high in either the presence or absence of isoproterenol. In NG 108-15 neroblastoma X glioma hybrid cells, there are multiple receptors, such as α_2-adrenergic, M_2-muscarinic, and δ-opioid receptors, all of which couple to adenylyl cyclase in an inhibitory fashion. These receptor-mediated decreases in cellular cAMP or inhibitions of adenylyl cyclase in membranes were no longer observable in cells or membranes that had been exposed to IAP to effect ADP-ribosylation of IAP-susceptible G proteins (Kurose et al., 1983).

Since then, many studies have revealed that membrane receptor-mediated inhibition of adenylyl cyclase is attenuated or abolished by the IAP-induced modification of G proteins in many types of cells (Murayama et al., 1983; Ui, 1984). Once ADP-ribosylated by IAP, G_i thus appeared to lose its ability to communicate between membrane-bound receptors and the adenylyl cyclase catalyst (see Section V). Thus, the substrate of IAP-catalyzed ADP-ribosylation proved to be G_i involved in adenylyl cyclase inhibition, an entity apparently different from G_s, which mediates the cyclase stimulation.

B. Purification and Characterization of IAP-Substrate G Proteins

An IAP-substrate G protein with a molecular weight of 41,000 was first identified and purified from rabbit liver membrane (Bokoch et al., 1983, 1984). The purified substrate, G_i protein, was proved to function as the inhibitory GTP-binding regulatory component of adenylyl cyclase in a reconstitution system (Katada et al., 1984a,b,c). Shortly thereafter, another IAP-substrate G protein, termed G_o, was identified in brain tissues in high abundance and was purified from the membranes (Sternweis and Robishaw, 1984; Neer et al., 1984; Milligan and Klee, 1985; Katada et al., 1986b). G_t (transducin), which mediates visual transduction in the rod and cone cells of vertebrate retinas (Stryer and Bourne, 1986), also served as the substrate of IAP-catalyzed ADP-ribosylation. All of the IAP-substrate G proteins acting as membrane signal transducers have been characterized by their heterotrimeric structures. They consist of a guanine nucleotide-binding α subunit (M_r of 39,000–41,000), which also contains the ADP-ribosylation site, a β subunit (35,000 or 36,000), and a γ subunit (6000–9000). Although G proteins are heterotrimers, they behave functionally as dimers; the $\beta\gamma$ components of G proteins are never resolved further unless they are denatured. As shown in Figure 8, the function of G proteins as signal transducers is regulated cyclically by (1) dissociation of bound GDP from α subunits, (2) association of GTP with α subunits, and (3) hydrolysis of GTP to GDP and P_i. The binding of GTP appears to result in dissociation of the

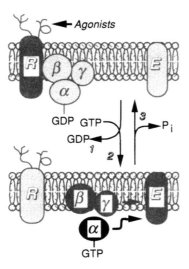

Fig. 8 Role of G proteins in membrane receptor-mediated signal transduction. See text for explanation.

trimeric G proteins into α and $\beta\gamma$ subunits, either of which consequently regulates the activity of appropriate effectors. Hydrolysis of GTP initiates the deactivation of G proteins. Thus, GTP-bound α subunits and/or $\beta\gamma$ subunits function as active components of G proteins for effector molecules.

The nucleotide binding to G proteins is influenced by Mg^{2+}, anions, and proteins that interact with α subunits (particularly $\beta\gamma$ subunits and receptors) (Katada et al., 1986b; Higashijima et al., 1987a,b,c,d). Binding of guanine nucleotide to $\alpha\beta\gamma$-trimeric G proteins or to their α subunits is clearly not a diffusion-controlled process, and it proceeds at a rate that is independent of nucleotide concentration. This anomaly is explainable by the fact that the proteins, as purified, contain stoichiometric amounts of GDP bound with high affinity (Ferguson et al., 1986). Dissociation of GDP obviously must precede binding of GTP, since there is only one site for nucleotide on G proteins. Although bound GTP is hydrolyzed to GDP by an intrinsic GTPase activity of α subunits, the rate of GTPase activity is remarkably low (approximately 1–5 min^{-1}) in comparison with activities catalyzed by the usual metabolic enzymes. Thus, the lifetime of GTP-bound G proteins would be many seconds. Agonist-receptor complex stimulates the GDP-GTP exchange reaction occurring on G proteins (Kurose et al., 1986). It also stimulates the steady-state rate of GTPase activity without affecting the actual catalytic rate (k_{cat}) (Haga et al., 1985). These effects are due exclusively to receptor-stimulated dissociation of GDP, subsequent association of GTP, and the resultant accumulation of significant levels of GTP-bound G proteins. Since the agonist-receptor complex functions catalytically, one receptor can interact with more than 10 molecules of G proteins over a period of a few seconds. Thus, the catalytic action of the agonist-receptor complex together with the relatively long lifetime of GTP-bound G proteins provides considerable amplification of the receptor-mediated signals. More detailed properties of G proteins have been extensively reviewed by Gilman (1987), Stryer and Bourne (1986), and more recently Kaziro et al. (1991).

C.   Molecular Heterogeneity of IAP-Sensitive G Proteins

Molecular cloning has revealed that there are at least seven complementary DNAs encoding the α subunits of IAP-substrate G proteins in mammalian cells—those of G_i-1, G_i-2, G_i-3, G_o-1, G_o-2, G_t-1, and G_t-2 (Kaziro et al., 1991). G_t-1 is localized in disk membranes of the rod outer segments in vertebrate retinas and communicates between light-activated rhodopsin and cyclic GMP-specific phosphodiesterase. G_t-2 is expressed selectively in the cones of the retina, where it may play an analogous role. Although the family of G_i and G_o is distributed with much less selectivity in many mammalian tissues, G_o is the major IAP substrate in mammalian brain. Immunohistochemical studies with the anti-G_o-α antibody reveal that G_o is localized not only in the central nervous system and retina but also in the nerve terminals in peripheral tissues and pancreatic islets (Terashima et al., 1987a,b,c, 1988a,b). Considerable amount of G_o were detected in neuroendocrine tumors but not in nonneuronal tumors by a specific radioimmunoassay (Kato et al., 1987).

The α subunits of two or more IAP-substrate G proteins are generally expressed in most mammalian tissues and cell lines. G_i-1 has proved to be the major G_i in brain (Asano et al., 1990), but G_i-2 and G_i-3 have also been detected and purified as minor components. On the other hand, G_i-2 is the major IAP substrate in a number of peripheral tissues, including neutrophils (and HL-60 cells) (Bokoch and Parkos, 1988; Dickey et al., 1987; Gierschik et al., 1986, 1987; Oinuma et al., 1987; Rotrosen et al., 1988; Uhing et al., 1987), human platelets (Crouch et al., 1989; Nagata et al., 1988), rat heart (Luetje et al., 1988), bovine lung (Kanaho et al., 1989; Morishita et al., 1988), a macrophage cell line (Backlund et al., 1988), and bovine chromaffin cells (Negishi et al., 1987, 1988; Toutant et al., 1987). G_i-1 or G_i-3 is the additional IAP substrate contained in these peripheral tissues.

The existence of multiple forms of the $\beta\gamma$ subunits of G proteins has also been demonstrated by biochemical, immunological, and molecular cloning techniques. Four different β-subunit cDNAs have been cloned from mammals; β_1, β_2, β_3, and β_4 (Amatruda et al., 1988; Gao et al., 1987; Levine et al., 1990; Simon et al., 1991). β_1, which is the major β subunit of G_t and encodes a 36-kDa polypeptide, appears to be expressed in all tissues including brain (Mumby et al., 1986). β_2 encodes a 35-kDa subunit that is absent from the retina but is expressed in many other tissues (Evans et al., 1987). The β_3 cDNA has been isolated from retinal cDNA libraries, and the corresponding mRNA and protein have been detected in retinal cones, brain and liver (Fung et al., 1992; Lee et al., 1992b; Levine et al., 1990). Different cDNAs for γ subunits (γ_1, γ_2, γ_3) have also been cloned (Hurley et al., 1984; Gautam et al., 1989, 1990). One subunit (γ_1) isolated from retina encodes the γ subunit of G_t, whereas γ_2 has been isolated from brain. The γ_2 mRNA is expressed in brain, adrenal, and liver tissues. The sequence of a fourth γ cDNA, γ_4, and that of a fifth γ cDNA have been established (Fisher and Aronson, 1992), and additional γ subtypes may exist (Tamir et al., 1991). Although these different β and γ subunits may form $\beta\gamma$ complexes with distinct characteristics (Kontani et al., 1992), their functional differences have been attributed to the γ subunits (Pronin and Gautam, 1992; Kleuss et al., 1992, 1993). The exact combinations of heterogeneous β and γ subunits under physiological conditions are still unclear.

IV.   SIGNAL TRANSDUCTION MEDIATED BY PERTUSSIS TOXIN-SENSITIVE G PROTEINS

A.   Adenylyl Cyclase and IAP-Sensitive G Proteins

Until the molecular heterogeneity of mammalian adenylyl cyclase became apparent, the activity of the cyclase had been supposed to be regulated simply by two G proteins, G_s

and G_i, responsible for stimulation and inhibition, respectively. However, the properties of the eukaryotic adenylyl cyclases that have been cloned to date (types I–V) are not identical (see review by Tang and Gilman, 1992). All of the mammalian enzymes are activated by the α subunit of G_s, and some (particularly the type I) by Ca^{2+}-calmodulin. Whereas the $\beta\gamma$ subunits of G proteins, which were initially proposed as a negative regulator for the cyclase catalyst directly or indirectly (Katada et al., 1984c, 1986a, 1987), apparently inhibit the type I cyclase when stimulated by either G_s or Ca^{2+}-calmodulin, the same $\beta\gamma$ greatly potentiates the stimulatory action of G_s-α on the type II or type IV cyclase (Tang and Gilman, 1991; Tang et al., 1991). Since the concentration of $\beta\gamma$ required for the regulation of cyclase activity is rather high, it is likely that the necessary $\beta\gamma$ fraction is supplied from IAP-substrate G proteins such as G_i and G_o, which are more abundant than G_s in most tissues or cells. Regulation of effectors by the $\beta\gamma$ subunits of G proteins in addition to their α subunits may well be a general phenomenon as described below.

Although the α subunits of IAP-substrate G proteins (G_i-1, G_i-2, and G_i-3) are demonstrably important for the inhibition of adenylyl cyclase in cells or membranes (Kobayashi et al., 1990; Wong et al., 1992), molecular mechanisms are still uncertain and may vary with the types of cyclase.

B. Regulation of Phospholipases by IAP-Sensitive G Proteins

As described above, IAP was initially developed as a probe for G proteins involved in the inhibition of adenylyl cyclase. However, the toxin exerted its influence on a variety of receptor-coupled signal transductions. The first example suggesting that IAP can interfere with a receptor–effector coupling other than the adenylyl cyclase system was reported by Nakamura and Ui (1983). They found that passive cutaneous anaphylaxis induced by an antigen challenge to antibody-sensitized rats was inhibited by prior treatment of the rats with IAP. The toxin-induced inhibition was due to inhibition of the antigen-induced histamine release from mast cells in these rats. Compound 48/80-mediated secretion of histamine from isolated mast cells was also prevented almost totally by prior treatment of the cells with IAP. The possiblity that the adenylyl cyclase system was involved in this IAP-sensitive signaling pathway in mast cells could be excluded, since the agonist-induced slight increase in cellular cAMP was inhibited rather than enhanced by IAP treatment of mast cells; cAMP accumulation generally attenuated histamine secretion in this cell type. The site of IAP action, or the site of an IAP-sensitive G-protein involvement, appeared to be receptor-coupled activation of phospholipase C in mast cells (Nakamura and Ui, 1984, 1985). Similar effects of IAP are also obtained for receptor-mediated generation of superoxide anion (O_2^-) in guinea pig neurophils (Okajima and Ui, 1984; Ohta et al., 1985; Okajima et al., 1985). Since the middle of the 1980s, many laboratories have published papers showing the involvement of G proteins in receptor-mediated regulation of phosphoinositide-specific phospholipase C (PLC) in various tissues and cells (see reviews by Ui, 1986, 1989).

During the course of these studies, the complexity of this signal transduction system has become apparent. In terms of IAP sensitivity, two distinct types of G proteins appear to couple receptors to PLC stimulation—an IAP-sensitive G protein, such as G_i and G_o, and an IAP-insensitive one, termed G_q (Smrcka et al., 1991; Taylor et al., 1991; Taylor and Exton, 1991). There are four distinct α subunits (G_q, G_{11}, G_{14}, and G_{16}) in the G_q family, and none of the four contains a site for pertussis toxin modification (Simon et al., 1991). Moreover, at least three types of PLC (PLC-β, PLC-γ, and PLC-δ) have been described; each type contains more than one subtype (see reviews by Rhee and Choi, 1992a,b). Recent studies with the purified proteins have clearly indicated that the α subunits of G_q and G_{11} specifically activate PLC-β1 but not PLC-γ1 or PLC-δ1 (Taylor et al., 1991; Blank et al., 1991; Conklin et al., 1992; Berstein et al., 1992a,b; Park et al., 1993; Wu et al., 1993); the α subunit of G_{16} activates most effectively PLC-β2 (Lee et al., 1992a). The receptors that activated PLC via the G_q family include those for thromboxane A_2, bradykinin, angiotensin, histamine, vasopressin, muscarinic acetylcholine, and α_1-adrenaline (Shenker et al., 1991; Wange et al., 1991; Gutowski et al., 1991; Berstein et al., 1992b; Wilkie et al., 1991; Mullaney et al., 1993; Wu et al., 1992a,b). More recently, in addition to the regulation of PLC by α subunits, it was reported that the $\beta\gamma$ subunits of G proteins can stimulate PLC-β, especially the β2 subtype (Blank et al., 1992; Boyer et al., 1992; Camps et al., 1992a,b; Carozzi et al., 1993; Katz et al., 1992). Thus, there appears to be specificity in the interaction of different members of the G_q subfamily or $\beta\gamma$ subunits with different PLC-β effectors.

Receptor activation of phospholipase A_2 also appears to be mediated by IAP-sensitive G proteins in certain types of cells. In Swiss 3T3 cells, thrombin or other agonist-induced stimulation of PLC is not affected, but stimulation of phospholipase A_2 is totally abolished by prior exposure of cells to IAP (Murayama and Ui, 1985). Since the first indication that an IAP-substrate G protein may couple certain receptors to phospholipase A_2 activation, an involvement of G proteins in phospholipase regulation has been suggested in several other types of cells (Axelrod et al., 1988; Silk et al., 1989; Narasimhan et al., 1990; Teitelbaum, 1990; Felder et al., 1991; Rubin et al., 1991; Murayama et al., 1990a; Ando et al., 1992; Xing and Mattera, 1992).

IAP-sensitive G proteins appear to be involved also in the activation of phospholipase D (PLD) in certain types of cells, such as neutrophils (and the differentiated HL-60 cells), and in rat 1 fibroblasts expressing α_2-C10 adrenergic receptor. Treatment of these types of cells with IAP prevents the agonist-stimulated PLD activation (Agwu et al., 1989; Kanaho et al., 1991; Pai et al., 1988; MacNulty et al., 1992). Thus, in analogy with PLC, the agonist activation of PLD is mediated by IAP-sensitive G proteins. In order to maximally activate PLD, however, another factor, Ca^{2+}, in addition to G proteins, may be required, since GTPγS and Ca^{2+} synergistically activate the enzyme in cell-free systems from HL-60 cells and human neutrophils (Anthes et al., 1989; Olson et al., 1991). In contrast, in other types of cells, including hepatocytes, rat mesangial cells, and rat pheochromocytoma PC12 cells, IAP has no effect on the agonist activation of PLD (Bocckino et al., 1987; Harris and Bursten, 1992; Kanoh et al., 1992). In these cells, IAP-insensitive G proteins appear to regulate PLD activity as GTPγS activates the enzyme in hepatocyte membranes, digitonin-permeabilized PC12 cells, and microsomes of rat mesangial cells. In common with PLC, PLD may be predicted to exist in multiple isoforms that may be regulated by IAP-sensitive and -insensitive G proteins.

C. Regulation of Ion Channels by IAP-Sensitive G Proteins

Cation channels are additional effectors that appear to be regulated directly by IAP-substrate G proteins (see review by Brown and Birnbaumer, 1990; Trautwein and Hescheler, 1990; Schultz et al., 1990). The best studied of these are cardiac K^+ channels. Not only the α subunit of G_i-3 (formerly named G_k in terms of its function) purified from human erythrocytes (Yatani et al., 1987) but also the recombinant α subunit (Mattera et al., 1989) was found to open K^+ channels in inside-out atrial membrane patches. Recombinant α

subunits of G_r-1 and G_i-2, however, were also effective in this regard, suggesting that any of the G_i subtypes may regulate more than one effector system (Yatani et al., 1988). Similarly, the α subunits of G_s were shown to stimulate both adenylyl cyclase and Ca^{2+} channels (Yatani et al., 1988; Mattera et al., 1989). In the case of voltage-dependent Ca^{2+} channels, IAP-substrate G proteins appear to be involved in channel gating in an inhibitory fashion; G_o is more potent than G_i in restoring opioid-induced inhibition of the channels in NG 108-15 cells (Rosenthal and Schultz, 1987; Rosenthal et al., 1988). Moreover, by intranuclear injection of antisense oligonucleotides into rat pituitary GH_3 cells, the essential role of the G_o-type G proteins in the Ca^{2+} channel inhibition has been elucidated; the subtypes G_o-1 and G_o-2 appear to mediate inhibition through the muscarinic and somatostatin receptors, respectively (Kleuss et al., 1991).

Atrial K^+ channels are activated not only by the α subunits of G_i or G_o but also by the $\beta\gamma$ subunits (Logothetis et al., 1987; Kurachi et al., 1989a). Although the two findings initially appeared to contradict each other (Birnbaumer, 1987; Bourne, 1987; Okabe et al., 1990), the $\beta\gamma$-induced K^+ channel activation has been proved (Kobayashi et al., 1990; Ito et al., 1991, 1992; Kurachi et al., 1992). Thus, both components of G protein subunits, α and $\beta\gamma$, may interact with K^+ channels under physiological conditions as has been observed in adenylyl cyclase regulation. As described above, the $\beta\gamma$ subunits of G proteins also activate phospholipase A_2, producing arachidonic acid. The metabolites of arachidonic acid, leukotrienes, may be secondarily responsible for receptor-coupled opening of K^+ channels in some cases (Kurachi et al., 1989b).

D. Other Cell Signals Affected by IAP

IAP-susceptible G proteins have been implicated in cell proliferation and differentiation (see references cited in reviews by Ui, 1986, 1989). The binding of growth factors to cell surface receptors causes rapid regulation of some of the effector systems, including phospholipases C and A_2, adenylyl cyclase, ion channels, and protein kinases. Intracellular signals generated by these effectors must trigger or form cascades of complex cellular events leading to cell proliferation or differentiation. Thus, it is reasonable to assume that G proteins, as signal transducers, couple certain growth factor receptors to some of these effector systems (Nishimoto et al., 1989; Crouch et al., 1990; Murayama et al., 1990b; Okamoto et al., 1990). In Swiss 3T3 fibroblasts, growth is triggered by combined addition of a competence factor and a progression factor. IAP added 3–6 h after the addition of the growth factors to serum-depleted 3T3 cells was still significantly effective in decreasing the cell proliferation (Murayama and Ui, 1987). A similar effect of IAP was observed on rat hepatocyte growth (Fujinaga et al., 1989). Moreover, retinoic acid-induced differentiation of human leukemic HL-60 cells was inhibited by prior exposure of the cells to IAP (Tohkin et al., 1989), and the toxin-induced inhibition was similarly observed with a delayed addition of IAP (Ohoka et al., unpublished observation). Thus, in addition to fulfilling their well-known roles as receptor-coupled signal transducers, IAP-substrate G proteins must be involved in the sustained duration of initial signals triggered via effectors or in the onset of other intracellular or autocrine signals that lag behind the initial signals responsible for cell proliferation and differentiation by an unknown mechanism.

Since the signal-coupling G proteins are mainly localized in the inner surface of plasma membranes of cells, all IAP-induced modifications are thought to be due to ADP-ribosylation of the plasma membrane-bound G proteins. However, it cannot be excluded that G proteins may play a postreceptor role in the slow cell responses. In this regard, there are several reports showing that IAP-substrate G proteins are present in cell cytosol (Takahashi et al., 1991), nuclei (Crouch, 1991; Takei et al., 1992), and Golgi (Barr et al., 1991; Stow et al., 1991; Burgoyne, 1992; Leyte et al., 1992) in addition to the plasma membranes.

Some nonmammalian cells also prosses G proteins that are readily ADP-ribosylated by IAP, and their cellular functions are modified by the toxin treatment (Arata et al., 1992; Kopf et al., 1986; Oinuma et al., 1986; Tsuda et al., 1986; Van Haastert et al., 1987; Tadenuma et al., 1991). An example is starfish oocytes (Shilling et al., 1989). In response to a meiosis-inducing hormone, 1-methyladenine (1-MA), starfish oocytes undergo reinitiation of meiosis with germinal vesicle breakdown (GVBD). The 1-MA-initiated GVBD was inhibited by prior microinjection of IAP into the oocytes, suggesting that an IAP-sensitive G protein is involved in the 1-MA-induced signal transduction. The G protein serving as the substrate of IAP has been purified from the plasma membranes of starfish oocytes as an $\alpha\beta\gamma$-trimeric structure consisting of 39-kDa, α, 37-kDa β, and 8-kDa γ subunits (Tadenuma et al., 1991). The purified starfish G protein displays high-affinity GTP binding and low GTPase activity. The toxin-catalyzed ADP-ribosylation of the 39-kDa α subunit is inhibited by its association with a nonhydrolyzable GTP analogue. Thus, the starfish G protein appears to be similar to mammalian G proteins at least in terms of structure and properties of nucleotide binding and modification by pertussis toxin (Chiba et al., 1992). There were apparently two forms of 1-MA receptors with high and low affinities in the membranes (Tadenuma et al., 1992). The high-affinity form was converted into the low-affinity form in the presence of a nonhydrolyzable analogue of GTP. The 39-kDa α subunit of the starfish G protein was also ADP-ribosylated by cholera toxin only when 1-MA was added to the membranes. The ADP-ribosylated 39-kDa α subunit could be immunoprecipitated with antibodies raised against the carboxy-terminal site of mammalian inhibitory G-α. Microinjection of the purified $\beta\gamma$ subunits into the oocytes induced GVBD as had been observed with 1-MA (Chiba et al., unpublished observation). These results indicate that 1-MA receptors are functionally coupled with the IAP-substrate G protein and that the $\beta\gamma$ complex resolved from the α subunit is an active molecule for the initiation of GVBD in starfish oocytes.

V. POSTTRANSLATIONAL MODIFICATIONS OF G PROTEINS

A. Characteristics of ADP-Ribosylation Reactions Catalyzed by Bacterial Toxins

Figure 9 shows ADP-ribosylation sites modified by IAP and cholera toxin in the α subunits of G proteins. The IAP site, cysteine, is the fourth amino acid away from the carboxy terminus of the α subunits of G_i, G_o, and G_t-α (West et al., 1985; Hoshino et al., 1990). The corresponding amino acid in G_s-α is tyrosine instead of cysteine. Therefore, IAP never ADP-ribosylates the α subunit of G_s. This domain is supposed to be an interacting region with receptor molecules. On the other hand, the cholera toxin site, arginine, is located on the middle of the α subunits of G_s and G_t (Van Dop et al., 1984). There are also quite similar sequences in the α subunits of G_i and G_o. However, G_i and G_o did not serve as the substrate for cholera toxin-induced modification under usual conditions (but see below). It has not been fully understood why G_i and G_o are poor substrates for cholera toxin. Abnormal G-α genes, termed *gsp* or *gip2*, have been discovered in several types of endocrine tumor tissues (Landis et al., 1989; Lyons et al., 1990). In some of these

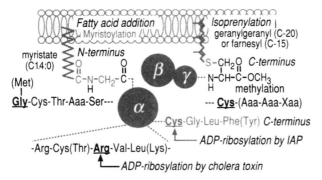

Fig. 9 Posttranslational modifications on the α and $\beta\gamma$ subunits of G proteins. See text for explanation.

oncogenes, the cholera toxin site, arginine, is replaced with cysteine or histidine, and properties of the mutated α subunits have been characterized (Pace et al., 1991; Wong et al., 1991).

These toxin-catalyzed ADP-ribosylation reactions appear to be dependent on the conformation of G protein structure. A GDP-bound $\alpha\beta\gamma$-trimeric form of G protein is the real substrate for IAP-catalyzed ADP-ribosylation; nonhydrolyzable GTP analogue-bound α subunits separated from $\beta\gamma$ never serve as the toxin substrate (Katada et al., 1986b). On the other hand, cholera toxin ADP-ribosylates G_s and activates the G proteins in membranes or cells. However, purified G_s does not serve as the substrate for cholera toxin-induced modification. A protein cofactor, termed ADP-ribosylation factor or ARF, is required for the toxin-induced ADP-ribosylation (Schleifer et al., 1982; Kahn and Gilman, 1984). ARF is also a GTP-binding protein with an M_r of 21,000, and binding of GTP or its analogues to ARF is necessary for ARF activity (Kahn and Gilman, 1986). The binding of the small GTP-binding protein to cholera toxin is GTP-dependent and activates the ADP-ribosyltransferase activity of the toxin (Bobak et al., 1990; Tsai et al., 1991). Studies have revealed that ARF is involved in vesicular trafficking between the endoplasmic reticulum and the cis-Golgi (Balch et al., 1992; Donaldson et al., 1992).

B. Modifications of the G-Protein Functions by ADP-Ribosylation

Although G_i and G_o do not serve as substrates of cholera toxin under usual conditions, it has been reported that the IAP-substrate G proteins could be ADP-ribosylated by cholera toxin when receptors coupled to the G proteins are stimulated by the agonists (Gierschik and Jakobs, 1987; Milligan, 1987, 1988; Milligan and McKenzie, 1988). The sites modified by the two toxins IAP and cholera toxin are clearly not identical, and the cholera toxin-induced ADP-ribosylation is dependent on GTP or its analogues mostly in a biphasic fashion (Iiri et al., 1989). IAP-substrate G proteins maintained in a certain unknown state in membranes as a result of stimulation of coupled receptors or the addition of GTP at the appropriate concentration are likely to serve as the substrate of cholera toxin. In this regard, there are reports that cholera toxin ADP-ribosylates the retinal G protein transducin only when it is bound to photoexcited rhodopsin and depleted of its nucleotide (Bornancin and Chabre, 1991). Since the IAP-substrate G proteins could be ADP-ribosylated by both pertussis and cholera toxins, these toxin-induced modifications have been investigated extensively with the purified proteins (Iiri et al., 1991, 1992; Bornancin et al., 1992).

Modifications of the G-protein functions by IAP- or cholera toxin-catalyzed ADP-ribosylation are summarized in Figure 10. The ADP-ribosylation by IAP results in uncoupling of the G proteins from receptors, and thus the receptor-mediated signal transduction is selectively abolished. There is no apparent change in other functions of G proteins; neither GDP-GTP exchange and GTPase reactions nor interaction of G proteins with the effectors, such as adenylyl cyclase, are modified by IAP (Katada et al., 1986a). On the other hand, cholera toxin-induced ADP-ribosylation exerts its influence on a variety of the functions of G proteins (Iiri et al., 1991, 1992; Bornancin et al., 1992). High-affinity agonist binding to receptors is stimulated, and GTPase activity of G proteins is markedly inhibited by the cholera toxin-induced modification. There is also an increase in the interaction of G proteins with the effector of phospholipase C. Moreover, the substrate activity of G proteins for IAP is still observable even after its ADP-ribosylation by cholera toxin, though there is a reduction in the substrate activity for IAP (Iiri et al., 1991). These properties imply that the cholera toxin-modified G protein would act as an efficient signal transducer between receptors and effectors.

C. Posttranslational Modifications of G Proteins Other Than ADP-Ribosylation

Figure 9 also illustrates posttranslational modifications other than ADP-ribosylation occurring on G-protein subunits. One of the recently well-characterized covalent modifications on G proteins is isoprenylation of the γ subunits. As has been observed in ras proteins (Finegold et al., 1990), the γ subunits of G proteins contain a Cys-A-A-X consensus sequence at the carboxy terminus (Gautam et al., 1989; Robishaw et al., 1989). Posttranslational addition of a polyisoprenoid (farnesyl or geranylgeranyl) group to the cysteine through a thioester linkage (Fukada et al., 1990; Lai et al., 1990; Maltese and Robishaw, 1990; Mumby et al., 1990a; Yamane et al., 1990; Sanford et al., 1991) appears to be involved in the association of γ subunits with plasma membranes or other protein components (Muntz et al., 1992; Ohguro et al., 1990, 1991).

On the other hand, the amino termini (NH_2-Met-Gly-X-X-X-Ser) of IAP-substrate G proteins, G_i and G_o, are myristoylated; the initial methionine is removed, exposing the α-amino group of the glycine, which is N-myristoylated (Buss et al., 1987; Duronio et al., 1991). The fatty acid addition appears to be essential for the attachment of α subunits

G_s	DLLRC**R**VLTS (Arg201)	RMHLRQYELL	
G_i-2	DVLRT**R**VKTT (Arg179)	KNNLKD**C**GLF (Cys352)	
G_o	DILRT**R**VKTT (Arg179)	ANNLRG**C**GLY (Cys351)	
G_t	DVLRS**R**VKTT (Arg173)	KENLKD**C**GLF (Cys347)	
	* *Cholera toxin-site*	** *IAP-site*	

Bacterial toxins	Functions of G proteins				
	Coupling with R	GDP-GTP exchange	GTPase	Interaction with E	Substrate for toxins
Cholera	↑	↓	↓	↑	↓ *for IAP*
IAP	*Uncoupling*	→	→	→	*No for CTX*

Fig. 10 ADP-ribosylation sites on the α subunits of G proteins and modifications of the functions of G proteins by ADP-ribosylation. See text for explanation.

to plasma membranes or the $\beta\gamma$ subunits (Jones et al., 1990; Mumby et al., 1990b; Linder et al., 1991). It has recently been reported that there is heterogeneous modification of the amino terminus of the α subunit of G_t with laurate ($C_{20:0}$) and unsaturated $C_{14:2}$ and $C_{14:1}$ fatty acids (Kokame et al., 1992).

VI. CONCLUDING REMARKS

Many studies are capable of taking advantage of pertussis toxin (IAP) as the best probe for G-protein involvement in membrane receptor-coupled regulation of effectors in a number of signal transduction systems at the level of purified proteins, isolated fractions intact cells, and experimental animals. IAP is the best agent to abolish or attenuate signal transduction of G proteins selectively when applied to animals or intact cells. The blockage of signaling that occurs upon application of the toxin provides conclusive evidence for the involvement of an IAP-susceptible G protein in the signaling. Although 10 years have passed since demonstration of the molecular mechanism of IAP action, the toxin is still used to analyze cell functions and to identify the signal-coupling G proteins.

Acknowledgments

The marjority of our studies were supported by research grants from the Scientific Research Fund of the Ministry of Education, Science, and Culture of Japan, and the Human Frontier Science Program.

REFERENCES

Agwu, D. E., McPhail, L. C., Chabot, M. C., Daniel, L. W., Wykle, R. L., and McCall, C. E. (1989). Choline-linked phosphoglycerides. A source of phosphatidic acid and diglycerides in stimulated neutrophils. *J. Biol. Chem. 264*: 1405–1413.

Amatruda, T. T. I., Gautam, N., Fong, H. K. W., Northup, J. K., and Simon, M. I. (1988). The 35- and 36-kDa β subunits of GTP-binding regulatory proteins are products of separate genes. *J. Biol. Chem. 263*: 5008–5011.

Ando, M., Furui, H., Suzuki, K., Taki, F., and Takagi, K. (1992). Direct activation of phospholipase A$_2$ by GTP-binding protein in human peripheral polymorphonuclear leukocytes. *Biochem. Biophys. Res. Commun. 183*: 708–713.

Anthes, J. C., Eckel, S., Siegel, M. I., Egan, R. W., and Billah, M. M. (1989). Phospholipase D in homogenates from HL-60 granulocytes: implications of calcium and G protein control. *Biochem. Biophys. Res. Commun. 163*: 657–664.

Arata, Y., Tada, S., and Ui, M. (1992). Probable occurrence of toxin-susceptible G proteins in the nematode *Caenorhabditis elegans*. *FEBS Lett. 300*: 73–76.

Asano, T., Shinohara, H., Morishita, R., and Kato, K. (1990). Immunochemical and immunohistochemical localization of the G protein Gil in rat central nervous tissues. *J. Biochem. 108*: 988–994.

Axelrod, J., Burch, R. M., and Jelsema, C. L. (1988). Receptor-mediated activation of phospholopase A$_2$ via GTP-binding proteins: arachidonic acid and its metabolites as second messengers. *Trends Neurosci. 11*: 117–123.

Backlund, P. S., Jr., Aksamit, R. R., Unson, C. G., Goldsmith, P., Spiegel, A. M., and Milligan, G. (1988). Immunochemical and electrophoretic characterization of the major pertussis toxin substrate of the RAW264 macrophage cell line. *Biochemistry 27*: 2040–2046.

Balch, W. E., Kahn, R. A., and Schwaninger, R. (1992). ADP-ribosylation factor is required for vesicular trafficking between the endoplasmic reticulum and the cis-Golgi compartment. *J. Biol. Chem. 267*: 13053–13061.

Barr, F. A., Leyte, A., Mollner, S., Pfeuffer, T., Tooze, S. A., and Huttner, W. B. (1991). Trimeric G-proteins of the trans-Golgi network are involved in the formation of constitutive secretory vesicles and immature secretory granules. *FEBS. Lett.* *294*: 239–243.

Berstein, G., Blank, J. L., Jhon, D. Y., Exton, J. H., Rhee, S. G., and Ross, E. M. (1992a). Phospholipase C-β1 is a GTPase-activating protein for Gq/11, its physiologic regulator. *Cell* *70*: 411–418.

Berstein, G., Blank, J. L., Smrcka, A. V., Higashijima, T., Sternweis, P. C., Exton, J. H., and Ross, E. M. (1992b). Reconstitution of agonist-stimulated phosphatidylinositol 4,5-bisphosphate hydrolysis using purified m1 muscarinic receptor, Gq/11, and phospholipase C-β1. *J. Biol. Chem.* *267*: 8081–8088.

Birnbaumer, L. (1987). Which G protein subunits are the active mediators in signal transduction? *Trends Pharmacol. Sci.* *8*: 209–211.

Blank, J. L., Ross, A. H., and Exton, J. H. (1991). Purification and charcterization of two G-proteins that activate the β1 isozyme of phosphoinositide-specific phospholipase C. Identification as members of the Gq class. *J. Biol. Chem.* *266*: 18206–18216.

Blank, J. L., Brattain, K. A., and Exton, J. H. (1992). Activation of cytosolic phosphoinositide phospholipase C by G-protein βγ subunits. *J. Biol. Chem.* *267*: 23069–23075.

Bobak, D. A., Bliziotes, M. M., Noda, M., Tsai, S. C., Adamik, R., and Moss, J. (1990). Mechanism of activation of cholera toxin by ADP-ribosylation factor (ARF): both low- and high-affinity interactions of ARF with guanine nucleotides promote toxin activation. *Biochemistry* *29*: 855–861.

Bocckino, S. B., Blackmore, P. F., Wilson, P. B., and Exton, J. H. (1987). Phosphatidate accumulation in hormone-treated hepatocytes via a phospholipase D mechanism. *J. Biol. Chem.* *262*: 15309–15315.

Bokoch, G. M., and Parkos C. A. (1988). Identification of novel GTP-binding proteins in the human neutrophil. *FEBS Lett.* *227*: 66–70.

Bokoch, G. M., Katada, T., Northup, J. K., Hewlett, E. L., and Gilman, A. G. (1983). Identification of the predominant substrate for ADP-ribosylation by islet activating protein. *J. Biol. Chem* *258*: 2072–2075.

Bokoch, G. M., Katada, T., Northup, J. K., Ui, M., and Gilman, A. G. (1984). Purification and properties of the inhibitory guanine nucleotide-binding regulatory component of adenylate cyclase. *J. Biol. Chem.* *259*: 3560–3577.

Bornancin, F., and Chabre, M. (1991). Cholera toxin ADP-ribosylates transducin only when it is bound to photoexcited rhodopsin and depleted of its nucleotide. *FEBS Lett.* *291*: 273–276.

Bornancin, F., Franco, M., Bigay, J., and Chabre, M. (1992). Functional modifications of transducin induced by cholera or pertussis-toxin-catalyzed ADP-ribosylation. *Eur. J. Biochem.* *210*: 33–44.

Bourne, H. R. (1987). "Wrong" subunit regulates cardiac potassium channels. *Nature 325*: 296–297.

Boyer, J. L., Waldo, G. L., and Harden, T. K. (1992). βγ-Subunit activation of G-protein-regulated phospholipase C. *J. Biol. Chem.* *267*: 25451–25456.

Brennan, M. J., David, J. K., Kenimer, J. G., and Manclark, C. R. (1988). Lectin-like binding of pertussis toxin to a 165-kilodalton Chinese hamster ovary cell glycoprotein. *J. Biol. Chem.* *263*: 4895–4899.

Brennan, M. J., Hannah, J. H., and Leininger, E. (1991). Adhesion of *Bordetella pertussis* to sulfatides and to the GalNAcβ4Gal sequence found in glycosphingolipids. *J. Biol. Chem.* *266*: 18827–18831.

Brown, A. M., and Birnbaumer, L. (1990). Ionic channels and their regulation by G protein subunits. *Annu. Rev. Physiol.* *52*: 197–213.

Burgoyne, R. D. (1992). Trimeric G proteins in Golgi transport. *Trends Biochem. Sci.* *17*: 87–88.

Burns, D. L., and Manclark, C. R. (1986). Adenine nucleotides promote dissociation of pertussis toxin subunits. *J. Biol. Chem.* *261*: 4324–4327.

Burns, D. L., and Manclark, C. R. (1989). Role of cysteine 41 of the A subunit of pertussis toxin. *J. Biol. Chem. 264*: 564–568.

Buss, J. E., Mumby, S. M., Casey, P. J., Gilman, A. G., and Sefton, B. M. (1987). Myristoylated α subunits of guanine nucleotide-binding regulatory proteins. *Proc. Natl. Acad. Sci. U.S.A. 84*: 7493–7497.

Camps, M., Carozzi, A., Schnabel, P., Scheer, A., Parker, P. J., and Gierschik, P. (1992a). Isozyme-selective stimulation of phospholipase C-β2 by G protein βγ-subunits. *Nature 360*: 684–686.

Camps, M., Hou, C., Sidiropoulos, D., Stock, J. B., Jakobs, K. H., and Gierschik, P. (1992b). Stimulation of phospholipase C by guanine-nucleotide-binding protein βγ subunits. *Eur. J. Biochem. 206*: 821–831.

Carozzi, A., Camps, M., Gierschik, P., and Parker, P. J. (1993). Activation of phosphatidylinositol lipid-specific phospholipase C-β3 by G-protein βγ subunits. *FEBS Lett. 315*: 340–342.

Chiba, K., Tadenuma, H., Matsumoto, M., Takahashi, K., Katada, T., and Hoshi, M. (1992). The primary structure of the alpha subunit of a starfish guanosine-nucleotide-binding regulatory protein involved in 1-methyladenine-induced oocyte maturation. *Eur. J. Biochem. 207*: 833–838.

Clark, C. G., and Armstrong, G. D. (1990). Lymphocyte receptors for pertussis toxin. *Infect. Immun. 58*: 3840–3846.

Conklin, B. R., Chabre, O., Wong, Y. H., Federman, A. D., and Bourne, H. R. (1992). REcombinant Gqα. Mutational activation and coupling to receptors and phospholipase C. *J. Biol. Chem. 267*: 31–34.

Cortina, G., and Barbieri, J. T. (1991). Localization of a region of the S1 subunit of pertussis toxin required for efficient ADP-ribosyltransferase activity. *J. Biol. Chem. 266*: 3022–3030.

Cortina, G., Krueger, K. M., and Barbieri, J. T. (1991). The carboxyl terminus of the S1 subunit of pertussis toxin confers high affinity binding to transducin. *J. Biol. Chem. 266*: 23810–23814.

Crouch, M. F. (1991). Growth factor-induced cell division is paralleled by translocation of G$_i$α to the nucleus. *FASEB J. 5*: 200–206.

Crouch, M. F., Winegar, D. A., and Lapetina, E. G. (1989). Epinephrine induces changes in the subcellular distribution of the inhibitory GTP-binding protein G$_{i2}$α amd a 38-kDa phosphorylated protein in the human platelet. *Proc. Natl. Acad. Sci. U.S.A. 86*: 1776–1780.

Crouch, M. F., Belford, D. A., Milburn, P. J., and Hendry, I. A. (1990). Pertussis toxin inhibits EGF-, phorbol ester- and insulin-stimulated DNA synthesis in BALB/c3T3 cells: evidence for post-receptor activation of G$_i$α. *Biochem. Biophys. Res. Commun. 167*: 1369–1376.

Dickey, B. F., Pyun, H. Y., Williamson, K. C., and Navarro, J. (1987). Identification and purification of a novel G protein from neurtophils. *FEBS Lett. 219*: 289–292.

Domenighini, M., Montecucco, C., Ripka, W. C., and Rappuoli, R. (1991). Computer modelling of the NAD binding site of ADP-ribosylating toxins: active-site structure and mechanism of NAD binding. *Mol. Microbiol. 5*: 23–31.

Donaldson, J. G., Cassel, D., Kahn, R. A., and Klausner, R. D. (1992). ADP-ribosylation factor, a small GTP-binding protein, is required for binding of the coatomer protein beta-COP to Golgi membranes. *Proc. Natl. Acad. Sci. U.S.A. 89*: 6408–6412.

Duronio, R. J., Rudnick, D. A., Adams, S. P., Towler, D. A., and Gordon, J. I. (1991). Analyzing the substrate specificity of *Saccharomyces cerevisiae* myristoyl-CoA:protein *N*-myristoyltransferase by co-expressing it with mammalian G protein α subunits in *Escherichia coli. J. Biol. Chem. 266*: 10498–10504.

Evans, T., Fawzi, A., Fraser, E. D., Brown, M. L., and Northup, J. K. (1987). Purification of a β35 form of the βγ complex common to G-proteins from human placental membranes. *J. Biol. Chem. 262*: 176–181.

Felder, C. C., Williams, H. L., and Axelrod, J. (1991). A transduction pathway associated with receptors coupled to the inhibitory guanine nucleotide binding protein Gi that amplifies ATP-mediated arachidonic acid release. *Proc. Natl. Acad. Sci. U.S.A. 88*: 6477–6480.

Ferguson, K. M., Higashijima, T., Smigel, M. D., and Gilman, A. G. (1986). The influence of bound GDP on the kinetics of guanine nucleotide binding to G proteins. *J. Biol. Chem.* *261*: 7393–7399.

Finegold, A. A., Schafer, W. R., Rine, J., Whiteway, M., and Tamanoi, F. (1990). Common modifications of trimeric G proteins and ras protein: involvement of polyisoprenylation. *Science 249*: 165–169.

Fisher, K. J., and Aronson, N. J. (1992). Characterization of the cDNA and genomic sequence of a G protein gamma subunit (gamma5). *Mol. Cell. Biol. 12*: 1585–1591.

Fujinaga, Y., Morozumi, N., Sato, K., Tokumitsu, Y., Fujinaga, K., Kondo, Y., Ui, M., and Okajima, F. (1989). A pertussis toxin-sensitive GTP-binding protein plays a role in the G_0-G_1 transition of rat hepatocytes following establishment in primary culture. *FEBS Lett. 245*: 117–121.

Fukada, Y., Takao, T., Ohguro, H., Yoshizawa, T., Akino, T., and Shimonishi, Y. (1990). Farnesylated gamma-subunit of photoreceptor G protein indispensable for GTP-binding. *Nature 346*: 658–660.

Fung, B. K., Lieberman, B. S., and Lee, R. H. (1992). A third form of the G protein beta subunit. 2. Purification and biochemical properties. *J. Biol. Chem. 267*: 24782–24788.

Gao, B., Gilman, A. G., and Robishaw, J. D. (1987). A second form of the β subunit of signal-transducing G proteins. *Proc. Natl. Acad. Sci. U.S.A. 84*: 6122–6215.

Gautam, N., Baetscher, M., Aebersold, R., and Simon, M. I. (1989). A G protein gamma subunit shares homology with ras proteins. *Science 244*: 971–974.

Gautam N., Northup, J., Tamir, H., and Simon, M. I. (1990). G protein diversity is increased by associations with a variety of γ subunits. *Proc. Natl. Acad. Sci. U.S.A. 87*: 7973–7977.

Gierschik, P., and Jakobs, K. H. (1987). Receptor-mediated ADP-ribosylation of a phospholipase C-stimulating G protein. *FEBS Lett. 224*: 219–223.

Gierschik, P., Falloon, J., Milligan, G., Pines, M., Gallin, J. I., and Spiegel, A. (1986). Immunochemical evidence for a novel pertussis toxin substrate in human neutrophils. *J. Biol. Chem. 261*: 8058–8062.

Gierschik, P., Sidiropoulos, D., Spiegel, A., and Jakobs, K. H. (1987). Purification and immunochemical characterization of the major pertussis toxin-sensitive guanine nucleotide-binding protein of bovine neutrophil membranes. *Eur. J. Biochem. 165*: 185–194.

Gill, D. M. (1978). Seven toxic peptides that cross cell membranes. In *Bacterial Toxins and Cell Membranes*, J. Jeljaszewicz and T. Wadström (Eds.), Academic, New York, pp. 291–332.

Gilman, A. G. (1987). G proteins: transducers of receptor-generated signals. *Annu. Rev. Biochem. 56*: 615–649.

Gutowski, S., Smrcka, A., Nowak, L., Wu, D., Simon, M., and Sternweis, P. C. (1991). Antibodies to the αq subfamily of guanine nucleotide-binding regulatory protein α subunits attenuate activation of phosphatidylinositol 4,5-bisphosphate hydrolysis by hormones. *J. Biol. Chem. 266*: 20519–20524.

Haga, K., Haga, T., Ichiyama, A., Katada, T., Kurose, H., and Ui, M. (1985). Functional reconstitution of purified muscarinic receptors and inhibitory guanine nucleotide regulatory protein. *Nature 316*: 731–733.

Harris, W. E., and Bursten, S. L. (1992). Lipid A stimulates phospholipase D activity in rat mesangial cells via a G-protein. *Biochem. J. 281*: 675–682.

Hausman, S. Z., and Burns, D. L. (1992). Interaction of pertussis toxin with cells and model membranes. *J. Biol. Chem. 267*: 13735–13739.

Hausman, S. Z., Manclark, C. R., and Burns, D. L. (1990). Binding of ATP by pertussis toxin and isolated toxin subunits. *Biochemistry 29*: 6128–6131.

Hazeki, O., and Ui, M. (1981). Modification by islet-activating protein of receptor-mediated regulation of cyclic AMP accumulation in isolated rat heart cells. *J. Biol. Chem. 256*: 2856–2862.

Heerze, L. D., Chong, P., and Armstrong, G. D. (1992). Investigation of the lectin-like binding domains in pertussis toxin using synthetic peptide sequences. Identification of a sialic acid binding site in the S2 subunit of the toxin. *J. Biol. Chem.* 267: 25810–25815.

Higashijima, T., Ferguson, K. M., Smigel, M. D., and Gilman, A. G. (1987a). The effect of GTP and magnesium on the GTPase activity and the fluorescent properties of G_o. *J. Biol. Chem.* 262: 757–761.

Higashijima, T., Ferguson, K. M., and Sternweis, P. C. (1987b). Regulation of hormone-sensitive GTP-dependent regulatory proteins by chloride. *J. Biol. Chem.* 262: 3597–3602.

Higashijima, T., Ferguson, K. M., and Sternweis, P. C., Ross, E. M., Smigel, M. D., and Gilman, A. G. (1987c). The effect of activating ligands on the intrinsic fluorescence of guanine nucleotide-binding regulatory proteins. *J. Biol. Chem.* 262:752–756.

Higashijima, T., Ferguson, K. M., and Sternweis, P. C., Smigel, M. D., and Gilman, A. G. (1987d). Effects of magnesium and the $\beta\gamma$-subunit complex on the interactions of guanine nuclotides with G proteins. *J. Biol. Chem.* 262: 762–766.

Hoshino, S., Kikkawa, S., Takahashi, K., Itoh, H., Kaziro, Y., Kawasaki, H., Suzuki, K., Katada, T., and Ui, M. (1990). Identification of sites for alkylation by N-ethylmaleimide and pertussis toxin-catalyzed ADP-ribosylation on GTP-binding proteins. *FEBS Lett.* 276: 227–231.

Hurley, J. B., Fong, H. K. W., Teplow, D. B., Dreyer, W. J., and Simon, M. I. (1984). Isolation and characterization of a cDNA clone for the γ subunit of bovine retinal transducin. *Proc. Natl. Acad. Sci. U.S.A.* 81: 6948–6952.

Iiri, T., Tohkin, M., Morishima, N., Ohoka, Y., Ui, M., and Katada, T. (1989). Chemotactic peptide receptor-supported ADP-ribosylation of a pertussis toxin-substrate GTP-binding protein by cholera toxin in neutrophil-type HL-60 cells. *J. Biol. Chem.* 264: 21394–21400.

Iiri, T., Ohoka, Y., Ui, M., and Katada, T. (1991). Functional modification by cholera-toxin-catalyzed ADP-ribosylation of a guanine-nucleotide-binding regulatory protein serving as the substrate of pertussis toxin. *Eur. J. Biochem.* 202: 635–641.

Iiri, T., Ohoka, Y., Ui, M., and Katada, T. (1992). Modification of the function of pertussis toxin substrate GTP-binding protein by cholera toxin-catalyzed ADP-ribosylation. *J. Biol. Chem.* 267: 1020–1026.

Ito, H., Sugimoto, T., Kobayashi, I., Takahashi, K., Katada, T., Ui, M., and Kurachi, Y. (1991). On the mechanism of basal and agonist-induced activation of the G protein-gated muscarinic K^+ channel in atrial myocytes of guinea pig heart. *J. Gen. Physiol.* 98: 517–533.

Ito, H., Tung, R. T., Sugimoto, T., Kobayashi, I., Takahashi, K., Katada, T., Ui, M., and Kurachi, Y. (1992). On the mechanism of G protein $\beta\gamma$ subunit activation of the muscarinic K^+ channel in guinea pig atrial cell membrane. Comparison with the ATP-sensitive K^+ channel. *J. Gen. Physiol.* 99: 961–983.

Jones, T., Simonds, W. F., Merendino, J. J., Brann, M. R., and Spiegel, A. M. (1990). Myristoylation of an inhibitory GTP-binding protein α subunit is essential for its membrane attachment. *Proc. Natl. Acad. Sci. U.S.A.* 87: 568–572.

Kahn, R. A., and Gilman, A. G. (1984). Purification of a protein cofactor required for ADP-ribosylation of the stimulatory regulatory component of adenylate cyclase by cholera toxin. *J. Biol. Chem.* 259: 6228–6234.

Kahn, R. A., and Gilman, A. G. (1986). The protein cofactor necessary for ADP-ribosylation of G_s by cholera toxin is itself a GTP binding protein. *J. Biol. Chem.* 261: 7906–7911.

Kanaho, Y., Crooke, S. T., and Stadel, J. M. (1989). Purification and characterization of predominant G-protein from bovine lung membranes: biochemical and immunochemical comparison with G_{i-1} and G_o purified from brain. *Biochem. J.* 259: 499–506.

Kanaho, Y., Kanoh, H., and Nozawa, Y. (1991). Activation of phospholipase D in rabbit neutrophils by fMet-Leu-Phe is mediated by a pertussis toxin-sensitive GTP-binding protein that may be distinct from a phospholipase C-regulating protein. *FEBS Lett.* 279: 249–252.

Kanoh, H., Kanaho, Y., and Nozawa, Y. (1992). Pertussis toxin-insensitive G protein mediates carbachol activation of phospholipase D in rat pheochromocytoma PC12 cells. *J. Neurochem.* 59: 1786–1794.

Kaslow, H. R., and Burns, D. L. (1992). Pertussis toxin and target eukaryotic cells: binding, entry, and activation. *FASEB J. 6*: 2684–2690.

Kaslow, H. R., Schlotterbeck, J. D., Mar, V. L., and Burnette, W. N. (1989). Alkylation of cysteine 41, but not cysteine 200, decreases the ADP-ribosyltransferase activity of the S1 subunit of pertussis toxin. *J. Biol. Chem. 264*: 6386–6390.

Katada, T., and Ui, M. (1979a). Effect of in vivo pretreatment of rats with a new protein purified from *Bordetella pertussis* on in vitro secretion of insulin: role of calcium. *Endocrinology 104*: 1822–1827.

Katada, T., and Ui, M. (1979b). Islet-activating protein: enhanced insulin secretion and cyclic AMP accumulation in pancreatic islets due to activation of native calcium ionophores. *J. Biol. Chem. 254*: 469–479.

Katada, T., and Ui, M. (1980). Slow interaction of islet-activating protein with pancreatic islets during primary culture to cause reversal of α-adrenergic inhibition of insulin secretion. *J. Biol. Chem. 255*: 9580–9588.

Katada, T., and Ui, M. (1981a). In vitro effects of islet-activating protein on secretion, adenosine $3':5'$-monophosphate accumulation and ^{45}Ca flux. *J. Biochem. 89*: 979–990.

Katada, T., and Ui, M. (1981b). Islet-activating protein: a modifier or receptor-mediated regulation of rat islet adenylate cyclase. *J. Biol. Chem. 256*: 8310–8317.

Katada, T., and Ui, M. (1982a). Direct modification of the membrane adenylate cyclase system by islet-activating protein due to ADP-ribosylation of a membrane protein. *Proc. Natl. Acad. Sci. U.S.A. 79*: 3129–3133.

Katada, T., and Ui, M. (1982b). ADP-ribosylation of the specific membrane protein of C6 cells by islet-activating protein as the mechanism for modification of membrane receptor–adenylate cyclase coupling. *J. Biol. Chem. 257*: 7210–7216.

Katada, T., Amano, T., and Ui, M. (1982). Modulation by islet-activating protein of adenylate cyclase activity in C6 glioma cells *J. Biol. Chem. 257*: 3739–3746.

Katada, T., Tamura, M., and Ui, M. (1983). The A-protomer of islet-activating protein, pertussis toxin, as an active peptide catalyzing ADP-ribosylation of a membrane protein. *Arch. Biochem. Biophys. 224*: 290–298.

Katada, T., Bokoch, G. M., Northup, J. K., Ui, M., and Gilman, A. G. (1984a). The inhibitory guanine nucleotide-binding regulatory component of adenylate cyclase: properties and function of the purified protein. *J. Biol. Chem. 259*: 3568–3577.

Katada, T., Bokoch, G. M., Smigel, M. D., Ui, M., and Gilman, A. G. (1984b). The inhibitory guanine nucleotide-binding regulatory component of adenylate cyclase: subunit dissociation and the inhibition of adenylate cyclase in S49 lymphoma cyc- and wild type membranes. *J. Biol. Chem. 259*: 3586–3595.

Katada, T., Northup, J. K., Bokoch, G. M., Ui, M., and Gilman, A. G. (1984c). The inhibitory guanine nucleotide-binding regulatory component of adenylate cyclase: subunit dissociation and guanine nucleotide-dependent hormonal inhibition. *J. Biol. Chem. 259*: 3578–3585.

Katada, T., Oinuma, M., and Ui, M. (1986a). Mechanisms for inhibition of the catalytic activity of adenylate cyclase by the guanine nucleotide-binding proteins serving as the substrate of islet-activating protein, pertussis toxin. *J. Biol. Chem. 261*: 5215–5221.

Katada, T., Oinuma, M., and Ui, M. (1986b). Two guanine nucleotide-binding proteins in rat brain serving as the specific substrate of islet-activating protein, pertussis toxin: interaction of the α-subunits with $\beta\gamma$-subunits in development of their biological activities. *J. Biol. Chem. 261*: 8182–8191.

Katada, T., Kusakabe, K., Oinuma, M., and Ui, M. (1987). A novel mechanism for the inhibitory GTP-binding proteins: calmodulin-dependent inhibition of the catalyst by the $\beta\gamma$-subunits of GTP-binding proteins. *J. Biol. Chem. 262*: 11897–11900.

Kato, K., Asano, T., Kamiya, N., Haimoto, H., Hosoda, S., Nagasaka, A., Ariyoshi, Y., and Ishiguro, Y. (1987). Production of the α subunit of guanine nucleotide-binding protein G_o by neuroendocrine tumors. *Cancer Res. 47*: 5800–5805.

Katz, A., Wu, D., and Simon, M. I. (1992). Subunits $\beta\gamma$ of heterotrimeric G protein activate $\beta2$ isoform of phospholipase C. *Nature 360*: 686–689.

Kaziro, Y., Itoh, H., Kozasa, T., Nakafuku, M., and Satoh, T. (1991). Structure and function of signal-transducing GTP-binding proteins. *Annu. Rev. Biochem. 60*: 349–400.

Kleuss, C., Hescheler, J., Ewel, C., Rosenthal, W., Schultz, G., and Wittig, B. (1991). Assignment of G-protein subtypes to specific receptors inducing inhibition of calcium currents. *Nature 353*: 43–48.

Kleuss, C., Scherübl, H., Hescheler, J., Schultz, G., and Wittig, B. (1992). Different β-subunits determine G-protein interaction wth transmembrane receptors. *Nature 358*: 424–426.

Kleuss, C., Scherübl, H., Hescheler, J., Schultz, G., and Wittig, B. (1993). Selectivity in signal transduction determined by γ subunits of heterotrimeric G proteins. *Science 259*: 832–834.

Kobayashi, I., Shibasaki, H., Takahashi, K., Tohyama, K., Kurachi, Y., Ito, H., Ui, M., and Katada, T. (1990). Purification and characterization of five different α subunits of guanine-nucleotide-binding proteins in bovine brain membranes: their physiological properties concerning the activities of adenylate cyclase and atrial muscarinic K^+ channels. *Eur. J. Biochem. 191*: 499–506.

Kokame, K., Fukada, Y., Yoshizawa, T., Takao, T., and Shimonishi, Y. (1992). Lipid modification at the N terminus of photoreceptor G-protein α-subunit. *Nature 359*: 749–752.

Kontani, K., Takahashi, K., Inanobe, A., Ui, M., and Katada, T. (1992). Molecular heterogeneity of the $\beta\gamma$-subunits of GTP-binding proteins in bovine brain membranes. *Arch. Biochem. Biophys. 294*: 527–533.

Kopf, G. S., Woolkalis, M. J., and Gerton, G. L. (1986). Evidence for a guanine nucleotide-binding regulatory protein in invertebrate and mammalian sperm. Identification by islet-activating protein-catalyzed ADP-ribosylation and immunochemical methods. *J. Biol. Chem. 261*: 7327–7331.

Kurachi, Y., Ito, H., Sugimoto, T., Katada, T., and Ui, M. (1989a). Activation of atrial muscarinic K^+ channels by low concentrations of $\beta\gamma$ subunits of rat brain G protein. *Pfülgers Arch. 413*: 325–327.

Kurachi, Y., Ito, H., Sugimoto, Shimizu, T., Miki, I., and Ui, M. (1989b). Arachidonic acid metabolites as intracellular modulators of the G protein-gated cardiac K^+ channel. *Nature 337*: 555–557.

Kurachi, Y., Tung, R. T., Ito, H., and Nakajima, T. (1992). G protein activation of cardiac muscarinic K^+ channels. *Prog. Neurobiol. 39*: 229–246.

Kurose, H., Katada, T., Amano, T., and Ui, M. (1983). Specific uncoupling by islet-activating protein, pertussis toxin, of negative signal transduction via α-adrenergic, cholinergic, and opiate receptors in neuroblastoma $\times$ glioma hybrid cells. *J. Biol. Chem. 258*: 4870–4875.

Kurose, H., Katada, T., Haga, T., Haga, K., Ichiyama, A., and Ui, M. (1986). Functional interaction of purified muscarinic receptors with purified inhibitory guanine nucleotide regulatory proteins reconstituted in phospholipid vesicles. *J. Biol. Chem. 261*: 6423–6428.

Lai, R. K., Perez, S. D., Cañada, F. J., and Rando, R. R. (1990). The gamma subunit of transducin is farnesylated. *Proc. Natl. Acad. Sci. U.S.A. 87*: 7673–7677.

Landis, C. A., Masters, S. B., Spada, A., Pace, A. M., Bourne, H. R., and Vallar, L. (1989). GTPase inhibiting mutations activate the α chain of G_s and stimulate adenylyl cyclase in human pituitary tumors. *Nature 340*: 692–696.

Lee, C. H., Park, D., Wu, D., Rhee, S. G., and Simon, M. I. (1992a). Members of the Gq α subunit gene family activate phospholipase C β isozymes. *J. Biol. Chem. 267*: 16044–16047.

Lee, R. H., Lieberman, B. S., Yamane, H. K., Bok, D., and Fung, B. K. (1992b). A third form of the G protein β subunit. 1. Immunochemical identification and localization to cone photoreceptors. *J. Biol. Chem. 267*: 24776–24781.

Levine, M. A., Smallwood, P. M., Moen, P. J., Helman, L. J., and Ahn, T. G. (1990). Molecular cloning of $\beta3$ subunit, a third form of the G protein β-subunit polypeptide. *Proc. Natl. Acad. Sci. U.S.A. 87*: 2329–2333.

Leyte, A., Barr, F. A., Kehlenbach, R. H., and Huttner, W. B. (1992). Multiple trimeric

G-proteins on the trans-Golgi network exert stimulatory and inhibitory effects on secretory vesicle formation. *EMBO J. 11*: 4795–4804.

Linder, M. E., Pang, I. H., Duronio, R. J., Gordon, J. I., Sternweis, P. C., and Gilman, A. G. (1991). Lipid modifications of G protein subunits. Myristoylation of $G_o\alpha$ increases its affinity for $\beta\gamma$. *J. Biol. Chem. 266*: 4654–4659.

Locht, C., and Keith, J. M. (1986). Pertussis toxin gene: nucleotide sequence and genetic organization. *Science 232*: 1258–1264.

Locht, C., Capiau, C., and Feron, C. (1989). Identification of amino acid residues essential for the enzymatic activities of pertussis toxin. *Proc. Natl. Acad. Sci. U.S.A. 86*: 3075–3079.

Locht, C., Lobet, Y., Feron, C., Cieplak, W., and Keith, J. M. (1990). The role of cysteine 41 in the enzymatic activities of the pertussis toxin S1 subunit as investigated by site-directed mutagenesis. *J. Biol. Chem. 265*: 4552–4559.

Logothetis, D. E., Kurachi, Y., Galper, J., Neer, E. J., and Clapham, D. E. (1987). The $\beta\gamma$ subunits of GTP-binding proteins activate the muscarinic K^+ channel in heart. *Nature 325*: 321–326.

Luetje, C. W., Tietje, K. M., Christian, J. L., and Nathanson, N. M. (1988). Differential tissue expression and developmental regulation of guanine nucleotide binding regulatory proteins and their messenger RNAs in rat heart. *J. Biol. Chem. 263*: 13357–13365.

Lyons, J., Landis, C. A., Harsh, G., Vallar, L., Grünewald, K., Feichtinger, H., Duh, Q. Y., Clark, O. H., Kawasaki, E., Bourne, H. R. and McCormick, F. (1990). Two G protein oncogenes in human endocrine tumors. *Science 249*: 655–659.

MacNulty, E. E.., McClue, S. J., Carr, I. C., Jess, T., Wakelam, M. J. C., and Milligan, G. (1992). α_2-C10 adrenergic receptors expressed in rat 1 fibroblasts can regulate both adenylyl cyclase and phospholipase D-mediated hydrolysis of phosphatidylcholine by interacting with pertussis toxin-sensitive guanine nucleotide-binding proteins. *J. Biol. Chem. 267*: 2149–2156.

Maltese, W. A., and Robishaw, J. D. (1990). Isoprenylation of C-terminal cysteine in a G-protein γ subunit. *J. Biol. Chem. 265*: 18071–18074.

Mattera, R., Graziano, M. P., Yatani, A., Zhou, Z., Graf, R., Codina, J., Birnbaumer, L., Gilman, A. G., and Brown, A. M. (1989). Splice variants of the α subunit of the G protein G_s activate both adenylyl cyclase and calcium channels. *Science 243*: 804–807.

Mattera, R., Yatani, A., Kirsch, G. E., Graf, R., Okabe, K., Olate, J., Codina, J., Brown, A. M., and Birnbaumer, L. (1989). Recombinant α_r-3 subunit of G protein activates G_k-gated potassium channels. *J. Biol. Chem. 264*: 465–471.

Milligan, G. (1987). Guanine nucleotide regulation of the pertussis and cholera toxin substrates of rat glioma C6 BU1 cells. *Biochem. Biophys. Acta 929*: 197–202.

Milligan, G. (1988). Fetal calf serum enhances cholera toxin-catalyzed ADP-ribosylation of the pertussis toxin-sensitive guanine nucleotide binding protein, G_i2, in rat glioma C6BU1 cells. *Cell. Signalling 1*: 65–74.

Milligan, G., and Klee, W. A. (1985). The inhibitory guanine nucleotide-binding protein (N_i) purified from bovine brain is a high affinity GTPase. *J. Biol. Chem. 260*: 2057–2063.

Milligan, G., and McKenzie, F. R. (1988). Opioid peptides promote cholera-toxin-catalyzed ADP-ribosylation of the inhibitory guanine-nucleotide-binding protein (G_i) in membranes of neuroblastoma × glioma hybrid cells. *Biochem. J. 252*: 369–373.

Morishita, R., Kato, K., and Asano, T. (1988). Major pertussis toxin-sensitive GTP-binding protein of bovine lung: purification, characterization and production of specific antibodies. *Eur. J. Biochem. 174*: 87–94.

Mullaney, I., Dodd, M. W., Buckley, N., and Milligan, G. (1993). Agonist activation of transfected human M1 muscarinic acetylcholine receptors in CHO cells results in down-regulation of both the receptor and the α subunit of the G-protein Gq. *Biochem. J. 289*: 125–131.

Mumby, S. M., Kahn, R. A., Manning, D. R., and Gilman, A. G. (1986). Antisera of designed specificity for subunits of guanine nucleotide-binding regulatory proteins. *Proc. Natl. Acad. Sci. U.S.A. 83*: 265–269.

Mumby, S. M., Casey, P. J., Gilman, A. G., Gutowski, S., and Sternweis, P. C. (1990a). G protein gamma subunits contain a 20-carbon isoprenoid. *Proc. Natl. Acad. Sci. U.S.A. 87*: 5873–5877.

Mumby, S. M., Heukeroth, R. O., Gordon, J. I., and Gilman, A. G. (1990b). G-protein α-subunit expression, myristoylation, and membrane association in COS cells. *Proc. Natl. Acad. Sci. U.S.A. 87*: 728–732.

Muntz, K. H., Sternweis, P. C., Gilman, A. G., and Mumby, S. M. (1992). Influence of γ subunit prenylation on association of guanine nucleotide-binding regulatory proteins with membranes. *Cell Regul. 3*: 49–61.

Murayama, T., and Ui, M. (1983). Loss of the inhibitory function of the guanine nucleotide regulatory component of adenylate cyclase due to its ADP-ribosylation by islet-activating protein, pertussis toxin, in adipocyte membrane. *J. Biol. Chem. 258*: 3319–3326.

Murayama, T., and Ui, M. (1984). [^{3}H]GDP release from rat and hamster adipocyte membranes independently linked to receptors involved in activation or inhibition of adenylate cyclase. Differential susceptibility to two bacterial toxins. *J. Biol. Chem. 259*: 761–769.

Murayama, T., and Ui, M. (1985). Receptor-mediated inhibition of adenylate cyclase and stimulation of archidonic acid release in 3T3 fibroblasts. Selective susceptibility to islet-activating protein, pertussis toxin. *J. Biol. Chem. 260*: 7226–7233.

Murayama, T., and Ui, M. (1987). Possible involvement of a GTP-binding protein, the substrate of islet-activating protein, in receptor-mediated signaling responsible for cell proliferation. *J. Biol. Chem. 262*: 12463–12467.

Murayama, T., Katada, T., and Ui, M. (1983). Guanine nucleotide activation and inhibition of adenylate cyclase as modified by islet-activating protein, pertussis toxin, in mouse 3T3 fibroblasts. *Arch. Biochem. Biophys. 221*: 381–390.

Murayama, T., Kajiyama, Y., and Nomura, Y. (1990a). Histamine-stimulated and GTP-binding proteins-mediated phospholipase A$_2$ activation in rabbit platelets. *J. Biol. Chem. 265*: 4290–4295.

Murayama, Y., Okamoto, T., Ogata, E., Asano, T., Iiri, T., Katada, T., Ui, M., Grubb, J. H., Sly, W. S., and Nishimoto, I. (1990b). Distinctive regulation of the functional linkage between the human cation-independent mannose 6-phosphate receptor and GTP-binding proteins by insulin-like growth factor II and mannose 6-phosphate. *J. Biol. Chem. 265*: 17456–17462.

Nagata, K., Katada, T., Tohkin, M., Itoh, H., Kaziro, Y., Ui, M., and Nozawa, Y. (1988). GTP-binding proteins in human platelet membranes serving as the specific substrate of islet-activating protein, pertussis toxin. *FEBS Lett. 237*: 113–117.

Nakamura, T., and Ui, M. (1983). Suppression of passive cutaneous anaphylaxis by pertussis toxin, islet-activating protein, as a result of histamine release from mast cells. *Biochem. Pharmacol. 32*: 3435–3441.

Nakamura, T., and Ui, M. (1984). Islet-activating protein, pertussis toxin, inhibits Ca^{2+}-induced and guanine nucleotide-dependent releases of histamine and arachidonic acid from rat mast cells. *FEBS Lett. 173*: 414–418.

Nakamura, T., and Ui, M. (1985). Simultaneous inhibitions of inositol phospholipid breakdown, arachidonic acid release, and histamine secretion in mast cells by islet-activating protein, pertussis toxin. A possible involvement of the toxin-specific substrate in Ca^{2+}-mobilizing receptor-mediated signal transduction. *J. Biol. Chem. 260*: 3584–3593.

Narasimhan, V., Holowka, D., and Baird, B. (1990). A guanine nucleotide-binding protein participates in IgE receptor-mediated activation of endogenous and reconstituted phospholipase A$_2$ in a permeabilized cell system. *J. Biol. Chem. 265*: 1459–1464.

Neer, E. J., Lok, J. M., and Wolf, L. G. (1984). Purification and properties of the inhibitory guanine nucleotide regulatory unit of brain adenylate cyclase. *J. Biol. Chem. 259*: 14222–14229.

Negishi, M., Ito, S., Tanaka, T., Yokohama, H., Hayashi, H., Katada, T., Ui, M., and Hayaishi, O. (1987). Covalent cross-linking of prostaglandin E receptor from bovine adrenal medulla with a pertussis toxin-sensitive guanine nucleotide-binding protein. *J. Biol. Chem. 262*: 12077–12084.

Negishi, M., Ito, S., Yokohama, H., Hayashi, H., Katada, T., Ui, M., and Hayaishi, O. (1988). Functional reconstitution of prostaglandin E receptor from bovine adrenal medulla with guanine nucleotide binding proteins. *J. Biol. Chem. 263*: 6893–6900.

Nicosia, A., Perugini, M., Franzini, C., Casagli, M. C., Borri, M. G., Antoni, G., Almoni, M., Neri, P., Ratti, G., and Rappuoli, R. (1986). Cloning and sequencing of the pertussis toxin genes: operon structure and gene duplication. *Proc. Natl. Acad. Sci. U.S.A. 83*: 4631–4635.

Nishimoto, I., Murayama, Y., Katada, T., Ui, M., and Ogata, E. (1989). Possible direct linkage of insulin-like growth factor-II receptor with guanine nucleotide-binding proteins. *J. Biol. Chem. 264*: 14029–14038.

Ohguro, H., Fukada, Y., Yoshizawa, T., Saito, T., and Akino, T. (1990). A specific $\beta\gamma$-subunit of transducin stimulates ADP-ribosylation of the α-subunit by pertussis toxin. *Biochem. Biophys. Res. Commun. 167*: 1235–1241.

Ohguro, H., Fukada, Y., Takao, T., Shimonishi, Y., Yoshizawa, T., and Akino, T. (1991). Carboxyl methylation and farnesylation of transducin γ-subunit synergistically enhance its coupling with metarhodopsin II. *EMBO J. 10*: 3669–3674.

Ohta, H., Okajima, F., and Ui, M. (1985). Inhibition by islet-activating protein of a chemotactic peptide-induced early breakdown of inositol phospholipids and Ca^{2+} mobilization in guinea pig neutrophils. *J. Biol. Chem. 260*: 15771–15780.

Oinuma, M., Katada, T., Yokosawa, H., and Ui, M. (1986). Guanine nucleotide-binding protein in sea urchin eggs serving as the specific substrate of islet-activating protein, pertussis toxin. *FEBS Lett. 207*: 28–34.

Oinuma, M., Katada, T., and Ui, M. (1987). A new GTP-binding protein in differentiated human leukemic (HL-60) cells serving as the specific substrate of islet-activating protein, pertussis toxin. *J. Biol. Chem. 262*: 8347–8353.

Okabe, K., Yatani, A., Evans, T., Ho, Y. K., Codina, J., Birnbaumer, L., and Brown, A. M. (1990). $\beta\gamma$ Dimers of G proteins inhibit atrial muscarinic K^+ channels. *J. Biol. Chem. 265*: 12854–12858.

Okajima, F., and Ui, M. (1984). ADP-ribosylation of the specific membrane protein by islet-activating protein, pertussis toxin, associated with inhibition of a chemotactic peptide-induced arachidonate release in neutrophils. A possible role of the toxin substrate in Ca^{2+}-mobilizing biosignaling. *J. Biol. Chem. 259*: 13863–13871.

Okajima, F., Katada, T., and Ui, M. (1985). Coupling of the guanine nucleotide regulatory protein to chemotactic peptide receptors in neutrophil membranes and its uncoupling by islet-activating protein, pertussis toxin. *J. Biol. Chem. 260*: 6761–6768.

Okamoto, T., Katada, T., Murayama, Y., Ui, M., Ogata, E., and Nishimoto, I. (1990). A simple structure encodes G protein-activating function of the IGF-II/mannose 6-phosphate receptor. *Cell 62*: 709–717.

Olson, S. C., Bowman, E. P., and Lambeth, J. D. (1991). Phospholipase D activation in a cell-free system from human neutrophils by phorbol 12-myristate 13-acetate and guanosine 5'-0-(3-thiotriphosphate). *J. Biol. Chem. 266*: 17236–17242.

Pace, A. M., Wong, Y. H., and Bourne, H. R. (1991). A mutant α subunit of G_i2 induces neoplastic transformation of Rat-1 cells. *Proc. Natl. Acad. Sci. U.S.A. 88*: 7031–7035.

Pai, J.-K., Siegel, M. I., Egan, R. W., and Billah, M. M. (1988). Phospholipase D catalyzes phospholipid metabolism in chemotactic peptide-stimulated HL-60 granulocytes. *J. Biol. Chem. 263*: 12472–12477.

Park, D., Jhon, D. Y., Lee, C. W., Ryu, S. H., and Rhee, S. G. (1993). Removal of the carboxyl-terminal region of phospholipase C-β1 by calpain abolishes activation by Gαq. *J. Biol. Chem. 268*: 3710–3714.

Pronin, A. N., and Gautam, N. (1992). Interaction between G-protein β and γ subunit types is selective. *Proc. Natl. Acad. Sci. U.S.A. 89*: 6220–6224.

Rhee, S. G., and Choi, K. D. (1992a). Regulation of inositol phospholipid-specific phospholipase C isozymes. *J. Biol. Chem. 267*: 12393–12396.

Rhee, S. G., and Choi, K. D. (1992b). Multiple forms of phospholipase C isozymes and their activation mechanisms. *Adv. Second Messenger Phosphoprotein Res. 26*: 35–61.

Robishaw, J. D., Kalman, V. K., Moomaw, C. R., and Slaughter, C. A. (1989). Existence of two γ subunits of the G proteins in brain. *J. Biol. Chem. 264*: 15758–15761.

Rosenthal, W., and Schultz, G. (1987). Modulations of voltage-dependent ion channels by extracellular signals. *Trends Pharmacol. Sci. 8*: 351–354.

Rosenthal, W., Hescheler, J., Trautwein, W., and Schultz, G. (1988). Control of voltage-dependent calcium channels by G protein-coupled receptors. *FASEB J 2*: 2784–2790.

Rotrosen, D., Gallin, J. I., Spiegel, A. M., and Malech, H. L. (1988). Subcellular localization of Gi α in human neutrophils. *J. Biol. Chem. 263*: 10958–10964.

Rubin, R. P., Withiam, L. M., and Laychock, S. G. (1991). Modulation of phospholipase A$_2$ activity in zymogen granule membranes by GTP[S]; evidence for GTP-binding protein regulation. *Biochem. Biophys. Res. Commun. 177*: 22–26.

Sanford, J., Codina, J., and Birnbaumer, L. (1991). γ-Subunits of G proteins, but not their α- or β-subunits, are polyisoprenylated. Studies on post-translational modifications using in vitro translation with rabbit reticulocyte lysates. *J. Biol. Chem. 266*: 9570–9579.

Saukkonen, K., Burnette, W. N., Mar, V. L., Masure, H. R., and Tuomanen, E. I. (1992). Pertussis toxin has eukaryotic-like carbohydrate recognition domains. *Proc. Natl. Acad. Sci. U.S.A. 89*: 118–122.

Schleifer, L. S., Kahn, R. A., Hanski, E., Northup, J. K., Sternweis, P. C., and Gilman, A. G. (1982). Requirements for cholera toxin-dependent ADP-ribosylation of the purified regulatory component of adenylate cyclase. *J. Biol. Chem. 257*: 20–23.

Schultz, G., Rosenthal, W., Hescheler, J., and Trautwein, W. (1990). Role of G proteins in calcium channel modulation. *Annu. Rev. Physiol. 52*: 275–292.

Shenker, A., Goldsmith, P., Unson, C. G., and Spiegel, A. M. (1991). The G protein coupled to the thromboxane A2 receptor in human platelets is a member of the novel Gq family. *J. Biol. Chem. 266*: 9309–9313.

Shilling, F., Chiba, K., Hoshi, M., Kishimoto, T., and Jaffe, L. A. (1989). Pertussis toxin inhibits 1-methyladenine-induced maturation in starfish oocytes. *Dev. Biol. 133*: 446–451.

Silk, S. T., Clejan, S., and Witkom, K. (1989). Evidence of GTP-binding protein regulation of phospholipase A2 activity in isolated human platelet membranes. *J. Biol. Chem. 264*: 21466–21469.

Simon, M. I., Strathmann, M. P., and Gautam, N. (1991). Diversity of G proteins in signal transduction. *Science 252*: 802–808.

Smrcka, A. V., Hepler, J. R., Brown, K. O., and Sternweis, P. C. (1991). Regulation of polyphosphoinositide-specific phospholipase C activity by purified Gq. *Science 251*: 804–807.

Sternweis, P. C., and Robishaw, J. D. (1984). Isolation of two proteins with high affinity for guanine nucleotides from membranes of bovine brain. *J. Biol. Chem. 259*: 13806–13813.

Stow, J. L., De, A. J., Narula, N., Holtzman, E. J., Ercolani, L., and Ausiello, D. A. (1991). A heterotrimeric G protein, G$_{\alpha i}$3, on Golgi membranes regulates the secretion of a heparan sulfate proteoglycan in LLC-PK1 epithelial cells. *J. Cell Biol. 114*: 1113–1124.

Stryer, L., and Bourne, H. R. (1986). G proteins: a family of signal transducers. *Annu. Rev. Cell Biol. 2*: 391–419.

Tadenuma, H., Chiba, K., Takahashi, K., Hoshi, M., and Katada, T. (1991). Purification and characterization of a GTP-binding protein serving as pertussis toxin substrate in starfish oocytes. *Arch. Biochem. Biophys. 290*: 411–417.

Tadenuma, H., Takahashi, K., Chiba, K., Hoshi, M., and Katada, T. (1992). Properties of 1-methyladenine receptors in starfish oocyte membranes: involvement of pertussis toxin-sensitive GTP-binding protein in the receptor-mediated signal transduction. *Biochem. Biophys. Res. Commun. 186*: 114–121.

Takahashi, S., Hashida, K., Yatsunami, K., Fukui, T., Negishi, M., Katada, T., Ui, M., Kanaho, Y., Asano, T., and Ichikawa, A. (1991). Characterization of cytosolic pertussis toxin-sensitive GTP-binding protein in mastocytoma P-815 cells. *Biochem. Biophys. Acta 1093*: 207–215.

Takei, Y., Kurosu, H., Takahashi, K., and Katada, T. (1992). A GTP-binding protein in rat liver nuclei serving as the specific substrate of pertussis toxin-catalyzed ADP-ribosylation. *J. Biol. Chem. 267*: 5085–5089.

Tamir, H., Fawzi, A. B., Tamir, A., Evans, T., and Northup, J. K. (1991). G-protein βγ forms: identify of β and diversity of γ subunits. *Biochemistry 30*: 3929–3936.

Tamura, M., Nogimori, K., Murai, S., Yajima, M., Ito, K., Katada, T., Ui, M., and Ishii, S. (1982). Subunit structure of islet-activating protein, pertussis toxin, in conformity with A-B model. *Biochemistry 21*: 5516–5522.

Tang, W. J., and Gilman, A. G. (1991). Type-specific regulation of adenylyl cyclase by G protein βγ subunits. *Science 254*: 1500–1503.

Tang, W. J., and Gilman, A. G. (1992). Adenylyl cyclases. *Cell 70*: 869–872.

Tang, W. J., Krupinski, J., and Gilman, A. G. (1991). Expression and characterization of calmodulin-activated (type I) adenylyl cyclase. *J. Biol. Chem. 266*: 8595–8603.

Taylor, S. J., and Exton, J. H. (1991). Two α subunits of the Gq class of G proteins stimulate phosphoinositide phospholipase C-β1 activity. *FEBS Lett. 286*: 214–216.

Taylor, S. J., Chae, H. Z., Rhee, S. G., and Exton, J. H. (1991). Activation of the β1 isozyme of phospholipase C by α subunits of the Gq class of G proteins. *Nature 350*: 516–518.

Teitelbaum, I. (1990). The epidermal growth factor receptor is coupled to a phospholipase A₂-specific pertussis toxin-inhibitable guanine nucleotide-binding regulatory protein in cultured rat inner medullary collecting tubule cells. *J. Biol. Chem. 265*: 4218–4222.

Terashima, T., Katada, T., Oinuma, M., Inoue, Y., and Ui, M. (1987a). Endocrine cells in pancreatic islets of Langerhans are immunoreactive to antibody against guanine nucleotide-binding protein (G₀) purified from rat brain. *Brain Res. 417*: 190–194.

Terashima, T., Katada, T., Oinuma, M., Inoue, Y., and Ui, M. (1987b). Immunohistochemical localization of guanine nucleotide-binding protein in rat retina. *Brain Res. 410*: 97–100.

Terashima, T., Katada, T., Okada, E., Ui, M., and Inoue, Y. (1987c). Light microscopy of GTP-binding protein (G₀) immunoreactivity within the retina of different vertebrates. *Brain Res. 436*: 384–389.

Terashima, T., Katada, T., Oinuma, M., Inoue, Y., and Ui, M. (1988a). Immunohistochemical analysis of the localization of guanine nucleotide-binding proteins in the mouse brain. *Brain Res. 442*: 305–311.

Terashima, T., Katada, T., Takayama, C., Ui, M., and Inoue, Y. (1988b). Immunohistochemical detection of GTP-binding regulatory protein (G₀) in the autonomic nervous system including the enteric nervous system, superior cervical ganglion and adrenal medualla. *Brain Res. 455*: 353–359.

Tohkin, M., Iiri, T., Ui, M., and Katada, T. (1989). Inhibition by islet-activating protein, pertussis toxin, of retinoic acid-induced differentiation of human leukemic (HL-60) cells. *FEBS Lett. 255*: 187–190.

Toutant, M., Aunis, D., Bockaert, J., Homburger, V., and Rouot, B. (1987). Presence of three pertussis toxin substrates and G₀α immunoreactivity in both plasma and granule membranes of chromaffin cells. *FEBS Lett. 215*: 339–344.

Trautwein, W., and Hescheler, J. (1990). Regulation of cardiac L-type calcium current by phosphorylation and G proteins. *Annu. Rev. Physiol. 52*: 257–274.

Tsai, S. C., Adamik, R., Moss, J., and Vaughan, M. (1991). Guanine nucleotide dependent formation of a complex between choleragen (cholera toxin) A subunit and bovine brain ADP-ribosylation factor. *Biochemistry 30*: 3697–3703.

Tsuda, M., Tsuda, T., Terayama, Y., Fukuda, Y., Akino, T., Yamanaka, G., Stryer, L., Katada, T., Ui, M., and Ebrey, T. (1986). Kinship of cephalopod photoreceptor G-protein with vertebrate transducin. *FEBS Lett. 198*: 5–10.

Uhing, R. J., Polakis, P. G., and Snyderman, R. (1987). Isolation of GTP-binding proteins from myeloid HL-60 cells: identification of two pertussis toxin substrates. *J. Biol. Chem. 262*: 15575–15579.

Ui, M. (1984). Islet-activating protein, pertussis toxin: a probe for functions of the inhibitory guanine nucleotide regulatory component of adenylate cyclase. *Trends Pharmacol. Sci. 5*: 227–279.

Ui, M. (1986). Pertussis toxin as a probe of receptor coupling to inositol lipid metabolism. In *Receptor Biochemistry and Methodology: Receptor and Phosphoinositides*, J. W. Putney (Ed.), Alan R. Liss, New York, pp. 163–195.

Ui, M. (1989). G proteins identified as pertussis toxin substrate. In *G Proteins and Calcium Mobilization*, P. H. Naccache (Ed.), CRC Press, Boca Raton, FL, pp. 1–24.

Ui, M. (1990). Pertussis toxin as a valuable probe for G-protein involvement in signal transduction. In *ADP-Ribosylating Toxins and G Proteins: Insights into Signal Transduction*, J. Moss and M. Vaughan (Eds.), American Society for Microbiology, Washington, DC, pp. 45–77.

Ui, M., Katada, T., Murayama, T., Kurose, H., Yajima, M., Tamura, M., Nakamura, T., and Nogimori, K. (1984). Islet-activating protein, pertussis toxin: a specific uncoupler of receptor-mediated inhibition of adenylate cyclase. *Adv. Cyclic Nucleotide Protein Phosphorylation Res. 17*: 145–151.

Van Dop, C., Tsubokawa, M., Bourne, H. R., and Ramachandran, J. (1984). Amino acid sequence of retinal transducin at the site ADP-ribosylated by cholera toxin. *J. Biol. Chem. 259*: 696–698.

Van Haastert, P. J. M., Snaar-Jagalska, B. E., and Janssens, P. M. W. (1987). The regulation of adenylate cyclase by guanine nucleotides in *Dictyostelium discoideum* membranes. *Eur. J. Biochem. 162*: 251–258.

Wange, R. L., Smrcka, A. V., Sternweis, P. C., and Exton, J. H. (1991). Photoaffinity labeling of two rat liver plasma membrane proteins with $[^{32}P]\gamma$-azidoanilido GTP in response to vasopressin. Immunologic identification as alpha subunits of the Gq class of G proteins. *J. Biol. Chem. 266*: 11409–11412.

West, R. E. J., Moss, J., Vaughan, M., Liu, T., and Liu, T. Y. (1985). Pertussis toxin-catalyzed ADP-ribosylation of transducin. Cysteine 347 is the ADP-ribose acceptor site. *J. Biol. Chem. 260*: 14428–14430.

Wilkie, T. M., Scherle, P. A., Strathmann, M. P., Slepak, V. Z., and Simon, M. I. (1991). Characterization of G-protein alpha subunits in the Gq class: expression in murine tissues and in stromal and hematopoietic cell lines. *Proc. Natl. Acad. Sci. U.S.A. 88*: 10049–10053.

Wong, Y. H., Federman, A., Pace, A. M., Zachary, I., Evans, T., Pouysségur, J., and Bourne, H. R. (1991). Mutant α subunits of Gi2 inhibit cyclic AMP accumulation. *Nature 351*: 63–65.

Wong, Y. H., Conklin, B. R., and Bourne, H. R. (1992). GZ-mediated hormonal inhibition of cyclic AMP accumulation. *Science 255*: 339–342.

Wu, D., Ho, L. C., Goo, R. S., and Simon, M. I. (1992a). Activation of phospholipase C by the α subunits of the Gq and G11 proteins in transfected Cos-7 cells. *J. Biol. Chem. 267*: 1811–1817.

Wu, D., Katz, A., Lee, C. H., and Simon, M. I. (1992b). Activation of phospholipase C by α1-adrenergic receptors in mediated by the alpha subunits of Gq family. *J. Biol. Chem. 267*: 25798–25802.

Wu, D., Jiang, H., Katz, A., and Simon, M. I. (1993). Identification of critical regions on phospholipase C-β1 required for activation by G-proteins. *J. Biol. Chem. 268*: 3704–3709.

Xing, M., and Mattera, R. (1992). Phosphorylation-dependent regulation of phospholipase A$_2$ by G-proteins and Ca^{2+} in HL60 granulocytes. *J. Biol. Chem.* 25967–25975.

Yamane, H. K., Fransworth, C. C., Xie, H., Howald, W., Fung, B. K., Clarke, S., Gelb, M. H., and Glomset, J. A. (1990). Brain G protein γ subunits contain an all-*trans*-geranylgeranyl-cysteine methyl ester at their carboxyl termini. *Proc. Natl. Acad. Sci. U.S.A. 87*: 5868–5872.

Yamazaki, S., Katada, T., and Ui, M. (1982). Alpha2-adrenergic inhibition of insulin secretion via interference with cyclic AMP generation in rat pancreatic islets. *Mol. Pharmacol. 21*: 648–653.

Yatani, A., Codina, J., Brown, A. M., and Birnbaumer, L. (1987). Direct activation of mammalian atrial muscarinic potassium channels by GTP regulatory protein G$_k$. *Science 235*: 207–211.

Yatani, A., Imoto, Y., Codina, J., Hamilton, S. L., Brown, A. M., and Birnbaumer, L. (1988). The stimulatory G protein of adenylyl cyclase, G$_s$, also stimulates dihydropyridine-sensitive calcium channels. Evidence for direct regulation independent of phosphorylation by cAMP-dependent protein kinase or stimulation by a dihydropyridine agonist. *J. Biol. Chem. 263*: 9887–2895.

Yatani, A., Mattera, R., Codina, J., Graf, R., Okabe, K., Padrell, E., Iyengar, R., Brown, A. M., and Birnbaumer, L. (1988). The G protein-gated atrial potassium channel is stimulated by three distinct G$_i\alpha$-subunits. *Nature 336*: 680–682.

21

Modification of Actin and Rho Proteins by Clostridial ADP-Ribosylating Toxins

Klaus Aktories and Gertrud Koch

Institut für Pharmakologie und Toxikologie der Universität des Saarlandes, Homburg, Germany

I. INTRODUCTION

ADP-ribosylation of eukaryotic regulatory proteins is a well-established mechanism by which various bacterial toxins affect the eukaryotic organism. These toxins are capable of transferring the ADP-ribose moiety of NAD to eukaryotic regulatory proteins, which are functionally altered upon introduction of this bulky group. Well-known members of this family of toxins are diphtheria toxin and *Pseudomonas* exotoxin A, which ADP-ribosylate elongation factor II, thereby inhibiting protein synthesis (for review see Collier, 1990; Wick and Iglewski, 1990; Pastan and FitzGerald, 1989; Perentesis et al., 1992). A second group of transferases is composed of pertussis toxin (for review see Ui, 1990; Gierschik, 1992), cholera toxin (Fishman, 1990; Gill, 1977; Spangler, 1992), and the largely homologous heat-labile *E. coli* enterotoxins (Spangler, 1992). This group of ADP-ribosyltransferases modify heterotrimeric G proteins, which are involved in signal transduction. During the last years several clostridial ADP-ribosyltransferases have been described. These toxins can be divided into two groups. One group, which is represented by *Clostridium botulinum* C2 toxin, *Clostridium perfringens* iota toxin, *Clostridium spiroforme* toxin, and an ADP-ribosyltransferase produced by *Clostridium difficile,* modify actin. The second group with *Clostridium botulinum* C3 transferase, *Clostridium limosum,* and the related transferases from *Bacillus cereus* and *Staphylococcus aureus* ADP-ribosyl-ate small GTP-binding proteins of the Rho family. In this chapter both classes of clostridial enzymes are described in more detail. During recent years several reviews about clostridial bacterial ADP-ribosyltransferases have been published that focus on actin-ADP-ribosylat-ing toxins (Aktories and Just, 1990; Aktories et al. 1992b; Aktories and Wegner, 1989, 1992; Considine and Simpson, 1991) or on Rho-ADP-ribosylating toxins (Aktories and Hall, 1989; Aktories et al., 1992a) or deal with both toxin families (Aktories, 1990, 1992; Aktories et al., 1993).

II. CLOSTRIDIAL ACTIN-ADP-RIBOSYLATING TOXINS

A. Introduction

The actin-ADP-ribosylating toxins are *Clostridium botulinum* C2 toxin, *Clostridium perfringens* iota toxin, *Clostridium spiroforme* toxin, and an ADP-ribosyltransferase from *Clostridium difficile*. These toxins are constructed according to the A-B model. One component (A) is biologically active and possesses transferase activity, whereas the B component is involved in binding the toxin to and transferring it into the target cell. This toxin structure is known for the *Pseudomonas* exotoxin A (Wick and Iglewski, 1990), diphtheria toxin (Collier, 1990), cholera toxin (Gill, 1977), and pertussis toxin (Tamura et al., 1982). In contrast to these toxins, the functionally distinct components of actin-ADP-ribosylating toxins are separate proteins that are linked by neither covalent nor noncovalent bonds (Ohishi et al., 1980; Ohishi, 1983). Thus, the actin-modifying toxins are binary in structure and are comparable to anthrax toxin (Leppla, 1982) and leukocidin (Noda et al., 1981).

The first toxin that was shown to ADP-ribosylate actin was *Clostridium botulinum* C2 toxin (Aktories et al., 1986b). Many *C. botulinum* type C and D strains produce the C2 toxin. Usually these strains also produce neurotoxins type C1 and D (Eklund and Poysky, 1972; Habermann and Dreyer, 1986; Eklund et al., 1989). Furthermore, these strains also produce ADP-ribosyltransferase C3 (Aktories et al., 1987). However, strains exist that produce neither neurotoxins nor transferase C3. The reason for this diversity is the fact that production of type C and D neurotoxins and transferase C3 is phage-mediated (Eklund et al., 1971, 1972, 1989; Popoff et al., 1990). In contrast, C2 toxin is not phage-encoded. Therefore, treatment of type C and D strains with acridine orange resulted in bacteria that produced only C2 toxin and were cured of neurotoxins and transferase C3 (Eklund et al., 1989).

B. Purification of *Clostridium botulinum* C2 Toxin

Purification of both components of C2 toxin form *Clostridium botulinum* type C strain 92-13 was achieved by ammonium sulfate precipitation, acid precipitation, DEAE-Sephadex, CM-Sephadex, and finally gel filtration with Sephadex G-100. The separation of toxin components depends on the high affinity of the binding component (C2II) for carboxymethyl-Sephadex. On the other hand, the enzyme component (C2I) binds to DEAE-Sephadex (Ohishi et al., 1980). The purification was improved by using hydroxy-apatite chromatography (Ohishi and DasGupta, 1987). In our laboratory, the enzyme component is purified by isoelectric focusing on a Westerberg-Svenson column.

C. The Binding Component of *C. botulinum* C2 Toxin

The purified binding component of C2 toxin has a molecular weight of about 100,000 (Ohishi et al., 1980). Quite early it was found that in contrast to botulinum neurotoxins, the C2 toxin gains full activity by trypsin treatment (Eklund and Poysky, 1972; Jansen and Knoetze, 1971). Ohishi studied the proteolytic activation of the toxins in more detail and found that the binding component is activated by trypsin (Oshishi, 1987), which cleaves the ~100-kDa component into a ~74-kDa active fragment (Fig. 1). Ohishi and coworkers described differences of the two components of C2 toxin produced by different strains of *Clostridium botulinum* types C and D (Ohishi and Akada, 1986; Ohishi and Hama, 1992). The binding components and the trypsin-activated fragments of these strains differ considerably in their M_r. It appears that the activated component C2II binds to the cell surface, thereby inducing a binding site for the enzyme component C2I (Ohishi and Yanagimoto, 1992; Ohishi and Miyake, 1985) (Fig. 2). Most likely, C2 toxin enters the target cell by receptor-mediated endocytosis (Simpson, 1989; Ohishi and Yanagimoto, 1992). For example, the membrane transit of the toxin is temperature-dependent. At low temperature the binding component (C2II) is still able to bind to the cell surface and to induce a binding site for the enzyme component (C2I); however, the components remain on the cell surface and membrane transfer is inhibited. Incubation of cells with C2II at 37°C causes internalization of the binding component, with the consequence that C2II is thereafter no longer available for the transport of C2I. Recently, the binding and internalization of the C2 toxins were studied with fluorescence-labeled components in Vero cells (Ohishi and Yanagimoto, 1992). It was shown that the trypsin-activated binding component first binds to the cell surface to form mosaic-like patterns. Thereafter, the binding component aggregates in clusters and enters the cell via endosomes. In contrast, the untrypsinized component also binds in a mosaic-like manner to the cell but is only inefficiently internalized by the cell, where it is found on the nuclear surface. By using a fluorescence-labeled enzyme component, it was demonstrated that C21 binds directly to the trypsinized but not to the untrypsinized binding component of C2 toxin (Ohishi and Yanagimoto, 1992).

It has been suggested that the activated 74-kDa binding component forms pentamers with hemagglutinating and hemolytic activity, whereas the trypsinized monomer

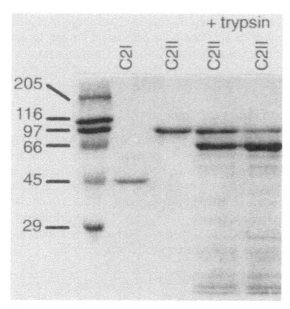

Fig. 1 *Clostridium botulinum* C2 toxin. Lane 1, M_r marker; lane 2, C2I (M_r ~45,000, 1 μg), the enzyme component possessing ADP-ribosyltransferase activity; lane 3, C2II (M_r ~100,000, 3 μg), the binding component of C2 toxin. The binding component C2II was activated by partial proteolysis with trypsin for 15 min (lane 4) and 30 min (lane 5) at 37°C; thereby an active protein fragment of ~70 kDa is released. (From Aktories et al., 1993, with permission.)

possesses only hemagglutinating properties (Ohishi, 1987). We have recently observed that the activated toxin forms cation-selective pores in artificial membranes (Schmid et al., 1994). These pores are selectively closed in the presence of anti-C2II antibody and are modified in the presence of enzyme component C2I, indicating an interaction of the enzyme component with the activated component. Its precise role is not known, but pore formation may have an important role in the entry of C2 toxin. It is worth noting that other toxins, such as diphtheria toxin, that enter cells by receptor-mediated endocytosis also form cation-selective pores (Zalman and Wisnieski, 1984; Falnes et al., 1992). Similarly, the binding component of anthrax toxin is a pore-forming agent (Koehler and Collier, 1991).

So far, the precise nature of the cell surface receptor for C2 toxin is not known. However, from inhibition studies with various glycoproteins and mono- and disaccharides that were performed with trypsin- and pronase-treated human erythrocytes, it has been suggested that the receptor for C2 toxin is a glycoprotein (Sugii and Kozaki, 1990).

D. The Enzyme Component of *C. botulinum* C2 Toxin

The enzyme component of C2 toxin has a molecular weight of about 45,000 (Ohishi et al., 1980). The complete sequences of the enzyme and binding components are not known. However, the amino acid composition of both toxin components clearly differed from that of *Clostridium botulinum* C1 neurotoxin (Takasawa et al., 1987). It was first reported by Simpson (1984) that C2 toxin possesses ADP-ribosyltransferase activity. In our

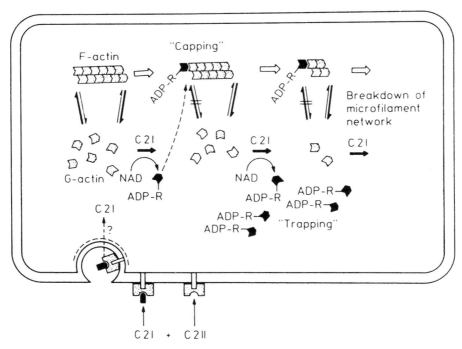

Fig. 2 Model of the cytopathic effects of actin ADP-ribosylating toxins. The activated binding component of the toxin (e.g., component C2II) binds to an acceptor located on the surface of the eukaryotic cell. This induces a binding site for the enzyme component (C2I). The toxin components most likely enter the cell by endocytosis. In the cell, the toxin ADP-ribosylates G-actin and disturbs the dynamic equilibrium between monomeric G-actin and filamentous F-actin. ADP-ribosylation of actin inhibits actin polymerization and traps actin in the monomeric form. Moreover, ADP-ribosylated actin binds to the barbed ends of filamentous F-actin in a capping-protein-like manner. Capping of F-actin inhibits further polymerization at the fast-growing ends of actin filaments. In contrast, ADP-ribosylated actin has no effect on the pointed end of F-actin, at which depolymerization occurs. The consequence of "trapping" G-actin and "capping" F-actin is the destruction of the microfilament network. But see also Figure 3. (From Aktories, 1990, with permission.)

laboratory the cellular pathophysiological substrate was identified as actin (Aktories et al., 1986b). The various ADP-ribosylating toxins can be classified according to the acceptor amino acid. Whereas diphtheria toxin and *Pseudomonas* exotoxin A ADP-ribosylate elongation factor II on diphthamide, a posttranslationally modified histidine (Van Ness et al., 1980), pertussis toxin (West et al., 1985), and cholera toxin (Moss and Vaughan, 1977) selectively attach the ADP-ribose moiety of NAD to cysteine and arginine, respectively. Like cholera toxin, all actin-ADP-ribosylating toxins modify arginine residues. Using protein chemistry and mutagenesis, arginine 177 has been identified as the amino acceptor acid in actin (Vandekerckhove et al., 1988; Just et al., 1993a) (see also below). The modification of actin was highly selective. G proteins, the substrates for cholera and pertussis toxins, were not modified. Other cytoskeletal components such as tubulin were not substrates for C2 toxin or *Clostridium perfringens* iota toxin. Like other bacterial ADP-ribosylating toxins, the actin-modifying toxins catalyze mono-ADP-ribosylation reactions. Accordingly the actin-bound ADP-ribose is cleaved by *Crotalus adamanteus* phosphodiesterase, thereby releasing 5'-AMP. The K_m for NAD for the ADP-ribosylation reaction is about 4 μM for C2 toxin and iota toxin (Aktories et al., 1986a; Schering et al., 1988). Like other ADP-ribosyltransferases, the actin-modifying toxins possess NAD glycohydrolase activity that cleaves NAD even in the absence of a protein substrate and releases nicotinamide and ADP-ribose (Ohishi, 1986). However, this activity is several times lower than the transferase activity, and a functional role is questionable.

E.   Other Actin-ADP-Ribosylating Toxins

The other actin-ADP-ribosylating toxins are *Clostridium perfringens* iota toxin. *Clostridium spiroforme* toxin, and an ADP-ribosyltransferase produced by *Clostridium difficile*. The latter is clearly distinct from *Clostridium difficile* toxins A and B, which have been etiologically linked with the antibiotic-associated pseudomembranous colitis (Popoff et al., 1988). All the actin-modifying toxins have similar structures and are binary toxins. Iota toxin has been purified from culture medium of *Clostridium perfringens* type E strain NCIB 10748; it comprises an enzyme component with M_r of 47,500 and a binding component with M_r of 71,500 that have isoelectric points of 5.2 and 4.2, respectively (Simpson et al., 1987; Stiles and Wilkens, 1986a,b). Although it has been reported that the toxicity of *Clostridium perfringens* iota toxin is increased by proteolysis, no effect of trypsin was observed with the purified components, most likely indicating the action of an endogenous protease (Stiles and Wilkins, 1986a,b). On the other hand, the activity of the binding component of *Clostridium spiroforme* toxin was increased by trypsin treatment. The enzyme component of *C. spiroforme* toxin is apparently heterogeneous with an M_r of 43,000–47,000. No binding component has been found to be produced by *Clostridium difficile* strains that synthesize the ADP-ribosyltransferase (Popoff et al., 1988). However, the binding components of iota toxin and spiroforme toxins can interchange with the *Clostridium difficile* transferase. In this respect it is noteworthy that apparently a subgroup of iotalike actin ADP-ribosylating toxins exists, because the binding and enzyme components of iota and spiroforme toxins are exchangeable (Simpson et al., 1989; Popoff and Boquet, 1988). This is not true for iota toxin and C2 toxin. Thus, the binding component of iota toxin is not able to translocate the enzyme component of C2 toxin, and, vice versa, C2I, the binding component of C2 toxin, cannot transfer the enzyme component of iota toxin into the host cell (Popoff and Boquet, 1988; Simpson et al. 1987, 1989). A further difference between C2 and iota toxins is their substrate specificity (Schering et al., 1988). Iota toxin is able to modify all actin isoforms, whereas C2 toxin is not (see below).

F.   Actin, the Eukaryotic Substrate for ADP-Ribosylation by Bacterial Toxins

Actin has been found in all eukaryotic cells. In fact, it is often the most abundant protein in non-muscle cells. Actin plays an important role in the organization of the microfilament network of the cytoskeleton (for review see Bershadsky and Vasiliev, 1988; Pollard and Cooper, 1986; Kabsch and Vandekerckhove, 1992). Moreover, it participates in various cellular motile functions such as migration, phagocytosis, secretion, and intracellular transport.

Actin is a single-chain peptide of 375 or 374 residues with a molecular weight of about 42,000 and an isoelectric point of about 5.4. It is a nucleotide-binding protein that binds ATP or ADP with high affinity [K_D(ATP) $\approx 10^{-10}$]. Moreover, actin has binding sites for divalent cations, which are most likely Mg^{2+} in intact cells. Recently, the atomic structure of actin has been reported, showing that actin consists of four subdomains (I–IV) with a cleft between domains I and II and domains III and IV, which contains the actin-bound nucleotide and the divalent cation (Kabsch et al., 1990; Kabsch and Vanderkerckhove, 1992).

The physiological role of actin is related to its ability to form filaments in a reversible manner. The filamentous F-actin is constructed as a double-stranded right-handed helix with 14 actin molecules in each strand per complete turn. Because actin filaments are polar structures, the interaction of monomeric actin (G-actin) with F-actin differs at the two F-actin ends (Holmes et al., 1990). The critical concentration of actin, that is, the minimal concentration of G-actin necessary for actin polymerization, is about one order of magnitude lower at the so-called plus or barbed end than at the minus or pointed end of actin filaments. Therefore, actin filaments tend to polymerize at their barbed end and to depolymerize at the pointed end, a phenomenon that has been called treadmilling (Selve and Wegner, 1986; Pollard and Cooper, 1986). In non-muscle cells, an equilibrium exists between G- and F-actin that is regulated by actin itself and by a still increasing number of actin-binding proteins such as gelsolin, profilin, and insertin (Hartwig and Kwiatkowski, 1991; Pollard and Cooper, 1986; Vandekerckhove, 1990).

G. Actin as Substrate for ADP-Ribosylation

At least six mammalian actin isoforms have been described. These are skeletal muscle α-actin, cardiac muscle α-actin, smooth muscle α- and γ-actin, and cytoplasmic β- and γ-actin (Vanderkerckhove and Weber, 1978b, 1979). The actin isoforms are classified as α-, β-, and γ-actin according to their isoelectric point, with α-actin the most acidic. All these isoforms are highly homologous, with the maximal difference in the primary structure about 7% and no sequence differences among various mammalian species. Similarly, all vertebrate actin isoforms are more than 93% identical (Bershadsky and Vasiliev, 1988; Vandekerckhove and Weber, 1978b, 1979). Moreover, even actin from the slime mold *Physarum polycephalum* differs in only about 4% of its primary structure from cytoplasmic actins (Vandekerckhove and Weber, 1978a). Thus, actin is a highly conserved protein, and differences appear to be function- rather than species-specific.

Clostridium botulinum C2 toxin can ADP-ribosylate cytoplasmic actin but not α-actin isoforms such as skeletal muscle α-actin, cardiac muscle α-actin, or smooth muscle α-actin (Aktories et al., 1986b; Mauss et al., 1990). Other substrates for ADP-ribosylation by C2 toxin are *Physarum polycephalum* actin (unpublished observation) and *Drosophila* indirect flight muscle actin (see below). In addition to *Drosophila* indirect flight muscle actin, arthrin, an actin-ubiquitin conjugate, is ADP-ribosylated by C2 toxin and iota toxin (Just et al., 1993a). In contrast to C2 toxin, *Clostridium perfringens* iota toxin ADP-ribosylates all actin isoforms studied (Mauss et al., 1990). That means also that α-actin isoforms, which are not substrates for C2 toxin, are modified by iota toxin. As mentioned above, the modification of actin by ADP-ribosylation occurs on Arg-177. This amino acid is also found in other actin isoforms (e.g., α-actin) that are not substrates for ADP-ribosylation. It has been suggested that the preceding amino acid 176, which is leucine in cytoplasmic actin and methionine in skeletal muscle actin, may be responsible for the substrate specificity of C2 toxin (Vandekerckhove et al., 1988). To test this hypothesis, the ADP-ribosylation of mutants of *Drosophila* indirect flight muscle actin was studied. Actin translated in vitro from indirect flight muscle specific gene *Act88F* was ADP-ribosylated by *Clostridium botulinum* C2 toxin and *Clostridium perfringens* iota toxin. When arginine 177 was mutated to glutamine there was no ADP-ribosylation by either toxin, confirming Arg-177 as the ADP-ribose acceptor. However, mutant L176M actin was modified by both toxins, indicating that amino acid 176 of actin does not define the substrate specificity of *Clostridium botulinum* C2 toxin (Just et al., 1993a). Because smooth muscle γ-actin is modified by C2 toxin and differs only in the N-terminal region from smooth muscle α-actin, it has been suggested that the N terminus defines the substrate specificities for ADP-ribosylation by C2 toxin (Mauss et al., 1990).

H. Properties of ADP-Ribosylated Actin

ADP-ribosylation of actin by bacterial toxins has gross functional consequences. First, ADP-ribosylated actin is unable to polymerize (Aktories et al., 1986a,b; Schering et al., 1988). Even in the presence of phalloidin, which markedly decreases the critical concentration for polymerization, ADP-ribosylated actin is not capable of forming actin filaments (Aktories et al., 1986a). Second, ADP-ribosylated actin still interacts with unmodified actin and binds like a capping protein to the fast-polymerizing end of actin filaments to inhibit further association or dissociation of unmodified actin monomers at this filament end (Wegner and Aktories, 1988; Weight et al., 1989). In contrast, ADP-ribosylated actin does not inhibit polymerization or depolymerization of unmodified actin at the pointed or slow-growing filament end (Wegner and Aktories, 1988). Third, ADP-ribosylation inhibits the actin ATPase activity (Geipel et al., 1989, 1990). As mentioned above, actin is an ATP-binding protein with ATPase activity. ATP is hydrolyzed both by filamentous polymeric actin and by monomeric G-actin to yield ADP. Inhibition of actin-catalyzed ATP hydrolysis is observed with actin at concentrations below the critical concentration for actin polymerization and at low concentrations of Mg^{2+} (<50 μM). ATPase activity is inhibited even in the quasi-monomeric actin-DNAse I complex after stimulation with cytochalasins (Geipel et al., 1990).

Reversal of ADP-ribosylation of actin by C2 toxin and iota toxin is possible in the presence of high concentrations of nicotinamide (30–50 mM) and in the absence of NAD (Just et al., 1990). The pH optima for both ADP-ribosylation and de-ADP-ribosylation was determined to be pH 5.5–6. Concomitant with the removal of ADP-ribose, the ability of actin to polymerize is restored and actin ATPase activity increases. Neither ADP-ribosylation nor removal of ADP-ribose is observed after treatment of actin with EDTA, indicating that the native structure of actin is required for both reactions. De-ADP-ribosylation of actin by either C2 toxin or iota toxin reflects the substrate specificity of the ADP-ribosylation reaction. Thus, ADP-ribosylation of platelet actin by iota toxin is reversed by C2 toxin and vice versa. However, ADP-ribosylation of skeletal muscle actin by iota toxin is not reversed by C2 toxin (Just et al., 1990).

Most of the findings reported on the ADP-ribosylation of actin and its functional consequences can be nicely explained in view of the recently published atomic models of G-actin and F-actin (Kabsch et al., 1990; Holmes et al., 1990). The acceptor amino acid, Arg-177, is located in domain III of actin. This specific area of the molecule is apparently a subunit contact site of F-actin. Thus, the cumbersome ADP-ribose group blocks the formation of F-actin by steric hindrance. This explains the inhibition of actin polymerization by ADP-ribosylation. Furthermore, this model makes clear that F-actin is not a substrate for ADP-ribosylation because the acceptor amino acid is not at the surface of the F-actin molecule. Moreover, there is only one place that accepts the ADP-ribose group, and this is the barbed end of the actin filament. Therefore, ADP-ribosylated actin acts like a capping protein at the fast-growing end of F-actin and not at the slow-growing, pointed end (Fig. 2). Finally, arginine 177 is located rather near the γ-phosphate of the actin-bound ATP. This might be the reason for the inhibitory effects of ADP-ribosylation on the actin-associated ATPase.

Wille et al. (1992) have shown that the gelsolin-actin complex is ADP-ribosylated by the toxins. These findings are of particular importance because the monomeric G-actin, which is a substrate for ADP-ribosylation, is not free in cells but bound to various regulatory actin-binding proteins such as profilin, β-thymosin, or gelsolin. Gelsolin is a calcium-dependent actin-binding protein, $M_r \approx 82,000$, that is universally found in vertebrate tissue (Pollard and Cooper, 1986). This actin-binding protein has at least three important functions. It severs F-actin and increases the number of short filaments (Harris and Weeds, 1984). It is a barbed-end-capping protein that inhibits actin polymerization at the fast-growing end. Moreover, it binds two actin monomers to form a 1:2 complex with nucleation activity for polymerization (Janmey et al., 1985). The two actin-binding sites are nonequivalent. At one binding site, actin binds with a moderate affinity and the complex is released by decreasing the Ca^{2+} concentration (this is called the Ca^{2+}-sensitive binding site of gelsolin). At the second site, actin binds with high affinity, and it is not removed by decreasing the Ca^{2+} concentration (this site is called the EGTA-resistant binding site of gelsolin). At this site, however, actin is released by interaction of the complex with phosphatidyl inositol phosphates (Janmey and Stossel, 1987). By using gel permeation chromatography and nondenaturing gels, we have shown that the 1:1 and 1:2 gelsolin-actin complexes are modified by ADP-ribosylation (Wille et al., 1992). *Clostridium perfringens* iota toxin ADP-ribosylates the 1:1 human platelet gelsolin-actin complex as effectively as skeletal muscle actin. The K_m for NAD ($\sim$4 μM) is almost identical for ADP-ribosylation of actin and of the gelsolin-actin complex. The ADP-ribosylation of the gelsolin-actin complex by C2 toxin is less effective than the ADP-ribosylation of non-muscle actin. This finding is in agreement with the view that C2 toxin has a higher substrate specificity. The ADP-ribosylated 1:1 gelsolin-actin complex (G-A$_r$) binds unmodified actin to form the 1:2 complex (G-A$_r$-A), and, vice versa, the unmodified G-A complex binds ADP-ribosylated actin to give the G-A-A$_r$ complex. ADP-ribosylated actin shows about a threefold higher affinity for the 1:1 gelsolin-actin complex than unmodified actin. Interestingly, the ADP-ribosylation of the various gelsolin-actin complexes exhibits different functional consequences. When ADP-ribosylated actin is bound to the Ca^{2+}-sensitive binding site of gelsolin, no change in the nucleation activity is observed. In contrast, the binding of ADP-ribosylated actin to the EGTA-resistant binding site inhibits the nucleation activity of the gelsolin-actin complex (Wille et al., 1992) (Fig. 3).

I. Pathophysiological Effects of Actin-ADP-Ribosylating Toxins

1. Cytopathic Effects

It has been demonstrated in various cell types and tissues that actin is the pathobiochemical substrate for ADP-ribosylation in intact cells (Reuner et al., 1987; Aktories et al., 1989; Norgauer et al., 1988; Suttorp et al., 1991). The most striking effect of the toxins on cultured cells is the destruction of the microfilament network (Reuner et al., 1987; Wiegers et al., 1991) (Fig. 4). The actin filaments depolymerize,

Gelsolin

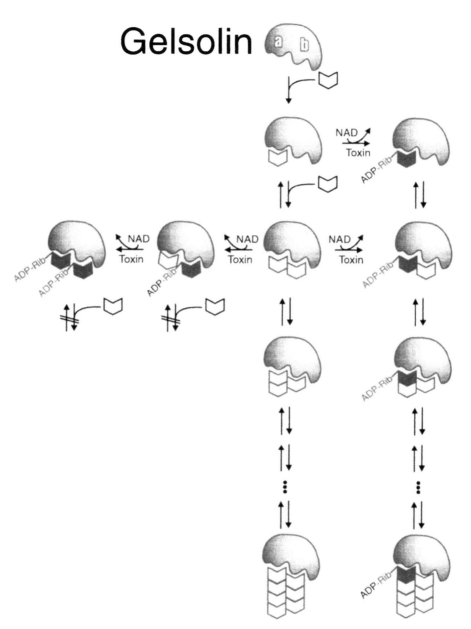

Fig. 3 ADP-ribosylation of the actin-gelsolin complex. In the presence of Ca^{2+}, Gelsolin binds to monomeric actin to form 1:1 and 1:2 complexes. The two actin-binding sites on gelsolin are referred to as EGTA-resistant (a) and Ca^{2+}-sensitive (b). Gelsolin-actin complexes can act as nuclei for actin polymerization. Actin bound to gelsolin is a substrate for ADP-ribosylation by toxins. ADP-ribosylation of actin bound to the Ca^{2+}-sensitive site causes inhibition of the nucleation activity of gelsolin-actin complex. ADP-ribosylation of actin bound to the EGTA-resistant binding site has no effect on nucleation. (Taken from Aktories and Wegner, 1992.)

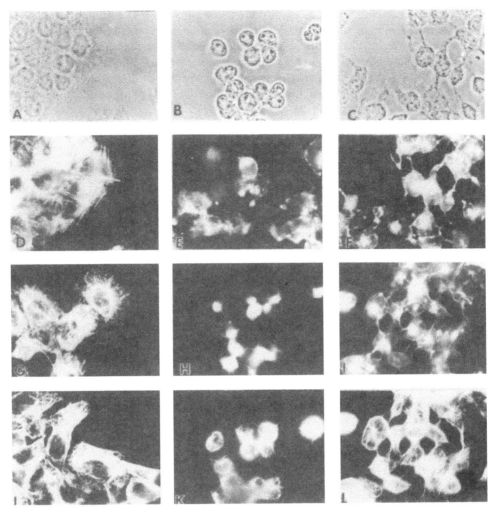

Fig. 4 Influence of the actin-ADP-ribosylating *C. botulinum* C2 toxin and the Rho-ADP-ribosylating *C. botulinum* C3 exoenzyme on the cytoskeleton of FAO cells. FAO cells were incubated with C2 toxin (C2I = 100 ng/mL and C2II = 400 ng/mL) for 3 h (B, E, H, K) or with C3 (10 μg/mL) for 48 h (C, F, I, L). The cells were labeled for actin (D, E, F), vimentin (G, H, I), and tubulin (J, K, L). A, B, C are phase-contrast micrographs. Bar 5 μm. Both toxins cause the redistribution of the actin cytoskeleton. The intermediate filaments are also affected by both transferases. However, these effects are most likely secondary (Wiegers et al., 1991). In contrast, the microtubule system is largely unaffected by the toxins. (Modified from Wiegers et al., 1991.)

and the G-actin content of the cells increases. This effect can be explained by the inhibition of the actin polymerization after ADP-ribosylation and by the barbed-end-capping effect of the modified actin, which blocks polymerization at the fast-growing end of actin filaments while the depolymerization at the pointed end is still possible (Fig. 2). Furthermore, it is feasible that not only the ADP-ribosylation of actin per se but also the modification of actin in complexes with actin-binding proteins is responsible for the cytopathic effects (Fig. 3). In this context it is noteworthy that the gelsolin actin complexes have been shown to be substrates for ADP-ribosylation in intact cells (Just et al., 1993c). After leukocytes loaded with [^{32}P]orthophosphate were intoxicated with C2-toxin, [^{32}P]ADP-ribosylated gelsolin-actin complex was precipitated with anti-gelsolin antibody from the cell lysate, indicating that the complex is substrate for ADP-ribosylation in intact cells. These data suggest that the modified functions of actin-actin-binding protein complexes may participate in the pathophysiological and pathobiochemical effects of actin-ADP-ribosylating toxins.

2. Effects of Clostridium botulinum C2 Toxin on Leukocytes

C2 toxin inhibits random migration and migration of human neutrophils induced by chemotactic agents such as formyl-methionine-leucine-phenylalanine (FMLP), complement C5a, and leukotriene B$_4$ (Norgauer et al., 1988). Treatment of human neutrophils with C2 toxin for 45 min increases superoxide anion production stimulated by FMLP 1.5–5-fold (Norgauer et al., 1988). Concomitantly, only a minor actin pool (about 20%) is ADP-ribosylated. C2 toxin also enhances the stimulation of superoxide anion production by concanavalin A and platelet-activating factor, but not by phorbol-12-myristate-13-acetate (PMA). C2 toxin increases the FMLP-stimulated release of elastase and N-acetylglucos-aminidase severalfold. Also the release of vitamin B$_{12}$-binding protein stimulated by FMLP is increased after C2 toxin treatment by about 150%. These effects of C2 toxin depend on toxin concentration, the presence of both components of the toxin, and incubation time. Norgauer et al. (1989) demonstrated that C2 toxin treatment (component I 400 ng/mL and component II 1600 ng/mL) of neutrophils for 30 min inhibits the FMLP-induced polymer-ization of G-actin and decreases the amount of F-actin in unstimulated neutrophils by about 30%. C2 toxin treatment of neutrophils for 60 min destroys about 75% of F-actin in un-stimulated neutrophils. Surprisingly, fluorescence flow cytometry and fluorospectrometric binding studies show little alteration in N-formyl peptide binding or dissociation dynamics in the toxin-treated cells. Endocytosis of the N-formyl peptide ligand-receptor complex is slower but still possible in degranulating neutrophils treated with C2 toxin for 60 min. The half-time of endocytosis is slowed from 4 min to 8 min in C2-treated neutrophils (Norgauer et al., 1989). Grimminger et al. (1991a) showed that C2 toxin augments ligand-elicited phosphoinositide hydrolysis with a major increase in inositol monophosphate. This is par-alleled by a severalfold amplification of diacylglycerol formation and a prolonged elevation of cytosolic calcium. Furthermore, it has been demonstrated that C2 toxin markedly affects the production of lipid mediators by neutrophils (Grimminger et al., 1991b). At rather low concentrations of C2 toxin (25–200 ng/mL C2I and 50–400 ng/mL and C2II), the ligand-induced formation of leukotriene B$_4$ and 5-hydroxyeicosatetraenoic acid (5-HETE) is am-plified. PAF generation in response to FMLP is increased greatly by C2 toxin at concentra-tions of 200–800 and 400–1600 ng/mL C2I and C2II, respectively. In contrast, the zymosan-effected generation of PAF and leukotrienes is inhibited by C2 toxin, a phenomenon that is paralleled by inhibition of migration and phagocytosis (Grimminger et al., 1991b). All these data indicate that the C2 toxin-induced redistribution of the actin cytoskeleton dramatically affects the activation of leukocytes by changing the temporal and spatial limitation of chemo-attractant-induced activation of neutrophils.

3. Effects of C. botulinum C2 Toxin on Rat Pheochromocytoma Cells

The influence of C2 toxin on [^{3}H]noradrenaline release has been studied with rat pheochromocytoma (PC12) cells (Matter et al., 1989). Noradrenaline release was stimulated 5–15-fold by carbachol or K^+ and 10–30-fold by ionophore A23187. Treatment of PC12 cells with C2 toxin for 4–8 h at 20°C increased carbachol-, K^+-, and A23187-induced catecholamine release 1.5–3-fold. Most important, the basal release was not affected by the toxin. Under these conditions about 75% of the cellular actin pool is ADP-ribosylated. Treatment of cells with C2 toxin for up to 1 h at 37°C also increases stimulated [³H]noradrenaline secretion. However, toxin treatment for more than 1 h at 37°C inhibited release stimulated by carbachol or K^+. In contrast, release induced by A23187 was not decreased after toxin treatment. Concomitant with toxin-induced stimulation of secretion, 20–50% of the cellular actin was ADP-ribosylated, whereas more than 60% of actin was ADP-ribosylated when exocytosis was decreased. These studies show again the importance of the actin cytoskeleton for exocytosis but indicate that C2 toxin acts on transmitter release by affecting different actin pools.

4. *Effects of C. botulinum C2 Toxin on Mast Cells*

Böttinger et al. (1987) studied the influence of C2 toxin on histamine release in rat mast cells. In this tissue, the effect of C2 toxin appears to differ from its effects on leukocytes or PC12 cells. Treatment of rat mast cells with C2 toxin for 4 h inhibited histamine release stimulated by compound 48/80 or MCD-peptide maximally by about 50%. Half-maximal and maximal inhibition occurred at about 30 and 450 ng/mL C2 toxin, respectively. Concomitantly, cellular actin was ADP-ribosylated by C2 toxin. C2 toxin also reduced histamine release stimulated by 12-*O*-tetradecanoylphorbol-13-acetate but not by A23187. Recent studies in our laboratory show that treatment of mast cells with C2 toxin for a shorter period (0.5–2 h) does not result in an increased exocytosis as observed with PC12 cells. Thus, it appears that in mast cells the intact actin cytoskeleton or the polymerizability of actin is not involved in limitation of transmitter release but is necessary for maximal exocytosis.

5. *Effects of C. botulinum C2 Toxin of Steroid Release*

Effects of C2 toxin on steroid release from Y-1 cells have been reported by Considine et al. (1992). Incubation of Y-1 cells with C2 toxin induced an increase in steroid release that was accompanied by rounding of the cell. Both effects were observed without any detectable increase in cyclic AMP, which is suggested to be the physiological mediator of ACTH-induced steroid release. Therefore, it appears that the enhanced release of steroids was caused by a mechanism related to toxin effects on the cellular microfilament network. However, in contrast to transmitter release from PC12 cells (described above), steroids are not released from vesicles. Therefore, a different mechanism appears to be involved in toxin action in Y-1 cells. A possible explanation is the increased transport of cholesterol from the membrane to the microsomal compartments observed after toxin treatment. Thus, destruction of the cytoskeleton may increase substrate availability for production of steroids (Considine et al., 1992) and finally increase steroid release.

6. *Effects of C. botulinum C2 Toxin on Endothelium Cells*

Suttorp et al. (1991) studied the influence of the C2 toxin-catalyzed ADP-ribosylation of actin on endothelial permeability. Exposure of endothelial cell monolayers to C2 toxin resulted in a concentration- and time-dependent increase in the hydraulic conductivity and decrease in the selectivity of the cell monolayers. The effects of C2 toxin were accompanied by a time- and concentration-dependent increase in ADP-ribosylation of G-actin. F-actin content decreased and the G-actin increased reciprocally in treated endothelial cells. In agreement with the view that C2 toxin causes depolymerization of

F-actin, phalloidin, which stabilizes F-actin, reduced the effects of C2 toxin on endothelial permeability. Importantly, C2 toxin induced the formation of interendothelial gaps. These effects occurred without overt cell damage and were not reversible within 2 h. These studies showed the importance of the microfilament network for regulation of endothelial permeability. Moreover, the toxin-induced breakdown of the vascular barrier function of the endothelium may be the reason for the massive edema and hypotonic effects observed after intoxication of intact animals with C2 toxin (Simpson, 1982).

7. Effects of C. botulinum C2 Toxin on Smooth Muscle Contraction

Simpson (1982) reported that C2 toxin had no effect on the neuromuscular transmission and neither caused relaxation of aortic strips nor antagonized the aortic contraction evoked by noradrenaline. However, Mauss et al. (1989) showed that treatment of the isolated longitudinal muscle of guinea pig ileum by botulinum C2 toxin impaired muscle contraction. C2 toxin inhibited the muscle contraction induced by electrical stimulation (60 V, 0.5 ms, 0.33 Hz) in a time- and concentration-dependent manner. The inhibitory effect occurred with a delay of about 1 h and required the presence of both toxin components. After 4 h with C2 toxin (C2I 1.7 μg/mL; C2II 6.7 μg/mL), smooth muscle contraction was inhibited by about 60%. At this time, about 55% of the modifiable smooth muscle actin was ADP-ribosylated. Similarly, contraction of the ileum smooth muscle stimulated by bradykinin or bethanechol is inhibited. Furthermore, the toxin inhibited muscle contraction induced by Ba^{2+} or by direct membrane depolarization (60 V, 10 ms, 0.33 Hz). Because C2 toxin modifies exclusively nomomeric actin, these findings were interpreted to indicate that a dynamic G/F-actin equilibrium is involved in the contraction of smooth muscle (Mauss et al., 1989). The puzzling finding that the ileum muscle contraction but not the contraction of vascular smooth muscle is affected by the toxin may be explained by the presence of different actin isoforms in the two tissues. Whereas vascular smooth muscle cells are especially rich in smooth muscle α-actin and poor in γ actin, the inverse relationship exits in visceral smooth muscle cells (Skalli et al., 1987). Because it has been shown that smooth muscle γ-actin is ADP-ribosylated by C2 toxin but α-actin is not (Mauss et al., 1990), the differences in the toxin effects on contraction appear to be plausible.

III. C3-LIKE ADP-RIBOSYLTRANSFERASES

A. Introduction

Some years ago it was shown that in addition to the seven serotypes of *Clostridium botulinum* neurotoxins and the actin ADP-ribosylating C2 toxin, some C and D strains of *Clostridium botulinum* produce exoenzyme C3, which ADP-ribosylates low molecular mass GTP-binding proteins of the Rho family. Meanwhile, Rho protein ADP-ribosylating transferases from other bacteria have been described, and it is quite clear that there is a large family of C3-like exoenzymes that are not restricted to clostridia. This family of C3-like transferases is described below.

B. Genetics of *Clostridium botulinum* C3 ADP-Ribosyltransferase

Clostridium botulinum C3 ADP-ribosyltransferase was detected serendipitously in our laboratory during screening for high producer strains of *Clostridium botulinum* C2 toxin. Using culture supernatant of *C. botulinum* type C, in addition to [^{32}P]ADP-ribosylation of actin by C2 toxin in human platelet membranes, [^{32}P]ADP-ribosylation of 20-kDa proteins was observed, which was apparently caused by a novel enzyme. The novel enzyme was termed C3 ADP-ribosyltransferase to distinguish it from *C. botulinum* neurotoxin C1 and from C2 toxin (Aktories et al., 1987).

ADP-ribosyltransferase C3 activity has been demonstrated in culture supernatants of various strains of *C. botulinum* types C and D. However, the characteristics of these ADP-ribosyltransferases are not identical. ADP-ribosyltransferase activity of C3 from strain C6813 was about 15-fold that of C3 from strain Stockholm (Moriishi et al., 1991). Furthermore, C3 ADP-ribosyltransferase from strain C003-9 (Nemoto et al., 1991) differs by about 40% in amino acid sequence from the C3 exoenzyme produced by strains C468 and D1873 (Popoff et al., 1991). Thus, the enzymes from various strains of *C. botulinum* most likely represent a group of isoenzymes.

Clostridium botulinum ADP-ribosyltransferase C3 is bacteriophage-encoded (Rubin et al., 1988) together with neurotoxins C1 and D (Popoff et al., 1991) and with the main component of hemagglutinin (Tsuzuki et al., 1990). The phage has a hexagonal head with a diameter of 80–100 nm and a 300–400-nm-long tail with a diameter of 10–15 nm. It has a molecular mass of $(70–100) \times 10^6$ and contains 111 kbp of double-stranded DNA (Fujii et al., 1988).

The C3 gene cloned and sequenced by Narumiya and coworkers from strain C003-9 encodes a protein of 244 amino acids with an M_r of 27,362 (Nemoto et al., 1991). During secretion, a putative signal peptide of 40 amino acids is cleaved, yielding a protein of 204 amino acids and a molecular mass of 23,119. Popoff and coworkers (Popoff et al., 1990, 1991) cloned and sequenced a gene from strains D1873 and C468 coding for a 251-amino-acid protein with a molecular mass of 27,823. After cleavage of the putative signal peptide, the mature C3 ADP-ribosyltransferase consists of 211 amino acids and has a molecular mass of 23,546. Features of the signal peptide are like those of other secreted proteins; the N terminus contains three basic amino acids, the core is hydrophobic, and the C terminus contains either serine or threonine residues (Nemoto et al., 1991; Popoff et al., 1991). The cleavage site of the signal peptide consists either of glycine and serine (Nemoto et al., 1991) or lysine and alanine (Popoff et al., 1991).

The organization of the C3 gene is similar to that of gram-positive bacteria. Like other clostridial genes it has a high A + T content. Upstream of the coding region a putative ribosomal binding site (Shine-Delgarno sequence) was found (Nemoto et al., 1991; Popoff et al., 1991). A −35 and a −10 consensus sequence with 16-bp (Nemoto et al., 1991) and 21-bp (Popoff et al., 1991) spacing, which is homologous to a gram-positive promoter, were identified. Downstream of the coding region a putative transcription terminator was identified that consisted of an inverted repeat that is able to form a hairpin structure (Nemoto et al., 1991; Popoff et al., 1991).

C. Purification of *C. botulinum* C3 ADP-Ribosyltransferase

Clostridium botulinum C3 exoenzyme is a basic, heat-stable (1 min, 95°C), and trypsin-resistant protein (Aktories et al., 1987, 1988). The C3 ADP-ribosyltransferase can be effectively purified by taking advantage of its basic structure (Aktories et al., 1987, 1988; Moriishi et al., 1991; Nemoto et al., 1991). After ammonium sulfate precipitation from the culture supernatant, ion-exchange chromatography is performed. Because exoenzyme C3 has an isoelectric point greater than pH 10, it can be recovered in the flow-through of anion-exchange columns. This step proved to be the most effective in the purification (Aktories et al., 1988). Purification to homogeneity is now performed in our laboratory by gel permeation chromatography with Superdex 75 (Fig. 5).

D. Exoenzyme C3 and *Clostridium botulinum* Neurotoxins

Concomitant with the description of *Clostridium botulinum* C3 ADP-ribosyltransferase by Aktories et al. (1988), several groups demonstrated ADP-ribosyltransferase activity of 20-kDa proteins in neurotoxin C1 and D preparations (Ohashi and Narumiya, 1987; Matsuoka et al., 1989). However, a large body of evidence has accumulated that the neurotoxin preparations used were contaminated by C3 ADP-ribosyltransferase. From this it is quite clear that inhibition of secretion and ADP-ribosylation by neurotoxin D are unrelated events. For example, C3 antiserum inhibited ADP-ribosylation of Rho protein in bovine chromaffin cells, but Ca^{2+}-dependent catecholamine secretion was unaffected (Adam-Vizi et al., 1988). Antiserum against exoenzyme C3 could not precipitate neurotoxic activity from a neurotoxin C1 preparation, but it could precipitate the ADP-ribosyltransferase activity (Morii et al., 1990). Further, C3-catalyzed ADP-ribosylation of 20-kDa platelet membrane proteins was about 1000 times stronger than ADP-ribosylation by neurotoxin C1 (Aktories and Frevert, 1987; Rösener et al., 1987). Finally, neurotoxic activity and ADP-ribosyltransferase activity in a neurotoxin D preparation were separated by hydroxyapatite chromatography (Moriishi et al., 1990).

E. ADP-Ribosylation by *C. botulinum* C3 ADP-Ribosyltransferase

Like other bacterial ADP-ribosylating exoenzymes such as cholera toxin, pertussis toxin, and *Clostridium botulinum* C2 toxin, C3 catalyzes a mono-ADP-ribosyltransferase reaction (Aktories et al., 1988). Phosphodiesterase treatment of [^{32}P]ADP-ribosylated Rho

Fig. 5 C3-like ADP-ribosyltransferases. SDS polyacrylamide gel electrophoresis of *C. botulinum* C3 exoenzyme (lane 1), *C. limosum* exoenzyme (lane 2), and *B. cereus* exoenzyme (lane 3) (1 μg of protein per lane).

protein released only [^{32}P]5′-AMP and not [^{32}P]phosphoribosyl-AMP, a cleavage product of poly(ADP-ribose) (Aktories et al., 1988; Rubin et al., 1988). ADP-ribosylation was not abolished by thymidine, which is known as an inhibitor of poly(ADP-ribose) polymerase, nor was it blocked by isonicotinic acid hydrazide, an inhibitor of NAD glycohydrolases (Rubin et al., 1988). Nonenzymatic transfer of an ADP-ribose moiety to Rho protein was excluded by the observation that ADP-ribosylation of Rho protein was not inhibited by ADP-ribose but was inhibited by NAD and nicotinamide (Aktories et al., 1988). Also, [^{32}P]ADP-ribose was not incorporated into Rho protein by C3 (Rubin et al., 1988). The K_m value was calculated to be about 0.5 μM (Just et al., 1992a). As known for other bacterial transferases, C3 possesses NAD glycohydrolase activity (Aktories et al., 1988), and the ADP-ribosylation reaction is reversed in the presence of nicotinamide and absence of NAD. The pH optima for ADP-ribosylation and de-ADP-ribosylation are 7.5 and 5.5, respectively (Habermann et al., 1991).

C3 ADP-ribosylation is influenced by various lipids and detergents. Sodium cholate (0.2%), SDS (0.01%), dimyristoylphosphatidylcholine (3 mM), and deoxycholate (0.1%) stimulate C3 ADP-ribosylation (Williamson et al., 1990; Maehama et al., 1991; Just et al., 1993b). CHAPS, Lubrol-PX, SDS (>0.03%), and various phospholipids inhibit ADP-ribosylation (Just et al., 1993b). Enhancement as well as inhibition of C3 NAD glycohydrolase by detergents has been reported (Just et al., 1993b). SDS (0.01%) decreased the K_m of exoenzyme C3 for NAD from 10 μM to 0.6 μM (Just et al., 1993b). Stimulation of C3 ADP-ribosylation by detergents is dependent on the type of Rho protein preparation used. Human platelet membrane Rho, recombinant Rho B (Just et al., 1993b), and Rho purified from bovine brain cytosol (Williamson et al., 1990) were not affected by detergents. However, stimulation of C3 ADP-ribosylation was observed using human platelet cytosol, recombinant Rho A (Just et al., 1993b), and Rho A purified from bovine brain cytosol (Williamson et al., 1990). Apparently, two different mechanisms are involved in these effects. On one site, the detergents and certain phospholipids may change the conformational state of Rho and/or induce a more lipophilic environment of Rho in order to render it more susceptible to C3 ADP-ribosylation. On the other hand, it has been reported that detergents (sodium dodecyl sulfate) or phospholipids split the complex of Rho proteins with the regulatory GDI protein in which complex Rho is not a substrate for ADP-ribosylation (see below) (Kikuchi et al., 1992).

Since magnesium and guanine nucleotides are ligands of small GTP-binding proteins, their effect on C3 ADP-ribosylation of Rho was studied by several groups. The C3 ADP-ribosylation of cytosolic Rho protein was enhanced by magnesium, calcium, barium, and manganese ions; cadmium and lanthanum ions reduced it (Narumiya et al., 1988a). Reports on the influence of guanine nucleotides on C3 ADP-ribosylation have been controversial. GTP, GTPγS, and GDP either stimulated or inhibited C3 ADP-ribosylation of Rho proteins in various preparations. However, compilation of the data obtained by different groups shows that without magnesium or at low magnesium concentrations these guanine nucleotides generally have stimulatory effects, most likely because in the absence of the divalent cation the nucleotide is released and the protein is susceptible to denaturing effects (Habermann et al., 1991). At higher magnesium concentrations the effects of nucleotides depend on the localization of Rho. ADP-ribosylation of membrane-located Rho was lower with GTP or GTPγS than with GDP. The same was true for recombinant RhoA protein (Habermann et al., 1991). On the other hand, ADP-ribosylation of cytosolic Rho appeared to be greater with GTP or GTPγS than with GDP (Williamson et al., 1990).

F.   Other C3-Like Exoenzymes

Besides the exoenzymes produced by *Clostridium botulinum*, several other bacterial exoenzymes have been described that modify the GTP-binding Rho proteins, forming a family of C3-like transferases. Members of this family are the *Clostridium limosum* exoenzyme, the recently described exoenzyme produced by *Bacillus cereus*, and a transferase from *Staphylococcus aureus* that is called EDIN (epidermal differentiation inhibitor). The exoenzyme from *Clostridium limosum* (Fig. 5) is closely related to C3, with about 70% identity on the amino acid level. The specific activity is 3.1 nmol/(mg min) with a K_m for NAD of 0.3 nM. In contrast to C3, *Clostridium limosum* exoenzyme is auto-ADP-ribosylated in the presence of 0.01% sodium dodecyl sulfate, forming an ADP-ribose protein bond highly stable toward hydroxylamine. EDIN, the exoenzyme from certain strains of *Staphylococcus aureus*, is produced as a 247-amino-acid precursor protein having a 35-amino-acid leader sequence (Inoue et al., 1991; Sugai et al., 1992). The mature protein (EDIN) of 212 amino acids shares about 35% amino acid identity with C3 (Inoue et al., 1991). Moreover, Just et al. (1992b) described a protein of about 25 kDa produced by *Bacillus cereus* that ADP-ribosylates Rho proteins (Fig. 5). Polyclonal anti-C3 or anti-*Clostridium limosum* exoenzyme antibody did not cross-react with the *Bacillus cereus* exoenzyme, indicating another more distantly related subtype of this transferase family. All these exoenzymes are basic proteins (PI > 9), have similar molecular masses of 25,000–28,000, and modify Rho proteins with high specificity, most likely on Asp-41 (Just et al., 1992a).

G.   The NAD-Binding Site of C3-Like Exoenzymes

The NAD-binding site of the Rho-ADP-ribosylating C3-like transferases was determined by Jung et al. (1993). By ultraviolet irradiation of *Clostridium limosum* exoenzyme in the presence of [carbonyl-^{14}C]NAD, 1 mole of label per mole of exoenzyme was incorporated into the transferase. Concomitantly, the transferase and NAD glycohydrolase activities were impaired. Subsequent proteolysis of the irradiated exoenzyme and analyses of the labeled peptides obtained by amino acid sequencing showed that the glutamic acid that corresponds to Glu-174 of *Clostridium botulinum* exoenzyme C3 was selectively labeled. These findings indicate that Glu-174 is part of the NAD-binding site of the catalytic center of the exoenzyme. Similar results were obtained with the exoenzyme from *Bacillus cereus*. Interestingly, the *B. cereus* exoenzyme has almost no sequence homology with the other transferases on the basis of available amino acid sequence data. In contrast, the region around the putative NAD-binding site is very similar in all C3-like exoenzymes. Furthermore, it appears that Glu-174 is equivalent to Glu-553 of *Pseudomonas aeruginosa* exotoxin A (Carroll and Collier, 1987), Glu-148 of diphtheria toxin (Carroll et al., 1985; Carroll and Collier, 1984), and Glu-129 of pertussis toxin (Barbieri et al., 1989). These data support the view that the ADP-ribosyltransferase reactions catalyzed by bacterial ADP-ribosyltransferases follow the same catalytic mechanism and that the C3-like exoenzymes possess an NAD-binding site that is common to all bacterial ADP-ribosyltransferases.

H.   Rho Proteins, Substrates for ADP-Ribosylation

The substrate of C3 ADP-ribosyltransferase was purified and subsequently identified as Rho protein (Yamamoto et al., 1988; Narumiya et al., 1988b; Kikuchi et al., 1988; Braun et al., 1989; Chardin et al., 1989). Originally Rho proteins (ras homologues) were described in the marine snail *Aplysia californica* (Madaule and Axel, 1985). In humans three isoforms were found and were termed RhoA, B, C (Chardin et al., 1988; Yeramian et al., 1987; Ogorochi et al., 1989). The Rho proteins consist of about 200 amino acids and have a molecular mass of 20–25 kDa. Like the ras proteins they have guanine nucleotide-binding and hydrolyzing activity. Rho proteins are activated by GTP binding and inactivated by subsequent hydrolysis of GTP to GDP (Hall, 1990, 1993).

As known for other low molecular mass GTP-binding proteins, several protein factors have been described that regulate the guanine nucleotide cycle of Rho (Hall, 1990). Proteins that stimulate or inhibit the dissociation of GDP were characterized (Mizuno et al., 1991; Ueda et al., 1990). However, the functions of these putative regulatory proteins are not completely clear. The guanine nucleotide dissociation inhibitor (RhoGDI) regulates the binding of Rho to membranes; that is, membrane dissociation is stimulated by RhoGDI (Hori et al., 1991). However, it appears that GDI acts not only with Rho but also with other GTP-binding proteins of the Rho family (see below) such as Rac (Hiraoka et al., 1992) and CDC42Hs (Leonard et al., 1992). A similar broad specificity has been described for the guanine nucleotide dissociation stimulator called GDS (Hiraoka et al., 1992), which also acts even on K-Ras, smg21, Rac, CDC42Hs, and other low molecular mass GTP-binding proteins. Thus, other more specific factors, which have not yet been described, may be more important for the regulation of Rho. The protein factor that stimulates Rho GTPase activity has been termed RhoGAP (Garrett et al., 1989; Morii et al., 1991). It has sequence homology with parts of the *bcr* (breakpoint cluster region) gene product chimaerin, the regulatory subunit of phosphatidylinositol 3-kinase, and p190, the RasGAP-binding protein (Hall, 1992). Subsequently RhoGAP activity of recombinant p190 was shown (Settleman et al., 1992). However, chimaerin and the *bcr* gene product apparently have higher RacGAP than RhoGAP activity (Diekmann et al., 1991).

I. Rho-Related Proteins

Several Rho-related proteins have been described that form the Rho family of low molecular mass GTP-binding proteins. Members of this family show 50–90% identity with each other (Vincent et al., 1992) and share about 35% amino acid homology with ras proteins (Drivas et al., 1990). Furthermore, all these proteins have asparagine, the amino acid acceptor for ADP-ribosylation by C3-like enzymes, at the equivalent position, which is Asp-41 in Rho. Rac1 and Rac2 (ras-related C3 botulinum toxin substrate), which were cloned from the human promyelocytic leukemia cell line HL60 (Didsbury et al., 1989), appear to be regulatory components of the superoxide-generating NADPH oxidase complex in neutrophils (Abo et al., 1991; Knaus et al., 1992). Although these GTP-binding proteins have been called C3 substrate, it was shown recently that Rac proteins are very poorly ADP-ribosylated by C3 and are most likely not physiological substrates of the transferase (Just et al., 1992a). CDC42Hs (G25K) is a 22-kDa GTP-binding protein that is the human homologue of a *Saccharomyces cerevisiae* yeast cell division cycle protein (Shinjo et al., 1990; Evans et al., 1986; Munemitsu et al., 1990; Hart et al., 1991). CDC42Hs is not modified by C3 (Just et al., 1992a). RhoG, which is more closely related to Rac than to Rho proteins, accumulates in late G_1 of the cell cycle and may be involved in cell shaping and growth (Vincent et al., 1992), and, finally, TC10 is expressed in a human teratocarcinoma cell line (Drivas et al., 1990). It is not known whether RhoG and TC10 are ADP-ribosylated by C3-like enzymes.

Furthermore, in yeast at least four *RHO* genes *(RHO1, 2, 3, 4)* have been described whose functions are still under investigation (Madaule et al., 1987; Matsui and Toh-e, 1992).

J. Functional Consequences of the ADP-Ribosylation of Rho Proteins

So far, the consequences of Rho ADP-ribosylation are best explained by a functional inactivation of the GTP-binding protein (see below) (Fig. 6). This proposed inactivation of Rho proteins by ADP-ribosylation is apparently not caused by an effect on the endogenous GTPase cycle of Rho proteins. The basal and GAP-stimulated GTPase single-cycle GTPase activity of Rho proteins is not affected by ADP-ribosylation (Paterson et al., 1990). However, ADP-ribosylation increased the steady-state GTPase activity of the recombinant RhoA and RhoB proteins in vitro about 50–100% (Mohr et al., 1992). This effect is rather small compared to the severalfold stimulation of GTPase activity by the RhoGAP protein (Garrett et al., 1989) and may be based on an increase in GDP/GTP exchange of the ADP-ribosylated Rho protein. Thus, further studies are necessary to clarify the significance of this finding.

The apparent biological inactivation of Rho by ADP-ribosylation might be explained by inhibition of the interaction of ADP-ribosylated Rho with a putative effector. The C3-like transferases modify Rho proteins on Asp-41 (Sekine et al., 1989). In Ras proteins

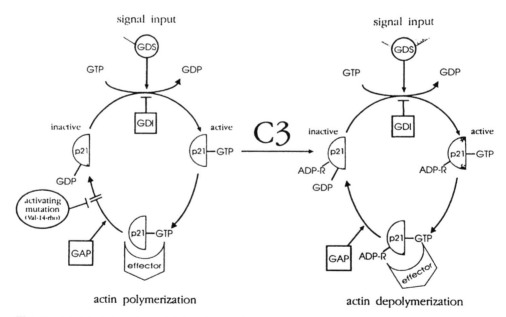

Fig. 6   Model of the action of *C. botulinum* C3 ADP-ribosyltransferase on the GTP-binding protein Rho. Rho is inactive in the GDP-bound form and active in the GTP-bound form. The active Rho is involved in the regulation of actin polymerization by an unknown mechanism. The GDP–GTP exchange is induced by guanine nucleotide dissociation stimulators (GDS) and prevented by guanine nucleotide dissociation inhibitors (GDI). The active state of Rho is terminated by hydrolysis of the bound GTP by an intrinsic GTPase activity, which is stimulated by GTPase-activating protein (GAP). ADP-ribosylation of Rho (ADP-R) occurs in the so-called effector region of the GTP-binding protein, which results in the functional inactivation of Rho and depolymerization of actin cytoskeleton. (Modified from Aktories, 1990.)

the equivalent amino acid forms part of an effector domain (Pai et al., 1989); that is, mutations in these positions render Ras biologically inactive. Provided that Ras and Rho are sterically homologous, steric hindrance of the interaction of Rho with its effector would be an explanation for the biological inactivation of Rho.

K.   Biological Effects of *C. botulinum* C3 ADP-Ribosyltransferase

Exoenzyme C3 injected into mice was nontoxic. No effect was observed after intraperitoneal injection of 100 μg per mouse, whereas for Clostridium *botulinum* neurotoxin C1 the median lethal dose (LD$_{50}$) was calculated to be 15 pg per mouse (Rösener et al., 1987; Aktories et al., 1988; Habermann and Dreyer, 1986). This lack of toxicity may be due to the difficulty of introducing C3 into intact cells. In contrast to ADP-ribosyltransferase C2, the enzyme component of which is incorporated into intact cells by its specific binding protein, C3 has no such protein. Therefore, C3 was introduced into cells by osmotic shock, electropermeabilization, or microinjection or by prolonged incubation for 1 day or more (Rubin et al., 1988; Stasia et al., 1991; Paterson et al., 1990; Wiegers et al., 1991).

Using osmotic shock, it was shown that C3 induces the rounding up of 3T3 cells, the formation of cell processes, and the occurrence of binucleated cells. After removal of exoenzyme C3, the fibroblasts reverted to normal cell shape (Rubin et al., 1988). It is noteworthy that similar morphological changes were observed after microinjection of RhoGDI into Swiss 3T3 cells (Miura et al., 1993). In PC12 cells, C3 causes the formation of neurite-like processes (Rubin et al., 1988). Morphological changes in Vero cells were accompanied by destruction of the microfilament network while the microtubule system was still intact (Chardin et al., 1989). In hepatoma cells (FAO), C3 depolymerized the microfilament cytoskeleton and additionally caused the redistribution of the intermediate filaments, an effect possibly secondary to microfilament destruction (Wiegers et al., 1991) (Fig. 4). Further insight into the role of Rho proteins and their modification by C3 was achieved by microinjection of Val-14 RhoA into Swiss 3T3 cells (Paterson et al., 1990; Ridley and Hall, 1992). This Rho mutant is persistently activated by inhibition of its endogenous GTPase activity. Microinjection of Val-14 Rho into nonconfluent Swiss 3T3 cells caused rapid changes in cell morphology with the formation of fingerlike processes that contained large amounts of microfilaments. Microinjection into confluent and serum-starved Swiss 3T3 cells, which contain a rather scarce microfilament cytoskeleton, induced the formation of focal adhesions and of stress fibers within 10–15 min. Prior ADP-ribosylation of the Rho protein prevented these effects (Paterson et al., 1990; Ridley and Hall, 1992). These data suggest that Rho proteins are involved in the regulation of the microfilament network and that ADP-ribosylation of Rho renders the GTP-binding protein biologically inactive. In agreement with this notion is the observation that C3 introduced into neutrophils by electropermeabilization inhibited chemotaxis (Stasia et al., 1991).

ADP-ribosylation of RhoA inhibited lymphocyte-mediated cytotoxicity (Lang et al., 1992), and C3 inhibited the PMA-induced, LFA-1-dependent aggregation of JY cells by ADP-ribosylation of Rho. These data suggest that RhoA protein works downstream of protein kinase C activation, thereby linking PMA stimulation to LFA-1 activation and aggregation of lymphocytes (Tominaga et al., 1993). In bovine retina membranes. ADP-ribosylation of Rho and/or Rho-like 21-kDa proteins by C3 in the presence of GTP[S] was reduced by exposure to light (Wieland et al., 1990a,b). These data were interpreted as indicating that the GTP-binding proteins that are substrates for C3 also interact with the G-protein-regulated rhodopsin receptor in the retina.

Whether the above-described effects are somehow related to the regulatory function of Rho in cytoskeleton organization is not clear. When C3 was added to the cultured Swiss 3T3 cells, it reduced cell growth rate and saturation density. Flow cytometric analysis of growing cells showed that the enzyme, in a concentration-dependent manner, caused the cells to accumulate in the G_1 phase of the cell cycle, suggesting that RhoA plays a critical role in G_1–S transition in Swiss 3T3 cells (Yamamoto et al., 1993).

IV. CONCLUDING REMARKS

From data accumulated during recent years, it is obvious that the discovery and subsequent characterization of the actin-ADP-ribosylating toxins resulted in significant progress in our knowledge about the pathogenetic mechanisms of cytotoxins. Moreover, the actin-modifying toxins are the most specific agents for modification of cellular actin. Thus, these toxins are very useful tools with which to study the physiological and pathophysiological functions of actin and of actin in complexes with actin-binding proteins. Further studies are necessary to determine the primary structure of the toxins and to elucidate the mechanisms and components involved in the cell-surface binding and transfer of the toxins into the cell.

Although the pathogenetic and pathophysiological functions of the C3-like exoenzymes are far from clear, this group of transferases are already valuable for studies on the role of small GTP-binding proteins of the Rho family. A disadvantage of the exoenzymes is their poor cell membrane penetration. Whether an unidentified binding component may exist is still under investigation. To circumvent this disadvantage, a chimeric fusion protein was recently constructed that consisted of the binding and transfer moiety of diphtheria toxin and the enzyme moiety of C3 (Aullo et al., 1993). Toxin chimeras, which are able to enter cells, will greatly extend the application of C3-like exoenzymes.

REFERENCES

Abo, A., Pick, E., Hall, A., Totty, N., Teaham, C. G., and Segal, A. W. (1991). Activation of the NADPH oxidase involves the small GTP-binding protein p21rac. *Nature 353*: 668–670.

Adam-Vizi, V., Rösener, S., Aktories, K., and Knight, D. E. (1988). Botulinum toxin-induced ADP-ribosylation and inhibition of exocytosis are unrelated events. *FEBS Lett. 238*: 277–280.

Aktories, K. (1990). Clostridial ADP-ribosyltransferases—modification of low molecular weight GTP-binding proteins and of actin by clostridial toxins. *Med. Microbiol. Immunol. 179*: 123–136.

Aktories, K. (1992). *Clostridium botulinum* C2 toxin and *C. botulinum* C3 ADP-ribosyltransferase. In: Handbook of Experimental Pharmacology, H. Herken and F. Hucho (Eds.), Springer-Verlag, Berlin, pp. 841–854.

Aktories, K., and Frevert, J. (1987). ADP-ribosylation of a 21–24 kDa eukaryotic protein(s) by C3, a novel botulinum ADP-ribosyltransferase, is regulated by guanine nucleotide. *Biochem. J. 247*: 363–368.

Aktories, K., and Hall, A. (1989). Botulinum ADP-ribosyltransferase C3: a new tool to study low molecular weight GTP-binding proteins. *Trends Pharmacol. Sci. 10*: 415–418.

Aktories, K., and Just, I. (1990). Botulinum C2 toxin. In: ADP-ribosylating toxins and G-proteins, J. Moss and M. Vaughan (Eds.), American Society for Microbiology, Washington, DC, pp. 79–95.

Aktories, K., and Wegner, A. (1989). ADP-ribosylation of actin by clostridial toxins. *J. Cell Biol. 109*: 1385–1387.

Aktories, K., and Wegner, A. (1992). Mechanisms of the cytopathic action of actin-ADP-ribosylating toxins. *Mol. Microbiol. 6*: 2905–2908.

Aktories, K., Ankenbauer, T., Schering, B., and Jakobs, K. H. (1986a). ADP-ribosylation of platelet actin by botulinum C2 toxin. *Eur. J. Biochem. 161*: 155–162.

Aktories, K. Bärmann, M., Ohishi, I., Tsuyama, S., Jakobs, K. H., and Habermann, E. (1986b). Botulinum C2 toxin ADP-ribosylates actin. *Nature 322*: 390–392.

Aktories, K., Weller, U., and Chhatwal, G. S. (1987). *Clostridium botulinum* type C produces a novel ADP-ribosyltransferase distinct from botulinum C2 toxin. *FEBS Lett. 212*: 109–113.

Aktories, K., Rösener, S., Blaschke, U., and Chhatwal, G. S. (1988). Botulinum ADP-ribosyltransferase C3. Purification of the enzyme and characterization of the ADP-ribosylation reaction in platelet membranes. *Eur. J. Biochem. 172*: 445–450.

Aktories, K., Reuner, K.-H., Presek, P., and Bärmann, M. (1989) Botulinum C2 toxin treatment increases the G-actin pool in intact chicken cells: a model for the cytopathic action of actin-ADP-ribosylating toxins. *Toxicon 27*: 989–993.

Aktories, K., Mohr, C., and Koch, G. (1992a). *Clostridium botulinum* C3 ADP-ribosyltransferase. *Curr. Top. Microbiol. Immunol. 175*: 115–131.

Aktories, K., Wille, M., and Just, I. (1992b). Clostridial actin-ADP-ribosylating toxins. *Curr. Top. Microbiol. Immunol. 175*: 97–113.

Aktories, K., Koch, G., and Just, I. (1993). Molecular biology of clostridial ADP-ribosyltransferases and their substrates. In: Genetics and Molecular Biology of Anaerobic Bacteria, M. Sebald (Ed.), Springer, New York, pp. 195–210.

Aullo, P., Giry, M., Olsnes, S., Popoff, M. R., Kocks, C., and Boquet, P. (1993). A chimeric toxin to study the role of the 21 kDa GTP binding protein rho in the control of actin microfilament assembly. *EMBO J. 12*: 921–931.

Barbieri, J. T., Mende-Mueller, M., Rappuoli, R., and Collier, R. J. (1989). Photolabeling of Glu-129 of the S-1 subunit of pertussis toxin with NAD. *Infect. Immun. 57*: 3549–3554.

Bershadsky, A. D., and Vasiliev, J. M. (1988). *Cytoskeleton*, Plenum, New York.

Böttinger, H., Reuner, K. H., and Aktories, K. (1987). Inhibition of histamine release from rat mast cells by botulinum C2 toxin. *Int. Arch. Allergy Appl. Immunol. 84*: 380–384.

Braun, U., Habermann, B., Just, I., Aktories, K., and Vandekerckhove, J. (1989). Purification of the 22 kDa protein substrate of botulinum ADP-ribosyltransferase C3 from porcine brain cytosol and its characterization as a GTP-binding protein highly homologous to the *rho* gene product. *FEBS Lett. 243*: 70–76.

Carroll, S. F., and Collier, R. J. (1984). NAD binding site of diphtheria toxin: identification of a residue within the nicotinamide subsite by photochemical modification with NAD. *Proc. Natl. Acad. Sci. U.S.A. 81*: 3307–3311.

Carroll, S. F., and Collier, R. J. (1987). Active site of *Pseudomonas aeruginosa* exotoxin A. Glutamic acid 553 is photolabeled by NAD and shows functional homology with glutamic acid 148 of diphtheria toxin. *J. Biol. Chem. 262*: 8707–8711.

Carroll, S. F., McCloskey, J. A., Crain, P. F., Oppenheimer, N. J., Marschner, T. M., and Collier, R. J. (1985). Photoaffinity labeling of diphtheria toxin fragment A with NAD: structure of the photoproduct at position 148. *Proc. Natl. Acad. Sci. U.S.A. 82*: 7237–7241.

Chardin, P., Madaule, P., and Tavitian, A. (1988). Coding sequence of human rho cDNAs clone 6 and clone 9. *Nucleic Acids Res. 16*: 2717.

Chardin, P., Boquet, P., Madaule, P., Popoff, M. R., Rubin, E. J., and Gill, D. M. (1989). The mammalian G protein rho C is ADP-ribosylated by *Clostridium botulinum* exoenzyme C3 and affects actin microfilament in Vero cells. *EMBO J. 8*: 1087–1092.

Collier, R. J. (1990). Diphtheria toxin: structure and function of a cytocidal protein. In: ADP-ribosylating toxins and G-proteins, J. Moss and M. Vaughan (Eds.), American Society for Microbiology, Washington, DC, pp. 3–19.

Considine, R. V., and Simpson, L. L. (1991). Cellular and molecular actions of binary toxins possessing ADP-ribosyltransferase activity. *Toxicon 29*: 913–936.

Considine, R. V., Simpson, L. L., and Sherwin, J. R. (1992). *Botulinum* C toxin and steroid production in adrenal Y-1 cells: the role of microfilaments in the toxin-induced increase in steroid release. *J. Pharmacol. Exp. Ther. 260*: 859–864.

Didsbury, J., Weber, R. F., Bokoch, G. M., Evans, T., and Snyderman, R. (1989). rac, a novel ras-related family of proteins that are botulinum toxin substrates. *J. Biol. Chem. 264*: 16378–16382.

Diekmann, D., Brill, S., Garrett, M. D., Totty, N., Hsuan, J., Monfries, C., Hall, C., Lim, L., and Hall, A. (1991). *Bcr* encodes a GTPase-activating protein for p21. *Nature 351*: 400–402.

Drivas, G. T., Shih, A., Coutavas, E., Rush, M. G., and D'Eustachio, P. (1990). Characterization of four novel *ras*-like genes expressed in a human teratocarcinoma cell line. *Mol. Cell. Biol. 10*: 1793–1798.

Eklund, M. W., and Poysky, F. T. (1972). Activation of a toxic component of *Clostridium botulinum* types C and D by trypsin. *Appl. Microbiol. 24*: 108–113.

Eklund, M. W., Poysky, F. T., Reed, S. M., and Smith, C. A. (1971). Bacteriophage and the toxigenicity of *Clostridium botulinum* type. *Science 172*: 480–482.

Eklund, M. W., Poysky, F. T., and Reed, S. M. (1972). Bacteriophage and the toxigenicity of *Clostridium botulinum* type D. *Nature New Biol. 235*: 16–17.

Eklund, M. W., Poysky, F. T., and Habig, W. H. (1989). Bacteriophages and plasmids in *Clostridium botulinum* and *Clostridium tetani* and their relationship to production of toxins. In: Botulinum neurotoxin and tetanus toxin, L. L. Simpson (Ed.), Academic, New York, pp. 25–51.

Evans, T., Brown, M. L., Fraser, E. D., and Northrup, J. K. (1986). Purification of the major GTP-binding proteins from human placental membranes. *J. Biol. Chem. 261*: 7052–7059.

Falnes, P. O., Madshus, I. H., Sandvig, K., and Olsnes, S. (1992). Replacement of negative by positive charges in the presumed membrane-inserted part of diphtheria toxin B fragment. *J. Biol. Chem. 267*(17): 12284–12290.

Fishman, P. H. (1990). Mechanism of action of cholera toxin. In: ADP-ribosylating toxins and G-proteins, J. Moss and M. Vaughan (Eds.), American Society for Microbiology, Washington, DC, pp. 127–140.

Fujii, N., Oguma, K., Yokosawa, N., Kimura, K., and Tsuzuki, K. (1988). Characterization of bacteriophage nucleic acids obtained from *Clostridium botulinum* types C and D. *Appl. Environ. Microbiol. 54*: 69–73.

Garrett, M. D., Self, A. J., v. Oers, C., and Hall, A. (1989). Identification of distinct cytoplasmic targets for ras, R-ras and rho regulatory proteins. *J. Biol. Chem. 264*: 10–13.

Geipel, U., Just, I., Schering, B., Haas, D., and Aktories, K. (1989). ADP-ribosylation of actin causes increase in the rate of ATP exchange and inhibition of ATP hydrolysis. *Eur. J. Biochem. 179*: 229–232.

Geipel, U., Just, I., and Aktories, K. (1990). Inhibition of cytochalasin D-stimulated G-actin ATPase by ADP-ribosylation with *Clostridium perfringens* iota toxin. *Biochem. J. 266*: 335–339.

Gierschik, P. (1992). ADP-ribosylation of signal-transducing guanine nucleotide-binding proteins by pertussis toxin. *Curr. Top. Microbiol. Immunol. 175*: 69–98.

Gill, D. M. (1977). Mechanism of action of cholera toxin. *Adv. Cyclic. Nucleotide Res. 8*: 85–118.

Grimminger, F., Sibelius, U., Aktories, K., Just, I., and Seeger, W. (1991a). Suppression of cytoskeletal rearrangement in activated human neutrophils by botulinum C toxin. Impact on cellular signal transduction. *J. Biol. Chem. 266*: 19276–19282.

Grimminger, F., Sibelius, U., Aktories, K., Suttorp, N., and Seeger, W. (1991b). Inhibition of cytoskeletal rearrangement by botulinum C2 toxin amplifies ligand-evoked lipid mediator generation in human neutrophils. *Mol. Pharmacol. 40*: 563–571.

Habermann, B., Mohr, C., Just, I., and Aktories, K. (1991). ADP-ribosylation and de-ADP-ribosylation of the *rho* protein by *Clostridium botulinum* exoenzyme C3. Regulation by EDTA, guanine nucleotides and pH. *Biochim. Biophys. Acta 1077*: 253–258.

Habermann, E., and Dreyer, F. (1986). Clostridial neurotoxins: handling and action at the cellular and molecular level. *Curr. Top. Microbiol. Immunol. 129*: 93–179.

Hall, A. (1990). The cellular functions of small GTP-binding proteins. *Science 249*: 635–640.

Hall, A. (1992). Signal transduction through small GTPases—a tale of two GAPs. *Cell 69*: 389–391.

Hall, A. (1993). Ras-related proteins. *Curr. Opin. Cell Biol. 5*: 265–268.

Harris, H. E., and Weeds, A. G. (1984). Plasma gelsolin caps and severs actin filaments. *FEBS Lett. 177*: 184–188.

Hart, M. J., Shinjo, K., Hall, A., Evans, T., and Cerione, R. A. (1991). Identification of the human platelet GTPase activating protein for the CDC42Hs protein. *J. Biol. Chem. 266*: 20840–20848.

Hartwig, J. H., and Kwiatkowski, D. J. (1991). Actin-binding proteins. *Curr. Opin. Cell Biol. 3*: 87–97.

Hiraoka, K., Kaibuchi, K., Andos, S. Musha, T., Takaishi, K., Mizuno, T., Asada, M., Ménard, L., Tomhave, E., Didsbury, J., Snyderman, R., and Takai, Y. (1992). Both stimulatory and inhibitory GDP/GTP exchange proteins, *smg* GDS and *rho* GDI, are active on multiple small GTP-binding proteins. *Biochem. Biophys. Res. Commun. 182*: 921–930.

Holmes, K. C., Popp, D., Gebhard, W., and Kabsch, W. (1990). Atomic model of the actin filament. *Nature 347*: 44–49.

Hori, Y., Kikuchi, A., Isomura, M., Katayama, M., Miura, Y., Fujioka, H., Kaibuchi, K., and Takai, Y. (1991). Post-translational modifications of the C-terminal region of the rho protein are important for its interaction with membranes and the stimulatory and inhibitory GDP/GTP exchange proteins. *Oncogene 6*: 515–522.

Inoue, S., Sugai, M., Murooka, Y., Paik, S.-Y., Hong, Y.-M., Ohgai, H., and Suginaka, H. (1991). Molecular cloning and sequencing of the epidermal cell differentiation inhibitor gene from *Staphylococcus aureus*. *Biochem. Biophys. Res. Commun. 174*: 459–464.

Janmey, P. A., and Stossel, T. P. (1987). Modulation of gelsolin function by hosphatidylinositol 4,5-bisphosphate. *Nature 325*: 362–364.

Janmey, P. A., Chaponnier, C., Lind, S. E., Zaner, K. S., Stossel, T. P., and Yin, H. L. (1985). Interactions of gelsolin and gelsolin-actin complexes with actin. Effects of calcium on actin nucleation, filament severing, and end blocking. *Biochemistry 24*: 3714–3723.

Jansen, B. C., and Knoetze, P. C. (1971). Tryptic activation of *Clostridium botulinum* type Cβ toxin. *Onderstepoort J. Vet. Res. 38*: 237–238.

Jung, M., Just, I., van Damme, J., Vanderkerckhove, J., and Aktories, K. (1993). NAD-binding site of the C3-like ADP-ribosyltransferase from *Clostridium limosum*. *J. Biol. Chem. 268*: 23215–23218.

Just, I., Geipel, U., Wegner, A., and Aktories, K. (1990). De-ADP-ribosylation of actin by *Clostridium perfringens* iota-toxin and *Clostridium botulinum* C2 toxin. *Eur. J. Biochem. 192*: 723–727.

Just, I., Mohr, C., Schallehn, G., Menard, L., Didsbury, J. R., Vandekerckhove, J., van Damme, J., and Aktories, K. (1992a). Purification and characterization of an ADP-ribosyltransferase produced by *Clostridium limosum*. *J. Biol. Chem. 267*: 10274–10280.

Just, I., Schallehn, G., and Aktories, K. (1992b). ADP-ribosylation of small GTP-binding proteins by *Bacillus cereus*. *Biochem. Biophys. Res. Commun. 183*: 931–936.

Just, I., Hennessey, E. S., Drummond, D. R., Aktories, K., and Sparrow, J. C. (1993a). ADP-ribosylation of *Drosophila* indirect flight muscle actin and arthrin by *Clostridium botulinum* C2 toxin and *Clostridium perfringens* iota toxin. *Biochem. J. 291*: 409–412.

Just, I., Mohr, C., Habermann, B., Koch, G., and Aktories, K. (1993b). Enhancement of *Clostridium botulinum* C3-catalyzed ADP-ribosylation of recombinant rhoA by sodium dodecyl sulfate. *Biochem. Pharmacol. 45*: 1409–1416.

Just, I., Wille, M., Chaponnier, C., and Aktories, K. (1993c). Gelsolin-actin complex is target for ADP-ribosylation by *Clostridium botulinum* C2 toxin in intact human neutrophils. *Eur. J. Pharmacol. 246:* 293–297.

Kabsch, W. and Vandekerckhove, J. (1992). Structure and function of actin. *Annu. Rev. Biophys. Biophys. Chem. 21:* 49–76.

Kabsch, W., Mannherz, H. G., Suck, D., Pai, E. F., and Holmes, K. C. (1990). Atomic structure of the actin: DNase I complex. *Nature 347:* 37–44.

Kikuchi, A., Yamamoto, K., Fujita, T., and Takai, Y. (1988). ADP-ribosylation of the bovine brain rho protein by botulinum toxin type C1. *J. Biol. Chem. 263:* 16303–16308.

Kikuchi, A., Kuroda, S., Sasaki, T., Kotani, K., Hirata, K., Katayama, M., and Takai, Y. (1992). Functional interactions of stimulatory and inhibitory GDP/GTP exchange proteins and their common substrate small GTP-binding protein. *J. Biol. Chem. 267:* 14611–14615.

Knaus, U. G., Heyworth, P. G., Kinsella, B. T., Curnutte, J. T., and Bokoch, G. M. (1992). Purification and characterization of Rac 2, a cytosolic GTP-binding protein that regulates human neutrophil NADPH oxidase. *J. Biol. Chem. 267:* 23575–23582.

Koehler, T. M., and Collier, R. J. (1991). Anthrax toxin protective antigen: low-pH-induced hydrophobicity and channel formation in liposomes. *Mol. Microbiol. 5:* 1501–1506.

Lang, P., Guizani, L., Vitté-Mony, I., Stancou, R., Dorseuil, O., Gacon, G., and Bertoglio, J. (1992). ADP-ribosylation of the *ras*-related, GTP-binding protein RhoA inhibits lymphocyte-mediated cytotoxicity. *J. Biol. Chem. 267:* 11677–11680.

Leonard, D., Hart, M. J., Platko, J. V., Eva, A., Henzel, W., Evans, T., and Cerione, R. A. (1992). The identification and characterization of a GDP-dissociation inhibitor (GDI) for the CDC42Hs protein. *J. Biol. Chem. 267:* 22860–22868.

Leppla, S. (1982). Anthrax toxin edema factor: a bacterial adenylate cyclase that increases cyclic AMP concentrations in eukaryotic cells. *Proc. Natl. Acad. Sci. U.S.A. 79:* 3162–3166.

Madaule, P., and Axel, R. (1985). A novel *ras*-related gene family. *Cell 41:* 31–40.

Madaule, P., Axel, R., and Myers, A. M. (1987). Characterization of two members of the *rho* gene family from the yeast *Saccharomyces cerevisiae*. *Proc. Natl. Acad. Sci. U.S.A. 84:* 779–783.

Maehama, T., Takahashi, K., Ohoka, Y., Ohtsuka, T., Ui, M., and Katada, T. (1991). Identification of a botulinum C3-like enzyme in bovine brain that catalyzes ADP-ribosylation of GTP-binding proteins. *J. Biol. Chem. 266:* 10062–10065.

Matsui, Y., and Toh-e, A. (1992). Isolation and characterization of two novel *ras* superfamily genes in *Saccharomyces cerevisiae*. *Gene 114:* 43–49.

Matsuoka, I., Sakuma, H., Syuto, B., Moriishi, K., Kubo, S., and Kurihara, K. (1989). ADP-ribosylation of 24–26-kDa GTP-binding proteins localized in neuronal and non-neuronal cells by botulinum neurotoxin D. *J. Biol. Chem. 264:* 706–712.

Matter, K., Dreyer, F., and Aktories, K. (1989). Actin involvement in exocytosis from PC12 cells: studies on the influence of botulinum C2 toxin on stimulated noradrenaline release. *J. Neurochem. 52:* 370–376.

Mauss, S., Koch, G., Kreye, V. A. W., and Aktories, K. (1989). Inhibition of the contraction of the isolated longitudinal muscle of the guinea-pig ileum by botulinum C2 toxin: evidence for a role of G/F-actin transition in smooth muscle contraction. *Naunyn-Schmiedebergs Arch. Pharmacol. 340:* 345–351.

Mauss, S., Chaponnier, C., Just, I., Aktories, K., and Gabbiani, G. (1990). ADP-ribosylation of actin isoforms by *Clostridium botulinum* C2 toxin and *Clostridium perfringens* iota toxin. *Eur. J. Biochem. 194:* 237–241.

Miura, Y., Kikuchi, A., Musha, T., Kuroda, S., Yaku, H., Sasaki, T., and Takai, Y. (1993). Regulation of morphology by *rho* p21 and its inhibitory GDP/GTP exchange protein (*rho* GDI) in Swiss 3T3 cells. *J. Biol. Chem. 268:* 510–515.

Mizuno, T., Kaibuchi, K., Yamamoto, T., Kawamura, M., Sakoda, T., Fujioka, H., Matsuura, Y., and Takai, Y. (1991). A stimulatory GDP/GTP exchange protein for smg p21 is active on the post-translationally processed form of c-Ki-ras p21 and rhoA p21. *Proc. Natl. Acad. Sci. U.S.A. 88*: 6442–6446.

Mohr, C., Koch, G., Just, I., and Aktories, K. (1992). ADP-ribosylation by *Clostridium botulinum* C3 exoenzyme increases steady state GTPase activities of recombinant rhoA and rhoB proteins. *FEBS Lett. 297*: 95–99.

Morii, N., Ohashi, Y., Nemoto, Y., Fujiwara, M., Ohnishi, Y., Nishiki, T., Kamata, Y., Kozaki, S., Narumiya, S., and Sakaguchi, G. (1990). Immunochemical identification of the ADP-ribosyltransferase in botulinum C1 neurotoxin as C3 exoenzyme-like molecule. *J. Biochem. 107*: 769–775.

Morii, N., Kawano, K., Sekine, A., Yamada, T., and Narumiya, S. (1991). Purification of GTPase-activating protein specific for the *rho* gene. *J. Biol. Chem. 266*: 7646–7650.

Moriishi, K., Syuto, B., Oguma, K., and Saito, M. (1990). Separation of toxic activity and ADP-ribosylation activity of botulinum neurotoxin D. *J. Biol. Chem. 265*: 16614–16616.

Moriishi, K., Syuto, B., Yokosawa, N., Oguma, K., and Saito, M. (1991). Purification and characterization of ADP-ribosyltransferases (exoenzyme C3) of *Clostridium botulinum* type C and D strains. *J. Bacteriol. 173*: 6025–6029.

Moss, J., and Vaughan, M. (1977). Mechanism of action of choleragen. Evidence for ADP-ribosyltransferase activity with arginine as an acceptor. *J. Biol. Chem. 252*: 2455–2457.

Munemitsu, S., Innis, M. A., Clark, R., McCormick, F., Ullrich, A., and Polakis, P. (1990). Molecular cloning and expression of a G25K cDNA, the human homolog of the yeast cell cycle gene CDC42. *Mol. Cell. Biol. 10*: 5977–5982.

Narumiya, S., Morii, N., Ohno, K., Ohashi, Y., and Fujiwara, M. (1988a). Subcellular distribution and isoelectric heterogeneity of the substrate for ADP-ribosyltransferase from *Clostridium botulinum. Biochem. Biophys. Res. Commun. 150*: 1122–1130.

Narumiya, S., Sekine, A., and Fujiwara, M. (1988b). Substrate for botulinum ADP-ribosyltransferase, Gb, has an amino acid sequence homologous to a putative *rho* gene product. *J. Biol. Chem. 263*: 17255–17257.

Nemoto, Y., Namba, T., Kozaki, S., and Narumiya, S. (1991). *Clostridium botulinum* C3 ADP-ribosyltransferase gene. *J. Biol. Chem. 266*: 19312–19319.

Noda, M., Kato, I., Matsuda, F., and Hirayama, T. (1981). Mode of action of staphylococcal leukocidin: relationship between binding of ^{125}I-labeled S and F components of leukocidin to rabbit polymorphonuclear leukocytes and leukocidin activity. *Infect. Immun. 34*: 362–367.

Norgauer, J., Kownatzki, E., Seifert, R., and Aktories, K. (1988). Botulinum C2 toxin ADP-ribosylates actin and enhances O_2 production and secretion but inhibits migration of activated human neutrophils. *J. Clin. Invest. 82*: 1376–1382.

Norgauer, J., Just, I., Aktories, K., and Sklar, L. A. (1989). Influence of botulinum C2 toxin on F-actin and *N*-formyl peptide receptor dynamics in human neutrophils. *J. Cell Biol. 109*: 1133–1140.

Ogorochi, T., Nemoto, Y., Nakajima, M., Nakamura, E., Fujiwara, M., and Narumiya, S. (1989). cDNA cloning of Gb, the substrate for botulinum ADP-ribosyltransferase from bovine adrenal gland and its identification as a *rho* gene product. *Biochem. Biophys. Res. Commun. 163*: 1175–1181.

Ohashi, Y., and Narumiya, S. (1987). ADP-ribosylation of an M_r 21,000 membrane protein by type D botulinum toxin. *J. Biol. Chem. 262*: 1430–1433.

Ohishi, I. (1983). Response of mouse intestinal loop to botulinum C2 toxin: enterotoxic activity induced by cooperation of nonlinked protein components. *Infect. Immun. 40*: 691–695.

Ohishi, I. (1986). NAD-glycohydrolase activity of botulinum C2 toxin: a possible role of component I in the mode of action of the toxin. *J. Biochem. (Tokyo) 100*: 407–413.

Ohishi, I. (1987). Activation of botulinum C2 toxin by trypsin. *Infect. Immun. 55*: 1461–1465.

Ohishi, I., and Akada, Y. (1986). Heterogeneities of two components of C2 toxin produced by *Clostridium botulinum* types C and D. *J. Gen. Microbiol. 132*: 125–131.

Ohishi, I, and DasGupta, B. R. (1987). Molecular structure and biological activities of *Clostridium botulinum* C2 toxin. In: *Avian Botulism*, M. W. Eklund, and V. R. Dowell (Eds.), Thomas, Springfield, pp. 223–247.

Ohishi, I., and Hama, Y. (1992). Purification and characterization of heterologous component IIs of botulinum C toxin. *Microbiol. Immunol. 36*(3): 221–229.

Ohishi, I., and Miyake, M. (1985). Binding of the two components of C2 toxin to epithelial cells and brush borders of mouse intestine. *Infect. Immun. 48*: 769–775.

Ohishi, I., and Yanagimoto, A. (1992). Visualizations of binding and internalization of two nonlinked protein components of botulinum C toxin in tissue culture cells. *Infect. Immun. 60*: 4648–4655.

Ohishi, I., Iwasaki, M., and Sakaguchi, G. (1980). Purification and characterization of two components of botulinum C2 toxin. *Infect. Immun. 30*: 668–673.

Pai, E. F., Kabsch, W., Krengel, U., Holmes, K. C., John, J., and Wittinghofer, A. (1989). Structure of the guanine-nucleotide-binding domain of the Ha-*ras* oncogene product p21 in the triphosphate conformation. *Nature 341*: 209–214.

Pastan, I., and FitzGerald, D. (1989). *Pseudomonas* exotoxin: chimeric toxins. *J. Biol. Chem. 264*: 15157–15160.

Paterson, H. F., Self, A. J., Garrett, M. D., Just, I., Aktories, K., and Hall, A. (1990). Microinjection of recombinant p21 induces rapid changes in cell morphology. *J. Cell Biol. 111*: 1001–1007.

Perentesis, J. P., Miller, S. P., and Bodley, J. W. (1992). Protein toxin inhibitors of protein synthesis. *BioFactors 3*: 173–184.

Pollard, T. D., and Cooper, J. A. (1986). Actin and actin-binding proteins. A critical evaluation of mechanisms and functions. *Ann. Rev. Biochem. 55*: 987–1035.

Popoff, M. R., and Boquet, P. (1988). *Clostridium spiroforme* toxin is a binary toxin which ADP-ribosylates cellular actin. *Biochem. Biophys. Res. Commun. 152*: 1361.

Popoff, M. R., Rubin, E. J., Gill, D. M., and Boquet, P. (1988). Actin-specific ADP-ribosyltransferase produced by a *Clostridium difficile* strain. *Infect. Immun. 56*: 2299–2306.

Popoff, M. R., Boquet, P., Gill, D. M., and Eklund, M. W. (1990). DNA sequence of exoenzyme C3, an ADP-ribosyltransferase encoded by *Clostridium botulinum* C and D phages. *Nucleic Acids Res. 18*: 1291.

Popoff, M. R., Hauser, D., Boquet, P., Eklund, M. W., and Gill, D. M. (1991). Characterization of the C3 gene of *Clostridium botulinum* types C and D and its expression in *Escherichia coli*. *Infect. Immun. 59*: 3673–3679.

Reuner, K. H., Presek, P., Boschek, C. B., and Aktories, K. (1987). Botulinum C2 toxin ADP-ribosylates actin and disorganizes the microfilament network in intact cells. *Eur. J. Cell Biol. 43*: 134–140.

Ridley, A. J., and Hall, A. (1992). The small GTP-binding protein rho regulates the assembly of focal adhesions and actin stress fibers in response to growth factors. *Cell 70*: 389–399.

Rösener, S., Chhatwal, G. S., and Aktories, K. (1987). Botulinum ADP-ribosyltransferase C3 but not botulinum neurotoxins C1 and D ADP-ribosylates low molecular mass GTP-binding proteins. *FEBS Lett. 224*: 38–42.

Rubin, E. J., Gill, D. M., Boquet, P., and Popoff, M. R. (1988). Functional modification of a 21-kilodalton G protein when ADP-ribosylated by exoenzyme C3 of *Clostridium botulinum*. *Mol. Cell. Biol. 8*: 418–426.

Schering, B., Bärmann, M., Chhatwal, G. S., Geipel, U., and Aktories, K. (1988). ADP-ribosylation of skeletal muscle and non-muscle actin by *Clostridium perfringens* iota toxin. *Eur. J. Biochem. 171*: 225–229.

Schmid, A., Benz, R., Just, I., and Aktories, K. (1994). Interaction of *Clostridium botulinum* C2 toxin with lipid bilayer membranes: formation of cation-selective channels and inhibition of channel function by chloroquine and peptides. *J. Biol. Chem. 269*: 16706–16711, 1994.

Sekine, A., Fujiwara, M., and Narumiya, S. (1989). Asparagine residue in the *rho* gene product is the modification site for botulinum ADP-ribosyltransferase. *J. Biol. Chem. 264*: 8602–8605.

Selve, N., and Wegner, A. (1986). Rate of treadmilling of actin filaments in vitro. *J. Mol. Biol. 187*: 627–631.

Settleman, J., Albright, C. F., Foster, L. C., and Weinberg, R. A. (1992). Association between GTPase activators for Rho and Ras families. *Nature 359*: 153–154.

Shinjo, K., Koland, J. G., Hart, M. J., Narasimhan, V., Johnson, D. I., Evans, T., and Cerione, R. A. (1990). Molecular cloning of the gene for the human placental GTP-binding protein Gp (G25K) identification of this GTP-binding protein as the human homolog of the yeast cell-division-cycle protein CDC42. *Proc. Natl. Acad. Sci. U.S.A. 87*: 9853–9857.

Simpson, L. L. (1982). A comparison of the pharmacological properties of *Clostridium botulinum* type C1 and C2 toxins. *J. Pharmacol. Exp. Ther. 223*: 695–701.

Simpson, L. L. (1984). Molecular basis for the pharmacological actions of *Clostridium botulinum* type C2 toxin. *J. Pharmacol. Exp. Ther. 230*: 665–669.

Simpson, L. L. (1989). The binary toxin produced by *Clostridium botulinum* enters cells by receptor-mediated endocytosis to exert its pharmacologic effects. *J. Pharmacol. Exp. Ther. 251*: 1223–1228.

Simpson, L. L., Stiles, B. G., Zapeda, H. H., and Wilkins, T. D. (1987). Molecular basis for the pathological actions of *Clostridium perfringens* iota toxin. *Infect. Immun. 55*: 118–122.

Simpson, L. L., Stiles, B. G., Zepeda, H., and Wilkins, T. D. (1989). Production by *Clostridium spiroforme* of an iotalike toxin that possesses mono(ADP-ribosyl)transferase activity: identification of a novel class of ADP-ribosyltransferase. *Infect. Immun. 57*: 255–261.

Skalli, O., Vandekerckhove, J., and Gabbiani, G. (1987). Actin-isoform pattern as a marker of normal or pathological smooth-muscle and fibroblastic tissues. *Differentiation 33*: 232–238.

Spangler, B. D. (1992). Structure and function of cholera toxin and the related *Escherichia coli* heat-labile enterotoxin. *Microbiol. Rev. 56*: 622–647.

Stasia, M.-J., Jouan, A., Bourmeyster, N., Boquet, P., and Vignais, P. V. (1991). ADP-ribosylation of a small size GTP-binding protein in bovine neurotrophils by the C3 exoenzyme of *Clostridium botulinum* and effect on the cell motility. *Biochem. Biophys. Res. Commun. 180*: 615–622.

Stiles, B. G., and Wilkens, T. D. (1986a). Purification and characterization of *Clostridium perfringens* iota toxin: dependence on two nonlinked proteins for biological activity. *Infect. Immun. 54*: 683–688.

Stiles, B. G., and Wilkins, T. D. (1986b). *Clostridium perfringens* iota toxin: synergism between two proteins. *Toxicon 24*: 767–773.

Sugai, M., Hashimoto, K., Kikuchi, A., Inoue, S., Okumura, H., Matsumota, K., Goto, Y., Ohgai, H., Moriishi, K., Syuto, B., Yoshikawa, K., Suginaka, H., and Takai, Y. (1992). Epidermal cell differentiation inhibitor ADP-ribosylates small GTP-binding proteins and induces hyperplasia of epidermis. *J. Biol. Chem. 267*: 2600–2604.

Sugii, S., and Kozaki, S. (1990). Hemagglutinating and binding properties of botulinum C2 toxin. *Biochem. Biophys. Acta 1034*: 176–179.

Suttorp, N., Polley, M., Seybold, J., Schnittler, H., Seeger, W., Grimminger, F., and Aktories, K. (1991). Adenosine diphosphate-ribosylation of G-actin by botulinum C2 toxin increases endothelial permeability in vitro. *J. Clin. Invest. 87*: 1575–1584.

Takasawa, T., Ohishi, I., and Shiokawa, H. (1987). Amino-acid composition of components I and II of botulinum C2 toxin. *FEMS Microbiol. Lett. 40*: 51–53.

Tamura, M., Nogimuri, K., Murai, S., Yajima, M., Ito, K., Katada, T., Ui, M., and Ishii, S. (1982). Subunit structure of islet-activating protein, pertussis toxin, in conformity with the A-B model. *Biochemistry 21*: 5516–5522.

Tominaga, T., Sugie, K., Hirata, M., Morii, N., Fukata, J., Uchida, A., Imura, H., and Narumiya, S. (1993). Inhibition of PMA-induced, LFA-1-dependent lymphocyte aggregation by ADP-ribosylation of the small molecular weight GTP binding protein, rho. *J. Cell Biol. 120*(6): 1529–1537.

Tsuzuki, K., Kimura, K., Fujii, N., Yokosawa, N., Indoh, T., Murakami, T., and Oguma, K. (1990). Cloning and complete nucleotide sequence of the gene for the main component of hemagglutinin produced by *Clostridium botulinum* type C. *Infect. Immun. 58*(10): 3173–3177.

Ueda, T., Kikuchi, A., Ohga, N., Yamamoto, J., and Takai, Y. (1990). Purification and characterization from bovine brain cytosol of a novel regulatory protein inhibiting the dissociation of GDP from and the subsequent binding of GTP to *rhoB* p20, a *ras* p21-like GTP-binding protein. *J. Biol. Chem. 265*: 9373–9380.

Ui, M. (1990). Pertussis toxin as a valuable probe for G-protein involvement in signal transduction. In: ADP-ribosylating toxins and G-proteins, J. Moss and M. Vaughan (Eds.), American Society for Microbiology, Washington, DC, pp. 45–77.

Van Ness, B. G., Howard, J. B., and Bodley, J. W. (1980). ADP-ribosylation of elongation factor 2 by diphtheria toxin. *J. Biol. Chem. 255*: 10717–10720.

Vandekerckhove, J. (1990). Actin-binding proteins. *Curr. Opin. Cell Biol. 2*: 41–50.

Vandekerckhove, J., and Weber, K. (1978a). The amino acid sequence of *Physarum* actin. *Nature 276*: 720–721.

Vandekerckhove, J., and Weber, K. (1978b). At least six different actins are expressed in a higher mammal: an analysis based on the amino acid sequence of the amino-terminal tryptic peptide. *J. Mol. Biol. 126*: 783–802.

Vandekerckhove, J., and Weber, K. (1979). The complete amino acid sequence of actins from bovine aorta, bovine heart, bovine fast skeletal muscle and rabbit slow skeletal muscle. *Differentiation 14*: 123–133.

Vandekerckhove, J., Schering, B., Bärmann, M., and Aktories, K. (1988). Botulinum C2 toxin ADP-ribosylates cytoplasmic β/γ-actin in arginine 177. *J. Biol. Chem. 263*: 696–700.

Vincent, S., Jeanteur, P., and Fort, P. (1992). Growth-regulated expression of *rhoG*, a new member of the *ras* homolog gene family. *Mol. Cell. Biol. 12*: 3138–3148.

Wegner, A., and Aktories, K. (1988). ADP-ribosylated actin caps the barbed ends of actin filaments. *J. Biol. Chem. 263*: 13739–13742.

Weigt, C., Just, I., Wegner, A., and Aktories, K. (1989). Nonmuscle actin ADP-ribosylated by botulinum C2 toxin caps actin filaments. *FEBS Lett. 246*: 181–184.

West, R. E., Moss, J., Vaughan, M., Liu, T., and Liu, T.-Y. (1985). Pertussis toxin-catalyzed ADP-ribosylation of transducin. *J. Biol. Chem. 260*: 14428–14430.

Wick, M. J., and B. H. Iglewski (1990). *Pseudomonas aeruginosa* exotoxin A. In: ADP-ribosylating toxins and G-proteins, J. Moss and M. Vaughan (Eds.), American Society for Microbiology, Washington, DC, pp. 31–43.

Wiegers, W., Just, I., Müller, H., Hellwig, A., Traub, P., and Aktories, K. (1991). Alteration of the cytoskeleton of mammalian cells cultured in vitro by *Clostridium botulinum* C2 toxin and C3 ADP-ribosyltransferase. *Eur. J. Cell Biol. 54*: 237–245.

Wieland, T., Ulibarri, I., Aktories, K., Gierschik, P., and Jakobs, K. H. (1990a). Interaction of small G proteins with photoexcited rhodopsin. *FEBS Lett. 263*: 195–198.

Wieland, T., Ulibarri, I., Gierschik, P., Hall, A., Aktories, K., and Jakobs, K. H. (1990b). Interaction of recombinant rho A GTP-binding proteins with photoexcited rhodopsin. *FEBS Lett. 274*: 111–114.

Wille, M., Just, I., Wegner, A., and Aktories, K. (1992). ADP-ribosylation of the gelsolin-actin complex by clostridial toxins. *J. Biol. Chem. 267*: 50–55.

Williamson, K. C., Smith, L. A., Moss, J., and Vaughan, M. (1990). Guanine nucleotide-dependent ADP-ribosylation of soluble rho catalyzed by *Clostridium botulinum* C3 ADP-ribosyl-transferase. *J. Biol. Chem. 265*: 20807–20812.

Yamamoto, K., Kondo, J., Hishida, T., Teranishi, Y., and Takai, Y. (1988). Purification and characterization of a GTP-binding protein with a molecular weight of 20,000 in bovine brain membranes. *J. Biol. Chem. 263*: 9926–9932.

Yamamoto, M., Marui, N., Sakai, T., Morii, N., Kozaki, S., Ikai, K., Imamura, S., and Narumiya, S. (1993). ADP-ribosylation of the *rhoA* gene product by botulinum C3 exoenzyme causes Swiss 3T3 cells to accumulate in the G phase of the cell cycle. *Oncogene 8*: 1449–1455.

Yeramian, P., Chardin, P., Madaule, P., and Tavitian, A. (1987). Nucleotide sequence of human rho cDNA clone 12. *Nucleic Acids Res. 15*: 1869.

Zalman, L. S., and Wisnieski, B. J. (1984). Mechanism of insertion of diphtheria toxin: peptide entry and pore size determinations. *Proc. Natl. Acad. Sci. U.S.A. 81*: 3341–3345.

22

Tetanus Toxin

Jane L. Halpern

Center for Biologics Evaluation and Research, Food and Drug Administration, Bethesda, Maryland

I. INTRODUCTION

Tetanus toxin is a protein neurotoxin produced by *Clostridium tetani*. This organism is a common environmental bacterium that has been identified in soil samples taken from all over the world (Hatheway, 1990). Unlike many pathogenic bacteria that produce a number of different virulence factors, all of the clinical effects seen in animals that have contracted tetanus can be attributed to this one toxin. Spores of *C. tetani* that are introduced into an animal through wounds will grow in a sufficiently anaerobic environment and produce toxin. After tetanus toxin is produced at a wound site, it is taken up by peripheral nerve endings and delivered to the central nervous system by the process of retrograde axonal transport and transsynaptic transport. In the central nervous system, tetanus toxin acts to inhibit the release of inhibitory neurotransmitters. The unopposed release of excitatory neurotransmitters results in the clinical symptoms seen in tetanus, such as muscle stiffness, trismus, opisthotonus, and reflex spasms. Symptoms of tetanus can become apparent as quickly as 48 h after exposure to *C. tetani* or may not appear for up to 3 weeks (see Bleck, 1989, for a review). The disease can remain localized to muscles near the entry site of the bacteria (localized tetanus) or affect the entire body (generalized tetanus). The effects of tetanus toxin on neuronal cells are extremely long-lasting, making recovery from tetanus a slow process. Recovery is thought to result at least in part from nerve sprouting and establishment of new synaptic contacts rather than the recovery of intoxicated cells.

The current vaccine against tetanus consists of partially purified tetanus toxin that has been chemically inactivated (tetanus toxoid). The effectiveness of vaccination against tetanus is apparent in the low incidence of disease seen in countries with immunization programs. In the United States, 117 cases of tetanus were reported during the period 1989–1990, for an annual incidence of 0.02/100,000 (CDC, 1992). Protection against the disease is conferred by the presence of serum antibodies to the toxin.

Although descriptions of tetanus are found in literature from very early times, a great deal of understanding of both the structure and mechanism of action of tetanus toxin has been gained only in recent years. The sequence of the structural gene has been determined, and recent studies have demonstrated that tetanus toxin is an enzyme possessing an endoprotease activity. This chapter focuses on current work regarding the structure and function of the toxin. It describes the structure of tetanus toxin and the gene encoding it and reviews what is known regarding the mechanism of the toxin's entry into and intoxication of cells.

The botulinum toxins, which are produced by *Clostridium botulinum*, are closely related to tetanus toxin in both structure and function. There are seven antigenically distinct botulinum toxins, A, B, C1, D, E, F, and G. Although these toxins are not the focus of this review, important similarities and differences between botulinum and tetanus toxins are mentioned.

II. STRUCTURE OF TETANUS TOXIN

Tetanus toxin is synthesized by *C. tetani* as a single polypeptide chain of 1315 amino acids (M_r 150,700). The toxin structure is shown schematically in Figure 1. It is released into the culture medium when the bacterial cells lyse and cleaved at one site by endogenous proteases (Helting et al., 1979) to yield two fragments, the heavy chain (H) and the light chain (L). The two chains are covalently linked through a single disulfide bond between

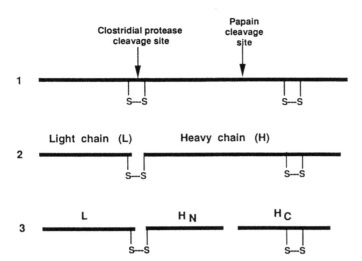

Fig. 1 A model of the structure of tetanus toxin. The single-chain holotoxin (1) is proteolytically cleaved to form the two-chain toxin (2). Proteolysis at one additional site in the heavy chain by papain or other enzymes leads to the formation of the L-H_N fragment and the H_C fragment (3).

Cys-438 on the light chain and Cys-466 on the heavy chain (Krieglstein et al., 1990). Except for the removal of the first methionine residue, this proteolysis appears to be the only posttranslational modification. Tetanus toxin contains 10 cysteine residues, which are present as six free sulfhydryl groups and two disulfide bonds. In addition to the disulfide bond between the heavy chain and the light chain, there is a second disulfide bond in the heavy chain formed by cysteines 1076 and 1092.

The heavy chain contains one additional site that is sensitive to cleavage with papain and certain other proteases (Helting and Zwisler, 1977). Cleavage at this site results in one fragment consisting of the light chain linked via a disulfide bond to the amino-terminal fragment of the heavy chain (L-H_N) and a second fragment consisting of the carboxy-terminal portion of the heavy chain (H_C).

The fragments of tetanus toxin that are produced by proteolysis have been referred to by a number of different names by various researchers. A scheme to simplify the nomenclature of the clostridial neurotoxins was proposed at the Fifth European Workshop on Bacterial Protein Toxins (Niemann, 1992) and is used in this review.

A. The Three-Domain Structure of Tetanus Toxin

Tetanus toxin appears to fit into the group of toxins referred to as A-B toxins, in which specific activities are a function of individual domains. The interaction of tetanus toxin with cells involves binding to a receptor, internalization of the toxin and transport of the light chain to its site of action, and the actual intoxication step. Each of these steps seems to be a specific function of one of the three structural domains defined by the protease-sensitive sites in tetanus toxin. The light chain has recently been demonstrated to be a zinc-binding endoprotease that inhibits neurotransmitter release by cleavage of synaptobrevin, a synaptic vesicle associated protein (Schiavo et al., 1992a; Link et al., 1992). H_C is known to retain the binding activity of the holotoxin and to undergo retrograde axonal transport, and H_N is thought to be important for translocation of the toxin across membranes. Although this model has been very useful in the design of experiments for characterization of tetanus toxin activity, it may be slightly inaccurate in that certain functions of the toxin appear to require interaction of multiple domains.

The three functional domains that have been identified in tetanus toxin provide a basis for explaining how tetanus toxin may act once it is present at the surface of target cells. A fourth activity of tetanus toxin is its ability to undergo retrograde and transsynaptic transport in order to reach its in vivo site of action in the spinal cord. Therefore, the structure of the toxin must also contain information that allows targeting to the appropriate trafficking pathway in addition to the three domains just described.

B. Proteolytic Activation of Tetanus Toxin

Tetanus toxin that has been purified from culture filtrates is primarily ($>90\%$) in the two-chain form, whereas single-chain toxin can be purified from intact bacteria. Cultures of *C. tetani* produce several different proteases that can cleave tetanus toxin (Helting et al., 1979), resulting in heterogeneous forms of the two-chain toxin. When the N-terminal amino acid of the different heavy chains was analyzed, it was determined that single-chain toxin is cleaved by clostridial proteases at two residues, Glu-449 and Ala-456 (Krieglstein et al., 1991). Single-chain toxin also was cleaved by a number of other proteases at residues within the loop formed by the disulfide bond between the heavy chain and light chain fragments. Two-chain toxins formed by all of these proteases were active, indicating that the exact site of cleavage is not critical. This is in contrast to diphtheria toxin, which also is cleaved within a disulfide loop to generate functional A and B chains. There are three arginines (residues 190, 192, and 193) in the disulfide loop; however, only cleavage at arginine 190 results in a toxin that is translocation-competent (Moskaug et al., 1989).

The conversion from the single-chain to the two-chain form of tetanus toxin is an activation step, as it has been demonstrated to generate a more potent toxin. In primary mouse spinal cord neurons, the potency of a preparation of two-chain toxin was compared to that of a preparation of toxin containing 10% two-chain and 90% single-chain forms (Bergey et al., 1989). Pure two-chain toxin was approximately tenfold more potent as measured by the lag time prior to the onset of paroxysmal depolarizing events. Two-chain toxin prepared by cleavage with a number of different enzymes (including clostridial proteases) was 5–11 times more potent than pure single-chain toxin in inhibiting the release of norepinephrine from rat brain homogenate (Krieglstein et al., 1991).

One important question concerns whether or not the single-chain toxin is active prior to cleavage or if the activity seen in these studies is due to proteolysis during the experiment. Experiments in permeabilized adrenal chromaffin cells suggest that some processing of the single-chain toxin is necessary for activity. In this system, tetanus light-chain or two-chain toxin was a potent inhibitor of calcium-stimulated secretion of catecholamines, while single-chain tetanus toxin had no effect (Ahnert-Hilger et al., 1989). The time course of these experiments is such that there is little opportunity for processing of the toxin by the cells.

C. Preparation of Isolated Heavy and Light Chains

The two chains of tetanus toxin cannot be readily separated after reduction of the interchain disulfide bond, indicating that noncovalent interactions must also exist between the two chains. In order to study specific functions associated with each domain, it was important to develop methods that would allow the purification of biologically active isolated chains. One group described the separation of the two chains by gel filtration chromatography after treatment with 4.0 M urea and 0.1 M dithiothreitol (DTT) (Matsuda and Yoneda, 1975). Removal of DTT and urea by dialysis allowed reassociation of the two chains and formation of active toxin. More recently, successful separation of the two chains has been achieved by the use of isoelectric focusing in the presence of 2 M urea and DTT (Weller et al., 1989). H and L chains prepared by this technique demonstrate good biological activity in a number of different assays.

D. Immunological Characterization of Tetanus Toxin

A number of different antibodies have been produced against tetanus toxin and have been useful in characterizing the structure and function of individual domains. Seventeen different monoclonal antibodies were produced against tetanus toxin using either tetanus toxoid or H_C as the immunogen, and binding of the antibodies was mapped to epitopes within each of the three functional domains (Kenimer et al., 1983). Preincubation of tetanus toxin with individual monoclonal antibodies against each domain blocked toxicity in mice, indicating that epitopes important for neutralization are distributed throughout the molecule. In competitive binding experiments using 57 different monoclonal antibodies prepared against tetanus toxoid, at least 20 different epitopes on tetanus toxin were identified, based on the ability of the antibodies to bind simultaneously to tetanus toxin in an ELISA (Volk et al., 1984).

Different overlapping fragments of tetanus toxin produced with in vitro expression systems also have been used to identify more precisely the epitopes recognized by a panel of monoclonal antibodies (Andersen-Beckh et al., 1989). In these studies, most of the antibodies were unable to recognize in vitro synthesized peptides, suggesting that the majority of the epitopes on tetanus toxin are conformational.

Several experiments from our laboratory would tend to support a model in which the majority of the antigenic determinants on tetanus toxin are conformational. We have attempted to identify the specific epitopes recognized by monoclonal antibodies by measuring the binding to synthetic peptides derived from tetanus toxin and proteolytic fragments of tetanus toxin. The monoclonal antibodies did not bind to any fragment smaller than L, H_N, or H_C, suggesting that these antibodies do not recognize a linear sequence. In a separate set of experiments, polyclonal antibodies were produced against a number of peptides 15–20 amino acids in length corresponding to different sequences of tetanus toxin (Halpern et al., 1989). With one exception (residues 1295–1314), none of the sera produced against these peptides were able to bind to native tetanus toxin.

III. MOLECULAR BIOLOGY OF TETANUS TOXIN

A. Structure of the Tetanus Toxin Gene

The sequence of the structural gene for tetanus toxin was reported in 1986 (Eisel et al., 1986; Fairweather and Lyness, 1986; Fairweather et al., 1986). As had been previously reported (Finn et al., 1984), the tetanus toxin gene is located extrachromosomally on a large plasmid. The nucleotide sequence of the tetanus toxin gene contained a single open reading frame coding for a mature protein of 1314 amino acids. Analysis of the upstream and downstream sequences from the coding sequence indicates that the toxin gene is not part of an operon. The coding region of the tetanus toxin gene was 72.1% A + T, which is typical for clostridial DNA, and the codon usage was comparable to that for other clostridial genes.

Tetanus toxin does not show significant homology to other known proteins except for other clostridial neurotoxins. The complete sequences of five botulinum toxins (A, B, C1, D, and E) have been reported (Binz et al., 1990a,b; Kimura et al., 1990; Hauser et al., 1990; Thompson et al., 1990; Whelan et al., 1992; Poulet et al., 1992) and can be compared to tetanus toxin. The overall homology between tetanus toxin and the different botulinum toxins is approximately 30%, which is seen as domains of highly conserved amino acids interrupted by areas with divergent sequences. The same degree of homology that is seen when comparing holotoxins is seen when the isolated heavy chains of the different toxins are examined. The light chain of tetanus toxin is considerably more homologous to some serotypes of botulinum toxin than to others. Tetanus toxin shares 50% homology with botulinum toxin type B and 40% homology with type E; homology with serotypes A, C1, and D is approximately 30%.

Because the mechanism by which the light chain blocks neurotransmitter release had not been established by previous experiments, one important finding from the sequence of the different neurotoxin genes was that each of the light-chain genes contained the sequence HELXH (residues 233–237 in tetanus toxin), which has been identified in a family of zinc-binding endoproteases. The finding that this sequence is conserved in the light chain of each clostridial neurotoxin prompted a number of workers to study the possibility that these toxins block neurotransmitter release by acting as a protease. The hypothesis has recently been confirmed by the demonstration that tetanus toxin cleaves synaptobrevin, a synaptic vesicle associated protein (Link et al., 1992; Schiavo et al., 1992a).

B. Expression of Tetanus Toxin

The production of tetanus toxin and immunogenic fragments of tetanus toxin in high-yield expression systems has been of interest because of the possibility of developing recombinant vaccines against tetanus toxin. Most of this work has focused on H_C because it is a nontoxic protein and animals immunized with H_C are protected against challenge with tetanus toxin. The initial expression of soluble H_C in E. coli reported yields of <5% of total cell protein (Halpern et al., 1990a; Makoff et al., 1989a). The relatively low yield of H_C was speculated to result from the presence of codons in the H_C sequence that are rare in E. coli. Replacement of certain codons in the gene for H_C with codons optimal for E. coli resulted in a three- to four-fold increase in the expression of H_C (Makoff et al., 1989b). High yields of H_C have been obtained in yeast (Saccharomyces cerevisiae and Pichia pastoris). In the methylotrophic yeast P. pastoris, multiple copies of the H_C expression cassette could be integrated into the chromosome to give yields of up to 27% of total cell protein (Clare et al., 1991). H_C has also been expressed in insect cells using a baculovirus expression vector (Charles et al., 1991). High-level expression of H_C was obtained in baculovirus using the native gene rather than a synthetic gene optimized for use in E. coli, demonstrating that codon usage does not appear to be a problem in this system. H_C has also been expressed in an attenuated (SL2361aroA) strain of Salmonella typhimurium (Fairweather et al., 1990). This strain was able to induce protective immunity in mice when administered either orally or intravenously as a live vaccine.

Identification and mutagenesis of critical residues in the L fragment has resulted in the production of mutants that lack toxic activity (Niemann, 1991). Since monoclonal antibodies against the light chain have been demonstrated to be protective, certain nontoxic light-chain mutants may have value as subunit vaccines. Further characterization of the conformation and immunogenicity of these mutants will be important for selecting suitable molecules.

IV. CELLULAR RECEPTORS FOR TETANUS TOXIN

The binding of tetanus toxin to receptors has been widely studied in a number of different experimental systems in order to describe this step (for reviews see Habermann and Dreyer, 1986; Wellhöner, 1992). Standard receptor-binding assays using radiolabeled tetanus toxin have been used to characterize the number and affinity of toxin-binding sites, and various studies employing light or electron microscopy have described the distribution of tetanus toxin bound to cell surfaces. Table 1 summarizes some of the binding studies that have attempted to characterize the tetanus toxin receptor in biological membranes. In spite of the extensive amount of work, no binding model has been developed that is consistent with all of the data. What seems to be most clear from these different studies is that under physiological conditions tetanus toxin does not appear to interact with a single class of high-affinity binding sites. Several difficulties have been reported in binding studies; these probably result from the presence of low-affinity or nonspecific binding sites for tetanus toxin on neurons. For example, it is difficult to demonstrate specific binding except at very low membrane concentrations. If membrane concentrations typical for ligand-binding assays are used, the level of nonspecific binding is greatly increased, suggesting the presence of a large number of nonspecific binding sites. A second observation that has

TABLE 1 Summary of Receptor Binding Studies

Preparation	Comments	Reference
Rat and bovine brain membranes	Single, high-affinity binding site in 50 mM Tris-acetate, pH 6.0 buffer	Rogers and Snyder, 1981
Rat brain membranes	Single high-affinity binding site in 25 mM Tris-acetate, pH 6.0 buffer	Goldberg et al., 1981
NGF-differentiated PC12 cells	Binding sites similar to those reported by Rogers and Snyder (1981)	Walton et al., 1988
Rat brain membranes	Single high-affinity binding site in 25mM Tris-acetate, pH 6.0, non-saturable binding in a physiological buffer	Pierce et al., 1986
Rat brain membranes	Saturable binding of both H_C and holo-toxin in a low-pH, low ionic strength buffer, saturable binding of holotoxin but not H_C in a physiological buffer.	Weller et al., 1986
Dibutyryl cyclic AMP-differentiated NG108-15 cells	Two binding sites in a physiological buffer	Wellhöner and Neville, 1987
N18-RE-105 cells	Receptor similar to that reported by Rogers and Snyder (1981)	Staub et al., 1986
Rat brain and thyroid membranes	Optimal binding at pH 6.0	Lee et al., 1979

been made is that specific, saturable binding of tetanus toxin has been difficult to measure under physiologic conditions (Pierce et al., 1986; Critchley et al., 1986). Tetanus toxin binding to membranes is enhanced in a low-pH, low ionic strength buffer (25 mM Tris, pH 6.0) relative to binding under more physiologic conditions, and many studies that report specific binding of tetanus toxin have used these conditions. Although some studies have demonstrated specific saturable binding under both conditions, it has also been reported that binding is saturable only when a low-pH, low ionic strength buffer is used. The presence of a large number of nonspecific binding sites, as suggested by these data, would make the characterization of a small number of high-affinity receptors for tetanus toxin difficult in binding assays.

A. Gangliosides as Receptors

The only components of neuronal membranes that have been demonstrated unambiguously to bind tetanus toxin are gangliosides. This interaction was first reported by van Heyningen, who demonstrated that gangliosides were able to neutralize tetanus toxin (van Heyningen and Miller, 1961; van Heyningen, 1974). The affinity of tetanus toxin was highest for gangliosides of the G_{1b} series, with significantly (200-fold) lower binding to G_{M1} amd G_{D1a}. The ability of tetanus toxin to bind to gangliosides and the relative affinity of tetanus toxin for different gangliosides has been confirmed by other investigators (Holmgren et al., 1980; Morris et al., 1980). The enhanced binding of tetanus toxin to cell membranes at pH 6.0, as noted above, can probably be attributed to gangliosides, as this effect is also seen when measuring binding to purified gangliosides.

The importance of gangliosides as the functional receptor for tetanus toxin is still unclear. One criticism of this model is that tetanus toxin does not have a pronounced specificity for one particular ganglioside. This is in contrast to cholera toxin, which exhibits high specificity for G_{M1} compared with other gangliosides. The affinity of cholera toxin for G_{M1} is at least 1000 times greater than for any other ganglioside.

Different studies have been carried out both to examine the importance of gangliosides and to determine if tetanus toxin may bind to other membrane components. Not all of these studies can be reviewed in this chapter, but a discussion of several provides good examples of the types of experiments and results that have been reported.

The expression of gangliosides and the level of tetanus toxin binding have been measured in PC12 cells grown under various conditions. Treatment with nerve growth factor (NGF) or growth of the cells at a high density resulted in an increase in the content of gangliosides (primarily G_{T1b}) that correlated with increased levels of tetanus toxin binding (Walton et al., 1988). However, cells differentiated with NGF were sensitive to tetanus toxin, whereas cells differentiated by growth at high density were not (Sandberg et al., 1989). This would suggest that the expression of gangliosides is not sufficient to mediate toxin binding and uptake. Alternatively, NGF differentiation may result in sensitivity to tetanus toxin as a result of expression of a protein important for another step in toxin action.

The importance of gangliosides for tetanus toxin action has also been demonstrated using primary cultures of bovine adrenal chromaffin cells, which had no detectable endogenous G_{T1b} or G_{D1b} (Marxen et al., 1989). Tetanus toxin had no effect on secretion from these cells. If the cells were preincubated with exogenous gangliosides, subsequent exposure to tetanus toxin resulted in a 40% inhibition of norepinephrine secretion. While these results do not address the presence of a second receptor, they suggest that gangliosides are required for tetanus toxin action.

Pierce et al. (1986) measured the binding of ^{125}I-labeled tetanus toxin to rat brain membranes in both a low-pH, low ionic strength buffer and a physiologic buffer (Krebs-Ringer buffer, pH 7.4). A small number of high-affinity sites for tetanus toxin were apparent when binding was measured at physiologic pH and ionic strength, but not under conditions that optimized binding to gangliosides. The treatment of rat brain membranes with trypsin reduced the binding of tetanus toxin by >80% at physiologic conditions, whereas the binding measured at pH 6.0 was insensitive to trypsin. These data were interpreted as evidence of a protein receptor to which tetanus toxin binds optimally under physiologic conditions. The trypsin-insensitive binding seen at pH 6.0, which also has been seen in other studies, was attributed to interaction with gangliosides.

Much of the binding data reported for tetanus toxin is difficult to interpret because it has not been correlated with a biological effect of tetanus toxin. The importance of binding at pH 6.0 versus pH 7.4 might be established by careful study of both tetanus toxin binding and inhibition of exocytosis at various doses in parallel in one model system.

B. Identification of Additional Receptors

Several approaches have been taken to determine if a protein receptor may play a role in the binding and uptake of tetanus toxin. Schiavo et al. (1991) demonstrated that ^{125}I-labeled tetanus toxin bound to the surface of NGF-differentiated PC12 cells could be covalently bound to a 14,000-Da protein in the presence of two different chemical cross-linkers. The cross-linking was blocked by excess unlabeled tetanus toxin, suggesting that tetanus toxin bound specifically and saturably to this protein.

Synaptosomal proteins separated by SDS-polyacrylamide gel electrophoresis and transferred to nitrocellulose paper have been demonstrated to bind ^{125}I-labeled tetanus toxin (Schengrund et al., 1992). Tetanus toxin bound to two proteins of 80,000 and 116,000 Da. This binding was significantly enhanced when tetanus toxin was preincubated with G_{T1b} ganglioside prior to incubation with the nitrocellulose blot, suggesting that binding to G_{T1b} enhances the affinity of tetanus toxin for these proteins.

A model that takes into account the binding of tetanus toxin to both gangliosides and a protein receptor has been proposed by Montecucco (1986). In this model, tetanus toxin initially binds to gangliosides present in neuronal cell membranes. This interaction does not lead to internalization but allows the toxin-ganglioside complex to bind to a protein receptor that is expressed on neuronal cells. Both the H_N domain and the H_C domain are proposed to be necessary for binding to this protein receptor. Formation of this complex between ganglioside, toxin molecule, and protein receptor would allow the internalization of tetanus toxin and intoxication of the target cell.

C. Structure of the Tetanus Toxin Binding Domain

Although the binding activity of tetanus toxin has been demonstrated to be associated with the H_C domain, other regions of tetanus toxin may be important for full expression of binding activity. In one study, the ability of H_C and tetanus toxin to compete with ^{125}I-labeled tetanus toxin for binding sites on rat brain membranes was measured in both a low-pH, low ionic strength buffer and an isotonic buffer at pH 7.4 (Weller et al., 1989). H_C was as potent as tetanus toxin in displacing ^{125}I-labeled tetanus toxin binding in the low-pH, low ionic strength buffer but was at least 1000-fold weaker at pH 7.4. These data are consistent with a model in which domains other than H_C are important for binding, possibly through multiple receptors as described above. The binding to ganglioside plus protein receptor is optimal under physiologic conditions, whereas binding to gangliosides alone is best seen under conditions of acidic pH.

Limited work has been done to define specific regions in the H_C domain that are involved in binding. We constructed a series of deletion mutants of H_C that were tested for binding to gangliosides and neuronal cells (Halpern and Loftus, 1993). Removal of 10 or more amino acids from the carboxy-terminal end of H_C resulted in a total loss of binding activity, whereas removal of up to 263 amino acids from the amino-terminal end did not affect binding. A peptide corresponding to the amino-terminal 20 amino acids did not bind to gangliosides or compete with H_C for binding. These data suggest that the binding of tetanus toxin to receptors requires a conformational domain that includes the carboxy-terminal 10 amino acids as well as another region from H_C. The involvement of the extreme COOH terminus in binding to target cells has been reported for other bacterial toxins such as diphtheria toxin (Rolf et al., 1990), anthrax toxin (Singh et al., 1991), and *Clostridium perfringens* exotoxin (Hanna et al., 1991).

V. ENTRY INTO TARGET CELLS

A. Internalization

Because the toxic effect of tetanus toxin requires delivery of the light chain to the cytosol, a requisite step in toxin action is internalization and translocation across a membrane. The internalization mechanism of tetanus toxin has been proposed to be uptake into acidic compartments via receptor-mediated endocytosis (see Simpson, 1986). In response to the decreased pH inside endocytic vesicles, the heavy chain undergoes a conformational change and inserts into membranes, resulting in the translocation of light chain into the cytosol. The H_N domain of tetanus toxin is thought to be responsible for translocation activity.

Other bacterial toxins that act in the cytosol, such as diphtheria toxin, have been demonstrated to be internalized via this pathway. A number of different studies have demonstrated that diphtheria toxin B chain undergoes conformational changes in response to low pH (reviewed by Sandvig and Olsnes, 1991). The ability of drugs that inhibit vesicle acidification to block the toxic effect of diphtheria toxin is also consistent with translocation occurring through acidic vesicles.

Similar pharmacologic and biochemical experiments have been done with tetanus toxin to characterize the mechanism of internalization. Simpson demonstrated that treatment of the neuromuscular junction with the lysosomotropic agents ammonium chloride and methylamine caused a significant delay in the time to intoxication by tetanus toxin (Simpson, 1983). More recently, monensin was shown to block the effects of tetanus toxin on secretion from primary murine spinal cord cultures (Williamson et al., 1992a). These data have been interpreted as evidence that tetanus toxin enters the cytoplasm through an acidic compartment.

Electron microscopy has been used to describe the internalization of tetanus toxin adsorbed to colloidal gold in primary cultures of mouse spinal cord neurons (Parton et al., 1987). Tetanus toxin binding to the cell surface appeared to preferentially localize to coated pits, although binding to noncoated regions was also seen. The toxin appeared to be rapidly internalized through coated pits at 37°C.

B. Translocation Across Membranes

Tetanus toxin must be transferred across the plasma membrane or an endocytic vesicle membrane in order to reach the cytosol. Studies examining the interaction of tetanus toxin with artificial membranes have provided evidence that tetanus toxin undergoes conformational changes and interacts with lipids in a pH-dependent manner. Liposomes containing photoreactive phospholipids were used to study the effect of pH on the interaction of tetanus toxin on membranes (Montecucco et al., 1986). Tetanus toxin incubated with these liposomes at pH 4.33 was covalently labeled with phospholipids, whereas no labeling occurred after an incubation at pH 7.4. The phospholipid probes specifically labeled the light chain and the H_N domain.

The susceptibility of tetanus toxin to proteolysis has been used to measure its interaction with lipid bilayers (Roa and Boquet, 1985). In these studies, tetanus toxin was incubated with lipid vesicles at either pH 7.2 or 3.0 and treated with various proteases. At pH 3.0, several fragments of tetanus toxin were protected against proteolysis by pepsin; this was not seen at pH 7.2. The protected fragments were from the H_N domain, suggesting that this region of the toxin interacts with lipid membranes at acidic pH.

In addition to artificial membranes, tetanus toxin has also been demonstrated to increase the permeability of synaptosomal membranes (Högy et al., 1992). The incubation of synaptosomes with heavy chain, but not holotoxin, at a physiological pH was shown to stimulate the release of lactate dehydrogenase and potassium ions. Intact heavy chain was required for this effect. The authors suggest that the holotoxin was inactive in this assay because the domains important for this activity are buried in the holotoxin.

The sequence of tetanus toxin has been analyzed for potential membrane-spanning domains with the Hopp–Wood algorithm (Eisel et al., 1986), and a potential membrane-spanning domain was identified in a region in H_N (Asn-660 through Ala-691). Identification and modification of critical amino acids in H_N that are important for membrane spanning and translocation will be important for additional characterization of this step.

C. Retrograde Axonal Transport

The binding of tetanus toxin to receptors present on peripheral nerves leads to internalization into axonal vesicles and movement of these vesicles into the central nervous system by retrograde axonal transport (Price et al., 1975). Retrograde axonal transport may be thought of as a specialized form of normal endocytosis. Both pinocytic vesicles containing nonspecific labels and vesicles derived from receptor-mediated endocytosis undergo retrograde axonal transport (Vallee and Bloom, 1991).

The initial steps in this process of binding, uptake, and retrograde axonal transport are not unique to tetanus toxin. Many proteins that bind to peripheral nerve endings have been demonstrated to be transported in a retrograde manner, including cholera toxin, certain lectins, horseradish peroxidase, and nerve growth factor. However, the retrograde transport of most of these other proteins does not result in transsynaptic transport. Instead, these proteins are incorporated into lysosomes and presumably are degraded (Schwab et al., 1979). In contrast to these other proteins, tetanus toxin is not degraded in the peripheral nerve cell body. A significant fraction of the transported toxin is released from the cell and taken up by presynaptic terminals. In this way, tetanus toxin can be transported through a chain of neurons to reach the site of action in the spinal cord. The transfer of tetanus toxin or H_C through a chain of neurons has been demonstrated in several experimental systems (Schwab and Thoenen, 1976; Cabot et al., 1991).

The mechanism of this trafficking pathway of tetanus toxin is unknown. Retrograde transport of tetanus toxin and several other proteins has been studied using electron microscopy. While intraaxonal vesicles containing cholera toxin, ricin, and lectins appeared to be targeted exclusively to lysosomes, vesicles containing tetanus toxin were able to fuse with the cell membranes and release the contents into the synaptic cleft (Schwab et al., 1979). This would suggest that tetanus toxin is internalized or segregated after internalization into a unique population of vesicles. These vesicles, unlike vesicles containing the other ligands, would contain a signal for targeting to the postsynaptic membrane. A number of different proteins that are important for the vectorial transport of vesicles in nonneuronal cells have been identified recently (Bourne, 1988). Biochemical characterization of different populations of axonal transport vesicles may provide important information on targeting mechanisms associated with retrograde axonal transport.

The domains of tetanus toxin that are important for retrograde axonal transport have not been studied extensively. H_C can clearly undergo retrograde axonal transport and transsynaptic transport and has been widely used by neurobiologists to study synaptic connections. However, other domains of tetanus toxin may also be important for retrograde transport (Weller et al., 1986). In a study in which the dose-response curves for retrograde axonal transport of H_C and tetanus toxin were compared, H_C was transported with approximately 50–100-fold less efficiency than holotoxin, suggesting that additional domains in the heavy chain are important for this activity. Retrograde transport of the intact heavy chain has not been studied. This large difference in retrograde axonal transport between H_C and holotoxin was not apparent in a second study (Morris et al., 1980). Although the two proteins were not compared quantitatively, the difference in transport appeared to be tenfold at most.

The unusual ability of tetanus toxin and H_C to enter the central nervous system by retrograde axonal transport has raised the question of whether H_C can be used as a nontoxic carrier to transport other molecules. Although the practical applications of this approach have not been realized, several reports have demonstrated that this is feasible. Habig et al. (1983) demonstrated that a noncovalent complex of H_C and a nonneutralizing antibody against H_C could be retrogradely transported in the rat sciatic nerve.

H_C has been used to reverse the phenotype of cells with a neuronal lysosomal enzyme defect (Dobrenis et al., 1992). Feline G_{M2} gangliosidosis is a model for Tay-Sachs and Sandhoff diseases, which result from defects in lysosomal storage. A defect in the enzyme β-N-acetylhexosaminidase leads to the buildup of undegraded G_{M2} gangliosides in lysosomes, resulting in CNS degeneration. H_C was chemically conjugated to β-N-acetylhexosaminidase, and the bivalent molecule was used to treat neuronal cell cultures with a G_{M2} gangliosidosis phenotype. Cultures treated with this conjugate for 3 days had no levels of G_{M2} detectable by immunofluorescence, whereas untreated cells were positive.

One problem with using H_C as a delivery system may be the presence of antibodies against H_C after vaccination against tetanus. Delivery of chimeric molecules into the nervous system rather than systemically may allow this approach to work.

VI. STRUCTURE AND FUNCTION OF THE LIGHT CHAIN

Several techniques that circumvent the need for binding and internalization have been used to demonstrate that the light chain of tetanus toxin contains the domain responsible for inhibition of neurotransmitter release. The plasma membrane of bovine chromaffin cells can be permeabilized by using a detergent (Bittner and Holz, 1988) or streptolysin O (Ahnert-Hilger et al., 1989) to introduce tetanus toxin directly into the cytoplasm. Permeabilized cells remain exocytosis-competent, and catecholamine release can be stimulated by increasing the concentration of calcium in the media. Light chain that was artificially introduced into the cytoplasm of bovine chromaffin cells permeabilized with digitonin or streptolysin O was able to inhibit secretion as effectively as holotoxin (Ahnert-Hilger et al., 1989; Bittner et al., 1989). The inhibition by light chain under these conditions was rapid, suggesting that much of the lag in toxin action may be in steps prior to the actual intoxication.

The marine mollusk *Aplysia californica* has also been used to study the effects of the light chain on exocytosis. The synaptic connections between different ganglia in *Aplysia* have been well characterized. It is possible to measure neurotransmitter release from a cell by stimulating it with a microelectrode and measuring the evoked activity in a known postsynaptic cell. The injection of either holotoxin or tetanus toxin light chain into different cells was shown to block neurotransmitter release (Mochida et al., 1989). Injection of light-chain mRNA also effectively blocked secretion (Mochida et al., 1990), indicating that the active light chain could be translated in the *Aplysia* neuron. This system has been used to identify regions of the light chain important for activity. A series of mutants of the light-chain gene with deletions at either the 5' or 3' end were generated, and the corresponding mRNAs were microinjected into a presynaptic cell and tested for activity (Kurazono et al., 1992). Up to eight amino acids could be removed from the amino terminus without losing activity; however, removing two more amino acids resulted in a nontoxic molecule. At the carboxy terminus, removing 65 amino acids had no effect, but the deletion of 68 amino acids caused loss of activity.

A. californica has been a very useful system for studying the clostridial neurotoxins; however, there seem to be important differences between it and mammalian systems in the functions of different domains of tetanus toxin. For example, the L-H_N fragment is relatively nontoxic in mammalian systems except at extremely high doses, presumably because of the lack of a receptor-binding domain. In *Aplysia*, L-H_N derived from either tetanus or botulinum toxin is as potent as holotoxin, indicating that H_C is not required for binding to receptors (Poulain et al., 1991). A different internalization pathway or expression of different receptors in *Aplysia* may account for these results.

A major advance in understanding the molecular basis of tetanus toxin action was the characterization of the enzymatic activity of tetanus toxin and the identification of a cellular substrate. The primary sequence of tetanus toxin and the botulinum toxins indicated that the light chain of each toxin contained a sequence motif (HELXH) common to zinc-binding metalloendoproteases. The crystal structures of several enzymes containing this sequence have been resolved, and the histidine residues are known to be important for binding zinc (reviewed in Vallee and Auld, 1990). In addition to the two histidines, a glutamic acid and a water molecule are involved in the coordination of zinc. The light chain of tetanus toxin was demonstrated to bind zinc, and this activity could be localized to a peptide (amino acids 225–243) containing the zinc-binding consensus motif (Wright et al., 1992). A synthetic peptide from botulinum toxin B that contained the zinc-binding motif also was able to bind zinc (Schiavo et al., 1992b). The importance of the histidines in zinc binding and light-chain activity has been confirmed further with both biochemical techniques and site-directed mutagenesis. The number of available histidines in both zinc-containing and zinc-depleted preparations of purified light chain was quantitated by titration with diethyl pyrocarbonate (Schiavo et al., 1992b). Light chain depleted of zinc had four reactive histidines compared with two in zinc-containing light chain.

Light-chain mRNAs with substitutions at each of the histidine residues were tested for activity in *Aplysia* (Niemann, 1991). Changing His-233 to leucine or valine or His-237 to glycine or valine had no effect; however, changing His-237 to proline resulted in a nontoxic molecule. Finally, specific inhibitors of zinc-binding endoproteases (Schiavo et al., 1992a; Sanders and Habermann, 1992) have been shown to block the effect of tetanus toxin.

The substrate of both tetanus toxin and botulinum toxin type B light chain has been identified as synaptobrevin 2, a 19,000-Da protein associated with small synaptic vesicles. Synaptobrevin is also referred to as vesicle-associated membrane protein (VAMP). This protein was isolated from both rat brain (Baumert et al., 1989) and *Torpedo californica* (Trimble et al., 1988) small synaptic vesicles and shown to be an integral membrane protein. It is present in small synaptic vesicles in cells of the nervous system and in certain types of neuroendocrine cells such as anterior pituitary cells and adrenal chromaffin cells. The DNA sequences for the coding region of synaptobrevin from a number of different species have been reported. Two forms of synaptobrevin, 1 and 2, have been identified in mammals; these isoforms are 77% homologous in humans (Archer et al., 1990). Each form is well conserved among species; bovine and human synaptobrevin 2 differ by only one amino acid. The sequence of synaptobrevin predicts a protein with a small intravesicular domain, one transmembrane segment, and a large cytosolic domain. The location of this protein suggests that it may play a role in the fusion of synaptic vesicles with the plasma membrane.

Treatment of highly purified synaptic vesicles from rat cerebral cortex with tetanus toxin or botulinum toxin type B resulted in the specific cleavage of synaptobrevin at one site to yield two fragments of 12,000 and 7000 Da (Schiavo et al., 1992c). The site of cleavage was determined to be between Gln-76 and Phe-77. Interestingly, this sequence is found only in synaptobrevin 2; synaptobrevin 1 has a valine at position 76 and is not cleaved by tetanus toxin. Treatment of partially purified synaptic vesicles with tetanus toxin resulted in complete degradation of synaptobrevin (Link et al., 1992) rather than the selective cleavage seen with highly purified vesicles. One explanation is that synaptobrevin may become a substrate for additional proteases after the initial cleavage by tetanus toxin.

Botulinum toxin type A did not cleave synaptobrevin or any other known small synaptic vesicle associated protein. This finding is in agreement with earlier electrophysiologic studies that suggested a different site of action for tetanus toxin and botulinum toxin type B than for botulinum toxin type A (Gansel et al., 1987; Dreyer et al., 1987).

VII. THE SPECIFICITY OF TETANUS TOXIN FOR INHIBITORY NEURONS

The preferential inhibition of inhibitory neurotransmitters by tetanus toxin results in the clinical symptoms that are seen in cases of clinical tetanus. A block in the release of inhibitory neurotransmitters results in excessive and unbalanced activity of excitatory neurons, causing the spastic paralysis typical of tetanus. While tetanus toxin inhibits secretion from many types of neuronal cells, the greater potency for inhibitory neurons is also apparent *in vitro*. When the effect of tetanus toxin on the release of multiple neurotransmitters was measured in rat brain homogenate, the release of inhibitory neurotransmitters was inhibited by lower concentrations of tetanus toxin than the release of excitatory neurotransmitters (Habermann, 1988). Recently, the effect of tetanus toxin on the release of glycine (an inhibitory neurotransmitter) and glutamate (an excitatory neurotransmitter) was measured in primary cultures of mouse spinal cord neurons (Williamson et al., 1992b). At a concentration of 0.6 nM tetanus toxin, glycine release was completely inhibited, whereas glutamate release was still apparent even after incubation with a tenfold higher concentration of tetanus toxin.

The electrical activity of tetanus toxin-treated spinal cord neurons also shows a greater sensitivity of inhibitory neurons relative to excitatory neurons (Bergey et al., 1987). The frequency of inhibitory postsynaptic potentials (IPSPs) was found to decrease and disappear much faster than that of excitatory postsynaptic potentials (EPSPs) measured in the same culture.

Data that have been obtained in *Aplysia* indicate that the greater potency of tetanus toxin for specific neurons is a function of cell surface receptors that recognize H_N. In *Aplysia*, tetanus toxin exhibits a greater potency at noncholinergic synapses than at cholinergic synapses. To determine which toxin domain was responsible for the relative potencies at different synapses, heterologous toxin molecules prepared with different chains from botulinum and tetanus toxin were tested for activity at noncholinergic and cholinergic neurons (Poulain et al., 1991). A hybrid toxin consisting of the heavy chain of botulinum toxin A and the light chain of tetanus toxin was equivalent in potency to botulinum toxin A at cholingeric synapses. These results suggest that the light chain of tetanus toxin is equivalent in potency to the light chain of botulinum toxin if delivered intracellularly. While these data support the hypothesis that the specificity of tetanus toxin for inhibitory neurons is due to binding domains, it is important to remember that the different neurotoxin domains may have somewhat different functions in *Aplysia* than in mammals. Because tetanus toxin and botulinum toxin A act on different substrates to inhibit exocytosis, the different light-chain activities may contribute to the differences in potency of tetanus toxin light chain in inhibitory versus excitatory synapses. Careful assessments of specific light chains in different types of neurons will help answer this question.

VIII. TETANUS TOXIN ACTION ON NONNEURONAL CELLS

Tetanus toxin also is known to inhibit calcium-stimulated secretion of neurotransmitters that are contained in large dense core vesicles (Janicki and Habermann, 1983; Halpern et al., 1990b). This type of secretory vesicle appears to be involved in the secretion of neuropeptides and hormones and is found in both neurons and other types of secretory cells (De Camilli and Jahn, 1990). When small synaptic vesicles and large dense core vesicles are present in the same type of cell (neuronal cells, for example), exocytosis may be stimulated independently for each type of vesicle. Large dense core vesicles are morphologically and biochemically different from small synaptic vesicles and do not appear to contain synaptobrevin and other proteins that have been identified as components of small synaptic vesicles (Navone et al., 1989). Because large dense core vesicles do not contain synaptobrevin, the mechanism by which tetanus toxin inhibits secretion may be different from the reported mechanism for small synaptic vesicles. Further characterization of large dense core vesicle associated proteins may help define the basis for tetanus toxin action in this type of vesicle.

Tetanus toxin has also been demonstrated to inhibit calcium-dependent exocytosis from nonneuronal secretory cells. Several different types of cells have been reported to be responsive to tetanus toxin. The release of lysozyme from human peripheral blood monocytes was inhibited by tetanus toxin in a time and dose-dependent manner (Ho and Klempner, 1985). Tetanus toxin also inhibited the release of lysozyme from γ-interferon-stimulated GG2EE cells, a mouse macrophage cell line (Pitzurra et al., 1989). However, the calcium-dependent release of amylase from permeabilized pancreatic acinar cells was not affected by tetanus toxin (Stecher et al., 1992). In our laboratory, the effect of tetanus toxin or tetanus toxin light chain on serotonin release from a rat basophil cell (RBL-2H3) line was measured. Holotoxin or light chain artificially introduced into cells by hypotonic lysis had no effect on secretion, while the release of norepinephrine from comparably treated PC12 cells was blocked (unpublished data).

Since the only identified activity of tetanus toxin is cleavage of a synaptic vesicle-specific protein, it is not clear how exocytosis is inhibited in cells that lack this protein. It is possible that tetanus toxin can block secretion by a mechanism unrelated to the cleavage of synaptobrevin. Alternatively, a synaptobrevin-like protein present on secretory vesicles may serve as substrate for tetanus toxin. Further elucidation of the intracellular activity of tetanus toxin will help clarify the specificity of tetanus toxin for different cell types.

IX. CONCLUSIONS

Research by a number of groups that has been discussed in this article has resulted in significant advances in understanding the mechanism of action of tetanus toxin. These studies have helped to confirm and extend the basic model of how tetanus toxin interacts with neuronal cells to inhibit exocytosis. A number of questions concerning the steps in this process remain to be answered by future work. Important details of tetanus toxin action that should be studied include (1) the identification of additional receptors for tetanus toxin, (2) how the toxin is correctly routed to its in vivo site of action, (3) characterization of the protease activity of the light chain, and (4) the mechanism by which exocytosis is inhibited in nonneuronal cells. Detailed information about these processes will hopefully provide important information on both the function of tetanus toxin and the cell biology of neurons.

REFERENCES

Ahnert-Hilger, G., Weller, U., Dauzenroth, M.-E., Habermann, E., and Gratzl, M. (1989). The tetanus toxin light chain inhibits exocytosis. *FEBS Lett. 242*: 245–248.

Andersen-Beckh, B., Binz, T., Kurazono, H., Mayer, T., Eisel, U., and Niemann, H. (1989). Expression of tetanus toxin subfragments in vitro and characterization of epitopes. *Infect. Immun. 57*: 3498–3505.

Archer, B. T., III, Özçelik, T., Jahn, R., Francke, U., and Südhof, T. C. (1990). Structures and chromosomal localizations of two human genes encoding synaptobrevins 1 and 2. *J. Biol. Chem. 265*: 17267–17273.

Baumert, M., Maycox, P. R., Navone, F., De Camilli, P., and Jahn, R. (1989). Synaptobrevin: an integral membrane protein of 18,000 daltons present in small synaptic vesicles of rat brain. *EMBO J. 8*: 379–384.

Bergey, G. K., Bigalke, H., and Nelson, P. G. (1987). Differential effects of tetanus toxin on inhibitory and excitatory synaptic transmission in mammalian spinal cord neurons in culture: a presynaptic locus of action for tetanus toxin. *J. Neurophysiol. 57*: 121–131.

Bergey, G. K., Habig, W. H., Bennett, J. I., and Lin, C. S. (1989). Proteolytic cleavage of tetanus toxin increases activity. *J. Neurochem. 53*: 155–161.

Binz, T., Kurazono, H., Wille, M., Frevert, J., Wernars, K., and Niemann, H. (1990a). The complete sequence of botulinum neurotoxin type A and comparison with other clostridial neurotoxins. *J. Biol. Chem. 265*: 9153–9158.

Binz, T., Kurazono, H., Popoff, M. R., Eklund, M. W., Sakaguchi, G., Kozaki, S., Krieglstein, K., Henschen, A., Gill, D. M., and Niemann, H. (1990b). Nucleotide sequence of the gene encoding *Clostridium botulinum* neurotoxin type D. *Nucleic Acids Res. 18*: 1291.

Bittner, M. A., and Holz, R. W. (1988). Effect of tetanus toxin on catecholamine release from intact and digitonin-permeabilized chromaffin cells. *J. Neurochem. 51*: 451–456.

Bittner, M. A., Habig, W. H., and Holz, R. W. (1989). Isolated light chain of tetanus toxin inhibits exocytosis: studies in digitonin-permeabilized cells. *J. Neurochem. 53*: 966–968.

Bleck, T. P. (1989). Clinical aspects of tetanus. In *Botulinum Neurotoxin and Tetanus Toxin*, L. L. Simpson (Ed.), Academic Press, San Diego, pp. 379–398.

Bourne, H. R. (1988). Do GTPases direct membrane traffic in secretion? *Cell 53*: 669–671.

Cabot, J. B., Mennone, A., Bogan, N., Caroll, J., Evinger, C., and Erichsen, J. T. (1991). Retrograde, transsynaptic and transneuronal transport of fragment C of tetanus toxin by sympathetic preganglionic neurons. *Neuroscience 40*: 805–823.

CDC. (1992). Centers for Disease Control *Surveillance Summaries*, December 11, 1992. *MMWR 41*: SS-8.

Charles, I. G., Rodgers, B. C., Makoff, A. J., Chatfield, S. N., Slater, D. E., and Fairweather, N. F. (1991). Synthesis of tetanus toxin fragment C in insect cells by use of a baculovirus expression system. *Infect. Immun. 59*: 1627–1632.

Clare, J. J., Rayment, F. B., Ballantine, S. P., Sreekrishna, and Romanos, M. A. (1991). High-level expression of tetanus toxin fragment C in *Pichia pastoris* strains containing multiple tandem integrations of the gene. *Biotechnology 9*: 455–460.

Critchley, D. R., Habig, W. H., and Fishman, P. H. (1986). Reevaluation of the role of gangliosides as receptors for tetanus toxin. *J. Neurochem. 47*: 213–222.

De Camilli, P., and Jahn, R. (1990). Pathways to regulated exocytosis in neurons. *Annu. Rev. Physiol. 52*: 625–645.

Dreyer, F., Rosenberg, F., Becker, C., Bigalke, H., and Penner, R. (1987). Differential effects of various secretagogues on quantal transmitter release from mouse motor nerve terminals treated with botulinum A and tetanus toxin. *Naunyn-Schmeideberg's Arch. Pharmacol. 335*: 1–7.

Dobrenis, K., Joseph, A., and Rattazzi, M. C. (1992). Neuronal lysosomal enzyme replacement using fragment C of tetanus toxin. *Proc. Natl. Acad. Sci. U.S.A. 89*: 2297–2301.

Eisel, U., Jarausch, W., Goretzki, K., Henschen, A., Engels, J., Weller, U., Hudel, M., Habermann, E., and Niemann, H. (1986). Tetanus toxin sequence: primary structure, expression in *E. coli*, and homology with botulinum toxin. *EMBO J. 5*: 2495–2502.

Fairweather, N. F., and Lyness, V. A. (1986). The complete amino acid sequence of tetanus toxin. *Nucleic Acids Res. 14*: 7809–7812.

Fairweather, N. F., Lyness, V. A., Pickard, D. J., Allen, G., and Thomson, R. O. (1986). Cloning, nucleotide sequencing and expression of tetanus toxin fragment C in *Escherichia coli*. *J. Bacteriol. 165*: 21–27.

Fairweather, N. F., Chatfield, S. N., Makoff, A. J., Strugnell, R. A., Bester, J., Maskell, D. J., and Dougan, G. (1990). Oral vaccination of mice against tetanus by use of a live attenuated *Salmonella* strain. *Infect. Immun. 58*: 132–136.

Finn, C. W., Jr., Silver, R. P., Habig, W. H., Hardegree, M. C., Zon, G., and Garon, C. F. (1984). The structural gene for tetanus neurotoxin is on a plasmid. *Science 224*: 881–884.

Gansel, M., Penner, R., and Dreyer, F. (1987). Distinct sites of action of clostridial neurotoxins revealed by double-poisoning of mouse motor nerve terminals. *Pflügers Arch. 409*: 533–539.

Goldberg, R. L., Costa, T., Habig, W. H., Kohn, L. D., and Hardegree, M. C. (1981).

Charcterization of fragment C and tetanus toxin binding to rat brain membranes. *Mol. Pharmacol. 20*: 565–570.

Habermann, E. (1988). Inhibition by tetanus and botulinum A toxin of the release of [^{3}H]noradrenaline and [^{3}H]GABA from rat brain homogenate. *Experientia 44*: 224–226.

Habermann, E., and Dreyer, F. (1986). Clostridial neurotoxins: handling and action at the cellular and molecular level. *Curr. Top. Microbiol. Immuno. 129*: 93–179.

Habig, W. H., Kenimer, J. G., and Hardegree, M. C. (1983). Retrograde axonal transport of tetanus toxin: toxin mediated antibody transport. In *Frontiers in Biochemical and Biophysical Studies of Proteins and Membranes*, T.-Y. Liu, S. Sakakibara, A. Schechter, K. Yagi, H. Yajima, and K. T. Yasunobu (Eds.), Elsevier, New York, pp. 463–473.

Halpern, J. L., and Loftus, A. (1993). Characterization of the receptor binding domain of tetanus toxin. *J. Biol. Chem.* in press.

Halpern, J. L., Smith, L. A., Seamon, K. B., Groover, K. A., and Habig, W. H. (1989). Sequence homology between tetanus and botulinum toxins detected by an antipeptide antibody. *Infect. Immun. 57*: 18–22.

Halpern, J. L., Habig, W. H., Neale, E. A., and Stibitz, S. (1990a). Cloning and expression of functional fragment C of tetanus toxin. *Infect. Immun. 58*: 1004–1009.

Halpern, J. L., Habig, W. H., Trenchard, H., and Russell, J. T. (1990b). Effect of tetanus toxin on oxytocin and vasopressin release from nerve endings of the neurohypophysis. *J. Neurochem. 55*: 2072–2078.

Hanna, P. C., Mietzner, T. A., Schoolnik, G. K., and McClane, B. A. (1991). Localization of the receptor-binding region of *Clostridium perfringens* enterotoxin utilizing cloned toxin fragments and synthetic peptides. *J. Biol. Chem. 266*: 11037–11043.

Hatheway, C. L. (1990). Toxigenic clostridia. *Clin. Microbiol. Rev.*: 66–98.

Hauser, D., Eklund, M. W., Kurazano, H., Binz, T., Niemann, H., Gill, D. M., Boquet, P., and Popoff, M. R. (1990). Nucleotide sequence of *Clostridium botulinum* C1 neurotoxin. *Nucleic Acids Res. 18*: 4924.

Helting, T. B., and Zwisler, O. (1977). Structure of tetanus toxin. I. Breakdown of the toxin molecule and discrimination between polypeptide fragments. *J. Biol. Chem. 252*: 187–193.

Helting, T. B., Parschat, S., and Engelhardt, H. (1979). Structure of tetanus toxin. Demonstration and separation of a specific enzyme converting intracellular tetanus toxin to the extracellular form. *J. Biol. Chem. 254*: 10728–10733.

Ho, J. L., and Klempner, M. S. (1985). Tetanus toxin inhibits secretion of lysosomal contents from human macrophages. *J. Infect. Dis. 152*: 922–929.

Högy, B., Dauzenroth, M. E., Hudel, M., Weller, U., and Habermann, E. (1992). Increase of permeability of synaptosomes and liposomes by the heavy chain of tetanus toxin. *Toxicon 30*: 63–76.

Holmgren, J., Elwing, H., Fredman, P., and Svennerholm, L. (1980). Polystyrene-adsorbed gangliosides for investigation of the structure of the tetanus toxin receptor. *Eur. J. Biochem. 106*: 371–379.

Janicki, P. K., and Habermann, E. (1983). Tetanus and botulinum toxins inhibit, and black widow spider venom stimulates the release of methionine-enkephalin-like material in vitro. *J. Neurochem. 41*: 395–402.

Kenimer, J. G., Habig, W. H., and Hardegree, M. C. (1983). Monoclonal antibodies as probes of tetanus toxin structure and function. *Infect. Immun. 42*: 942–948.

Kimura, K., Fujii, N., Tsuzuki, K., Murakami, T., Indoh, T., Yokosawa, N., Takeshi, K., Syuto, B., and Oguma, K. (1990). The complete nucleotide sequence of the gene coding for botulinum type C1 toxin in the c-st phage genome. *Biochem. Biophys. Res. Commun. 171*: 1304–1311.

Krieglstein, K., Henschen, A., Weller, U., and Habermann, E. (1990). Arrangement of disulfide bridges and positions of sulfhydryl groups in tetanus toxin. *Eur. J. Biochem. 188*: 39–45.

Krieglstein, K. G., Henschen, A. G., Weller, U., and Habermann, E. (1991). Limited proteolysis of tetanus toxin. *Eur. J. Biochem. 202*: 41–51.

Kurazano, H., Mochida, S., Binz, T., Eisel, U., Quanz, M., Grebensteim, O., Wernars, K., Poulain, B., Tauc, L., and Niemann, H. (1992). Minimal essential domains specifying toxicity of tetanus toxin and botulinum neurotoxin type A. *J. Biol. Chem. 267*: 14721–14729.

Lee, G., Groll, E. F., Dyer, S., Beguinot, F., Kohn, L. D., Habig, W. H., and Hardegree, M. C. (1979). Tetanus toxin and thyrotropin interactions with rat brain membranes. *J. Biol. Chem. 254*: 3826–3832.

Link, E., Edelmann, L., Chou, J. H., Binz, T., Yamasaki, S., Eisel, U., Baumert, M., Südhof, T. C., Niemann, H., and Jahn, R. (1992). Tetanus toxin action: inhibition of neurotransmitter release linked to synaptobrevin proteolysis. *Biochem-Biophys. Res. Commun. 189*: 1017–1023.

Makoff, A. J., Ballantine, S. P., Smallwood, A. E., and Fairweather, N. F. (1989a). Expression of tetanus toxin fragment C in *E. coli*: its purification and potential use as a vaccine. *Biotechnology 7*: 1043–1046.

Makoff, A. J., Oxer, M. D., Romanos, M. A., Fairweather, N. F., and Ballantine, S. (1989b). Expression of tetanus toxin fragment C in *E. coli*: high level expression by removing rare codons. *Nucleic Acids Res. 17*: 10191–10202.

Marxen, P., Fuhrmann, U., and Bigalke, H. (1989). Gangliosides mediate inhibitory effects of tetanus and botulinum A neurotoxins on exocytosis in chromaffin cells. *Toxicon 27*: 849–859.

Matsuda, M., and Yoneda, M. (1975). Isolation and purification of two antigenically active, "complementary" polypeptide fragments of tetanus neurotoxin. *Infect. Immun. 12*: 1147–1153.

Mochida, S., Poulain, B., Weller, U., Habermann, E., and Tauc, L. (1989). Light chain of tetanus toxin intracellularly inhibits acetylcholine release at neuro-neuronal synapses, and its internalization is mediated by heavy chain. *FEBS Lett. 253*: 47–51.

Mochida, S., Poulain, B., Eisel, E., Binz, T., Kurazano, H., Niemann, H., and Tauc, L. (1990). Exogenous mRNA encoding tetanus or botulinum neurotoxins expressed in *Aplysia* neurons. *Proc. Natl. Acad. Sci. U.S.A. 87*: 7844–7848.

Montecucco, C. (1986). How do tetanus toxin and botulinum toxins bind to neuronal membranes? *Trends Biochem. Sci. 11*: 314–317.

Montecucco, C., Schiavo, G., Brunner, J., Duflot, E., Boquet, P., and Roa, M. (1986). Tetanus toxin is labeled with photoactivatable phospholipids at low pH. *Biochemistry 25*: 919–924.

Morris, N. P., Consiglio, E., Kohn, L. D., Habig, W. H., Hardegree, M. C., and Helting, T. B. (1980). Interaction of fragments B and C of tetanus toxin with neural and thyroid membranes and with gangliosides. *J. Biol. Chem. 255*: 6071–6076.

Moskaug, J. Ø., Sletten, K., Sandvig, K., and Olsnes, S. (1989). Translocation of diphtheria toxin A-fragment to the cytosol. *J. Biol. Chem. 264*: 15709–15713.

Navone, F., Di Gioa, G., Jahn, R., Browning, M., Greengard, P., and De Camilli, P. (1989). Microvesicles of the neurohypophysis are biochemically related to small synaptic vesicles of presynaptic nerve terminals. *J. Cell Biol. 109*: 3425–3433.

Niemann, H. (1991). Molecular biology of clostridial neurotoxins. In *Sourcebook of Bacterial Protein Toxins*, J. E. Alouf and J. H. Freer (Eds.), Academic, San Diego, pp. 301–348.

Niemann, H. (1992). Clostridial neurotoxins—proposal of a common nomenclature. *Toxicon 30*: 223–225.

Parton, R. G., Ockleford, C. D., and Critchley, D. R. (1987). A study of the mechanism of internalization of tetanus toxin by primary mouse spinal cord cultures. *J. Neurochem. 49*: 1057–1068.

Pierce, E. J., Davison, M. D., Parton, R. G., Habig, W. H., and Critchley, D. R. (1986). Characterization of tetanus toxin binding to rat brain membranes. *Biochem. J. 236*: 845–852.

Pitzurra, L., Marconi, P., Bistoni, F., and Blasi, E. (1989). Selective inhibition of cytokine-induced lysozyme activity by tetanus toxin in the GG2EE macrophage cell line. *Infect. Immun. 57*: 2452–2456.

Poulain, B., Mochida, S., Weller, U., Högy, B., Habermann, E., Wadsworth, J. D. F., Shone, C. C., Dolly, J. O., and Tauc, L. (1991). Heterologous combinations of heavy and light chains from botulinum neurotoxin A and tetanus toxin inhibit neurotransmitter release in *Aplysia. J. Biol. Chem. 266*: 9580–9585.

Poulet, S., Hauser, D., Quanz, M., Niemann, H., and Popoff, M. R. (1992). Sequences of the botulinal neurotoxin E derived from *Clostridium botulinum* type E (strain beluga) and *Clostridium butyricum* (strains ATCC 43181 and ATCC 43755). *Biochem. Biophys. Res. Commun. 183*: 107–113.

Price, D. L., Griffin, J., Young, A., Peck, K., and Stocks, A. (1975). Tetanus toxin: direct evidence for retrograde intraaxonal transport. *Science 188*: 945–947.

Roa, M., and Boquet, P. (1985). Interaction of tetanus toxin with lipid vesicles at low pH. *J. Biol. Chem. 260*: 6827–6835.

Rogers, T. B., and Snyder, S. H. (1981). High affinity binding of tetanus toxin to mammalian brain membranes. *J. Biol. Chem. 256*: 2402–2407.

Rolf, J. M., Gaudin, H. M., and Eidels, L. (1990). Localization of the diphtheria toxin receptor-binding domain to the carboxyl-terminal M_r ~6000 region of the toxin. *J. Biol. Chem. 265*: 7331–7337.

Sandberg, K., Berry, C. J., and Rogers, T. B. (1989). Studies on the intoxication pathway of tetanus toxin in the rat pheochromocytoma (PC12) cell line. *J. Biol. Chem. 264*: 5679–5686.

Sanders, D., and Habermann, E. (1992). Evidence for a link between specific proteolysis and inhibition of [^{3}H]noradrenaline release by the light chain of tetanus toxin. *Naunyn-Schmeideberg's Arch. Pharmacol. 346*: 358–361.

Sandvig, K., and Olsnes, S. (1991). Membrane translocation of diphtheria toxin. In *Sourcebook of Bacterial Protein Toxins*, J. E. Alouf and J. H. Freer (Eds.), Academic, San Diego, pp. 57–67.

Schengrund, C.-L., Ringler, N. J., and DasGupta, B. R. (1992). Adherence of botulinum and tetanus neurotoxins to synaptosomal proteins. *29*: 917–924.

Schiavo, G.,Ferrari, G., Rossetto, O., and Montecucco, C. (1991). Tetanus toxin receptor. Specific cross-linking of tetanus toxin to a protein of NGF-differentiated PC12 cells. *FEBS Lett. 290*: 227–230.

Schiavo, G., Poulain, B., Rossetto, O., Benfenati, F., Tauc, L., and Montecucco, C.(1992a). Tetanus toxin is a zinc protein and its inhibition of neurotransmitter release and protease activity depend on zinc. *EMBO J. 11*: 3577–3583.

Schiavo, G., Rossetto, O., Santucci, A., DasGupta, B. R., and Montecucco, C. (1992b), Botulinum toxins are zinc proteins. *J. Biol. Chem. 267*: 23479–23484.

Schiavo, G., Benfenati, F., Poulain, B., Rossetto, O., Polverino de Laureto, P., DasGupta, B. R., and Montecucco, C. (1992C). Tetanus and botulinum B neurotoxins block neurotransmitter release by proteolytic cleavage of synaptobrevin. *Nature 359*: 832–835.

Schwab, M. E., and Thoenen, H. (1976). Electron microscopic evidence for a transsynaptic migration of tetanus toxin in spinal cord motoneurons: an autoradiographic and morphometric study. *Brain Res. 105*: 213–227.

Schwab, M. E., Suda, K., and Thoenen, H. (1979). Selective retrograde transsynaptic transfer of a protein, tetanus toxin, subsequent to its retrograde axonal transport. *J. Cell Biol. 82*: 798–810.

Simpson, L. L. (1983). Ammonium chloride and methylamine hydrochloride antagonize clostridial neurotoxins. *J. Pharmacol. Exp. Ther. 255*: 546–552.

Simpson, L. L. (1986). Molecular pharmacology of botulinum toxin and tetanus toxin. *Ann. Rev. Pharmacol. Toxicol. 26*: 427–453.

Singh, Y., Klimpel, K. R., Quinn, C. P., Chaudhary, V. K., and Leppla, S. H. (1991). The carboxyl-terminal end of protective antigen is required for receptor binding and anthrax toxin activity. *J. Biol. Chem. 266*: 15493–15497.

Staub, G. C., Walton, K. M., Schnaar, R. L., Nichols, T., Baichwal, R., Sandberg, K., and Rogers, T. B. (1986). Characterization of the binding and internalization of tetanus toxin in a neuroblastoma hybrid cell line. *J. Neurosci. 6*: 1443–1451.

Stecher, B., Ahnert-Hilger, G., Weller, U, Kemmer, T. P., and Gratzl, T. (1992). Amylase release from streptolysin O-permeabilized pancreatic acinar cells. *Biochem. J. 283*: 899–904.

Thompson, D. E., Brehm, J. K., Oultram, J. D., Swinfield, T.-J., Shone, C. C., Atkinson, T.,

Melling, J., and Minton, N. P. (1990). The complete amino acid sequence of the *Clostridium botulinum* type A neurotoxin, deduced by nucleotide sequence analysis of the encoding gene. *Eur. J. Biochem. 189*: 73–81.

Trimble, W. S., Cowan, D. W., and Scheller, R. H. (1988). VAMP-1: a synaptic vesicle associated integral membrane protein. *Proc. Natl. Acad. Sci. U.S.A. 85*: 4538–4542.

Vallee, B. L., and Auld, D. S. (1990). Zinc coordination, function, and structure of zinc enzymes and other proteins. *Biochemistry 29*: 5647–5659.

Vallee, R. B., and Bloom G. S. (1991). Mechanisms of fast and slow axonal transport. *Annu. Rev. Neurosci. 14*: 59–92.

van Heyningen, W. E. (1974). Gangliosides as membrane receptors for tetanus toxin, cholera toxin and serotonin. *Nature 249*: 415–417.

van Heyningen, W. E., and Miller, P. (1961). The fixation of tetanus toxin by ganglioside. *J. Gen Microbiol. 24*: 107–119.

Volk, W. A., Bizzini, B., Snyder, R. M., Bernhard, E., and Wagner, R. R. (1984). Neutralization of tetanus toxin by distinct monoclonal antibodies binding to multiple epitopes on the toxin molecule. *Infect. Immun. 45*: 604–609.

Walton, K. M., Sandberg, K., Rogers, T. B., and Schnaar, R. L. (1988). Complex ganglioside expression and tetanus toxin binding by PC12 pheochromocytoma cells. *J. Biol. Chem. 263*: 2055–2063.

Weller, U., Taylor, C. F., and Habermann, E. (1986). Quantitative comparison between tetanus toxin, some fragments and toxoid for binding and axonal transport in the rat. *Toxicon 24*: 1055–1063.

Weller, U., Dauzenroth, M.-E., Meyer Zu Heringdorf, D., and Habermann, E. (1989). Chains and fragments of tetanus toxin. Separation, reassociation and pharmacological properties. *Eur. J. Biochem. 182*: 649–656.

Wellhöner, H. H. (1992). Tetanus and botulinum neurotoxins. *Handb. Exp. Pharmacol. 102*: 356–417.

Whelan, S. M., Elmore, M. J., Bodsworth, N. J., Atkinson, T., and Minton, N. P. (1992). The complete amino acid sequence of the *Clostridium botulinum* type E neurotoxin, derived by nucleotide sequence analysis of the encoding gene. *Eur. J. Biochem. 204*: 657–667.

Williamson, L. C., Clarke, W. Y., Fitzgerald, L. C., and Neale, E. A. (1992a). Tetanus toxin enters neurons through acidic endosomes. *Soc. Neurosci. Abstr.* 671.11.

Williamson, L. C., Fitzgerald, S. C., and Neale, E. A. (1992b). Differential effects of tetanus toxin on inhibitory and excitatory release from mammalian spinal cord cells in culture. *J. Neurochem. 59*: 2148–2157.

Wright, J. F., Pernollet, M., Reboul, A., Aude, C., and Colomb, M. G. (1992). Identification and partial characterization of a low affinity metal-binding site in the light chain of tetanus toxin. *J. Biol. Chem. 267*: 9053–9058.

23

Anthrax Toxins

Stephen H. Leppla

National Institute of Dental Research, National Institutes of Health, Bethesda, Maryland

I. INTRODUCTION

In attempting to explain the disease-causing properties of *Bacillus anthracis*, Smith and colleagues (Smith and Keppie, 1954) discovered that sterile plasma from guinea pigs dying of *B. anthracis* infection was toxic when injected into other guinea pigs. This was the first description of anthrax toxin. Subsequently, it was shown that this toxin also caused edema when injected intradermally (Smith et al., 1955). By growing *B. anthracis* in liquid medium and treating the culture supernate in various ways, it was found that two distinct components had to be present to produce edema in skin (Harris-Smith et al., 1958). It was soon recognized (Strange and Thorne, 1958; Thorne et al., 1960) that one of these components, factor II, was the "protective antigen" that had been identified by Gladstone more than 10 years earlier as the material present in *B. anthracis* culture supernatants that induces immunity to infection. This component is still referred to as protective antigen (PA). The component needed to produce edema in skin was separated from PA by adsorption to glass filters (Stanley and Smith, 1961; Beall et al., 1962); its original designation as factor I has been superseded by the more descriptive term edema factor (EF). The ability of the anthrax toxin mixture to kill guinea pigs and rats was then shown to require the presence of a third component, factor III (Stanley and Smith, 1961; Smith and Stanley, 1962; Beall et al., 1962), now designated lethal factor (LF). More recently, an extension of the nomenclature (Friedlander, 1986) has come into use that designates the combination of PA and LF as lethal toxin (LT) and the combination of PA and EF as edema toxin (ET).

Although the work completed prior to 1965 firmly established that anthrax toxin contains three components, it proved difficult to obtain pure material, and biochemical characterization of the proteins proceeded slowly. A symposium held in 1967 summarized progress up to that time (Berry, 1967) but also marked the beginning of a period of about 10 years during which little work on the toxin occurred. A period of more active work began about 1980, by which time the success in analyzing other bacterial protein toxins (diphtheria, tetanus, cholera, etc.) had provided a new conceptual framework as well as greatly improved experimental tools. Work with these other toxins showed that the most potent toxins gain access to the cytosol of eukaryotic cells and act there by catalytic mechanisms (Gill, 1978). Many of these toxins contain separable domains, B and A, having receptor-binding and catalytic activities, respectively. It was evident that anthrax toxin fits this pattern if the PA protein is considered the B moiety and the LF and EF proteins are alternate A moieties. The implication that LF and EF are enzymes was not confirmed in an early search for catalytic activities (Stanley and Smith, 1963). However, the similarity of the skin edema caused by the combination of PA and EF to that induced by cholera toxin led directly to the demonstration that EF is an adenylyl cyclase (Leppla, 1982).

Subsequent basic research on anthrax toxin has been stimulated in part by interest in how the three components of anthrax toxin interact to enter cells and gain access to the cytosol. The expectation that LF is, like EF, an enzyme has only recently been confirmed by evidence that LF is a metalloprotease (discussed below). This chapter describes current knowledge about the structure and function of the three toxin components. Other reviews are available that emphasize aspects of microbial physiology and genetics (Thorne, 1993), protein structure (Turnbull, 1986; Leppla, 1988, 1991a,b), interaction with cells (Friedlander, 1990), pathogenesis (Stephen, 1981, 1986), and vaccine development (Hambleton et al., 1984; Turnbull, 1992; Hambleton and Turnbull, 1990). The published proceedings of a symposium held in 1989 includes concise summaries of progress up to that time (Turnbull, 1990b).

II. ROLE OF ANTHRAX TOXIN IN DISEASE

Anthrax is a toxigenic disease in the classical sense of diphtheria and tetanus because injection of the toxin produces symptoms nearly identical to those accompanying an infection (Turnbull, 1990a; Stephen, 1986). Furthermore, as in diphtheria and tetanus, antibodies to anthrax toxin protect against both toxin and bacterial infection (Hambleton et al., 1984). It was noted above that the PA component of anthrax toxin was first identified during searches for a vaccine that would protect against infection (Gladstone, 1946). All subsequent efforts to develop anthrax vaccines have focused on the central role of PA in immunity (Turnbull, 1992). Finally, genetic analysis shows that only those *B. anthracis* strains possessing the plasmid-borne toxin genes are virulent (Mikesell et al., 1983; Uchida et al., 1986; Thorne, 1985).

B. *anthracis* is fully virulent only if it produces, in addition to the three-component protein toxin, a second virulence factor consisting of a gamma-linked, poly-D-glutamic acid capsule. Evidence suggests that the capsule plays an essential role during initial stages of an infection by preventing opsonization and phagocytosis (Keppie et al., 1963; Thorne, 1960; Welkos, 1991). The acidic capsule may also bind and inhibit lysozyme, a strongly basic protein, although *B. anthracis* is already resistant to lysozyme due to deacetylation of the peptidoglycan (Zipperle et al., 1984). Edema toxin (ET, the mixture of PA+EF) probably also plays its most important role early in infection by inhibiting phagocytosis (O'Brien et al., 1985; Keppie et al., 1963) in the same way as the adenylate cyclase of *Bordetella pertussis* (Confer and Eaton, 1982). Lethal toxin (LT, the mixture of PA+LF) may play its major role late in infection, because the symptoms induced by LT resemble those observed in animals just prior to death from infection. However, it is notable that the death of certain animal species does not occur until the concentrations of bacteria in the blood exceed 10^7–10^9 mL^{-1} (Turnbull, 1990c), suggesting that LT is either not highly toxic to host cells in those species or is slow-acting.

A definitive analysis of the roles of the individual toxin components in virulence was recently obtained by genetic methods. Insertional mutagenesis was used to individually inactivate each of the three toxin genes in a strain unable to produce capsule (Cataldi et al., 1990; Pezard et al., 1991, 1993). Virulence for mice is reduced more than 1000-fold in the mutants lacking either PA or LF and about 10-fold in the mutant lacking EF. All three mutant strains have a greatly reduced ability to induce inflammation in mouse foot pads compared to the parental strain. Because all the mutant strains survive within the footpad tissue as well as the parental strain, their loss of inflammatory activity suggests that both LT and ET contribute to inflammation in this model.

III. STRUCTURAL ANALYSIS OF THE ANTHRAX TOXIN COMPONENTS

Analysis of the structure of the three protein components of anthrax toxin has been greatly facilitated by the relative ease with which they can be purified from culture supernatants of avirulent, noncapsulated, Sterne-type strains (Puziss et al., 1963; Ristroph and Ivins, 1983; Leppla, 1988). Procedures developed for growth in fermentors produce yields from each liter of culture of 10 mg PA, 2 mg LF, and 0.5 mg EF (Leppla, 1988, 1991b). The availability of purified components allowed characterization of their structure and function and made possible the cloning and sequencing of the genes, from which further knowledge about structure was deduced. These analyses are discussed in the following sections.

A. Protective Antigen: The Receptor-Binding and Translocation Component

The protective antigen (PA) protein is the central component of the toxin and is required for internalization of the LF and EF proteins into eukaryotic cells. PA is the component that has received the most study, largely due to its role as the essential ingredient in cell-free anthrax vaccines. It was the first component to be identified (Gladstone, 1946), the first to be cloned (Vodkin and Leppla, 1983) and sequenced (Welkos et al., 1988), and it remains the one most easily produced in large amounts (Wright and Angelety, 1971; Leppla, 1991b; Quinn et al., 1991).

1. Sequence Analysis of the Protective Antigen Gene

Cloning of the PA gene, *pag*, was facilitated by evidence (Mikesell et al., 1983) that the gene is located on the large *B. anthracis* plasmid, pXO1 (originally designated pBA1). A library of pXO1 restriction fragments was constructed in pBR322, and the resulting *Escherichia coli* transformants were screened with PA-specific antiserum. Two immunoreactive colonies were identified, and these were shown to produce biologically active PA, even though the bacteria produced only 5–10 ng of PA per milliliter of culture (Vodkin and Leppla, 1983). Subsequently, the gene was subcloned and the DNA sequence determined for a 4235-bp *Hin*dIII-*Bam*HI fragment containing *pag* (Welkos et al., 1988). The sequence is deposited in GenBank with accession number M22589.

The DNA sequence contains an open reading frame (ORF) of 764 codons, of which 29 correspond to a signal peptide. Thus, the mature protein contains 735 amino acids and has a mass of 82.7 kDa. The amino acid composition and amino-terminal sequence of the deduced protein matched those determined for PA purified from *B. anthracis*. The presence of a signal peptide was expected from evidence that PA, LF, and EF are secreted proteins. PA is notable for the absence of cysteine residues. LF and EF also lack cysteine (discussed below). An absence of cysteine is characteristic of a number of secreted bacterial proteins, perhaps reflecting the disadvantage of having an easily oxidized group in proteins that must exist in an oxidizing extracellular environment. The fact that many extracellular proteins lack cysteine was first noted when the amount of sequence data was still quite limited (Pollack and Richmond, 1962).

When the sequence of PA was first obtained, it was compared to protein databases in hopes of gaining insights as to its evolutionary origin and function. No significant similarities to known proteins were found. Recently, we became aware of two sequences that are similar to that of PA. The first of these is a theoretical protein encoded by DNA sequences upstream of *pag*. This 192–codon ORF was noted in the original report on the PA sequence (Welkos et al., 1988). However, the strong similarity of the sequence of this ORF to the carboxyterminal region of PA was not reported. The sequence similarity is most evident when comparing the translations of nucleotides 374–877 in the upstream ORF and nucleotides 3607–4095 within the PA structural gene. These 168 amino acids have a 40% identity and 63% similarity. The regions of amino acid homology end approximately at the termination codon of *pag* (bp 4095), and at the point (bp 877) in the upstream ORF where a putative hydrophobic membrane-spanning domain begins. The upstream ORF does not contain a consensus ribosome-binding site, and it is not known whether this region is transcribed or translated. Clearly, the similarity of the upstream ORF and PA sequences imply that a gene duplication has occurred in this region. Recognition of this duplication may aid in tracing the evolution of the toxin genes.

The second protein that has recently been shown to be related to PA is the *Clostridium perfringens* iota toxin. The iota toxin is one of a group of several clostridial toxins that share with anthrax toxin the unusual property of containing separate proteins that must interact to intoxicate cells. The best-studied member of the group, *Clostridium botulinum* C_2 toxin (which is not related to the more potent botulinum neurotoxins), consists of proteins of 55 and 105 kDa, called components I and II, respectively (Ohishi et al., 1980). Component I is an ADP-ribosyltransferase that modifies cellular actin (Aktories et al., 1986), and component II is the receptor-binding moiety. Component II must be activated by trypsin before it becomes able to bind component I; activation is accompanied by aggregation of component I to a form of about 365 kDa (Ohishi, 1987). The *C. perfringens* iota toxin and the *C. spiroforme* toxin also contain an ADP-ribosyltransferase subunit and a cell-binding subunit that is activated proteolytically. Recently, Popoff and colleagues (Perelle et al., 1993) cloned and sequenced the iota toxin genes. They found that the cell-binding subunit of iota toxin, designated Ib, has extensive sequence similarity to residues 170–590 of the anthrax toxin PA. In PA, this region is implicated in membrane translocation and in binding of the catalytic subunit. The regions that are similar in both toxins begin at the sites at which the activating protease cleavage occurs. In PA, an RKKR^{167} sequence (R = Arg, K = Lys) is cleaved by furin, as discussed in a later section. In iota Ib, the activated protein probably begins at Ala-212 which corresponds to Gly-172 in PA. The region in iota Ib at which cleavage occurs contains a single Arg residue and two closely spaced occurrences of a Phe-Phe sequence. Interestingly, a Phe-Phe sequence at residues 312–313 in anthrax toxin PA is highly susceptible to cleavage by chymotrypsin and by bacterial metalloproteases such as thermolysin (discussed in following section). This suggests that iota toxin may normally be activated by clostridial proteases with specificities like that of thermolysin rather than proteases of the target eukaryotic cells. The fact that iota toxin and anthrax toxin PA have both incorporated a similar translocation domain suggests that this domain must be a functionally efficient motif for protein toxins.

2. Structural and Functional Domains of Protective Antigen

PA Domains Identified by Analysis of Large Peptide Fragments. PA has two sites that are susceptible to cleavage by low concentrations of proteases. These sites define three fragments that have distinct functions (Fig. 1). Treatment with 1 μg/mL chymotrypsin produces fragments of 37 and 47 kDa by specific cleavage on the carboxyl side of F314 in the sequence ASFFD^{315} (Leppla, 1991a). Thermolysin cleaves in the same region, between S312 and F313. The 37- and 47-kDa fragments produced by chymotrypsin and thermolysin are the same size as the major degradation products found in *B. anthracis* culture supernatants, as expected from the similarity of the major protease secreted by *B. anthracis* to thermolysin, a protease isolated from *Bacillus thermoproteolyticus*. The fragments produced by cleavage near residues 313–314 do not immediately dissociate; the "nicked" PA behaves like uncleaved PA during gel filtration or ion-exchange chromatography. However, the nicked PA is biologically inactive (Leppla, 1991a; Novak et al., 1992). The functional defect in chymotrypsin-nicked PA is discussed in a later section. When nicked PA preparations are radioiodinated and incubated with cultured cells at 4°C, the 47-kDa fragment binds to the cells, whereas the 37-kDa fragment does not. This is one of several types of data that show that the region from the carboxy terminal to residue 314 contains the receptor-binding domain.

The other site in PA that is highly sensitive to proteases is the sequence RKKR^{167}. Treatment with 0.1 μg/mL trypsin causes cleavage at one or more of the basic residues in this sequence (Fig. 1) to produce fragments of 20 and 63 kDa (Leppla et al., 1988; Blaustein et al., 1989; Leppla, 1991a). The fragments remain tightly associated in the nicked toxin but can be separated by chromatography on MonoQ resin at pH 9.0. Specific cleavage at the same site can be achieved with 1 μg/mL clostripain, which is specific for

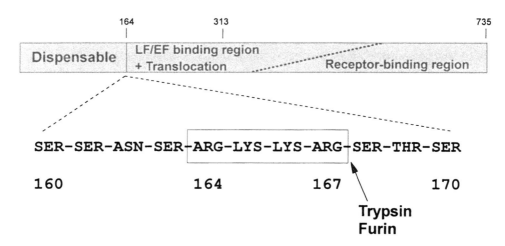

Fig. 1   Domain structure of anthrax toxin protective antigen.

arginine. Many different types of evidence show that cleavage at this site is a normal and obligatory step in the process of toxin action. In studies designed to measure the possible binding of PA to LF in solution, using sedimentation equilibrium to measure apparent molecular weights, it was noted that only a small fraction of PA was able to bind LF. PA repurified on MonoQ resins in a manner shown to remove molecules nicked at the RKKR site did not bind LF. This led to the demonstration that the 63-kDa fragment (PA63) binds tightly to LF, whereas intact PA and the 20-kDa fragment do not. The interaction of LF with the PA fragments can also be studied using polyacrylamide gel electrophoresis at pH 8.5 in buffers containing nondenaturing detergents (Leppla, 1991a). PA63 migrates in these gels as an aggregate of about 350 kDa, suggesting a pentamer or hexamer. Radiolabeled LF binds to the PA63 oligomer to produce a set of bands suggesting that five to six LF molecules bind to the PA63 oligomer. The sharp bands observed in this system indicate that the protein–protein interactions are very strong, so that little dissociation occurs during the course of the electrophoresis experiment.

Cleavage at the RKKR site was found to be biologically relevant when it was discovered that PA bound to eukaryotic cells at 4°C becomes nicked at the same site by a cellular protease (Leppla et al., 1988; Singh et al., 1989a; Novak et al., 1992), recently identified as furin (details presented in a later section). PA63 formed from receptor-bound PA remains bound to the cells, whereas the 20-kDa fragment is released to the medium. This result agrees with that obtained using chymotrypsin-nicked PA in showing that the receptor-binding domain lies in the carboxy-terminal region. It was then shown that PA63 can substitute for PA in combining with LF or EF to cause toxicity (Leppla et al., 1988; Leppla, 1991a). This result proves that all the essential receptor-binding and translocation functions of PA are located in the domains carboxyterminal to residue 167.

Important data implicating the PA63 fragment in the translocation function were obtained using artificial lipid membranes (Blaustein et al., 1989; Koehler and Collier, 1991). As had been shown for other toxins that cross membranes (e.g., diphtheria, tetanus, and botulinum), addition of purified toxin to planar phospholipid bilayers under appropriate conditions leads to formation of ion-conductive channels. Trypsin-nicked PA and purified PA63 produce channels; native PA and the 20-kDa fragment do not (Blaustein et al., 1989). In buffers free of a chelating agent, channels form rapidly at both pH 7.4 and 6.8. In contrast, in the presence of 1 mM EDTA and 100 mM KCl, channel formation is very slow at the higher pH. It was suggested that the rapid channel formation at pH 7.4 in the absence of EDTA is due to heavy metal binding to histidine residues on PA63, giving them a positive charge equivalent to that obtained by lowering the pH to the pK of histidine. Detailed electrophysiological analysis of the channels shows they are highly cation-selective, having a 20-fold preference for K^+ over Cl^-. Quaternary ammonium ions bind to a single site in the lumen of the channel, thereby decreasing conductance, but are driven through the channel by higher voltages (Blaustein and Finkelstein, 1990b; Blaustein et al., 1990). Comparison of permeability to tetrahexylammonium and tetra-heptylammonium ions indicates that the membrane channels are large, with a pore diameter of about 12 Å (Blaustein et al., 1990; Blaustein and Finkelstein, 1990a). The channels formed by diphtheria, tetanus, and botulinum toxins are also about 12 Å, supporting the suggestion that these channels constitute the structures by which the catalytic subunits of the toxins are translocated through endosomal membranes to the cytosol.

Channel formation and membrane interaction by PA63 also occurs in liposomes (Koehler and Collier, 1991; Gordon et al., 1989) and in intact cultured cells (Milne and Collier, 1993). PA63 causes release of K^+ but not calcein from single-walled asolectin vesicles; release requires a pH gradient with the external, toxin-containing solution below pH 6.0 (Koehler and Collier, 1991) Retention of calcein shows that PA63 does not disrupt the integrity of the vesicle membrane and that the K^+-conductive channels formed have diameters of <12 Å (Koehler and Collier, 1991). Diphtheria toxin caused release of calcein from the same vesicle preparation, suggesting differences in the properties of the channels produced by the two toxins. Multilamellar liposomes containing 4-methyl-umbelliferone phosphate release this marker when exposed to hemolytically active preparations of the *B. pertussis* adenylyl cyclase but not when treated with PA (Gordon et al., 1989). Another indication of interaction with membranes is evidence that PA and PA63 partition into the hydrophobic phase of aqueous Triton X-114 systems, especially at low pH (Koehler and Collier, 1991). However, partitioning is not highly selective for PA63 compared to intact PA, consistent with the absence of any strongly hydrophobic domains in the deduced amino acid sequence.

The difficulties in detecting formation in liposomes of channels like those observed in artificial lipid bilayers suggests that membrane insertion of PA63 may be highly dependent on the exact lipid content of the membrane, in contrast to the situation with several toxins previously studied. Although the behavior of these artificial systems is of uncertain physiological significance, the relevance of channel formation is confirmed by evidence that PA63 forms channels in cultured Chinese hamster ovary cells (Milne and Collier, 1993). Channels that allow release of ^{86}Rb are formed in these cells at pH <5.0; the channels are blocked by the quaternary ammonium salts previously shown to decrease conductance in the artificial lipid bilayers (Blaustein et al., 1990).

Site-Specific Mutagenesis to Characterize Functionally Important Residues. Biochemical evidence that cleavage at the sequence RKKR[167] by a cellular protease is essential for toxicity was confirmed by analysis of a mutated PA lacking residues 163–168 (Singh et al., 1989a). The altered PA was constructed by a mutagenic polymerase chain reaction (PCR) procedure and was expressed from the protease-deficient DB104 strain of *Bacillus subtilis* (Kawamura and Doi, 1984). The mutated PA, PA Δ163–168, has the properties predicted from the other evidence that cleavage at the RKKR[167] sequence is essential for toxicity. Thus, the mutated PA is not susceptible to cleavage by a low concentration of trypsin, is not cleaved into fragments of 20 and 63 kDa when bound to cells at 4°C, does not support binding of radiolabeled LF to cells, and is not toxic to cells when combined with LF (Singh et al., 1989a). The mutated PA blocks toxicity of normal LT for macrophages in a competitive manner, and a Schild plot of the data shows that the mutated PA binds with a K_d of 3×10^{-9} M. The mutated PA also blocks the action of LT in Fischer 344 rats, implying that this material has at least theoretical potential as an antitoxic therapy. In an extension of the analysis on the role of the RKKR site, a large number of mutants were made containing substitutions of these four residues. This work led to identification of furin as the cellular protease that cleaves PA, as described in a later section.

Evidence that PA is cleaved specifically at residues 313–314 by low concentrations of chymotrypsin and thermolysin (described in an earlier section) suggests that these residues are on the surface of the protein. Because these highly hydrophobic residues would normally be buried in the interior of the protein, their apparent presence on the surface of PA suggests that they may have a role in initiating the membrane insertion process. This hypothesis is consistent with data showing that PA nicked at this site is inactive (Leppla, 1991a; Novak et al., 1992). To further characterize this site, it was altered by site-specific mutagenesis. Substitutions were made using a PCR procedure, and the mutated proteins were produced from *B. subtilis* DB104 as described above (Y. Singh et al., unpublished). Mutated PA proteins in which residues 313–315, FFD, are replaced with the sequences FFA or AAA, are as toxic as native toxin indicated that hydrophobic residues at this site are not essential. However, a mutated PA in which the two Phe residues are deleted is nontoxic. PA with a CFD sequence retains approximately 20% of normal activity. These data support the view that this region is important but do not provide insights into its functional role.

PA proteins containing amino acid substitutions at one or both of the protease-sensitive sites have several potential uses. Mutant PA proteins resistant to cleavage are expected to be obtained more easily in a homogeneous form, free of degradation products. Also, total yields may be increased if protein is not lost to degradation. Based on this argument, a PA mutant altered at both sites, PA (RKKR)164-167(SNSS), (FFD)313-315(--D), has been proposed for development as the major component of an improved anthrax vaccine. This altered PA should retain nearly all the epitopes of native PA while being totally nontoxic and refractory to cleavage at either site and therefore more easily produced. Its lack of toxic activity would make it possible to include LF or a mutated LF in the vaccine so as to obtain antibodies to both PA and LF.

Two studies in which alterations were made to the carboxy terminus of PA provide evidence that this region contains structures responsible for binding to the cellular receptor (Singh et al., 1991; Little and Lowe, 1991). PA proteins truncated at the carboxyl terminus by three, five, or seven residues show progressively greater decreases in toxicity, up to 20-fold for the deletion of seven residues. Competition assays show that the loss in toxicity correlates with decreased affinity for receptor. Proteins with these small truncations are not significantly more sensitive to proteases than is native PA, indicating that their structures are nearly normal. In contrast, truncation by 12 or 14 residues makes the protein much more susceptible to proteases and causes a complete loss of activity (Singh et al., 1991). A separate study exploited the prior demonstration that several monoclonal antibodies to PA, 14B7 and 3B6, react with a site on PA so as to block its binding to receptor (Little et al., 1988). These antibodies react with PA proteins truncated at the carboxy terminus by 30 residues but not with those truncated by 155 residues (Little and Lowe, 1991), consistent with the view that regions near the carboxy terminus are involved in receptor binding. However, differences in reactivity of the constructs may result from differences in the extent to which refolding to a native conformations occurs, because the epitopes recognized by these monoclonal antibodies are at least partially conformation-dependent (Singh et al., 1991).

A separate effort using site-specific mutagenesis exploited the fact that PA does not contain any cysteine. A set of 12 PA proteins was constructed in which single Ser or Thr residues were replaced by Cys (K. R. Klimpel and S. H. Leppla, unpublished work). This conservative substitution is not expected to alter the physical properties of the mutant PA. In fact, two of the mutated PA proteins were inactive. The single sulfhydryl group located at known sites in the protein provides a highly reactive side chain to which a variety of reagents can be attached. For example, sulfhydryl-specific cross-linking reagents can be used to measure the proximity of an added Cys to other proteins bound to the mutated PA or PA63, such as the PA receptor or LF.

Although biochemical analyses and site-specific mutagenesis provide important information about the structure of PA, a full understanding requires solving the PA structure by X-ray diffraction. Crystals of PA of adequate quality are available (Allured et al., 1985), and work is progressing on the structure (R. C. Liddington, personal communication). Several of the mutants of PA described above in which single Ser residues are replaced by Cys have yielded heavy metal derivatives that will aid in solving the structure.

B. Edema Factor: An "Invasive" Adenylate Cyclase

The edema factor (EF) protein was the first component of anthrax toxin shown to have an enzymatic activity, that of an adenylyl cyclase (Leppla, 1982). Because cAMP is an important regulatory molecule in both prokaryotes and eukaryotes (but not in bacilli), many concepts and methods developed for study of adenylyl cyclases in other organisms have been available for use in studying the *B. anthracis* enzyme. In particular, it has proved useful to compare EF to the related adenylyl cyclase from *B. pertussis* (Hanski and Coote, 1991). Both EF and the *B. pertussis* cyclase have the unique property of requiring a eukaryotic cofactor, the calcium-binding protein calmodulin. Although EF is the least abundant of the three anthrax toxin proteins, it is still possible to purify milligram amounts of the enzyme (Leppla, 1988, 1991b).

1. Sequence Analysis of the Edema Factor Gene

As in the case of the PA gene, cloning of the EF gene, *cya*, was facilitated by evidence (Mikesell et al., 1983) that the gene is located on plasmid pXO1. Two separate groups independently cloned the gene (Mock et al., 1988; Robertson, 1988). One group screened cloned restriction fragments of pXO1 for the ability to complement an adenylyl cyclase-deficient mutant of *E. coli* (Mock et al., 1988). A compatible plasmid that encoded calmodulin was present in the *E. coli* to provide the cofactor required for EF cyclase activity. The other group cloned the gene by screening colonies with a specific antiserum and then confirmed the clone using oligonucleotide probes based on the amino-terminal sequence of the purified protein (Tippetts and Robertson, 1988). The sequences of the cloned *cya* genes were determined by each group (Robertson et al., 1988; Escuyer et al., 1988) and are deposited as GenBank accessions M23179 and M24074. A few differences between the published sequences were resolved when it was concluded that the sequence presented by Escuyer et al. (1988) is the correct one, as stated in revised annotations to the GenBank sequences. The DNA sequence contains an ORF of 800 codons, of which 33 correspond to a signal peptide (Leppla, 1991a). Thus the mature protein contains 767 amino acids and has a mass of 88.8 kDa.

Comparison of the EF amino acid sequence to protein databases identifies two related proteins, anthrax toxin LF and *B. pertussis* adenylyl cyclase (BPCYA) (Robertson, 1988; Escuyer et al., 1988). The similarity to LF resides in residues 1–250. Because both LF and EF interact with PA, and do so competitively, the PA binding site must lie within this region of sequence similarity. The homologous regions of LF and EF also have very similar hydrophilicity profiles, suggesting that the secondary and tertiary structures of these domains are nearly identical (Robertson, 1988). The region having sequence similarity to the BPCYA is approximately residues 265–570 of EF, which align with residues 10–323 of the pertussis cyclase (Escuyer et al., 1988; Robertson, 1988). These homologous regions contain the adenylyl cyclase catalytic domains of the proteins.

2. The Structure and Functional Domains of Edema Factor

Edema Factor (EF) is a very efficient adenylyl cyclase (Leppla, 1984, 1991a; Labruyere et al., 1990), especially compared to the eukaryotic enzymes that are the subject of more extensive study. Kinetic analysis shows that the K_m for Mg^{2+}-ATP is 0.16 mM and V_{max} is 1.2 mmol cAMP/(min mg enzyme). Similar values have been obtained with the BPCYA (Glaser et al., 1989; Bouhss et al., 1993). The catalytic activity is absolutely dependent on the presence of calmodulin, and the stimulatory activity of calmodulin is in turn highly dependent on calcium. Thus, the concentrations of calmodulin needed to obtain half-maximal stimulation of activity are 2 nM and 5000 nM in the presence and absence of calcium, respectively. Nucleotides such as 2′d3′AMP that are known to inhibit eukaryotic adenylyl cyclases by binding at a distinct "P" site also inhibit EF but appear to do so by competing with ATP at the substrate-binding site (Johnson and Shoshani, 1990). The high catalytic activity of EF suggests that it could find use in the synthesis of cAMP. If the enzyme can be shown to accept ATP analogues as substrates, then certain cAMP analogues could also be prepared.

Additional studies of the catalytic properties of EF have been performed using a recombinant protein lacking residues 1–261 (Labruyere et al., 1990). The 62-kDa product, designated CYA62, was produced in *E. coli*; a 20-L fermentor culture yielded 44 mg of purified protein. Kinetic analysis showed that CYA62 has catalytic properties like those of the full-size protein, with $K_m = 0.25$ mM for ATP, and an absolute requirement for calmodulin. The K_d for calmodulin activation is 23 nM. Cross-linking studies indicate that calmodulin interacts with the carboxyterminal 150 residues of CYA62. Truncation of the carboxyterminal 127 amino acids of CYA62 reduces the catalytic activity to 0.1% of the original level. High-affinity binding of ATP, measured with a fluorescent ATP analogue, 3′-anthraniloyl-2′-deoxy-ATP (Sarfati et al., 1990), is dependent on the binding of calmodulin, suggesting that binding of this cofactor induces a functional conformation in the EF protein. The ability to produce large amounts of this smaller, catalytically active form of EF may facilitate determination of the structure by X-ray diffraction.

Functional roles for several portions of the EF protein are known. Residues 314–321 of EF match the consensus sequence GxxxxGKS found in nucleoside triphosphate-binding proteins. This consensus sequence is also found in BPCYA (Fig. 2). This region is inferred to contain at least part of the ATP-binding site. All catalytic activity is lost when the Lys

```
consensus                 GxxxxGKS
                          |    |||
EF           309    GVATKGLNVHGKSSDW   324
                    ||||||| || |||||
BPCYA         54    GVATKGLGVHAKSSDW    71
```

Fig. 2 Sequence similarity of anthrax toxin edema factor (EF) and the *Bordetella pertussis* adenylyl cyclase (BPCYA). Residue numbers are those of the mature proteins.

residue within the consensus sequence is replaced by Met, as in the construct EF K320M (Xia and Storm, 1990). Homologous mutations in the pertussis cyclase K65 residue have the same effect (Glaser et al., 1989; Au et al., 1989). The Lys residue preceding the consensus sequence, Lys-313 in EF, is also implicated in ATP binding, because the construct EF K313Q is catalytically inactive (Labruyere et al., 1991). (Residue numbers used here are corrected to those in the mature EF protein, whereas the cited report assigned numbers from the beginning of the 33-residue signal peptide.) Replacement of His-63 in BPCYA with any of several residues causes more than a 100-fold decrease in activity (Munier et al., 1992); kinetic analysis indicates that the His serves as an essential acid/base catalyst in the ATP cyclization. Involvement of additional residues in the ATP binding site is suggested by results of mutagenesis of several Asp (D) residues in BPCYA. Substitutions for either D188 or D190 inactivate the enzyme and greatly decrease binding of the ATP analogue, suggesting that the homologous residues in EF, D458 and D460, may constitute part of the ATP-binding site (Glaser et al., 1991).

Identification of the EF sequences making up the site to which calmodulin binds has been more difficult. Much of what is known is inferred from study of the BPCYA. The requirement for recombining two protease fragments of BPCYA to regain affinity for calmodulin implies that the binding site may be spread over a large surface (Ladant, 1988). Certain recombinant fragments of the BPCYA that include residues 189–267 bind to calmodulin with nearly the same binding energy as the full-length enzyme, localizing the binding site to this region (Bouhss et al., 1993). Further implicating this region is the large decrease in calmodulin binding caused by mutagenesis of Trp-242 (Glaser et al., 1989). Insertion of a dipeptide after Ala-247 has the same effect, whereas dipeptide insertions in 14 other locations in the protein have little effect (Ladant et al., 1992). A synthetic peptide having the sequence of EF residues 499–532, corresponding approximately to residues 239–273 of BPCYA, binds tightly to calmodulin (Munier et al., 1993), indicating that the calmodulin-binding sites in EF and BPCYA are in homologous locations. This peptide has properties of an amphipathic α helix, which is characteristic of many other calmodulin-binding domains.

C. Lethal Factor: A Metalloprotease?

Lethal factor plays a major role in pathogenesis, as proved by the decrease in virulence when the LF gene is insertionally inactivated (Pezard et al., 1991). LT causes dramatic effects in the two most responsive biological systems, the Fischer 344 rat and cultured mouse macrophages. Intravenous injection of large doses of LT causes death of the rats in exactly 38 min from an overwhelming pulmonary edema (Beall et al., 1962; Ezzell et al., 1984). Treatment of susceptible mouse macrophages with LT leads to complete lysis in about 90 min (Friedlander, 1986; Hanna et al., 1992). The latter response provides a sensitive and convenient bioassay and is the system favored by researchers seeking to identify the molecular mechanism of action of LF. As discussed below, recent studies in macrophages provide convincing evidence that LF is a metalloprotease.

1. DNA Sequence Analysis of Lethal Factor Gene

The gene encoding LF, designated *lef*, was cloned from a library of pXO1 DNA by an immunochemical screening procedure, and positive colonies were confirmed with oligonucleotide probes based on amino-terminal protein sequence (Robertson and Leppla, 1986). The DNA sequence was determined, GenBank accession M29081(Bragg and Robertson, 1989), and confirmed independently by another group, Genbank accession M30210. The DNA sequence contains an ORF of 809 codons, of which 33 residues constitute a signal peptide (Leppla, 1991a). Thus the mature protein contains 776 amino acids and has a mass of 90.2 kDa.

For several years after the sequence was determined, databases were searched for proteins having sequences similar to LF. The only sequence found was that of EF. LF and EF are very similar in residues 1–250, as discussed above. The other notable feature of the LF amino acid sequence is the presence of five imperfect repeats of 19 amino acids (Bragg and Robertson, 1989). Although the original sequence report showed the five repeats in a dot matrix plot, our previous discussions have stated that there are only four repeats (Leppla, 1991a; Quinn et al., 1991; Arora and Leppla, 1993). The first repeat was overlooked when the earlier alignments were prepared because it is separated from the other four contiguous repeats by seven amino acids. The repeat region in rich in charged residues.

2. The Structure and Functional Domains of Lethal Factor

The similarity in the amino-terminal regions of LF and EF strongly implies that this is the region that binds to PA63. This view is confirmed by results from a mutagenesis study in which oligonucleotide cassettes encoding two amino acids, either RV, TR, or AY, were inserted into LF in a semirandom manner (Quinn et al., 1991). Of four insertions within residues 1–250, two completely inactivate LF and the other two cause decreases in activity to 1 or 4% of control levels. In each case, the decreases in activity correlate to decreases in binding of the mutated LF proteins to PA63 bound on cells, as measured in a competitive binding protocol. The same mutagenesis procedure also provided information on the function of the other portions of LF. Of four insertions in the repeat region, within residues 293–376, three destabilize the protein so severely that no protein can be recovered, whereas the other insert causes only a two- to threefold decrease in toxicity. These data suggest that the repeat region may be highly ordered, so that minor disruptions make it susceptible to proteolysis. However, no insights are provided as to the possible function of the repeat region. The three insertions within residues 429–491 are all well tolerated, causing no measurable loss in toxicity. The four insertions located closest to the carboxy terminus, distal to residue 594, totally inactivate LF but do not decrease the binding of LF to PA63 as measured in the competitive assay described above. This is persuasive evidence that this region contains the catalytic domain of LF.

The sequence similarities of the amino-terminal regions of LF and EF are not so extensive as to induce substantial amounts of cross-reacting antibodies. It has been a general finding that polyclonal sera to LF do not react with EF, or vice versa. This is surprising in view of the extensive evidence that both proteins bind to the same site on PA63. Monoclonal antibodies are available that react with both LF and EF (Little et al., 1990). Of 61 monoclonal antibodies obtained to LF examined, only three reacted with EF. Two of these cross-reactive antibodies blocked binding of LF to PA63 and neutralized

LF action on cells. Identification of the epitopes recognized by these antibodies may help to explain the high-affinity interaction of LF and EF with PA63.

Additional evidence delineating essential regions of LF derives from studies of fusion proteins containing portions of LF combined with the ADP-ribosylation domain of *Pseudomonas* exotoxin A (PE), contained within residues 362–613. These fusion proteins are highly toxic when combined with PA and administered to cells having PA receptors (Arora et al., 1992; Arora and Leppla, 1993). Concentrations as low as 2 pM kill Chinese hamster ovary (CHO) cells. Pharmacological evidence shows that the LF-PE fusion proteins are internalized by endocytosis and trafficked in the same way as LF. Constructs containing residues 1–254 of LF fused to PE362–613 are as toxic as analogous constructs containing the entire LF sequence. Structures essential to LF activity lie within residues 1–41 and 198–254, because deletion of these regions from LF or the fusion proteins renders them inactive.

3. Catalytic Properties of Lethal Factor

Even prior to the demonstration that EF is an enzyme, it was suspected that LF had catalytic activity. Continuing efforts over several years to find sequences in protein databases that have similarity to LF were finally successful with the identification of similarity to metalloproteases. The similarity to metalloproteases is restricted to a small sequence at the active site (Fig. 3). In metalloproteases, this sequence, HExxH, contains two His (H) residues that chelate a zinc ion and a Glu (E) residue that acts as the nucleophile during bond cleavage. This small region of sequence similarity would have been dismissed as insignificant except for the fact that pharmacological evidence had implicated aminopeptidases in the action of LF, and certain aminopeptidases share with metalloproteases the same zinc-binding active-site sequence (Vallee and Auld, 1990). The pharmacological evidence referred to was data that certain peptide derivatives protect mouse macrophages from LT action. Protective compounds include bestatin, leucinamide, phenylalaninamide, and chloromethyl ketones of leucine and phenylalanine. Di- and tripeptides of Phe and Leu are also protective if the carboxyl groups are blocked as amides. Bestatin, (2S,3R)-(3-amino-2-hydroxy-4-phenylbutanoyl)-L-leucine, is obtained from *Streptomyces olivoreticuli* (Umezawa et al., 1976). It inhibits leucine aminopeptidase and aminopeptidases B and M, stimulates certain immune responses, and has been tested in humans as an anticancer agent. Addition of 0.2 mM bestatin decreases LT action on mouse macrophages by 50%. Although it remains uncertain that the protective action of these compounds results from inhibition of a target cell aminopeptidase, recognition of their action was important in directing attention to the possible role of zinc-binding enzymes in LT action.

The hypothesis that LF is a zinc-binding peptidase is made more credible by the recent discovery that the clostridial neurotoxins have this activity (Schiavo et al., 1992). All the botulinum and tetanus neurotoxins have the HExxH sequence characteristic of

Lethal Factor

```
686   HEFGHAVDDY.AGYLLDKNQSDLVTNSKKFIDIF
      ||:.|||.||  ||  |:  .|:|:  :  |  .  :  |||
374   HELTHAVTDYTAG.LVYQNESGAI.N.EAMSDIF
```

Thermolysin

Fig. 3 Sequence similarity of anthrax toxin lethal factor (LF) to thermolysin.

zinc-binding peptidases, a fact noted in 1989 (Jongeneel et al., 1989) but apparently overlooked for several years by workers in this field. It was recently shown that the neurotoxins cleave nerve cell proteins that are involved in docking and fusion of exocytic vesicles (Schiavo et al., 1992; Südhof et al., 1993). Proteins cleaved at unique sites by one or more of the neurotoxins include synaptobrevin, SNAP-25, and syntaxin. Inactivation of these proteins provides a reasonable and satisfying molecular explanation for the physiological effects of the neurotoxins.

Characterization of mutant LF proteins altered in and near the HEFGH[690] sequence by site-specific mutagenesis supports the hypothesis that LF is a zinc-binding peptidase (Klimpel et al., 1993). Mutated LF proteins having the single amino acid replacements H686A, E687C, or H690A are totally inactive (Klimpel et al., 1994). Substitutions E720A and E721A made to residues suspected of contributing another chelating group for the zinc ion have no effect on LF activity. These data strongly support the hypothesis that LF is a zinc-dependent enzyme. Analogous studies of the role of the two His residues in the zinc-binding site of *Streptococcus sanguis* IgA protease show that replacement of either His inactivated the enzyme (Gilbert et al., 1991). Replacement of the Glu (E) residue in the sequence HELIH inactivates tetanus light chain (McMahon et al., 1993).

Protease and peptidase activities have not yet been detected in purified LF preparations, in spite of the use of sensitive assays for these activities. If LF does have peptidase activity, it is likely to be restricted to specific cellular substrates, as in the case of the clostridial neurotoxins. Proving that LF acts as a metalloprotease, and identifying the critical cellular substrate(s) it modifies, would fill one of the major gaps in our understanding of anthrax toxin action.

IV. BIOSYNTHESIS OF ANTHRAX TOXIN

A. Genetic Control of Toxin Synthesis

1. Genetics of Anthrax Virulence Factors

DNA hybridization demonstrates that *B. anthracis* is most closely related to *Bacillus thuringiensis* and *Bacillus cereus* (Kaneko et al., 1978). The similarity of these species was confirmed by showing that the sequences of their ribosomal RNAs are >99% identical (Ash et al., 1991; Ash and Collins, 1992). Methods developed for genetic analysis of these species and particularly for the more distantly related *B. subtilis* have been useful in characterizing the genes of *B. anthracis* (Youngman, 1990; Thorne, 1993).

The two recognized virulence factors of *B. anthracis* are encoded by separate large plasmids. The three proteins of the toxin are encoded by plasmid pXO1 (184 kb), and the polyglutamate capsule is encoded by pXO2 (97 kb). Comparison of parental and plasmid-cured strains proved that these plasmids encode toxin (Mikesell et al., 1983) and capsule (Green et al., 1985; Uchida et al., 1985), respectively, and that they are required for virulence (Uchida et al., 1986; Mikesell et al., 1983). Both plasmids have been physically characterized and mapped (Kaspar and Robertson, 1987; Robertson et al., 1990; Uchida et al., 1987). The genes for all three toxin components and for two regulatory genes (discussed below) are located in a 20-kb region of pXO1 (Fig. 4). The large size of the plasmids suggests the presence of other genes involved in pathogenesis, but none have been identified other than the several discussed below.

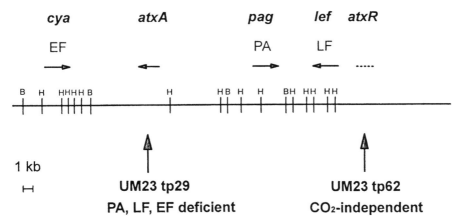

Fig. 4 Toxin gene region of plasmid pXO1. Genes encoding EF, PA, LF are *cya*, *pag*, and *lef*, respectively. Restriction enzyme sites are B, BamHI; H, HindIII.

2. Regulation of Toxin Gene Expression

Early studies showed that the material secreted by *B. anthracis* that induces protective immunity, now known to be PA, is produced only when bicarbonate is added to culture media (Gladstone, 1946). It was subsequently shown that regulation by bicarbonate occurs at the transcriptional level and that a trans-acting gene required for toxin synthesis is present on pXO1 (Bartkus and Leppla, 1989; Cataldi et al., 1992). The genes involved in regulation (Fig. 4) were localized to pXO1 by transposon mutagenesis (Hornung and Thorne, 1991). One type of Tn*917* mutant obtained, UM23 tp29, is deficient in synthesis of all three toxin components, whereas a second type of mutant, UM23 tp62, produces toxin in the absence of bicarbonate. The gene inactivated in UM23 tp29 was cloned and sequenced (Uchida et al., 1993). Complementation analysis identified a single open reading frame (ORF) of 476 codons, designated *atxA*, as a trans-acting, positive regulator required for transcription of each of the three toxin genes. The putative positive regulatory protein does not have significant sequence similarity to regulatory proteins of other bacteria, and it may therefore represent a new class of regulator. Recent studies have shown that the *AtxA* protein product binds to DNA sequences upstream of the PA structural gene (Uchida and Leppla, unpublished work).

The requirement for bicarbonate is lost in mutants like UM23 tp62. This is considered evidence that the gene disrupted in this mutant encodes a repressor for which bicarbonate is the corepressor. The region of pXO1 containing the inserted Tn*917* has been sequenced, and a gene has been identified that is designated *atxR* (Uchida and Leppla, unpublished work). The cloned gene complements the mutation in UM23 tp62, restoring the requirement for bicarbonate.

The effects of *AtxA* and *AtxR* on toxin production are most evident when these genes are present at low copy number on pXO1 and when *B. anthracis* is grown in a synthetic medium. Under other circumstances, *AtxA* is not absolutely required for transcription of the toxin genes. For example, good expression of PA has been obtained from recombinant plasmids containing the PA gene under its own promoter in *B. anthracis* (Singh et al., 1989a) and in *B. subtilis* (Ivins and Welkos, 1986) in the absence of *AtxA* and *AtxR* (discussed in a later section). In this case, expression is best in a rich medium, quite poor in the synthetic medium, and unresponsive to addition of bicarbonate. However, the plasmids used in these studies, such as pYS5 (Singh et al., 1989a), contain only about 200 bp of DNA upstream of the structural gene and probably lack the sequences to which regulatory factors bind.

B. Anthrax Toxin Production and Purification

1. Toxin Production from B. anthracis and B. subtilis

Current methods for growth of the Sterne strain of *B. anthracis* to produce the anthrax toxin components are not fundamentally different from those developed many years ago for production of the acellular PA-based anthrax vaccine (Puziss et al., 1963; Wright and Angelety, 1971). Toxin is usually produced from Sterne-type strains, such as the protease-deficient strain V770-NP1-R (ATCC #14185), although a rifampicin-resistant mutant, SRI-1, is reported to secrete 50–75% more toxin (Leppla, 1988). Anaerobic growth promotes PA production, and media compositions were therefore optimized for anaerobic conditions (Puziss et al., 1963). The defined "R medium" proposed in 1983 (Ristroph and Ivins, 1983) was derived from a previously used medium (Haines et al., 1965) by replacement of the casamino acid mixture with a mixture of purified amino acids. The toxin components may be purified by ion-exchange and hydrophobic interaction chromatography (Wilkie and Ward, 1967; Quinn et al., 1988) or with monoclonal antibody immunoadsorbent columns (Larson et al., 1988). Detailed protocols are available for fermentation of the Sterne strain and for recovery and purification of the toxin components (Leppla, 1988, 1991b).

B. *anthracis* strains in which one or more of the toxin components have been inactivated (Pezard et al., 1993) can be used to produce subsets of the toxin components. Because these strains are only slightly altered from their Sterne-type parents, the growth conditions developed for Sterne support good expression of the toxin proteins. The absence of LT, ET, or both from the culture supernatants would allay some concerns about safety, as well as providing immunogens for production of antisera free of antibodies to other toxin components.

Current work on toxin structure and function requires that mutated toxin proteins be expressed and purified. The most successful systems for expression of recombinant forms of the toxin genes are based on plasmid pUB110, originally isolated from *Staphylococcus aureus*. The PA gene inserted into pUB110 is expressed well in *B. subtilis*, yielding 20 μg PA per milliliter (Ivins and Welkos, 1986), an amount equal to that obtained from *B. anthracis* Sterne grown under optimal conditions. A shuttle plasmid, pYS5, derived from pUB110 and pVC8f(+)T offers the advantage that it is able to replicate in both *Bacillus* species and in *E. coli* (Singh et al., 1989a). It also contains the f1 phage origin, allowing production of the single-stranded DNA needed in certain mutagenesis protocols. The pYS5 plasmids also function well in *B. anthracis*. However, expression of toxin from these pUB110-based plasmids occurs at low levels in the anaerobic R medium cultures that support good production from the Sterne strain (Quinn et al., 1991; Singh et al., 1989a; Uchida et al., 1993). Instead, production of toxin from pYS5 plasmids in *B. anthracis* and *B. subtilis* requires a rich medium and vigorous aeration. A rich medium containing tryptone and yeast extract (FA medium) gives yields of PA exceeding 50 μg/mL culture supernatant (Singh et al., 1989a).

One approach to improving production of the toxin proteins from *B. subtilis* and *B. anthracis* is the use of protease-deficient strains. Effective genetic tools are available to manipulate *B. subtilis* (Youngman, 1990), making possible a systematic approach to identification and inactivation of its extracellular proteases. This effort yielded strain DB104, lacking two proteases (Kawamura and Doi, 1984), and more recently strain WB600, lacking six proteases (Wu et al., 1991). More than half of the PA expressed from a pUB110 vector in a parental *B. subtilis* strain was cleaved to fragments of 47 and 37 kDa, characteristic of cleavage at residue 313. Less degradation is evident when DB104 is used as the host (Singh et al., 1989a; Ivins et al., 1990). Preliminary studies with WB600 in this laboratory confirm the claims that proteolysis in culture supernatants of this strain is negligible.

B. anthracis has an inherent advantage as an expression host because it appears to produce less extracellular protease than *B. subtilis*. More than half the protease activity in the supernatant is due to a single metalloprotease, one that is probably very similar to the well-characterized *B. cereus* metalloprotease (Stark et al., 1992) and to thermolysin. We have further improved the utility of *B. anthracis* as an expression host by inactivating the predominant extracellular protease. Strain UM44-1C9 was mutagenized with Tn*917* to obtain strain BH441, which produces a zone of clearing on casein agar that is much smaller than that of the parent strain. DNA sequencing confirms that Tn*917* is inserted into a protease gene (K. Klimpel et al., manuscript in preparation). Recombinant proteins secreted from this host appear to suffer less degradation, and this is particularly evident for proteins containing alterations expected to destabilize their native conformations.

It has previously been difficult to produce milligram amounts of recombinant LF proteins (Quinn et al., 1988). Production of LF has been achieved by taking advantage of the efficient expression of PA from pYS5-type plasmids. Fusion proteins were constructed that contain residues 1–164 of PA and the entire LF protein, joined by a sequence that can be cleaved by the factor Xa protease. These fusion proteins are expressed from plasmids of the pYS5 type at concentrations of 50 µg/mL (Klimpel et al., 1994). Use of the protease-deficient *B. anthracis* mutant BH441 as host improves the integrity of the secreted proteins. The fusion proteins are purified and then cleaved by the factor Xa protease to yield the desired native or mutant LF proteins.

2. Toxin Production from Baculovirus and Vaccinia Virus Vectors

In a separate approach to producing PA, the PA gene was transferred into a baculovirus vector that was used to infect insect cells. The PA gene was also recombined into vaccinia virus (Iacono-Connors et al., 1990). Expression of PA in both systems was confirmed by immunoblots. Injection of the infected insect cells or infection with the recombinant vaccinia virus induced antibodies to PA. Subsequently (Iacono-Connors et al., 1991), it was shown that immunization with the vaccinia virus partially protected mice and guinea pigs against infection by virulent *B. anthracis*, as did PA purified from the baculovirus-infected insect cells. Only small amounts of purified PA were obtained in these initial trials; extensive efforts were not made to optimize yields. Although the baculovirus system has unique properties, it is not evident that it has advantages over *Bacillus* hosts for the routine production of toxin proteins.

3. Toxin Production from E. coli Hosts

Many different vector systems are available for expression of heterologous proteins in *E. coli*. In this laboratory, vectors containing the T7 promoter have been used to produce native and mutated PA and LF proteins (Singh et al., 1991; Arora et al., 1992). Fusion proteins of LF connected to PE or other polypeptides have been produced in the same vectors (Arora et al., 1992; Arora and Leppla, 1993), although it has recently been found that better expression is obtained as fusions to the glutathione-*S*-transferase (GST) (Guan and Dixon, 1991). This system allows affinity purification of the tripartite GST-LF-PE fusions (Arora and Leppla, 1994).

V. ANTHRAX TOXIN INTERACTION WITH CELLS

Interest in how the separate anthrax toxin proteins interact to cause toxic effects arose as soon as researchers recognized the requirement that components be recombined to obtain toxicity. An early study involving sequential injections into Fischer 344 rats indicated that PA binds to host tissues, whereas LF remains in the circulation, where it is available to combine with PA added several hours later (Molnar and Altenbern, 1963). Identification of mouse macrophages as a cell system that is rapidly and dramatically responsive to LT (Friedlander, 1986) was accompanied by demonstration that internalization is an acid-dependent process, analogous to that of diphtheria toxin. These and other results led to a model that described the overall internalization process (Leppla et al., 1990) and provided a basis on which to experimentally examine each step in detail.

A. Characterization of the Cellular Receptor for PA

Internalization of PA into cells begins with its binding to a specific cell surface receptor (Escuyer and Collier, 1991; Singh et al., 1990; Leppla, 1991a). Nearly all types of cells possess receptors. The presence of receptors on a particular type of cell can be inferred in the absence of direct measurements by showing that cells are sensitive to PA combined with an LF-PE fusion (described above). Trypsin treatment greatly reduces the sensitivity of cells, as does treatment for several hours with cycloheximide (Gordon et al., 1989), suggesting that the receptor is a protein. Studies using ^{125}I-labeled PA show that binding is highly specific and saturable; the data are most consistent with there being a single, homogeneous class of receptors having a K_d of about 1 nM (Friedlander et al., 1993; Escuyer and Collier, 1991). Maximal binding requires the presence of Ca^{2+} (Bhatnagar et al., 1989). CHO cells contain 10,000 receptors per cell, with a K_d of 1 nM (Escuyer and Collier, 1991), while primary mouse macrophages have 30,000 receptors per cell and a K_d of 0.5 nM (Friedlander et al., 1993). The cell type with the highest density of receptors so far observed is the rat myoblast cell line L-6, which has about 50,000 receptors per cell. Cross-linking of radiolabeled PA bound on CHO cells identifies a cellular protein of 85–90 kDa that may be the PA receptor (Escuyer and Collier, 1991).

B. Activation of PA by Cellular Proteases to Produce PA63

Proteolytic activation of PA is required for binding of LF and EF (Leppla et al., 1988; Singh et al., 1989a; Leppla, 1991a). This is most clearly shown by the lack of toxicity of a mutated PA in which residues 163–168 are deleted (Singh et al., 1989a). When it was recognized that proteolysis is normally performed by a cellular enzyme, it became important to identify the protease(s) involved. Cassette mutagenesis produced a set of PA mutants in which residues 164–167 are replaced by a random sequence containing Arg, Lys, Ser, or Asn (Klimpel et al., 1992b). Of the 19 mutated proteins initially characterized, only those having Arg residues at both positions 164 and 167 are toxic to cells. Mutated proteins having one or more Arg or Lys residues within this sequence, but not having Arg at both positions 164 and 167, become active if treated with trypsin. Additional mutated PA proteins constructed to have defined sequences confirm that only an RAAR sequence confers rapid toxicity, whereas AKKR and RKKA sequences do not.

The substrate specificity of the cellular protease deduced from these data matches that of furin, a recently identified eukaryotic protease responsible for posttranslational processing of certain protein precursors (van de Ven et al., 1990; Steiner et al., 1992). Direct evidence that PA is activated by furin comes from studies of the nicking of ^{125}I-PA that occurs on the surface of cells incubated at 4°C or fixed by chemical cross-linking. Nicking requires >200 μM Ca^{2+}, is blocked by dithiothreitol and *p*-chloromercuribenzoate, and is resistant to moderate concentrations of serine protease inhibitors such as phenyl methanesulfonyl chloride (Klimpel et al., 1992b). All these responses match those of purified furin as measured with synthetic substrates. Finally, a purified recombinant form of furin cleaves PA very efficiently after Arg-167 (Molloy et al., 1992), and the susceptibility of mutated PA proteins to furin correlates to the relative toxicities of the proteins, which both decrease in the order RKKR > RAAR >> AKKR, RKKA.

Although furin is located principally in the trans-Golgi region, the demonstration that PA cleavage can occur on the cell surface implies the presence there of some enzyme. Recent evidence suggests that furin may cycle to the surface and also that cleavage of furin may separate the catalytic domain from the transmembrane anchor, thereby producing a soluble form of furin. This latter process could account for the presence in plasma of a Ca^{2+}-dependent protease able to cleave PA (Ezzell and Abshire, 1992).

Identification of furin as the protease that activates PA is mirrored in work with other toxins having cleavage sites containing multiple Arg residues. Mutagenesis of the putative cleavage sites of PE (Ogata et al., 1992) and of diphtheria toxin-IL-2 fusion proteins (Williams et al., 1990) clearly demonstrate a requirement for Arg residues at positions -1 and -4 relative to the bonds cleaved. Diphtheria toxin is efficiently cleaved by furin (Klimpel et al., 1992a). Very convincing evidence that furin is needed to activate PE derives from work with a PE-resistant mutant CHO cell, RPE.40, that is also resistant to enveloped viruses due to failure to proteolytically process viral envelope proteins (Inocencio et al., 1993). Complementation with a furin cDNA restores both phenotypes to that of the parental CHO cell (Moehring et al., 1993), indicating that the mutant cell lacks furin and that this explains the resistance to PE.

Although furin is the cellular enzyme that accounts for most of the activation of PA, other enzymes can activate PA under certain circumstances. PA mutants with cleavage site sequences of SSRR retain some toxicity. Inhibition of this toxicity by leupeptin implies the participation of a protease that recognizes a pair of basic residues (V. M. Gordon et al., unpublished studies). Furthermore, furin-deficient CHO cell mutants isolated in this laboratory and considered to be equivalent to RPE.40 retain some sensitivity to diphtheria toxin and native PA while being totally resistant to PE and mutated PA proteins with an RAAR cleavage site. Again, this points to activation by proteases that recognize multiple basic residues. Studies with additional mutant CHO cells and mutationally altered toxins are under way to identify the other proteases involved in alternative activation pathways.

C. Binding of LF and EF to PA63

Receptor-bound PA that has been cleaved by furin releases the amino-terminal 20-kDa fragment (Leppla et al., 1988). Release appears to be facilitated by binding to receptor, because PA nicked in solution by trypsin does not readily dissociate. Removal of the 20-kDa fragment exposes a site to which LF and EF bind with high affinity. LF binding can be measured conveniently using cells preincubated at 4°C to allow prior nicking of PA without internalization; chemically fixed cells provide an alternative system that can be used at 23° or 37° (Klimpel et al., 1992b). LF and EF bind competitively to PA63, an effect that was detected in animals in early studies and confirmed in cell culture systems (Leppla, 1982; Leppla et al., 1985). No measurements of the stoichiometry of the binding of PA63 to LF have been reported. However, evidence from nondenaturing gels (described above) and the analyses of binding curves suggest that each PA63 binds one LF. This view should be reexamined because PA63 is likely to exist as a multimer in the ion-conductive channels suggested by studies in artificial lipid membranes.

D. Endocytosis and Translocation of LF and EF to the Cytosol

Translocation of LF or EF to the cytosol follows endocytosis of the complex of PA63 and LF or EF and passage through an acidified compartment, as shown in a number of studies modeled on prior work with other toxins and protein ligands (Sandvig and Olsnes, 1981). Increases in cAMP in response to ET begin only 10 min after toxin addition at 30°C, a lag interpreted as the time required for passage through endosomes (Gordon et al., 1989). Cytochalasin D, an inhibitor of endocytosis, protects cells from ET (Gordon et al., 1988). Agents that block acidification of endosomes, such as 30 mM ammonium chloride, 0.1 mM chloroquine, and 1 μM monensin, also protect cells from ET (Gordon et al., 1988) and LT (Friedlander, 1986; Arora et al., 1992). Acidification of the toxin within endosomes is believed to lead to membrane insertion and translocation of the catalytic domain (LF or EF) to the cytosol. This event has not been measured biochemically. Instead, translocation is inferred to occur by analogy with other toxins such as diphtheria. In a process thought to mimic that occurring in endosomes, LT toxin bound on the surface of macrophages can be "acid-shocked" into the cells; toxicity induced in this manner is not blocked by chloroquine (Friedlander, 1986; Gordon et al., 1988).

The membrane translocation event is likely to involve portions of both PA63 and LF or EF. Evidence that PA63 alone can form ion-conductive channels was presented in an earlier section. Speculation that the amino terminus of LF initiates translocation by entering a protein-conductive pore formed by PA63 (Arora et al., 1992) requires modification based on data from a larger group of LF fusion proteins (Arora and Leppla, 1994). Fusions in which the amino terminus of LF is extended by 10 amino acids introduced by recombinant manipulations or, more dramatically, by the entire ADP-ribosylation domain of diphtheria toxin (amino acids 1–189), are not substantially decreased in toxicity relative to the original fusions. Therefore, translocation does not require the presence of a particular sequence at the amino terminus, implying that membrane insertion is initiated by an interior peptide loop. This view suggests a process like that recently proposed for membrane insertion of diphtheria toxin, based on features evident from the complete structure of that protein (Choe et al., 1992).

E. Toxin Component Action in the Cytosol and on Cells and Tissues

1. Edema Factor Action

The predictable and observed result of introducing EF into the cytosol of cells is an increase in cAMP concentration (Leppla, 1982). The increases observed in some cell types are very dramatic, with cAMP concentrations rising 1000-fold to reach 2000 μmol per milligram of cell protein (Gordon et al., 1988, 1989). This represents conversion of 20–50% of the ATP present. Cell types differ greatly in the amount of cAMP produced after ET treatment, and this does not correlate in a simple way with receptor numbers. Human polymorphonuclear neutrophils (PMNs) demonstrate only a five- to eightfold increase in cAMP levels (O'Brien et al., 1985; Wade et al., 1985). These differences could reflect cell-specific differences in compartmentalization, cAMP phosphodiesterase content, EF stability, etc. In CHO cells, EF appears to have a half-life of less than 2 h (Leppla, 1982).

Increases in cAMP concentration have many effects on cellular metabolism, all of which are believed to be mediated through cAMP-dependent protein kinase. Even very large increases in cAMP are not cytotoxic to cells. The alterations in eukaryotic cells caused by cAMP that have proven relevance to bacterial virulence are those seen in phagocytic cells (Confer and Eaton, 1982). Treatment of human PMNs with ET inhibits phagocytosis of opsonized vegetative *B. anthracis* (O'Brien et al., 1985), an effect that is reversible upon toxin removal. ET stimulates the chemotactic response of these cells to *N*-formyl-Met-Leu-Phe (Wade et al., 1985). Surprisingly, LT has the same effect, a response that has not been explained. Subsequently, it was shown that ET blocks priming of PMNs by muramyl dipeptide or lipopolysaccharide, decreasing the release of superoxide anion caused by later stimulation with the chemotactic peptide (Wright et al., 1988).

2. LF Action

Although a large body of data exists on the effects of anthrax toxin on animals (Stephen, 1986), this did not lead to productive hypotheses about the intracellular mechanism of LF. Mouse macrophages provide a useful system in which to identify the early changes occurring after toxin treatment (Friedlander, 1986), and it is in this system that important progress has been achieved (Bhatnagar et al., 1989; Hanna et al., 1992, 1993). The first changes detected in RAW264.7 cells are increases in permeability to K^+ and Rb^+ ions, beginning 45 min after LT addition (Hanna et al., 1992). Beginning 60 min after LT, ATP levels drop dramatically and permeability to other ions increases. At 75 min, gross changes in morphology are evident, and at 90 min the cells begin to lyse. The events beginning at 60 min require extracellular calcium (Bhatnagar et al., 1989) and are blocked by osmotic stabilizing agents such as 0.3 M sucrose (Hanna et al., 1992). It was suggested that the increases in permeability to Na^+ and water cause depletion of ATP, which in turn leads to an influx of Ca^{2+} that produces further damage leading to lysis. This view suggests that LF acts in the cytosol to damage ion pumps or to deregulate ion channels. In view of the evidence that LF is a metalloprotease, it is interesting to note that certain voltage-gated K^+ channels can be rendered permanently open by proteolysis that cleaves the peptide "chain" in the "ball-and-chain" description of the gate (Miller, 1991). However, the evidence that PA63 alone causes increases in ion permeability in cells (Milne and Collier, 1993) suggests that future work will be needed to clearly distinguish the effects of LT from those of PA63.

Evidence that macrophages play a key role in LT action in animals was obtained from studies in BALB/c mice (Hanna et al., 1993). Depletion of macrophages by repeated injection of silica renders mice resistant to LT. Sensitivity is restored by injection of 10^8 RAW264.7 cells into the mice along with the LT. Administration of the same number of cells of the LT-resistant mouse macrophage line IC-21 (Singh et al., 1989b) does not restore sensitivity. Because the symptoms of animals dying of anthrax toxin resemble those of septic shock, measurements were made of cytokine induction. Remarkably low concentrations of LT, 10^{-9} μg/mL LF with 0.1 μg/mL PA, induced synthesis of high levels of tumor necrosis factor (TNF). Based on the hypothesis that death of the mice was due to high levels of TNF and interleukin-1, the effectiveness of antisera to these cytokines was tested. Animals were partially protected by each antiserum alone, and completely protected when the two antisera were combined. This provides convincing evidence that it is induction and/or release of excessive amounts of IL-1 and TNF from macrophages that causes the death of toxin-treated animals.

The increased understanding of the response of macrophages to LT may help to explain differences in toxin sensitivity between different animal species. Resistance to infection appears to be inversely correlated among animal species with resistance to toxin. If resistance to infection is due largely to the number or availability of macrophages, then that protective mechanism may provide a larger target for a toxin that is specific for macrophages (Hanna et al., 1993). Because the Fischer 344 rat is uniquely sensitive to LT, even compared to other rat strains (Beall et al., 1962), it will be interesting to test the macrophages of this strain.

Some primary and immortalized mouse macrophage cells are totally resistant to LT (Singh et al., 1989b; Friedlander et al., 1993). Macrophages from A/J mice are totally resistant, whereas macrophages from C3H/HeJ mice are very sensitive (Friedlander et al., 1993). Similarly, the established line IC-21 is resistant, whereas the J774A.1, RAW264.7, and other lines are sensitive. Resistant cells have a normal number of receptors, activate PA, and bind and internalize LF and EF. Introduction of LF into the cytosol by osmotic lysis of pinosomes does not lead to cell death. Therefore, the resistance of the A/J and IC-21 cells appears to result from an absence of the cytosolic target of LF. If will be important to compare sensitive and resistant cells for their content of any proteins tentatively identified as intracellular targets of LF.

VI. SUMMARY AND FUTURE DIRECTIONS

Rapid progress has occurred in characterizing the structure and function of anthrax toxin, and this will continue as more investigators become interested in this unique set of proteins. Identification of the enzymatic mechanism of action of LF, which can be expected to occur in the near future, can be viewed as bringing to a close the first phase of research on this toxin. Some researchers will then wish to focus on characterizing in detail the binding interactions of the proteins, and others will seek to identify the cellular receptor for PA.

A detailed understanding of structural domains of the toxin provides an opportunity to exploit these domains in the construction of therapeutic agents, as has been done with PE, diphtheria toxin, ricin, and other protein toxins (Pastan et al., 1992). The LF fusion proteins described above provide a promising starting point in this work. Efforts will be needed to redirect anthrax toxin from its native receptor to other specific receptors to obtain cell-type-specific cytotoxins. The extensive work with other toxin-based therapeutic agents suggests that this can be achieved.

Acknowledgments

I thank the members of the Laboratory of Microbial Ecology, National Institute of Dental Research, for insights and stimulating discussions that have contributed to the preparation of this review. I thank Jerry Keith for his continuing support and encouragement.

REFERENCES

Aktories, K., Barmann, M., Ohishi, I., Tsuyama, S., Jakobs, K. H., and Habermann, E. (1986). Botulinum C_2 toxin ADP-ribosylates actin. *Nature 322*: 390–392.

Allured, V. S., Case, L. M., Leppla, S. H., and McKay, D. B. (1985). Crystallization of the protective antigen protein of *Bacillus anthracis*. *J. Biol. Chem. 260*: 5012–5013.

Arora, N., and Leppla, S. H. (1993). Residues 1–254 of anthrax toxin lethal factor are sufficient to cause cellular uptake of fused polypeptides. *J. Biol. Chem. 268*: 3334–3341.

Arora, N., and Leppla, S. H. (1994). Fusions of anthrax toxin lethal factor with Shiga toxin and diphtheria toxin enzymatic domains are toxic to mammalian cells. *Infect Immun. 62*: 4955–4961.

Arora, N., Klimpel, K. R., Singh, Y., and Leppla, S. H. (1992). Fusions of anthrax toxin lethal factor to the ADP-ribosylation domain of *Pseudomonas* exotoxin A are potent cytotoxins which are translocated to the cytosol of mammalian cells. *J. Biol. Chem. 267*: 15542–15548.

Ash, C., and Collins, M. D. (1992). Comparative analysis of 23S ribosomal RNA gene sequences of *Bacillus anthracis* and emetic *Bacillus cereus* determined by PCR direct sequencing. *FEMS Microbiol. Lett. 73*: 75–80.

Ash, C., Farrow, J. A., Dorsch, M., Stackebrandt, E., and Collins, M. D. (1991). Comparative analysis of *Bacillus anthracis*, *Bacillus cereus*, and related species on the basis of reverse transcriptase sequencing of 16S rRNA. *Int. J. Syst. Bacteriol. 41*: 343–346.

Au, D. C., Masure, H. R., and Storm, D. R. (1989). Site directed mutagenesis of lysine 58 in a putative ATP-binding domain of the calmodulin-sensitive adenylate cyclase from *Bordetella pertussis* abolishes catalytic activity. *Biochemistry 28*: 2772–2776.

Bartkus, J. M., and Leppla, S. H. (1989). Transcriptional regulation of the protective antigen gene of *Bacillus anthracis*. *Infect. Immun. 57*: 2295–2300.

Beall, F. A., Taylor, M. J., and Thorne, C. B. (1962). Rapid lethal effect in rats of a third component found upon fractionating the toxin of *Bacillus anthracis*. *J. Bacteriol. 83*: 1274–1280.

Berry, L. J. (1967). General summary of the anthrax conference. *Fed. Proc. 26*: 1569–1570.

Bhatnagar, R., Singh, Y., Leppla, S. H., and Friedlander, A. M. (1989). Calcium is required for the expression of anthrax lethal toxin activity in the macrophagelike cell line J774A.1. *Infect. Immun. 57*: 2107–2114.

Blaustein, R. O., and Finkelstein, A. (1990a). Voltage-dependent block of anthrax toxin channels in planar phospholipid bilayer membranes by symmetric tetraalkylammonium ions. Effects on macroscopic conductance. *J. Gen. Physiol. 96*: 905–919.

Blaustein, R. O., and Finkelstein, A. (1990b). Diffusion limitation in the block by symmetric tetraalkylammonium ions on anthrax toxin channels in planar phospholipid bilayer membranes. *J. Gen. Physiol. 96*: 943–957.

Blaustein, R. O., Koehler, T. M., Collier, R. J., and Finkelstein, A. (1989). Anthrax toxin: channel-forming activity of protective antigen in planar phospholipid bilayers. *Proc. Natl. Acad. Sci. U.S.A. 86*: 2209–2213.

Blaustein, R. O., Lea, E. J., and Finkelstein, A. (1990). Voltage-dependent block of anthrax toxin channels in planar phospholipid bilayer membranes by symmetric tetraalkylammonium ions. Single-channel analysis. *J. Gen. Physiol. 96*: 921–942.

Bouhss, A., Krin, E., Munier, H., Gilles, A. M., Danchin, A., Glaser, P. and Bârzu, O. (1993). Cooperative phenomena in binding and activation of *Bordetella pertussis* adenylate cyclase by calmodulin. *J. Biol. Chem. 268*: 1690–1694.

Bragg, T. S., and Robertson, D. L. (1989). Nucleotide sequence and analysis of the lethal factor gene (*lef*) from *Bacillus anthracis*. *Gene 81*: 45–54.

Cataldi, A., Labruyere, E., and Mock, M. (1990). Construction and characterization of a protective antigen-deficient *Bacillus anthracis* strain. *Mol. Microbiol. 4*: 1111–1117.

Cataldi, A., Fouet, A., and Mock, M. (1992). Regulation of *pag* gene expression in *Bacillus anthracis*—use of a *pag-lacZ* transcriptional fusion. *FEMS Microbiol. Lett. 98*: 89–93.

Choe, S., Bennett, M. J., Fujii, G., Curmi, P. M., Kantardjieff, K. A., Collier, R. J., and Eisenberg, D. (1992). The crystal structure of diphtheria toxin. *Nature 357*: 216–222.

Confer, D. L., and Eaton, J. W. (1982) Phagocyte impotence caused by the invasive bacterial adenylate cyclase. *Science 217*: 948–950.

Escuyer, V., and Collier, R. J. (1991). Anthrax protective antigen interacts with a specific receptor on the surface of CHO-K1 cells. *Infect. Immun. 59*: 3381–3386.

Escuyer, V., Duflot, E., Sezer, O., Danchin, A., and Mock, M. (1988). Structural homology between virulence-associated bacterial adenylate cyclases. *Gene 71*: 293–298.

Ezzell, J. W., Jr., and Abshire, T. G. (1992). Serum protease cleavage of *Bacillus anthracis* protective antigen. *J. Gen. Microbiol. 138*: 543–549.

Ezzell, J. W., Ivins, B. E., and Leppla, S. H. (1984). Immunoelectrophoretic analysis, toxicity, and kinetics of in vitro production of the protective antigen and lethal factor components of *Bacillus anthracis* toxin. *Infect. Immun. 45*: 761–767.

Friedlander, A. M. (1986). Macrophages are sensitive to anthrax lethal toxin through an acid-dependent process. *J. Biol. Chem. 261*: 7123–7126.

Friedlander, A. M. (1990). Anthrax toxins. In *Trafficking of Bacterial Toxins*, C. B. Saelinger (Ed.), CRC Press, Boca Raton, FL, pp. 121–128.

Friedlander, A. M., Bhatnagar, R., Leppla, S. H., Johnson, L., and Singh, Y. (1993). Characterization of macrophage sensitivity and resistance to anthrax lethal toxin. *Infect. Immun. 61*: 245–252.

Gilbert, J. V., Plaut, A. G., and Wright A. (1991). Analysis of the immunoglobulin A protease gene of *Streptococcus sanguis*. *Infect. Immun. 59*: 7–17.

Gill, D. M. (1978). Seven toxin peptides that cross cell membranes. In *Bacterial Toxins and Cell Membranes*, J. Jeljaszewicz and T. Wadstrom (Eds.), Academic, New York, pp. 291–332.

Gladstone, G. P. (1946). Immunity to anthrax. Protective antigen present in cell-free culture filtrates, *Br. J. Exp. Pathol. 27*: 349–418.

Glaser, P., Elmaoglou-Lazaridou, A., Krin, E., Ladant, D., Bârzu, O., and Danchin, A. (1989). Identification of residues essential for catalysis and binding of calmodulin in *Bordetella pertussis* adenylate cyclase by site-directed mutagenesis. *EMBO J. 8*: 967–972.

Glaser, P., Munier, H., Gilles, A. M., Krin, E., Porumb, T., Bârzu, O., Sarfati, R., Pellecuer, C., and Danchin, A. (1991). Functional consequences of single amino acid substitutions in calmodulin-activated adenylate cyclase of *Bordetella pertussis*. *EMBO J. 10*: 1683–1688.

Gordon, V. M., Leppla, S. H., and Hewlett, E. L. (1988). Inhibitors of receptor-mediated endocytosis block the entry of *Bacillus anthracis* adenylate cyclase toxin but not that of *Bordetella pertussis* adenylate cyclase toxin. *Infect. Immun. 56*: 1066–1069.

Gordon, V. M., Young, W. W., Jr., Lechler, S. M., Gray, M. C., Leppla, S. H., and Hewlett, E. L. (1989). Adenylate cyclase toxins from *Bacillus anthracis* and *Bordetella pertussis*. Different processes for interaction with and entry into target cells. *J. Biol. Chem. 264*: 14792–14796.

Green, B. D., Battisti, L., Koehler, T. M., Thorne, C. B., and Ivins, B. E. (1985). Demonstration of a capsule plasmid in *Bacillus anthracis*. *Infect. Immun. 49*: 291–297.

Guan, K. L., and Dixon, J. E. (1991). Eukaryotic proteins expressed in *Escherichia coli*: an improved thrombin cleavage and purification procedure of fusion proteins with glutathione S-transferase. *Anal. Biochem. 192*: 262–267.

Haines, B. W., Klein, F., and Lincoln, R. E. (1965). Quantitative assay for crude anthrax toxins. *J. Bacteriol. 89*: 74–83.

Hambleton, P., and Turnbull, P. C. (1990). Anthrax vaccine development: a continuing story. *Adv. Biotechnol. Processes 13*: 105–122.

Hambleton, P., Carman, J. A., and Melling, J. (1984). Anthrax: the disease in relation to vaccines. *Vaccine 2*: 125–132.

Hanna, P. C., Kouchi, S., and Collier, R. J. (1992). Biochemical and physiological changes induced by anthrax lethal toxin in J774 macrophage-like cells. *Mol. Biol. Cell 3*: 1269–1277.

Hanna, P. C., Acosta, D., and Collier, R. J. (1993). On the rol of macrophages in anthrax. *Proc. Natl. Acad. Sci. U.S.A. 90*: 10198–10201.

Hanski, E., and Coote, J. G. (1991). *Bordetella pertussis* adenylate cyclase toxin. In *Sourcebook of Bacterial Protein Toxins*, J. E. Alouf and J. H. Freer (Eds.), Academic, London, pp. 349–366.

Harris-Smith, P. W., Smith, H., and Keppie, J. (1958). Production *in vitro* of the toxin of *Bacillus anthracis* previously recognized *in vivo*. *J. Gen. Microbiol. 19*: 91–103.

Hornung, J. M., and Thorne, C. B. (1991). Insertion mutations affecting pXO1-associated toxin production in *Bacillus anthracis*, Abstr. *91st Annu. Meet. Am. Soc. Microbiol. 98*, p. 98, Abstr. D-121.

Iacono-Connors, L. C., Schmaljohn, C. S., and Dalrymple, J. M. (1990). Expression of the *Bacillus anthracis* protective antigen gene by baculovirus and vaccinia virus recombinants. *Infect. Immun. 58*: 366–372.

Iacono-Connors, L. C., Welkos, S. L., Ivins, B. E., and Dalrymple, J. M. (1991). Protection against anthrax with recombinant virus-expressed protective antigen in experimental animals. *Infect. Immun. 59*: 1961–1965.

Inocencio, N. M., Moehring, J. M., and Moehring, T. J. (1993). A mutant CHO-K1 strain with resistance to *Pseudomonas* exotoxin A is unable to process the precursor fusion glycoprotein of Newcastle disease virus. *J. Virol. 67*: 593–595.

Ivins, B. E., and Welkos, S. L. (1986). Cloning and expression of the *Bacillus anthracis* protective antigen gene in *Bacillus subtilis*. *Infect. Immun. 54*: 537–542.

Ivins, B. E., Welkos, S. L., Knudson, G. B., and Little, S. F. (1990). Immunization against anthrax with aromatic compound-dependent (Aro-) mutants of *Bacillus anthracis* and with recombinant strains of *Bacillus subtilis* that produce anthrax protective antigen. Infect. Immun. 58: 303–308.

Johnson, R. A., and Shoshani, I. (1990). Inhibition of *Bordetella pertussis* and *Bacillus anthracis* adenylyl cyclases by polyadenylate and "P"-site agonists. *J. Biol. Chem. 265*: 19035–19039.

Jongeneel, C. V., Bouvier, J., and Bairoch, A. (1989). A unique signature identifies a family of zinc-dependent metallopeptidases. *FEBS Lett. 242*: 211–214.

Kaneko, T., Nozaki, R., and Aizawa, K. (1978). Deoxyribonucleic acid relatedness between *Bacillus anthracis*, *Bacillus cereus* and *Bacillus thuringiensis*. *Microbiol. Immunol. 22*: 639–641.

Kaspar, R. L., and Robertson, D. L. (1987). Purification and physical analysis of *Bacillus anthracis* plasmids pXO1 and pXO2. *Biochem. Biophys. Res. Commun. 149*: 362–368.

Kawamura F., and Doi, R. H. (1984). Construction of a *Bacillus subtilis* double mutant deficient in extracellular alkaline and neutral proteases. *J. Bacteriol. 160*: 442–444.

Keppie, J., Harris-Smith, P. W., and Smith, H. (1963). The chemical basis of the virulence of *Bacillus anthracis*. IX. Its aggressins and their mode of action. *Br. J. Exp. Pathol. 44*: 446–453.

Klimpel, K. R., Molloy, S. S., Bresnahan, P. A., Thomas, G., and Leppla, S. H. (1992a). Cleavage of diphtheria toxin and the protective antigen of *Bacillus anthracis* by the eukaryotic protease furin. *Abstr. 92nd Annu. Meet. AM. Soc. Microbiol. 31*, p. 31, Abstr. B-32.

Klimpel, K. R., Molloy, S. S., Thomas, G., and Leppla, S. H. (1992b). Anthrax toxin protective antigen is activated by a cell-surface protease with the sequence specificity and catalytic properties of furin. *Proc. Natl. Acad. Sci. U.S.A. 89*: 10277–10281.

Klimpel, K. R., Arora, N., and Leppla, S. H. (1993). Anthrax toxin lethal factor has homology to the thermolysin-like proteases and displays proteolytic activity. *Annu. Meet. Am. Soc. Microbiol. 45*, Abstr. B-111.

Klimpel, K. R., Arora, N., and Leppla, S. H. (1994). Anthrax toxin lethal factor contains a zinc metalloprotease consensus sequence which is required for lethal toxin activity. *Mol. Microbiol. 13:* 1093–1100.

Koehler, T. M., and Collier, R. J. (1991). Anthrax toxin protective antigen: low-pH-induced hydrophobicity and channel formation in liposomes. *Mol. Microbiol. 5*: 1501–1506.

Labruyere, E., Mock, M., Ladant, D., Michelson, S., Gilles, A. M., Laoide, B., and Bârzu, O. (1990). Characterization of ATP and calmodulin-binding properties of a truncated form of *Bacillus anthracis* adenylate cyclase. *Biochemistry 29*: 4922–4928.

Labruyere, E., Mock, M., Surewicz, W. K., Mantsch, H. H., Rose, T., Munier, H., Sarfati, R. S., and Bârzu, O. (1991). Structural and ligand-binding properties of a truncated form of *Bacillus anthracis* adenylate cyclase and of a catalytically inactive variant in which glutamine substitutes for lysine-346. *Biochemistry 30*: 2619–2624.

Ladant, D. (1988). Interaction of *Bordetella pertussis* adenylate cyclase with calmodulin: identification of two separate calmodulin-binding domains. *J. Biol. Chem. 263*: 2612–2618.

Ladant, D., Glaser, P., and Ullmann, A. (1992). Insertional mutagenesis of *Bordetella pertussis* adenylate cyclase. *J. Biol. Chem. 267*: 2244–2250.

Larson, D. K., Calton, G. J., Little, S. F., Leppla, S. H., and Burnett, J. W. (1988). Separation of three exotoxic factors of *Bacillus anthracis* by sequential immunosorbent chromatography. *Toxicon 26*: 913–921.

Leppla, S. H. (1982). Anthrax toxin edema factor: a bacterial adenylate cyclase that increases cyclic AMP concentrations of eukaryotic cells. *Proc. Natl. Acad. Sci. U.S.A. 79*: 3162–3166.

Leppla, S. H. (1984). *Bacillus anthracis* calmodulin-dependent adenylate cyclase: chemical and enzymatic properties and interactions with eucaryotic cells. *Adv. Cyclic Nucleotide Protein Phosphorylation Res. 17*: 189–198.

Leppla, S. H. (1988). Production and purification of anthrax toxin. *Methods Enzymol. 165*: 103–116.

Leppla, S. H. (1991a). The anthrax toxin complex. In *Sourcebook of Bacterial Protein Toxins*, J. E. Alouf and J. H. Freer (Eds.), Academic, London, pp. 277–302.

Leppla, S. H. (1991b). Purification and characterization of adenylyl cyclase from *Bacillus anthracis*. *Methods Enzymol. 195*: 153–168.

Leppla, S. H., Ivins, B. E., and Ezzell, J. W., Jr. (1985). Anthrax toxin. In *Microbiology—1985*, L. Leive, P. F. Bonventre, J. A. Morello, S. Schlessinger, S. D. Silver, and H. C. Wu (Eds.), Am. Soc. Microbiology, Washington, DC, pp. 63–66.

Leppla, S. H., Friedlander, A. M., and Cora, E. M. (1988). Proteolytic activation of anthrax toxin bound to cellular receptors. In *Bacterial Protein Toxins* F. J. Fehrenbach, J. E. Alouf, P. Falmagne, W. Goebel, J. Jeljaszewicz, D. Jurgen, and R. Rappuoli (Eds.), Gustav Fischer, New York, pp. 111–112.

Leppla, S. H., Friedlander, A. M., Singh, Y., Cora, E. M., and Bhatnagar, R. (1990). A model for anthrax toxin action at the cellular level. *Salisbury Med. Bull. 68 (Spec. Suppl.)*: 41–43.

Little, S. F., and Lowe, J. R. (1991). Location of receptor-binding region of protective antigen from *Bacillus anthracis*. *Biochem. Biophys. Res. Commun. 180*: 531–537.

Little, S. F., Leppla, S. H., and Cora, E. (1988). Production and characterization of monoclonal antibodies to the protective antigen component of *Bacillus anthracis* toxin. *Infect. Immun. 56*: 1807–1813.

Little, S. F., Leppla, S. H., and Friedlander, A. M. (1990). Production and characterization of monoclonal antibodies against the lethal factor component of *Bacillus anthracis* lethal toxin. *Infect. Immun. 58*: 1606–1613.

McMahon, H. T., Ushkaryov, Y. A., Edelmann, L., Link, E., Binz. T., Niemann, H., Jahn, R., and Südhof, T. C. (1993). Cellubrevin is a ubiquitous tetanus-toxin substrate homologous to a putative synaptic vesicle fusion protein. *Nature 364*: 346–349.

Mikesell, P., Ivins, B. E., Ristroph, J. D., and Dreier, T. M. (1983). Evidence for plasmid-mediated toxin production in *Bacillus anthracis*. *Infect. Immun. 39*: 371–376.

Miller, C. (1991). 1990: annus mirabilis of potassium channels. *Science 252*: 1092–1096.

Milne, J. C., and Collier, R. J. (1993). pH-dependent permeabilization of the plasma membrane of mammalian cells by anthrax toxin protective antigen. *Mol. Microbiol. 10*: 647–653.

Mock, M., Labruyere, E., Glaser, P., Danchin, A., and Ullmann, A. (1988). Cloning and expression of the calmodulin-sensitive *Bacillus anthracis* adenylate cyclase in *Escherichia coli. Gene 64*: 277–284.

Moehring, J. M., Inocencio, N. M., Robertson, B. J., and Moehring, T. J. (1993). Expression of mouse furin in a Chinese hamster cell resistant to *Pseudomonas* exotoxin A and viruses complements the genetic lesion. *J. Biol. Chem. 268*: 2590–2594.

Molloy, S. S., Bresnahan, P. A., Leppla, S. H., Klimpel, K. R., and Thomas, G. (1992). Human furin is a calcium-dependent serine endoprotease that recognizes the sequence Arg-X-X-Arg and efficiently cleaves anthrax toxin protective antigen. *J. Biol. Chem. 267*: 16396–16402.

Molnar, D. M., and Altenbern, R. A. (1963). Alterations in the biological activity of protective antigen of *Bacillus anthracis* toxin. *Proc. Soc. Exp. Biol. Med. 114*: 294–297.

Munier, H., Bouhss, A., Krin, E., Danchin, A., Gilles, A. M., Glaser, P., and Bârzu, O. (1992). The role of histidine 63 in the catalytic mechanism of *Bordetella pertussis* adenylate cyclase. *J. Biol. Chem. 267*: 9816–9820.

Munier, H., Blanco, F. J. Prêcheur, B., Diesis, E., Nieto, J. L., Craescu, C. T., and Bârzu, O. (1993). Characterization of a synthetic calmodulin-binding peptide derived from *Bacillus anthracis* adenylate cyclase. *J. Biol. Chem. 268*: 1695–1701.

Novak, J. M., Stein, M. P., Little, S. F., Leppla, S. H., and Friedlander, A. M. (1992). Functional characterization of protease-treated *Bacillus anthracis* protective antigen. *J. Biol. Chem. 267*: 17186–17193.

O'Brien, J., Friedlander, A., Dreier, T., Ezzell, J., and Leppla, S. (1985). Effects of anthrax toxin components on human neutrophils. *Infect. Immun. 47*: 306–310.

Ogata, M., Fryling, C. M., Pastan, I., and FitzGerald, D. J. (1992). Cell-mediated cleavage of *Pseudomonas* exotoxin between Arg279 and Gly280 generates the enzymatically active fragment which translocates to the cytosol. *J. Biol. Chem. 267*: 25396–25401.

Ohishi, I. (1987). Activation of botulinum C_2 toxin by trypsin. *Infect. Immun. 55*: 1461–1465.

Ohishi, I., Iwasaki, M., and Sakaguchi, G. (1980). Purification and characterization of two components of botulinum C_2 toxin. *Infect. Immun. 30*: 668–673.

Pastan, I., Chaudhary, V., and FitzGerald, D. J. (1992). Recombinant toxins as novel therapeutic agents. *Annu. Rev. Biochem. 61*: 331–354.

Perelle, S., Gibert, M., Boquet, P., and Popoff, M. R. (1993). Characterization of *Clostridium perfringens* iota-toxin genes and expression in *Escherichia coli. Infect. Immun. 61*: 5147–5156.

Pezard, C., Berche, P., and Mock, M. (1991). Contribution of individual toxin components to virulence of *Bacillus anthracis. Infect. Immun. 59* 3472–3477.

Pezard, C., Duflot, E., and Mock, M. (1993). Construction of *Bacillus anthracis* mutant strains producing a single toxin component. *J. Gen. Microbiol. 139*: 2459–2463.

Pollack, M. R., and Richmond, M. H. (1962). Low cysteine content of bacterial extracellular proteins: its possible physiological significance. *Nature 194*: 446–449.

Puziss, M., Manning, L. C., Lynch, J. W., Barclay, E., Abelow, I., and Wright, G. G. (1963). Large-scale production of protective antigen of *Bacillus anthracis* in anaerobic cultures. *Appl. Microbiol. 11*: 330–334.

Quinn, C. P., Shone, C. C., Turnbull, P. C., and Melling, J. (1988). Purification of anthrax-toxin components by high-performance anion-exchange, gel-filtration and hydrophobic-interaction chromatography. *Biochem. J. 252*: 753–758.

Quinn, C. P., Singh, Y., Klimpel, K. R., and Leppla, S. H. (1991). Functional mapping of anthrax toxin lethal factor by in-frame insertion mutagenesis. *J. Biol. Chem. 266*: 20124–20130.

Ristroph, J. D., and Ivins, B. E. (1983). Elaboration of *Bacillus anthracis* antigens in a new, defined culture medium. *Infect. Immun. 39*: 483–486.

Robertson, D. L. (1988). Relationships between the calmodulin-dependent adenylate cyclases produced by *Bacillus anthracis* and *Bordetella pertussis. Biochem. Biophys. Res. Commun. 157*: 1027–1032.

Robertson, D. L., and Leppla, S. H. (1986). Molecular cloning and expression in *Escherichia coli* of the lethal factor gene of *Bacillus anthracis*. *Gene 44*: 71–78.

Robertson, D. L., Tippetts, M. T., and Leppla, S. H. (1988). Nucleotide sequence of the *Bacillus anthracis* edema factor gene (*cya*): a calmodulin-dependent adenylate cyclase. *Gene 73*: 363–371.

Robertson, D. L., Bragg, T. S., Simpson, S., Kaspar, R., Xie, W., and Tippetts, M. T. (1990). Mapping and characterization of *Bacillus anthracis* plasmids pXO1 and pXO2. *Salisbury Med. Bull. 68 (Spec. Suppl.)*: 55–58.

Sandvig, K., and Olsnes, S. (1981). Rapid entry of nicked diphtheria toxin into cells at low pH. Characterization of the entry process and effects of low pH on the toxin molecule. *J. Biol. Chem. 256*: 9068–9076.

Sarfati, R. S., Kansal, V. K., Munier, H., Glaser, P., Gilles, A. M., Labruyere, E., Mock, M., Danchin, A., and Bârzu, O. (1990). Binding of 3′-anthraniloyl-2′-deoxy-ATP to calmodulin-activated adenylate cyclase from *Bordetella pertussis* and *Bacillus anthracis*. *J. Biol. Chem. 265*: 18902–18906.

Schiavo, G., Benfenati, F., Poulain, B., Rossetto, O., Polverino de Laureto, P., Dasgupta, B. R., and Montecucco, C. (1992). Tetanus and botulinum-B neurotoxins block neurotransmitter release by proteolytic cleavage of synaptobrevin. *Nature 359*: 832–835.

Singh, Y., Chaudhary, V. K., and Leppla, S. H. (1989a). A deleted variant of *Bacillus anthracis* protective antigen is non-toxic and blocks anthrax toxin action *in vivo*. *J. Biol. Chem. 264*: 19103–19107.

Singh, Y., Leppla, S. H., Bhatnagar, R., and Friedlander, A. M. (1989b). Internalization and processing of *Bacillus anthracis* lethal toxin by toxin-sensitive and -resistant cells. *J. Biol. Chem. 264*: 11099–11102.

Singh, Y., Leppla, S. H., Bhatnagar, R., and Freidlander, A. M. (1990). Basis of cellular sensitivity and resistance to anthrax lethal toxin. *Salisbury Med. Bull. 68*: 46–48.

Singh, Y., Klimpel, K. R., Quinn, C. P., Chaudhary, V. K., and Leppla, S. H. (1991). The carboxyl-terminal end of protective antigen is required for receptor binding and anthrax toxin activity. *J. Biol. Chem. 266*: 15493–15497.

Smith, H., and Keppie, J. (1954). Observations on experimental anthrax: demonstration of a specific lethal factor produced *in vivo* by *Bacillus anthracis*. *Nature 173*: 869–870.

Smith, H., and Stanley, J. L. (1962). Purification of the third factor of anthrax toxin. *J. Gen. Microbiol. 29*: 517–521.

Smith, H., Keppie, J., and Stanley, J. L. (1955). The chemical basis of the virulence of *Bacillus anthracis*. V. The specific toxin produced by *B. anthracis* in vivo. *Br. J. Exp. Pathol. 36*: 460–472.

Stanley, J. L., and Smith, H. (1961). Purification of factor I and recognition of a third factor of anthrax toxin. *J. Gen. Microbiol. 26*: 49–66.

Stanley, J. L., and Smith, H. (1963). The three factors of anthrax toxin: their immunogenicity and lack of demonstrable enzymic activity. *J. Gen. Microbiol. 31*: 329–337.

Stark, W., Pauptit, R. A., Wilson, K. S., and Jansonius, J. N. (1992). The structure of neutral protease from *Bacillus cereus* at 0.2-nm resolution. *Eur. J. Biochem. 207*: 781–791.

Steiner, D. F., Smeekens, S. P., Ohagi, S., and Chan, S. J. (1992). The new enzymology of precursor processing endoproteases. *J. Biol. Chem. 267*: 23435–23438.

Stephen, J. (1981). Anthrax toxin. *Pharmacol. Ther. 12*: 501–513.

Stephen, J. (1986). Anthrax toxin. In *Pharmacology of Bacterial Toxins*, F. Dorner and J. Drews (Eds.), Pergamon, Oxford, pp. 381–395.

Strange, R. E., and Thorne, C. B. (1958). Further purification of the protective antigen of *Bacillus anthracis* produced *in vitro*. *J. Bacteriol. 76*: 192–201.

Südhof, T. C., De Camilli, P., Niemann, H., and Jahn, R. (1993). Membrane fusion machinery: insights from synaptic proteins. *Cell 75*: 1–4.

Thorne, C. B. (1960). Biochemical properties of virulent and avirulent strains of *Bacillus anthracis*. *Ann. N.Y. Acad. Sci. 88*: 1024–1033.

Thorne, C. B. (1985). Genetics of *Bacillus anthracis*. In *Microbiology—1985*, L. Lieve, P. F. Bonventre, J. A. Morello, S. Schlessinger, S. D. Silver, and H. C. Wu (Eds.), Am. Soc. Microbiology, Washington, DC, pp. 56–62.

Thorne, C. B. (1993). *Bacillus anthracis*. In *Bacillus subtilis and Other Gram-Positive Bacteria*, A. B. Sonenshein, J. A. Hoch, and R. Losick (Eds.), Am. Soc. Microbiology, Washington, DC, pp. 113–124.

Thorne, C. B., Molnar, D. M., and Strange, R. E. (1960). Production of toxin *in vitro* by *Bacillus anthracis* and its separation into two components. *J. Bacteriol. 79*: 450–455.

Tippetts, M. T., and Robertson, D. L. (1988). Molecular cloning and expression of the *Bacillus anthracis* edema factor toxin gene: a calmodulin-dependent adenylate cyclase. *J. Bacteriol. 170*: 2263–2266.

Turnbull, P. C. B. (1986). Thoroughly modern anthrax. *Abstr. Hyg. Trop. Med. 61*: R1–R13.

Turnbull, P. (1990a). Anthrax. In *Topley and Wilson's Principles of Bacteriology, Virology and Immunity*, Vol. 3, *Bacterial Diseases*, G. R. Smith and C. R. Easmon (Eds.), Edward Arnold, Sevenoaks, Kent, United Kingdom, pp. 364–377.

Turnbull, P. C. B. (1990b). Proceedings of the International Workshop on Anthrax. *Salisbury Med. Bull. 68 (Spec. Suppl.)*

Turnbull, P. C. B. (1990c). Terminal bacterial and toxin levels in the blood of guinea pigs dying of anthrax. *Salisbury Med. Bull. 68*: 53–55.

Turnbull, P. C. B. (1992). Anthrax vaccines: past, present and future. *Vaccine 9*: 533–539.

Uchida I., Sekizaki, T., Hashimoto, K., and Terakado, N. (1985). Association of the encapsulation of *Bacillus anthracis* with a 60-megadalton plasmid. *J. Gen. Microbiol. 131*: 363–367.

Uchida, I., Hashimoto, K., and Terakado, N. (1986). Virulence and immunogenicity in experimental animals of *Bacillus anthracis* strains harbouring or lacking 110 MDa and 60 MDa plasmids. *J. Gen. Microbiol. 132*: 557–559.

Uchida, I., Hashimoto, K., Makino, S., Sasakawa, C., Yoshikawa, M., and Teradado, N. (1987). Restriction map of a capsule plasmid of *Bacillus anthracis*. *Plasmid 18*: 178–181.

Uchida, I., Hornung, J. M., Thorne, C. B., Klimpel, K. R., and Leppla, S. H. (1993). Cloning and characterization of a gene whose product is a transactivator of anthrax toxin synthesis. *J. Bacteriol. 175*: 5329–5338.

Umezawa, H., Aoyagi, T., Suda, H., Hamada, M., and Takeuchi, T. (1976). Bestatin, an inhibitor of aminopeptidase B, produced by actinomycetes. *J. Antibiot. 29*: 97–99.

Vallee, B. L., and Auld, D. S. (1990). Zinc coordination, function, and structure of zinc enzymes and other proteins. *Biochemistry 29*: 5647–5659.

van de Ven, W. J., Voorberg, J., Fontijn, R., Pannekoek, H., van den Ouweland, A. M., van Duijnhoven, H. L., Roebroek, A. J., and Siezen, R. J. (1990). Furin is a subtilisin-like proprotein processing enzyme in higher eukaryotes, *Mol. Biol. Rep. 14*: 265–275.

Vodkin, M. H., and Leppla, S. H. (1983). Cloning of the protective antigen gene of *Bacillus anthracis*. *Cell 34*: 693–697.

Wade, B. H., Wright, G. G., Hewlett, E. L., Leppla, S. H., and Mandell, G. L. (1985). Anthrax toxin components stimulate chemotaxis of human polymorphonuclear neutrophils. *Proc. Soc. Exp. Biol. Med. 179*: 159–162.

Welkos, S. L. (1991). Plasmid-associated virulence factors of non-toxigenic (pXO1-) *Bacillus anthracis*. *Microb. Pathog. 10*: 183–198.

Welkos, S. L., Lowe, J. R., Eden-McCutchan, F., Vodkin, M., Leppla, S. H., and Schmidt, J. J. (1988). Sequence and analysis of the DNA encoding protective antigen of *Bacillus anthracis*. *Gene 69*: 287–300.

Wilkie, M. H., and Ward, M. K. (1967). Characterization of anthrax toxin. *Fed. Proc. 26*: 1527–1531.

Williams, D. P., Wen, Z., Watson, R. S., Boyd, J., Strom, T. B., and Murphy, J. R. (1990). Cellular processing of the interleukin-2 fusion toxin DAB486-IL-2 and efficient delivery of diphtheria fragment A to the cytosol of target cells requires Arg194. *J. Biol. Chem. 265*: 20673–20677.

Wright, G. G., and Angelety, L. H. (1971). Effect of the method of agitation on the accumulation of protective antigen in cultures of *Bacillus anthracis*. *Appl. Microbiol. 22*: 135–136.

Wright, G. G., Read, P. W., and Mandell, G. L. (1988). Lipopolysaccharide releases a priming substance from platelets that augments the oxidative response of polymorphonuclear neutrophils to chemotactic peptide. *J. Infect. Dis. 157*: 690–696.

Wu, X. C., Lee, W., Tran, L., and Wong, S. L. (1991). Engineering a *Bacillus subtilis* expression-secretion system with a strain deficient in six extracellular proteases. *J. Bacteriol. 173*: 4952–4958.

Xia, Z. G., and Storm, D. R. (1990). A-type ATP binding consensus sequences are critical for the catalytic activity of the calmodulin-sensitive adenylyl cyclase from *Bacillus anthracis*. *J. Biol. Chem. 265*: 6517–6520.

Youngman, P. (1990). Use of transposons and integrational vectors for mutagenesis and construction of gene fusions in *Bacillus* species. In *Molecular Biological Methods for Bacillus*, C. R. Harwood and S. M. Cutting (Eds.), Wiley, Chichester, pp. 221–266.

Zipperle, G. F., Jr., Ezzell, J. W., Jr., and Doyle, R. J. (1984). Glucosamine substitution and muramidase susceptibility in *Bacillus anthracis*. *Can J. Microbiol. 30*: 553–559.

24

Leukocidins

Masatoshi Noda

Chiba University, Chiba, Japan

I. INTRODUCTION

Leukocidins produced by *Staphylococcus aureus* and *Pseudomonas aeruginosa* that are cytotoxic to polymorphonuclear leukocytes and macrophages but not to erythrocytes and other cells are called leukocidal toxins. These toxins are clearly distinct from various cytotoxins that possess cytotoxicity not only to phagocytes but also to other cells. Leukocidal toxins are known to be important in the pathogenicity to staphylococcal and pseudomonal infections.

    Staphylococcal leukocidin consists of two protein components (S and F) that act synergistically to induce cytotoxic changes in human and rabbit polymorphonuclear leukocytes (Woodin, 1960; Soboll et al., 1973, Noda et al., 1980a). These components when tested individually are not cytotoxic. The S and F components are preferentially bound and inactivated by G_{M1} ganglioside and phosphatidylcholine, respectively (Noda et al., 1980b). Specific binding of the S component to G_{M1} ganglioside on the cell membrane of leukocidin-sensitive cells induces activation of phospholipid methyl-transferases (Noda et al., 1985) and phospholipase A_2 (Noda et al., 1982) and an increase in the number of F-component-binding sites (Noda et al., 1981). This increased number of binding sites for F component is dependent on the specific enzymatic action of S component and is correlated with increased phospholipid methylation and phospholipase A_2 activation. F component, bound to leukocidin-sensitive cells in the presence of S component, rapidly caused degradation on the cell membrane, resulting in cell lysis. This degradation of cell membrane is associated with stimulation of ouabain-insensitive Na^+,K^+-ATPase activity and inhibition of cyclic AMP-dependent protein kinase.

Pseudomonas aeruginosa often causes fatal infections in burned and other hospitalized patients. The importance of leukocidin in the pathogenesis of *P. aeruginosa* infections is becoming recognized, since polymorphonuclear leukocytes have been known to be one of the most important cells in the host defense mechanisms against pseudomonal infection.

It has been reported that two types of pseudomonal leukocidins exist whose molecular weights are around 25,000 and 42,500. The toxin with low molecular weight was initially designated leukocidin because of its strong cytotoxic activity on the leukocytes, but it was later renamed cytotoxin because it has a wide range of cytotoxic activity to various cells. The toxin of higher molecular weight is called pseudomonal leukocidin because of its specific cytotoxicity to polymorphonuclear leukocytes and macrophages but not to other cells. Most of our knowledge concerning pseudomonal leukocidin comes from the studies of Hirayama et al. (1983). Hirayama and Kato (1984) reported that leukocidin stimulated phosphatidylinositol (PI) metabolism. Although the leukocidin-stimulated PI metabolism is thought to be important for its cytotoxicity to leukocytes, little is known about the mode of action of this leukocidin at molecular level.

II. STAPHYLOCOCCAL LEUKOCIDIN

A. Purification of Leukocidin[1]

The V8 strain of *Staphylococcus aureus* (American Type Culture Collection No. 27733) is believed to produce more leukocidin than most staphylococcal strains. The cocci are grown at 37°C for 22 h on an enriched medium (Table 1) (Gladstone and van Heyningen, 1957) in a reciprocating shaker (120 cycles/min). The culture is then centrifuged at 11,000 × *g* for 20 min at 4°C, and the clear supernatant fluid (5000 mL, step 1, see Scheme 1) is purified at 4°C by the following procedure. $ZnCl_2$ (3.7 M) is added dropwise to the supernatant (pH 6.5) until a final concentration of 75 mM is reached. After 30 min at 4°C, a precipitate is formed, which is collected by centrifugation at 8000 × *g* for 15 min. The pellet is dissolved gradually in 0.4 M sodium phosphate buffer (pH 6.5). To remove metal ions, the solution is dialyzed against 0.05 M sodium acetate buffer (pH 5.2) containing 0.2 M NaCl (buffer A). Solid $(NH_4)_2SO_4$ is then added until saturation is

---

[1] This section is taken from Noda et al., 1980a.

TABLE 1 Composition of Growth Medium
(pH 7.4)

Yeast extract	25 g
Casamino acid	20 g
Sodium glycerophosphate	20 g
$Na_2HPO_4 \cdot 12H_2O$	6.25 g
KH_2PO_4	400 mg
$MgSO_4 \cdot 7H_2O$	20 mg
$MnSO_4 \cdot 4H_2O$	10 mg
Sodium lactate, 50%	19.8 mL
$FeSO_4 \cdot 7H_2O$, 0.32%(w/v), plus citric acid, 0.32%(w/v)	2 mL
Distilled water to total	1 L

reached, and the solution is allowed to stand overnight. The precipitate is collected by centrifugation at 15,000 $\times$ g for 20 min, dissolved in a small volume of buffer A, and then dialyzed against buffer A. The dialysate (315 $\times$ 10^5 U, step 2) is applied to a column (5 $\times$ 90 cm) of carboxymethyl-Sephadex C-50 equilibrated with buffer A. The fractions containing the highest leukocidin activity are eluted with 0.05 M sodium acetate buffer (pH 5.2) containing 1.2 M NaCl. The active fractions are pooled (254 $\times$ 10^5 U, step 3), saturated with $(NH_4)_2SO_4$ by addition of the salt, and allowed to stand overnight. The precipitate is collected by centrifugation, dissolved in buffer A, and then dialyzed against buffer A.

Further purification is achieved by applying the leukocidin preparation to a column (2.5 $\times$ 85 cm) of carboxymethyl-Sephadex C-50 equilibrated with buffer A and elution with a linear gradient from 0.2 to 1.2 M NaCl in 0.05 M sodium acetate buffer (pH 5.2). The fractions with F (fast) component activity, which eluted at about 0.45 M NaCl (230 $\times$ 10^5 U, step 4), and those with S component activity, which eluted at about 0.9 M NaCl (235 $\times$ 10^5 U, step 4) (Fig. 1), are pooled and concentrated by dialysis against saturated $(NH_4)_2SO_4$.

Each concentrated solution is applied to a column (2.5 $\times$ 85 cm) of Sephadex G-100 equilibrated with buffer A. The fractions with F or S component activity (F component, 217 $\times$ 10^5 U, or S component, 231 $\times$ 10^5 U, step 5) are concentrated by dialysis against saturated $(NH_4)_2SO_4$ and then dialyzed against 0.05 M veronal buffer (pH 8.6). After dialysis each component is purified by zone electrophoresis on starch [12 mA for 12 h (vessel, 1.5 $\times$ 3 $\times$ 40 cm)] using 0.05 M veronal buffer (pH 8.6).

Both the F and S fractions show a single protein peak associated with leukocidin activity. Fractions with the highest activity (F component, 210 $\times$ 10^5 U, or S component, 209 $\times$ 10^5 U, step 6) are dialyzed against buffer A and pooled for crystallization. The recovery and degree of purification at each step are shown in Table 2. The yields of the highly purified F and S components of leukocidin are about 2 mg/L of culture filtrate.

B. Crystallization of the S and F Components of Leukocidin

The S and F components of leukocidin are crystallized by dialysis against a saturated $(NH_4)_2SO_4$ solution, pH 7.0, at 4°C for 16 h. Slight opalescence appears in 6 h, along with a white precipitate at the bottom of the cellulose tube. The crystalline precipitate

1. Culture filtrate (step 1, Table 2)

2. ZnCl₂ precipitation at 4°C, pH 6.5
       | Centrifugation (8000 × g, 15 min)
    ┌──────┴──────┐
    ppt   sup

    Dissolved in 0.4 m sodium phosphate buffer (pH 6.5)
       | Centrifugation (10.000 × g. 20 min)
    ┌──────┴──────┐
    ppt   sup
                  |
                Dialyzed against 0.05 M acetate buffer (pH 5.2)
                containing 0.2 M NaCl (buffer A)
                  |
                Concentrated by saturated (NH₄)₂SO₄ solution
                (pH 7.0) (SAS)
                   | Centrifugation (15.000 × g. 20 min)
            ┌──────┴──────┐
            ppt   sup

    Dialyzed against buffer A (step 2, Table 2)

3. Carboxymethyl-Sephadex C-50 column chromatography
       | Eluted stepwise
    ├────────Effluent with buffer A
    Effluent with 0.05M acetate buffer (pH 5.2) containing 1.2 M NaCl
    (step 3, Table 2)

    Solution concentrated by SAS

4. Carboxymethyl-Sephadex C-50 column chromatography
       | Eluted with linear gradient (0.2–1.2 M NaCl in 0.05 M
       | acetate buffer, pH 5.2)
    ┌────────────────────────┐
    F fraction (eluted with about   S fraction (eluted with about
    0.45 M NaCl concentration)   0.9 M NaCl concentration
    (step 4, Table 2)   (step 4, Table 2)

5.   Concentrated by dialysis against SAS

            Sephadex G-100 gel filtration

            Eluted with buffer A (step 5, Table 2)

            Solution concentrated by dialysis against SAS

            Dialysis against 0.05 M veronal buffer (pH 8.6)

6. Starch zone electrophoresis in 0.05 M veronal buffer (pH 8.6)

        purified F   Purified S
        component   component
        (step 6, Table 2)   (step 6, Table 2)

Scheme 1 Purification procedures for leukocidin.

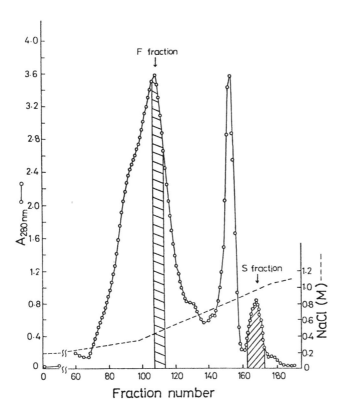

Fig. 1 The gradient elution pattern of leukocidin from CM-Sephadex C-50. CM-Sephadex C-50 column (2.5 × 85 cm) was equilibrated with 0.05 M sodium acetate buffer (pH 5.2) containing 0.2 M NaCl. The F and S fractions of leukocidin were eluted by 0.05 M sodium acetate buffer (pH 5.2) containing a linear gradient of NaCl concentration (0.2–1.2 M). Leukocidin activity of the F and S fractions was determined by the microscopic slide adhesion method. Aliquots of 3 mL were measured for A_{280} mm (○) and NaCl concentration (——) from each fraction (8 mL). The individual diagonally hatched areas of the F fraction (fractions 107–113) and S fraction (fractions 163–173) were applied to the gel filtration in step 5.

formed is collected by gentle centrifugation at 4°C and dissolved with the same volume of chilled 0.05 M sodium acetate buffer (pH 5.2) containing 0.2 M NaCl. The crystals are highly soluble in the buffer. Recrystallization of the two components is repeated twice more by dialyzing the component solution against a large volume of a 95% saturated $(NH_4)_2SO_4$ solution at 4°C. Microscopic examinations of the white precipitates of the purified S and F components of leukocidin in the dialysis bag reveals crystals in the form of very fine needles and square plates, respectively (Fig. 2).

C. Determination of Leukocidin Activity

Polymorphonuclear leukocytes are prepared from rabbit peripheral blood on an isokinetic gradient of Ficoll (Pretlow and Luberoff, 1973) and washed with Hanks' solution. The purity of polymorphonuclear leukocyte suspensions averages 98% as judged by exami-

TABLE 2 Preparation of Staphylococcal Leukocidin

Step	Volume (mL)	Total protein (mg)	Total activity (1×10^{-5} U)	Specific activity (1×10^{-5} U/mg protein)	Recovery of activity (%)	Recovery of protein (%)
1. Culture filtrate	5000	13,840	346	0.025	100	100
2. ZnCl$_2$ precipitation	300	6,330	315	0.050	91.1	45.7
3. 1.2 M NaCl effluent	1500	1,420	254	0.180	73.4	10.3
4. Gradient elution						
F component	50	138	230[a]	1.67	66.5	1.0
S component	80	47.1	235[a]	4.99	67.9	0.34
5. Gel filtration						
F component	45	39.9	217[a]	5.44	62.7	0.29
S component	40	42.0	231[a]	5.50	66.8	0.30
6. Zone electrophoresis						
F component	15	12.0	210[a]	17.50	60.7	0.09
S component	10	11.0	209[a]	19.00	60.4	0.08

[a]For the determination of the activity of each fraction, the F (or S) component (6 ng) was added into the serially diluted S (or F) fraction solution in 20 μL of phosphate-buffered saline (pH 7.2) containing 0.5% gelatin, since each fraction itself has no or little leukocidin activity.

nation of Giemsa-stained smears. Viability averages 99% as assayed by nigrosine dye exclusion. Aliquots of 10^6 cells (10 μL) are placed on glass slides. Since the F and S fractions individually have little or no leukocidin activity, to quantify the components of leukocidin, 6 ng of the S (or F) component of leukocidin is added to serially diluted fractions containing the other component.

To standardize leukocidin, the microscope slide adhesion method (Gladstone and van Heyningen, 1957) is used. Serial dilutions (10 μL) of each component of leukocidin in 0.1 M phosphate-buffered saline (pH 7.2) containing 0.5% gelatin are incubated at 37°C for 10 min with the other component (6 ng) on a glass slide (total volume, 20 μL) in a moist chamber. After incubation, morphological changes in a slide field (about 1000 cells) are observed by phase-contrast microscopy. The end point is the smallest amount of leukocidin causing about 100% morphological changes in a standard polymorphonuclear leukocyte suspension (10^6 cells), the number of units in a leukocidin preparation being numerically equal to the dilution at the end point. When 500 pg of purified F component and various amounts (100–700 pg) of purified S component are incubated with suspensions of polymorphonuclear leukocytes (10^6 cells) on the glass slide (total volume, 20 μL), optimal destruction of all polymorphonuclear leukocytes is obtained at 500 pg of S component (Fig. 3). A linear relationship between polymorphonuclear leukocyte destruction (percentage) and protein concentration is obtained in the range of 300–500 pg.

Results similar to those shown in Figure 3 are obtained when 500 pg of S component is incubated with various amounts (100–700 pg) of F component and polymorphonuclear leukocytes (10^6 cells). Optimal destruction of all polymorphonuclear leukocytes is obtained at 500 pg of F component. One unit of each component of leukocidin is obtained at 500 pg. When a suspension of rabbit polymorphonuclear leukocytes (10^6 cells) is incubated with 500 pg of both components of leukocidin, it is revealed by light microscopy that, after 2 min of the incubation, the polymorphonuclear leukocytes become round and

(a)

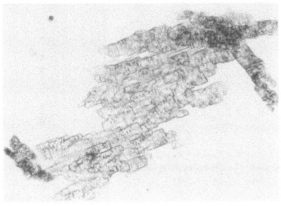

(b)

Fig. 2 Crystals of **(a)** S component and **(b)** F component after recrystallization three times at 4°C. Magnification × 200.

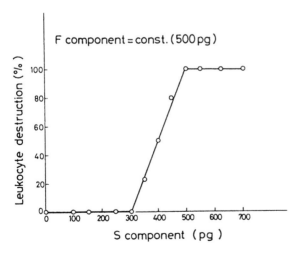

Fig. 3 Titration of the optimal leukocytolytic doses of the S and F components. Mixtures of 500 pg of F component and varying amounts (100–700 pg) of S component were incubated at 37°C for 10 min with suspensions of leukocytes (1×10^6 cells) in 20 μL of 0.1 M phosphate-buffered saline (pH 7.2) containing 0.5% gelatin.

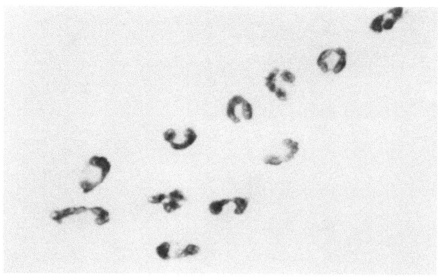

(a)

Fig. 4   Morphological changes in rabbit polymorphonuclear leukocytes caused by staphylococcal leukocidin. **(a)** Before exposure to leukocidin; **(b)** after exposure to leukocidin for 2 min; **(c)** after exposure to leukocidin for 10 min. Cells were stained with Giemsa.

somewhat swollen; the lobulated nuclei eventually become spherical. The terminal event in leukocidin action is cytoplasmic degranulation, rupture of nuclei and complete cell lysis. These effects are fully developed after 10 min of incubation (Fig. 4).

Leukocidin activity is determined also by assaying for ^{86}Rb released from ^{86}Rb-labeled polymorphonuclear leukocytes. Similar data are obtained by examining morphological changes of polymorphonuclear leukocytes. The study of leukocidin effects is confined to the first 10 min of intoxication; maximal effect results from treating 10^6 polymorphonuclear leukocytes per 20 μL with 500 pg of each component of leukocidin.

Chromium-release assay is also reported (Loeffler et al., 1986).

D. Criteria of Purity

The F and S components crystallize in the form of plates and needles, respectively. The amino-terminal residue of both components is alanine. The crystallized components migrate as single bands on the SDS-polyacrylamide gels. Molecular weights of crystallized S and F components of leukocidin determined from relative mobilities of marker proteins are 31,000 and 32,000, respectively. To further establish purity, the isoelectric point (pI) of each is determined. A single protein is evident for both S and F components with isoelectric points of 9.39 ± 0.05 and 9.08 ± 0.05, respectively.

E. Storage of Leukocidin

The purified F and S components of leukocidin stored at –80°C in 0.1 M phosphate-buffered saline (pH 7.2) containing 0.5% gelatin are stable for at least 3 months.

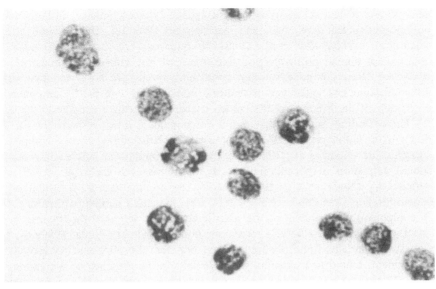

(b)

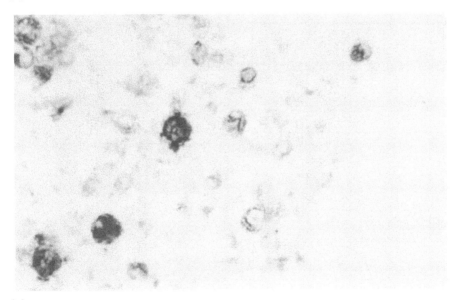

(c)

F. Physicochemical Properties of Leukocidin

No detectable half-cystine and free sulfhydryl groups are found in either component.

G. Mode of Action of Leukocidin

Staphylococcal leukocidin, known to be important in the pathogenicity of certain staphylococcal diseases (Woodin, 1970; Wenk and Blobel, 1970), consists of two protein components (S and F) that act synergistically to induce cytotoxic changes in human and rabbit polymorphonuclear leukocytes and macrophages (Woodin, 1960; Soboll et al., 1973; Noda et al., 1980a; Kato et al., 1988). No other cell type has been found to be susceptible except human promyelocytic leukemia cell line HL-60 (Morinaga et al., 1987a,b, 1988). These components when tested individually are not cytotoxic. When rabbit polymorphonuclear leukocytes have been preincubated with the S component at 37°C for 10 min, if the F component is added it causes immediate damage to the cells (Fig. 5). These findings suggest that the S component is more responsible for the interaction with the leukocytes than the F component. Specific binding of the S component to G_{M1} ganglioside (Noda et al., 1980b) on the cell membrane of leukocidin-sensitive cells induces activation of phospholipid methyltransferases (Noda et al., 1985) and phospholipase A_2 (Noda et al., 1982), leading to an increase in the number of F-component-binding sites (Noda et al., 1981). This increased number of binding sites for the F component is dependent on the specific enzymatic action of the S component, that is, ADP-ribosylation of a 37-kDa membrane protein (Kato and Noda, 1989a,b), and is correlated with increased phospholipid methylation and phospholipase A_2 activation. The F component, bound to leukocidin-sensitive cells in the presence of S component, rapidly causes degradation of the cell membrane, resulting in cell lysis. The F component

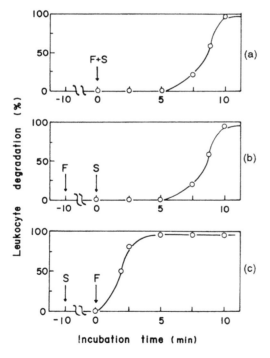

Fig. 5   Synergistic effect of the S and F components of leukocidin on cytolysis of rabbit leukocytes. (a) A 20-μL mixture containing 500 pg of each component was incubated at 37°C for 10 min with 1×10^6 rabbit leukocytes in 0.1 M phosphate-buffered saline (pH 7.2) containing 0.5% gelatin. (b) The F component alone (500 pg) was incubated with the cells, and after 10 min the S component (500 pg) was added. (c) The S component alone (500 pg) was incubated with the cells, and after 10 min the F component (500 pg) was added.

also possesses ADP-ribosyltransferase activity and ADP-ribosylates a 41-kDa membrane protein (Kato and Noda, 1989a,b) and stimulates phosphatidylinositol metabolism of leukocidin-sensitive cells (Kato and Noda, 1989a,b; Wang et al., 1990).

Morinaga et al. (1993) also reported that 6 h exposure of human promyelocytic leukemia HL-60 cells to 12-O-tetradecanoylphorbol 13-acetate (TPA) made the cells much more resistant to staphylococcal leukocidin than original HL-60 cells. These cells treated with TPA could bind S component of leukocidin to the same degree as the original HL-60 cells, suggesting that TPA treatment does not interfere with the S component binding to G_{M1} ganglioside on the cells, whereas in these cells leukocidin-induced calcium influx, phospholipase A_2 activity, and hydrolysis of PIP and PIP_2, which is a result of phospholipase C activity, were not enhanced. These data suggest that a signal transduction system that takes effect after binding of S component of leukocidin to the cell surface receptor is suppressed in the cells incubated with TPA for a short term.

These phenomena were observed in the cells treated for 6 h with the concentration of more than 2 ng/mL of TPA and at a stage when no differentiation of the cells was observed morphologically and functionally. Although further experiments attempting to clarify the mechanism of TPA suppressing these leukocidin-induced enzymatic activities transiently are in progress, it was clarified directly that leukocidin-induced calcium influx and phospholipase A_2 and C activities are important to induce leukocytolysis by using these transiently resistant cells. The degradation of the cell membrane is associated with stimulation of phosphatidylinositol metabolism and ouabain-insensitive Na^+,K^+-ATPase activity and inhibition of cyclic AMP-dependent protein kinase (Noda et al., 1982).

H. Nucleotide Sequence of Leukocidin Gene

The nucleotide sequence of *lukS* gene encoding the S component of staphylococcal leukocidin from methicillin-resistant *Staphylococcus aureus* (MRSA) has been determined (Rahman et al., 1991). The structural gene of *lukS* consists of 857 base pairs. An open reading frame that could encode a 35,556-Da polypeptide consisting of 315 amino acids is assigned. The molecular size of the polypeptide predicted from the amino acid composition is close to the value of pro-S component determined in DNA-directed transcription/translation system. Inspection of the amino acid sequence deduced from the nucleotide sequence of *lukS* and that from the S component of leukocidin clarifies that pro-S component contains a typical signal sequence at the NH_2 terminus. The amino acid sequence of predicted S component correlates exactly with the known N-terminal 50-amino-acid sequence of S component from MRSA and *S. aureus* V8. The molecular size of the predicted matured protein is also close to the value of S component determined in both MRSA and *S. aureus* V8. The nucleotide sequence of the 5' flanking region shows the presence of the consensus sequence of ribosome binding site, Pribnow box, and the RNA polymerase recognition site in *Escherichia coli*.

A *lukF* gene encoding the F component of staphylococcal leukocidin from MRSA has been cloned, and the nucleotide sequence of *lukF* gene has been determined (Rahman et al., 1992). The sequence data revealed an open reading frame that encodes a polypeptide with 323 amino acid residues. Inspection of the amino acid sequence deduced from the nucleotide sequence of *lukF* and that from the F component of leukocidin from *S. aureus* V8 clarifies that pro-F component contains a typical signal peptide at the NH_2 terminus and that the ATG starting codon for pro-F component is present one base downstream from the TGA that is the translation termination codon for the S component of leukocidin.

The nucleotide sequence of the 5′ flanking region of *lukF* shows the presence of the consensus sequence of ribosome binding site in the internal region of the structural gene of the S component. The *lukF* gene is transcribed in the same direction as *lukS*. No Pribnow box can be discerned in the intercistronic region between the *lukS* and *lukF* genes. The amino acid sequence homology between the S and F components is 31%. The F component was expressed in *Escherichia coli* DH5α harboring plasmid pFRK92 that contained *lukF* gene.

III. PSEUDOMONAL LEUKOCIDIN

A.   Purification and Crystallization of Leukocidin

Pseudomonas aeruginosa 158 is grown on a trypticase soy broth medium containing 0.5% glucose with a reciprocal shaker (120 cycles/min) at 30°C for 12 h. The culture fluid (10 liters per batch) is then centrifuged at 11,000 × *g* for 15 min at 4°C, and the bacteria are washed and suspended in 1400 mL of phosphate-buffered saline (pH 7.4) containing 0.137 M NaCl, 2.68 mM KCl, 0.5 mM MgCl$_2$, 1.47 mM KH$_2$PO$_4$, and 2.9 mM Na$_2$HPO$_4$. Autolysis of the bacteria is carried out by shaking incubation at 37°C for 48 h, rotating at 120 cycles/min. The supernatant (1612 mL) after centrifugation of the autolysate at 20,000 × *g* for 20 min is dialyzed against 0.05 M sodium phosphate buffer (pH 7.2). The dialysate is applied to a DEAE-Sephadex A-50 column (5 × 100 cm) equilibrated with 0.05 M sodium phosphate buffer (pH 7.2). The dialysate is applied to a DEAE-Sephadex A-50 column (5 × 100 cm) equilibrated with 0.05 M sodium phosphate buffer (pH 7.2). After washing the column with the equilibration buffer (3700 mL), fractions with leukocytotoxic activity are eluted with 0.5 M sodium phosphate buffer (pH 7.2) (Fig. 6) and concentrated to 80 mL by an Amicon ultrafiltration apparatus equipped with a PM-10 membrane. The resulting solution is applied to a Sephadex G-100 column (5 × 100 cm) that was equilibrated with 0.05 M sodium phosphate buffer (pH 7.2). Protein is eluted with the same buffer and assayed for leukocytotoxic activity. Fractions containing leukocytotoxic activity are collected and concentrated to 6 mL. This solution, equilibrated with 0.05 M sodium phosphate buffer (pH 8.0), is applied to zone electrophoresis on Pevikon C-870 using the same buffer at 20 mA for 24 h. Active fractions from this final purification step are collected, concentrated by membrane filtration (Diaflo PM-10), crystallized by means of dialysis against ammonium sulfate (Noda et al., 1980a), and stored at −80°C. The recovery and degree of purification at each step are shown in Table 3.

B.   Determination of Leukocidin Activity

Polymorphonuclear leukocytes are prepared from rabbit peripheral blood on isokinetic Ficoll gradients (Pretlow and Luberoff, 1973). Leukocytotoxic activity is assayed by the microscopic slide adhesion method as described previously (Noda et al., 1980a) except for the use of Ca^{2+}-Dulbecco's phosphate-buffered saline (PBS; 0.01 M sodium potassium phosphate buffer, pH 7.4, 0.15 M NaCl, 2.5 mM KCl, and 1 mM CaCl$_2$). End points are determined as the smallest amount of the toxin causing morphological changes in 50% of leukocytes in a standard assay suspension (1 × 10^6 cells/20 μL). The leukocytotoxic activity is expressed in units, the number of units in a toxin preparation being numerically equal to the dilution (in mL) at the point.

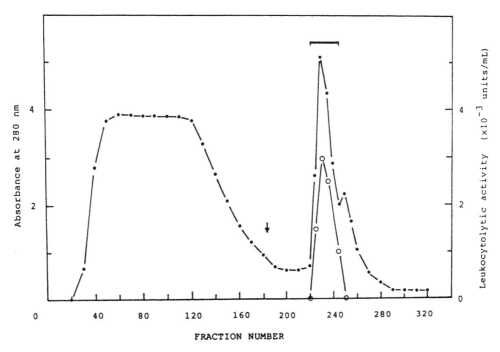

Fig. 6 DEAE-Sephadex A-50 column chromatography. The dialysate of bacterial autolysate was applied to the column and equilibrated with 0.05 M sodium phosphate buffer (pH 7.2). The buffer was replaced by 0.5 M sodium phosphate buffer (pH 7.2) at 10°C for elution of leukocidin (indicated by arrow). Fractions were measured for absorbance at 280 nm (●) and leukocytotoxic activity (○) by the microscopic adhesion method. Fractions that were pooled are indicated by a bar.

C. Physicochemical Properties of Leukocidin

The molecular weight of the leukocidin is estimated to be 42,500 by SDS-polyacrylamide gel electrophoresis, 40,000 by gel filtration, and 44,700 by sucrose density gradient centrifugation. The isoelectric point of the leukocidin is estimated to be 6.3 by isoelectrofocusing. Three residues of half-cystine are detected. The (Glx + Asx) / (Lys + Arg) ratio is 0.5, in agreement with the acidic nature of the toxin (Hirayama et al., 1983).

D. Mode of Action of Leukocidin

The minimum required doses of leukocidin for destruction of 1×10^6 rabbit leukocytes, rabbit blood lymphocytes, or rabbit erythrocytes are 13 ng, 390 ng, and 480 μg, respectively. No morphological changes are observed in rabbit platelets. Hirayama and Kato (1984) reported that the pseudomonal leukocidin specifically bound to a 50-kDa membrane protein of rabbit leukocytes and that the subsequent stimulation of the leukocytes by the toxin resulted in a rapid increase in the incorporation of $[^{32}P]P_i$ into phosphatidyl-4-phosphate (PIP), phosphatidyl-4,5-bisphosphate (PIP$_2$), and phosphatidic acid (PA). These findings suggested that the leukocidin stimulated the phospholipase C-induced hydrolysis of the inositol phospholipid pathway, which is the general process involved in signal transduction, leading to both activation of protein kinase C and enhancement of Ca^{2+} mobilization. In fact, Hirayama and Kato (1982, 1983) showed

TABLE 3 Purification of Leukocidin from the Autolysate of *P. aeruginosa* 158

Step	Volume (mL)	Protein (μg/mL)	Total protein (mg)	Total activity ($\times 10^{-3}$ U)	Specific activity (U/mg protein)	Recovery of activity (%)
1. Supernatant of bacterial autolysate	1,612	806	1,300	806	620	100
2. DEAE-Sephadex A-50 chromatography	520	1,530	796	780	980	96.8
Concentration by ultrafiltration	80	8,060	645	600	930	74.4
3. Sephadex G-100 chromatography	162	679	110	486	4,420	60.3
Concentration by ultrafiltration	6	16,000	96	420	4,380	52.1
4. Pevikon zone electrophoresis	12	292	3.5	360	103,000	44.7

that leukocidal activity of the leukocidin was observed only in the presence of Ca^{2+} and that leukocidin stimulated the activity of Ca^{2+}-dependent protein kinase C, resulting in the phosphorylation of a 28-kDa protein in the membranes of cell surface and lysosome particles from which the leakage of lysosomal enzymes occurred at almost the same time as the leukocyte destruction. Electron microscopic experiments (Hirayama et al., 1983) regarding morphological changes of leukocidin-treated leukocytes indicated an apparent increase in vacuoles resulting from the loss of cytosolic granule contents before leukocyte enlargement.

Pseudomonal leukocidin induces the ADP-ribosylation of 18- and 41-kDa membrane proteins of rabbit polymorphonuclear leukocytes (Noda et al., 1991). The ADP-ribosylation of these proteins is observed to be dependent on the incubation time and toxin dose. The addition of agmatine has no effect on the ADP-ribosylation of the 18- and 41-kDa proteins. These results suggest that pseudomonal leukocidin differs in catalytic properties from cholera toxin and is not an NAD:agmatine ADP-ribosyltransferase. It is known that staphylococcal leukocidin ADP-ribosylates 37- and 41-kDa proteins of rabbit polymorphonuclear leukocytes (Kato and Noda, 1989a,b). The staphylococcal toxin also stimulates PI metabolism (Wang et al., 1990). The 41-kDa membrane protein that is ADP-ribosylated by staphylococcal leukocidin is thought to regulate PI-specific phospholipase C. As pseudomonal leukocidin also ADP-ribosylates the 41-kDa membrane protein, it is tempting to speculate that this ADP-ribosylation activates PI-specific phospholipase C, resulting in phosphorylation of the 28-kDa protein by the activation of protein kinase C.

REFERENCES

Gladstone, G. P., and van Heyningen, W. E. (1957). Staphylococcal leukocidin. *Br. J. Exp. Pathol. 38*:123–137.

Hirayama, T., and Kato, I. (1982). Two types of Ca^{2+}-dependent phosphorylation in rabbit leukocytes. *FEBS Lett. 146*:209–212.

Hirayama, T., and Kato, I. (1983). Rapid stimulation of phosphatidylinositol metabolism in rabbit leukocytes by pseudomonal leukocidin. *FEBS Lett. 157*:46–50.

Hirayama, T., and Kato, I. (1984). Mode of cytotoxic action of pseudomonal leukocidin on phosphatidylinositol metabolism and activation of lysosomal enzyme in rabbit leukocytes. *Infect. Immun. 43*:21–27.

Hirayama, T., and Kato, I., Matsuda, F., and Noda, M. (1983). Crystallization and some properties of leukocidin from *Pseudomonas aeruginosa. Microbiol. Immunol. 27*:575–588.

Kato, I., and Noda, M. (1989a). ADP-ribosylation of cell membrane proteins by staphylococcal α-toxin and leukocidin in rabbit erythrocytes and polymorphonuclear leukocytes. *FEBS Lett. 255*:59–62.

Kato, I., and Noda, M. (1989b). ADP-ribosyltransferase activities of staphylococcal cytolytic toxins. In *ADP-Ribosyl Toxins and Target Molecules*, I. Kato and T. Uchida (Eds.), Saikon Publishers, Tokyo, p. 135–161.

Kato, I., Morinaga, N., and Muneto, R. (1988). Non-thiol-activated cytolytic bacterial toxin: current status. *Microbiol. Sci. 5*:53–57.

Loeffler, D. A., Schat, K. A., and Norcross, N. L. (1986). Use of ^{51}Cr release to measure the cytotoxic effects of staphylococcal leukocidin and toxin neutralization on bovine leukocytes. *J. Clin. Microbiol. 23*:416–420.

Morinaga, N., Nagamori, M., and Kato, I. (1987a). Suppressive effect of calcium on the cytotoxicity of staphylococcal leukocidin for HL-60 cells. *FEMS Microbiol. Lett. 42*:259–264.

Morinaga, N., Nagamori, M., and Kato, I. (1987b). Stimulation of Ca^{2+}-dependent protein phosphorylation in HL-60 cells by staphylococcal leukocidin. *FEMS Microbiol. Lett. 44*:431–434.

Morinaga, N., Nagamori, M., and Kato, I. (1988). Changes in binding of staphylococcal leukocidin to HL-60 cells during differentiation induced by dimethyl sulfoxide. *Infect. Immun.* 56:2479–2483.

Morinaga, N., Kato, I., and Noda, M. (1993). Changes in the susceptibility of TPA-treated HL-60 cells to staphylococcal leukocidin. *Microbiol. Immunol.* 37:537–541.

Noda, M., Hirayama, T., Kato, I., and Matsuda, F. (1980a). Crystallization and properties of staphylococcal leukocidin. *Biochim. Biophys. Acta* 633:33–44.

Noda, M., Kato, I., Hirayama, T., and Matsuda, F. (1980b). Fixation and inactivation of staphylococcal leukocidin by phosphatidylcholine and ganglioside G_{M1} in rabbit polymorphonuclear leukocytes. *Infect. Immun.* 29:678–684.

Noda, M., Kato, I., Matsuda, F., and Hirayama, T. (1981). Mode of action of staphylococcal leukocidin: relationship between binding of ^{125}I-labeled S and F components of leukocidin to rabbit polymorphonuclear leukocytes and leukocidin activity. *Infect. Immun.* 34:362–367.

Noda, M., Kato, I., Hirayama, T., and Matsuda, F. (1982). Mode of action of staphylococcal leukocidin: effects of the S and F components on the activities of membrane-associated enzymes of rabbit polymorphonuclear leukocytes. *Infect. Immun.* 35:38–45.

Noda, M., Hirayama, T., Matsuda, F., and Kato, I. (1985). An early effect of the S component of staphylococcal leukocidin on methylation of phospholipid in various leukocytes. *Infect. Immun.* 50:142–145.

Noda, M., Kato, I., Wang, X., and Hirayama, T. (1991). ADP-ribosylation and activation of phosphatidylinositol-specific phospholipase C by pseudomonal leukocidin. In *Pseudomonas aeruginosa in Human Diseases* (*Antibiot. Chemother.*, Vol. 44), J. Y. Homma, H. Tanimoto, I. A. Holder, N. Høiby, and G. Döring (Eds.), Karger, Basel, p. 59–62.

Pretlow, T. G., II and Luberoff, D. E. (1973). A new method for separating lymphocytes and granulocytes from human peripheral blood using programmed gradient sedimentation in an isokinetic gradient. *Immunology* 24:85–92.

Rahman, A., Izaki, K., Kato, I., and Kamio, Y. (1991). Nucleotide sequence of leukocidin S-component gene (*lukS*) from methicillin resistant *Staphylococcus aureus*. *BBRC 181*:138–144.

Rahman, A., Nariya, H., Izaki, K., Kato, I., and Kamio, Y. (1992). Molecular cloning and nucleotide sequence of leukocidin F-component gene (*lukF*) from methicillin resistant *Staphylococcus aureus*. *BBRC 184*:640–646.

Soboll, H., Ito, A., Schaeg, W., and Blobel, H. (1973). Leukozidin von Staphylokokken verschiedener Herkunft. *Zbl. Bakt. Parasitenkd. Infekt. Hyg. Abt. 1 Orig. Reihe A224*:184–193.

Wang, X., Noda, M., and Kato, I. (1990). Stimulatory effect of staphylococcal leukocidin on phosphoinositide metabolism in rabbit polymorphonuclear leukocytes. *Infect. Immun.* 58:2745–2749.

Wenk, K., and Blobel, H. (1970). Untersuchungen an "Leukozidinen" von Staphylokokken Verschiedener Herkunft. *Bakteriol. Parasitenkd. Infektionskr. Hyg. Abt. 1 Orig.* 213:479–487.

Woodin, A. M. (1960). Purification of two components of leukocidin from *Staphylococcus aureus*. *Biochem. J.* 75:158–165.

Woodin, A. M. (1970). Staphylococcal leukocidin. In *Microbial Toxins*, Vol. 3, T. C. Montie, S. Kadis, and S. J. Ajl (Eds.), Academic Press, New York, p. 327–355.

25

Endotoxin: A Mediator of and Potential Therapeutic Target for Septic Shock

Robert L. Danner and Charles Natanson

National Institutes of Health, Bethesda, Maryland

I. INTRODUCTION

Over the past 50 years, researchers have used purified endotoxin in animals and human volunteers to study the pathophysiology of septic shock. During this time, scientists have speculated about the role of endotoxin in producing human septic shock. Some believed that endotoxin was the agent that could induce all forms of shock (Fine, 1954; Ravin et al., 1960; van Deventer et al., 1988). Others were convinced that endotoxin was only one of several mediators that could cause the septic syndrome (Zweifach et al., 1958; Danner et al., 1989a).

Recently, large clinical trials were conducted in septic patients to evaluate the effectiveness of therapies directed against endotoxin. In these studies, the new treatment was combined with standard antibiotic and supportive therapy (Greenman et al., 1991; Ziegler et al., 1991). The results of these studies were inconclusive and generated much controversy about the role of endotoxin in bacterial infection and septic shock. To clarify the role of this bacterial toxin in septic shock, we review its biochemistry, its effects when administered to animals and humans, and its relation to clinical therapies.

II. THE BIOCHEMISTRY OF ENDOTOXIN

The term "endotoxin" refers to a structurally heterogeneous class of lipopolysaccharides (LPS) found in the outer membrane of gram-negative bacteria that have similar biological activities (Danner et al., 1989a). Compared to other bacterial products with toxic properties, endotoxin has a number of unique features. Although some bacterial toxins may convey a survival advantage to a particular strain or species of bacteria, they are not essential. In contrast, endotoxin is a necessary component of virtually all gram-negative bacteria. Failure to synthesize the minimally required endotoxin structure and assemble an outer membrane is a lethal event for gram-negative bacteria (Raetz et al., 1991). Further, most bacterial toxins directly injure eukaryotic cells by disrupting vital functions such as membrane integrity, signal transduction, or protein synthesis. In contrast, it has become clear over the past 10 years that endotoxin, in physiologically relevant concentrations is not a direct cellular toxin or poison. The harmful effects of endotoxin are largely caused by activation of the host immune system (Corriveau and Danner, 1993).

The mammalian reaction to endotoxin may in fact represent an adaptive defense mechanism that has evolved to protect the host from gram-negative bacteria. By this view, the host's response to endotoxin is useful under most circumstances because it creates a hostile environment for invading microorganisms. However, in the setting of overwhelming bacteremia, this vigorous response may initiate a cascade of events that lead to shock, organ failure, and death (Danner et al., 1989a). These harmful actions of endotoxin are mediated by the release of host-derived substances that have potent proinflammatory activity such as the cytokines tumor necrosis factor (TNF) and interleukin-1 (IL-1) (Tracey et al., 1988; Okusawa et al., 1988).

The effects of endotoxin in vivo are further complicated because endotoxin also induces the release of anti-inflammatory mediators, including transforming growth factor β (Dinarello, 1991), IL-1 receptor antagonist (Granowitz et al., 1991), IL-4 (te Velde et al., 1990), and IL-10 (Fiorentino et al., 1991). In addition, endotoxin activates poorly understood mechanisms that produce a form of tachyphylaxis to endotoxin called tolerance (Zuckerman et al., 1991). Therefore, the clinical manifestations of endotoxin "toxicity" may actually represent conditions where there is a deleterious imbalance between the proinflammatory and anti-inflammatory effects of endotoxin. Self-injury occurs when the resultant immunologic response is in excess of that necessary to control the initiating infection. Conversely, an inadequate inflammatory response would fail to eradicate the causative organism and result in a persistent infection with ongoing tissue injury.

Recent advances have shown that the mammalian response to endotoxin is complex and highly regulated. It relies on both humoral and cell membrane bound recognition molecules that specifically interact with lipid A (Wright, 1991). In the remainder of this section, the structure of endotoxin and the interaction between endotoxin and the host immune system are discussed.

A. Endotoxin Structure

For discussion of the immunochemical properties of endotoxin, it has been useful to divide the molecule into three domains: the O-polysaccharide side chain, core oligosaccharide region, and lipid A (Fig. 1). The O-polysaccharide side chain is composed of a series of repeating oligosaccharides that vary among bacteria in composition and length (Westphal et al., 1983). This domain is also referred to as the O-antigen or O-specific chain and is responsible for the serotypic specificity of bacterial strains. Notably, the O-polysaccharide side chain is important in protecting gram-negative bacteria from environmental conditions such as detergents and enables them to establish infections but is not absolutely required for bacterial growth or survival (Raetz et al., 1988). Although this region of the molecule can directly activate the alternative pathway of complement (Morrison and Kline, 1977), overall, in the absence of specific antibody, it helps prevent serum-mediated lysis of the bacterium.

The core oligosaccharide domain contains the unique sugar 2-keto-3-deoxyoctonate (KDO). This region is more conserved than the highly variable O-antigens (Brade et al., 1988) but is buried within the LPS molecule (Fig. 1) and probably interacts only minimally with the host immune system because of steric hindrance (Gigliotti and Shenep, 1985; Pollack et al., 1989; Heumann et al., 1991). Bacteria that lack O-antigens and have exposed core sugars are called rough mutants; they do not cause infections because they are readily lysed by complement. At least part of the core structure, that containing KDO, is necessary for bacterial viability (Raetz et al., 1991).

KDO links the core region to lipid A, which is the most structurally conserved moiety of the endotoxin molecule. Lipid A is a D-glucosamine-based, $\beta(1\rightarrow6)$-linked disaccharide phospholipid (Brozek and Raetz, 1990) that can activate complement via the classical pathway (Morrison and Kline, 1977). Importantly, lipid A alone fully conveys endotoxin toxicity and is capable of initiating a broad inflammatory response through immune cell activation and the release of cytokines (Galanos et al., 1985). Although the signal transduction pathway through which endotoxin causes its biological effects is still incompletely understood, a number of the recognition molecules that regulate these events have been identified (Raetz et al., 1991; Wright, 1991).

B. Endotoxin Binding Proteins in Plasma

The serum of animals and humans contain circulating proteins that appear to interact with endotoxin in biologically relevant ways (Fig. 2). These proteins may have other, unrelated functions, whereas the only known purpose for some is modulating the interaction between endotoxin and the host. The effect of these serum factors can be to either inhibit or enhance the immunological response to endotoxin.

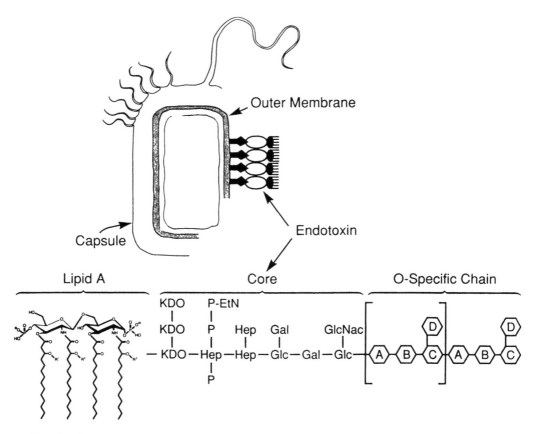

Fig. 1 Location and structure of endotoxin in gram-negative bacteria. See text for details. In brief, on top is shown a gram-negative bacterium with capsule, flagella, and outer membrane. On the bottom is a schematic of the tripartite structure of endotoxin with three domains: lipid A, core, and O-specific chain. On top to the right are shown a series of black arrows, open circles, and black combs that indicate the position of lipid A, core, and the O-specific chain of endotoxin, respectively, in the outer bacterial membrane. KDO = 2-keto-3-deoxyoctonate; P = phosphate; Hep = heptose; GAL = glucosamine; ETN = ethanolamine. (Modified from Young et al., 1977 with permission.)

The plasma lipoproteins that transport triglycerides and cholesterol and regulate their metabolism may also be a natural defense system against endotoxin. Plasma lipoproteins in vitro and in vivo can bind endotoxin, reduce its toxicity, and promote its clearance from the circulation (Warren et al., 1988; Harris et al., 1990). Reconstituted high-density lipoproteins are discussed in Section IV as a treatment for septic shock (Quezado et al., 1993).

In contrast to the antiendotoxin activity of serum lipoproteins, another protein produced by the liver binds to the lipid A portion of endotoxin and actually increases its ability to activate phagocytic cells (Tobias et al., 1989). Lipopolysaccharide binding protein (LBP) is a 60-kDa glycoprotein that acts as an opsonin by promoting the binding of gram-negative bacteria and endotoxin to phagocytes (Tobias et al., 1988). This increased binding of LPS to macrophages is observed with different types and sources of

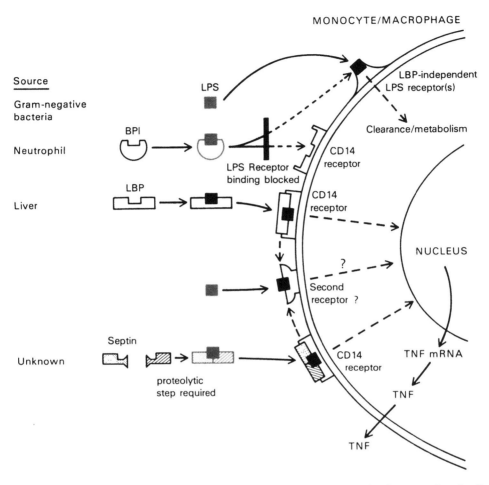

Fig. 2 Schematic of monocyte and macrophage activation by LPS. See text for details. (Reproduced from Corriveau and Danner, 1993, with permission.)

endotoxin and causes a marked increase in cell activation (Schumann et al., 1990). For example, the addition of small amounts of LBP to cultured macrophages increases 100-fold the ability of endotoxin to induce TNF production (Schumann et al., 1990). In healthy humans, LBP is present in plasma at a concentration of <0.5 μg/mL but can rise within 24 h to levels of 50 μg/mL during an acute-phase response (Schumann et al., 1990).

A proteolytic cascade distinct from the coagulation and complement systems has been described (Wright et al., 1992) in human plasma that yields a product like LBP, which binds endotoxin and mediates the recognition of endotoxin by phagocytes. This plasma activity, given the name "septin," is not blocked by LBP-neutralizing antibodies but is blocked by the addition of protease inhibitors. Further, fractionation of plasma by chromatography has shown that at least two different components of plasma must be combined to observe this "septin" activity (Wright et al., 1992).

Two other proteins that bind LPS have been described, but their function in the circulation is currently less well understood. One is a neutrophil product called bactericidal/permeability-increasing protein (BPI). BPI is an inducible cationic protein with a molecular weight of 55,000. BPI is closely related to LBP and can bind and neutralize endotoxins from a variety of bacteria (Weiss et al., 1982; Marra et al., 1990). BPI appears to function primarily as an endogenous antimicrobial agent important in the nonoxidative killing of ingested gram-negative bacteria (Weiss et al., 1982). However, recent evidence suggests that BPI can also be found on the surface neutrophils, indicating that it may be able to detoxify endotoxin in the circulation (Weersink et al., 1992). Recombinant forms of BPI are discussed in Section IV as a possible treatment for septic shock.

The other protein that binds LPS, although its function in the circulation is not well understood, is the soluble form of the CD14 receptor (sCD14). CD14 is the only cell surface receptor identified that has been linked to the LPS-induced activation of monocytes and macrophages for cytokine production (Wright, 1991). Of note, 99% of CD14 in whole blood is present as sCD14 (Frey et al., 1992). Although several studies have suggested that sCD14 can block LPS-induced cell activation (Schütt et al., 1991), recent evidence indicates that complexes of endotoxin, LBP, and sCD14 are capable of directly activating cells, in particular those that do not naturally express CD14 receptors (Frey et al., 1992). These data suggest that sCD14, secreted or shed from phagocytes, may mediate the interaction of endotoxin with a wide variety of other cells and tissues.

C. Endotoxin–Cell Interactions

The CD14 receptor was first described as a monocyte/macrophage cell differentiation marker (Wright, 1991). It is a 55-kDa cell surface glycoprotein that is attached to the membrane by a phosphoinositol anchor and lacks a cytoplasmic domain (Wright, 1991). Both LBP and septin, described above, increase the bioactivity of endotoxin through the binding of LBP-endotoxin complexes (Wright et al., 1990) and septin-endotoxin complexes (Wright et al., 1992) to CD14 receptors on monocytes, macrophages, and neutrophils. However, the precise events leading to signal transduction after the binding of these LPS-protein complexes to CD14 are uncertain. Notably, LPS in relatively high concentrations can activate cells in the absence of serum factors such as LBP and septin (Raetz et al., 1991; Wright, 1991). Further, cells depleted or naturally devoid of CD14 receptors can still be activated by LPS to produce cytokines (Wright et al., 1990). One proposed explanation for these findings is that CD14 merely acts to facilitate the delivery of endotoxin to a second definitive binding site that is directly linked to signal transduction (Raetz et al., 1991). Although this second binding site involved in CD14-mediated cell activation by LPS has not been described, similar two-receptor systems have been demonstrated for other ligands such as IL-6. In support of this receptor model for LPS-induced cell activation, it has been shown that lipid IVa, a lipid A analogue that antagonizes LPS-induced TNF production, does not block the binding of endotoxin to CD14 (Kitchens et al., 1992). However, lipid IVa does inhibit an LPS-induced increase in the nuclear binding activity of NF-kβ, a cytoplasmic factor involved in TNF gene transcription. These data suggest that lipid IVa, a structural analogue of endotoxin, blocks LPS-stimulated cell activation at a binding site distinct from the CD14 receptor.

A number of other cellular binding sites for LPS have been described, but these receptors have either an uncertain role or no role in the LPS-induced secretory response of monocytes and macrophages. The CD18 antigen (Wright, 1991) and the scavenger receptor (Hampton et al., 1991) both bind LPS in the absence of serum factors but do not initiate signal transduction. These cell surface proteins may be important as endotoxin clearance pathways that do not lead to cell activation (Wright, 1991; Hampton et al., 1991). Another LPS receptor has been described on murine lymphocytes through the use of a radioiodinated derivatized LPS probe that forms a covalent bond upon photoactivation (Lei et al., 1991). This 73-kDa receptor has also been found on human monocytes and neutrophils (Halling et al., 1992). Further, a monoclonal antibody to this receptor appears to protect mice against the lethal effects of endotoxin, but this antibody may actually activate its targeted receptor rather than block it (Morrison et al., 1990). The precise relationship between LPS-induced cell activation and the 73-kDa receptor remains to be defined.

In the following sections, the importance of endotoxin in septic shock is further evaluated using data from animal models, human volunteers challenged with endotoxin, and patients. Further, endotoxin is explored as a target for therapeutic intervention. Ultimately, antiendotoxic agents developed for the treatment of septic shock should be capable of effectively blocking the LPS–host interactions reviewed in this section, although the suitability of endotoxin-directed treatment strategies remains controversial.

III. ADMINISTRATION OF ENDOTOXIN TO AND MEASUREMENT OF ENDOTOXEMIA IN HUMANS AND ANIMALS: INVESTIGATING THE ROLE OF ENDOTOXIN AS A POTENTIAL THERAPEUTIC TARGET IN SEPTIC SHOCK

A. Data Supporting Endotoxin as an Important Toxin in Septic Shock

Investigators have explored the clinical importance of endotoxin by giving challenges both to animals (Shwartzman, 1928; Brigham et al., 1979; Esbenshade et al., 1982; Kikeri et al., 1986; Zager, 1986; Zager and Prior, 1986; Natanson et al., 1989a; Fink et al., 1991) and in safe, low doses to human volunteers (O'Dwyer et al., 1988; Suffredini et al., 1989a,b). More than half a century ago, Shwartzman showed that injections of endotoxin into small animals produced abnormal coagulation and necrotic skin lesions (Shwartzman, 1928). Since that time, depending on the model used, endotoxin challenges in animals have been shown to induce lung injury (Brigham et al., 1979; Esbenshade et al., 1982), renal dysfunction (Kikeri et al., 1986; Zager, 1986; Zager and Prior, 1986), gastrointestinal mucosal injury (Fink et al., 1991), and cardiovascular dysfunction similar to that seen in patients with septic shock (Natanson et al., 1989a).

In normal volunteers, small safe doses of endotoxin (4 ng/kg) result in release of cytokines and a cardiovascular response that, although much milder, is remarkably similar to that of patients with septic shock (Suffredini et al., 1989a). In other human studies, endotoxin in small doses induced release of plasminogen activator inhibitor (Suffredini et al., 1989b) and increased intestinal permeability to lactulose and mannitol (O'Dwyer et al., 1988). Inadvertently transfused blood products contaminated with cold-growing gram-negative bacteria (Stevens et al., 1953) and crude endotoxin preparations for the treatment of cancers can produce a fulminant syndrome indistinguishable from septic shock (Brues and Shear, 1944). More recently, an otherwise healthy cancer patient self-administered a high dose of purified endotoxin that reproduced the clinical features of septic shock including a high cardiac output form of hypotension requiring vasopressor therapy over days, disseminated intravascular coagulation, and multiple organ dysfunction

(Taveira da Silva et al., 1993). Thus, endotoxin challenges in humans and animals can produce most if not all of the abnormalities of de novo septic shock in patients.

Another area of endotoxin research has examined levels of circulating endotoxin during human and animal septic shock. In these studies, endotoxin was measured using a *Limulus* amebocyte lysate (LAL) assay that relies on proteins obtained from the lysate of horseshoe crab amebocytes (Levin and Bang, 1965). Endotoxin activates a proenzyme in the lysate to produce a gel (Levin and Bang, 1965; Levin et al., 1970; Reinhold and Fine, 1971). There is no specific biochemical assay of endotoxin to validate this assay. Nonetheless, using methods to neutralize interfering proteins and a quantitative chromogenic substrate to increase its sensitivity (lower limit is 0.5 pg/mL) (Iwanaga et al., 1978; Thomas et al., 1981, 1984), this assay is relatively specific and has been used productively in animal models and humans to study the role of endotoxin in septic shock (Elin and Wolff, 1973; Elin et al., 1976; Jorgensen, 1986).

Notably in a canine model of septic shock that used intraperitoneal challenges of live *Escherichia coli* in fibrin clots, endotoxin levels that developed in the blood over time were correlated with important outcomes. Nonsurviving animals, compared with survivors, had higher levels of endotoxemia, which predicted a decreased survival time (Natanson et al., 1990). Consistent with these canine data, during meningococcemia in humans, levels of endotoxemia have correlated with the occurrence of renal failure, lung injury, shock, and death (Brandtzaeg et al., 1989). In humans with septic shock at the National Institutes of Health (NIH), serial LAL tests were done in 100 consecutive patients (Danner et al., 1991). Forty-three percent of patients had endotoxemia, which was associated with more severe episodes of renal, pulmonary, and cardiac dysfunction. Thus, endotoxin challenges reproduce many of the abnormalities of septic shock, and the presence of endotoxemia during septic shock is correlated with more severe disease, suggesting that this bacterial product has an important role in septic shock.

B. Data Not Supporting Endotoxin as an Important Therapeutic Target in Septic Shock

As a potential therapeutic modality, animals and humans have been injected with small safe doses of endotoxin, which produce minimal toxicity and induce tolerance to endotoxin in the hope of alleviating the toxicities of subsequent gram-negative bacterial infections. Normal human volunteers made tolerant to endotoxin had no benefit from the febrile toxic course of experimental typhoid fever or tularemia (Greisman et al., 1969). Some mice (C3H/HeJ) are genetically resistant to endotoxin. Macrophages from these mice produce abnormally low amounts of tumor necrosis factor (a cytokine that is believed to be directly involved in the pathophysiology of septic and endotoxic shock). However, these endotoxin-resistant mice are actually more susceptible to bacterial challenge with saprophytes than normal mice (Weinstein et al., 1986; Ayala et al., 1992). In other studies, normal mice, C3H/HeJ mice, and mice injected with BCG differed by 5000-fold in their sensitivity to endotoxin; however, in these different groups of mice, infection with gram-negative bacteria produced similar outcomes and organ bacterial counts (McCabe and Olans, 1981). Thus, induced tolerance, genetic resistance, or increased sensitivity to endotoxin does not alter the course of gram-negative infections in humans or animals. In addition, a normal response to endotoxin in some circumstances might be a necessary protective mechanism against bacterial invasion as shown by studies with C3H/HeJ mice (Weinstein et al., 1986; Ayala et al., 1992).

Other studies in animals and humans that have measured endotoxin levels during infection have not supported the hypothesis that this molecule is the primary bacterial toxin responsible for septic shock. In the canine model discussed above, the abilities of different microorganisms to produce cardiovascular toxicity and elevate levels of LAL positivity in the blood were compared during septic shock (Hoffman et al., 1989, 1990a,b; Natanson et al., 1989b; Danner et al., 1990). In one set of experiments, comparable challenges (in number of colony-forming units) of *Pseudomonas aeruginosa* and *E. coli* were implanted into the peritoneal cavity (Danner et al., 1990). Those animals infected with *P. aeruginosa* demonstrated greater cardiovascular dysfunction and lethality but markedly lower levels of endotoxemia than those infected with *E. coli*. In similar experiments, a virulent *E. coli* strain with serum resistance, an antiphagocytic capsule, and the ability to produce hemolysins was compared with a nonvirulent strain of *E. coli* without these properties (Hoffman et al., 1990a). The virulent strain produced greater cardiovascular dysfunction and lethality but less endotoxemia (Hoffman et al., 1989, 1990a). However, the endotoxins from these two *E. coli* strains were equipotent (Hoffman et al., 1989, 1990a,b). In yet another set of experiments, animals infected with *Staphylococcus aureus* (an organism without endotoxin in its cell wall) were compared with animals infected with *E. coli* (Natanson et al., 1989b). *S. aureus* produced the same qualitative pattern of cardiovascular changes as did *E. coli*. However, *S. aureus* implanted intraperitoneally at lower doses produced greater morbidity and mortality than *E. coli* in the absence of endotoxemia. Thus, no correlation was found between clinical outcome and level of endotoxemia when different gram-negative organisms were compared. Further, endotoxin and endotoxemia were not necessary to produce septic shock.

In the study of 100 consecutive patients with septic shock at NIH cited earlier (Danner et al., 1991), it was found that LAL positivity did not correlate with the type of infecting organism. Surprisingly, 8 of 14 patients with gram-positive bacteremia had endotoxemia, as well as 7 of 7 patients with fungemia and 19 of 63 culture-negative patients; however, 8 of 18 patients with gram-negative bacteremia had no endotoxemia. Other investigations also have not found that LAL positivity correlates with the presence of gram-negative bacteremia (Elin, 1979; McCartney et al., 1987). Further, in the NIH study, there was no association with mortality (by multivariate analysis), and amounts of endotoxin (levels of endotoxemia) found in the blood did not correlate with severity of illness (Danner et al., 1991). No other clinical reports have convincingly correlated the degree of LAL positivity with the outcome of a group of diverse patients with septic shock of various etiologies.

In summary, endotoxin is probably one among many bacterial products that can elicit the septic shock response. It is not clear whether it is the most important or the only bacterial stimulus even among gram-negative bacteria. Tolerance or decreased sensitivity to endotoxin did not benefit animals or humans subsequently infected with gram-negative bacteria in several investigations, suggesting that treatment strategies aimed at this toxin may not provide protection in sepsis. Although endotoxemia may be a marker of disease, there is no direct evidence that it is the putative agent or persistent mediator of toxicity in septic shock.

IV. STUDIES USING ENDOTOXIN AS A THERAPEUTIC TARGET IN EXPERIMENTAL AND CLINICAL SEPTIC SHOCK

Endotoxin is an integral part of all gram-negative bacteria. This diverse group of microorganisms causes a substantial percentage (30–85%) of all cases of septic shock

(Glauser et al., 1991). Endotoxin elicits an intense inflammatory response from the host, and, as discussed above, challenges with endotoxin in animals can produce a pattern of organ damage and cardiovascular changes similar to those observed in human septicemia (Shwartzman, 1928; Brigham et al., 1979; Esbenshade et al., 1982; Kikeri et al., 1986; Zager, 1986; Zager and Prior, 1986; Natanson et al., 1989a; Fink et al., 1991). Further, endotoxemia when present in clinical sepsis is associated with worse organ damage (Brandtzaeg et al., 1989; Danner et al., 1991). Thus, scientists have developed several antiendotoxin approaches to treat septic shock. There are antiendotoxin antibodies that bind to the different parts of the endotoxin molecular (O-antigen, core, lipid A), nonantibody agents that bind endotoxin [cationic peptides (polymyxin B), cationic proteins (bactericidal/permeability-increasing protein), and lipoproteins (high-density lipoprotein)], analogues of lipid A that either induce tolerance (monophosphorylated lipid A) or are direct endotoxin antagonists (lipid X), and plasma detoxification schemes to remove endotoxin directly (plasmapheresis).

To date, no antiendotoxin therapy has been approved for clinical use by the Food and Drug Administration (FDA). However, antibodies directed against endotoxin have undergone several multicenter, randomized clinical trials, and one monoclonal antibody directed against the lipid A portion of endotoxin (HA-1A) was approved in Europe (Ziegler et al., 1991). The use of this antibody in the United States has been controversial, and recently, because of excess mortality in the treatment arm of a large clinical trial, HA-1A was removed from the European market (McCloskey et al., 1994; Piercey, 1993). These clinical studies have cast doubt on the existence of broadly cross-reactive antibodies directed at conserved epitopes on the endotoxin molecule and renewed the debate about the importance of endotoxin per se in human septic shock. Therefore, we will review the development of antiendotoxin antibodies in detail and then discuss other nonantibody approaches to neutralizing endotoxin in septic shock.

A. Endotoxin Antibodies Directed at *O*-Side Chain

Animals challenged with endotoxin or bacteria have been shown repeatedly to be protected by antibodies directed at the specific homologous *O*-polysaccharide of the endotoxin with which they were challenged (Greisman et al., 1979; van Dijk et al., 1981; Cross et al., 1984; Baumgartner et al., 1990). These antibodies are opsonic and have complement-dependent bactericidal activity (van Dijk et al., 1981; Kaufman et al., 1986; Sagawa et al., 1990; Oishi et al., 1992). However, because the *O*-polysaccharide is antigenically distinct for each strain or serotype of gram-negative bacteria and many different serotypes of gram-negative bacteria cause infection, *O*-polysaccharide-specific antibodies are difficult to use for the empirical treatment of septic shock. One would need an arsenal of many different O-serotype antibodies and rapid methods both to obtain culture results and to identify the O-serotype of the infecting pathogen. Alternatively, acutely ill patients with septic shock could be given a combination of O-specific antibodies that are effective against the most common pathogens before the organism is identified (Baumgartner, 1990; Baumgartner and Glauser, 1990). In part because of these problems, researchers have focused on developing antibodies directed at conserved epitopes on the endotoxin molecule such as the core sugars and the lipid A moiety with the hope of finding broadly "cross-protective" antibodies to a wide variety of heterogeneous gram-negative bacteria (Chedid et al., 1968; Braude and Douglas, 1972; McCabe and Greely, 1972; McCabe et al., 1972; Braude et al., 1973).

B. Polyclonal Antibodies Directed at the Endotoxin Core and Lipid A

1. Laboratory Studies

Much of the research into antibodies directed against shared epitopes in the core and lipid A regions of endotoxin has used two strains of bacteria, the J5 mutant of *E. coli* O111:B4 and *Salmonella minnesota* R595. These organisms produce endotoxins that lack O-antigens and possess only core elements and Lipid A. These are referred to as "rough" strains, as opposed to those with O-antigens, which are called "smooth" forms. Antisera developed from these rough strains have been reported to protect animals from challenges of bacteria of the smooth type and purified heterologous endotoxins (Ziegler et al., 1973; Young et al., 1975; Braude et al., 1977; McCabe et al., 1988). These results have been ascribed to the direct effects of core/lipid A-reactive antibodies and their presumed ability to bind and neutralize endotoxin, but these results have not been consistently reproducible (Greisman et al., 1979), particularly when serum obtained before immunization was used as a control (Greisman and Johnston, 1988). Several theories have been put forth to try to reconcile these differences: test antisera may have contained O-specific antibodies (Siber et al., 1985; Greisman and Johnston, 1988; Baumgartner, 1991); pretreatment was a part of most experiments, and inadvertent endotoxin contamination with the induction of tolerance could have occurred (Baumgartner, 1991).

2. Human Clinical Trials

Six clinical trials have used core-specific antiserum or core-specific intravenous immunoglobulin either as prophylaxis or for the treatment of patients with gram-negative septic shock (Table 1) (Ziegler et al., 1982; Baumgartner et al., 1985; McCutchan et al., 1983; Calandra et al., 1988; The Intravenous Immunoglobulin Collaborative Study Group, 1992; The J5 Study Group, 1992), but, like the animal data (Greisman et al., 1979), their findings were inconsistent. The first such human clinical trial studied patients with gram-negative sepsis treated with J5 antiserum produced from firemen immunized with heat-killed J5 *E. coli* (Ziegler et al., 1982). Preimmunization serum was used as the control. A total of 304 patients were entered into this randomized, blinded, controlled trial, and 212 were considered to have gram-negative bacteremia. A curious feature of this study is that 39 patients were found to have localized gram-negative infections without bacteremia, and 21 of these were included in the bacteremic group because they were on appropriate antibiotics at the time blood cultures were obtained. Overall mortality rates in the gram-negative bacteremia group, including the 21 additional patients, were reduced from 39% in control subjects to 22% in treated patients ($p = 0.011$). In the subgroup of septic patients requiring vasoactive drugs for an extended period of time (>6 h), mortality rates were 77% and 44%, respectively ($p = 0.003$). The impact of J5 antisera on all-cause mortality or on the major treatment group (i.e., all patients with gram-negative infection) was not reported, however, which complicated interpretation of these findings. Since the study was published, scientists have questioned the extremely high mortality rate in the control group, which may indicate an imbalance in pretreatment randomization (Baumgartner, 1990, 1991; Baumgartner and Glauser, 1990). In addition, no significant relation was found between anti-J5 antibody levels and improved outcome (Ziegler et al., 1982; Baumgartner, 1991). Consequently, it is not known whether factors other than cross-protection contributed to the reported differences between treatment groups.

Five subsequent clinical trials evaluating polyclonal core-reactive antiserum or immunoglobulin have not confirmed the beneficial effects (by standard two-tailed tests)

TABLE 1 Summary of Six Clinical Trials with LPS-Core-Directed Antiserum or Plasma

Therapy	Reported efficacy	Reported *p* value (survival)	Methodologic concerns about reported results	Target group	*p* value (survival)
J5 antiserum (Ziegler et al., 1982)	GNB + (GNI and appropriate anti-biotic therapy)	0.011	High mortality in controls	GNS	Not reported
J5 antiserum (McCutchan et al., 1983)	None	None	None	Prophylaxis of GNS in patients with pro-longed neutropenia	NS
J5 immune plasma (Baumgartner et al., 1985)	GNS (shock)	0.046	One-tailed test for efficacy	Prophylaxis of GNS in high-risk surgical patients	NS
J5 immune plasma (J5 Study Group, 1992)	None	None	None	Infectious purpura in children (shock)	NS
J5 IVIG (Calandra et al., 1988)	None	None	None	GNS (shock)	NS
J5 IVIG (IVIG Group, 1992)	None	None	None	Prophylaxis of high-risk surgical patients	NS

GNB = gram-negative bacteremia; GNS = gram-negative sepsis; GNI = gram-negative infection; LPS = lipopolysaccharide.

noted in this initial clinical trial (Table 1) (McCutchan et al., 1983; Baumgartner et al., 1985; Calandra et al., 1988; The Intravenous Immunoglobulin Collaborative Study Group, 1992; The J5 Study Group, 1992). In these studies, more than 800 patients were treated with either J5 antiserum, J5 immune plasma, or J5 intravenous immune globulin (IVIG). Three of the five studies were prophylactic (McCutchan et al., 1983; Baumgartner et al., 1985; The Intravenous Immunoglobulin Collaborative Study Group, 1992), and two were treatment of established infection (Calandra et al., 1988; The J5 Study Group, 1992). Scientists have suggested that sample sizes and study designs could explain the lack of protection in these studies. Additionally, some studies used immune globulin of the IgG class. Several investigators maintain that only IgM class immune globulin is protective (McCabe et al., 1988); however, this has not been convincingly proved. As a group, these studies raise serious doubts about the efficacy of polyclonal core-specific anti-endotoxin antibodies.

C. Monoclonal Antibodies Directed Against the Endotoxin Core and Lipid A

Despite problems with animal and clinical studies of polyclonal antibodies directed against core structures of endotoxin, many investigators still maintain that J5 antiserum contains antibodies against core/lipid A structures of endotoxin that cross-protect against diverse gram-negative infections. With the advent of hybridoma technology in the late 1970s, monoclonal antibodies against core and lipid A structures were developed in the hopes of finding more specific cross-protection therapy than that provided by polyclonal antiserum. Of clinical importance, monoclonal antibody therapy circumvents problems in the production of J5 polyclonal antiserum such as transmission of infection from serum donors to patients, variable antibody content of antiserum preparations, and potential toxicity produced during vaccination of serum donors (Smith et al., 1992).

To date, almost all of the laboratory and clinical data on monoclonal antibodies against endotoxin have been derived from two such IgM class antibodies, HA-1A (Centocor, Malvern, PA) and a human monoclonal antibody, E5 (XOMA, Berkeley, CA), of murine origin (Table 2). It may be unfortunate that these two antibodies have been used to study cross-reactive anticore monoclonal antibodies, because investigations using them have been inconclusive and cast further doubt on the cross-protection hypothesis.

1. Laboratory Studies of HA-1A and E5

An important requirement of these cross-protective, core/lipid A-reactive monoclonal antibodies is that they bind to endotoxin and cross-react with the endotoxins from a variety of gram-negative bacteria. Initial reports showed that both E5 and HA-1A fulfilled these criteria (Teng et al., 1985; Gazzono-Santoro et al., 1991). More recently, however, the binding specificity of HA-1A has been questioned, and descriptions of its binding characteristics have differed substantially from initial reports (Baumgartner, 1991; Warren et al., 1993). Of particular concern is that even if these antibodies do bind to endotoxin, it has not been convincingly shown that they neutralize endotoxin or are opsonic or bactericidal (Chia et al., 1989; Baumgartner et al., 1990; Warren et al., 1993). Thus, if they have an antiendotoxic effect, the mechanism is unknown.

Animal studies using these two IgM class antibodies have produced conflicting results. In initial reports, HA-1A protected mice from lethal bacteremia (Teng et al., 1985). In a rabbit model of sepsis, HA-1A prevented the dermal Shwartzman reaction (Teng et al.,

TABLE 2 Summary of Four Clinical Trials with LPS-Core-Directed Monoclonal Antibodies

Therapy	Reported efficacy	Reported p value (survival)	Methodologic concerns about reported results	Target group	p value (survival)
E5 (Greenway, 1991)	GNS (no shock)	0.01	Retrospective	GNS	NS
E5 (Wenzel et al., 1991)	None	None	None	GNS (no shock)	NS
HA-1A (Ziegler et al., 1991)	GNB (shock, 28 days all cause mortality)	0.017	Revised analytic plan rejected	GNS (14-day mortality)	NS
HA-1A (McCloskey et al., 1994)	Early stopping criteria made for excess mortality in the HA-1A group.				

GNB = gram-negative bacteremia; GNS = gram-negative sepsis; GNI = gram-negative infection; LPS = lipopolysaccharide.

1985), and in another rabbit model, decreased mortality (Ziegler et al., 1987). E5 also protected mice challenged with bacteria (Young et al., 1989). However, in studies of E5 in sheep challenged with endotoxin there was no meaningful protection (Wheeler et al., 1990). Other researchers have been unable to reproduce the beneficial effects of HA-1A found in mice and rabbits (Baumgartner, 1991; Warren et al., 1992). In one study, HA-1A did not protect mice from bacterial or endotoxic challenge (Baumgartner et al., 1990). It is disturbing that in a blinded, controlled study using a canine model of septic shock, HA-1A was harmful (Quezado et al., 1991, 1992, 1993a). All canines were infected with an intraperitoneal clot containing *E. coli* O111:B4, the smooth, parent form of the *E. coli* J5 organism used to generate the heteromyeloma cells that produce HA-1A. Among animals treated with HA-1A (10 mg/kg) there were significantly more deaths and worse cardiovascular dysfunction and organ injury. Furthermore, HA-1A had no effect on clearing endotoxemia or bacteremia. Despite conflicting laboratory data, both of these antibodies have been tested in two large, multicenter, randomized clinical trials.

2. Human Clinical Trials Using HA-1A and E5

HA-1A and E5 have each undergone two placebo-controlled, randomized multicenter clinical trials in patients with presumed gram-negative sepsis (Table 2) (Greenman et al., 1991; Wenzel et al., 1991; Ziegler et al., 1991; McCloskey et al., 1994). In the first E5 trial, 468 patients were enrolled, and 316 had documented gram-negative infection (Greenman et al., 1991). E5 had no effect in these patients, but in a retrospectively identified, unusual subset of patients ($n = 137$) with gram-negative infection who were not in refractory shock, a statistically significant increase in survival was reported (70% vs. 50%, $p = 0.01$). Results from the second E5 trial have not been fully published (Wenzel et al., 1991). In brief, 847 patients were enrolled, of whom 530 had gram-negative infections without refractory shock. The study failed to confirm the favorable effect in the subgroup without refractory shock from the original study but did confirm that E5 had no effect on mortality from gram-negative infections. XOMA Corporation, using meta-analysis and combining these two clinical trials, found that E5 significantly decreased the time to recovery from organ dysfunction and improved survival in a subgroup of patients with gram-negative infection and organ dysfunction without refractory shock. At present, a third clinical trial is enrolling patients. (Xoma corporation, personal communication).

In the first clinical trial using HA-1A, 543 patients were enrolled, and the authors reported that HA-1A had no effect in the overall group randomized or in patients with gram-negative infection (Table 2) (Ziegler et al., 1991). In the subgroup of patients (*n* = 200) with gram-negative bacteremia, HA-1A significantly decreased mortality at 28 days compared with placebo (30% vs. 49%, *p* = 0.014). This effect was even more significant in the subset of patients with gram-negative bacteremia and shock. The disclosure of methods and the analysis of the data was published in the *New England Journal of Medicine* article (Ziegler et al., 1991) caused considerable debate among the FDA (Siegel, 1990; Pierce, 1993), the *New England Journal of Medicine* (Relman, 1990), the scientific community (Warren et al., 1992; Pierce, 1993), the published article's author, and Centocor Corporation (Ziegler and Smith, 1992; Pierce, 1993).

At an FDA advisory committee meeting on September 4, 1991, the presentation of HA-1A data differed substantially from that in the published article (Warren et al., 1992; Piercey, 1993). The analytic plan presented to the FDA contained three overlapping patient subgroups, two types of mortality (septic and all-cause), and two end points for survival (14 and 28 days). There is no standard approach to adjust *p* values for these multiple overlapping subgroups, and it was concluded that if HA-1A did have an effect, it was marginal (Warren et al., 1992). Another criticism was that randomization of patients into HA-1A and control groups was unbalanced, such that sicker patients and patients not given effective antibiotic therapy may have been assigned more often to the placebo group (Warren et al., 1992). The reason the published report differed from the FDA analysis has not been stated. Of note, Dr. J. Siegel (who oversaw the HA-1A clinical trial at the FDA) and Dr. A. Relman (then editor of the *New England Journal of Medicine*), just before publication of this article, debated in the *NEJM* correspondence section whether prospective analytic plans should be submitted with manuscripts. Dr. Relman concluded that this was not necessary because

> Reports of clinical trials are supposed to describe the initial protocol adequately and mention any subsequent modifications. Failure to do so constitutes a breach of scientific conduct. The rare scientist . . . would do so whether or not they submit their original protocol.

The FDA has subsequently determined that this clinical trial did not conclusively establish the efficacy of HA-IA. The HA-1A clinical trial had two separate analytic plans, and at the FDA meeting the (second) revised analytic plan was presented (Siegel et al., 1992; Pierce, 1993). The FDA believed that the revised analytic plan had been submitted without knowledge of preliminary data from the clinical trial. The FDA subsequently learned that data from the clinical trial had become available to Centocor staff members before the revised analytic plan was submitted, so bias could not be excluded. Analysis of the study using the original analytic plan and the original prospective mortality end point of 14 days showed that HA-1A did not significantly improve survival in patients with gram-negative bacteremia (Siegel et al., 1992; Piercey, 1993).

Accordingly, the FDA did not approve HA-1A for clinical use. A second clinical trial was performed to assess whether HA-1A therapy is beneficial in patients with gram-negative bacteremia. Almost a year after HA-1A had been reported to be harmful in canines, this second clinical trial was stopped because interim safety monitoring showed excess mortality in the group without gram-negative bacteremia receiving HA-1A (Piercey, 1993; McCloskey et al., 1994). Further this second clinical trial showed no benefit of HAIA in patients with gram-negative bacteria (McCloskey et al., 1994).

Thus, neither clinical nor animal data support the use of these antibodies (E5, HA-1A) to treat septic shock. To date, the cross-protective hypothesis has not been proved, and a mechanism of therapeutic action has not been convincingly demonstrated, for core/lipid A-reactive antibodies. The increased mortality shown in animal and human data with one monoclonal antibody (HA-1A) emphasizes the need to understand how core-reactive antibodies work before they are widely tested in humans with severe infections (Quezado et al., 1991; Quezado and Hoffman, 1993; Hoffman and Natanson, 1993a,b).

V. NONANTIBODY THERAPIES DIRECTED AT ENDOTOXIN IN SEPTIC SHOCK

The controversial results of monoclonal antibody therapies directed against core endotoxin structures in septic shock should not be used to judge all therapies directed at this toxin. Newer agents, such as polymyxin B nonapeptide, bactericidal/permeability-increasing protein, and tachyplesin, have been developed that are known to bind as well as neutralize endotoxin. Importantly, these agents may help determine whether binding and neutralizing endotoxin, independent of antibacterial therapy, is an appropriate therapeutic strategy to treat septic shock. This final section discusses these nonantibody antiendotoxin therapies according to their therapeutic mechanisms.

A. Nonantibody Agents That Bind Endotoxin

1. Cationic Polypeptides

Polymyxin B and colistin are polycationic antibiotics that have antiendotoxin properties (Rifkind and Palmer, 1966; Craig et al., 1974; Bannatyne et al., 1977; Vaara and Vaara, 1983; Warren et al., 1985; Danner et al., 1989b; Stokes et al., 1989). These agents bind tightly to the lipid A region of endotoxin by both ionic and hydrophobic interactions that neutralize endotoxic properties (Morrison and Jacobs, 1976). In small-animal models of endotoxic shock, Polymyxin B lowered mortality rates if given prophylactically, simultaneously, or temporally close to an endotoxin challenge (Rifkind, 1967; Craig et al., 1974; Flynn et al., 1987). In a large-animal model, however, in doses protective in mice, polymyxin B was toxic and did not improve survival rates (Craig et al., 1974). In bacterial challenge models using an appropriate nonpolycationic antibiotic and polymyxin B, although morbidity was improved (amelioration of hypotension and acidosis), survival was unaffected (Flynn et al., 1987).

Because of the untoward side effects of polymyxin B (nephrotoxicity, seizures, neuromuscular blockade, and confusion) (Craig et al., 1974; Corrigan and Kiernat, 1979) and limited therapeutic success (Rifkind, 1967; Craig et al., 1974; Flynn et al., 1987), less toxic derivatives have been developed. One such derivative, polymyxin B nonapeptide, possesses antiendotoxin properties in vitro and lacks the in vivo toxicities of the parent compound. Investigators have found, however, that polymyxin B nonapeptide, like its parent compound, may need to be given before or simultaneously with endotoxin to be effective (Danner et al., 1989b). To reduce toxicity and increase potential beneficial effects, polymyxin B has been immobilized on a filter to use with extracorporeal hemoperfusion. This treatment was protective in large animals challenged with gram-negative bacteria (Hanasawa et al., 1988). Recently, a polymyxin B-dextran 70 conjugate with reduced toxicity and good antiendotoxin activity was made (Handly and Lake, 1992). Toxicity, dosage, and timing of therapy are being established for the possible evaluation of this compound in a human clinical trial.

2. Cationic Proteins

Bactericidal/permeability-increasing protein (BPI), a 55-kDa protein produced by human neutrophils (Weis et al., 1978) discussed in Section II, is being actively studied because it binds and neutralizes endotoxin (Weis et al., 1978; Marra et al., 1990, 1992; Ooi et al., 1991; Gazzano-Santoro et al., 1992). BPI binds to the lipid A moiety of a wide variety of gram-negative bacteria (Gazzano-Santoro et al., 1992) and can inhibit endotoxin-mediated effects such as tumor necrosis factor production and neutrophil priming (Ooi et al., 1991). BPI and a recombinant amino-terminal fragment of BPI (rBPI$_{23}$) also being developed to treat septic shock are protective in mice models of sepsis (Marra et al., 1990, 1992; Ooi et al., 1991). Another protein, tachyplesin, with 102 amino acids, inhibits endotoxin-induced coagulation in horseshoe crabs. This coagulation cascade forms the basis of the *Limulus* amebocyte lysate (LAL) test used to detect endotoxin (Ohashi et al., 1984). Further studies are being done to evaluate cationic proteins to treat septic shock in the hopes of bringing these agents into human clinical trials in the near future.

3. Lipoproteins

As noted in Section II, lipoproteins from normal serum can bind, neutralize, and clear endotoxins from the circulation (Ulevitch et al., 1979; Warren et al., 1988; Harris et al., 1990). Reconstituted high-density lipoproteins made from either purified or recombinant material have been shown to protect small animals from endotoxic challenge (Harris et al., 1990). Problems with dosages, purification, and untoward side effects are presently being addressed prior to bringing these agents into human clinical trials (Quezado et al., 1994).

B. Analogues of Lipid A

Another area of active research during the past 10 years relates to compounds that are structurally similar to the toxic moiety of endotoxin, lipid A. Analogues are being sought that either maintain the beneficial immunostimulatory properties of endotoxin without the toxicities or directly antagonize the effects of endotoxin. Several lipid A analogues have been investigated: deacylated endotoxin, lipid X, and monophosphoryl lipid A (Takayama et al., 1983; Galanos et al., 1984; Kiso et al., 1984; Ray et al., 1984; Tanamoto et al., 1984; Munford and Hall, 1985; Astiz et al., 1989).

1. Deacylated Endotoxins

Deacylated endotoxins are formed from the hydrolysis of acyloxyacyl bonds in lipid A (Munford and Hall, 1985, 1986). Acyloxyacyl hydrolases that break these bonds are found in neutrophils and may be part of the natural host system for the detoxification of endotoxin (Munford and Hall, 1986; Riedo et al., 1990). Deacylated endotoxin maintains some of its potentially beneficial immunostimulatory properties while having greatly reduced tissue toxicity (dermal Shwartzman reaction) (Munford and Hall, 1986). Further, deacylated endotoxin inhibits the ability of endotoxin to augment endothelial cell–neutrophil adherence, prostaglandin production, and plasminogen activator inhibitor-1 expression (Pohlman et al., 1987; Riedo et al., 1990). Deacylated endotoxins have not yet been evaluated in animal models of sepsis to determine whether these agents have protective effects in vivo.

2. Lipid X

Lipid X is a monosaccharide that is a precursor of lipid A (Takayama et al., 1983; Raetz, 1984; Ray et al., 1984; Danner, 1990). Lipid X, depending on dose and time of incubation, prevents endotoxin-induced enhancement of superoxide production in an in vitro neutrophil system (Danner et al., 1987). The pattern of inhibition and the structural similarity between lipid X and lipid A suggests that lipid X may be a competitive inhibitor of some endotoxin–cell interactions (Danner et al., 1987). Early studies in animals using lipid X have been difficult to interpret because of contaminating disaccharide derivatives, which may have affected experimental results (Burhop et al., 1985; Golenbock et al., 1987; Aschauer et al., 1990; Lam et al., 1991; Corriveau and Danner, 1993). Studies in mice using pure lipid X demonstrate protection from a lethal endotoxic challenge (Lam et al., 1991); however, this protection was not seen in neutropenic mice infected with *Pseudomonas aeruginosa* (Lam et al., 1991) or in antibiotic-treated dogs infected with *E. coli* (Danner et al., 1993). Thus, lipid X may have antiendotoxin activity, but this may not be beneficial in actual gram-negative infections. More recently, other lipid A derivatives, such as diphosphoryl lipid A of *Rhodopseudomonas sphaeroides*, have been found to block endotoxin-induced tumor necrosis factor production by monocytes/macrophages, an effect not seen with lipid X (Takayama et al., 1989). Newer agents in this class of antiendotoxin therapy are being actively investigated (Takayama et al., 1989; Kovach et al., 1990; Van Dervort et al., 1992).

3. Monophosphoryl Lipid A

Monophosphoryl lipid A has been isolated from a mutant strain of *Salmonella* (Astiz et al., 1989). In vitro, it can inhibit endotoxin-induced enhancement of neutrophil superoxide production (Heiman et al., 1990). In rats, monophosphoryl lipid A did not produce the cardiovascular abnormalities seen with endotoxin (Astiz et al., 1989). Monophosphoryl lipid A given to endotoxin-challenged animals attenuated the adverse cardiovascular effects and improved short-term survival (Rackow et al., 1989; Carpati et al., 1990). In normal human volunteers, monophosphoryl lipid A caused release of cytokines into the circulation. However, a second dose of monophosphoryl lipid A given 24 h after the first produced an attenuated cytokine response (Von Eschen et al., 1992). These lipid A analogues are now being used in a human clinical trial of sepsis. (Ribi Company, personal communication).

C. Plasma Detoxification

Activated charcoal, bentonite, and Kaopectate are all efficient at absorbing endotoxin from plasma in vitro (Bysani et al., 1990). Extracorporeal activated charcoal hemoperfusion can remove almost all the circulating endotoxin infused into animals within an hour (Bende and Bertók, 1986). Several experimental therapies have focused on removing harmful products such as endotoxin from the blood less selectively. Case reports have suggested that exchange whole blood transfusion in neonates affects survival by clearing endotoxin (Togari et al., 1983). In infants and children with severe sepsis, case reports suggest that exchange transfusion mitigates the abnormalities of disseminated intravascular coagulation and thus improves survival (Gross and Melhorn, 1971; Töllner et al., 1977; Vain et al., 1980). There are similar anecdotal reports of benefit in severe meningococcal and pneumococcal sepsis treated with plasmapheresis (Scharfman et al., 1979; Bjorvatn et al., 1984; Brandtzaeg et al., 1985). In a controlled clinical trial in an animal bacteremic shock model, however, plasmapheresis was found to be harmful. In that study, plasma exchange worsened cardiovascular abnormalities and decreased survival (Natanson et al., 1993). A similarly designed study using continuous arteriovenous hemofiltration also showed no benefit (Freeman et al., 1994). These animal data suggest that nonselective plasma detoxification schemes (exchange transfusion, hemoperfusion, plasmapheresis, continuous arteriovenous hemofiltration), in addition to removing bacterial toxins, may also remove host mediators to the extent that various beneficial effects may be compromised. However, highly specific or selective resins to detoxify plasma may still prove to be useful in the treatment of septic shock.

VI. CONCLUSION

In summary, endotoxin is a complex molecule that elicits a broad nonspecific host response. Although essential to protect the host from infection, when excessive this inflammatory response may produce organ damage and death. Endotoxin is not, however, the only mediator of septic shock, and the relationship between circulating endotoxin and outcome is not well defined. Clinical trials using antibodies against endotoxin have been unsuccessful, and it is still unknown whether endotoxin is an appropriate target for the treatment of septic shock. New nonantibody agents are being developed that, in contrast to core-directed antiendotoxin antibodies, actually neutralize or inhibit endotoxin. Future studies may help clarify whether therapeutic strategies using these agents may be effective for treating patients with septic shock.

REFERENCES

Aschauer, H., Grob, A., Hildebrandt, J., Schuetze, E., and Stuetz, P. (1990). Highly purified lipid X is devoid of immunostimulatory activity. *J. Biol. Chem. 265*: 9159–9164.

Astiz, M. E., Rackow, E. C., Kim, Y. B., and Weil, M. H. (1989). Hemodynamic effects of monophosphoryl lipid A compared to endotoxin. *Circ. Shock 27*: 193.

Ayala, A., Kisala, J. M., Felt, J. A., Perrin, M. M., and Chaudry, I. H. (1992). Does endotoxin tolerance prevent the release of inflammatory monokines (interleukin a, interleukin 6, or tumor necrosis factor) during sepsis? *Arch. Surg. 127*: 191–196.

Bannatyne, R. M., Harnett, N. M., Lee, K. Y., and Biggar, W. D. (1977). Inhibition of the biologic effects of endotoxin on neutrophils by polymyxin B sulfate. *J. Infect. Dis. 136*: 469.

Baumgartner, J. D. (1990). Monoclonal anti-endotoxin antibodies for the treatment of gram-negative bacteremia and septic shock. *Eur. J. Clin. Microbiol. Infect. Dis. 9*: 711–716.

Baumgartner, J. D. (1991). Immunotherapy with antibodies to core lipopolysaccharide: a critical appraisal. *Infect. Dis. Clin. North Am. 5*: 915–927.

Baumgartner, J. D., and Glauser, M. P. (1990). Immunotherapy of gram-negative septic shock. In *Update in Intensive Care and Emergency Medicine, Vol. 10*, J. L. Vincent (Ed.), Springer-Verlag, Frankfurt, pp. 107–120.

Baumgartner, J. D., Glauser, M. D., McCutchan, J. A., Ziegler, E. J., van Melle, G., Klauber, M. R., Vogt, M., Muehlen, E., Lüthy, R., Chiolero, R., and Geroulanos, S. (1985). Prevention of gram-negative shock and death in surgical patients by prophylactic antibody to endotoxin core glycolipid. *Lancet 2*: 59–63.

Baumgartner, J. D., Heumann, D., Gerain, J., Weinbreck, P., Grau, G. E., and Glauser, M. P. (1990). Association between protective efficacy of anti-lipopolysaccharide (LPS) antibodies and suppression of LPS-induced tumor necrosis factor α and interleukin 6. *J. Exp. Med.* *171*: 889–896.

Bende, S., and Bertók, L. (1986). Elimination of endotoxin from the blood by extracorporeal activated charcoal hemoperfusion in experimental canine endotoxin shock. *Circ. Shock 19*: 239.

Bjorvatn, B., Bjertnaes, L., Fadnes, H. O., Flaegstad, T., Gutteberg, T. J., Kristiansen, B. E., Pape, J., Rekvig, O. P., Osterud, B., and Aanderud, L. (1984). Meningococcal septicaemia treated with combined plasmapheresis and leucapheresis or with blood exchange. *Br. Med. J. 288*: 439.

Brade, H., Brade, I., Schade, U., Zähringer, U., Holst, O., Kuhn, H. M., Rozalski, A., Röhrscheidt, E., and Rietschel, E. T. (1988). Structure, endotoxicity, immunogenicity and antigenicity of bacterial lipopolysaccharides (endotoxins, O-antigens). *Prog. Clin. Biol. Res.* *272*: 17–45.

Brandtzaeg, P., Sirnes, K., Folsland, B., Godal, H. C., Kierulf, P., Bruun, J. N., and Dobloug, J. (1985). Plasmapheresis in the treatment of severe meningococcal or pneumococcal septicemia with DIC and fibrinolysis. *Scand. J. Clin. Lab. Invest. 45*: 53.

Brandtzaeg, P., Kierulf, P., Gaustad, P., Skulberg, A., Bruun, J. N., Halvorsen, S., and Sorensen, E. (1989). Plasma endotoxin as a predictor of multiple organ failure and death in systemic meningococcal disease. *J. Infect. Dis. 159*: 195–204.

Braude, A. I., and Douglas, H. (1972). Passive immunization against the local Shwartzman reaction. *J. Immunol. 108*: 505–512.

Braude, A. I., Douglas, H., and Davis, C. E. (1973). Treatment and prevention of intravascular coagulation with antiserum to endotoxin. *J. Infect. Dis. 128*(Suppl): S157–S164.

Braude, A. I., Ziegler, E. J., Douglas, H., and McCutchan, J. A. (1977). Antibody to cell wall glycolipid of gram-negative bacteria: induction of immunity to bacteremia and endotoxemia. *J. Infect. Dis. 136*: S167–S173.

Brigham, K. L., Bowers, R. E., and Haynes, J. (1979). Increased sheep lung vascular permeability caused by *Escherichia coli* endotoxin. *Circ. Res. 45*: 292–297.

Brozek, K. A., and Raetz, C. R. H. (1990). Biosynthesis of lipid A in *Escherichia coli*. *J. Biol. Chem. 265*: 15410–15417.

Brues, A. M., and Shear, M. J. (1944). Chemical treatment of tumors. X. Reactions of four patients with advanced malignant tumors to injection of a polysaccharide from *Serratia marcescens* culture filtration. *J. Natl. Cancer Inst. 5*: 195–208.

Burhop, K. E., Proctor, R. A., Helgerson, R. B., Raetz, C. R. H., Starling, J. R., and Will, J. A. (1985). Pulmonary pathophysiological changes in sheep caused by endotoxin precursor, lipid X. *J. Appl. Physiol. 59*: 1726–1732.

Bysani, G. K., Shenep, J. L., Hildner, W. K., Stidham, G. L., and Roberson, P. K. (1990). Detoxification of plasma containing lipopolysaccharide by adsorption. *Crit. Care Med. 18*: 67.

Calandra, T., Glauser, M. P., Schellekens, J., Verhoef, J., and the Swiss-Dutch J5 Immunoglobulin Study Group. (1988). Treatment of gram-negative septic shock with human IgG antibody to *Escherichia coli* J5: a prospective, double-blind, randomized study. *J. Infect. Dis. 158*: 312–319.

Carpati, C., Astiz, M. E., Rackow, E. C., Kim, J. W., Kim, Y. B., and Weil, M. H. (1990). Monophosphoryl lipid A attenuates septic shock in pigs. *Crit. Care Med. 18*: S260 (Abstract).

Chedid, L., Parant, M., Parant, F., and Boyer, F. (1968). A proposed mechanism for natural immunity to enterobacterial pathogens. *J. Immunol. 100*: 292–301.

Chia, J. K. S., Pollack, M., Guelde, G., Koles, N. L., Miller, M., and Evans, M. E. (1989). Lipopolysaccharide (LPS)-reactive monoclonal antibodies fail to inhibit LPS-induced tumor necrosis factor secretion by mouse-derived macrophages. *J. Infect. Dis. 159*: 872–880.

Corrigan, J. J., Jr., and Kiernat, J. F. (1979). Effect of polymyxin B sulfate on endotoxin activity in a gram-negative septicemia model. *Pediatr. Res. 13*: 48.

Corriveau, C. C., and Danner, R. L. (1993). Endotoxin as a therapeutic target in septic shock. *Infect. Agents Dis. 2*: 35–43.

Craig, W. A., Turner, J. H., and Kunin, C. M. (1974). Prevention of the generalized Shwartzman reaction and endotoxin lethality by polymyxin B localized in tissues. *Infect. Immun. 10*: 287.

Cross, A. S., Gemski, P., Sadoff, J. C., Orskov, F., and Orskov, I. (1984). The importance of the K1 capsule in invasive infections caused by *Escherichia coli. J. Infect. Dis. 149*: 184–193.

Danner, R. L. (1990). Mediators and endotoxin inhibitors. *Ann. Intern. Med. 113*: 227.

Danner, R. L., Joiner, K. A., and Parrillo, J. E. (1987). Inhibition of endotoxin-induced priming of human neutrophils by lipid X and 3-aza-lipid X. *J. Clin. Invest. 80*: 605.

Danner, R. L., Suffredini, A. F., Natanson, C., and Parrillo, J. E. (1989a). Microbial toxins: role in the pathogenesis of septic shock and multiple organ failure. In *Multiple Organ Failure*, D. J. Bihari and F. B. Cerra (Eds.), Society of Critical Care Medicine, Fullerton, CA, pp. 151–191.

Danner, R. L., Joiner, K. A., Rubin, M., Patterson, W. H., Johnson, N., Ayers, K. M., and Parrillo, J. E. (1989b). Purification, toxicity, and antiendotoxin activity of polymyxin B nonapeptide. *Antimicrob. Agents Chemother. 33*: 1428.

Danner, R. L., Natanson, C., Elin, R. J., Hosseini, J. M., Banks, S., MacVittie, T. J., and Parrillo, J. E. (1990). *Pseudomonas aeruginosa* compared with *Escherichia coli* produces less endotoxemia but more cardiovascular dysfunction and mortality in a canine model of septic shock. *Chest 98*: 1480–1487.

Danner, R. L., Elin, R. J., Hosseini, J. M., Wesley, R. A., and Parrillo, J. E. (1991). Endotoxemia in human septic shock. *Chest 99*: 169–176.

Danner, R. L., Doerfler, M. E., Eichacker, P. Q., Reilly, J. M., Ratica, D., Wilson, J., MacVittie, T. J., Stuetz, P., Parrillo, J. E., and Natanson, C. (1993). A therapeutic trial of lipid X in a canine model of septic shock. *J. Infect. Dis. 167*: 378–384.

Dinarello, C. A. (1991). The proinflammatory cytokines interleukin-1 and tumor necrosis factor and treatment of the septic shock syndrome. *J. Infect. Dis. 163*: 1177–1184.

Elin, R. J. (1979). Clinical utility of the limulus test with blood, CSF, and synovial fluid. In *Biomedical Applications of the Horseshoe Crab (Limulidae)*, E. Cohen (Ed.), Alan R. Liss, New York, pp. 279–292.

Elin, R. J., and Wolff, S. M. (1973). Nonspecificity of the *Limulus* amebocyte lysate test: positive reactions with polynucleotides and proteins. *J. Infect. Dis. 128*: 349.

Elin, R. J., Sandberg, A. L., and Rosenstreich, D. L. (1976). Comparison of the pyrogenicity, limulus activity, mitogenicity and complement reactivity of several bacterial antiendotoxins and related compounds. *J. Immunol. 117*: 1238.

Esbenshade, A. M., Newman, J. H., Lams, P. M., Jolles, H., and Brigham, K. L. (1982). Respiratory failure after endotoxin infusion in sheep: lung mechanics and lung fluid balance. *J. Appl. Physiol. 53*: 967–976.

Fine, J. (1954). The bacterial factor in traumatic shock. In *American Lecture Series: Monograph in American Lectures in Circulation*, I. H. Page and A. C. Corcoran (Eds.), Publ. No. 219, Thomas, Springfield, IL.

Fink, M. P., Antonsson, J. B., Wang, H. L., and Rothschild, H. R. (1991). Increased intestinal permeability in endotoxin pigs: mesenteric hypoperfusion as an etiologic factor. *Arch. Surg. 126*: 211–218.

Fiorentino, D. F., Zlotnik, A., Mosmann, T. R., Howard, M., and O'Garra, A. (1991). IL-10 inhibits cytokine production by activated macrophages. *J. Immunol. 147*: 3815–3822.

Flynn, P. M., Shenep, J. L., Stokes, D. C., Fairclough, D., and Hildner, W. K. (1987). Polymyxin B moderates acidosis and hypotension in established, experimental gram-negative septicemia. *J. Infect. Dis. 156*: 706.

Freeman, B. D., Yatsiv, I., Natanson, C., Solomon, M. A., Quezado, Z. M. N., Danner, R. L., Banks, S. M., and Hoffman, W. D. (1994). Continuous ateriovenous hemofiltration does not improve survival in a canine model of septic shock. *J. Pharmacol. Exp. Therap.* In press.

Frey, E. A., Miller, D. S., Jahr, T. G., Sundan, A., Bazil, V., Espevik, T., Finlay, B. B., and Wright, S. D. (1992). Soluble CD14 participates in the response of cells to lipopolysaccharide. *J. Exp. Med. 176*: 1665–1671.

Galanos, C., Lehmann, V., Lüderitz, O., Rietschel, E. T., Westphal, O., Brade, H., Brade, L., Freudlenberg, M. A., Hansen-Hagge, T., and Lüderitz, T. (1984). Endotoxic properties of chemically synthesized lipid A part structures. *Eur. J. Biochem. 140*: 221.

Galanos, C., Luderitz, O., Reitchel, E., Westphal, O., Brade, H., Brade, L., Freudenberg, M., Schade, U., Imoto, M., and Yoshimura, H. (1985). Synthetic and natural *Escherichia coli* free lipid A express identical endotoxin activities. *Eur. J. Biochem. 148*: 1–5.

Gazzono-Santoro, H., Parant, J. B., Wood, O. M., Lim, E., Pruyne, D. T., Troun, P. W., and Conlon, P. J. (1991). Reactivity of E5 monoclonal antibody to smooth lipopolysaccharides. In *Program and Abstracts of the 31st Interscience Conference on Antimicrobial Agents and Chemotherapy*, Chicago, Sept. 29–Oct. 2, 1991. Am. Soc. Microbiology, Washington, DC, p. 230 (Abstract).

Gazzano-Santoro, H., Parent, J. P., Grinna, L., Horwitz, A., Parsons, T., Theofan, G., Elsbach, P., Weiss, J., and Conlon, P. J. (1992). High-affinity binding of the bactericidal/permeability-increasing protein and a recombinant amino-terminal fragment to the lipid A region of lipopolysaccharide. *Infect. Immun. 60*: 1–6.

Gigliotti, F., and Shenep, J. L. (1985). Failure of monoclonal antibodies to core glycolipid to bind intact strains of *Escherichia coli*. *J. Infect. Dis. 151*: 1005–1011.

Glauser, M. P., Zanetti, G., Baumgartner, J. D., and Cohen, J. (1991). Septic shock pathogenesis. *Lancet 338*: 732–739.

Golenbock, D. T., Will, J. A., Raetz, C. R. H., and Proctor, R. A. (1987). Lipid X ameliorates pulmonary hypertension and protects sheep from death due to endotoxin. *Infect. Immun. 55*: 2471–2476.

Granowitz, E. V., Santos, A. A., Poutsiaka, D. D., Cannon, J. G., Wilmore, D. W., Wolff, S. M., and Dinarello, C. A. (1991). Production of interleukin-1-receptor antagonist during experimental endotoxaemia. *Lancet 338*: 1423–1424.

Greenman, R. L., Schein, R. M., Martin, M. A., Wenzel, R. P., MacIntyre, N. R., Emmanuel, G., Chmel, H., Kohler, R. B., McCarthy, M., Plouffe, J., Russell, J. A., and the XOMA Sepsis Study Group. (1991). A controlled clinical trial of E5 murine monoclonal IgM antibody to endotoxin in the treatment of gram-negative sepsis. *J. Am. Med. Assoc. 266*: 1097–1102.

Greisman, S. E., and Johnston, C. A. (1988). Failure of antisera to J5 and R595 rough mutants to reduce endotoxemic lethality. *J. Infect. Dis. 157*: 54–64.

Greisman, S. E., Hornick, R. B., Wagner, H. N., Jr., Woodward, W. E., and Woodward, T. E. (1969). The role of endotoxin during typhoid fever and tularemia in man. IV. The integrity of the endotoxin tolerance mechanisms during infection. *J. Clin. Invest. 48*: 613–629.

Greisman, S. E., DuBuy, J. B., and Woodward, C. L. (1979). Experimental gram-negative bacterial sepsis: prevention of mortality not preventable by antibiotics alone. *Infect. Immun. 25*: 538–557.

Gross, S., and Melhorn, D. K. (1971). Exchange transfusion with citrated whole blood for disseminated intravascular coagulation. *J. Pediatr. 78*: 415.

Halling, J. L., Hamill, D. R., Lei, M.-G., and Morrison, D. C. (1992). Identification and characterization of lipopolysaccharide-binding proteins on human peripheral blood cell populations. *Infect. Immun. 60*: 845–852.

Hampton, R. Y., Golenbock, D. T., Penman, M., Krieger, M., and Raetz, C. R. H. (1991). Recognition and plasma clearance of endotoxin by scavenger receptors. *Nature 352*: 342–344.

Hanasawa, K., Tani, T., Oka, T., Yoshioka, T., Aoki, H., Endo, Y., and Kodama, M. (1988). Selective removal of endotoxin from the blood by extracorporeal hemoperfusion with polymyxin B immobilized fiber. *Proc. Clin. Biol. Res. 264*: 337.

Handly, D. A., and Lake, P. (1992). Results with polymyxin B conjugates as anti-endotoxin agents. Presented at 2nd Annual Meeting on Advances in Prevention and Treatment of Endotoxemia and Sepsis, June 22–23, Philadelphia, PA, (Abstract).

Harris, H. W., Grunfeld, C., Feingold, K. R., and Rapp, J. R. (1990). Human VLDL and chylomicrons can protect against endotoxin-induced death in mice. *J. Clin. Invest. 86*: 696–792.

Heiman, D. F., Astiz, M. E., Rackow, E. C., Rhein, D., Kim, Y. B., and Weil, M. H. (1990). Monophosphoryl lipid A inhibits neutrophil priming by lipopolysaccharide. *J. Lab. Clin. Med. 116*: 237.

Heumann, D., Baumgartner, J. D., Jacot-Guillarmod, H., and Glauser, M. P. (1991). Antibodies to core lipopolysaccharide determinants: absence of cross-reactivity with heterologous lipopolysaccharides. *J. Infect. Dis. 163*: 762–768.

Hoffman, W. D., and Natanson, C. (1993a). Endotoxin in septic shock. *Anesth. Analg. 77*: 13–24.

Hoffman, W. D., and Natanson, C. (1993b). The role of endotoxin in bacterial septic shock. In *Update in Intensive Care and Emergency Medicine*, Vol. 13, S. L. Vincent (Ed.), Springer-Verlag, Frankfurt.

Hoffman, W. D., Natanson, C., Danner, R. L., Koev, L., Banks, S. M., Elin, R. J., Hosseini, J. M., and Parrillo, J. E. (1989). Bacterial organism virulence factors may be more important than endotoxemia in determining cardiovascular (CV) dysfunction and mortality in canine septic shock. *Clin. Res. 37*: 344 (Abstract).

Hoffman, W. D., Danner, R. L., Koev, L., Banks, S. M., Walker, L. D., Elin, R. J., Hosseini, J. M., Dolan, D. P., and Natanson, C. (1990a). Ability of endotoxin and heat stable components of two strains of *E. coli* to produce lethality in canine septic shock. *Clin. Res. 38*: 454 (Abstract).

Hoffman, W. D., Danner, R. L., Koev, L. A., Banks, S. M., Elin, R. J., Hosseini, J. M., Walker, L. D., Dolan, D. P., and Natanson, C. (1990b). Ability of endotoxin and heat stable components of two strains of *E. coli* to produce lethality in canine septic shock. *Crit. Care Med. 18*: S212 (Abstract).

The Intravenous Immunoglobulin Collaborative Study Group. (1992). Prophylactic intravenous administration of standard immune globin as compared with core-lipopolysaccharide immunoglobulin in patients at high risk of post surgical infections. *N. Engl. J. Med. 327*: 234–240.

Iwanaga, S., Morita, T., and Harada, T. (1978). Chromogenic substrate for horseshoe crab clotting enzyme: its application for the assay of bacterial endotoxins. *Haemostasis 7*: 183–188.

The J5 Study Group. (1992). Treatment of severe infectious purpura in children with human plasma from donors immunized with *Escherichia coli* J5: a prospective, double-blind study. *J. Infect. Dis. 165*: 695–701.

Jorgensen, J. H. (1986). Clinical applications of the *Limulus* amebocyte lysate test. In *Handbook of Endotoxin*, Vol. 4, Clinical Aspects of Endotoxin Shock, R. A. Proctor (Ed.), Elsevier, Amsterdam, pp. 127–160.

Kaufman, B. M., Cross, A. S., Futrovsky, S. L., Sidberry, H. F., and Sadoff, J. C. (1986). Monoclonal antibodies reactive with K1-encapsulated *Escherichia coli* lipopolysaccharide are opsonic and protect mice against lethal challenge. *Infect. Immun. 52*: 617–619.

Kikeri, D., Pennell, J. P., Hwang, K. H., Jacob, A. I., Richman, A. V., and Bourgoignie, J. J. (1986). Endotoxemic acute renal failure in awake rats. *Am. J. Physiol. 250*(6, Pt. 2): F1098–F1106.

Kiso, M., Ishida, H., and Hasegawa, A. (1984). Synthesis of biologically active, novel monosaccharide analogs of lipid A. *Agric. Biol. Chem. 48*: 251.

Kitchens, R. L., Ulevitch, R. J., and Munford, R. S. (1992). Lipopolysaccharide (LPS) partial structures inhibit responses to LPS in a human macrophage cell line without inhibiting LPS uptake by a CD14-mediated pathway. *J. Exp. Med. 176*: 485–494.

Kovach, N. L., Yee, E., Munford, R. S., Raetz, C. R. H., and Harlan, J. M. (1990). Lipid IV$_A$ inhibits synthesis and release of tumor necrosis factor induced by lipopolysaccharide in human whole blood *ex vivo*. *J. Exp. Med. 172*: 77–84.

Lam, C., Hildebrandt, J., Schütze, E., Rosenwirth, B., Proctor, R. A., Liehl, E., and Stütz, P. (1991). Immunostimulatory, but not antiendotoxin, activity of lipid X is due to small amounts of contaminating *N,O*-acylated disaccharide-1-phosphate: *in vitro* and *in vivo* reevaluation of the biological activity of synthetic lipid X. *Infect. Immun. 59*: 2351–2358.

Lei, M.-G., Stimpson, S. A., and Morrison, D. C. (1991). Specific endotoxic lipopolysaccharide-binding receptors on murine splenocytes. III. Binding specificity and characterization. *J. Immunol. 147*: 1925–1932.

Levin, J., and Bang, F. (1965). The role of endotoxin in the extracellular coagulation of *Limulus* blood. *Johns Hopkins Med. J. 115*: 265–274.

Levin, J., Poore, T. E., Zauger, N. P., and Oser, R. S. (1970). Detection of endotoxin in the blood of patients with sepsis due to gram-negative bacteria. *N. Engl. J. Med. 283*: 1313–1316.

McCloskey, R. V., Strobe, R. C., Sanders, C., Smith, S. M., Smith, C. R. (1994). Treatment of septic shock with human monoclonal antibody HA-1A *Ann. Intern. Med. 121*: 1–5.

McCabe, W. R., and Greely, A. (1972). Immunization with R mutants of *S. minnesota*. I. Protection against challenge with heterologous gram-negative bacilli. *J. Immunol. 108*: 601–610.

McCabe, W. R., and Olans, R. N. (1981). Shock in gram-negative bacteremia: predisposing factors, pathophysiology, and treatment. In *Current Clinical Topics in Infectious Diseases*, Vol. 2, J. S. Remington and M. N. Swartz (Eds.), McGraw-Hill, New York, pp. 121–150.

McCabe, W. R., Kreger, B. E., and Johns, M. (1972). Type-specific and cross-reactive antibodies in gram-negative bacteremia. *N. Engl. J. Med. 287*: 261–267.

McCabe, W. R., DeMaria, A., Jr., Berberich, H., and Johns, M. A. (1988). Immunization with rough mutants of *Salmonella minnesota*: protective activity of IgM and IgG antibody to the R595 (Re chemotype) mutant. *J. Infect. Dis. 158*: 291–300.

McCartney, A. C., Robertson, M. R. I., Piotrowicz, B. I., and Lucie, N. P. (1987). Endotoxaemia, fever, and clinical status in immunosuppressed patients: a preliminary study. *J. Infect. 15*: 201–206.

McCutchan, J. A., Wolf, J. L., Ziegler, E. J., and Braude, A. I. (1983). Ineffectiveness of single-dose human antiserum to core glycolipid (*Escherichia coli* J5) for prophylaxis of bacteremic, gram-negative infection in patients with prolonged neutropenia. *Schweiz, Med. Wochenschr. 113* (14, Suppl.): 40–55.

Marra, M. N., Wilde, C. G., Griffith, J. E., Snable, J. L., and Scott, R. W. (1990). Bactericidal/permeability-increasing protein has endotoxin-neutralizing activity. *J. Immunol. 144*: 662–666.

Marra, M. N., Thornton, M. B., Opal, S., Fisher, C., and Scott, R. W. (1992). Endotoxin binding and neutralizing activities of bactericidal/permeability-increasing protein *in vivo* and *in vitro*. Presented at 2nd Annual Meeting on Advances in Prevention and Treatment of Endotoxemia and Sepsis, June 22–23, Philadelphia, PA (Abstract).

Morrison, D. C., and Jacobs, D. M. (1976). Binding of polymyxin B to the lipid A portion of bacterial lipopolysaccharides. *Immunochemistry 13*: 813–818.

Morrison, D. C., and Kline, L. F. (1977). Activation of the classical and properdin pathways of complement by bacterial lipopolysaccharides (LPS). *J. Immunol. 118*: 362–368.

Morrison, D. C., Silverstein, R., Bright, S. W., Chen, T.-Y., Flebbe, L. M., and Lei, M.-G. (1990). Monoclonal antibody to mouse lipopolysaccharide receptor protects mice against the lethal effects of endotoxin. *J. Infect. Dis. 162*: 1063–1068.

Munford, R. S., and Hall, C. L. (1985). Uptake and deacylation of bacterial lipopolysaccharides by macrophages from normal and endotoxin-hyporesponsive mice. *Infect. Immun. 48*: 464.

Munford, R. S., and Hall, C. L. (1986). Detoxification of bacterial lipopolysaccharides (endotoxins) by a human neutrophil enzyme. *Science 234*: 203.

Natanson, C., Eichenholz, P. W., Danner, R. L., Eichacker, P. Q., Hoffman, W. D., Kuo, G. C., Banks, S. M., MacVittie, T. J., and Parrillo, J. E. (1989a). Endotoxin and tumor necrosis factor challenges in dogs simulate the cardiovascular profile of human septic shock. *J. Exp. Med. 169*: 823–832.

Natanson, C., Danner, R. L., Elin, R. J., Hosseini, J. M., Peart, K. W., Banks, S. M., MacVittie, T. J., Walker, R. I., and Parrillo, J. E. (1989b). The role of endotoxemia in cardiovascular dysfunction and mortality: *E. coli* and *S. aureus* challenges in a canine model of human septic shock. *J. Clin. Invest. 83*: 243–251.

Natanson, C., Danner, R. L., Reilly, J. M., Doerfler, M. L., Hoffman, W. D., Akin, G. L., Hosseini, J. M., Banks, S. M., Elin, R. J., MacVittie, T. J., and Parillo, J. E. (1990). Antibiotics versus cardiovascular support in a canine model of human septic shock. *Am. J. Physiol. 259*: H1440–H1447.

Natanson, C., Hoffman, W. D., Koev, L. A., Dolan, D. P., Banks, S. M., Bacher, J., Danner, R. L., Klein, H., and Parrillo, J. E. (1993). A controlled trial of plasmapheresis fails to improve outcome in an antibiotic treated canine model of human septic shock. *Transfusion Med. 33*: 243–248.

O'Dwyer, S. T., Michie, H. R., Ziegler, T. R., Revhaug, A., Smith, R. J., and Wilmore, D. W. (1988). A single dose of endotoxin increases intestinal permeability in healthy humans. *Arch. Surg. 123*: 1459–1464.

Ohashi, K., Niwa, M., Nakamura, T., Morita, T., and Iwanaga, S. (1984). Anti-LPS factor in the horseshoe crab *Tachypleus tridentatus. FEBS Lett. 176*: 207–210.

Oishi, K., Koles, N. L., Guelde, G., and Pollack, M. (1992). Antibacterial and protective properties of monoclonal antibodies reactive with *Escherichia coli* O111:B4 lipopolysaccharide: relation to antibody isotype and complement-fixing activity. *J. Infect. Dis. 165*: 34–45.

Okusawa, S., Gelfand, J. A., Ikejima, T., Connolly, R. J., and Dinarello, C. H. (1988). Interleukin 1 induces shock-like state in rabbits. *J. Clin. Invest. 81*: 1162–1172.

Ooi, C. H., Weiss, J., Doerfler, M. E., and Elsbach, P. (1991). Endotoxin-neutralizing properties of the 25 kD N-terminal fragment and a newly isolated 30 kD C-terminal fragment of the 55–60 kD bactericidal/permeability-increasing protein of human neutrophils. *J. Exp. Med. 174*: 649–655.

Piercey, L. (1993). HA-1A has a checkered past. *Biol. World Financial Watch*, January 25, pp. 1–2.

Pohlman, T. H., Munford, R. S., and Harlan, J. M. (1987). Deacylated lipopolysaccharide inhibits neutrophil adherence to endothelium induced by lipopolysaccharide *in vitro. J. Exp. Med. 165*: 1393.

Pollack, M., Chia, J. K. S., Koles, N. L., and Miller, M. (1989). Specificity and cross-reactivity of monoclonal antibodies reactive with the core and lipid A regions of bacterial lipopolysaccharide. *J. Infect. Dis. 159*: 168–188.

Quezado, Z. M. N., and Hoffman, W. D. (1993). Therapies directed against endotoxin: has the time come? *West. J. Med. 158*(4): 424–425.

Quezado, Z. M. N., Natanson, C., Banks, S. M., Alling, D. W., Koev, C. A., Danner, R. L., Elin, R. J., Hosseini, J. M., Bacher, J. D., Dolan, D. P., and Hoffman, W. D. (1991). Physiologic results from a controlled trial of human IgM monoclonal antibody against lipid A (HA-1A) in a canine model of gram-negative septic shock. In *Program and Abstracts of the 31st Interscience Conference on Antimicrobial Agents and Chemotherapy*, Chicago, Sept. 19–Oct. 2, 1991, Am. Soc. Microbiology, Washington, DC, p. 230 (Abstract).

Quezado, Z. M. N., Natanson, C., Banks, S. M., Alling, D. W., Koev, C. A., Danner, R. L., Elin, R. J., Hosseini, J. M., Pollack, M., Bacher, J. D., Dolan, D. P., and Hoffman, W. D. (1992). A human monoclonal IgM antibody (MAb) against endotoxin (HA-1A) decreased survival in a canine model of gram-negative bacterial septic shock. *Clin. Res. 20*: 286 (Abstract).

Quezado, Z. M. N., Natanson, C., Alling, D. W., Banks, S. L., Koev, C. A., Elin, R. L., Hosseini, J. M., Bacher, J. D., Danner, R. L., and Hoffman, W. O. (1993). A controlled trial of human lipid A-reactive monoclonal antibody HA-1A in a canine model of septic shock. *J. Am. Med. Assoc. 269*: 2221–2227.

Quezado, Z. M. N., Natanson, C., Banks, S. M., Alling, D. W., Koev, C. A., Danner, R. L., Elin, R. J., Hosseini, J. M., Parker, T. S., Levine, D. M., Rubin, A. L., and Hoffman, W. D. (1994). Therapeutic trial of reconstituted human high density lipoprotein in a canine model of gram-negative septic shock: preliminary results. *J. Pharmacol. Exp. Therap*. In Press.

Rackow, E. C., Astiz, M. E., Kim, Y. B., and Wil, M. H. (1989). Monophosphoryl lipid A blocks hemodynamic effects of lethal endotoxemia. *J. Lab. Clin. Med.* *113*: 112.

Raetz, C. R. H. (1984). The enzymatic synthesis of lipid A: molecular structure and biologic function of monosaccharide precursors. *Rev. Infect. Dis.* *6*: 463.

Raetz, C. R. H., Brozek, K. A., Clementz, T., Coleman, J. D., Galloway, S. M., Golenbock, D. T., and Hampton, R. Y. (1988). Gram-negative endotoxin: a biologically active lipid. *Cold Spring Harbor Symp. Quant. Biol.* *53*: 973–982.

Raetz, C. R. H., Ulevitch, R. J., Wright, S. D., Sibley, C. H., Ding, A., and Nathan, C. F. (1991). Gram-negative endotoxin: an extraordinary lipid with profound effects on eukaryotic signal transduction. *FASEB J.* *5*: 2652–2660.

Ravin, H. A., Rowley, D., Jenkins, C., and Fine, J. (1960). On the absorption of bacterial endotoxin from the gastrointestinal tract of normal and shocked animals. *J. Exp. Med.* *112*: 783–792.

Ray, B. L., Painter, G., and Raetz, C. R. H. (1984). The biosynthesis of gram-negative endotoxin: formation of lipid A disaccharides from monosaccharide precursors in extracts of *Escherichia coli*. *J. Biol. Chem.* *259*: 4852–4859.

Reinhold, R., and Fine, J. (1971). A technique for quantitative measurement of endotoxin in human plasma. *Proc. Soc. Exp. Biol. Med.* *137*: 334–340.

Relman, A. S. (1990). Editorial review of protocols for clinical trials [Letter]. *N. Engl. J. Med.* *323*: 1355.

Riedo, F. X., Munford, R. S., Campbell, W. B., Reisch, J. S., Chien, K. A., and Gerard, R. D. (1990). Deacylated lipopolysaccharide inhibits plasminogen activator inhibitor-1, prostacyclin, and prostaglandin E$_2$ induction by lipopolysaccharide but not by tumor necrosis factor α. *J. Immunol.* *144*: 3506.

Rifkind, D. (1967). Prevention by polymyxin B of endotoxin lethality in mice. *J. Bacteriol.* *93*: 1463.

Rifkind, D., and Palmer, J. D. (1966). Neutralization of endotoxin toxicity in chick embryos by antibiotics. *J. Bacteriol.* *815*: 92.

Sagawa, T., Hitsumoto, Y., Kanoh, M., Utsumi, S., and Kimura, S. (1990). Mechanisms of neutralization of endotoxin by monoclonal antibodies to O and R determinants of lipopolysaccharide. *Adv. Exp. Med. Biol.* *256*: 341–344.

Scharfman, W. B., Tillotson, J. R., Taft, E. G., and Wright, E. (1979). Plasmapheresis for meningococcemia with disseminated intravascular coagulation [Letter]. *N. Engl. J. Med.* *300*: 1277.

Schumann, R. R., Leong, S. R., Flaggs, G. W., Gray, P. W., Wright, S. D., Mathison, J. C., Tobias, P. S., and Ulevitch, R. J. (1990). LPS binding protein. *Science 249*: 1429–1433.

Schütt, C., Schilling, Th., and Krüger, C. (1991). sCD14 prevents endotoxin inducible oxidative burst response of human monocytes. *Allerg. Immunol.* *37*: 159–164.

Shwartzman, G. (1928). Studies on *Bacillus typhosus* toxic substances. I. Phenomenon of local skin reactivity to *B. typhosus* culture filtrate. *J. Exp. Med.* *38*: 247–268.

Siber, G. R., Kania, S. A., and Warren, H. S. (1985). Cross-reactivity of rabbit antibodies to lipopolysaccharides of *Escherichia coli* J5 and other gram-negative bacteria. *J. Infect. Dis.* *152*: 954–964.

Siegel, J. P. (1990). Editorial review of protocols for clinical trials [Letter]. *N. Engl. J. Med.* *323*: 1355.

Siegel, J. P., Stein, K. E., and Zoon, K. C. (1992). Antiendotoxin monoclonal antibodies [Letter]. *N. Engl. J. Med. 327*: 890–891.

Smith, C. R., Straube, R. C., and Ziegler, E. J. (1992). HA-1A: a human monoclonal antibody for the treatment of gram-negative sepsis. *Infect. Dis. Clin. North Am.* *6*: 253–266.

Stevens, A. R., Legg, J. S., Henry, B. S., Dille, J. M., Kirby, W. M. M., and Finch, C. A. (1953). Fatal transfusion reactions from contamination of stored blood by cold growing bacteria. *Ann. Intern. Med.* *39*: 1228–1239.

Stokes, D. C., Shenep, J. L., Fishman, M., Hildner, W. K., Bysani, G. K., and Rufus, K. (1989). Polymyxin B prevents lipopolysaccharide-induced release of tumor necrosis factor α from alveolar macrophages. *J. Infect. Dis. 160*: 52.

Suffredini, A. F., Fromm, R. E., Parker, M. M., Brenner, M., Kovacs, J. A., Wesley, R. A., and Parrillo, J. E. (1989a). The cardiovascular response of normal humans to administration of endotoxin. *N. Engl. J. Med. 321*: 280–287.

Suffredini, A. F., Harpel, P. C., and Parrillo, J. E. (1989b). Promotion and subsequent inhibition of plasminogen activation after administration of intravenous endotoxin to normal subjects. *N. Engl. J. Med. 320*: 1165–1172.

Takayama, K., Qureshi, N., Mascagni, P., Nashed, M. A., Anderson, L., and Raetz, C. R. (1983). Fatty acyl derivatives of glucosamine-1-phosphate in *Escherichia coli* and their relation to lipid A. *Biol. Chem. 258*: 7379.

Takayama, K., Qureshi, N., Beutler, B., and Kirkland, T. N. (1989). Diphosphoryl lipid A from *Rhodopseudomonas sphaeroides* ATCC 17023 blocks induction of cachetin in macrophages by lipopolysaccharide. *Infect. Immun. 57*: 1336–1338.

Tanamoto, K., Zähringer, U., McKenzie, G. R., Galanos, C., Reitschel, E. T., Lüderitz, O., Kusumoto, S., and Shiba, T. (1984). Biological activities of synthetic lipid A analogs: pyrogenicity, lethal toxicity, anticomplement activity, and induction of gelation of *Limulus* amebocyte lysate. *Infect. Immun. 44*: 421.

Taveira da Silva, A. M., Kaulbach, H., Chuidian, F., Lambert, D., Suffredini, A. F., and Danner, R. L. (1993). Shock and multiple organ dysfunction caused by self-administration of R595 *Salmonella minnesota* endotoxin. *N. Engl. J. Med.* in press.

Teng, N. N. H., Kaplan, H. S., and Hebert, J. M. (1985). Protection against gram-negative bacteremia and endotoxemia with human monoclonal IgM antibodies. *Proc. Natl. Acad. Sci. U.S.A. 82*: 1790–1794.

te Velde, A. A., Huijibens, R. J. F., Heije, K., de Vries, J. E., and Figdor, C. G. (1990). Interleukin-4 (IL-4) inhibits secretion of IL-1β, tumor necrosis factor α, and IL-6 by human monocytes. *Blood 76*: 1392–1397.

Thomas, L. L., Sturk, A., Buller, H. R., Ten Cate, J. W., Spijker, R. E., and Ten Cate, H. (1984). Comparative investigation of a quantitative chromogenic endotoxin assay and blood cultures. *Am. J. Clin. Pathol. 82*: 203–206.

Thomas, L. L. M., Sturk, A., Kahle, L. H., and Ten Cate, J. W. (1981). Quantitative endotoxin determination in blood with a chromogenic substrate. *Clin. Chem. Acta 116*: 63–68.

Tobias, P. S., Mathison, J. C., and Ulevitch, R. I. (1988). A family of lipopolysaccharide binding proteins involved in responses to gram-negative sepsis. *J. Biol. Chem. 263*: 13479–13481.

Tobias, P. S., Soldau, K., and Ulevitch, R. I. (1989). Identification of a lipid A binding site in the acute phase reactant lipopolysaccharide binding protein. *J. Biol. Chem. 264*: 10867–10871.

Togari, H., Mikawa, M., Iwanaga, T., Matsumoto, N., Kawase, A., Hagisawa, M., Ogino, T., Goto, R., Watanabe, I., Kito, H., Ogawa, Y., and Wada, Y. (1983). Endotoxin clearance by exchange blood transfusion in septic shock neonates. *Acta Paediatr. Scand. 72*: 87.

Töllner, U., Pohlandt, T. F., Heinze, F., and Henrichs, I. (1977). Treatment of septicaemia in the newborn infant: choice of initial antimicrobial drugs and the role of exchange transfusion. *Acta Paediatr. Scand. 66*: 605.

Tracey, R. J., Lowry, S. F., and Cerami, A. (1988). Cachectin: a hormone that triggers acute shock and chronic cachexia. *J. Infect. Dis. 3*: 413–420.

Ulevitch, R. J., Johnston, A. R., and Weinstein, D. B. (1979). New function for high density lipoproteins: their participation in intravascular reactions of bacterial lipopolysaccharides. *J. Clin. Invest. 64*: 1516–1524.

Vaara, M., and Vaara, T. (1983). Polycations sensitize enteric bacteria to antibiotics. *Antimicrob. Agents Chemother. 24*: 107.

Vain, N. E., Mazlumian, J. R., Swarner, O. W., and Cha, C. C. (1980). Role of exchange transfusion in the treatment of severe septicemia. *Pediatrics 66*: 693.

Van Dervort, A. L., Doerfler, M. E., Stuetz, P., and Danner, R. L. (1992). Antagonism of lipopolysaccharide-induced priming of human neutrophils by lipid A analogs. *J. Immunol. 149*: 359–366.

van Deventer, S. J. H., Ten Cate, J. W., and Ty Gat, G. N. J. (1988). Intestinal endotoxemia: clinical significance. *Gastroenterology 94*: 825–831.

van Dijk, W. C., Verbrugh, H. A., van Erne-van der Tol, M. E., Peters, R., and Verhoef, J. (1981). *Escherichia coli* antibodies in opsonization and protection against infection. *J. Med. Microbiol. 14*: 381–389.

Von Eschen, K. B., Howell, S., Stern, W., and Vargas, R. (1992). Pretreatment of humans with MPL monophosphoryl lipid A immunostimulant induces hyporesponsiveness to bacterial endotoxin. Presented at 2nd Annual Meeting on Advances in Prevention and Treatment of Endotoxemia and Sepsis, June 22–23, Philadelphia, PA (Abstract).

Warren, H. S., Kania, S. A., and Siber, G. R. (1985). Binding and neutralization of bacterial lipopolysaccharide by colistin nonapeptide. *Antimicrob. Agents Chemother. 28*: 107.

Warren, H. S., Riveau, G. R., DeDeckker, F. A., and Chedid, L. A. (1988). Control of endotoxin activity and interleukin-1 production through regulation of lipopolysaccharide-lipoprotein binding by a macrophage factor. *Infect. Immun. 56*: 204–212.

Warren, H. S., Danner, R. L., and Munford, R. S. (1992). Antiendotoxin monoclonal antibodies. *N. Engl. J. Med. 326*: 1153–1156.

Warren, H. S., Amato, S. F., Fitting, C., Black, K. M., Loiselle, P. M., Pasternack, M. S., and Cavaillon, J. M. (1993). Assessment of ability of murine and human antilipid A monoclonal antibodies to bind and neutralize lipopolysaccharide. *J. Exp. Med. 177*: 89–97.

Weersink, A. J. L., van Kessel, K. P. M., van den Tol, M. E., van Strijp, J. A. G., Torensma, R., Verhoef, J., Elsbach, P., and Weiss, J. (1992). Human granulocytes express a 55-kDa lipopolysaccharide-binding protein on the cell surface that is identical to the bactericidal/permeability-increasing protein. *J. Immunol. 150*: 253–263.

Weinstein, D. L., Lissner, C. R., Swanson, R. N., and O'Brien, A. D. (1986). Macrophage defect and inflammatory cell recruitment dysfunction in *Salmonella* susceptible C3H/HeJ mice. *Cell. Immunol. 102*: 68–77.

Weis, J. P., Elsbach, P., Olsson, I., and Odeberg, H. (1978). Purification and characterization of a potent bactericidal and membrane-active protein from the granules of human polymorphonuclear leukocytes. *J. Biol. Chem. 253*: 2664–2672.

Weiss, J., Victor, M., Stendhal, O., and Elsbach, P. (1982). Killing of gram-negative bacteria by polymorphonuclear leukocytes. Role of an O_2-independent bactericidal system. *J. Clin. Invest. 69*: 959–970.

Wenzel, R., Bone, R., Fein, A., Quenzer, R., Schentag, J., Gorelick, K. J., Wedel, N. I., and Perl, T. (1991). Results of a second double-blind, randomized, controlled trial of antiendotoxin antibody E5 in gram-negative sepsis. In *Program and Abstracts of the 31st Interscience Conference on Antimicrobial Agents and Chemotherapy*, Sept. 19–Oct. 2, 1991, Am. Soc. Microbiology, Chicago, p. 294 (Abstract).

Westphal, O., Hann, K., and Himmelspach, K. (1983). Chemistry and immunochemistry of bacterial lipopolysaccharide as cell wall antigens and endotoxin. *Prog. Allergy 33*: 9–39.

Wheeler, A. P., Hardie, W. D., and Bernard, G. (1990). Studies of an antiendotoxin antibody in preventing the physiologic changes of endotoxemia in awake sheep. *Am. Rev. Respir. Dis. 142*: 775–781.

Wright, S. D. (1991). Multiple receptors for endotoxin. *Curr. Opinion Immunol. 3*: 83–90.

Wright, S. D., Ramos, R. A., Tobias, P. S., Ulevitch, R. J., and Mathison, J. C. (1990). CD14, a receptor for complexes of lipopolysaccharide (LPS) and LPS binding protein. *Science 249*: 1431–1433.

Wright, S. D., Ramos, R. A., Patel, M., and Miller, D. S. (1992). Septin: a factor in plasma that opsonizes lipopolysaccharide-bearing particles for recognition by CD14 on phagocytes. *J. Exp. Med. 176*: 719–727.

Young, L. S., Martin, W. J., Meyer, R. D., Weinstein, R. J., and Anderson, E. T. (1977). Gram-negative bacteremia: Microbiologic immunologic, and therapeutic considerations, *Ann. Int. Med. 86*: 456–471.

Young, L. S., Stevens, P., and Ingram, I. (1975). Functional role of antibody against "core" glycolipid of Enterobacteriaceae. *J. Clin. Invest. 56*: 850–861.

Young, L. S., Gascon, R., Alam, S., and Bermudez, L. E. M. (1989). Monoclonal antibodies for treatment of gram-negative infections. *Rev. Infect. Dis. 11*(Suppl): S1564–S1571.

Zager, R. A. (1986). *Escherichia coli* endotoxin injections potentiate experimental ischemic renal injury. *Am. J. Physiol. 251*(6, Pt. 2): F988–F994.

Zager, R. A., and Prior, R. B. (1986). Gentamicin and gram-negative bacteremia: a synergism for the development of experimental nephrotoxic acute renal failure. *J. Clin. Invest. 78*: 196–204.

Ziegler, E. J., and Smith, C. R. (1992). Antiendotoxin monoclonal antibodies [Letter]. *N. Engl. J. Med. 326*: 1165.

Ziegler, E. J., Douglas, H., Sherman, J. E., Davis, C. E., and Braude, A. I. (1973). Treatment of *E. coli* and *Klebsiella* bacteremia in agranulocytic animals with antiserum to a UDP-Gal epimerase-deficient mutant. *J. Immunol. 111*: 433–438.

Ziegler, E. J., McCutchan, J. A., Fierer, J., Glauser, M. P., Sadoff, J. C., Douglas, H., and Braude, A. I. (1982). Treatment of gram-negative bacteremia and shock with human antiserum to a mutant *Escherichia coli*. *N. Engl. J. Med. 307*: 1225–1230.

Ziegler, E. J., Teng, N. N. H., Douglas, H., Wunderlich, A., Berger, H. J., and Bolmer, S. D. (1987). Treatment of *Pseudomonas* bacteremia in neutropenic rabbits with human monoclonal antibody against *E. coli* lipid A. *Clin. Res. 35*: 619 (Abstract).

Ziegler, E. J., Fisher, C. J., Sprung, C. L., Straube, R. C., Sadoff, J. C., Foulke, G. E., Wortel, C. H., Fink, M. P., Dellinger, R. P., Teng, N. N. H., Allen, I. E., Berger, H. J., Knatterud, G. L., LoBuglio, A. F., Smith, C. R., and the HA-1A Sepsis Study Group. (1991). Treatment of gram-negative bacteremia and septic shock with HA-1A human monoclonal antibody against endotoxin. *N. Engl. J. Med. 324*: 429–436.

Zuckerman, S. H., Evans, G. F., and Butler, L. D. (1991). Endotoxin tolerance: independent regulation of interleukin-1 and tumor necrosis factor expression. *Infect. Immun. 59*: 2774–2780.

Zweifach, B. W., Gordon, H. A., Wagner, M., and Reyniere, J. A. (1958). Irreversible hemorrhagic shock in germ-free rats. *J. Exp. Med. 107*: 437–450.

Index

Milton Keynes UK
Ingram Content Group UK Ltd.
UKHW052030071024
449327UK00027B/2503